Lecture Notes in Networks and Systems

Volume 362

The series "Lecture Notes in Networks and Systems" publishes the latest developments in Networks and Systems—quickly, informally and with high quality. Original research reported in proceedings and post-proceedings represents the core of LNNS.

Volumes published in LNNS embrace all aspects and subfields of, as well as new challenges in, Networks and Systems.

The series contains proceedings and edited volumes in systems and networks, spanning the areas of Cyber-Physical Systems, Autonomous Systems, Sensor Networks, Control Systems, Energy Systems, Automotive Systems, Biological Systems, Vehicular Networking and Connected Vehicles, Aerospace Systems, Automation, Manufacturing, Smart Grids, Nonlinear Systems, Power Systems, Robotics, Social Systems, Economic Systems and other. Of particular value to both the contributors and the readership are the short publication timeframe and the world-wide distribution and exposure which enable both a wide and rapid dissemination of research output.

The series covers the theory, applications, and perspectives on the state of the art and future developments relevant to systems and networks, decision making, control, complex processes and related areas, as embedded in the fields of interdisciplinary and applied sciences, engineering, computer science, physics, economics, social, and life sciences, as well as the paradigms and methodologies behind them.

Indexed by SCOPUS, INSPEC, WTI Frankfurt eG, zbMATH, SCImago.

All books published in the series are submitted for consideration in Web of Science.

More information about this series at https://link.springer.com/bookseries/15179

Rafik A. Aliev · Janusz Kacprzyk ·
Witold Pedrycz · Mo Jamshidi ·
Mustafa Babanli · Fahreddin M. Sadikoglu
Editors

11th International Conference on Theory and Application of Soft Computing, Computing with Words and Perceptions and Artificial Intelligence - ICSCCW-2021

Springer

Editors
Rafik A. Aliev
Department of Control Systems
Azerbaijan State Oil and Industry University
Baku, Azerbaijan

Janusz Kacprzyk
System Research Institute
Polish Academy of Sciences
Warsaw, Poland

Witold Pedrycz
Department of Electrical and Computer Engineering
University of Alberta
Edmonton, AB, Canada

Mo Jamshidi
Department of Electrical and Computer Engineering
University of Texas at San Antonio
San Antonio, TX, USA

Mustafa Babanli
Azerbaijan State Oil and Industry University
Baku, Azerbaijan

Fahreddin M. Sadikoglu
Department of Mechatronics
Near East University
North Cyprus, Turkey

ISSN 2367-3370 ISSN 2367-3389 (electronic)
Lecture Notes in Networks and Systems
ISBN 978-3-030-92126-2 ISBN 978-3-030-92127-9 (eBook)
https://doi.org/10.1007/978-3-030-92127-9

This Springer imprint is published by the registered company Springer Nature Switzerland AG
The registered company address is: Gewerbestrasse 11, 6330 Cham, Switzerland

Preface

The 11th International Conference on Theory and Applications Soft Computing, Computing with Words and Perceptions and Artificial Intelligence (ICSCCW-2021) is the premier international conference organized by Azerbaijan Association of "Zadeh's Legacy and Artificial Intelligence" (Azerbaijan), Azerbaijan State Oil and Industry University (Azerbaijan), University of Siegen (Siegen, Germany), BISC-Berkeley Initiative in Soft Computing (USA), University of Texas (San Antonio (USA), Georgia State University (Atlanta, USA), University of Alberta (Canada), University of Toronto (Toronto, Canada), System Research Institute of Polish Academy of Sciences, Near East University (North Cyprus).

This volume presents the proceedings of the 11th International Conference on Theory and Applications Soft Computing, Computing with Words and Perceptions and Artificial Intelligence, ICSCCW-2021 held in Antalya, Turkey, on August 23–24, 2021. It includes contributions from diverse areas of soft computing and computing with words such as uncertain computation, decision making under imperfect information, neuro-fuzzy approaches, deep learning, natural language processing, and others. The topics of the papers include theory and application of soft computing, information granulation, computing with words, computing with perceptions, image processing with soft computing, probabilistic reasoning, intelligent control, machine learning, fuzzy logic in data analytics and data mining, evolutionary computing, chaotic systems, soft computing in business, economics, earth sciences, engineering, material sciences, biomedical engineering and health care.

The volume covers new ideas from theories of soft computing and computing with words and their applications in economics, business, industry, education, medicine, earth sciences, and other fields. This volume will be a useful guide for academics, practitioners, and graduates in fields of soft computing and computing with words. It will allow for increasing of interest in development and applying of these paradigms in various real-life fields.

August 2021

Rafik Aliev
Chairman of ICSCCW-2021

Organization

Chairman

R. A. Aliev, Azerbaijan

Co-Chairmen and Guest Editors

J. Kacprzyk, Poland
W. Pedrycz, Canada
M. Jamshidi, USA
M. B. Babanli, Azerbaijan
F. S. Sadikoglu, North Cyprus

International Program Committee

A. Averkin, Russia
A. Musayev, Azerbaijan
A. Piegat, Poland
B. Fazlollahi, USA
B. Kang, China
B. Turksen, Canada, Turkey
C. Kahraman, Turkey
D. Dubois, France
D. Enke, USA
D. Kumar Jana, India
E. Babaei, Iran
F. Aminzadeh, USA
G. Imanov, Azerbaijan
H. Berenji, USA
H. Hamdan, France, UK
H. Prade, France
H. Roth, Germany
I. Batyrshin, Mexico
I. G. Akperov, Russia
I. Perfilieva, Czech Republic
K. Atanassov, Bulgary
K. Bonfig, Germany
K. Takahashi, Japan
M. Gupta, Canada
M. Nikravesh, USA
N. Yusupbekov, Uzbekistan
O. Huseynov, Azerbaijan
O. Kaynak, Turkey
R. R. Aliev, North Cyprus
R. Abiyev, North Cyprus
R. O. Kabaoglu, Turkey
R. Yager, USA
S. Ulyanov, Russia
T. Allahviranloo, Turkey

T. Fukuda, Japan
T. Takagi, Japan
V. Kreinovich, USA
V. Loia, Italy
V. Niskanen, Finland
V. Nourani, Iran
V. Novak, Czech Republic

Organizing Committee

Chairman

U. Eberhardt, Germany

Co-Chairmen

T. Abdullayev, Azerbaijan
L. Gardashova, Azerbaijan

Members

N. E. Adilova, Azerbaijan
A. Alizadeh, Azerbaijan
B. Guirimov, Azerbaijan
A. Guliyev, Azerbaijan
M. M. M. Elamin, North Cyprus
M. A. Salahli, Turkey

Conference Organizing Secretariat

Azadlig Ave. 20, AZ 1010 Baku, Azerbaijan
Phone: +99 412 493 45 38, Fax: +99 412 498 45 09
E-mail: icsccw2021.org@gmail.com, latsham@yandex.ru, adilovanigarr@gmail.com

Contents

Multistage Decision Making in Bimodal Information Environment 1
R. A. Aliev

Interactive Solution of Difficult Choice and Decision Making Problems: Effective and Efficient but not Always Easy 2
Janusz Kacprzyk

About Valery Borisovich TARASOV (16.02.1955–22.07.2021) 3
R. A. Aliev and Alexey Averkin

Designing and Evaluating Interpretable Rule-Based Architecture Under Privacy Constraints: A Framework of Granular Computing ... 6
Witold Pedrycz

Fuzzy Classification-Based Alloy Selection from a Large Dataset...... 7
M. B. Babanli

Consistent Z-preferences Formulation in Decision Analysis 14
Rafig R. Aliyev

New High Step-Up DC-DC Converter in PV System: Performance and Analysis.. 19
Senol Bektas, Farzad Mohammadzadeh Shahir, and Ebrahim Babaei

Investigation of the Quality of Fuzzy IF-THEN Model for a Control System 28
Nigar E. Adilova

Circular Intuitionistic Fuzzy Sets in Multi Criteria Decision Making... 34
Esra Çakır, Mehmet Ali Taş, and Ziya Ulukan

Z^+ - Number Based Alternatives Selection in Investment Problem..... 43
Shamil A. Ahmadov

Fuzzy Interval-Valued Temporal Automated Planning and Scheduling Problem . . . 51
Janusz Kacprzyk, Margarita Knyazeva, and Alexander Bozhenyuk

Z-numbers Based Preference of Expert Opinions on Social Capital . . . 59
G. Imanov, M. Murtuzaeva, and A. Aliyev

Quality Metrics of LSB Image Steganography Technique for Color Space HSI . . . 67
Yucel Inan

Z-Set Based Inference Using ALI-2 Implication for Control System Design . . . 75
Latafat A. Gardashova

Z-Numbers-Based Approach to Hotel Service Quality Assessment . . . 85
Aziz Nuriyev and Ahmet Bahadir Baysal

Predicting Stock Prices Using Random Forest and Logistic Regression Algorithms . . . 95
Aysel Aliyeva

Explainable Artificial Intelligence: Rules Extraction from Neural Networks . . . 102
Alexey Averkin and Sergey Yarushev

Application of Emotional Neural Network in Modeling Evaporation . . . 110
Fahreddin Sadikoglu, Vahid Nourani, Nardin Jabbarian Paknezhad, and Soheil Emamalipour

Malaria Detection Using Convolutional Neural Network . . . 116
Khaled Almezhghwi

Selection of Information - Measuring Components on the Basis of Layout Diagram of Flexible Manufacturing Cell . . . 124
I. R. Aliyev, J. F. Mammadov, and Sh. R. Rakhimov

Minimizing Handover Process and Wireless Propagation Lose by Using Multilayer Perception Neural Network . . . 132
Jamal Fathi

Automatic Control of Reactive Power in the Load Node of the Power Supply System Based on Fuzzy Logic . . . 140
Z. I. Farkhadov and R. Z. Azizov

Fuzzy Aided Generalized Probability Distribution Function for Wireless Wearable Medical Sensors . . . 148
Bülent Bilgehan and Fahreddin Sadikoglu

Use of Decision Tree and Fuzzy Logic Methods to Predict Academic Achievement of University Freshmen 156
M. A. Salahli, T. Gasimzade, V. Salahli, F. Alasgarova, and A. Guliyev

Monthly Prediction of Reference Evapotranspiration in Northcentral Nigeria Using Artificial Intelligence Tools: A Comparative Study 165
Jazuli Abdullahi and Gozen Elkiran

Estimation of Benchmarking Influence in Buyer's Decision-Making Process by Using Fuzzy AHP 173
Khatira J. Dovlatova

Attitudes of Families of Children with Special Needs towards Technology 183
Rojda Kılınçaslan Akdağ, Şahin Akdağ, Çağla Kılınç, Yücehan Yücesoy, and Başak Bağlama

Some Aspects of Applying Fuzzy Logic Theory in Steel Metallurgy. . . . 190
R. I. Karimov

Imperfect Knowledge Base Self-organization in Robotic Intelligent Cognitive Control: Quantum Supremacy 196
S. V. Ulyanov, A. V. Shevchenko, A. A. Shevchenko, and A. G. Reshetnikov

Solution of IT Investment Problem Using Fuzzy Logic-Based Approach 205
J. M. Babanli

Application of ELECTRE Method to Decision Making Under Z-number-valued Information 214
K. I. Jabbarova and Aynur I. Jabbarova

Fuzzy Logic Evaluation of Teacher Competencies Through Students Perceptions 221
Saide Sadikoglu and Konul Memmedova

Graph-Based Load Balancing Model for Exascale Computing Systems 229
Araz R. Aliev and Nigar T. Ismayilova

The Impact of Store Environment on Purchase Intention in Supermarkets 237
Gunay Sadikoglu, Nigar E. Adilova, and Promise Isioma Anene

Simulation of Electrodynamic Processes in the Cylindrical-Rectangular Microwave Waveguide Systems Transmitting Information 246
Islam J. Islamov, Mehman H. Hasanov, and Mikael H. Abbasov

Ranking Integration Factors Using Fuzzy TOPSIS Method 254
Tural Suleymanli

Evaluation of HCV Infection Laboratory Test Results Using Machine Learning Methods 261
Hamit Altıparmak, Şerife Kaba, and Meliz Yuvalı

Fuzzy Logic Approach to the Amount of Medication Taken During Breathing with an Inhalator 270
Aynur J. Jabiyeva

Fuzzy Modeling of the Relationship Between Tax Administration Efficiency and Tax Obligations Fulfillment 278
A. F. Musayev and M. Kh. Gazanfarli

Project Selection Under U-Number-Valued Information 286
K. I. Jabbarova

Downscaling Precipitation from GCM Parameters Using Recurrent Neural Networks 294
Vahid Nourani, Fahreddin Sadikoglu, Kasra Khodkar, and Arash Shahli

Using Z-Number-Based Information in Personnel Selection Problem . . . 302
Latafat A. Gardashova and Salman Salmanov

Portfolio Selection Model Using Z-Numbers Theory 308
Leyla Hasanova

Predicting the Mechanical Power of a New-Style Savonius Wind Turbine Using Machine Learning Techniques and Multiple Linear Regression: Comparative Study 316
Youssef Kassem, Hüseyin Çamur, and Mohamed Almojtba Hamid Ali Abdalla

Analysis of Consistency of Pairwise Comparison Matrix with Fuzzy Type-2 Elements 324
Aygul Dadasheva

Analysis and Assessment of Accuracy of the Oil Recovery Factor Calculations by Using the Fuzzy Clustering Algorithm 331
B. N. Koilybayev, M. K. Karazhanova, G. M. Efendiyev, A. S. Strekov, and L. B. Zhetekova

Using Fuzzy Linear Regressions by the Estimation of Expected Changes in the Oil and Gas Saturation's Capacity at Great Depths . . . 340
Adalat B. Hasanov and Mubarize S. Ibrahimli

Fuzzy Control of Mechanical Ventilation System 347
Gorkem Say, Nurullah Akkaya, Ersin Aytac, Sanan Abizada, Tolga Yirtici, Kemal Ruso, Irfan Gunsel, Murat Tuzunkan, and Rahib H. Abiyev

Modeling of Relationship Between the Quality Indicators
of Plastic Details 355
Djahid A. Kerimov

Fuzzy-Multiple Model of Gamified Mobile Application "Smart House"
Based on Bartle Classification of Psychotypes of Players 365
S. Alexeev Anton, A. Batishcheva Galina, A. Kiyaschenko Tatyana,
V. Lukyanova Galina, and A. Ovsyannikov Victor

Decision on Facility Location on Base of Interval Data 371
Kamala Aliyeva

Fuzzy Gain Scheduling Controller for Quadrotor 375
Rahib H. Abiyev, Nurullah Akkaya, Ersin Aytac, and Sanan Abizada

Application of Fuzzy Logic Model for Optimal Solution of Light
Reflection Value in Lighting Calculations 384
Tarlan Abdullayev, Rahib Imamguluyev, and Niyar Umarova

Fuzzy-Multiple Model of Adjustment of the "Smart House"
Mobile Application Under the User Level 392
Ansari Mustafa, Alexander N. Kuzminov, Michael B. Stryukov,
and Oleg A. Ternovsky

Generation of Fuzzy-Chaotic Behavior in Secure Communication
System 401
K. M. Babanli

Exploring of the Problematic Industry 4.0 and Platform-Based
Economic Development 412
Nadirbek Yusupbekov, Shuhrat Gulyamov, Nargiza Usmanova,
and Jamshid Khoshimov

Toward an Analysis of Pairwise Comparison Matrices
with Z-number-valued Elements 420
R. R. Aliyev, O. H. Huseynov, and Javanshir Guliyev

Development of an Automatic Parking Algorithm Based
on Fuzzy Logic 428
A. B. Sultanova

Staff Selection with a Fuzzy Analytical Hierarchy Process in the
Tourism Sector 437
Ahmad Valiyev, Rahib Imamguluyev, and Ilkin Gahramanov

Type-2 Fuzzy Logic Approach for Port Selection 445
Kamala Aliyeva

Developing the PP Balance Model Scale and Fuzzy Logic Evaluation
of the Latent Variable 450
Konul Memmedova and Guldal Kan Şebnem

Application of ANFIS to Predict Adhesive Wear Rate in Different Fiber-Reinforced HDPE Matrix Composite Materials 459
M. B. Babanli and K. H. Ismayilova

Some Aspects of Fuzzy Decision Making in Digital Marketing Analysis 465
Gunay E. Imanova and Gunel Imanova

Modeling of Number of Small and Medium Enterprises in Rostov Region Using a Modified Model Fuzzy Time Series 474
Taras Bogachev, Tamara Alekseychik, Natalia Vovchenko, and Yana Kurinova

Decision Making on Firm's Software Selection with Type-2 Fuzzy Data 480
Nihad Mehdiyev

Mathematical Model of Production and Exchange: Bifurcation Analysis and Management 488
V. Bratishchev Alexander, A. Batishcheva Galina, I. Zhuravleva Maria, and Rogozhin Sergei

Identification of the Multiconnection Uncertainty Dynamic Objects Based on Neural Networks 496
S. M. Jafarov and A. S. Aliyeva

Application of Digital Twin Theory for Improvement of Natural Gas Treatment Unit 505
Nodirbek Yusupbekov, Farukh Adilov, and Arsen Ivanyan

Determination of the Boundaries of Solid Solutions in the $MnTe\text{-}Sb_2Te_3$ and $SnTe\text{-}Sb_2Te_3$ Systems 513
E. N. Orujlu, A. E. Seidzade, A. N. Mammadov, D. B. Tagiev, and M. B. Babanly

Numerical Determination of Gas and Oil Reserves 522
I. Y. Bayramov, A. N. Gurbanov, O. M. Mirzayev, and I. Z. Sardarova

Development of the Geoinformation Space Soft Model Considering Its Intelligent Properties 529
Vladimir V. Khramov

Use of Defuzzification Technique in a Pulse Echo Method for Cable Fault Locating 537
Salahaddin I. Yusifov, Farid H. Aghayev, and Rza S. Safarov

Application of the Fuzzy Cobb-Douglas Model to Predict the Final Product of the Region 545
V. J. Akhundov

Applying Multi-layers Feature Fusion in SSD for Detection of Small-Scale Objects 552
Elbrus Imanov and Zubair Shah

Fuzzy Information Granulation Methodology for Identification of Class-Conscious Speech 560
Farida Huseynova

Possibility Optimal Risk Management in Magistral Gas and Oil Transportation Systems 567
S. A. Agammedova and I. A. Habibov

Modern Approaches to the Formation of Management Methodology in the Conditions of Digital Transformation 575
Akperov Imran Gurru Ogly

Synthesis of the Polymer Bitumen Compositions for Pavement Using Neural Network 580
D. S. Mamed Hasan-zade, A. I. Babayev, and G. S. Hasanov

Assessment of the University Competitiveness in the Paradigm of the Humanistic System Behavior 591
Misir Mardanov, Ramin Rzayev, and Zeynal Jamalov

Fuzzy Model of Estimating the Specific Weight of Shadow Economy Being a Manifestation of Economic Crime on the Basis of the Utility Function 599
Mirali S. Kazimov, Nazim K. Aliyev, and Shahzada G. Madatova

Optimal Regimes of Cooling Processes of Plastic Details 607
Djahid A. Kerimov

The Problem of Measuring Human Capacity Using Ahangyol (Way to Harmony) Methods and the Need for Fuzzy Evaluation 615
Ahmad Gashamoglu

The Use of Mathematical Methods in the Analysis and Forecasting of Stock Indices 621
I. V. Fedorova, L. K. Popova, T. N. Sharypova, and E. N. Lozina

EFL Teachers' Competencies According to Student Opinions 630
Saide Sadıkoğlu, Şahin Akdağ, Murat Tezer, and Rashad Khalilov

On a View of Decision Making Under Z-information 636
Akif V. Alizadeh

Assessment of the Economic Efficiency of Investment Projects Using a Fuzzy Inference System 645
Naila Kalantarly, Inara Rzayeva, and Vasif Hasanov

Stable Algorithms for Solving the Problem of Determining the Weighting Coefficients of Neural Networks with Radial-Basis Activation Functions . . . 654
Husan Igamberdiev, Azizbek Yusupbekov, Uktam Mamirov, and Inomjon Abdukaxxarov

Conversion of Volatile Time Series into a Fuzzy Time Series by the Example of the Dow Jones Index Dynamics . . . 662
Parvin Alizada

Multi-criteria Evaluation of Investment Projects Using the Fuzzy Method of Weighted Maximin Convolution . . . 671
Elchin Aliyev, Ramin Rzayev, and Adila Ali

Investigating of Some Macroeconomic Indicators of the Republic of Azerbaijan in the Post-oil Period by Using Interval Analysis . . . 680
Revana I. Davudova

Problems of Fuzzy – Multiple Control of the "Smart House" System (Review) . . . 689
G. Akperov Imran and Sakharova Lyudmila

A Fuzzy Logic Approach for Evaluating the Effectiveness of the Use of Information Technology in an Industrial Enterprise . . . 697
Nusrat Muradov

Development of Methods for Processing Acoustic Emission Signals of Sensors for the Compressor-Pump Station's Control . . . 704
Valeh I. Bakhshali, Nail T. Mardanov, Aygun A. Bekirova, and Ismayil A. Ismayil

Financial Performance Analysis with Intuitive Fuzzy Logic and Entropy-Based Multi-criteria Decision Making Method . . . 711
Vuqar Salahli and Abil Suleymanov

Selection of Data Science Software by Using Choquet Integral . . . 719
V. H. Salimov

Artificial Intelligence and Fuzzy Set Theory in the Methodology for Studying the Dynamics of the Financial and Economic State 'of the IT Industry . . . 724
N. G. Kuznetsov, N. Rodionova, Kh. Abdurahman, and M. Vatolina

Method for Diagnostics and Selection of the Type of Market Structure of Commodity Markets Based on Fuzzy Set Theory . . . 732
Rena Mikayilova and Ramiz Alekperov

Decision Making on Tourism by Using Natural Language Processing . . . 741
Nigar F. Huseynova

Solving the Problem of Building an Automatic Control System for the Process of Water Chemical Treatment Using Fuzzy Logic 748
K. A. Mammadova and E. N. Aliyeva

Fuzzy Multi-criteria Decision Making Approach for Choosing the Best Subcontractor Company 757
Rashad M. Suleymanli

Determination of Radial Strains in Sealing Elements with Rubber Matrix Based on Fuzzy Sets 765
J. N. Aslanov, A. B. Sultanova, Z. S. Huseynli, and F. F. Mustafayev

Mobile Robot Navigation with Preference-Based Fuzzy Behaviors 774
L. M. Zeinalova and B. O. Jafarov

Author Index 783

Multistage Decision Making in Bimodal Information Environment

R. A. Aliev[1,2,3](✉)

[1] Georgia State University, Atlanta, USA
raliev@asoa.edu.az
[2] Azerbaijan State Oil and Industry University, 20 Azadlig Avenue, AZ1010 Baku, Azerbaijan
[3] Department of Information Systems Engineering, Near East University, Lefkosa, North Cyprus

Abstract. Multistage decision making finds its important applications in various fields. A series of works devoted to solving deterministic, stochastic and fuzzy multistage problems by using dynamic programming exist. It is needed to account for the fact that real-world multistage problems are characterized by partially reliable information. Unfortunately, up to day, no works on multistage decision making problems under bimodal information exist. In this work, we consider multistage problem under Z-number-valued information. States, control actions and transition law are described by Z-number-valued restrictions and relations. An optimal solution is found as one which allows to arrive to a predefined Z-number-valued goal. A numerical example is considered for analysis of validity of the proposed approach.

R. A. Aliev et al. (Eds.): ICSCCW 2021, LNNS 362, p. 1, 2022.
https://doi.org/10.1007/978-3-030-92127-9_1

Interactive Solution of Difficult Choice and Decision Making Problems: Effective and Efficient but not Always Easy

Janusz Kacprzyk(✉)

Systems Research Institute, Polish Academy of Sciences,
Ul. Newelska 6, 01-447 Warsaw, Poland
kacprzyk@ibspan.waw.pl

Abstract. The present world is characterized by a huge and growing complexity of processes and systems involved, amplified by many other characteristic features that even increase the complexity exemplified by the human centricity, time criticality, emergent behaviors, a need for trustworthiness, etc.

Therefore, the most efficient effective and efficient paradigms for the solution of today's complex problems for now, maybe also for the foreseeable future, is not the "full automation", the so-called automated decision making (ADM) but rather the so called quasi-automated decision making (Quasi-ADM) which boils down to a synergistic combination of human and computer capabilities, mainly related to human capabilities to solve "delicate" problems, with a sheer number crunching ability of the computer. This is basically the essence of an interactive problem solving which is advocated by many experts, for instance, Geoff Hinton, one of the founders of deep neural networks.

We consider difficult decision making and choice problems which involve optimization that can be solved either by a strict optimization procedure or a metaheuristic. First, we look at what can make an optimization based problem difficult. We propose an interactive approach via the so-called decision aid paradigm. We have a decision maker, called in this context a judge, who knows his/her area but not necessarily tools and techniques for solving resulting decision making and choice problems. So, the judge commissions an analyst, called here an advisor, who need not know the area but knows how to solve those resulting decision making and choice problems. Clearly, the advisor provides advice, information and suggestions, while the judge makes the final decision.

We discuss various issues related to how the 3 stakeholder (judge, advisor and the pair "judge-advisor") cooperate and collaborate, and in which relations they enter. This is considered, mainly in the context of advice giving (by the advisor) and advice taking (by the judge), and possible different intentions and interests of all parties. In particular, we focus on the so-called advice discounting related to an insufficient use of advice by the judge, unintentionally and intentionally, and analyze various cognitive and psychological reasons, and some ways out. We also discuss some economic aspects. Suggestions for the design and implementation of interactive solution procedure are proposed.

R. A. Aliev et al. (Eds.): ICSCCW 2021, LNNS 362, p. 2, 2022.
https://doi.org/10.1007/978-3-030-92127-9_2

About Valery Borisovich TARASOV (16.02.1955–22.07.2021)

R. A. Aliev[1,2,3](✉) and Alexey Averkin[4,5]

[1] Georgia State University, Atlanta, USA
raliev@asoa.edu.az
[2] Azerbaijan State Oil and Industry University, 20 Azadlig Ave., AZ1010 Baku, Azerbaijan
[3] Department of Information Systems Engineering, Near East University, Lefkosa, North Cyprus
[4] Federal Research Centre of Informatics and Computer Science of RAS, Moscow, Vavilova, 42, Moscow, Russia
[5] Educational and Scientific Laboratory of Artificial Intelligence, Neuro-Technologies and Business Analytics, Plekhanov Russian University of Economics, Moscow, Russia

July 22, 2021 died Valery Borisovich Tarasov, associate professor at Bauman Moscow State Technical University, Candidate of Technical Sciences, Associate Professor.

He was one of the leading national experts in the field of AI, distinguished by his encyclopedic knowledge in this field. Valeriy Borisovich obtained fundamental results in the field of AI methodology, system approach in AI and synergetic AI, semiotic modeling, agent theory and multi-agent systems, fuzzy sets theory, soft computing and measurement. He actively developed and implemented such areas as intelligent systems of computer-aided design and combined development, computer-integrated manufacturing and virtual enterprises, organizational design, enterprise modeling and re-engineering, knowledge management, methods for creating intelligent productions, networked and intelligent enterprises. He has developed a number of innovations in the field of engineering education and professional training, primarily on the basis of intelligent and networked technologies, open education and virtual departments.

Valery Borisovich was born in Moscow on February 16, 1955. Graduated in 1978 from Bauman Moscow Higher Technical School (now Bauman Moscow State Technical University). Since then he had been working at MVTU for more than 40 years, where he rose from a postgraduate and junior research assistant to a Candidate of Science, Associate Professor, Deputy Head of the Department of Computer Systems of Industrial Automation.

B. Mr. Tarasov participated in the founding congress of Soviet Association for Artificial Intelligence (now RAIA). Was one of its founders. From 1992–2000 he was Vice-President of Association, from 2000 - Vice-President. - From 1992 till 2000 Mr. Tarasov was Vice-president of the Association and from 2000 - member of Scientific Council of RAIA. V. Tarasov was also one of the founders of Russian Fuzzy Systems Association (RANS), a subsidiary of RAII, at the Founding Congress of RANS in Kazan in 1991. He was also one of the initiators of the 2005 congress in Kolomna, when RANS was reorganized to become the Russian Association of Fuzzy Systems and Soft Computing (RANSMV). From 2019 to 2021 V. B. Tarasov was president of this association for one year. Thus, he can rightfully be counted among the cohort of "founding fathers" of these Russian scientific communities. In the first two decades of the 21st century V. B. Tarasov

R. A. Aliev et al. (Eds.): ICSCCW 2021, LNNS 362, pp. 3–5, 2022.
https://doi.org/10.1007/978-3-030-92127-9_3

initiated and actively participated in the organization of the representative scientific conferences: "Integrated Models and Soft Computing in Artificial Intelligence" (IMMV) in Kolomna (10 conferences organized from 2001 to 2021), "Intellectual Systems and Computer-Integrated Production" (ISICP) in MSTU named after N. E. Bauman (Moscow State Technical University, Moscow). N.E. Bauman Moscow State Technical University (3 conferences were held), "Hybrid and Synergetic Intelligent Systems" (since 2012, 4 conferences were held).

B. B. Tarasov was Academician of International Academy of System Research, member of Central House of Scientists of RAS and Corresponding Member of International Academy of Informatization. Since 1992 he was a member of the editorial board, and since 1998 he has been executive secretary of the editorial board of the Artificial Intelligence News and a member of the editorial board of the Artificial Intelligence and Decision Making journals, Software Products and Systems, and Soft Measurements and Computing. He initiated the book series "Artificial Sciences" and was the executive secretary of its editorial board. He was a laureate of the 2002 RAIA Prize for his monograph "From Multi-Agent Systems to Intelligent Organizations".

His activity in the field of science popularization is worth mentioning separately. He repeatedly spoke at the "Pospelov Readings", at the Summer Schools on AI, at the "Cybernetics" section of the CDU, with brilliant lectures, dedicated to the review of scientific achievements and unique biographical facts about the life of D. A. Pospelov, G. S. Pospelov and L. Zadeh, with whom he maintained friendly relations for many years. His lectures and articles on the history of development of systemology, theory of fuzzy sets and artificial intelligence in the USSR and Russia, describing the stages of life and scientific heritage of outstanding Russian scientists - A. A. Bogdanov, E. V. Popov, V. V. The authors are also grateful to the authors of the book.

B. B. Tarasov has significantly contributed to the development of international scientific cooperation. From 1995 till 1997 he worked as a visiting professor at the University of Valenciennes and Heno-Cambresi (France). He has repeatedly presented papers at international conferences and symposiums in the USA, France, Belgium, Germany, Mexico, Brazil, Turkey, Czech Republic, Slovakia, Uzbekistan, Azerbaijan and other countries.

He was the author of more than 250 scientific works, including monographs "Fuzzy sets in control models and artificial intelligence" (co-authors: A. N. Averkin, I. Z. Batyrshin, A. F. Blishun, V. B. Silov), "Intellectual learning systems and virtual learning organizations" (co-authors: N. A. Golenkov, N. A. Gulkina, V. A. Emelyanov, etc.). (co-authors: V. V. Golenkov, N. A. Gulyakina, V. V. Emelyanov, etc.), "From Multi-Agent Systems to Intelligent Organizations", "Fuzzy Hybrid Systems. Theory and Practice" (co-authors Batyrshin I. Z., Nedosekin A. O., Stetsko A. A., Yazenin A. V., Yarushkina N.G.), "Methods and languages for ontological modeling" (co-authors: G.S. Plesnevich, B.S. Karabekov, Nguyen Thi Minh Vu).

Valery Borisovich's life style was accurately and succinctly characterized by two words: "collector" and "traveler. A collector of books (he had a very large home library), a collector of "scientific lands," who did not recognize rigid boundaries between different scientific disciplines and sought to find "flowers on the neutral strip," a collector of people of science at conferences and symposia, where there was always a welcoming and friendly atmosphere inherent in his spirit.

And an easy-going traveler, both literally and figuratively. For him it was necessary and extremely exciting to travel from one scientific area to another: from engineering to psychology, from psychology to information technology and artificial intelligence, from computer science, knowledge engineering and network theory - to economics, and from there back to the new technologies of the fourth industrial revolution, embodied the ideas of hybrid and synergetic intelligence. And his trips as a speaker to national and international conferences: from Kaliningrad to Vladivostok, from Lake Baikal to Smolensk and Sochi, from Rostov-on-Don and Taganrog to Kazan and Ulyanovsk, from Tashkent to Cancun, from Budapest and Bratislava to Lisbon and Rio de Janeiro for a new charge of creative energy.

In the nearest plans was a trip to Turkey with a plenary report at the 11th International Conference on Theory and Applications of Soft Computing, Computing with Words, Comprehension and Artificial Intelligence in August this year to Prof. R.A. Aliev at 11th INTERNATIONAL CONFERENCE ON THEORY AND APPLICATION OF SOFT COMPUTING, COMPUTING WITH WORDS, PERCEPTION AND ARTIFICIAL INTELLIGENCE).

Bright memory of Valery Borisovich Tarasov, a brilliant scientist, unique erudite, cheerful, kind and hospitable man will forever remain in our hearts.

Designing and Evaluating Interpretable Rule-Based Architecture Under Privacy Constraints: A Framework of Granular Computing

Witold Pedrycz(✉)

Department of Electrical and Computer Engineering, University of Alberta, Edmonton, Canada
wpedrycz@ualberta.ca

Abstract. In data analytics, system modeling, and decision-making models, the aspects of interpretability and explainability are of paramount relevance as emphasized in numerous studies on explainable Artificial Intelligence (XAI). Those requirements are especially timely when the design of models has to be realized when considering strict requirements of privacy and security.

We advocate that to efficiently address these challenges, it becomes beneficial to engage the fundamental framework of Granular Computing. It is demonstrated that a conceptualization of information granules can be conveniently carried out with the use of information granules (for example, fuzzy sets, sets, rough sets, and alike).

We cover a comprehensive discussion of information granules-oriented design of rule-based architectures. A way of forming condition parts of the rules through unsupervised federated learning is discussed along with algorithmic developments. Strategies of joint and separate learning of condition parts and conclusion parts are outlined. A granular characterization of the model formed by the server vis-a-vis data located at individual clients is presented. It is demonstrated that the quality of the rules at the client's end is described in terms of granular parameters and subsequently the global model becomes represented as a granular model with parameters in the form of information granules of type-2.

R. A. Aliev et al. (Eds.): ICSCCW 2021, LNNS 362, p. 6, 2022.
https://doi.org/10.1007/978-3-030-92127-9_4

Fuzzy Classification-Based Alloy Selection from a Large Dataset

M. B. Babanli(✉)

Azerbaijan State Oil and Industry University, 20 Azadlig Avenue, AZ1010 Baku, Azerbaijan

Abstract. Nowadays, large datasets of alloys designed for various purposes exist. This allows using computer-guided methods for alloy synthesis and selection instead of costly experiments. Modern alloy selection problems are characterized by a large space of alternatives with complex properties. An application of classical methods to such problems may not be adequate. In turn, fuzzy logic-based methods may be effective due to good abilities of summarization and robustness to imprecision.

In this work, fuzzy classification-based alloy selection from a complex dataset is considered. This allows partitioning of a large space of alternatives to typical classes. An optimal alloy may then be found within most relevant classes. Such technique reduces search space under well interpretability and provide satisfactory results. An example of an alloy selection from a large dataset of alloy mechanical properties is used to illustrate effectiveness of the approach.

Keywords: Alloy selection · Decision making · Fuzzy set · Fuzzy classification · Mechanical properties

1 Introduction

Modern alloy selection problems are characterized by a large space of alternatives with complex properties. The reason is that numerous alloys designed for various purposes exist. Thus, computer-guided methods for alloy selection from large datasets may alternate costly experiments. For such complex problems, an application of classical methods may not be adequate. In turn, fuzzy logic and soft computing-based methods may be effective due to good abilities of summarization and robustness to imprecision [1–5]. An application of fuzzy logic and soft computing techniques in alloys modelling and decision making is considered in [3, 4, 6–14]. In [3] they apply fuzzy logic to describe relationship between chemical composition and yield strength for A356 alloys. The obtained results are close to those of application of neural networks. Modeling of surface roughness is considered in [4]. [10] is devoted to application of fuzzy sets and TOPSIS method for selection of an optimal alloy in corrosive environment. Book [8] is devoted to an application of fuzzy logic in alloy selection and design. An overview of existing works in this realm is given. Solving of various problems by using FL are considered.

R. A. Aliev et al. (Eds.): ICSCCW 2021, LNNS 362, pp. 7–13, 2022.
https://doi.org/10.1007/978-3-030-92127-9_5

One of the directions of fuzzy logic related to data summarization is fuzzy classification [11, 15–19]. Fuzzy classification allows to partition complex datasets into interpretable fuzzy classes. Further, related linguistic interpretations help a user to reason with this information. Indeed, decision makers (DMs) often use linguistic terms to describe desired values of alloy properties for practical problems. Such terms are soft constraints used to reflect degrees to which alloys satisfy desired requirements. In this work, fuzzy classification-based alloy selection from a complex dataset is considered. This allows partitioning of a large space of alternatives to typical classes by using linguistic evaluation of properties values. Optimal alloys can then be considered as those belonging to the best classes. Such technique is characterized by low computational complexity and good explanation capability. An example of an alloy selection from a large dataset of alloy mechanical properties is used to illustrate effectiveness of the approach.

2 Problem Definition

Let us consider a problem of decision making on alloy selection from a large data set. Assume the following data set of alloys properties is given:

Table 1. A large dataset on alloy properties

Alloy	Property P_1	Property P_2	...	Property P_n
a_1	a_{11}	a_{12}	...	a_{1n}
.	.	.	...	.
.	.	.		.
.	.	.		.
a_m	a_{m1}	a_{m2}	...	a_{mn}

In Table 1, $a_{il}, i = 1, ..., m, l = 1, ..., n$ denotes a value of property l for alloy i.

Due to the complexity of the dataset, let us formulate the considered problem as a fuzzy classification problem. The set of alternatives a_i (alloys) is partitioned to fuzzy classes $C_k,\ k = 1, ..., K$, according to fuzzy values of properties $P_1, P_2, ..., P_m$. Without loss of generality, assume that the lower index k is used for a better class and C_k, and class C_K contains infeasible alternatives. An optimal alternative a^* is determined one with a maximum membership degree to the best class C^* (non-empty class with the lowest index):

$$\mu_{C^*}(a^*) = \max_{i=1,...,m} \mu_{C^*}(a_i)$$

In the next section we outline an approach to solving this problem.

3 Solution Approach

The problem is solved as follows.

Stage 1. Forming a fuzzy partition of properties values. For the sake of simplicity, the following type of a fuzzy partition for the material properties is used:

$$T_1 = Low(L) = (0, 0, 0.5),\ T_2 = Medium(M) = (0, 0.5, 1),$$
$$T_3 = High(H) = (0.5, 1, 1).$$

On this basis, the following fuzzy classes may be formed:

Class 1. The alloys with the values of the most important properties P_l described by term "H" are considered.
Class 2. The alloys with all properties evaluated as at least "M" (excluding those in Class 1).
Class 3: The alloys with at least one property evaluated as "Low".

As feasible alternatives, we consider those assigned to Class 1 and Class 2. Each class C_k includes cases:

$$Case_{ks} = (T_1, T_2, T_3), s = 1, ..., S_k; T_1, T_2, T_3 \in \{L, M, H\}.$$

For example, case (*H,L,M*) implies that value of P_1 is H, value of P_2 is L, value of P_3 is M. A general description of fuzzy classes structure is given in Table 2.

Table 2. Fuzzy classes structure description

			Most important properties			Less Important Properties		
	Class	Case	P_1	...	P_j	P_{j+1}	...	P_n
Feasible	C_1	$Case_{11}$	*H*	...	*H*	*M*	...	*M*
		...	...	...	...	...	...	...
		$Case_{1S_1}$	*H*	...	*H*	*H*	...	*M*
	C_2	$Case_{21}$	*M*	...	*M*	*M*	...	*H*
		...	...	...	...	...	...	...
		$Case_{2S_2}$	*M*	...	*H*	*H*	...	*M*
Infeasible	C_3	$Case_{11}$	*M*	...	*L*	*L*	...	*H*
		...	...	...	...	...	...	...
		$Case_{1S_3}$	*L*	...	*M*	*M*	...	*M*

Stage 2. A membership degree of a_i to C_k is computed as a maximum of membership degrees $\mu_{Case_{ks}}(a_i)$, $s = 1, ..., S_k$ to all the cases:

$$\mu_{Class\,k}(a_i) = \max_{s=1,...,S_k} \mu_{Case_{ks}}(a_i). \tag{1}$$

$\mu_{Case_{ks}}(a_i)$ is a minimum of membership degrees to the terms of $Case_{ks}$:

$$\mu_{Case_{ks}}(a_i) = \min_{l=1,\ldots,3} \mu_T^l(a_i),$$

where μ_T^l is a membership function of a term for property l.

Stage 3. For each a_i, its class is determined as one to which it belongs more than to the others:

Determine $C_{k'}$, such that

$$\mu_{C_{k'}}(a_i) = \max_{k=1,\ldots,3} \mu_{C_k}(a_i) \quad (2)$$

Stage 4. An optimal alternative a^* is found as one with the highest membership degree to the best class:

Determine a^*, such that

$$\mu_{C^*}(a^*) = \max_i \mu_{C^*}(a_i) \quad (3)$$

$$\text{s.t. } a_i \in \bigcup_{k=1,\ldots,K-1} C_k \quad (4)$$

Constraint (4) describes feasibility condition.

4 An Example. Alloy Selection Based on Mechanical Properties

Consider a choice of an optimal alloy from a large dataset containing information on alloys mechanical properties (Table 3).

Table 3. A fragment of an alloy data set

Alloy	Yield stress (MPa), P_1	Fatigue strength (MPa), P_2	Elongation %, P_3
a_1	407	776	59
a_2	338	415	20
a_3	140	380	45
. . .	...	...	...
a_m	255	470	28

Let us use the described fuzzy classification-based technique for solving of the problem.

At Stage 1, we form fuzzy classes considering that P_1, P_3 are the most important properties as it is described below.

Class 1. The alloys with properties P_1, P_3 evaluated as "H" are considered. This includes two cases:

$$Case_{11} = (H,\ H,\ H),\ Case_{12} = (H,\ M,\ H).$$

Class 2. The alloys with all properties evaluated as at least "M" are considered (excluding those in Class 1). This class includes 6 cases:

$$Case_{21} = (M,\ M,\ M),\ ....,\ Case_{26} = (H,\ H, M).$$

Class 3. The alloys with at least one property evaluated as "Low". 19 cases are included.

At Stage 2, we compute membership degrees of the alternatives to the classes. For example, consider a_2. The membership degrees to the fuzzy terms of the properties:

$$\mu_L^1(a_2) = 0,\ \mu_M^1(a_2) = 0.48,\ \mu_H^1(a_2) = 0.52;$$
$$\mu_L^2(a_2) = 0.39,\ \mu_M^2(a_2) = 0.61,\ \mu_H^2(a_2) = 1;$$
$$\mu_L^3(a_2) = 0.47,\ \mu_M^3(a_2) = 0.53,\ \mu_H^3(a_2) = 0.$$

Thus, the membership degrees to the cases of Class 2 are as follows:

$$\mu_{Case\,21}(a_2) = \min(\mu_M^1(a_2), \mu_M^2(a_2), \mu_M^3(a_2)) = 0.48,$$

.

.

.

$$\mu_{Case26}(a_2) = \min(\mu_H^1(a_2), \mu_H^2(a_2), \mu_M^3(a_2)) = 0.53.$$

The membership degree of a_2 to Class 2 computed by using (1):

$$\mu_{C_2}(a_2) = \max_{s=1,...,6} \mu_{Case_{2s}}(a_2) = \max(\mu_{Case_{21}}(a_2), ..., \mu_{Case_{26}}(a_2)) = 0.53$$

Analogously, we computed the membership degree of a_2 to Class 1:

$$\mu_{C_1}(a_2) = 0.$$

At Stage 3, we assign the alloys to the classes, see (2). For example, the class of a_2 is Class 2:

$$\mu_{C_{k'}}(a_2) = \max_{k=1,...,3} \mu_{C_k}(a_i) = \mu_{C_2}(a_2) = 0.53$$

The feasible alternatives assigned to fuzzy classes are shown in Table 4:

Table 4. Alternatives assigned to the fuzzy classes

a_i	μ_{C_1}	μ_{C_2}	Class
a_1	1.00	0.00	Class 1
a_2	0.00	0.52	Class 2
a_6	0.00	0.54	Class 2
a_8	0.09	0.53	Class 2
a_{11}	0.00	0.83	Class 2

At Stage 4, we determine an optimal alloy by using (3)–(4). An optimal alternative is a_1 as it is an only one that belongs to the best class (Class 1):

$$\mu_{C_1}(a_1) = 1.$$

Thus, a_1 dominates all the other alternatives.

5 Conclusion

Alloy selection problems may be characterized by a large space of alternatives and complex choice criteria. In this work, we proposed to use fuzzy classification technique to cope with these complexities. This allows to obtain a satisfactory solution under low computational complexity. An example of alloy selection based on mechanical properties from a large dataset is considered. By using fuzzy partition of the properties values, three fuzzy classes of alloys are formed. The best class is considered as a class of alloys with high values of the most important properties. An optimal alloy is found as one with the highest membership degree to the best class.

References

1. Zadeh, L.A.: Fuzzy sets. Inf. Control **8**, 338–353 (1965). https://doi.org/10.1016/S0019-9958(65)90241-X
2. Aliev, R.A., Fazlollahi, B., Aliev, R.R.: Soft Computing and Its Applications in Business and Economics. Springer, Heidelberg (2004). https://doi.org/10.1007/978-3-540-44429-9
3. Tarasov, V., Tan, H., Jarfors, A.E.W., Seifeddine, S.: Fuzzy logic-based modelling of yield strength of as-cast A356 alloy. Neural Comput. Appl. **32**(10), 5833–5844 (2019). https://doi.org/10.1007/s00521-019-04056-5
4. Barzani, M.M., Zalnezhad, E., Sarhan, A.A., Farahany, S., Ramesh, S.: Fuzzy logic based model for predicting surface roughness of machined Al–Si–Cu–Fe die casting alloy using different additives-turning. Measurement **61**, 150–161 (2015). https://doi.org/10.1016/j.measurement.2014.10.003
5. Babanli, M.B., Babanli, J.M.: Fuzzy decision method based on Zadeh's data aggregation approach. In: 14th International Conference on Theory and Application of Fuzzy Systems and Soft Computing. Springer 1306, Cham (2021). https://doi.org/10.1007/978-3-030-64058-3_5

6. Babanli, M.B.: Fuzzy Logic and Fuzzy Expert System-Based Material Synthesis Methods (2019). IntechOpen. https://www.intechopen.com/online-first/fuzzy-logic-and-fuzzy-expert-system-based-materialsynthesis-methods. https://doi.org/10.5772/intechopen.84493
7. Babanli, M.B., Huseynov, V.M.: Z-number-based alloy selection problem. Procedia Comput. Sci. **102**, 183–189 (2016). https://doi.org/10.1016/j.procs.2016.09.387
8. Babanli, M.B.: Fuzzy Logic-Based Material Selection and Synthesis. World Scientific, Singapore (2019). https://doi.org/10.1142/11164
9. Babanli, M.B.: Artificial ıntelligence-based new material design. In: 10th International Conference on Theory and Application of Soft Computing, Computing with Words and Perceptions. Springer 1095, Cham (2020). https://doi.org/10.1007/978-3-030-35249-3_2
10. Bansod, A.V., Patil, A.P., Kalita, K., Deshmukh, B.D., Khobragade, N.: Fuzzy multicriteria decision-making-based optimal Zn–Al alloy selection in corrosive environment. Int. J. Mater. Res. **111**, 953–963 (2020). https://doi.org/10.3139/146.111957
11. Scherer, R.: Multiple Fuzzy Classification Systems. Springer, Heidelberg (2012). https://doi.org/10.1007/978-3-642-30604-4
12. Chen, M.-Y., Wang, Q., Yang, Y.: A hybrid knowledge-based neural-fuzzy network model with application to alloy property prediction. In: Liu, D., Fei, S., Hou, Z.-G., Zhang, H., Sun, C. (eds.) ISNN 2007. LNCS, vol. 4491, pp. 528–535. Springer, Heidelberg (2007). https://doi.org/10.1007/978-3-540-72383-7_63
13. Lesiak, B., Zemek, J., Jiricek, P., Jóźwik, A.: Investigation of CoPd alloys by XPS and EPES using the pattern recognition method. J. Alloy. Compd. **428**, 190–196 (2007). https://doi.org/10.1016/j.jallcom.2006.01.108
14. Kluska-Nawarecka, S., Górny, Z., Mrzygłód, B., Wilk-Kołodziejczyk, D.: Methods of development fuzzy logic driven decision support models in copper alloys processing. Arch. Foundry Eng. **10**, 23–28 (2010)
15. Roubos, J., Setnes, M., Abonyi, J.: Learning fuzzy classification rules from data. Inf. Sci. **150**, 77–93 (2003). https://doi.org/10.1016/S0020-0255(02)00369-9
16. Kuncheva, L.I.: Fuzzy Classifier Design. Studies in Fuzziness and Soft Computing, Heidelberg (2000). https://doi.org/10.1007/978-3-7908-1850-5
17. Friedman, M., Kandel, A.: Introduction to Pattern Recognition. Series in Machine Perception and Artificial Intelligence. Springer, World Scientific (1999). https://doi.org/10.1142/3641
18. Puchkov, A., Dli, M., Kireyenkova, M.: Fuzzy classification on the base of convolutional neural networks. In: Hu, Z., Petoukhov, S.V., He, M. (eds.) AIMEE2018 2018. AISC, vol. 902, pp. 379–391. Springer, Cham (2020). https://doi.org/10.1007/978-3-030-12082-5_35
19. Ravikumar, P., Susheela, D.V.: Fuzzy classification of time series data. In: International Conference on Fuzzy Systems, Hyderabad. pp. 1–6 (2013). https://doi.org/10.1109/FUZZ-IEEE.2013.6622571

Consistent Z-preferences Formulation in Decision Analysis

Rafig R. Aliyev(✉)

Research Laboratory of Intelligent Control and Decision-Making Systems in Industry and Economics, Azerbaijan State Oil, and Industry University, Azadlyg Avenue, 20, AZ1010 Baku, Azerbaijan

Abstract. Comprehensive critical analysis of existing decision theories and tools has shown that effective research results on decision making methods under imprecise and partially reliable decision information are very scarce. In particular, in scientific literature there are no works on formulation of more adequate decision preferences characterized by bimodal information, i.e. information containing synergy of fuzzy and probabilistic uncertainties. The concept of Z-number - the Z-extension of fuzzy logic is a formal construct to describe such kind of information. In this paper we try to create method for construction of consistent and adequate preferences of a DM characterized by imprecise and partially reliable knowledge.

Keywords: Z-number · Z-valued PCM · Consistency-driven preferences · Consistency index

1 Introduction

Analyzing the famous existing decision theories [1, 2] we arrived at conclusion that these theories are developed for a decision environment characterized by well-described information on alternatives, states of nature, probabilities and outcomes.

It is very well known that validity and effectiveness of decision analysis mainly is related with consistent formulation of decision maker's (DM) preferences. In real-world problems, DM's knowledge is inherently associated with imprecision and partial reliability. This involves combination of fuzzy and probabilistic information. The concept of a Z-number is a formal construct to describe such kind of information. In this study, we formulate the concepts of Z-number-valued eigenvalue and eigenvector for matrices components of which are Z-numbers. Consequently, there is need to investigate consistency of a DM's preference knowledge which is related to eigenvalues and eigenvectors of decision matrices.

The notion of consistency is used to estimate the quality of preference knowledge and its stability for reliable evaluation of decision alternatives. In the famous AHP method there is a set of strict consistency conditions used to keep the rationality of preference intensities between compared elements. These requirements are not achievable in the real situations when DM has limited rationality and partially reliable preferences. A new

R. A. Aliev et al. (Eds.): ICSCCW 2021, LNNS 362, pp. 14–18, 2022.
https://doi.org/10.1007/978-3-030-92127-9_6

approach to deriving consistency-driven preference degrees for such kind of situations is a research problem.

Construction of consistency-driven decision preferences, characterized by imprecision and partial reliability by using eigensolutions of preference matrices.

The paper is structured as follows. In Sect. 2, necessary definitions including those of a Z-number, a Z-number valued pairwise comparisons matrix and others are provided. Section 3 is devoted to construction of consistency driven preferences of DM. The numerical example for the considered problem is described in Sect. 4. Section 5 is conclusion.

2 Z-Preference Related Some Preliminary Information

Definition 1. ***Continuous Z-number*** [3–5]. A continuous Z-number is an ordered pair $Z = (A, B)$ where A is a continuous fuzzy number playing a role of a fuzzy constraint on values that a random variable X may take:

$$X \text{ is } A,$$

B is a continuous fuzzy number with a membership function $\mu_B : [0, 1] \to [0, 1]$, $[0, 1]$, playing a role of a fuzzy constraint on the probability measure of A:

$$P(A) = \int_R \mu_A(x)p(x)dx \text{ is } B.$$

Definition 2. ***A Discrete Z-number*** [3, 4]. A discrete Z-number is an ordered pair $Z = (A, B)$ where A is a discrete fuzzy number which describes a fuzzy constraint on values that a random variable X may take – "X *is* A", and B is a discrete fuzzy number with a membership function $\mu_B : \{b_1, \ldots, b_n\} \to [0, 1]$, $\{b_1, \ldots, b_n\} \subset [0, 1]$, which describes a fuzzy constraint on the probability measure of A :

$$P(A) \text{ is } B$$

Definition 3. ***Ranking of Z-numbers*** [6]. For Z-numbers Z, Z' it holds:

$$Z \le Z'Z \le Z' \text{ iff } D(Z, (1, 1)) \ge D(Z', (1, 1))$$

where D is distance defined above, (1,1) is a fuzzy singletons-based Z-number.

One can easily prove that $\le\le$ is a partial order as it poses the following properties:

$Z \le Z,\ Z \le Z$ (reflexivity)

if $Z \le Z'\ Z \le Z'$ and $Z' \le Z\ Z' \le Z$ then $Z = Z'$ (antisimmetry)

if $Z \le Z'\ Z \le Z'$ and $Z' \le Z''\ Z' \le Z''$ then $Z \le Z''\ Z \le Z''$ (transitivity).

pij = p11...p1n.....pn1...pnn

Definition 4. ***Z-valued PCM*** [6]. A Z-number-valued PCM (Z_{ij}) Zij is a square matrix of Z-numbers:

$$(Z_{ij} = (A_{ij}, B_{ij})) = \begin{pmatrix} Z_{11} = (A_{11}, B_{11}) \; ... \; Z_{1n} = (A_{1n}, B_{1n}) \\ . \quad ... \quad . \\ Z_{n1} = (A_{n1}, B_{n1}) \; ... \; Z_{nn} = (A_{nn}, B_{nn}) \end{pmatrix} \text{Xij, i, j} = 1, \ldots, \text{n}.$$

A Z-number $Z_{ij} = (A_{ij}, B_{ij}),\ \ i, j = 1, ..., n$ describes partially reliable information on degree of preference for *i-th* alternative (criterion) against *j*-th one.

Definition 5. ***An inconsistency Index for Z-number-valued PCM*** [7]. An inconsistency index K for Z-number-valued PCM (Z_{ij}) is defined as follows:

$$K((Z_{ij})) = \max_{i<j<k} \min \left\{ D\left(Z(1), \left(\frac{z_{ik}}{z_{ij} z_{jk}} \right) \right) D\left(Z(1), \left(\frac{z_{ij} z_{jk}}{z_{ik}} \right) \right) \right\}, \tag{1}$$

3 Consistency-Driven Preferences

Decision making is based on preferences over alternatives and choice criteria. A DM's preference may be formally described by a pairwise comparison matrix (PCM) (a_{ij}), where an a_{ij} denotes a degree to which an *i-th* alternative (criterion) is preferred to *j*-th one. Natural conditions used for a_{ij} are $a_{ii} = 1$ and $a_{ji} = 1/a_{ij}$ (reciprocity), $\forall i, j = 1, ..., n$. Traditionally, consistency of (a_{ij}) is based on multiplicative transitivity condition (though different constructs are also used):

$$a_{ij} a_{jk} = a_{ik}, \forall i, j, k$$

Let us consider a problem of generation of consistent PCM (Z'_{ij}) most similar to a given inconsistent PCM (Z_{ij}). The elements of inconsistent Z-matrix (Z_{ij}) will be considered as a perturbation of the elements of matrix (Z'_{ij}) for which reciprocity and consistency are verified. We have to change elements of (Z_{ij}) in order to arrive at (Z'_{ij}). The problem is formulated as follows.

$$J = \sum_{i=1}^{n} \sum_{j=1}^{n} D(Z_{ij}, Z'_{ij}) \to \max \tag{2}$$

s.t. multiplicative reciprocity:

$$Z'_{ij} Z'_{ji} = Z(1) \tag{3}$$

multiplicative transitivity:

$$Z'_{ij} Z'_{jk} = Z'_{ik} \tag{4}$$

non-negativity:

$$Z'_{ij} \geq Z(0), \quad i, j = 1, \ldots, n \tag{5}$$

Problem (2–5) is a non-linear optimization problem characterized by fuzzy and probabilistic uncertainties. Taking into account these features, it is needed to develop a solution approach relying on differential evolution (DE) optimization technique [8].

4 Example

Assume that information on DM preferences can be formalized by a 3×3 matrix of Z-numbers with TFNs-based components:

$$\begin{pmatrix} Z_{11} = ((0.93, 0.95, 0.97), (0.95, 0.98, 1)) & Z_{12} = ((0.327, 0.333, 0.34), (0.7, 0.8, 0.9)) & Z_{13} = ((3.92, 4, 4.08), (0.7, 0.8, 0.9)) \\ Z_{21} = ((2.94, 3, 3.06), (0.7, 0.8, 0.9)) & Z_{22} = ((0.93, 0.95, 0.97), (0.95, 0.98, 1)) & Z_{23} = ((8.82, 9, 9.02), (0.6, 0.7, 0.8)) \\ Z_{31} = ((0.245, 0.25, 0.255), (0.7, 0.8, 0.9)) & Z_{32} = ((0.1108, 0.111, 0.113), (0.6, 0.7, 0.8)) & Z_{33} = ((0.93, 0.95, 0.97), (0.95, 0.98, 1)) \end{pmatrix}$$

Zij denotes a Z-valued degree to which the *i*-th criterion is preferred to the *j*-th one. For example, $Z_{21} = ((2.94, 3, 3.06), (0.7, 0.8, 0.9))$ **Error! Digit expected.** is a Z-valued degree to which C_2 is preferred to C_1C1. The value of computed inconsistency index K in accordance with (1) for the considered matrix (Z_{ij}) is $K((Z_{ij})) = 0.31$. Let us consider extraction of consistent Z-valued matrix (Z'_{ij}) closest to the initial one. For this purpose, optimization problem (2–5) will be solved by using the proposed method.

Optimal Z-valued matrix $(Z'_{ij}) = MP(V'_{best})$ is retrieved:

$$\begin{pmatrix} Z_{11} = ((1.000761, 1.002326, 1.002326), (0.9996, 0.9998, 0.9998)) & Z_{12} = ((0.402933, 0.402933, 0.402935), (0.72, 0.995931, 0.995931)) & Z_{13} = ((3.650363, 3.651220, 3.651235), (0.996759, 0.997332, 0.997332)) \\ Z_{21} = ((2.482609, 2.482835, 2.482835), (0.707104, 0.707104, 0.991447)) & Z_{22} = ((0.995452, 1, 1), (0.95, 0.98, 1)) & Z_{23} = ((9.060410, 9.060411, 9.064086), (0.994743, 0.994743, 0.995316)) \\ Z_{31} = ((0.273967, 0.273967, 0.273967), (0.007870, 0.008223, 0.502359)) & Z_{32} = ((0.110189, 0.110424, 0.110424), (0.499450, 0.999329, 0.999329)) & Z_{33} = ((0.998922, 1.002763, 1.002763), (0.99867, 0.99986, 0.999886)) \end{pmatrix}$$

At the final step, we have to verify whether the value of K for the obtained (Z'_{ij}) exceeds a predefined threshold $\theta_K = 0.1$. The computed value of K is $K((Z'_{ij})) = 0.003$ which does not exceed θ_K. Thus, the obtained matrix can be considered as consistent.

5 Conclusion

Decision making is based on preferences over alternatives and choice criteria. The notion of consistency in this case is used to estimate the quality of preference knowledge and its stability for reliable evaluation of decision alternatives. Existing works have a set of strict consistence conditions used to keep the rationality of preference intensities between compared elements. In this study we have proposed an approach to deriving consistency-driven preference for such kind of situation. A preference degree is described by a Z-number to reflect imprecision and partial reliability of preference knowledge. An optimization problem with Z-number-valued variables is used to design consistent preferences.

References

1. Neumann, J.V., Morgenstern, O.: Theory of Games and Economic Behaviour. Princeton University Press, USA (1944). https://doi.org/10.2307/2280142
2. Kahneman, D., Tversky, A.: Prospect theory: an analysis of decision under uncertainty. Econometrica **47**, 263–291 (1979). https://doi.org/10.1142/9789814417358_0006
3. Zadeh, L.A.: A note on Z-numbers. Inform. Sci. **181**(14), 2923–2932 (2011). https://doi.org/10.1016/j.ins.2011.02.022
4. Aliev, R.A., Huseynov, O.H., Aliyev, R.R., Alizadeh, A.V.: The Arithmetic of Z-Numbers: Theory and Applications, World Scientific, Singapore (2015). https://doi.org/10.1142/9575
5. Aliev, R.A., Huseynov, O.H., Zeinalova, L.M.: The arithmetic of continuous Z-numbers. Inform Sci. **373**, 441–460 (2016). https://doi.org/10.1016/j.ins.2016.08.078
6. Aliev, R.A., Guirimov, B.G., Huseynov, O.H., Aliyev, R.R.: Z-relation equation-based decision making. Expert Syst. Appl. **184**, 115387 (2021). https://doi.org/10.1016/j.eswa.2021.115387
7. Brunelli, M.: A survey of inconsistency indices for pairwise comparisons. Int. J. General Syst. **47**(8), 751–771 (2018). https://doi.org/10.1080/03081079.2018.1523156
8. Storn, R., Price, K.: Differential Evolution - A simple and efficient adaptive scheme for global optimization over continuous spaces. J. Global Optim. **11**, 341–359 (1997)

New High Step-Up DC-DC Converter in PV System: Performance and Analysis

Senol Bektas[1], Farzad Mohammadzadeh Shahir[2(✉)], and Ebrahim Babaei[2]

[1] Engineering Faculty, Near East University, Nicosia, North Cyprus, Mersin 10, 99138 Mersin, Turkey
senol.bektas@neu.esu.tr

[2] Faculty of Electrical and Computer Engineering, University of Tabriz, Tabriz, Iran
e-babaei@tabrizu.ac.ir

Abstract. These days, solar energy can replace fossil fuels in many energy applications. One of the most important advantages of this type of new energy is no air pollution and their lack of completion. Using Photovoltaic (PV) can be one solution. DC-DC converters have main role in PV systems where the output voltage is higher than the voltage of input side. This paper presents a new topology using boost DC-DC converter for providing maximum power load achievement from PV via the Perturb-and-Observe (P&O) technique. The considered framework is comprised like battery, bidirectional and proposed DC-DC converters. The different operating conditions are tested in this paper. In the proposed system, a storage system is connected to PV system via a bidirectional DC-DC converter in order to improve efficiency of the proposed system. Also, the new controlling scheme is proposed for this new PV system. The MATLAB/Simulink software results are used to verify theoretical concepts.

Keywords: DC-DC converter · Photovoltaic · MPPT · PV arrays

1 Introduction

These days, solar energy can replace fossil fuels in many energy applications. One of the most important advantages of this type of new energy is no air pollution and their lack of completion. However, these types of energies have disadvantages such as the amount of variable power according to some parameters and low output voltage. Maximum power point tracking (MPPT) is one of the tasks used to solve the disadvantages of this type of energy [1]. P&O method [2–4], adjustable conductance [5], adding ANN and Fuzzy system [6, 7], partial open-circuit-voltage, other part as partial short-circuit-current, while the most widely used is the RCC method in MPPT techniques. In [8, 9], the conventional DC-DC boost converter is used in order to provide higher voltage at the output while resulting in high conductivity of the used switch, in this case, the used converter's losses are increased to result in the decreased of the efficiency. Therefore, one solution to overcome this problem is to improve the boost converter.

In this paper, MPPT by P&O techniques are considered for PV system with its arrays as firstly, and secondly, high-voltage DC-DC converter and battery-storage system. The

R. A. Aliev et al. (Eds.): ICSCCW 2021, LNNS 362, pp. 19–27, 2022.
https://doi.org/10.1007/978-3-030-92127-9_7

modeling and analyzing of the proposed system are done and the validity of theoretical concepts are recorded by MATLAB/Simulink software.

2 Description of the Proposed Structure in PV System

Figure 1 shows the investigated system including high-gain-boost DC-DC. Taking into consideration that the providing voltage of PV is depends on both temperature and radiation. The high gain boost DC-DC is installed for two reasons, firstly to obtain high voltage, while secondly which uses battery to stabilize the DC-link, as much as stabilizing the power flowing in bi-directional DC-DC converter.

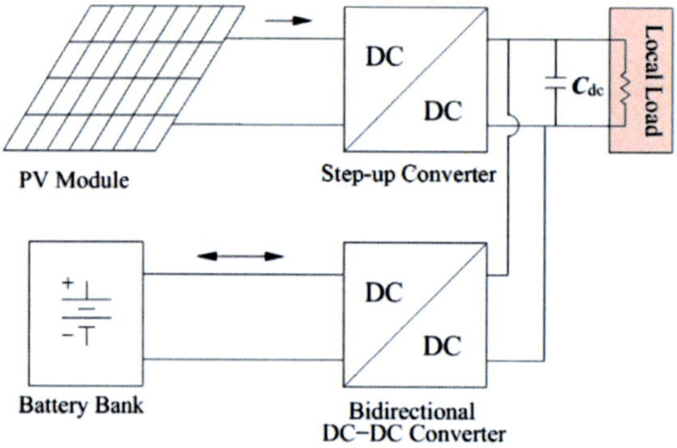

Fig. 1. The studied system.

2.1 PV Model and Algorithm of MPPT

PV model base on equivalent circuit technique [10–12] is described in Fig. 2. According to Fig. 2, we have:

$$I = I_{pv} - I_o\left[\exp(\frac{V + R_s I}{aV_t}) - 1)\right] - \frac{V + R_s I}{R_p} \tag{1}$$

Then:

$$I_{pv} = (I_{pv,n} + K_I \Delta_T)\frac{G}{G_n} \tag{2}$$

$$I_o = I_{o,n}(\frac{T_o}{T})^3 \exp\left[\frac{qE_g}{ak}\left(\frac{1}{T_n} - \frac{1}{T}\right)\right] \tag{3}$$

$$I_{o,n} = \frac{I_{sc,n}}{\exp(V_{ov,n}/aV_{t,n}) - 1} \tag{4}$$

The applied MPPT algorithm is based on P&O technique. The steps of P&O technique are shown in Fig. 3. The features of this technique are lower cost, few measurable parameters and easy implementation.

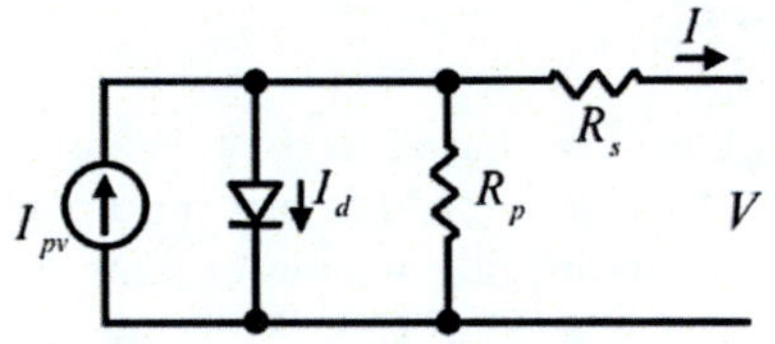

Fig. 2. PV modules circuit

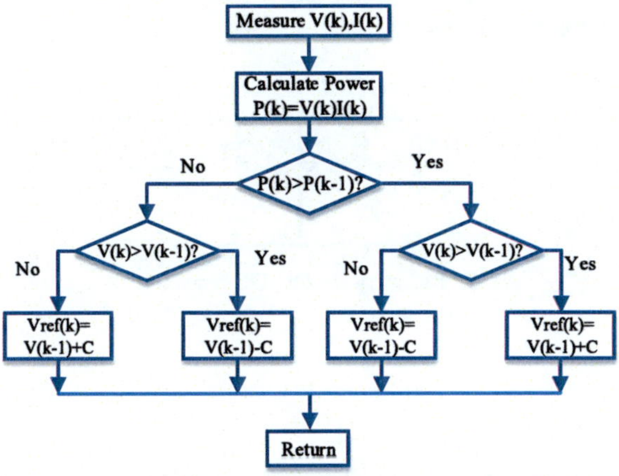

Fig. 3. P&O algorithm flowchart

2.2 High Gain DC-DC Converter

Figure 4 represents the proposed dc-dc converter [13]. Proposed DC-DC converter includes 3-inductors, 4-diodes, 3-capacitors and 2-switches. The S_1 and S_2 switches are controlled by PWM technique. Proposed converter's voltage gain is

$$\frac{V_o}{V_i} = \frac{2}{D(1-D)} \quad (5)$$

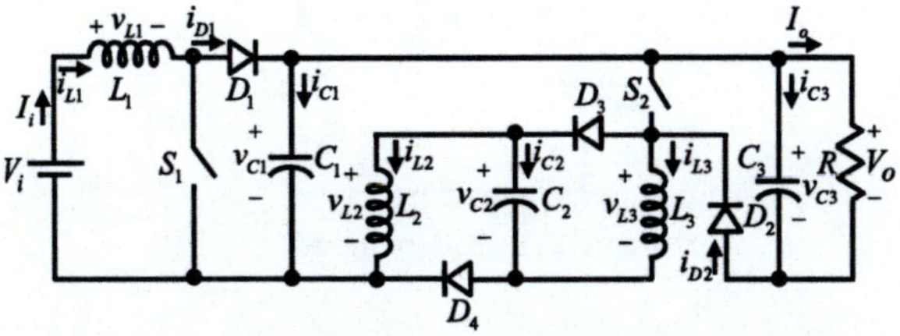

Fig. 4. Structure non-isolated boost DC-DC converter.

2.3 Bidirectional DC-DC Converter

To increase the reliability, a battery is used to store excess energy of the solar cell. During the periods of reduced solar cell production, it can provide part of system power. Thus, due to the charging and discharging and power transmission in both directions, a bidirectional dc-dc converter is needed. (Fig. 5). Dc-dc converter is a buck-boost converter output and is supposed to stabilize the output voltage of proposed converter. By applying pulse to S_3, the converter operates in boost mode and power will be delivered from the battery to the dc bus; by applying pulse to S_4 converter works in buck mode and extra power is stored in the battery.

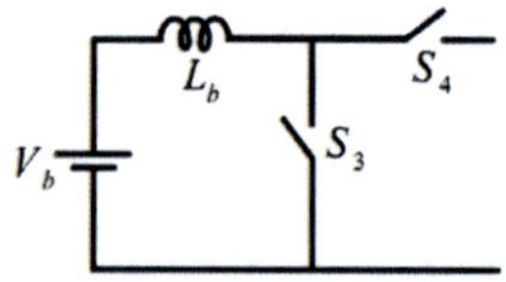

Fig. 5. Bidirectional DC-DC converter.

2.4 The Battery Model

Figure 6 shows the battery model using Thevenin method [14].

$$E_{bat} = E_o - k\frac{Q}{Q - it}i^* - k\frac{Q}{Q - it}it \tag{6}$$

$$E_{bat} = E_o - k\frac{Q}{it + 0.1Q}i^* - k\frac{Q}{Q - it}it \tag{7}$$

In (6) and (7), E_o and E_{bat} are the constant voltage and open circuit voltage of the battery, respectively, k is polarity resistance, Q is the maximum capacity of the battery and i is battery current. Integral blocks are used to calculate it and passing the current of battery from low-pass filter i^* can be obtained.

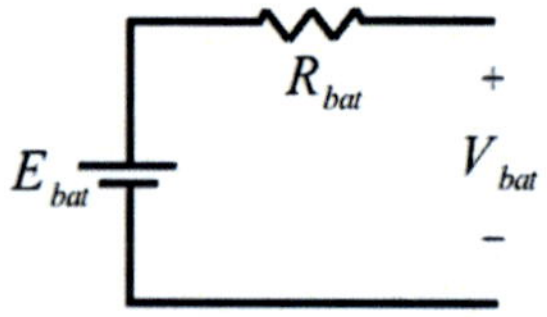

Fig. 6. The battery model.

3 Description of the Control Scheme

The proposed controlling scheme is predicted in Fig. 7. As shown, the proposed controlling scheme has two different parts. First part is allocated for providing pulse for S_1 and S_2 by comparison between PI controller and carrier wave. The second part has stabilizes duty of dc- link voltage via generating pulse for S_3 and S_4. This aim is based on second PI controller, controller 3 and reference current.

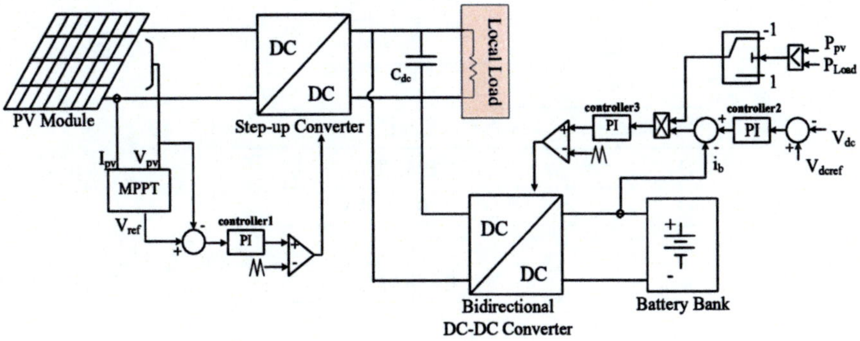

Fig. 7. Proposed system's scheme.

4 Simulation Results

16 PV modules (2-sets of 8 arrays), 2mH for inductor values of the dc-dc converter and 40 kHz for switching frequency, Tables 1, 2 and 3 are considered to evaluated system by MATLAB/Simulink. Two different scenarios are considered: first, in 25 °C, while radiation of sun changed from 500 W/m^2 to 1000 W/m^2 at 0.7 s, second, in 900 W/m^2, temperature changed from 25 °C to 60 °C in 0.8 s.

Table 1. Photovoltaic profile in solar radiation at 1000 W/m^2 and temperature of 25 °C

Characteristic	Rating
Maximum Power (P_{max})	200 W
Voltage at Pmax (V_{max})	26.3 V
Current at Pmax (I_{max})	7.61 A
Short-circuit current (I_{sc})	8.21 A
Open-circuit voltage (V_{sc})	32.9 V
Solar cells	54 cells in series

Table 2. Parameters of non-isolated high step-up dc-dc converter

Parameters	Value
$L_1 = L_2 = L_3$	2 mH
$C_1 = C_2$	100 μF
C_3	47 μF
f_s	40 kHz

Table 3. Battery parameters

Parameters	Value
E_o	196.37 V
Q	104.16 Ah
k	0.057
R_{bat}	0.0192 Ω

4.1 First Scenario

The first scenarios results are discussed in this section. As shown in Fig. 8(c), the reference voltage is constant at 500 V for 100 Ω and the load power is equal to 250 W. During in this scenario, using MPPT algorithm resulted in 52 V and 30A as in Fig. 8(d) whereas the produced power is 1560 W. Then, 970 W is provided by battery and 30W is the loss power of the studied system. At 0.7 s, PV voltage and current by MPPT algorithm change to 53V and 60.4A Fig. 8(d). 3200 W, 640 W and 60 W are the produced power, stored power in battery and loss power of the system, respectively. Figure 8(b) shows the battery SOC which delays that after 0.7 s, direct proportional between PV capacity, battery charges and SOC.

4.2 Second Scenario

The second scenario's results are presented in this section. Assuming 900 W/m^2 and 25 °C, 287 W, 52 V, and 454.6 A are the PV output, voltage and current, respectively Fig. 9(d). From 377 W consumption power, 312 W stored in battery and 65 W is loss Fig. 9(a). At 0.8 s, 2384 W, 43.7 V and 54.5 A are the PV output, voltage and current, respectively, where 34 W provided by battery. It is noted that SOC decreased in this condition.

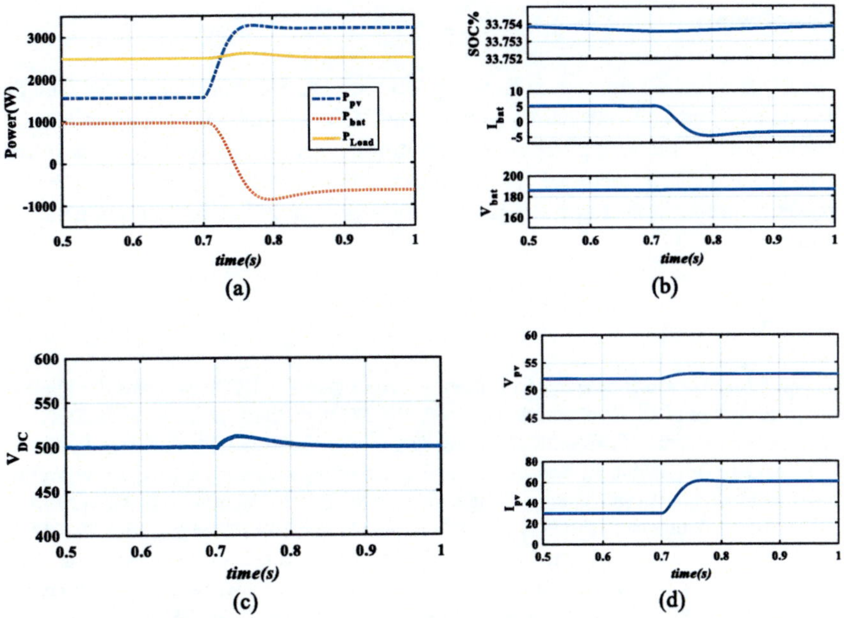

Fig. 8. Simulation's results using first scenario, (a) power flow response, (b) battery, (c) dc link, (d) PV voltage and current.

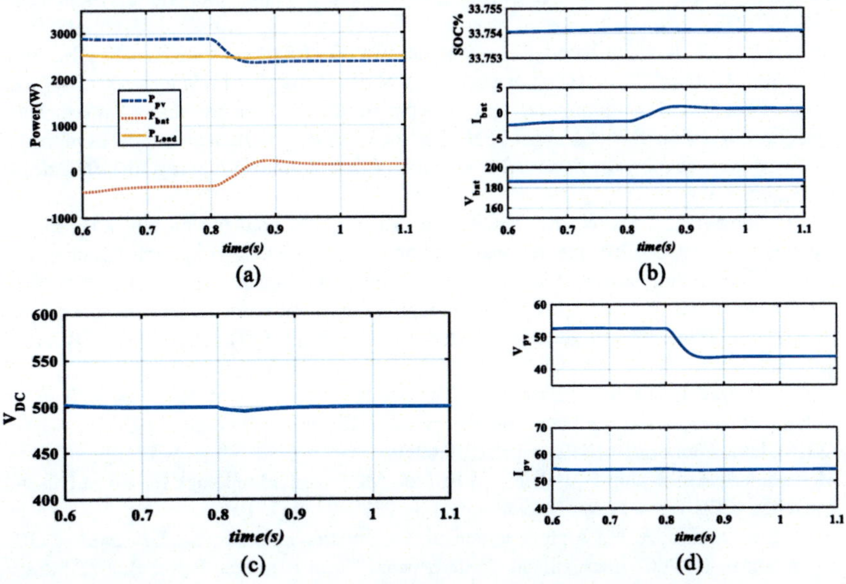

Fig. 9. Second scenario, (a) power flow response, (b) battery, (c) DC link, (d) PV voltage and current.

5 Conclusion

This paper presents a new non-isolated topology of DC-DC boost converter including PV, battery and bidirectional converter. The P&O technique is applied to MPPT. The mentioned technique applied via new controlling scheme by different controllers and comparators. From recorded results, it can mention that the controlling scheme and presented dc-dc converter have good performance for radiations and temperature variations.

References

1. Shahir, F.M., Aberoumandazar, M., Babaei, E.: High gain DC-DC boost converter applied in hybrid system of photovoltaic and battery. In: 2021 International Symposium on Devices, Circuits and Systems (ISDCS) (2021). https://doi.org/10.1109/isdcs52006.2021.9397922
2. Mohammadzadeh Shahir, F., Babaei, E.: Application of high voltage gain DC-DC converter in photovoltaic system with energy storage. In: 2017 8th Power Electronics, Drive Systems & Technologies Conference (PEDSTC) (2017). https://doi.org/10.1109/pedstc.2017.7910334
3. Femia, N., Petrone, G., Spagnuolo, G., Vitelli, M.: Optimizing duty-cycle perturbation of P&O MPPT technique. In: 2004 IEEE 35th Annual Power Electronics Specialists Conference (IEEE Cat. No.04CH37551) (n.d.). https://doi.org/10.1109/pesc.2004.1355414
4. Li, X., Hui, D., Lai, X.: Battery energy storage station (BESS)-based smoothing control of photovoltaic (PV) and wind power generation fluctuations. IEEE Trans. Sustain. Energ. **4**(2), 464–473 (2013). https://doi.org/10.1109/tste.2013.2247428
5. Aberoumandazar, M., Shahir, F.M.: Stability analysis of Luo converter using state space modeling. J. Circ. Syst. Comput. **30**(07), 2150127 (2020). https://doi.org/10.1142/s0218126621501279
6. Campbell, R. C.: A circuit-based photovoltaic array model for power system studies. In: 2007 39th North American Power Symposium (2007). https://doi.org/10.1109/naps.2007.4402293
7. Wong, Y.C., Mak, O.C., Ioinovici, A.: Development of boost converter based on switched-capacitor circuits. In: Proceedings of TENCON '93. IEEE Region 10 International Conference on Computers, Communications and Automation (n.d.). doi:https://doi.org/10.1109/tencon.1993.320696
8. Zhu, G., Ioinovici, A.: Switched-capacitor power supplies: DC voltage ratio, efficiency, ripple, regulation. In: 1996 IEEE International Symposium on Circuits and Systems. Circuits and Systems Connecting the World. ISCAS 96 (n.d.). https://doi.org/10.1109/iscas.1996.540007
9. Zhu, G., Ioinovici, A.: DC-to-DC converter with no magnetic elements and enhanced regulation. IEEE Trans. Aerosp. Electron. Syst. **33**(2), 499–506 (1997). https://doi.org/10.1109/7.575889
10. Mohammadzadeh Shahir, F., Babaei, E., Farsadi, M.: Analysis and design of voltage-lift technique-based non-isolated boost dc–dc converter. IET Power Electron. **11**(6), 1083–1091 (2018). https://doi.org/10.1049/iet-pel.2017.0259
11. Mohammadzadeh Shahir, F., Babaei, E., Farsadi, M.: Extanded topology for Boost DC-DC converter. IEEE Trans. Power Electeronic **34**(3), 2375–2384 (2019)
12. Altas, I. H., Sharaf, A. M.: A photovoltaic array simulation model for matlab-simulink GUI environment. In: 2007 International Conference on Clean Electrical Power (2007). https://doi.org/10.1109/iccep.2007.384234

13. Mohammadzadeh Shahir, F., Babaei, E., Sabahi, M., Laali, S.: A new DC–DC converter based on voltage-lift technique. Trans. Electr. Energ. Syst. (2015). https://doi.org/10.1002/etep.2133
14. Tremblay, O., Dessaint, L.-A.: Experimental validation of a battery dynamic model for EV applications. World Electr. Veh. J. **3**(2), 289–298 (2009). https://doi.org/10.3390/wevj3020289

Investigation of the Quality of Fuzzy IF-THEN Model for a Control System

Nigar E. Adilova(✉)

Azerbaijan State Oil and Industry University, 20 Azadlyg Avenue, Baku AZ1010, Azerbaijan
nigar.adilova@asoiu.edu.az

Abstract. The main facet of fuzzy logic is relational facet which is associated with IF-THEN rules. They are being increasingly applied to a variety problem as a model of different systems. Up to date research results on investigation of the quality of fuzzy rule-based systems (specificity, transparency, inconsistency) are very scars. In this paper author suggests a new approach to increase quality of IF-THEN rule for a control system.

Keywords: Fuzzy If-Then rules · Specificity · Interpretability · Inconsistency · Pareto optimality

1 Introduction

It is a known fact that fuzzy IF-THEN model of control systems is a foundational means to provide analysis of complex processes in decision making. The consideration of specificity and other quality criteria of fuzzy rule base is one of the important factors to develop a model. The specificity of fuzzy rule base can be widely used to determine the level of concentration of information granules and a degree of informativeness of consequents according to given IF-THEN model. Considering the perspective of informativeness measure in a rule-based system, large number of works are devoted to the investigation and calculation of specificity [1–5].

It is possible to solve several problems in knowledge extraction, decision-making, and support systems with the assistance of interpretability measure, which is considered as a quality criterion for fuzzy rule-based systems. The interpretability, and, particularly, transparency is an important factor of understanding of fuzzy rules by users. It is worth noting that lots of works on the investigation of interpretability measure exist [6–8].

The other quality criteria, inconsistency, directly reveals anomalies in fuzzy IF-THEN rules. Initial experiments on the investigation of irregularity of fuzzy rule-based system reflect the manifestation of potential inconsistencies in validating uncertain knowledge bases [9].

In addition to the specificity of fuzzy rule-based systems, noticing other quality criteria such as complexity, coverage, partition, and inconsistency has a positive effect on an efficiency of IF-THEN model for control systems. A discussion of the quantification of specificity and other quality criteria is still open and there is a serious theoretical

R. A. Aliev et al. (Eds.): ICSCCW 2021, LNNS 362, pp. 28–33, 2022.
https://doi.org/10.1007/978-3-030-92127-9_8

and practical requirement for a comprehensive approach to this problem. A remarkable trade-off among quality criteria of fuzzy rules bases was discussed in [10].

This paper is devoted to explore and improve specificity and other quality criteria for the IF-THEN model of a control system.

The paper is structured as follows. Section 2 presents some quality measures of IF-THEN model. Section 3 is devoted to the statement of the problem of analysis of quality measures and synthesis of a rule base. The application of the proposed method is illustrated in Sect. 4. Eventually, Sect. 5 covers concluding marks and future works.

2 Preliminaries

Definition 1. If-Then model of a control system. [11–15]: IF-THEN model of a control system, which characterizes fuzzy conditional statements is identified for multi-input case as follows:

$$\text{If } (X_1 \text{is } A_1),\ (X_2 \text{is } A_2), \ldots,\ (X_n \text{is } A_n) \text{ Then } Y \text{ is } B$$

where A_i describes information granules for input of a control system.

Definition 2. Main criteria of quality IF-THEN rules:Interpretability [6]. Classic interpretability is determined by the product of 3 quality criteria of fuzzy rules.

$$I = comp \times \overline{cov} \times \overline{part} \tag{1}$$

where, *comp*, *cov*, *part* are the indices of complexity, coverage, partition respectively [14].

Inconsistency [9]. Inconsistency measure is computed as.

$$f_{incons} = \sum_{i=1}^{N} Incons(i), \tag{2}$$

$$Incons(i) = \sum_{\substack{1 \le k \le N \\ k \ne i}} [1.0 - Cons(R^1(i), R^1(k))] + \sum_{\substack{1 \le l \le L \\ i=1,2,\ldots,N}} [1.0 - Cons(R^1(i), R^2(l))],$$

where $Incons(i)$ is inconsistency index for the ith rule.

Definition 3. Specificity [3]: The degree of specificity for fuzzy variables in a control system can be calculated by the following formula:

$$Sp(A) = \int_0^{hgt(A)} \frac{1}{|A^\mu|} d\mu \tag{3}$$

where $|A^\mu|$ is cardinality of μ-cut of fuzzy number A.

Definition 4.Pareto-optimality [16–19]. For any two candidate points, such as $A_1, A_2 \in \mathcal{A}$; A_1 is considered to dominate A_2 in the Pareto sense for only the given conditions.

$$f_i(A_1) \le f_i(A_2) \quad for\ all\ \ i \in \{1, 2, ..., M\}, \tag{4}$$

$$f_j(A_1) < f_j(A_2) \quad for\ at\ least\ one\ \ j \in \{1, 2, ..., M\} \tag{5}$$

$A^* \in \mathcal{A}$ is PO if there is no $A \in \mathcal{A}$ such that A P-dominates A^*.

3 Statement of the Problem

For simplicity, suppose that IF-THEN model of a control system is constructed based on the journey schedule as follows:

1. IF *Distance* is close, *vehicle's speed* is fast THEN *duration* of the journey is short.
2. IF *Distance* is far, *vehicle's speed* is slow THEN *duration* of the journey is long.
3. IF *Distance* is too far, *vehicle's speed* is very slow THEN *duration* of the journey is too long.

The model is organized with 2 input variables (Distance and Vehicle's speed) and 1 Output variable (Duration of the journey) generally including the triangular fuzzy values for the linguistic terms given below.

For Distance: *too far* = (0; 0.13; 0.25); *far* = (0.3; 0.35; 0.45); *close* = (0.75; 0.8; 0.88);

For Vehicle's speed: *very slow* = (0.05; 0.15; 0.3); *slow* = (0.25; 0.4; 0.45); *fast* = (0.65; 0.72; 0.8);

For Duration: *too long* = (−0.99; − 0.85; − 0.7); *long* = (−0.6; −0.55; −0.45); *short* = (0.5; 0.65; 0.75);

Figure 1 characterizes the description of membership functions.

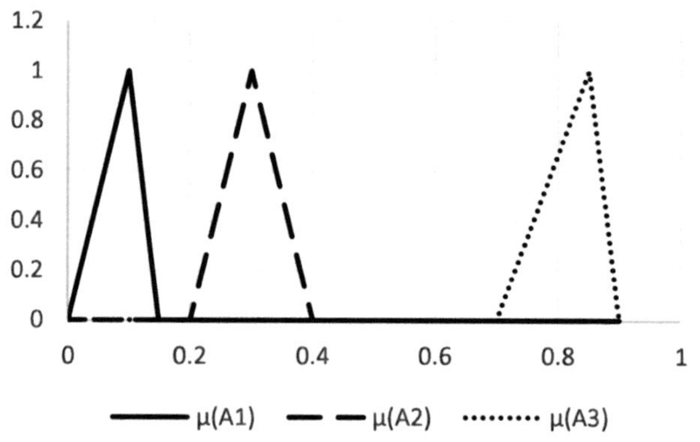

Fig. 1. Description of membership functions

The purpose here is to calculate specificity and other quality criteria of the model and find a near optimal trade among them.

4 Solution of the Problem

To demonstrate the suggested approach to solving the problem, experiments are focused on the calculation of specificity (as a dominant criterion) and the other quality criteria

Table 1. Measures of quality criteria for the initial IF-THEN model

Criteria of IF-THEN model	Values
Complexity	0.5
Coverage	0.298
Partition	0.2
Inconsistency	0.124
Specificity	0.934

for fuzzy IF-THEN model. The computational result for the mentioned quality criteria and specificity is stratified in Table 1 by using (1)–(3).

Consequently, interpretability index based on the product of first 3 criteria will be $I = 0.0298$.

For the next step, investigation will be expanded by changing the degrees of membership functions (given below) and gaining a better result (Table 2).

For Distance: *too far* = (0; 0.2; 0.35); *far* = (0.2; 0.35; 0.5); *close* = (0.5; 0.75; 1);

For Vehicle's speed: *very slow* = (0; 0.1; 0.15); *slow* = (0.2; 0.3; 0.4); *fast* = (0.7; 0.85; 0.9);

For Duration: *too long* = (−1; −0.7; −0.55); *long* = (−0.55; −0.4; −0.15); *short* = (0.2; 0.35; 0.55);

Table 2. Measures of quality criteria for the second IF-THEN model

Criteria of IF-THEN model	Values
Complexity	0.5
Coverage	0.446
Partition	0.2
Inconsistency	0.087
Specificity	0.870

Clearly seems from the experimental outcome that the interpretability measure is $I = 0.0446$.

Final step of the experimental analysis is related to creating a balance among criteria with the assistance of the following goal functions:

$$com \rightarrow \max, \mathrm{cov} \rightarrow \max, part_j \rightarrow \max, 1 - f_{incons} \rightarrow \max, Sp(A) \rightarrow \max$$

One of the most important directions in this regard is to create a balance for inconsistency degree which is considered competitive parameter for the other criteria. Trade-off among criteria makes is considered in terms of Pareto optimality (Def. 4). These degrees of optimality computed by using the fuzzy Pareto optimality approach [16] are given in Table 3:

Table 3. Degrees of optimality (*do*) for 2 cases

Alternatives	*Do*
Case-1	0.1811
Case-2	1

The obtained results show that the initial fuzzy IF-THEN model for the considered control system is not perfect. The new rule-base, Case-2, represents a better model for the control system.

5 Conclusion and Discussion

Experimental analysis of the proposed method has been carried out by involving the use of quality criteria of IF-THEN model for the control system. The results have revealed that the better trade-off. Indeed, the modification of the values of membership functions has improved some of the main factors.

In the current implementation, fuzzy Pareto optimality approach was used for solving of multi-criteria optimization problem. However, it would be very interesting to investigate the quality of Z-extension of IF-THEN model for a control system. Therefore, a piece of further research will be on the exploration of quality criteria over Z-based IF-THEN model.

References

1. Kacprzyk, J.: On measuring the specificity of IF-THEN rules. Int. J. Approx. Reason. **11**, 29–53 (1994). https://doi.org/10.1016/0888-613x(94)90018-3
2. Dubois, D., Prade, H.: A note on measures of specificity for fuzzy sets. Int. J. Gen. Syst. **10**(17), 279–283 (1995). https://doi.org/10.1080/03081078508934893
3. Yager, R.R.: Measures of specificity. In: Kaynak, O., Zadeh, L.A., Türkşen, B., Rudas, I.J. (eds.) Computational Intelligence: Soft Computing and Fuzzy-Neuro Integration with Applications. NATO ASI Series (Series F: Computer and Systems Sciences), vol. 162. Springer (1998). https://doi.org/10.1007/978-3-642-58930-0_6
4. Aliev, R.A.: Operations on Z-numbers with acceptable degree of specificity. Procedia Comput. Sci. **120**, 9–15 (2017)
5. Adilova, N.E.: Construction of fuzzy control system rule-base with predefined specificity. In: 13th International Conference on Theory and Application of Fuzzy Systems and Soft Computing, ICAFS-2018, Warsaw: Advantages of Intelligence System Computing, Springer, 901–904 (2018). https://doi.org/10.1007/978-3-030-04164-9_119
6. Nauck, D.D.: Measuring interpretability in rule-based classification systems, USA: FUZZ-IEEE, **1**, 196–201 (2003). https://doi.org/10.1109/FUZZ.2003.1209361
7. Gacto, M.J., Alcala, R., Herrera, F.: Interpretability of linguistic fuzzy rule-based systems: an overview of interpretability measure. Inform. Sci. **181**, 4340–4360 (2011). https://doi.org/10.1016/j.ins.2011.02.021
8. Razak, T.R., Garibaldi, J.M., Wagner, C., Pourabdollah, A., Soria, D.: Interpretability indices for hierarchical fuzzy systems. In: IEEE International Conference Fuzzy Systems (FUZZ-IEEE), Naples, Italy. IEEE, 1–6 (2017). https://doi.org/10.1109/FUZZ-IEEE.2017.8015616

9. Yager, R.R.: On discovering potential inconsistencies in validating uncertain knowledge bases by reflecting on the input, IEEE Trans. Syst., Man, Cybern., **21**, 790–901 (1991)
10. Huseynov, O.H., Adilova, N.E.: Multi criterial optimization problem for fuzzy IF-THEN rules. In: 14th International Conference Theory and Application of Fuzzy Systems and Soft Computing–ICAFS-2020, Advantage of Intelligence System Computing, Springer, **1306**, 80–88 (2021). https://doi.org/10.1007/978-3-030-64058-3_10
11. Zadeh, L.A., Aliev, R.A.: Fuzzy logic theory and applications. Part I Part II, Singapore World Sci., p. 612 (2019)
12. Aliev, R.A., Aliev, F., Babaev, M.: Fuzzy process control and knowledge engineering in petrochemical and robotic manufacturing, Verlag, Germany, p. 146 (1991)
13. Mirzhakhanov, V.E., Gardashova, L.A.: The incrementality issue in the wu-mendel approach for linguistic summarization using IF-THEN Rules. 13th International Conference Theory and Application of Fuzzy Systems and Soft Computing-ICAFS-2018, 293–300 (2018). https://doi.org/10.1007/978-3-030-04164-9_40
14. Adilova, N.E.: Quality criteria of fuzzy IF-THEN rules and their calculations. In: 14th International Conference Theory and Application of Fuzzy Systems and Soft Computing-ICAFS-2020, Budva, Advantage of Intelligence System Computing, Springer, **1306**, pp. 55–62 (2021). https://doi.org/10.1007/978-3-030-64058-3_7
15. Aliev, R.A., Gardashova, L.A.: Z-set based approach to control system design. In: 14th International Confernece Theory and Application of Fuzzy Systems and Soft Computing-ICAFS-2020, Advantages of Intelligence System Computing, **1306**, 10–21 (2021)
16. Farina, M., Amato, P.: A fuzzy definition of "optimality" for many-criteria optimization problem. IEEE Trans. Syst., Man, Cybern. Part a: Syst. Humans, **34**(3), (2004). https://doi.org/10.1109/TSMCA.2004.824873
17. Alizadch, A.V.: Application of the fuzzy optimality concept to decision making. In: 10th International Conference Theory and Application of Soft Computing, Computing with Words and Perceptions, 373–382 (2020). https://doi.org/10.1007/978-3-030-35249-3_69
18. Jabbarova, K.I., Alizadeh, A.V.: Z-number-based decision making in human resources department. In: 14th International Conference Theory and Application of Fuzzy Systems and Soft Computing–ICAFS-2020, Advanatages of Intelligence System Computing, **1306**, 498–507 (2021)
19. Jabbarova, K.I., Huseynov, O.H.: Solution of a decision making problem under risk with Z-Information. In: 14th International Conference Theory and Application of Fuzzy Systems and Soft Computing – ICAFS-2020, Advantages of Intelligence System Computing, **1306**, 781–786 (2021). https://doi.org/10.1007/978-3-030-64058-3_98

Circular Intuitionistic Fuzzy Sets in Multi Criteria Decision Making

Esra Çakır[1] (✉), Mehmet Ali Taş[2], and Ziya Ulukan[1]

[1] Department of Industrial Engineering, Galatasaray University, 34349 Ortakoy/Istanbul, Turkey
{ecakir,zulukan}@gsu.edu.tr

[2] Department of Industrial Engineering, Turkish-German University, 34820 Beykoz/Istanbul, Turkey
mehmetali.tas@tau.edu.tr

Abstract. Collaborating businesses are expected to be mutually beneficial and act as a single entity. It is possible to achieve the goals thanks to the businesses that come together in an industrial symbiosis network. Two (or more) businesses benefit in an industrial symbiosis relationship. The selection of enterprises in which industrial symbiosis can be established is a difficult task due to the different criteria that must be considered. In this study, the assessment of companies for industrial symbiosis is examined. The alternatives are evaluated with circular intuitionistic fuzzy numbers according to the criteria by experts, and a new circular intuitionistic fuzzy MCDM approach is developed. In addition to the existing literature, this study contributes to the circular intuitionistic fuzzy sets by proposing a new score function. This application is intended to guide future circular intuitionistic fuzzy MCDM approaches.

Keywords: Circular intuitionistic fuzzy set · Multi criteria decision making · Industrial symbiosis

1 Introduction

Industrial symbiosis simply refers to organizations working together to establish long-term associations aimed at improving environmental performance and competitiveness [1]. The basic dynamic in ensuring this is that the waste of one enterprise in the symbiosis relationship is evaluated as an input material by another enterprise [2]. In this manner, by reducing the amounts of carbon dioxide emissions, wastewater, and solid waste, the negative effects on air and water resources can be eliminated and their quality can be increased. Although the ecological benefits of being in industrial symbiosis activities are generally emphasized, it also offers widely positive effects in economic and social areas for enterprises and society [3]. Industrial symbiosis contributes to the sustainability of the circular economy [4]. In order to establish an industrial symbiosis relationship, the steps of examining the current situation, identifying potential flow materials and businesses and starting symbiosis studies are followed respectively [5]. In these application steps, techniques such as Life Cycle Assessment and Material Flow Analysis as well as multi-criteria decision making (MCDM) methods can be used [6].

R. A. Aliev et al. (Eds.): ICSCCW 2021, LNNS 362, pp. 34–42, 2022.
https://doi.org/10.1007/978-3-030-92127-9_9

In the literature, there are some studies in which various MCDM methods are used in the context of industrial symbiosis. Zhao et al. [7] examined the benefits of six sample eco-parks in China according to 26 benefits with fuzzy VIKOR. Alakaş et al. [8] evaluated criteria in terms of sustainability in industrial symbiosis implementations using Analytic Network Process (ANP). Vimal et al. [9] studied industrial symbiosis applications based on organizational outputs with Analytic Hierarchy Process (AHP). Jayakrishna et al. [10] ranked the eco-innovation factors of the industrial symbiosis relationship in the determined sectors in India with DEMATEL method. Li and Pinto [11] chose the AHP method to rank the success factors of industrial symbiosis applications in China. These methods can be diversified by using different approaches, and the fuzzy set theory can be implemented to express expert opinions in MCDM methods.

Fuzzy set theory was introduced by Zadeh in order to express uncertainty [12]. Based on this theory, various extensions have been proposed such as neutrosophic fuzzy sets [13], hesitant fuzzy sets [14], and so on. One of the commonly used one in the literature is the intuitionistic fuzzy sets (IFS), which is introduced by Atanassov [15]. In 2020, Atanassov also proposed a new extension of IFS called circular intuitionistic fuzzy sets (C-IFS) [16]. This new fuzzy set is a development of the IFS and differs by including a circle of the number consisting of membership and non-membership degrees [17].

In this study, the determination of the most suitable alternative for the industrial symbiosis network of a wood product manufacturer in Turkey is discussed. Thus, it is desired to contribute to the literature on industrial symbiosis. Another contribution of the study is the introduction of score functions for circular IFS. It is aimed to widen the use of circular intuitionistic fuzzy sets in MCDM methods. It is thought that this study fills the gaps in the literature in the mentioned areas.

The rest of the paper is organized as follows. Section 2 introduces circular intuitionistic fuzzy sets with a new defined score function. The circular intuitionistic fuzzy MCDM methodology is given step by step in Sect. 3. Proposed approach is applied on enterprise selection for industrial symbiosis in Sect. 4. In the last section, the paper is concluded with future perspectives.

2 Circular Intuitionistic Fuzzy Set

This section gives the preliminaries and definitions of circular intuitionistic fuzzy (C-IFS) set.

Definition 1: [15] Let $X = \{x_1, x_2, \ldots, x_n\}$ be a universe of discourse, an IFS A in X is given by

$$A = \{< x,\ u_A(x),\ v_A(x) > x \in X\} \tag{1}$$

where "$u_A : X \to [0, 1]$" and "$v_A : X \to [0, 1]$" with the conditions $0 \leq u_A(x) + v_A(x) \leq 1,\ \forall x \in X$. These numbers are the membership degree "$u_A(x)$" and "$v_A(x)$" non-membership degree of the element x to the set A, respectively.

Given an element x of X, the pair "$< u_A(x), v_A(x) >$" is called an IF value (IFV) [18]. For convenience, it can be denoted as $\tilde{a} =< u_{\tilde{a}}, v_{\tilde{a}} >$ such that $u_{\tilde{a}} \in [0, 1]$, $v_{\tilde{a}} \in [0, 1]$ and $0 \leq u_{\tilde{a}} + v_{\tilde{a}} \leq 1$. The indeterminacy degree is denoted by $\pi_{\tilde{a}}$, with the conditions of $\pi_{\tilde{a}} \in [0, 1]$ and $\pi_{\tilde{a}} = 1 - u_{\tilde{a}} - v_{\tilde{a}}$.

Definition 2: [16] Let $\tilde{a} = (u_{\tilde{a}}, v_{\tilde{a}})$ be an IFS, a score function S and an accuracy function of the IFV $\tilde{a}$ is defined as follows:

$$S(\tilde{a}) = u_{\tilde{a}} - v_{\tilde{a}} \text{ where } S(\tilde{a}) \in [-1, 1] \quad (2)$$

$$H(\tilde{a}) = u_{\tilde{a}} + v_{\tilde{a}} \text{ where } H(\tilde{a}) \in [0, 1] \quad (3)$$

Definition 3: [19] Let $\tilde{a}_i = \left(u_{\tilde{a}_i}, v_{\tilde{a}_i}\right)$ (i = 1,2,…,n) be an IFS, then their aggregated value by using the IF weighted averaging (IFWA) operator is also an IF value:

$$\text{IFWA}(\tilde{a}_1, \tilde{a}_2, \ldots, \tilde{a}_n) = \left(1 - \prod\nolimits_{j=1}^{n} \left(1 - u_{\tilde{a}_j}\right)^{w_j}, \prod\nolimits_{j=1}^{n} v_{\tilde{a}_j}^{w_j}\right) \quad (4)$$

where $\mathrm{w} = \{\mathrm{w}_1, \mathrm{w}_2, \ldots, \mathrm{w}_n\}$ is the weighting vector of $\tilde{a}_i$(i = 1,2,…,n) with $\mathrm{w}_j \in [0, 1]$ and $\sum_{\mathrm{j}=1}^{\mathrm{n}} \mathrm{w}_\mathrm{j} = 1$.

The circular intuitionistic fuzzy (C-IFS) set is introduced to the literature in 2020 by Atanassov [16]. This new fuzzy set is the extension of the IFS and differs from IFS by including a circle of the number consisting of membership and non-membership degrees [17].

Definition 4: Let E be a fixed universe, and a generic element a C-IFS C_r in E is denoted by x; $C_r = \{< \mathrm{x} : u_\mathrm{C}(x), v_\mathrm{C}(x); \mathrm{r} >, \mathrm{x} \in E\}$ is the form of an object that is the circular intuitionistic fuzzy set (C-IFS), where the functions "u, v: E → [0, 1]" define respectively the membership function and the non-membership function of the element $x \in E$ to the set C-IFS with condition:

$$0 \le u_\mathrm{C}(x) + v_\mathrm{C}(x) \le 1 \text{ and } \mathrm{r} \in \left[0, \sqrt{2}\right] \quad (5)$$

where r is the radius of the circle around each element $x \in E$ [17]. The indeterminacy function can be also defined as $\pi_\mathrm{C}(x) = 1 - u_\mathrm{C}(x) - v_\mathrm{C}(x)$. When r = 0, a C-IFS is reduced to a standard IFS.

Definition 5: Let $\left\{< m_{\mathrm{i},1}, n_{\mathrm{i},1} >, < m_{\mathrm{i},2}, n_{\mathrm{i},2} >, \ldots\right\}$ is a set of IF pairs. The C-IFS C_i is calculated from pairs where i is the number of the IF set and k_i is the number of IF pairs in each set. The arithmetic average of the set is as follows:

$$C_\mathrm{i} =< u_\mathrm{C}(C_\mathrm{i}), v_\mathrm{C}(C_\mathrm{i}) >=< \frac{\sum_{j=1}^{k_i} m_{\mathrm{i,j}}}{k_i}, \frac{\sum_{j=1}^{k_i} n_{\mathrm{i,j}}}{k_i} > \quad (6)$$

The radius r_i of the C_i is obtained by the maximum of the Euclidean distances as follows:

$$r_\mathrm{i} = \max_{1 \le j \le k_i} \sqrt{\left(u_\mathrm{C}(C_\mathrm{i}) - m_{\mathrm{i,j}}\right)^2 + \left(v_\mathrm{C}(xC_\mathrm{i}) - n_{\mathrm{i,j}}\right)^2} \quad (7)$$

Definition 6: Let $L^* = \{< a, b > | a, b \in [0, 1] \& a + b \le 1\}$. Therefore C_r can be rewritten in the form $C_r^* = \{< \mathrm{x} : O_r(u_\mathrm{C}(x), v_\mathrm{C}(x))) >, \mathrm{x} \in E\}$ where "$O_r(u_\mathrm{C}(x), v_\mathrm{C}(x)) = \left\{< a, b > | a, b \in [0, 1], \sqrt{(u_\mathrm{C}(x) - a)^2 + (v_\mathrm{C}(x) - b)^2} \le r, a + b \le 1\right\}$" is a function of circle representation. The C-IFs are illustrated in Fig. 1.

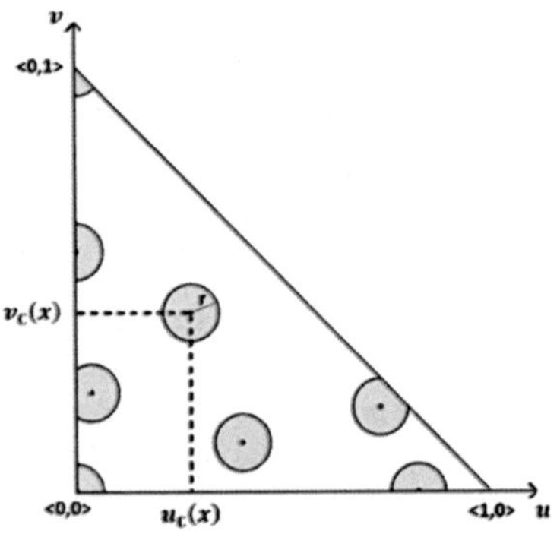

Fig. 1. Geometrical representation of circular intuitionistic fuzzy numbers.

Definition 7: Here, a new score and accuracy functions for C-IFSs are defined. Let $c = (u_c, v_c; r)$ be an C-IFV, a score function S_c and an accuracy function H_c of the C-IFV c are defined as follows with respect to the decision maker's (or manager's) preference information $\lambda \in [0, 1]$:

$$S_c(c) = (u_c - v_c + \sqrt{2}r(2\lambda - 1))/3 \tag{8}$$

$$H_c(c) = u_c + v_c \text{ where } H_c(c) \in [0, 1] \tag{9}$$

λ reflects the decision maker's perspective to the model. If λ is equal to zero, it shows the full pessimistic point of view, while λ is equal to one, it shows the optimistic point of view. In general acceptance, $\lambda \in [0, 0.5)$ indicates a pessimistic point of view, and $\lambda \in (0.5, 1]$ indicates an optimistic point of view. $\lambda = 0.5$ is the reflect indifferent attitude of decision maker.

Definition 8: Let $c_1 = \left(u_{c_1}, v_{c_1}; r\right)$ and $c_2 = \left(u_{c_2}, v_{c_2}; r\right)$ be two C-IFSs. Then, the ranking rule is defined as follows:

If $S_c(c_1) \succ S_c(c_2)$, then $c_1 \succ c_2$. If $S_c(c_1) = S_c(c_2)$, then;
If $H_c(c_1) \succ H_c(c_2)$, then $c_1 \succ c_2$. If $H_c(c_1) = H_c(c_2)$, then $c_1 = c_2$.

3 Methodology

This section introduces the new circular intuitionistic fuzzy MCDM method by proposing a new defuzzification (or score) function on C-IFS. The methodology aims to determine the rank order of the alternatives according to the criteria and the reviews of DMs. The steps of proposed C-IFS MCDM methodology are explained as follows.

Step 1: Define the case. Consider "$A = \{A_1, A_2, \ldots, A_m\}$" is the alternative set, "$C = \{C_1, C_2, \ldots, C_n\}$" is the criteria set and "$D = \{D_1, D_2, \ldots, D_k\}$" is the set of DMs. Let "$W_C = \left\{W_{C_1}, W_{C_2}, \ldots, W_{C_n}\right\}$" is the weight vector of criteria where $W_{C_i} \geq 0$ and $\Sigma W_{C_i} = 1$. These weight vectors are determined by DMs.

Step 2: Collect intuitionistic fuzzy decision matrices from DMs.

Step 3: Obtain the aggregated IF decision matrix using the IFWA operator by Eq. (4).

Step 4: According to the criteria weights, calculate aggregated decision $\tilde{D}$ of alternatives using the IFWA operator by Eq. (4).

Step 5: Calculate the maximum radius lengths of each aggregated decision $\tilde{D}$ by Eq. (6–7) from aggregated IF decision matrix and revise the aggregated decision $\tilde{D}$ with radius (C-IFS).

Step 6: Determine the λ value related to the case from a pessimistic or optimistic view.

Step 7: Calculate their score values by Eq. (8–9) and rank the alternatives.

Step 8: Evaluate the full optimistic ($\lambda = 1$) and full pessimistic ($\lambda = 0$) scenario for the case. Perform the sensitivity analysis by solving the problem for each value of $\lambda \in [0, 1]$.

4 Illustrative Example

A wood product manufacturer in Turkey wants to manage and minimize the waste generated during its processes. As a strategic goal, it was decided to create a holistic industrial ecology by combining investments in clean production technologies with industrial symbiotic relations. For this purpose, the existing activities and the environment are examined and four alternative enterprises A1, $A2$, $A3$, $A4$ that could establish a symbiotic relationship are determined. The alternatives are located in the same region and their distances from the manufacturer are identical. In this case, management wants to choose the best one to collaborate, therefore the steps of the proposed methodology are followed in order:

Step 1: Selection criteria considered for industrial symbiosis are: C_1: usage of renewable of energy [20], C_2: savings in raw material cost [21], C_3: water consumption control [8], C_4: environmental awareness [22] with the weights of "$W_{C_1} = 0.25$, $W_{C_2} = 0.23$, $W_{C_3} = 0.32$, $W_{C_4} = 0.20$." Three DMs "D_1, D_2, D_3" are selected by manufacturer.

Step 2: Three equally weighted DMs evaluate the alternative enterprises according to the criteria using IFSs as follows:

$$DM1 = \begin{matrix} A_1 \\ A_2 \\ A_3 \\ A_4 \end{matrix} \begin{bmatrix} \langle 0.3, 0.4\rangle & \langle 0.5, 0.1\rangle & \langle 0.6, 0.2\rangle & \langle 0.4, 0.5\rangle \\ \langle 0.4, 0.2\rangle & \langle 0.8, 0.1\rangle & \langle 0.3, 0.6\rangle & \langle 0.5, 0.4\rangle \\ \langle 0.5, 0.1\rangle & \langle 0.6, 0.3\rangle & \langle 0.7, 0.1\rangle & \langle 0.2, 0.3\rangle \\ \langle 0.8, 0.1\rangle & \langle 0.3, 0.4\rangle & \langle 0.5, 0.3\rangle & \langle 0.4, 0.2\rangle \end{bmatrix},$$

$$DM2 = \begin{matrix} A_1 \\ A_2 \\ A_3 \\ A_4 \end{matrix} \begin{bmatrix} \langle 0.6, 0.2\rangle & \langle 0.4, 0.3\rangle & \langle 0.5, 0.1\rangle & \langle 0.2, 0.5\rangle \\ \langle 0.8, 0.1\rangle & \langle 0.2, 0.7\rangle & \langle 0.8, 0.2\rangle & \langle 0.5, 0.4\rangle \\ \langle 0.4, 0.5\rangle & \langle 0.6, 0.3\rangle & \langle 0.1, 0.5\rangle & \langle 0.2, 0.6\rangle \\ \langle 0.6, 0.3\rangle & \langle 0.7, 0.2\rangle & \langle 0.5, 0.5\rangle & \langle 0.4, 0.2\rangle \end{bmatrix}$$

$$DM3 = \begin{matrix} A_1 \\ A_2 \\ A_3 \\ A_4 \end{matrix} \begin{bmatrix} \langle 0.4, 0.3\rangle & \langle 0.5, 0.1\rangle & \langle 0.2, 0.4\rangle & \langle 0.6, 0.3\rangle \\ \langle 0.3, 0.4\rangle & \langle 0.6, 0.1\rangle & \langle 0.6, 0.3\rangle & \langle 0.7, 0.1\rangle \\ \langle 0.2, 0.1\rangle & \langle 0.5, 0.3\rangle & \langle 0.4, 0.6\rangle & \langle 0.5, 0.4\rangle \\ \langle 0.6, 0.2\rangle & \langle 0.1, 0.4\rangle & \langle 0.6, 0.4\rangle & \langle 0.2, 0.5\rangle \end{bmatrix}.$$

Step 3: Aggregated IF decision matrix of three intuitionistic fuzzy decision matrices by IFWA operator is as follows:

$$Aggregated\ DM = \begin{matrix} & C_1 & C_2 & C_3 & C_4 \\ A_1 & \langle 0.584, 0.288\rangle & \langle 0.536, 0.144\rangle & \langle 0.609, 0.200\rangle & \langle 0.637, 0.422\rangle \\ A_2 & \langle 0.542, 0.200\rangle & \langle 0.542, 0.191\rangle & \langle 0.476, 0.330\rangle & \langle 0.441, 0.252\rangle \\ A_3 & \langle 0.658, 0.171\rangle & \langle 0.435, 0.300\rangle & \langle 0.696, 0.311\rangle & \langle 0.729, 0.416\rangle \\ A_4 & \langle 0.340, 0.182\rangle & \langle 0.724, 0.317\rangle & \langle 0.469, 0.391\rangle & \langle 0.683, 0.271\rangle \end{matrix}$$

Step 4: The aggregated decisions of alternatives are obtained by IFWA operator according to the criteria weights specified in Step 1.

$$\tilde{D} = \begin{matrix} A_1 \\ A_2 \\ A_3 \\ A_4 \end{matrix} \begin{bmatrix} \langle 0.5927,\ 0.2360\rangle \\ \langle 0.5023,\ 0.2434\rangle \\ \langle 0.6472,\ 0.2814\rangle \\ \langle 0.5647,\ 0.2862\rangle \end{bmatrix}$$

Step 5: For each aggregated decision, maximum radius lengths are calculated by Eqs. (6) and (7) with the aid of aggregated intuitionistic fuzzy decision matrix from Step 3. The aggregated decisions with radius (C-IFS) is calculated as follows:

$$\tilde{D} = \begin{matrix} A_1 \\ A_2 \\ A_3 \\ A_4 \end{matrix} \begin{bmatrix} \langle 0.5927,\ 0.2360;\ 0.240\rangle \\ \langle 0.5023,\ 0.2434;\ 0.194\rangle \\ \langle 0.6472,\ 0.2814;\ 0.132\rangle \\ \langle 0.5647,\ 0.2862;\ 0.164\rangle \end{bmatrix}$$

Step 6: The manager office wants to determine their optimistic point of view as $\lambda = 0.65$.

Step 7: By proposed score function in Eq. (8), the scores of the alternatives are $A_1 = 0.255, A_2 = 0.219, A_3 = 0.263, A_4 = 0.240$.

Therefore, the ranking of the alternatives are $A_3 \succ A_1 \succ A_4 \succ A_2$.

Step 8: For each value of $\lambda \in [0, 1]$, the values and ranking results are given in Table 1.

The full pessimistic case occurs when $\lambda = 0$, and the ranking result is $A_3 \succ A_4 \succ A_1 \succ A_2$. The full optimistic case occurs when $\lambda = 1$, and the ranking result is $A_1 \succ A_3 \succ A_4 \succ A_2$. Examining the sequences from pessimistic to optimistic, the best alternative changes from A_3 to A_1, but the worst alternative A_2 does not changes. The alternative A_4 goes backwards in the ranking, while alternative A_1 starts to rise in the ranking. Thus, the reflection of the λ value given by the manager to the score and ranking can be mentioned.

Table 1. Sensitivity analysis for λ value.

Score	A_1	A_2	A_3	A_4	Ranking
$\lambda = 0$	0.006	−0.005	0.060	0.015	$A_3 \succ A_4 \succ A_1 \succ A_2$
$\lambda = 0.1$	0.044	0.029	0.091	0.050	$A_3 \succ A_4 \succ A_1 \succ A_2$
$\lambda = 0.2$	0.082	0.064	0.122	0.084	$A_3 \succ A_4 \succ A_1 \succ A_2$
$\lambda = 0.3$	0.121	0.098	0.153	0.119	$A_3 \succ A_1 \succ A_4 \succ A_2$
$\lambda = 0.4$	0.159	0.133	0.185	0.154	$A_3 \succ A_1 \succ A_4 \succ A_2$
$\lambda = 0.5$	0.198	0.167	0.216	0.188	$A_3 \succ A_1 \succ A_4 \succ A_2$
$\lambda = 0.6$	0.236	0.202	0.247	0.223	$A_3 \succ A_1 \succ A_4 \succ A_2$
$\lambda = 0.7$	0.274	0.237	0.278	0.257	$A_3 \succ A_1 \succ A_4 \succ A_2$
$\lambda = 0.8$	0.313	0.271	0.309	0.292	$A_1 \succ A_3 \succ A_4 \succ A_2$
$\lambda = 0.9$	0.351	0.306	0.341	0.327	$A_1 \succ A_3 \succ A_4 \succ A_2$
$\lambda = 1$	0.389	0.340	0.372	0.361	$A_1 \succ A_3 \succ A_4 \succ A_2$

5 Conclusion

In order to continue their activities and increase their benefits, enterprises establish industrial relations. The ability to act as a single organization by establishing industrial symbiotic relationships is essential for increasing market share by competitiveness. In addition to the economic dimension, environmental benefits are also achieved, such as reducing waste, reducing resource consumption and improving quality. This situation has a multiplier effect with the selection of the appropriate enterprise. Therefore, the selection of enterprises in the industrial symbiosis network is crucial. This study proposes a new methodology on circular fuzzy MCDM to evaluate alternatives based on criteria. A new score function is proposed to rank C-IFS, which is a new extension of intuitionistic fuzzy. The method is illustrated in companies' evaluation of alternatives for industrial symbiosis. Along with the sensitivity analysis, the effect of the variability in the parameter λ is investigated.

For future research, in decision making process, new score and accuracy functions can be defined. The MCDM methodology should be enriched by adapting new operators. The real-life cases should be applied with circular intuitionistic MCDM procedures.

Acknowledgements. "This work has been supported by the Scientific Research Projects Commission of Galatasaray University under grant number # FBA-2020-1036."

References

1. Chertow, M.R.: "Uncovering" industrial symbiosis. J. Ind. Ecol. **11**, 11–30 (2007). https://doi.org/10.1162/jiec.2007.1110

2. Costa, I., Massard, G., Agarwal, A.: Waste management policies for industrial symbiosis development: case studies in European countries. J. Clean. Prod. **18**, 815–822 (2010). https://doi.org/10.1016/j.jclepro.2009.12.019
3. Neves, A., Godina, R., Azevedo, S.G., Pimentel, C., Matias, J.C.O.: The potential of industrial symbiosis: case analysis and main drivers and barriers to its implementation. Sustainability **11**, 7095 (2019). https://doi.org/10.3390/su11247095
4. Domenech, T., Bleischwitz, R., Doranova, A., Panayotopoulos, D., Roman, L.: Mapping Industrial symbiosis development in Europe_ typologies of networks, characteristics, performance and contribution to the circular economy. Resour. Conserv. Recy. **141**, 76–98 (2019). https://doi.org/10.1016/j.resconrec.2018.09.016
5. Felicio, M., Amaral, D., Esposto, K., Durany, X.G.: Industrial symbiosis indicators to manage eco-industrial parks as dynamic systems. J. Clean. Prod. **118**, 54–64 (2016). https://doi.org/10.1016/j.jclepro.2016.01.031
6. Özkan, A., Günkaya, Z., Özdemir, A., Banar, M.: Sanayide Temiz Üretim Ve Döngüsel Ekonomiye Geçişte Endüstriyel Simbiyoz Yaklaşimi: Bir Değerlendirme. Anadolu Univ. J. Sci. Technol. B-Theor. Sci., **6**, 84–97 (2018). https://doi.org/10.20290/aubtdb.332377
7. Zhao, H., Zhao, H., Guo, S.: Evaluating the comprehensive benefit of eco-industrial parks by employing multi-criteria decision making approach for circular economy. J. Clean. Prod. **142**, 2262–2276 (2017). https://doi.org/10.1016/j.jclepro.2016.11.041
8. Alakaş, H.M., Gür, Ş., Özcan, E., Eren, T.: Ranking of sustainability criteria for industrial symbiosis applications based on ANP. J. Environ. Eng. and Landsc. Manag. **28**, 192–201 (2020). https://doi.org/10.3846/jeelm.2020.13689
9. Vimal, K.E.K., Jayakrishna, K., Ameen, T., Afridhi, S.S., Vasudevan, V.: An investigation on the impact of industrial symbiosis implementation on organizational performance using analytical hierarchical approach. Benchmarking Int. J. **27**, 886–911 (2020). https://doi.org/10.1108/BIJ-12-2018-0423
10. Jayakrishna, K., Kek, V., Vibha, M., Jain, S., Kulatunga, A.K.: Fostering systematic eco-innovation in an industrial symbiosis environment using DEMATEL. Int. J. Ind. Syst. Eng. **34**, 20–42 (2020). https://doi.org/10.1504/IJISE.2020.104314
11. Li, Y., Pinto, M.C.B.: Analyzing the critical success factors for industrial symbiosis—a Chinese perspective. In: Phanden, R.K., Mathiyazhagan, K., Kumar, R., Paulo Davim, J. (eds.) Advances in Industrial and Production Engineering. LNME, pp. 23–33. Springer, Singapore (2021). https://doi.org/10.1007/978-981-33-4320-7_3
12. Zadeh, L.A.: Fuzzy sets. Inf. Control. **8**, 338–353 (1965). https://doi.org/10.1016/S0019-9958(65)90241-X
13. Smarandache, F.: Neutrosophy: neutrosophic probability, set, and logic: analytic synthesis and synthetic analysis (1998)
14. Torra, V.: Hesitant fuzzy sets. Int. J. Intell. Syst. **25**, 529–539 (2010). https://doi.org/10.1002/int.20418
15. Atanassov, K.: Intuitionistic fuzzy sets. Fuzzy Set. Syst. **20**, 87–96 (1986). https://doi.org/10.1016/S0165-0114(86)80034-3
16. Atanassov, K.: Circular intuitionistic fuzzy sets. J. Intell. Fuzzy Syst. **39**, 5981–5986 (2020). https://doi.org/10.3233/JIFS-189072
17. Kahraman, C., Alkan, N.: Circular intuitionistic fuzzy TOPSIS method with vague membership functions: supplier selection application context. Notes Intuitionistic Fuzzy Set **27**, 24–52 (2021). https://doi.org/10.7546/nifs.2021.27.1.24-52
18. Chen, S., Tan, J.: Handling multi criteria fuzzy decision making problems based on vague set theory. Fuzzy Set Syst. **67**, 163–172 (1994). https://doi.org/10.1016/0165-0114(94)90084-1
19. Xu, Z.: Intuitionistic fuzzy aggregation operators. IEEE Trans. Fuzzy Syst. **15**, 1179–1187 (2007). https://doi.org/10.1109/TFUZZ.2006.890678

20. Geng, Y., Liu, Z., Xue, B., Dong, H., Fujita, T., Chiu, A.: Energy-based assessment on industrial symbiosis: a case of Shenyang economic and technological development zone. Environ. Sci. Pollut. Res. **21**(23), 13572–13587 (2014). https://doi.org/10.1007/s11356-014-3287-8
21. Wahrlich, J., Simioni, F.J.: Industrial symbiosis in the forestry sector: a case study in southern Brazil. J. Ind. Ecol. **23**, 1470–1482 (2019). https://doi.org/10.1111/jiec.12927
22. Martin, M.: Quantifying the environmental performance of an industrial symbiosis network of biofuel producers. J. Clean. Prod. **102**, 202–212 (2015). https://doi.org/10.1016/j.jclepro.2015.04.063

Z^+ - Number Based Alternatives Selection in Investment Problem

Shamil A. Ahmadov[1,2](✉)

[1] Azerbaijan State Oil and Industry University, 20 Azadliq Avenue, Az 1010, Baku, Azerbaijan
ahmadov.shamil@asoiu.edu.az
[2] French-Azerbaijani University, 183 Nizami Street, Baku, Azerbaijan

Abstract. Investment problem is one of the most interesting and important multi-attribute decision making problem in real-world. This problem is difficult for solving when information are characterized by fuzzy and probability type uncertainty. There are several works related to analyzing partially reliable information based decision making problem. Researchers are implemented solution of this problem by different approaches. In this study we are analyzing the investment problem described by Z^+ number. The purpose of this paper is to define best alternatives using Z information based approach in investment problem. The considered problem is demonstrated by a numerical example. Solution of the problem is implemented using information described Z^+ number, distributions, probability measure and utility function. The given method is used to solve investment problem under uncertain information.

Keywords: Investment problem · Partially reliable information · Z^+ number · Distributions · Probability measure · Utility function · Ranking Z numbers

1 Introduction

We make many decisions in our daily life based on natural language. Our decisions can be more or less correct or wrong, optimal or non-optimal. The decisions are based on experience, intuition, events, representation of information. Intuitive decisions can be uninformative due to incorrect assessment of the elements of the situation. This is the issue that researchers focus on when solving MADM problems.

In MADM problems the state of nature, outcomes mainly are characterized by probability and fuzzy uncertainty. For example, investment problem is such MADM problem. Authors of [1] proposed a dynamic programming approach to solve investment problem including data in the interval type fuzzy numbers. This approach has been used to evaluate interval- valued fuzzy return. It is clear that in this study not considered a probability -based uncertainty.

Author of paper [2] solved an investment problem under uncertainty by using a dynamic programming approach.

In [3, 22] researchers introduced a methodology for solving investment problem. They used expert's evaluations and Chaos Environment for solving investment problem.

R. A. Aliev et al. (Eds.): ICSCCW 2021, LNNS 362, pp. 43–50, 2022.
https://doi.org/10.1007/978-3-030-92127-9_10

In [4] authors studied investment problem with rough interval data. For obtaining the best return they used dynamic programming. Also authors in [5] for solving investment problem used trapezoidal fuzzy numbers.

In [6] an extension of the Merton optimal investment problem is discussed using probability based uncertainty.

Authors of paper [7] analyzed the investment problem by using probability based preferences.

There are several approaches in interval and fuzzy MADM but, minority of scientific works are about multi-attribute decision making under Z-information [8, 23]. Author of the paper [8] is proposed a new approach for MADM based on decision relevant information which characterized by partial reliability and fuzzy uncertainty. Here author used a Continuous Z-number and a distance between Z-number-valued vectors, because Z-numbers is effective tool to express human knowledge, which consist of combination "constraint" and "reliability". There are a number of research works on the use of Z-numbers in decision-making.

In literature existing Z number based studies are characterized by a number of shortcomings, such as the set of probability distributions is often not considered [9–12]. The second part of Z-numbers is not considered as a measure of probability and often Z-numbers are converted to fuzzy numbers [13–16]. It results to loss of reliability information and probabilistic information. It is crystal clear that, analyzing an impact of partial reliability on ranking of alternatives is important.

In this study, proposed approach based on application expected utility using Z^+ information. In investment, it applied for solving decision making problem.

The rest of the paper is as follows. Section 2 is about the preliminaries. Section 3 gives information about a statement of the problem and its solution. Finally, Sect. 4 concludes the paper.

2 Preliminaries

Definition 1 [17]. Z^+-number. Basically, a Z^+-number, Z^+, is a combination of a fuzzy number, A, and a random number, R, written as an ordered pair $Z^+ \equiv (A; R)$. In this pair, A plays the same role as it does in a Z-number, and R is the probability distribution of a random number.

R may be viewed as the probability distribution of U in the Z-valuation (U, A, B) [18]. A Z^+-number may be expressed as (A, p_U) or, equivalently, (μ_A, p_X), where μ_A is the membership function of A. Then a Z^+-valuation is expressed as (U, A, p_U) (or as (U, μ_A, p_X), where p_U is the probability density over U. A Z^+-number is associated with a bimodal distribution as a combination the possibility and probability distributions over U. Then compatibility of these distributions should be considered which in [18] is understood as

$$\int_R u p_U(u) du = \frac{\int_R u \mu_A(u) du}{\int_R \mu_A(u) du}.$$

An adequate in some cases and an effective from computational point assumption is to consider p_U as a parametric distribution, for example, normal distributions with parameters m_U and σ_U. In this case, given $\mu_A = trapmf(u, [a_1, a_2,, a_3, a_4])$ and $p_U = normpdf(u, m, \sigma)$, compatibility conditions requires [18] $gmv(A) = m$.

The probability measure P_A of A is a scalar product of μ_A and p_U, $\mu_A \cdot p_U$:

$$\mu_A \cdot p_U = P_A = \int_R \mu_A(u) p_U(u) du.$$

This connects the concepts of a Z-number a Z^+-number as follows:

$$Z(A, B) = Z^+(A, \mu_A \cdot p_X \text{ is } B).$$

Definition 2 [19]. Expected utility. Let $P : S \to R$ be any probability measure on a set of states S such that P(s) > 0 for all $s \in S$. For each $s \in S$ define $v_s : Z \to R$. Then

$$U(f) = \sum_{s \in S} P(s) v_s(f(s))$$

Where f is an act, $z = f(s)$ is an outcome, $v_s(f(s))$ is a utility in state s and $U(f)$ is the expected value of utility.

Definition 3 [20]. Sum of two normal random variables. If X_1 and X_2 are normal random variable then $Z = f(X_1, X_2) = X_1 + X_2$.

$$\mu_Z = \mu_{X_1} + \mu_{X_2}, \sigma_Z^2 = \sigma_{X_1}^2 + \sigma_{X_2}^2$$

Normal distribution is represented as [21]:

$$p(x) \frac{1}{\sigma\sqrt{2\pi}} e^{\left(\frac{(x-\mu)^2}{2\sigma^2}\right)}$$

The parameter μ is the mean or expectation of the distribution, while the parameter σ is its standard deviation.

3 Investment Problem

Assume that $A = \{A_1, A_2, ..., A_n\}$ is a set of alternatives and $S = \{(s_1), (s_2), ..., (s_j)\}$ is a set of state of nature. All states of the nature are characterized by weight w_j which defined by decision maker. Decision environment is characterized by Z^+-information, the characteristic of the alternative $A_i, i = \overline{1, n}$ on weight w_j $(j = 1...m)$ is described by the crisp data

$$A_i = \{Z(A_{i1}, P_{i1}), Z(A_{i2}, P_{i2}, \ldots, Z(A_{ij}, P_{ij})\}$$

where Z (A_{i1}, P_{i1}) is value of A_i with respect to a state of the nature j. Evaluation of the information and weights of criteria are usually estimated by expert or decision maker

Table 1. Decision matrix

	s_1	s_2	s_3	s_4
A_1	$Z_{11}(A_{11},(\mu_{11},\sigma_{11}))$	$Z_{12}(A_{12},(\mu_{12},\sigma_{12}))$	$Z_{13}(A_{13},(\mu_{13},\sigma_{13}))$	$Z_{14}(A_{14},(\mu_{14},\sigma_{14}))$
A_2	$Z_{21}(A_{21},(\mu_{21},\sigma_{21}))$	$Z_{22}(A_{22},(\mu_{22},\sigma_{22}))$	$Z_{23}(A_{23},(\mu_{23},\sigma_{23}))$	$Z_{24}(A_{24},(\mu_{24},\sigma_{24}))$
A_3	$Z_{31}(A_{31},(\mu_{31},\sigma_{31}))$	$Z_{32}(A_{32},(\mu_{32},\sigma_{32}))$	$Z_{33}(A_{33},(\mu_{33},\sigma_{33}))$	$Z_{34}(A_{34},(\mu_{34},\sigma_{34}))$

and are characterized with partial reliability. Here, A_{ij} is constraint, as fuzzy number, P_{ij} is described as normal distribution (μ_{ij}, σ_{ij}).Hence, if n = 3 and m = 4 then decision matrix as follow (Table 1):

The values of probabilities of states of nature are P(s_1), P(s_2), P(s_3), P(s_4). Main aim in given problem is to select the best alternative.

Numerical example [22].

Suppose, that we need to choose one of three actions: buy bonds (a_1), buy shares of the company (a_2), or put money in the bank for a deposit (a_3). Each of the actions depends on four (m = 4) possible states of nature, which are states of the economy during one year: rapid economic growth (w_1), average economic growth (w_2), stable state of the economy (w_3), economic decline (w_4). The utility function characterizing the rate of return as a percentage of the invested amount is presented in Table 2.

Table 2. The utility values of actions

Alternatives	Rapid Economic growth P(s_1) = w_1 = 0.4	Average economic growth P(s_2) = w_2 = 0.2	Stable state of the economy P(s_3) = w_3 = 0.3	Economic decline P(s_4) = w_4 = 0.1
Bonds	((4,4.8,5.6), (12,4))	((6.5, 8,9.5)(8, 3))	((5,6,7),(6,2))	((2.4,3, 3.6)(3,2))
Share	((12,15,18),(15,5))	((6,7,8),(7, 4))	((2.4,3,3.6),(3, 2))	((–2.4,–2,–1.6)(–2,1))
Deposit	((6,7,8)(7,3))	((6,7,8)(7,2))	((6,7,8),(7,4))	((6,7,8),(7,5))

The expected utility value for A_1 A_2, A_3 are defined as follow:

$$Z_{U(A_1)} = Z_{11}(A_{11},(\mu_{11},\sigma_{11})) * w_1 + Z_{12}(A_{12},(\mu_{12},\sigma_{12}) * w_2 + Z_{13}(A_{13},(\mu_{13},\sigma_{13}) * w_3 + Z_{14}(A_{14},(\mu_{14},\sigma_{14})) * w_4$$

$$Z_{U(A_2)} = Z_{21}(A_{21},(\mu_{21},\sigma_{21})) * w_1 + Z_{22}(A_{22},(\mu_{22},\sigma_{22}) * w_2 + Z_{23}(A_{23},(\mu_{23},\sigma_{23}) * w_3 + Z_{24}(A_{24},(\mu_{24},\sigma_{24})) * w_4$$

$$Z_{U(A_3)} = Z_{31}(A_{31},(\mu_{31},\sigma_{31})) * w_1 + Z_{32}(A_{32},(\mu_{32},\sigma_{32}) * w_2 + Z_{33}(A_{33},(\mu_{33},\sigma_{33}) * w_3 + Z_{34}(A_{34},(\mu_{34},\sigma_{34})) * w_4$$

Using given data and proposed decision making method the expected values of utility for acts A_1, A_2, A_3 are calculated. Obtained results as follow (Figs. 1, 2 and 3):

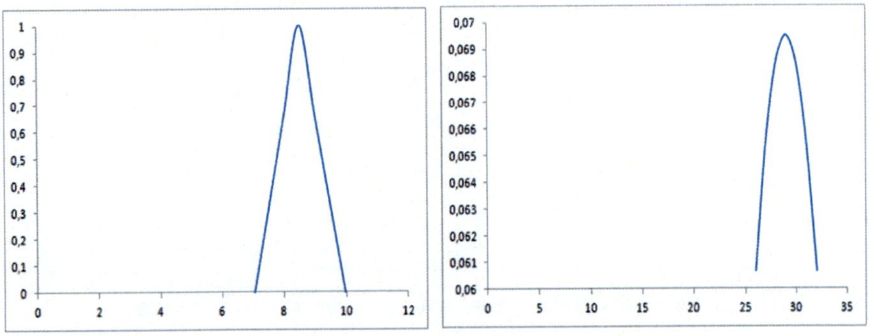

Fig. 1. The expected utility value for A_1 first alternative as Z^+ number

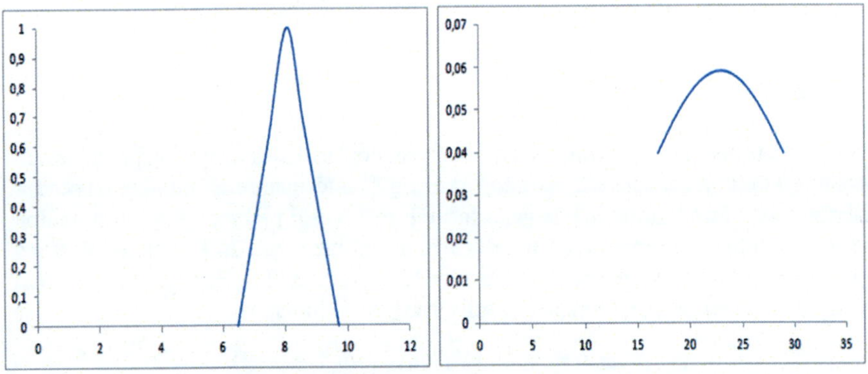

Fig. 2. The expected utility value for A_2 second alternative as Z^+ number

$$Z_{U(A_1)} = ((7,04; 8,5; 9,96),\ (29; 5,744563))$$

$$Z_{U(A_2)} = ((6,48; 8,1; 9,72),\ (23; 6,78233))$$

$$Z_{U(A_3)} = ((6; 7; 8), (28; 7,348469))$$

Probability measure for each alternative is calculated by using obtained data:

$P_{(A_1)} = 0{,}204826$
$P_{(A_2)} = 0{,}171155$
$P_{(A_3)} = 0{,}155075$

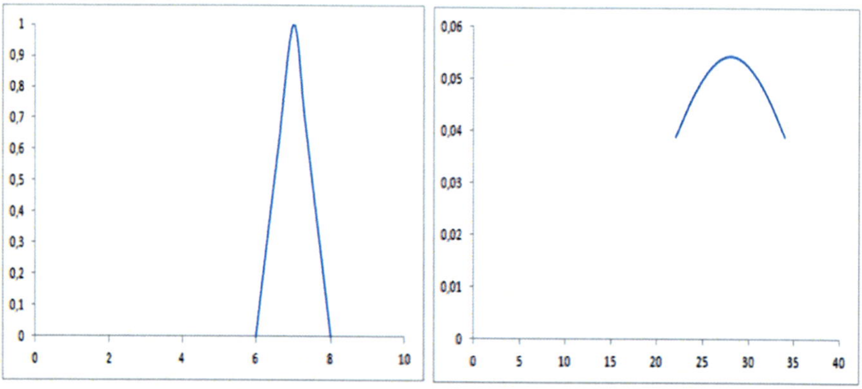

Fig. 3. The expected utility value for A_3 third alternative as Z^+ number

Ranking of Z^+ number of utilizes gives a preference to the first alternative, i.e.

$A_1 > A_2 > A_3$.

4 Conclusion

Mostly, reliability of the information are not considered in decision making processes. In this study we eliminated this shortcoming by using Z^+ information. Z^+ number is used for describing real information in adequate form in real-world problem. Real information and its reliability are represented usually by natural language. In this study obtained results using Z^+ number are compared result which given in [22]. Results described that proposed approach is true. Obtained results using Z^+ number are:

$$Z_{U(A_1)} = ((7,04; 8,5; 9,96),\ (29; 5,744563))$$

$$Z_{U(A_2)} = ((6,48; 8,1; 9,72),\ (23; 6,78233))$$

$$Z_{U(A_3)} = ((6; 7; 8),\ (28; 7,348469))$$

Using Wald criteria and expected maximum utility method solution of this problem is given in [24, p. 224] and in [24, p.227] as 1-st alternative is the best.

Proposed method considers the reliability of the information in alternative selection problem.The proposed algorithm can be used to analyze various real-life problems in investment and other fields.

References

1. Hussah, M.A.: On solving fuzzy investment problem using dynamic programming. Int. J. Comput. Appl. **181**(22), 14–20 (2018). https://doi.org/10.5120/ijca2018917923

2. Abel, A.B., Eberly, J.C.: A unified model of investment under uncertainty. Am. Econ. Rev. **84**(5), 1369–1384 (1994)
3. Sirbilaze, G., Khutsishvili, I., Ghvaberidze, B.: Multistage decision- making fuzzy methodology for optimal investments based on expert's evaluations. Europ. J. Operat. Res. **232**(1), 169–177 (2014). https://doi.org/10.1016/j.ejor.2013.06.035
4. Ammar, E.E., Khalifa, H.A.: On rough interval quadratic programming approach for minimizing the total variability in the future payments to portfolio selection problem. Int. J. Math. Arch. **6**(1), 67–75 (2015)
5. Ammar, E., Khalifa, H.A.: Characterization of optimal solutions of uncertainty investment problem. Appl. Math. Comput. **160**(1), 111–124 (2005). https://doi.org/10.1016/j.amc.2003.08.151
6. Tahar, B.I., Soner, M.H., Touzi, N.: The dynamic programming equation for the problem of optimal investment- under capital gains taxes. Soc. Ind. Appl. Math. **46**, 1779–1801 (2007). https://doi.org/10.1137/050646044
7. Shin, Y.H., Roh, K.-H.: An optimal consumption and investment problem with stochastic hyperbolic discounting. Adv. Dif. Equ. **2019**(1), 1–7 (2019). https://doi.org/10.1186/s13662-019-2144-y
8. Aliyev, R.R.: Multi-attribute decision making based on z-valuation. Proc. Comput. Sci. **102**, 218–222 (2016)
9. Gardashova, L.A.: Application of operational approaches to solving decision making problem using Z-numbers. Appl. Math. **05**(09), 1323–1334 (2014). https://doi.org/10.4236/am.2014.59125
10. Wang, F., Mao, J.: Approach to multicriteria group decision making with Z-Numbers based on TOPSIS and power aggregation operators. Math. Prob. Eng. **2019**, 1–18 (2019). https://doi.org/10.1155/2019/3014387
11. Zhang, S.-S.: A model of based on Z-number and fuzzy analytic hierarchy process. J. Math. Inf. **7**, 63–71 (2017). https://doi.org/10.22457/jmi.v7a8
12. Aliev, R.A., Huseynov, O.H., Serdaroglu, R.: Ranking of Z-numbers and its application in decision making. Int. J. Inform. Tech. Decision Making. **15**(6), 1503–1519 (2016). https://doi.org/10.1142/s0219622016500310
13. Mohamad, D., Shaharani, S.A., Kamis, N.H.: A Z-number-based decision making procedure with ranking fuzzy numbers method. AIP Conf. Proc. **1635**(1), 160–166 (2014). https://doi.org/10.1063/1.4903578
14. Božanić, D., Tešić, D., Milić, A.: Multicriteria decision making model with Z-Numbers based on FUCOM and MABAC model. Dec. Making Appl. Manag. Eng. **3**(2), 19–36 (2020). https://doi.org/10.31181/dmame2003019d
15. Kang, B., Wei, D., Li, Y., Deng, Y.: Decision making using Z-numbers under uncertain environment. J. Computat. Inf. Syst. **8**, 2807–2814 (2012)
16. Kang, B., Wei, D., Li, Y., Deng, Y.: A method of converting Z-number to classical fuzzy number. J. Inf. Comput. Sci. **9**(3), 703–709 (2012)
17. Zadeh, L.A.: A note on a Z-number. Inf. Sci. **181**, 2923–2932 (2011). https://doi.org/10.1016/j.ins.2011.02.022
18. Aliev, R.R., Mraiziq, D.A.T., Huseynov, O.H.: Expected utility based decision making under Z-information and its application. Comput. Intell. Neurosci. **2015**, 1–11 (2015). https://doi.org/10.1155/2015/364512
19. Paul A., Prasanta P., Clemens P.: The Handbook of Rational and Social Choice, Oxford Scholarship (2009)
20. Williamson R.C.: Ph.D. dissertation, University of Queensland, Australia. https://www.ime.usp.br-jstern/miscellanea/general/Williamson87.pdf
21. Weisstein, E.W.: Normal Distribution. mathworld.wolfram.com. August 15, 2020. https://mathworld.wolfram.com/NormalDistribution.htmls

22. Hamiden, K.: A study on investment problem in chaos environment. J. Appl. Res. Ind. Eng. **6**(3), 177–183 (2019). https://doi.org/10.22105/jarie.2019.195913.1099
23. Gardashova L.A.: Z-Number based TOPSIS method in multi-criteria decision making. In: Aliev R., Kacprzyk J., Pedrycz W., Jamshidi M., Sadikoglu F. (eds.) 13th International Conference on Theory and Application of Fuzzy Systems and Soft Computing — ICAFS-2018. Advances in Intelligent Systems and Computing. vol. 896, 42–50 (2018). Springer, Cham. https://doi.org/10.1007/978-3-030-04164-9_10
24. Utkin, L.V.: Risk analysis and decision-making with incomplete information. SPB, Science (2007). (In Russian)

Fuzzy Interval-Valued Temporal Automated Planning and Scheduling Problem

Janusz Kacprzyk[1], Margarita Knyazeva[2(✉)], and Alexander Bozhenyuk[2]

[1] Systems Research Institute Polish Academy of Sciences, Newelska 6, 01-447 Warsaw, Poland
janusz.kacprzyk@ibspan.waw.pl

[2] Southern Federal University, Nekrasovskiy Street, 44, 347928 Taganrog, Russia

Abstract. In this paper the descriptive state-oriented and temporal-oriented models are introduced to treat the assumptions and uncertainty for the automated planning and scheduling problem. The way how to estimate which state or set of possible states may result from performing an operation is discussed. Fuzzy graph interval-valued representation of scheduling problem and state-transition system is introduced. Fuzzy temporal model operates on fuzzy intervals; and both qualitative (precedence relations between operations, finish-to-start relations) and quantitative (interval-valued time durations) constraints are handled by it. The idea of temporal-ordered partial schedule associated with the planning state of the system is discussed. And the finite state machine model (automata) for the planning system under uncertainty is suggested.

Keywords: Automated planning · Fuzzy graph · Interval order · Scheduling · Decision-making · Temporal model · Temporal reasoning

1 Introduction

Decision-making in automated planning and scheduling problems usually presuppose decisions about actions or operations to be performed to achieve an objective. It refers to a reasoning process: what is the sequence of operations and their estimated durations; how to perform operations according to the time and resource restrictions; what is the final goal? Moreover, such kind of problem formulations is significant in the given context and changes in in the environment caused by uncertainty. So different types of assumptions may happen due to unpredictable dynamics of the environment, observability of events, uncertainty managed by such mathematical models and finally in the way how the temporal variables and concurrency are handled.

Dynamics of the environment can be handled by discrete fuzzy interval-valued models. Observability of the environment can be treated as multi-mode case of operation execution, so that some parameters may be always known, and the others may stay as observable in time. Uncertainty in decision-making problem presupposes the current state estimation of the model during the execution stage. And finally, the temporal aspect and concurrency may be handled as topological ordering of the events in time meeting

R. A. Aliev et al. (Eds.): ICSCCW 2021, LNNS 362, pp. 51–58, 2022.
https://doi.org/10.1007/978-3-030-92127-9_11

deadlines, precedence relations and resource consumption needed to perform an action or operation.

Temporal reasoning is common for the artificial intelligence problems, for example, recurrent neural network (RNN) where connections between nodes in a directed graph and a temporal sequence allow investigating temporal dynamic behavior. A number of works have been devoted to temporal reasoning (changing environment, events, operation execution), for example, qualitative approach was presented by Allen [1–3], first-order temporal logic was developed by Drew McDermott [4] and allowed to consider events and plans with causality, changes in quantities and relations between elements. A number of such works have laid the foundation for the implementation of a temporal-inference machine, which keeps in memory different "maps" of the timeline, one per discrete time moment. The Handbook of Temporal Reasoning in Artificial Intelligence raised the actual topics on time granularity, modal varieties of temporal logic and computational complexity of temporal constraint problems [5]. A formal discussion devoted to temporal networks was introduced by Ghallab et al. in Chapter 13 [6] and the book of Bartak et al. [7]. Dechter et al. proposed the framework, called temporal constraint satisfaction problem (TCSP), in which the variables represented time points and the temporal knowledge was presented by constraints or permitted intervals; the authors also investigated path consistency algorithms [8].

The paper is organized as follows. Section 2 introduces state-variable representation for the scheduling problem and discusses the applicability of planning domain approach to automated planning. Section 3 introduces temporal-oriented representation, fuzzy intervals and fuzzy relations for the scheduling problem. And finally Sect. 4 discusses fuzzy graph interval-valued representation of scheduling problem and state-transition system, which allows estimating current planning stage and partial plans during the operational stage.

2 State-Variable Representation of the Scheduling Problem

In automated planning it is common to use some formal notations, or planning domain to introduce the descriptive model. Descriptive model uses the formal apparatus of a finite-state automata (or finite-state machine) and classical planning assumptions, such as: requiring the sets of states and operations to be finite; discrete sequence of states and operations to be scheduled; the certain prediction about what state will be the system in if an action a is performed in a state s. Finite-state machine operates with the states, rules for the transition from one state to another and depends on the applied input and output.

Let us consider a state-transition system formulation model and some computational aspects of using it for scheduling problem.

Definition 1. *A State-transition system or planning domain* is a 4-tuple $\Sigma = (S, O, \gamma, cost)$ where:

S – is a finite set of states for the system, or a time-oriented partial plans;

O – is a finite set of operations to be scheduled and performed;

γ – is a state-transition function, that gives the next state, or possible next states, after an action or event, $S \times A \rightarrow S$ with $\gamma(s,a)$ being a predicted outcome.

Cost – is a partial function, $S \times A \rightarrow [0, \infty)$ having the same domain as γ. The cost function may represent monetary cost, time, or other resources to be minimized.

It is also common to represent each state $s \in S$ for the planning system as a set of features (properties) and to define a set of operators that can be used to compute transitions between the states. The way to estimate each state and specify the members of the states for the set *S* is to formulate a set of *consistency constraints*, so this set plays the role of restrictions over the possible combinations of variable assignments and check the state-transition function.

Definition 2. Let **R** and **X** be **sets of relations** and **state variables** over a set of objects B, and **S** be a **state-variable state space** over X. An *action template* (planning operators or action schemas) for S is a tuple $\alpha = (head(\alpha); pre(\alpha); eff(\alpha); cost(\alpha))$, where $head(\alpha)$ is a syntactic expression (planning operators or action schemas); $pre(\alpha)\{p_1, \ldots, p_m\}$ is a set of preconditions, $eff(\alpha)\{e_1, \ldots, e_n\}$ is a set of effects for the state; $cost(\alpha)$ denotes the cost of the applying action [9].

Definition 3. *A plan or a partial schedule* is a certain finite temporal-ordered sequence of actions or operations to be executed as follows: $\pi = o_{1t}, o_{2t}, \ldots, o_{nt}$ and the total cost of such plan is $cost(\pi) = \sum_{i=1}^{n} cost(o_{it})$.

Definition 4. A temporal-ordered partial schedule for the number of operations $\pi = o_{1t}, o_{2t}, \ldots, o_{nt}$ is *applicable* to estimate the state s_0 if there are states $s_1, s_2, \ldots, s_n$ such that: $\gamma(s_{i-1}, o_i) = s_i$ for $i = 1, \ldots, n$.

So we can define a number of states associated with the temporal-ordered partial schedules as follows:

$$\gamma(s_0, \pi) = s_n; \tag{1}$$

$$\hat{\gamma}(s_0, \pi) = s_0, s_1, \ldots, s_n. \tag{2}$$

In terms of discrete mathematics $\hat{\gamma}$ is called the transitive closure of γ. If a certain operation *o* isn't applicable in a state *s* in discrete time moment *t*, then $\gamma(s, o)$ is undefined. If an operation *o* is applicable in state *s* in discrete time moment *t*, then:

$$\gamma(s_t, o)(x) = \begin{cases} w, \text{if } eff(o) \text{contains an effect } x \leftarrow w, \\ s_t(x), \text{otherwise}, \end{cases} \tag{3}$$

where *x* – is a state variable from the set *X*.

In Resource-Constrained Scheduling Problems (RCSP) temporal modelling sub-problem usually presuppose variables, constraints and objective function to be minimized. Operations need to be planned and expected to occur during discrete time periods (or fuzzy intervals) and are to be located in time with respect to deadlines and resource constraints. A decision-maker uses a predictive model to decide what operations to execute and how to do this with respect to time-cost trade-offs. Such kind of goals define final solutions for partial or full temporal-scheduled plans, they are called *goal states*.

Definition 5. A *state-variable scheduling problem* is a triple $P = (s_0, g)$, where Σ is a state-variable planning domain, s_0 is an initial planning state of the system, and g is the set of goals.

Any feasible solution (partial temporal-scheduled plan) should satisfy the following condition: a solution for P is any execution plan for the operations that are to be located in timeline $\pi = o_{1t}, o_{2t}, \ldots, o_{nt}$ such that the state $\gamma(s_0, \pi) = s_n$ should satisfy goal g and $\gamma(s_0, \pi) \in S_g$, where S_g – is the set of goal states for the planning system. Any solution can be minimal according to the number of operations, temporal-shortest or cost-optimal. Operations execution may also overlap in time, if their conditions, execution modes and resource-effects are independent, but the goals are usually limited by deadlines and cost of the project.

Temporal modelling for the scheduling problems implies qualitative approach to handle operation allocation or topological ordering, as well as quantitative approach to handle uncertain durations, start-to-finish relations between operations with respect to numerous parameters. A formal statement of the problem with temporal variables for resource-constrained scheduling problem can be both *state-oriented* with the transitions between states and *temporal-oriented* within the timeline.

The next Sect. 3 discusses temporal-oriented method for modelling which apply intervals as temporal primitives, and take into consideration both qualitative and quantitative relations between operations in discrete time moments.

3 Temporal-Oriented Representation of the Scheduling Problem. Fuzzy Intervals and Fuzzy Relations

Let's consider a quantitative discrete model of time described by a set of temporal variables, for example, $t_1, t_2, \ldots, t_n$ and each variable designates a time point. An interval-valued variable is characterized by a pair $[t_1, t_2]$ such that $t_1 < t_2$ and its duration is positive number.

Definition 6. *A fuzzy interval* $\tilde{I}(\cdot)$ and the membership function $\mu_I(\cdot)$ represent a fuzzy subset, if the following condition is true:$\forall(x, y, z) \in \Re$, $z \in [x, y]$, and $\mu_I(z) \geq \min(\mu_I(x), \mu_I(y))$ [10, 11]. So fuzzy interval in scheduling problem is a fuzzy set in the real timeline whose left-right (L-R) α-cuts are intervals, and each state variable *x(t)* is a function of time interval.

A fuzzy interval can be also specified by an ordered pair of L-R intervals $\left(\tilde{l}^-, \tilde{r}^+\right)$, where $\tilde{l}^-$ is called the fuzzy lower bound and $\tilde{r}^+$ is called the fuzzy upper bound, and $A_{\tilde{M}}$ is an assignment function [12].

From the time-oriented point of view the state variable x can be characterized by its permanency (or persistence) and change within the fuzzy temporal assertion.

A *persistence* of $\left[\tilde{t}_1, \tilde{t}_2\right]x = v$ denotes that $x\left(\tilde{t}\right) = v$ for every $\tilde{t}$ in the temporal fuzzy interval $\tilde{t}_1 \leq \tilde{t} \leq \tilde{t}_2$.

A *change* $\left[\tilde{t}_1, \tilde{t}_2\right]x : (v_1, v_2)$ shows that the value of state variable x change over the fuzzy interval $\left[\tilde{t}_1, \tilde{t}_2\right]$ as follows: from $x\left(\tilde{t}_1\right) = v_1$ to $x\left(\tilde{t}_2\right) = v_2$ with $v_1 \neq v_2$.

Figure 1 illustrates a timeline for state variable x, where the persistence can be characterized by the fixed membership function $\mu_I(\cdot)$ for the interval $\left[\tilde{t}_2, \tilde{t}_3\right]$.

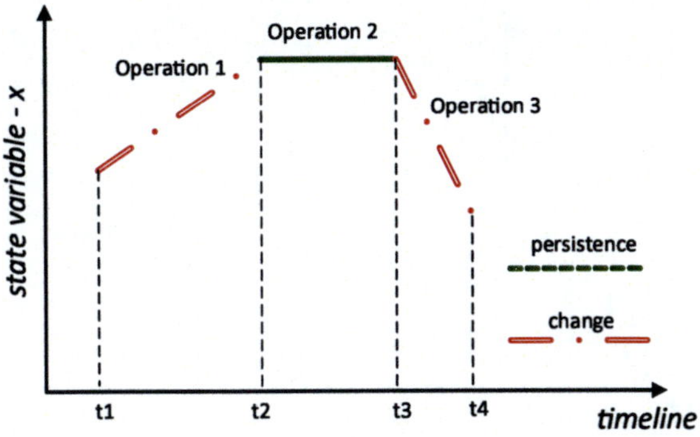

Fig. 1. A timeline for the state variable x in scheduling problem.

For example, assertion $[\tilde{t}_1, \tilde{t}_2]x$: (*operation*1, *operation*2) illustrates that state variable x (resources profile needed to execute operations) may change over time. The precise crisp moments of this change are not fixed, but the Fig. 2 shows that the values of state variable x for *operation 1* and time interval $[\tilde{t}_1, \tilde{t}_2]$ change and can be initialized only at acting stage. This kind of uncertainty can be handled by fuzzy intervals as well. Temporal statements in a timeline and their instances make sense in the corresponding planning domains and reflect possible evolution of the state variable x.

Definition 7. *An interval ordering* for a number of fuzzy temporal intervals $[\tilde{t}_i, \tilde{t}_j]$ on a real timeline is the partial order (schedule) or a sequence of temporal interval-valued operations to be planned with respect to precedence relations, cost function and finish-to-start relations.

Definition 8. *A fuzzy relation* $\tilde{R}$ is a mapping from the Cartesian space X_1 x X_2 to the interval [0,1], where the strength of the mapping is expressed by the membership function of the relation $\mu(x, y)$. Fuzzy temporal relation shows a degree of the presence or absence of interaction between the elements of two or more fuzzy interval sets.

The most traditional crisp method for formalizing the qualitative relations between time-dependent intervals is Allen's Interval Algebra, or Interval Algebra (IA) [2]. Allen introduced 13 temporal-depended relations between start-to-finish crisp intervals that can be applied in scheduling precedence relations between operations as follows: before, after, meets (is met by), overlaps (is overlapped), during (includes), starts (is started by), finish (is finished by) and equals [12].

Let's consider the following example. There are two sets: $X_1 = \{o_1, o_2, \ldots, o_5\}$ of crisp operations and $\tilde{X}_2 = \{\tilde{d}_1, \tilde{d}_2, \ldots, \tilde{d}_5\}$ of fuzzy temporal durations (they are presented as intervals), and the relation $R(X_1, \tilde{X}_2) = \{(o_1, \tilde{d}_2), (o_3, \tilde{d}_4), (o_4, \tilde{d}_1), (o_5, \tilde{d}_3)\}$ between them. This kind of relation is binary and is composed by the pairs $(o_i, \tilde{d}_j)$; and with the following membership function $\mu_R(o_i, \tilde{d}_j)$ this kind of fuzzy relation can be represented in three-dimensional space (Fig. 2).

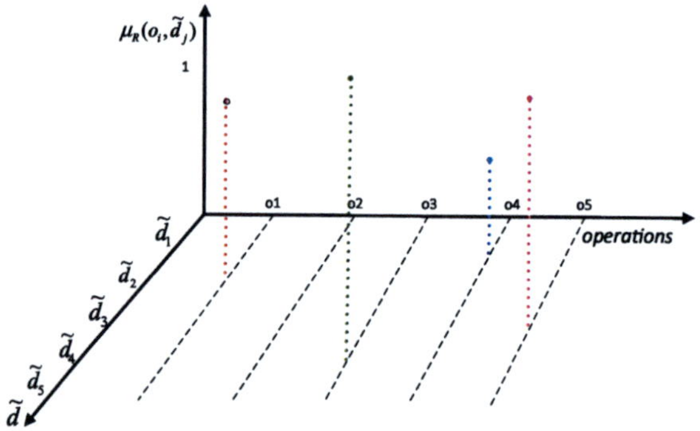

Fig. 2. A three-dimensional representation of fuzzy relations between operations and fuzzy durations.

A fuzzy *n*-ary relation $\tilde{R}$ defined on the domain $X = X_1 \times \ldots \times X_n$ is an ordered set of tuples of n-elements as follows:

$$\tilde{R} = \{((x_1, \ldots, x_n), \mu_R(x_1, \ldots, x_n)) | (x_1, \ldots, x_n) \in X\}, \tag{4}$$

where $\mu_R(x_1, \ldots, x_n) : X_1 \times \ldots \times X_n \to [0, 1]$ is a membership function of relation R, which maps the domain to a continuous interval [0,1].

An interval ordering problem in scheduling and planning can be treated as a special case of *interval distance geometry problem (iDGP)*, where distances are not given precisely. The main distance geometry problem as a decision-making problem is to estimate measurements for the pairwise distances between elements in some set and to calculate the coordinates of those points.

The next Sect. 4 discusses fuzzy graph interval-valued representation of scheduling problem and state-transition system, which allows estimating current planning stage and partial plan during the operational stage.

4 Fuzzy Graph Representation of the Interval-Valued Scheduling Problem

Let's consider a connected fuzzy graph $G = \left(V, E, \tilde{d}\right)$, where $\tilde{d} : E \to \mathbb{R}_+$ is positive fuzzy edge weights and the goal is to find embedding of the vertices V into Euclidean space $V \to \mathbb{R}^K$, so that for every edge $\{u, v\}$ the equation is true:

$$\forall \{u, v\} \in E \|x_u - x_v\| = \tilde{d}_{uv}. \tag{5}$$

In automated planning and scheduling problems durations of operations are usually uncertain due to the nature of the planning process and interval-valued problem can be presented as follows. Given a positive integer K and interval-valued weighted fuzzy graph $G = \left(V, E, \tilde{d}\right)$, where fuzzy durations maps edges $\{u, v\} \in E$ to positive interval weights $\left[\underline{d}(\{u, v\}), \overline{d}(\{u, v\})\right]$, the distance geometry problem is the problem of constructing the graph G, mapping $x : V \to \mathbb{R}^K$ and satisfying the interval distance constrains with respect to precedence relations:

$$\underline{d}(\{u, v\}) \leq \|x_u - x_v\| \leq \overline{d}(\{u, v\}), \forall \{u, v\} \in E, \tag{6}$$

where $\|\cdot\|$ denotes the Euclidean norm.

Each discrete moment is associated with partial schedule (the number of operations to be executed) and corresponds to a certain state S^t of the planning system, see Fig. 3.

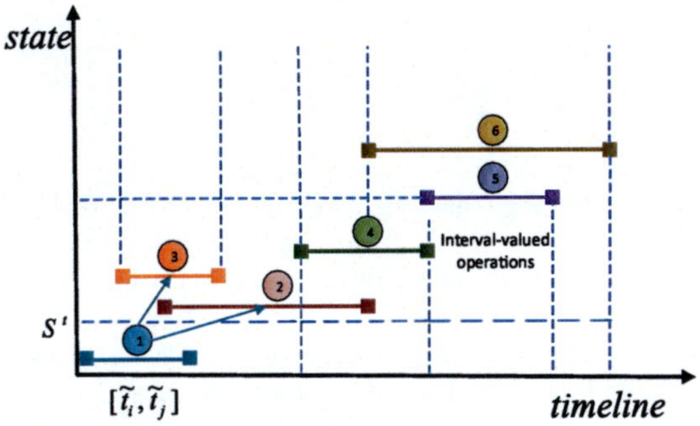

Fig. 3. Interval-valued representation of fuzzy precedence relations between operations, durations and states of the planning system.

The finite state machine model (automata) for the planning system under uncertainty can be stated as follows.

Definition 9. *A finite automata* is a 5-tuple $M = (S, \sigma, s_0, F)$, where S – is a finite set of possible states, Σ - is a finite set called alphabet, $\sigma : S\times \to Pos(S)$ is the transition function from state to possible state with respect to precedence relations of operations, $s_0 \in S$ is the start state, and $F \subseteq S$ is the set of accept states or final states of the planning system [13].

The definition above was given for the deterministic finite automata, but in practice of planning several ways may exist for the next state at any discrete moment *t*. Nondeterministic computation then is a tree with the root corresponds to the start of the planning process. Every branching point in the tree corresponds to a point in the computation at which the machine has multiple choices to select. In paper [12] we introduced the approach based on planning computational graph. Having a discrete sequence of states, time constraints, precedence relations between operations the scheduling problem can be solved by the computational planning algorithm for the s-states. For each state of enumeration scheme this algorithm evaluates all possible partial schedules and stores the current solution until all the operations are scheduled [12].

5 Conclusion

The paper discusses the descriptive state-oriented and temporal-oriented models to treat the constraints and uncertainty for the automated planning and scheduling problem. The main idea of the paper is to show how temporal reasoning and the finite state machine model (automata) can be applied for the planning system within qualitative and quantitative approach. Fuzzy graph interval-valued representation of scheduling problem and the state-transition system are introduced in this paper.

Acknowledgments. The reported study was funded by the Russian Foundation for Basic Research according to the research project #20-01-00197.

References

1. Allen, J., Kautz, H., Pelavin, R., Tenenberg, J.: Reasoning about Plans. Morgan Kaufmann, San Mateo (2014)
2. Allen, J.F.: Maintaining knowledge about temporal intervals. Commun. ACM **21**(11), 832–843 (1983)
3. Allen, J.: Planning as temporal reasoning. In: Proceedings of the International Conference on Principles of Knowledge Representation and Reasoning (KR), pp. 3–14. Morgan Kaufmann, San Mateo (1991)
4. McDermott, D.: A temporal logic for reasoning about processes and plans. Cogn. Sci. **6**, 101–155 (1982). https://doi.org/10.1016/S0364-0213(82)90003-9
5. Fisher, M., Gabbay, D.M., Vila, L.: Handbook of Temporal Reasoning in Artificial Intelligence, vol. 1. Elsevier (2005)
6. Ghallab, M., Nau, D.S., Traverso, P.: Automated Planning: Theory and Practice. The Morgan Kaufmann Series in Artificial Intelligence, Morgan Kaufmann, Amsterdam (2004)
7. Bartak, R., Morris, R., Venable, B.: An Introduction to Constraint-Based Temporal Reasoning. Morgan Claypool (2014). https://doi.org/10.2200/S00557ED1V01Y201312AIM026
8. Dechter, R., Meiri, I., Pearl, J.: Temporal constraint networks. Artif. Intell. **49**, 61–95 (1991). https://doi.org/10.1016/0004-3702(91)90006-6
9. Ghallab, M., Nau, D., Traverso, P.: Automated Planning and Acting. Cambridge University Press (2016)
10. Fortin, J., Dubois, D., Fargier, H.: Gradual numbers and their application to fuzzy interval analysis. IEEE Trans. Fuzzy Syst. **16**(2), 388–402 (2008)
11. Dubois, D., Kerre, E., Mesiar, R., Prade, H.: Fuzzy interval analysis. In: Dubois, D., Prade, H. (eds.) Fundamentals of Fuzzy Sets, pp. 483–581, The Handbooks of Fuzzy Sets Series, vol. 7. Springer, Boston (2000). https://doi.org/10.1007/978-1-4615-4429-6_11
12. Knyazeva, M., Bozhenyuk, A., Kaymak, U.: Fuzzy temporal graphs and sequence modelling in scheduling problem. In: Lesot, M.-J., et al. (eds.) IPMU 2020. CCIS, vol. 1239, pp. 539–550. Springer, Cham (2020). https://doi.org/10.1007/978-3-030-50153-2_40
13. Sipser, M.: Introduction to the Theory of Computation. Cengage Learning (2012)

Z-numbers Based Preference of Expert Opinions on Social Capital

G. Imanov, M. Murtuzaeva, and A. Aliyev(✉)

Control Systems Institute of the Azerbaijan National Academy of Sciences, Baku 1141, Azerbaijan

Abstract. This article sets out the methodology for description of preference of expert opinions (PEO) on social capital (SC) indicators. For this purpose, the theory and instruments of fuzzy Z-numbers are employed. The Z-number theory opened new horizons for many applications, especially in the areas of computation with probabilities and events stated in natural language. The generation of z-numbers is an effective tool to make choices among expert evaluations. With the objective to make PEO, the experimental researches on modified generation of Z-numbers based on ordered weighted average and maximum entropy is implemented. Firstly, on the basis of three probability distribution functions, corresponding normative Z-numbers are constructed. Then, in order to establish the most reliable version of expert opinions, the distances between the obtained probability distributions, and consequently the degrees of similarity and credibility are calculated. As a result, the rating of the expert opinions is assessed, in which the preferred one illustrates the level of SC in Azerbaijan.

Keywords: Social capital · Expert opinions · Z-numbers · Weights · Maximum entropy

1 Introduction

Social capital - made up of social commitments ("connections"), which is transformable, in peculiar conditions, into economic capital and may be established in the form of a nobility title.

SC is the accumulation of the current or prospective resources that ensures each of its members with the funding of the commonly - owned capital, a "credential" which entitles them to credit, in the multiple senses of the word [1].

Diverse concepts of the SC are specified as the institutions, relationships, and values that conduct communications among people and facilitate the economic and social progress [2].

SC is established by the OECD as "network comprising shared norms, values and perceptions that facilitate cooperation within or between groups". In this conceptualization, we can articulate this networks as real-life bonds inside of groups or between individuals. Consider the networks of friends, networks of families, networks of ex colleges, and so on. Our common norms, values and perceptions are less exact than our social networks.

R. A. Aliev et al. (Eds.): ICSCCW 2021, LNNS 362, pp. 59–66, 2022.
https://doi.org/10.1007/978-3-030-92127-9_12

Sociologists occasionally refer to norms, as society's tacit and unarguable rules. Values may be more open to question; actually societies often discuss whether their values are changing. But still, values – such as regard for people's safety and security – is a central focal point in every social group. Bring together, these networks and perceptions generate trust and so incite people to interact and cooperate [3].

Solability Swiss-Korean joint venture outlined the SC of a nation as the convergence of social stability and the well-being of the entire population. SC engenders social solidarity, which in turn facilitates a stable climate for the economy, and restrains natural resources from being exploited extensively. SC is not a tangible value, so it is difficult to quantify and measure in numeric values [4]. Definitions disagree but in general converge to those networks of relationships of people, who show faith and solidarity mutually, capacitating the society to cooperate and function productively.

In the course of evaluations of expert opinions and making PEO the pilot study of Tian Y. and Kang B. [5] is used. As compared to earlier work [6], in later some disadvantages were eliminated and generating z-numbers was modified.

2 Expert Evaluations of Social Capital Indicators

In this paper we evaluate Azerbaijan SC indicators proposed by two experts. In an effort to develop social capital indicators (SISC) presented by UN Basel Institute of Commons and Economics [3, 4], the Healthcare and Corruption indicators were also added to the list:

1. Social climate (psychological environment, social conditions) is characteristically outlined as the concepts of a social atmosphere that is bound to be shared by the members of the society – **SC**
2. The interpersonal trust among the people – to expect that someone is righteous and trustworthy and will not mistreat you – **TR**
3. Interest of the people to take individual strict measures with a view to fund public goods such as: safety, healthcare, education, environmental challenges, infrastructure, social allowances, media, culture – **PG**
4. Interest of the people to be involved in paying more taxes and making contributions to fund public goods such as: safety, healthcare, education, environmental challenges, infrastructure, social allowances, media, culture – **PT**
5. Willingness of the people for adding up to national and regional level investments such as: purchasing of cooperative shares, acquisition of national and local stocks, purchasing of SME shares (small and medium enterprises), foundation of private or family business – **IE**
6. Helpfulness of the people – the character of offering necessary help, and (2) friendliness evinced by warm and gracious attitude – **HE**
7. Friendliness of the people – The character of a person to be kind, and having affection for people – **FR**
8. Hospitality of the people – welcoming, entertaining, and generous treatment of guests, visitors or strangers - **HO**

9. Healthcare - Health care is the entire societal commitment, organized or not, either private or public, that aims to maintain, provide, fund, and improve health – **HL** [4]
10. Corruption is a severe crime that deflowers the social and economic growth and destroys the constitution of contemporary societies – **CO** [7]

The expert opinions are expressed in zero to ten scale [0–10] respectively, where 0 – very bad and 10 – very high, which are represented in Table 1.

Table 1. Expert opinions in crisp values

	SC	TR	PG	PT	IE	HE	FR	HO	HL	CO
Expert. Opinion 1	6	6	5	5	7	9	8	9	6	7.5
Expert. Opinion 2	9	9	5	6	6	10	10	10	7	9

With a view to evaluate expert opinions on SC in Azerbaijan modified method of generating Z-numbers is applied, which proposed by Ye Tian and B. Kang [6], which mainly advantaged from the ideas of Zadeh L. [8], Aliev R.A. [9], and Aliev R.A., Huseynov O.H., Zeinalova L.M. [10].

3 Estimation of the Values of Generalized Z-Numbers

For every version, experts were proposed three scenarios of weight distributions:

normative - $W^1[0.1, 0.1, 0.1, 0.1, 0.1, 0.1, 0.1, 0.1, 0.1, 0.1]$,
intuitionistic - $W^2[0.04, 0.04, 0.15, 0.15, 0.19, 0.07, 0.17, 0.07, 0.04, 0.08]$ is taken from work [11],
optimistic - $W^3[1, 0, 0, 0, 0, 0, 0, 0, 0, 0]$.

The ORNESS measure that evaluates each expert's attitude (W), which proposed by Yager, Kacprzyk, [12] is expressed by the following formula (1):

$$ORNESS(W) = \alpha = \frac{\sum_{i=1}^{n}(n-i) * w_i}{n-1} \tag{1}$$

where w_i (i = 1,2,..,10) are components of the vector W^i (i = 1,2,3).

For normative distribution ORNESS α_1 is calculated as:

$$\alpha_1 = \frac{1}{9}\sum_{i=1}^{10}(n-i) * w_i = \frac{1}{9} * \frac{1}{10}\sum_{i=1}^{10}(10-i) = \frac{45}{90} = 0.5$$

Following the same rule α_2, α_3 are calculated:

$$\alpha_2 = 0.505, \quad \alpha_3 = 1$$

Using the maximum entropy formula, the vector of the probability distribution of the options accepted by the expert can be calculated, that is, using α_i $(i = 1, 2, 3)$ the following optimization problem (2) is solved:

$$\max H(W) = -\sum_{i=1}^{n} w_i * \ln(w_i)$$

$$\text{s.t.} \quad \begin{cases} \alpha = \frac{\sum_{i=1}^{n}(n-i)*w_i}{n-1} \\ \sum_{i=1}^{n} w_i = 1 \\ 0 \le w_i \le 1 \end{cases} \tag{2}$$

Afterwards, the probability for the constraint part A is denoted as:

$$p_X(x_i) = w_{n-i+1}$$

As a result, referring to papers [5, 6] the following outcomes are obtained:

$$P_N = [0.1, 0.1, 0.1, 0.1, 0.1, 0.1, 0.1, 0.1, 0.1, 0.1],$$
$$P_i = [0.04, 0.04, 0.15, 0.15, 0.19, 0.07, 0.17, 0.07, 0.04, 0.08],$$
$$P_O = [0, 0, 0, 0, 0, 0, 0, 0, 0, 1]$$

Expert opinions on SC indicators which proposed by expert I and II given in crisp values are fuzzified, using fuzzy triangle membership function (where a = 6, b = 5.5, c = 10 is pre-set), which is given in Table 2.

Table 2. Membership function values of expert opinions on SC indicators

	SC	TR	PG	PT	IE	HE	FR	HO	HL	CO
I Expert Opinion	0.89	0.89	0.89	0.89	0.67	0.22	0.44	0.22	0.89	0.55
II Expert Opinion	0.67	0.67	0.89	0.89	0.89	0.00	0.00	0.00	0.67	0.67

By utilizing indicators from Table 2, the constraint part (A_i) of Z-number for normative attitude is calculated as below:

$$A_1 = \frac{\mu_{A_1}}{x} = \frac{0.89}{6} + \frac{0.89}{6} + \frac{0.89}{5} + \frac{0.89}{5} + \frac{0.67}{7} + \frac{0.22}{9} + \frac{0.44}{8} + \frac{0.22}{9} + \frac{0.89}{6} + \frac{0.55}{7.5}$$
$$A_2 = \frac{\mu_{A_2}}{x} = \frac{0.67}{9} + \frac{0.67}{9} + \frac{0.89}{5} + \frac{0.89}{5} + \frac{0.89}{5} + \frac{0.00}{10} + \frac{0.00}{10} + \frac{0.00}{10} + \frac{0.67}{7} + \frac{0.67}{9}$$

Fuzzy probability of the argument X for the constraint part A_1, A_2, are calculated:

$$\frac{P_x(x_i) * \mu_{A_1}(x_i)}{x_i} = \frac{0.89 * 0.1}{6} + \frac{0.89 * 0.1}{6} + \frac{0.89 * 0.1}{5} + \frac{0.89 * 0.1}{5} + \frac{0.67 * 0.1}{7} + \frac{0.22 * 0.1}{9} + \frac{0.44 * 0.1}{8} + \frac{0.22 * 0.1}{9} + \frac{0.89 * 0.1}{6} + \frac{0.55 * 0.1}{7.5}$$

$$\frac{P_X(x_i) * \mu_{A_2}(x_i)}{x_i} = \frac{0.67 * 01}{9} + \frac{0.67 * 0.1}{9} + \frac{0.89 * 0.1}{5} + \frac{0.89 * 0.1}{5} + \frac{0.89 * 0.1}{5} + \frac{0.00 * 0.1}{10} + \frac{0.00 * 0.1}{10} + \frac{0.00 * 0.1}{10} + \frac{0.67 * 0.1}{7} + \frac{0.67 * 0.1}{9}$$

4 Obtaining the Reliability Values of Expert Opinions

On the base of calculation $P(A)$, the quantity of b of the reliability part (B) is obtained by formula (3):

$$b = p(A) = \int_R \mu_A(x) p_X(x) dx,\ b \in B \tag{3}$$

Herein, R is the domain of real numbers.

The quantity of reliability part of B for expert I and II are:

$$b_1^1 = P\left(A_1^1\right) = 0.655 \quad b_1^2 = P\left(A_1^2\right) = 0.535$$
$$b_2^1 = P\left(A_2^1\right) = 0.651 \quad b_2^2 = P\left(A_2^2\right) = 0.565$$
$$b_3^1 = P\left(A_3^1\right) = 0.55 \quad b_3^2 = P\left(A_3^2\right) = 0.67$$

Here, $i = 1, 2, 3$ (subscripts) denote weight distribution versions, $j = 1, 2$ (superscripts) denote experts.

As evidenced from calculations the degrees of fuzzy membership functions equal to the value of *ORNESS*:

$$\mu_B(x) = \mu_{p_A}(p_A) = k\alpha_1$$
$$\mu_{B_1} = \mu_{PA1}(P_{A1}) = \alpha_1 = 0.5;$$
$$\mu_{B_2} = \mu_{PA2}(P_{A2}) = \alpha_2 = 0.505$$
$$\mu_{B_3} = \mu_{PA3}(P_{A3}) = \alpha_3 = 1$$

To establish the most reliable version of expert opinions, the distance measures between the obtained probability distributions, the degrees of similarity and credibility have to be calculated.

The Hellinger distance for discrete probability distributions is calculated by the following formula (4):

$$D_H(P, Q) = \frac{1}{\sqrt{2}} \sqrt{\sum\nolimits_{i=1}^{n} \left(\sqrt{p_i} - \sqrt{q_i}\right)^2} \tag{4}$$

All possible distances between distributions are also calculated and the following values are obtained:

$$D_H(P_1, P_2) = 0.185, \quad D_H(P_1, P_3) = 0.805, \quad D_H(P_2, P_3) = 0.843$$

Constructed elements of similarly matrix between three probability distribution (5) defined as:

$$Sim(p_i, p_j) = 1 - D_H(p_i, p_j) \tag{5}$$

The obtained values are:

$$SimP_{12} = 0.815, \quad SimP_{13} = 0.195, \quad SimP_{23} = 0.157$$

On the base of results obtained above the similarity measure matrix (SMM) is constructed as follows:

$$\text{SMM} = \begin{pmatrix} 1.00 & 0.815 & 0.195 \\ 0.815 & 1.00 & 0.157 \\ 0.195 & 0.157 & 1.00 \end{pmatrix}$$

The support degree $Sup(P_j)$, is defined by the following formula (6):

$$Sup(P_j) = \sum_{\substack{i=1 \\ i \neq j}}^{n} Sim(P_i, P_j) \tag{6}$$

As a result:

$$Sup(P_1) = 0.815 + 0.195 = 1.01$$
$$Sup(P_2) = 0.815 + 0.157 = 0.972$$
$$Sup(P_3) = 0.195 + 0.157 = 0.352$$
$$\sum\nolimits_{i=1}^{3} Sup(P_i) = 2.334$$

Creditability degree of probability distributions is defined by formula (7):

$$crd_j = \frac{Sup(P_j)}{\sum_{j=1}^{n} Sup(P_j)}, \ (i = 1, 2, 3) \tag{7}$$

That can be employed as weight and transformed into discount coefficient to get the membership degree of the reliability part B. The outcomes appear as follows:

$$crd_1 = \frac{1,01}{2,334} = 0,433$$
$$crd_2 = \frac{0,972}{2,334} = 0,416$$
$$crd_3 = \frac{0,352}{2,334} = 0,151$$

The membership degree for reliability part (B) of Z-number is defined as (8):

$$\mu_B^j(x) = \frac{k * \alpha_j * crd_j}{max(k * \alpha_j * crd_j)} \quad j = \overline{(1, , n)} \tag{8}$$

In this research, for the sake of simplicity, the relation coefficient k between orness measure and membership function of B is set: $k = 1$

$\alpha_1 = 0.5, \quad \alpha_2 = 0.505, \quad \alpha_3 = 1$
$k * \alpha_1 * crd_1 = 0.439 * 0.5 = 0.2165 \quad \mu_{B_1}(x) = \frac{0.5 * 0.433}{0.2165} = 1$

$$k * \alpha_2 * crd_2 = 0.505 * 0.416 = 0.2101 \quad \mu_{B_2}(x) = \frac{0.505 * 0.416}{0.2165} = 0.9703$$

$$k * \alpha_3 * crd_3 = 1 * 0.151 = 0.151 \quad \mu_{B_3}(x) = \frac{0.151 * 1}{0.2165} = 0.692$$

In the end, the generated Z-numbers have been developed all-inclusive. This allows one to choose the most reliable option with the maximum value of the B part of the Z-number, which corresponds to the normative option. For the normative attitude, the obtained values of the Z-numbers for both experts are as followings:

$$B_1^1 = 0.5/0.655 \quad B_1^2 = 0.5/0.535$$

It is obvious that both expert opinions have equal membership degree for the normative option. But, the opinion of the I expert is preferred due to the greatest reliability.

5 Conclusion

In this paper, we made PEO with the application of the theory and instruments of Z-numbers. Concept of Z-numbers is suitable to make PEO, on that account we applied Kang's modified generalized Z-number method based on ordered weighted average and maximum entropy. Taking into consideration the reliability part of Z-numbers results in more reasonable solution that contemplates human judgements, and is an advantage over the traditional fuzzy methods in making PEO. The proposed approach can be a piece of help for decision-makers in evaluation of different aspects of socio-economic systems.

References

1. Bourdieu, P.: Economical Capital, Cultural Capital, Social Capital. Otto Shartz & Co., Goettingen (1983)
2. The Sustainable Competitiveness Report. Methodology, Calculation, & Report Compilation by SolAbility, 8th edn, p. 49 (2019)
3. Verbeek, J., Dill, A: The forgotten dimension of the SDG indicators – Social Capital. World Bank, 15 Aug 2017
4. Donev, D., Kovacic, L., Laaser, U.: The role and organization of health systems. Forum for Public Health in South Eastern Europe (2013)
5. Tian, Y., Kang, B.: A modified method of generating Z-number based on OWA weights and maximum entropy. Soft. Comput. **24**(20), 15841–15852 (2020). https://doi.org/10.1007/s00500-020-04914-8
6. Kang, B., Deng, Y., Hewage, K., Sadiq, R.: Generating Z-number based on OWA weights and maximum entropy. Int. J. Intell. Syst. **33**(8), 1745–1755 (2018). https://doi.org/10.1002/int.21995
7. Jain, A.K.: Corruption: a review. J. Econ. Surv. **15**(1), 71–121 (2001). https://doi.org/10.1111/1467-6419.00133

8. Zadeh, L.A.: A note on z-numbers. Inf. Sci. **181**(14), 2923–2932 (2011). https://doi.org/10.1016/j.ins.2011.02.022
9. Aliev, R.A., Alizadeh, A.V., Huseynov, O.H.: The arithmetic of discrete z-numbers. Inf. Sci. **290**, 134–155 (2015). https://doi.org/10.1016/j.ins.2014.08.024
10. Aliev, R.A., Huseynov, O.H., Zeinalova, L.M.: The arithmetic of continuous z-numbers. Inf. Sci. **373**, 441–460 (2016). https://doi.org/10.1016/j.ins.2016.08.078
11. Imanov, G.J.: Quality estimation level of country's development, Real Academia de Ciencias Economicas y Financieras, La vejes conocimiento, vivencia y experiencias. In: XV Acto Internacional Real Academia de Ciencias Economicas y Financieras – Barcelona, Spain, pp. 253–276 (2021)
12. Yager, R.R., Kacprzyk, J.: The Ordered Weighted Averaging Operators: Theory and Applications. Springer, New York (2012)

Quality Metrics of LSB Image Steganography Technique for Color Space HSI

Yucel Inan(✉)

Near East University, Nicosia, TRNC, Mersin, Turkey
yucel.inan@neu.edu.tr

Abstract. The aim of the paper is to find the distortion rates of the images by using Steganography. The cover and message images are scaled to 256 by 256 pixels. RGB level images are converted into HSI color space. HSI color space based metric establishes a better relationship with human perception for the processing. For the 'message hiding', the LSB Steganography method is utilized. In the image application, the least significant bit (LSB) data embedding is performed on the image. The effect of changing LSBs up to the 7th bit plane on the image has been examined with various quality metrics values like PSNR, SSIM, and MSE. Here we focus on the degradation level of the tested image to evaluate the effect of the RGB level image converted to the HSI color space, the altered effect of the original image on the data embedded image and calculate the image quality criteria.

Keywords: Steganography · Color space · Image quality criteria · LSB · HIS

1 Introduction

Steganography is the art of hiding a confidential data by embedding. The word steganography is derived from the Greek alphabet. Hidden means "στεγανό-ς" and covered "γραφ-ει" text [1]. The purpose of Steganography is to hide the presence of confidential message or information. The message to be sent is kept in an innocent-looking environment, preventing third parties from being aware of the presence of the message. Steganography is divided into two parts. Linguistics is Steganography and Technical Steganography. Linguistics Steganography: Steganography part with carrier data (text). Technical Steganography: Invisible ink, hidden places, Microdots, Computer based methods. Computer based methods: Text, audio, video, image are data hiding areas using Steganography files. There are several methods in Image Steganography that includes hiding information into image files [2]: Adding the least significant bit (LSB), Masking and filtering, Algorithms and transforms. In Steganography image application, the most insignificant bit (LSB) data embedding is performed on the image. In order to evaluate the effect of the RGB level image transformed into HSI color space, the changing effect of the original image on the data embedded image, and to calculate the image quality criteria, the distortion level of the image tested here is focused on PSNR, MSE, and SSIM. Although many mathematical measurements are also simplicity, this method is not well related to perceptual quality. By incorporating HSI features, the performance of

R. A. Aliev et al. (Eds.): ICSCCW 2021, LNNS 362, pp. 67–74, 2022.
https://doi.org/10.1007/978-3-030-92127-9_13

Metric measurements can be improved. An attempt was made to model HSI in a mathematical form and to normalize the distortion rates according to visibility. HSI based metric establishes a better relationship with human perception. The difference between the test images of HSI human perception is normalized according to visibility.

2 LSB Method 24-Bit Color Image Application

In this method, every bit of data to be hidden is written to the last bit of one byte of the image data. In 24-bit images, 3 bytes are used per pixel. The color of each pixel is obtained from three main colors: R (red), G (green), B (blue). This is called the RGB value of the pixel. By changing the last bit in each byte, 3 bits of information can be hide in a pixel [3].

$$\text{Bytes} = \text{Total pixels color depth}$$
$$8\ \text{Bit}\ [1\ \text{Byte}]$$

In other words, it has 24 bit depth (256 × 26 × 24)/8 = 196.608 bit (24.576 byte) [1]. Below has the ASCII 65 (decimal) code with the letter 'A' which corresponds to 01000001 in the binary system. Three pixels are required for a 24-bit image to embed 'A' information. Suppose Pixel values are as follows Table 1.

Table 1. MSB, LSB, and RGB values of example pixels.

Pixel values	MSB LSB	MSB LSB	MSB LSB	RGB values
Pixel1	1001 0101	0000 1101	1100 1001	149,13,201
Pixel2	1001 0110	0000 1111	11001010	150,15,202
Pixel3	1001 1111	0001 0000	1100 1011	159,16,234

The new pixel values that occur when the bit "01000001" information to be embedded in the 3 pixel given as the original image bits is hidden is as in Table 2 below.

Table 2. Altered values of example pixels in HSI format.

Pixel values	MSB LSB	MSB LSB	MSB LSB	HSI values
Pixel1	1001 01**00**	0000 1101	1100 100**0**	**148**,13,**200**
Pixel2	1001 0110	0000 111**0**	11001010	150,**14**,202
Pixel3	1001 111**0**	0001 000**1**	1100 1011	**158**,**17**,234

Changes were made in the last bits. If the bit to be embedded is the same as the next bit, the result will not change. The data will be embedded in the desired channel, as there are three channels in red, green, and blue in the color pictures. In Table 2 above, only 5 bits are changed, and the information is hidden. It is important to go to the result by making the least change in this method.

3 Color Spaces

Color spaces are mathematical models used to describe colors. Color spaces are generally divided into two groups, device dependent and device independent color spaces. In Fig. 1 some commonly used color spaces are seen [4].

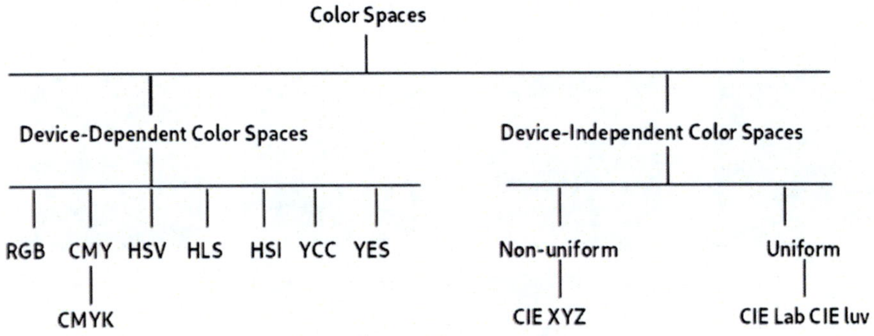

Fig. 1. Commonly used Color spaces

3.1 RGB Color Model

RGB consists of a combination of three basic colors, red, green and blue. The colors that make up this model are basic colors and they are not obtained with a mixture of two colors. Since the colors obtained by the combination are structurally lighter in red, green and blue, white is formed by the combination of all of them. By mixing these colors together, cygan, magenta and yellow are obtained. It is white from 100% mixture of three, and black is obtained when it is reduced to 0%. Televisions, computer monitors and scanners use the RGB color model. RGB color model is in the color receiving structure of the human eye. These models are difficult to use when recognizing or comparing colors in the human brain. Therefore, in this type of study, the HSI color model is used.

3.2 HSI Color Model

Detecting and replacing light intensity and saturation in color images requires a very intensive mathematical operation in RGB color space. This increases workload and directs the researcher to find simpler solutions in color image applications. For this purpose, HSI color spaces have been developed to make color selection easier and intuitive easier. In HSI color space, colors are defined as color essence, hue (H), saturation (S) and density (I), unlike RGB color space. For this reason, light changes can be recognized in the HSI color space and difficulties during image processing are eliminated [5]. With the combination of R (Red), 2nd layer G (Green), 3rd layer B (Blue), and three of the RGB layers of a color image, a colorful picture with RGB components emerges. Transition from RGB color space to HSI color space Three components of H (hue), S (saturation), I (Intensity) are shown separately in Fig. 2.

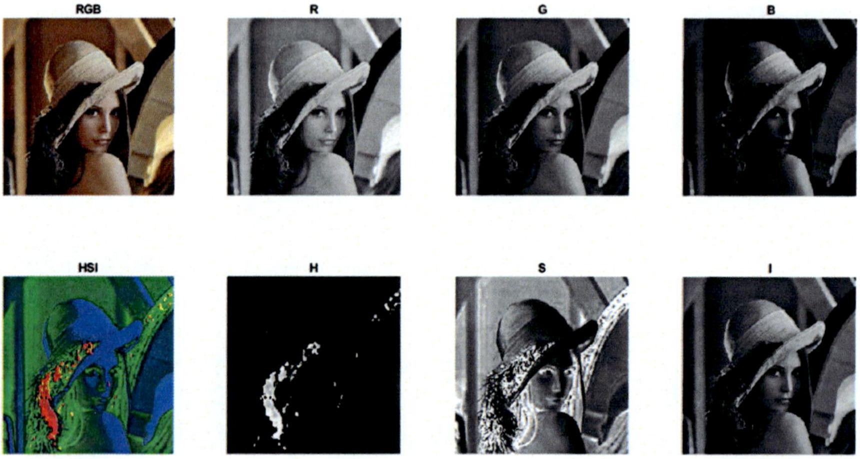

Fig. 2. RGB and HSI Color spaces components.

3.2.1 Used Formulas

The steps below are mathematically formulated for conversion of the RGB color space to the HSI color space [6].

$$r = \frac{R}{R+G+B} g = \frac{G}{R+G+B} b = \frac{B}{R+G+B} \tag{1}$$

$$\Theta = \cos^{-1}\left[\frac{\frac{1}{2}x[(r-g)+(r-b)]}{[(r-g)^2+(r-b)(g-b)]^{\frac{1}{2}}}\right] h(Hue) = \left\{\begin{matrix} \Theta \text{ } if \text{ } b \le g \\ 360 \text{ } if \text{ } b > g \end{matrix}\right\} \tag{2}$$

$$S(\text{Saturation}) = 1 - \frac{3}{(\text{R+G+B})}[\min(\text{R,G,B})] \tag{3}$$

$$I(\text{Intensity}) = \frac{1}{3}(\text{R+G+B}) \tag{4}$$

4 Image Quality Measures

When embed processing an Image is applied, it may result in information loss or quality loss. Objective and subjective methods are used to evaluate the quality of this image [7].

* Subjective Methods: It works according to the human vision system.
* Objective Methods: Based on comparisons using numerical criteria.

In this application, we analyzed image quality criteria SSIM, PSNR, MSE measures.

4.1 Structural Similarity Index (SSIM)

It is a quality criterion used to measure the similarity between the original image and the image embedded in it. SSIM is designed for image distortion, information loss, brightness distortion factors. This human visual system (HSV) is thought to be close to quality perception. In this approach, they take the value in the value range [0…1]. SSIM is calculated by the formula mentioned below.

$$SSIM(X, Y) = \frac{((2\mu_x\mu_y + c_1)(2\sigma_{xy} + c_2))}{\left(\left(\mu_x^2\mu_y^2 + c_1\right)\left(\sigma_x^2 + \sigma_y^2 + c_2\right)\right)} \quad (5)$$

4.2 Peak Signal-to-Noise Ratio (PSNR)

It is the ratio between the noise-canceling power that affects the highest value of the signal. It is most often used in image reconstruction as a measure of its quality. In this case, it calculates the quality of similarity between the original image and the image containing confidential data. As a result of the calculation, the high PSNR value means that the quality is also high, that is, the effect of the change on the image on the picture is low. PSNR is calculated with the formula specified below.

$$\text{PSNR(db)} = 20\log_{10}\frac{\text{MAX}(X^2)_f}{\sqrt{\text{MSE}}} \quad (6)$$

4.3 Mean Square Error (MSE)

Measures the average of squares of errors. Values close to zero are better. MSE is calculated by the formula mentioned below.

$$\sigma^2 = \frac{1}{N}\sum_{n=1}^{N}(x_n + y_n)^2 \quad (7)$$

5 Experimental Results

Lenna picture taken from the web was tested on two jpg images of 256 × 256 pixel size. After converting the picture from RGB color level to HSI color space, secret information embedding was performed on the picture using LSB technique. In this test, there are no noticeable objects on the image up to the 4th bit plane. But there are clues that the image has changed. With the embedded image, the file size has increased and the examination of the images shows that it consists of two colors. There are changes in histogram test, hidden information embedded image and original image. The differences between these histograms can be easily perceived by the human eye. Changes made to the digital image cause distortion on the image. These distortions cause visual differences in the picture. Even if the differences are not visible with the human eye, they can be detected through analysis to be performed in digital media. There are basically two different approaches for detecting deterioration in quality measurements. HSV human vision system is the second quality measurement metrics. Below are the original pictures and data embedded images by LSB bit embedding method. The images and histograms of the 7th bit plane from the 0th bit plane of the HSI levels from the RGB level of the pictures and the differences between these histograms can be seen in Fig. 3 In the next part, the results obtained after the above experiments are given on the chart and table. Based on the experiment, the quality measurement values were taken after embedding in a picture (see Table 3) The cover image shows the numerical values of the file sizes varying in the Bit plane of the Stego noise.

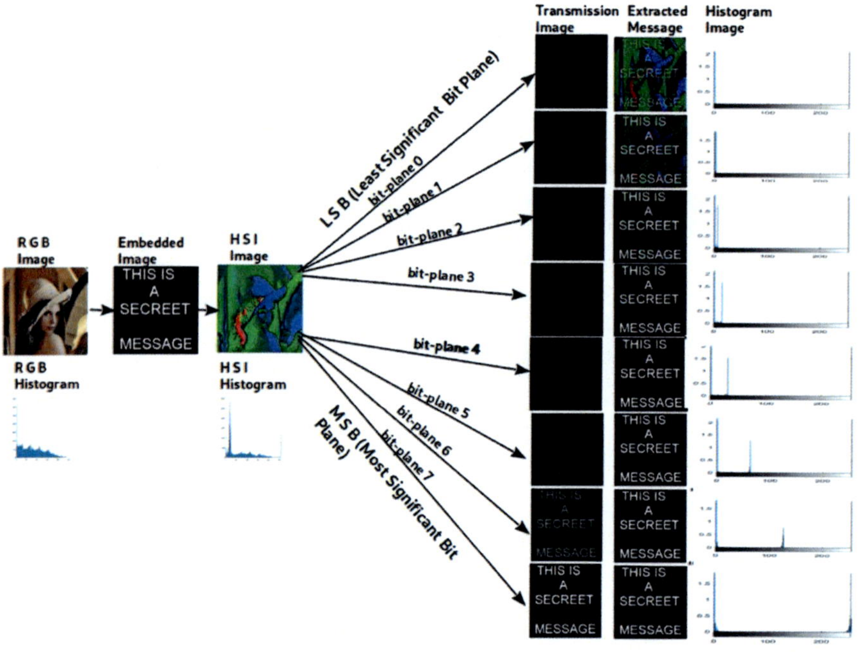

Fig. 3. RGB and HSI image, Stego image and Histograms

There are no noticeable symbols on the stego image in this test, up to the 4th Bit plane. In fact, there is a lossy compression here. Distortion ratio is important in lossy compression. The main thing to be tested when making a comparison is not how much it is compressed but how high quality it compresses. Here, a successful, high quality compression was achieved. In planes after the 5th bit plane, the hidden information in the image is noticeable and an increase in file size is observed. In Table 4 it was observed how the quality criteria related to each distortion affect the values of the distortion in order to evaluate the sensitivity of the image to the distortions. PSNR and SSIM seem to be sensitive to noise jpg compression. In the discrimination of the image embedded up to the 4th.bit plane, the PSNR indicates that the SSIM value provides higher image quality. Values close to 0th are more successful than MSE, PSNR.

Table 3. The cover image and Stego image.

Cover image dimension	Message image dimension	Size of cover	Size of message	Bit plane1	Size of stego
256 × 256	256 × 256	14.0 Kb	11.6 Kb	0.Bit-plane	3.4 Kb
				1.Bit-plane	4.0 Kb
				2.Bit-plane	5.0 Kb
				3.Bit-plane	6.5 Kb
				4.Bit-plane	8.7Kb
				5.Bit-plane	11.2 Kb
				6.Bit-plane	14.4 Kb
				7.Bit-plane	18.5 Kb

Table 4. Bit-plane quality metrics

Bit I-plane	PSNR	SSIM	MSE
0st-Bit-plane	59.224	0.994	0.078
1st-Bit-plane	49.225	0.952	0.777
2nd-Bit-plane	41.593	0.881	4.506
3rd-Bit-plane	34.861	0.815	21.234
4th-Bit-plane	28.511	0.765	91.610
5th-Bit-plane	22.345	0.725	378.965
6th-Bit-plane	16.268	0.692	1535.745
7th-Bit-plane	10.222	0.664	6178.369

6 Conclusions

The study was coded in Matlab program environment and applied on the classical Lenna picture taken from the web. Images 256 × 256 pixel sized blocks were converted from RGB color space to HSI color space. The subsequent LSB image was hidden using the method of embedding confidential data into the image using the Steganography method. The changing effect of the hidden information embedded image on the Bit-plane was tested using quality criteria metrics on the image using numerical criteria. The results are expressed on numerical values on how image distortions affect the picture. The aim is to reveal the differences and similarities of SSIM, PSNR, MSE quality criteria in general separations by embedding confidential information with LSB method on the picture. PSNR appears to be more sensitive in jpeq compression. SSIM performed well in distinguishing structural content in images. However, the color value remained black in the cover image. Despite the noisy distortion, MSE and PSNR were observed to be good at evaluating the quality of the images.

References

1. Watheq, R.A., Almasalha, F., Qutqut, M.H.: A new steganography technique using JPEG images. Int. J. Adv. Comput. Sci. Appl. **9**, 751–760 (2018). https://doi.org/10.14569/IJACSA.2018.0911107
2. Inan, Y.: Comparing image distortion of LSB. In: Aliev, R.A., Kacprzyk, J., Pedrycz, W., Jamshidi, Mo., Sadikoglu, F.M. (eds.) ICAFS 2018. AISC, vol. 896, pp. 82–90. Springer, Cham (2019). https://doi.org/10.1007/978-3-030-04164-9_14
3. Swain, G., Lenka, S.K.: A better RGB channel based image steganography technique. In: Krishna, P.V., Babu, M.R., Ariwa, E. (eds.) ObCom 2011. CCIS, vol. 270, pp. 470–478. Springer, Heidelberg (2012). https://doi.org/10.1007/978-3-642-29216-3_51
4. Poornima, R., Iswarya, R.J.: An overview of digital image steganography. Int. J. Comput. Sci. Eng. Surv. **4**(1), 23–31 (2013). https://doi.org/10.5121/ijcses.2013.4102
5. Jurio, A., Pagola, M., Galar, M., Lopez-Molina, C., Paternain, D.: A comparison study of different color spaces in clustering based image segmentation. In: Hüllermeier, E., Kruse, R., Hoffmann, F. (eds.) IPMU 2010. CCIS, vol. 81, pp. 532–541. Springer, Heidelberg (2010). https://doi.org/10.1007/978-3-642-14058-7_55
6. Khan, M., Ahmad, J., Farman, H., Zubair, M.: A novel image steganographic approach for hiding text in color images using HSI color model. Middle-East J. Sci. Res. **22**(5), 647–654 (2014). https://doi.org/10.5829/idosi.mejsr.2014.22.05.21946
7. Hore, A., Ziou, D.: Image quality metrics: PSNR vs. SSIM. Int. Conf. Pattern Recogn. 2366–2368 (2010). https://doi.org/10.1109/ICPR.2010.579

Z-Set Based Inference Using ALI-2 Implication for Control System Design

Latafat A. Gardashova(✉)

Azerbaijan State Oil and Industry University, Azadlig 35, Nasimi, Baku, Azerbaijan
l.qardashova@asoiu.edu.az

Abstract. Knowledge based systems are characterized with bimodal information. There are a lot of works in scientific literature related on fuzzy and probabilistic reasoning. Fuzzy reasoning and probabilistic reasoning have a wide application, in both theoretical and applied works. It is lucid that used implication impacts to result of reasoning. In this work, Z-set based inference using Ali-2 implication for control system design is discussed. Ali 2 logic satisfies the relation $\mu_A(x)\hat{}(\mu_A(x) \to \mu_B(x)) \leq \mu_B(x)$ which is essential for progress of fuzzy conditional inference rules. It gives them possibility to utilize it for the creation of developed rules of fuzzy conditional inference and for the modeling of different systems. For clarification of the situation, new approach for approximate reasoning is proposed by author. Proposed approach is based on processing imperfect information.

Keywords: Ali-2 fuzzy implication · Z-set · Probabilistic implication · Z-If-Then rules bimodal information

1 Introduction

Fuzzy implication has a very important role in fuzzy control, multivalued mathematical logic, approximate reasoning and data analysis. It is a kernel of approximate reasoning and fuzzy logic. Inference process is realizing through fuzzy implication in many fuzzy rule-based systems. An important problem in implication is their representation and it impacts to the result of inference. This question is discussed in a lot of published works [1–5].

Review about Fuzzy Implication Functions is given in [6, 7]. Different problems related to Construction Methods of Fuzzy Implications are discussed in this study.

[8] is dedicated to Yager's implications and their characterization. Their basic properties is discussed in this research. Using this idea authors have characterized well-known fuzzy implications -Goguen, Reichenbach implication.

There are several different fuzzy implications used in fuzzy control system. For instance, Ali-1, Ali-2, Ali-3, KD. Advantages of these logic are proved by examples in [9, 10]. Ali-2 is fuzzy logic, which truthiness of the implication depends on the degrees of truthiness of the antecedent and consequent. This logic able to implement the conditional

R. A. Aliev et al. (Eds.): ICSCCW 2021, LNNS 362, pp. 75–84, 2022.
https://doi.org/10.1007/978-3-030-92127-9_14

inference rule, that satisfies criteria of Masumoto [11] and Modus Ponens [12]. Ali-2 logic has own logical connectives [12].

In this work we propose using Ali-2 implication. This implication and its comparative semantic analysis are given in [13, 14].

Also, Ali2 and Ali3 implications satisfy above given inequality too.

The study [15] is devoted to the uncertainty management. Discussed Zadeh approach give possibility operate different type uncertainty. However, combining two probability measures is the problem of judgment, it is not a problem of mathematics. Probabilistic implication-based reasoning is described in [16]. Conditional copulas is a bridge between fuzzy logic and probability theory.

There are several works related on modeling problem of combination fuzzy and probabilistic uncertainties. They are belief networks, fuzzy and probabilistic information combination, belief rule-based systems, and etc. [17–24]. Nowadays imperfect information can be represented by different effective tools such as Z-numbers and Z-sets.

Authors [25–27] used the Z-number but didn't use the Z-set. A new method for approximate reasoning using Ali-2 implication and Z-set based rules is proposed.

The structure of the paper consists of five sections. In the second section basic definitions related on Z-set based reasoning are given. A statement of the problem is presented in Sect. 3. The basic steps of the proposed approach for reasoning with Z-rules are considered in Sect. 4. An example is illustrated in last section.

2 Preliminaries

Definition 1 [26]. Z IF-THEN Rules. If X_1 is $Z_{X_1,n} = (A_{X_1,n}, B_{X_1,n})$ and, …, and X_m is $Z_{X_m,n} = (A_{X_m,n}, B_{X_m,n})$ then Y is $Z_Y = (A_{Y,n}, B_{Y,n})$.

$Z_{X_m,n} = (A_{X_m,n}, B_{X_m,n})$ is the Z-sets [26] based inputs and $Z_Y = (A_{Y,n}, B_{Y,n})$ is Z sets based outputs of the rues.

Definition 2. AlI-2 implication [12].

$$\mu_s(u, e) = \begin{cases} 1, & \mu_{E_i}(e) \le \mu_{U_j}(u) \\ (1 - \mu_{E_i}(e))^{\mu}_{U_j}(u), & \mu_{E_i}(e) > \mu_{U_j}(u) \end{cases}$$

$$i,\ j,\ s = \overline{1, n}.$$

Logical connective conjunction on Ali 2 implication is given below:

$$\mu_s(u, e) = \begin{cases} \mu_{E_i}(e), & \mu_{E_i}(e) + \mu_{U_j}(u) < 1 \\ 0, & \mu_{E_i}(e) + \mu_{U_j}(u) = 1 \\ \mu_{U_j}(u), & \mu_{E_i}(e) + \mu_{U_j}(u) > 1 \end{cases}$$

Definition 3 [28]. Aggregation distributions. Probability aggregation method a weighted sum as follow:

$$P_G(p) = \sum_{i-1}^{n} w_i P_i(p).$$

Definition 4 [29]. The compositional rule of fuzzy inference is:

$$\mu_U(u) = \max_u \min[\mu_E(e^c_{ic}), \mu_R(u, e)]$$

3 Statement of the Problem

Implemented control system is described using seven rules which antecedent and consequent is represented using Z-set. The goal is to determine output according to given input.

Z-set based rules are as follow [28]:

1. If the error *e* is [negative big (NB), small sure] THEN the control action *u* is [negative big (NB), small sure].
2. If the error *e* is [negative medium (NM), sure] THEN the control action *u* is [negative medium (NM), sure].
3. If the error *e* is [negative small (NS), very sure] THEN the control action *u* is [negative small (NS), very sure].
4. If the error *e* is [zero (ZE), medium] THEN the control action *u* is [zero, medium].
5. If the error *e* is [positive small (PS), sure] THEN the control action *u* is [positive small (PS), medium].
6. If the error *e* is [positive medium (PM), very sure] THEN the control action *u* is [positive medium (PM), sure].
7. If the error *e* is [positive big (PB), medium] THEN the control action *u* is [positive big (PB), sure].

Here, variable *e* shows the errors and *u* describes the control action. The value of the linguistic variables is described as follows:

$$\mathit{medium} = 0.2/0.5 + 1/0.6 + 0.4/0.8;\ \mathit{small\ sure} = 0.3/0.7 + 1/0.73 + 0.2/0.75;$$
$$\mathit{very\ sure} = 0.2/0.8 + 1/0.9 + 0.2/1;\ \mathit{sure} = 0.1/0.7 + 1/0.8 + 0.3/0.9.$$

Rule 1

$$\mu_{E1}(e) = 1.00/-10 + 0.73/-7 + 0.34/-3 + 0.20/0 + 0.13/3 + 0.08/7 + 0.06/10$$

$$\begin{aligned}\mu_{U1}(u) = &1.00/-1 + 0.92/-0.7 + 0.67/-0.3 + 0.50/0 + 0.37/0.3\\ &+ 0.26/0.7 + 0.20/1\end{aligned}$$

...

Rule 7

$$\mu_{E7}(e) = 0.06/-10 + 0.08/-7 + 0.13/-3 + 0.20/0 + 0.34/3 + 0.73/7 + 1.00/10$$

$$\begin{aligned}\mu_{U7}(u) = &0.20/-1 + 0.26/-0.7 + 0.37/-0.3 + 0.50/0 + 0.67/0.3\\ &+ 0.92/0.7 + 1.00/1\end{aligned}$$

The aim is to accomplish Z-inference using Ali-2 implication and define result according to new input, which described below:

New input: $(\mu_{\mathrm{E}}(e_{ic}^{c}) = 0.11/-10 + 0.17/-7 + 0.34/-3 + 0.61/0 + 0.96/3 + 0.73/7 + 0.41/10,\ \text{sure})$.

4 Solution Method

Algorithm of solution method as follow.

1. Calculation ***A*** part of the Z set using fuzzy implication. The value of the output determined utilizing the compositional rule of inference.
2. Calculation ***B*** part of the Z-set.
   Determination set of probaility distribition ***G*** over all rules using point of support ***B*** Calculation copula is based on distributions of inputs and outputs on rules.
3. Aggregation probability distribition set is performed by aggregation method.
4. Calculation set of probability distributions using points of ***B*** on new input.
5. Calculation distance between two vectors such as one of them probability vector on new input and second aggregated probability vector over inputs of the rule.
6. Calculation similarity measure.
7. Definition the support points of output ***B*** using ***A***, similarity measure and probability distribution.

Example. Control system is described by seven rules [30].

First step is related on creation relations over rules [31] by using ALI-2 implication over 7 rules. Intersection is determined by using logical connective of Ali 2, obtained composed fuzzy relation is described in Table 1.

Table 1. Obtained fuzzy relation

$R = \cup R_S,\ s = 1, \ldots, n$						
1	0	0	0	0	0	0
0	1	0	0	0	0	0
0	0	1	0	0	0	0
0	0	0	1	0	0	0
0	0	0	0	1	0	0
0	0	0	0	0	1	0
0	0	0	0	0	0	1

The value of the control action is defined as follows:$\mu_{Eout}(e) = 0.11/-1+0.17/-0.7+0.34/-0.3+0.61/0+0.96/0.3+0.73/0.7+0.41/1.$

The next step consists calculation of probability distributions on input and output of the rules. For instance, result on the seventh rules determined according to [28] is given in Table 2.

Table 2. Probability distributions on input and output of the rules

Distributions on seventh rule inputs			Distributions on seventh rule outputs		
0.5	***0.6***	***0.8***	***0.7***	***0.8***	***0.9***
0.0	0.0	0.21	0.0	0.20888355	0.21
0.0	0.0	0.0	0.06	0.0	0.0
0.0	0.0	0.0	0.14	0.0	0.0
.00.0	0.0	0.0	0.17	0.0	0.0
0.77	0.55	0.0	0.19101044	0.0	0.0
0.0	0.13	0.0	0.203559604	0.79111645	0.79
0.240	0.32	0.79	0.229758495	0.0	0.0

Result of the probability aggregation operation over set of probability distribution on antecedent as follows (Table 3):

Table 3. Result of the aggregation

P1	P2	P3
0.0439	0.0481	0.1501
0.0309	0.0240	0.0189
0.0468	0.0513	0.0480
0.0506	0.0640	0.0680
0.4910	0.3424	0.0949
0.1791	0.2983	0.2195
0.1578	0.1719	0.4006

Using the distributions on the input and outpust calculate a copula for every rule (see Table 4).

Table 4. Result over the points (0.75,0.75)

Copula						
0.214	0.178	0.009	0.000	0.000	0.000	0.151
0.090	0.074	0.004	0.000	0.000	0.000	0.063
0.022	0.018	0.001	0.000	0.000	0.000	0.016
0.000	0.000	0.000	0.000	0.000	0.000	0.000
0.000	0.000	0.000	0.000	0.000	0.000	0.000
0.003	0.003	0.000	0.000	0.000	0.000	0.002
0.059	0.049	0.003	0.000	0.000	0.000	0.041

Using probability implication and copula the distributions are determined. For points (0.75,0.75) over seventh rule is probability distribution is defined: p = (0.388148,0.321775,0.016603,0,0,0,0.273474) (Table 5).

Table 5. Probability distributions on 7-th rule

Points			Probability distributions					
(0.5,0.7)	0	0.06059	0.140805	0.174276	0.191011	0.20356	0.229759	
(0.5,0.8)	0.208884	0	0	0	0	0.791117	0	
(0.5,0.9)	0.21	0	0	0	0	0.790001	0	
(0.6,0.7)	0	0.06059	0.140805	0.1742765	0.191011	0.20356	0.229759	
(0.6,0.8)	0.208884	0	0	0	0	0.791118	0	
(0.6,0.9)	0.21	0	0	0	0	0.790002	0	
(0.8,0.7)	0	0.06059	0.140805	0.1742761	0.19101	0.20356	0.229758	
(0.8,0.8)	0.208884	0	0	0	0	0.791116	0	
(0.8,0.9)	0.21	0		0	0	0.79	0	

Obtained 63 distributions for seven rules are aggregated and result is described as vector in Table 6:

Table 6. Distributions

Probability distributions						
0.121668	0.065734	0.11557	0.129142	0.128281	0.117438	0.322169
0.182069	0.03164	0.048667	0.064212	0.083381	0.519071	0.070961
0.132204	0.029034	0.046529	0.058941	0.284119	0.427881	0.021293
0.121668	0.065734	0.11557	0.129142	0.128281	0.117438	0.322169
0.182069	0.03164	0.048667	0.064212	0.083381	0.519071	0.070961
0.131722	0.024096	0.057288	0.05587	0.282213	0.427144	0.021669
0.138369	0.084436	0.122166	0.139029	0.146407	0.117438	0.252156
0.182069	0.03164	0.048667	0.064212	0.083381	0.51907	0.070961
0.131721	0.024096	0.057288	0.05587	0.282213	0.427143	0.021669

First part of the Z set or *A* part of the Z-set is determined as follows:

$$\mu_{Eout}(e) = 0.11/-1 + 0.17/-0.7 + 0.34/-0.3 + 0.61/0 + 0.96/0.3 + 0.73/0.7 + 0.41/1$$

The next step related on calculation similarity measure between current and aggregated input vector of probability distributions.

Probabilty distributions on new input are given in Table 7.

Table 7. Probabilty distributions on new input

Distributions		
0.7	***0.8***	***0.9***
0.0	0.0	0.0
0.0425	0.0	0.0
0.1162	0.0389	0.0
0.1994	0.274	0.1548
0.3345	0.5132	0.8199
0.2641	0.1732	0.0252
0.0431	0.0	0.0

Find the distance between the new vector and aggregated probability vector. Determined result as follows:

$$ZHD(P_{Z_1}, P_{Z_2}) = \min_{P_{zi} \in G}\left(\frac{1}{\sqrt{2}}\left(\sum_{j=1}^{k}(p_{z1_j}^{1/2} - p_{z2_j}^{1/2})\right)^{1/2}\right) = 0,263482$$

Obtained similarity measure between a new vector and aggregated input is:

$$sm = 1/(1 + ZHD) = 0.787851$$

Find probability distributions in output part utilizing similarity degree. Determined A as first part of Z-set, probability distributions and similarity measure.

For defining B we use minimal value of all input and output which make P_i, $i = \overline{1,9}$. For example,

$$P_1 = \left(\begin{array}{l}\left(\frac{0.3}{0.7}, \frac{0.3}{0.7}\right), \left(\frac{0.1}{0.7}, \frac{0.1}{0.7}\right), \left(\frac{0.2}{0.8}, \frac{0.2}{0.8}\right), \left(\frac{0.2}{0.5}, \frac{0.2}{0.5}\right), \left(\frac{1}{0.7}, \frac{0.2}{0.5}\right), \\ \left(\frac{0.2}{0.8}, \frac{1}{0.7}\right), \left(\frac{0.2}{0.5}, \frac{1}{0.7}\right)\end{array}\right)$$

$$P_1 = \left(\begin{array}{l}\left(\frac{0.3}{0.7}, \frac{0.3}{0.7}\right), \left(\frac{0.1}{0.7}, \frac{0.1}{0.7}\right), \left(\frac{0.2}{0.8}, \frac{0.2}{0.8}\right), \left(\frac{0.2}{0.5}, \frac{0.2}{0.5}\right), \left(\frac{1}{0.7}, \frac{0.2}{0.5}\right), \\ \left(\frac{0.2}{0.8}, \frac{1}{0.7}\right), \left(\frac{0.2}{0.5}, \frac{1}{0.7}\right)\end{array}\right)$$

...

$$P_5 = \left(\begin{array}{l}\left(\frac{1}{0.73}, \frac{1}{0.73}\right), \left(\frac{1}{0.8}, \frac{1}{0.8}\right), \left(\frac{1}{0.9}, \frac{1}{0.9}\right), \left(\frac{1}{0.6}, \frac{1}{0.6}\right), \left(\frac{1}{0.8}, \frac{1}{0.8}\right), \\ \left(\frac{1}{0.9}, \frac{1}{0.8}\right), \left(\frac{1}{0.6}, \frac{1}{0.8}\right)\end{array}\right)$$

...

$$P_9 = \left(\begin{array}{l}\left(\frac{0.2}{0.75}, \frac{0.2}{0.75}\right), \left(\frac{0.3}{0.9}, \frac{0.3}{0.9}\right), \left(\frac{0.2}{1}, \frac{0.2}{1}\right), \left(\frac{0.4}{0.8}, \frac{0.4}{0.8}\right), \left(\frac{0.3}{0.9}, \frac{0.4}{0.8}\right), \\ \left(\frac{0.2}{1}, \frac{0.3}{0.9}\right), \left(\frac{0.4}{0.8}, \frac{0.3}{0.9}\right)\end{array}\right)$$

Obtained result as follow:

A = 0.11/−1 + 0.17/−0.7 + 0.34/−0.3 + 0.61/0 + 0.96/0.3 + 0.73/0.7 + 0.41/1

B = 0.1/0,36753 + 0.1/0,36753 + 0.1/0,36904 + 0.2/0,43258 + 0.2/0,43258 + 1/0,43258 +

+ 0.2/0,50446341 + 0.2/0,50446386 + 0.1/0,50546814

Determined result is represented in Fig. 1.

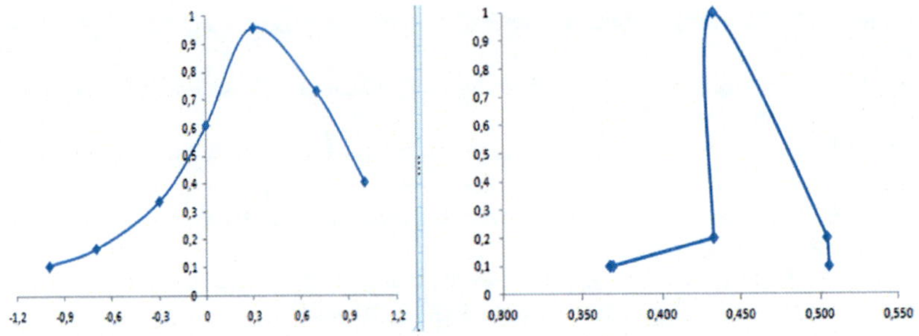

Fig. 1. Determined result

5 Conclusion

In given research, a new inference method based on Ali-2 implication and Z-set based antecedents and consequents is given. Proposed approach has ability to operate fuzzy and probabilistic uncertainties. The priority of the suggested approach is to keep main properties of data, such as information characterized by probability and fuzzy uncertainty. Discussed method related to definitions, such as probabilistic and fuzzy implication, fuzzy and probability aggregations, similarity measure, a distance. It is clear that we can apply this approach in different decision-making problems when the information represented using probability and membership function.

References

1. Massanet, S., Joan, T.: The law of importation versus the exchange principle on fuzzy implications. Fuzzy Sets Syst. **168**(1), 47–69 (2011). https://doi.org/10.1016/j.fss.2010.12.012
2. Mas, M., Monserrat, M., Torrens, J.: A characterization of (U, N), RU, QL and D-implications derived from uninorms satisfying the law of importation. Fuzzy Sets Syst. **161**(10), 1369–1387 (2010). https://doi.org/10.1016/j.fss.2009.10.022
3. Mas, M., Massanet, S., Torrens, J.: The law of importation for discrete implications. Inf. Sci. **179**(24), 4208–4218 (2009). https://doi.org/10.1016/j.ins.2009.08.028
4. Balasubramaniam, J.: Yager's new class of implications *J* and some classical tautologies. Inf. Sci. Int. J. **177**(3), 930–946 (2007). https://doi.org/10.1016/j.ins.2006.08.006
5. Baczyński, M., Jayaram, B.: (U, N)-implications and their characterizations. Fuzzy Sets Syst. **160**, 2049–2062 (2009)
6. Mas, M., Monserrat, M., Torrens, J., Trillas, E.: A survey on fuzzy implication functions. IEEE Trans. Fuzzy Syst. **15**(6), 1107–1121 (2007). https://doi.org/10.1109/TFUZZ.2007.896304
7. Baczyński, M., Beliakov, G., Bustince, S., Pradera, H.: A. Advances in fuzzy implication functions (2013).https://doi.org/10.1007/978-3-642-35677-3
8. Massanet, S., Torrens, J.: On the characterization of Yager's implications. Inf. Sci. **201**(15), 1–18 (2012). https://doi.org/10.1016/j.ins.2012.03.008
9. Potiupkin, A.Yu.: Solving of the problem of fuzzy systems identification. News Acad. Sci. Russia **4**, 40–46 (1996). (in Russian)
10. Potiupkin, A.Yu.: Estimation technical state of an objects described by fuzzy relational equations. News Acad. Sci. Russia **4**, 111–119 (1999). (in Russian)

11. Mizumoto, M.: Fuzzy conditional inference under max-composition. Inf. Sci. **27**(2), 183–209 (1982)
12. Aliev, R.A., Aliev, R.R.: Soft Computing and Its Application. World Scientific, New Jersey, London, Singapore, Hong Kong (2001)
13. Aliev, R.A.: Fuzzy Knowledge Based Intelligent Robots. Moscow: Radio i svyaz (1995). (in Russian)
14. Aliev, R.A., Mamedova, G.A., Tserkovny, A.E.: Fuzzy control systems. Energoatomizdat, Moscow (1991). (in Russian)
15. Zadeh, L.A.: Outline of a new approach to the analysis of complex systems and decision processes. IEEE Trans. Syst. Man Cyberns. SMC-3, 28–44 (1973)
16. Przemysław, G.: Probabilistic implications. Fuzzy Sets Syst. **226**, 53–66 (2013). https://doi.org/10.1016/j.fss.2013.01.003
17. Touazi, F., Cayrol, C., Dubois, D.: Possibilistic reasoning with partially ordered beliefs. J. Appl. Log. **13**(4), 770–798 (2015)
18. Dubois, D., Durrieu, C., Prade, H., Rico, A., Ferro, Y.: Extracting decision rules from qualitative data using Sugeno integral: a case-study. In: Destercke, S., Denoeux, T. (eds.) ECSQARU 2015. LNCS (LNAI), vol. 9161, pp. 14–24. Springer, Cham (2015). https://doi.org/10.1007/978-3-319-20807-7_2
19. Couso, I., Dubois, D.: Statistical reasoning with set-valued information: ontic vs. epistemic views. Int. J. Approx. Reason **55**(7), 1502–1518 (2014)
20. Jiao, L., Pana, Q., Denœux, T., Liang, Y., Feng, X.: Belief rule-based classification system: extension of FRBCS in belief functions framework. Inform. Sci. **309**, 26–49 (2015)
21. Couso, I., Dubois, D.: A perspective on the extension of stochastic orderings to fuzzy random variables. In: 16th World Congress of the International Fuzzy Systems Association (IFSA), Gijon, pp. 1486–1492 (2015)
22. Zhou, S., Chen, Q., Wang, X.: Fuzzy deep belief networks for semi-supervised sentiment classification. Neurocomputing **131**, 312–322 (2014). https://doi.org/10.1016/j.neucom2013.10.011
23. Dubois, D., Prade, H.: Possibility theory and its applications: where do we stand? In: Kacprzyk, J., Pedrycz, W. (eds.) Springer Handbook of Computational Intelligence, pp. 31–60. Springer, Heidelberg (2015). https://doi.org/10.1007/978-3-662-43505-2_3
24. Dubois, D., Liu, W., Ma, J., Prade, H.: The basic principles of uncertain information fusion. An organized review of merging rules in different representation frameworks. Inform. Fusion. **32**, 2–39 (2016). https://doi.org/10.1016/j.inffus.2016.02.006
25. Aliev, R.A., Huseynov, O.H., Zulfugarova, R.X.: Z-distance based IF-THEN rules. Sci. World J. **2016**, 1673537 (2016). https://doi.org/10.1155/2016/1673537
26. Aliev, R.A., Pedrycz, W., Huseynov, O.H., Eyupoglu, S.Z.: Approximate reasoning on a basis of Z-number valued If-Then rules. IEEE Trans. Fuzzy Syst. **25**(6), 1589–1600 (2016). https://doi.org/10.1109/tfuzz.2016.2612303
27. Jiang, W., Cao, Y., Deng, X.: A novel Z-network model based on Bayesian network and Z-number. IEEE Trans. Fuzzy Syst. **28**(8), 1585–1599 (2020). https://doi.org/10.1109/TFUZZ.2019.2918999
28. Aliev, R.A., Gardashova, L.A.: Z-set based approach to control system design. In: Aliev, R.A., Kacprzyk, J., Pedrycz, W., Jamshidi, Mo., Babanli, M., Sadikoglu, F.M. (eds.) ICAFS 2020. AISC, vol. 1306, pp. 10–21. Springer, Cham (2021). https://doi.org/10.1007/978-3-030-64058-3_2
29. Zadeh, L.A.: Fuzzy logic and approximate reasoning. Synthese **30**, 407–428 (1975)
30. Aliev, R., Mamedova, G., Aliev, R.: Fuzzy Set Theory and its Application, Tabriz, Iran (1993)
31. Dadasheva, A.N.: Analysis of consistency of pairwise comparison matrix with fuzzy type-2 elements. In: 11th International Conference on Theory and Application of Soft Computing, Computing with Words, Perception and Artificial Intelligence ICSCCW 2021, in press (2021)

Z-Numbers-Based Approach to Hotel Service Quality Assessment

Aziz Nuriyev[1,2](✉) and Ahmet Bahadir Baysal[2]

[1] Azerbaijan State Oil and Industry University, Azadlyg Avenue, 20, AZ1010 Baku, Azerbaijan
[2] Khazar University, Mehsety Street 41, AZ1069 Baku, Azerbaijan
bbaysal@khazar.org

Abstract. In this study, we are analyzing the possibility of using Z-numbers for measuring the service quality and decision-making for quality improvement in the hotel industry. Techniques used for these purposes are based on consumer evaluations - expectations and perceptions. As a rule, these evaluations are expressed in crisp numbers (Likert scale) or fuzzy estimates. However, descriptions of the respondent opinions based on crisp or fuzzy numbers formalism not in all cases are relevant. The existing methods do not take into account the degree of confidence of respondents in their assessments. A fuzzy approach better describes the uncertainties associated with human perceptions and expectations. Linguistic values are more acceptable than crisp numbers. To consider the subjective natures of both service quality estimates and confidence degree in them, the two-component Z-numbers $Z = (A, B)$ were used. Z-numbers express more adequately the opinion of consumers. The proposed and computationally efficient approach (Z-SERVQUAL, Z-IPA) allows to determine the quality of services and identify the factors that required improvement and the areas for further development. The suggested method was applied to evaluate the service quality in small and medium-sized hotels in Turkey and Azerbaijan, illustrated by the example.

Keywords: Hotel service quality · Z-SERVQUAL · Z-IPA · Z-number · Fuzzy number

1 Introduction

The leading role of tourism in the economic development of the various countries in the post-pandemic era will significantly increase competition in the industry. In such conditions, the well-informed decisions and the justified decision-making approaches for the service quality become one of the most important issues for gaining an advantageous position in such a highly competitive environment as the hotel industry.

Customer satisfaction is one of the primary research issues in the service quality in tourism [1]. There are several methods for measuring the service quality from the consumer's point of view - SERVQUAL, SERVPREF, Importance-Performance analysis (IPA), Critical Incident Technique, Mystery Shopper [2]. Other methods used include Kano's Model, Failure Mode and Effects Analysis (FMEA), Multi-Criteria Decision-Making Method (MCDM) [3, 4].

R. A. Aliev et al. (Eds.): ICSCCW 2021, LNNS 362, pp. 85–94, 2022.
https://doi.org/10.1007/978-3-030-92127-9_15

Developed in 1988 and currently widely used, the SERVQUAL methodology [5] assesses the differences between the consumer's perception of the service quality at the time of its provision and the expected level of quality before it is provided. This methodology includes the evaluation of the service quality in five dimensions (tangibles, reliability, responsiveness, assurance, empathy) according to 22 criteria. It was originally used for banks, long-distance/international telephone calls, repairs, and maintenance. Numerous studies have shown the applicability of the technique in other areas of service. In 1990, the authors expanded the methodology by adding an indicator of relative importance for each measurement [6]. Despite discussions about the applicability in certain areas and the need to increase or decrease the number of dimensions and criteria, the SERVQUAL technique keeps the leading positions. In the tourism sector, this technique has been utilized since the nineties [7] and successfully applied till today. In this methodology, Likert scale is used to determine the consumer expectations and perceptions regarding the service quality. There are fuzzy modifications of the method [8, 9] as well.

In all these approaches, the degree of the respondent's confidence in their evaluations is not considered. A person's reasoning about their preferences is often vague, and it is generally difficult to assess them with a crisp number. In addition, a consumer often cannot be completely confident in their assessments and perceptions.

Z-numbers, proposed by L. Zadeh [10], allow us to describe a customer's perception and expectation in a linguistic form and consider the degree of confidence in the assessments. In our work, we studied the applicability of Z-numbers for evaluation of the service quality in the tourism sector using the small and medium-sized four hotels in Azerbaijan and five hotels in Turkey. An integrated approach to assessing the quality of service based on Z-numbers is proposed. At the first stage, based on the prepared Z-SERVQUAL questionnaire, the differences between the expected and perceived quality of service in the hotel is determined, at the next stage, by using the Z-Importance-Performance Analysis (Z-IPA), attributes or indicators of service that need to be improved are determined.

2 Method

2.1 Hotel Service Quality Analysis Methodology

The utilization of the SERVQUAL methodology in the tourism sector is widely described in the scientific literature and, therefore, we shall not dwell on the one. It is necessary to underline that Likert scale, widely used by researchers, has limited capacity for describing the customer's subjective expectation and perceptions in a highly uncertain subject area like the hotel industry.

A fuzzy approach better describes the uncertainties associated with a person's subjective perception and expectations. Linguistic assessments are more acceptable than clear-cut ones. For evaluation, a term-set of linguistic meanings can be established that consumers can use for expressing their opinions. Assessment of the service quality based on linguistic values is preferable since most people cannot use accurate numerical estimations for describing perceptions or expressing feelings. Examples of a fuzzy approach to assessing the quality of service in hotels include such works as [11, 12].

Despite the development of various techniques based on SERVQUAL [13, 14] the basic dimensions remain. The number of items may vary, because some of the 22 items of the basic methodology have a general interpretation, and it is necessary to use specific subitems for their effective application in tourism. With the preservation of 5 basic dimensions, in [15] a questionnaire consisting of 38 items was used. The modified methodology for the hospitality industry HOLSERV [13], contains 5 new items in addition to the basic ones.

Since the application of the SERVQUAL methodology is a time-consuming process, after calculation of the differences between expectations and perceptions, it is also advisable to define the next steps to improve the service quality. A proven toolkit for determining a quality improvement strategy Importance-Performance Analysis (ISA) is used in conjunction with SERVQUAL. After calculations of gaps, a place for each item of the service in the space of possible future decisions is determined - in one of the following quadrants - 1- "focus on this"; 2- "successful result"; 3- "low priority"; 4- "possible revaluation" [15, 16]. An example of using the integrated approach with fuzzy numbers is described in [17].

2.2 Importance of Z-Numbers for Service Quality Analysis

Working with subjective perceptions and expectations involves the use of such a formalism that could adequately reflect the evaluations provided by consumers. The simple and widely used Likert scale allows a wider range of responses to be captured than a simple binary response. However, when using this scale, one may encounter inaccuracies because its perception depends on the respondent [18]. Moreover, subjective perception as well as the necessity of mental transformation of preferences into scores can lead to distortion or change in preferences [4], forcing customers to make the score too high or too low. Vague information is perceived by different consumers in different ways. And if the assessment process takes place in a fuzzy environment, then the use of linguistic values is preferable instead of the use numerical ones.

Traditional SERVQUAL and IPA methods based on crisp or fuzzy numbers do not consider the degree of respondent's confidence in the assessments. Moreover, the direct utilization of the exact numbers or defuzzified values of the fuzzy numbers in the assessment of the service quality leads to the paradox of operating with exact values obtained from various sources of the non-numerical nature with different levels of confidence.

Until 2011, the researchers did not have tools for processing uncertain information with varying degrees of reliability. The Z-numbers proposed in 2011 allow us to consider the reliability of information or confidence in it when measuring the service quality. The proposed two-component Z-numbers Z = (A, B) allow to evaluate the factors of service and their relevance. Parts A and B, described by fuzzy numbers, are more adequate in expressing people's evaluations. Working with consumers is a key aspect of analyzing the quality of the services provided. Consumers rarely express their assessments in precise numbers, more often in the form of *"very sure that is good"* or *"sure that is bad"*. In addition, the information they use is not 100% accurate. They can make an assessment based on their own perception, the perception of their relatives and the opinions of other people. Consumers also can be influenced in their assessments by advertising or other one-time exposure.

From this point of view, the utilization of the Z-numbers allows to adequately reflect the opinion and assessments of consumers in the service quality measurement. By applying Z-numbers, it is possible to obtain the values of the estimated items of service and the degree of confidence in these values. The developed computation methods with Z-numbers [19] make it possible to process individual evaluations, obtain generalized assessments, and rank them.

3 Results

We are considering the usage of Z-SERVQUAL and Z-IPA to estimate the service quality and identify the improvement strategy based on an example of one of the above-mentioned hotels. The suggested approach was realized as shown below.

Step 1. Z-SERVQUAL questionnaire has been compiled, where 21 items of the basic SERVQUAL scale corresponded to 1–4 subitems. In the questionnaire (a total 38 questions related to 21 item), travelers rated their expectations and perceptions by using Z-numbers. Expectations were evaluated by applying the terms *Very bad, bad, average, good, very good.* For evaluation of the degree of confidence, the terms *Not sure, not quite sure, quite sure, very sure, extremely sure* were used. In our case, 52 people took part in the survey.

Step 2. After completing the survey, the results were summarized for each of the 38 questions. The aggregated Z-values of perception and expectation for each item are calculated by averaging of all Z-evaluations of expectations and service perception.

$$Zaver = \frac{n_1}{N} Z_1 + \frac{n_2}{N} Z_2 + \cdots + \frac{n_m}{N} Z_m \tag{1}$$

Here– nm – number of respondents whose evaluation is Zm.

N – total number of respondents, where $N = \sum_1^m n_m$.

To calculate Z_{aver}, the operations of the multiplication of Z-number and scalar, as well as the addition of Z-numbers are used. The arithmetic mean had been used since the opinion of each hotel guest was equally important. If more than 75% of Z-evaluations of the perceptions (expectations) have the same value, then to reduce the calculations, the value of this item can be generalized (aggregated). The 75% threshold can also be changed based on the results obtained.

In our case, the parts of Z-numbers are expressed by triangular fuzzy numbers (TFN) with values of the triangular membership functions (TMF) shown in Table 1.

For the 1st subitem's expectation *("The hotel has modern and comfortable furniture")* of the 1st item *("Contemporary/new equipment")* –10 respondents gave a Z-evaluation as *(good, very sure)*, 29 respondents rated it as *(good, quite sure)*, 13 rated it as *(very good, quite sure)*.

Then the Z-evaluation of the expectation for this is calculated as follows:

$$Z_{E1} = 10 \cdot ((3\ 4\ 5)(0.75\ 1\ 1))/52 + 29 \cdot ((3\ 4\ 5)(0.5\ 0.75\ 1))/52$$
$$+13 \cdot ((4\ 5\ 5)(0.5\ 0.75\ 1))/52$$

Table 1. Codebook of A and B parts

Variable	A (level)	A (TMF value)	B (level)	B (TMF value)
Expectation,	Very bad	1/0, 1/1, 0/0	Not sure (NS)	1/0, 1/0.05, 0/0.25
gap,	Bad	0/1, 1/2, 0/3	Not very sure (NVS)	0/0.05, 1/0.25, 0/0.5
Perception	Average	0/2, 1/3, 0/4	Sure (S)	0/0.25, 1/0.5, 0/0.75
	Good	0/3, 1/4, 0/5	Very sure (VS)	0/0.5, 1/0.75, 0/1
	Very good	0/4, 1/5, 1/5	Extremely sure (ES)	0/0.75, 1/1, 1/1

$$Z_{E1} = (3.25\ 4.25\ 4.99)\ (0.4\ 0.67\ 0.86)$$

For the perception of the 1st subitem of the 1st item 80% of respondents gave a Z-evaluation *(good, very sure)*. So

$$Z_{P1} = (3\ 4\ 5)\ (0.5\ 0.75\ 1)$$

Similarly, for each subitem, the Z-evaluations of the expectation and perception of the service were calculated.

Step 3. It is necessary to define a gap for each SERVQUAL item. If the SERVQUAL item corresponds to several subitems, then, depending on the item, either the average value (if the questions related to the indicator are of the same importance) or the weighted average (if the questions have different importance) should be calculated. The significance of a question can be expressed either as a crisp number from 0 to 1, or as a Z-number.

In our case, the 1st item *"Contemporary/new equipment"* corresponds to three subitems: *"Modern and comfortable furniture"*, *"Modern air conditioning and lighting systems"*, *"Conveniences for people with disabilities"*. Let us assign the weight of importance to each question. For clarity, we used normalized weights - crisp numbers in the range from 0 to 1, giving a total of 1. For the 1st subitem, we set the weight indicator to 0.2, for the 2nd - 0.4, and for the 3rd - 0.4. For the first SERVQUAL item, the Table 2 of the obtained Z-evaluations is as follows:

The gap is determined by the following formula:

$$Z_{gapj} = \sum\nolimits_{1}^{n} w_i (Z_{pi} - Z_{ei}) \tag{2}$$

$i = 1, m$, where m – number of subitems in j-th SERVQUAL item
w_i – weight of importance of i-th subitem within framework of j-th item
If item has only one subitem then formula (2) is

$$Z_{gapj} = Z_{pi} - Z_{ei} \tag{3}$$

here, $j = 1, 2, \ldots, 21$ (items of SERVQUAL);
$\sum w_i \cdot Z_{pi}$ and $\sum w_i \cdot Z_{ei}$ - are Z-evaluations of expectation and perception for each item correspondingly.

Table 2. Subitems of 1st item, Z-evaluations, and importance weights (IW)

Item	Subitem	Expectation		Perception		IW
		Z-value	Confidence	Z-value	Confidence	
1	1.1	3.25 4.25 4.99	0.4 0.67 0.86	3 4 5	0.5 0.75 1	0.2
	1.2	3.1 4.1 4.8	0.18 0.4 0.67	3.12 4.16 4.98	0.41 0.7 0.86	0.4
	1.3	2.84 3.88 4.66	0.03 0.18 0.41	3.19 4.82 4.98	0.39 0.67 0.88	0.4

here 1.1. - Modern and comfortable furniture; 1.2 - Modern air-conditioning system, lightening 1.3 - Conveniences for people with disabilities.

According to the formula (2)

$$Z_{gap1} = (-1.65\ 0.35\ 1.95)\ (0.01\ 0.08\ 0.29)$$

Then, in the same way, we calculated the Z-evaluation of expectation and perception, as well as the gap for each indicator. If Z_{ei} and Z_{pi} have the same values, then the difference between them is considered equal to zero, and we consider this in formulas (2) and (3).

We can set the tolerance threshold for differences for fuzzy triangular numbers. For example, if the values of the left or right side differ by 0.01, then such numbers can be considered the same.

Further, according to the methodology proposed in [19–22], the ranking of Z_{expec} and Z_{gap} was carried out. As a result, we have the following Z-SERVQUAL Table 3:

Table 3. Ranked Z-SERVQUAL values of perceptions and expectations

N	SERVQUAL items	Z-Expectation	Z_{exp} rank	Z-Perception	Z-gaps	Z_{gap} rank
1	Contemporary/new equipment	(3.03 4.04 4.78) (0.01 0.11 0.34)	4	(3.12 4.39 4.98) (0.16 0.49 0.78)	(−1.65 0.35 1.95) (0.01 0.08 0.29)	8
2	Visually attractive physical facilities	(1.6 2.7 3.8) (0.12 0.21 0.5)	19	(1.75 2.75 3.75) (0.18 0.44 0.75)	(−1.36 0.64 2.64) (0.04 0.18 0.43)	6
3	Personnel is elegantly dressed and neat	(2.81 3.77 4.82) (0.21 0.47 0.75)	3	(2.81 3.77 4.82) (0.21 0.47 0.75)	No gap	10–17

(*continued*)

Table 3. (*continued*)

N	SERVQUAL items	Z-Expectation	Z_{exp} rank	Z-Perception	Z-gaps	Z_{gap} rank
4	Facilities have been keeping according to the types of services provided	(1.63 2.64 3.44) (0.23 0.44 0.63)	18	(2.15 3.15 3.9) (0.39 0.54 0.75)	(−1.29 0.51 2.27) (0.14 0.3 0.55)	3
5	Hotel staff do promised job on pledged time	(2.19 3.19 4.19) (0.44 0.7 0.88)	8	(2.19 3.19 4.19) (0.44 0.7 0.88)	No gap	10–17
	…	…	…	…	…	…
17	Individual attention for guests	(1.19 2.19 3.2) (0.32 0.64 0.98)	17	(1.96 3.01 4.02) (0.29 0.45 0.57)	(−1.24 0.82 2.83) (0.19 0.39 0.57)	1
18	Personal attention for you	(2.23 3.23 4.15) (0.26 0.38 0.69)	10	(2.23 3.22 4.15) (0.26 0.38 0.69)	No gap	10–17
19	Personnel foresee your needs	(2.32 3.33 4.21) (0.12 0.32 0.63)	11	(1.35 2.35 3.32) (0.32 0.64 0.81)	(−2.86 −0.98 1) (0.8 0.28 0.58)	21
20	Hotel is taking care about your interest	(2.01 3.01 4.01) (0.05 0.25 0.5)	15	(1.5 2.5 3.5) (0.5 0.75 1)	(−2.51 −0.51 1.49) (0.26 0.38 0.56)	18
21	Convenient operating time	(2.01 3.01 4.02) (0.17 0.5 0.73)	14	(2 3.01 4.02) (0.17 0.5 0.73)	No gap	10–17

Step 4. After the ranking, we conducted an Importance-Performance Analysis to measure the quality of service and determine those indicators of quality improvement that the service provider should pay attention to. Usually, to construct quadrants, the average of the expectation values and the gap values are determined [16, 17]. We are using the median values from Z_{exp} and Z_{gap}.

Median values for Z_{gap} can be interpreted as the Low Service Gap, and median values for Z_{exp} as the High Expectations. As a result, we get the following Z-IPA model (Fig. 1):

Step 5. After building the Z-IPA model, we determined the problematic indicators of the service quality that require the improvement as "*Deliver service at pledged time*", "*Personnel is always ready to help guests*", "*Personnel foresee your needs*", "*Hotel is taking care about your interest*".

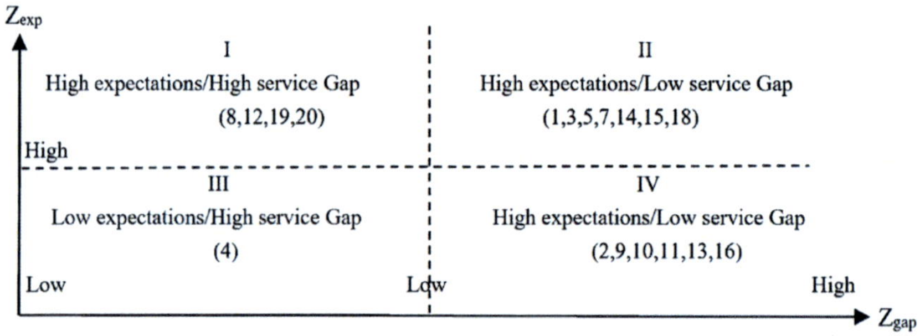

Fig. 1. Z-IPA model

4 Conclusion

In our study, we investigated the applicability of Z-numbers for the estimation of the service quality and the identification of the service improvement strategies in the tourism sector.

Measurement of the service quality encounters uncertainties associated with subjective perception, subjective assessment, and the degree of confidence in assessment.

The paper illustrates the possibility of constructing Z-SERVQUAL tables and Z-IPA models based on Z-numbers. Using Z-numbers, the respondent expresses his evaluations and degrees of confidence in them in a natural language. In this case, there is no need for intermediate mental transformations into crisp numbers. Recent achievements in Z-number calculation techniques potentiate to process of verbal information that considers both the value of the uncertain variable and the reliability of this value. The proposed methodology allows to measure the service quality effectively and does not depend on the number of items and dimensions.

The Z-numbers based approach is applicable not only to the hospitality sector, but also to other fields of the tourism and service sector.

References

1. Park, N.K., Jeong, E.B.: Service quality in tourism: a systematic literature review and keyword network analysis. Sustainability. **11**, 3665 (2019). https://doi.org/10.3390/su11133665
2. Ingaldi, M.: Overview of the main methods of service quality analysis. Prod. Eng. Archiv. **18**, 54–59 (2017). https://doi.org/10.30657/pea.2018.18.10
3. Tang, L.-L., Chen, S.-H., Lin, C.-C.: Integrating FMEA and the Kano model to improve the service quality of logistics centers. Processes **9**, 51 (2021). https://doi.org/10.3390/pr9010051
4. Chang, T.-Y., Yen, C.-H.: The evaluation of airline service quality by fuzzy MCDM. Tour. Manage. **23**, 107–115 (2002). https://doi.org/10.1016/S0261-5177(01)00050-4
5. Parasuraman, A.P., Zeithaml, V., Berry, L.: SERVQUAL: a multiple- item scale for measuring consumer perceptions of service quality. J. Retail. (1988)
6. Parasuraman, A.P., Zeithaml, V., Berry, L.: Refinement and reassessment of the SERVQUAL scale. J. Retail. **67**, 420–450 (1991)
7. Saleh, F., Ryan, C.: Analysing service quality in the hospitality industry using the SERVQUAL model. Serv. Ind. J. **11**(3), 324–345 (1991)
8. Haryadi, S.: Costumers' satisfaction analysis at Kartika Hotel International in East Java using fuzzy servqual approach. Int. Bus. Manag. **8**, 159–167 (2014)
9. Firdaus, I., Farikhin, B.S.: Application of fuzzy Servqual method to measure user satisfaction of Mooc service quality. E3S Web Conf. **202**, 14007 (2020)
10. Zadeh, L.A.: A note on Z-numbers. Inf. Sci. **181**(14), 2923–2932 (2011). https://doi.org/10.1016/j.ins.2011.02.022
11. Stefano, N.M., Casarotto Filho, N., Barichello, R., Sohn, A.P.: A fuzzy SERVQUAL based method for evaluated of service quality in the hotel industry. Procedia CIRP **30**, 433–438 (2015). https://doi.org/10.1016/j.procir.2015.02.140
12. Pérez-Campdesuñer, R., Garcia Vidal, G., Martínez-Vivar, R., Sánchez-Rodríguez, A., Akhunjonov, U., Obrenovic, B.: Application of the fuzzy logic tool to evaluate customer satisfaction in Hotels. J. Int. Bus. Res. Market. **3**, 24–29 (2018). https://doi.org/10.18775/jibrm.1849-8558.2015.32.3004
13. Wong, A., Dean, A., White, C.: Analysing service quality in the hospitality industry. Manag. Serv. Qual. **9**, 136–143 (1999). https://doi.org/10.1108/09604529910257920
14. Choi, H., Ann, S., Lee, Kwang-Woo., Park, Duk-Byeong.: Measuring service quality of rural accommodations. Sustainability **10**(2), 443 (2018)
15. Ho, L.-H., Peng, T.-F., Feng, S.-Y., Yen, T.-M.: Integration of Kano's model and SERVQUAL for enhancing standard hotel customer satisfaction. Afr. J. Bus. Manage. **7**(23), 2257–2265 (2013). https://doi.org/10.5897/AJBM11.1421
16. Hamzah, Z., Purwati, A.A., Suryani, F.: Quality improvement strategy of islamic banking services in Indonesia through the integration of Servqual and Importance Performance Analysis (IPA). Rev. ESPACIOS **40**(30) (2019)
17. Wijayanti, E., Noya, S.: Integrating fuzzy-servqual into importance performance analysis and quality function deployment for improve Ksp kusuma artha lestari service quality. J. Ilmiah Teknik Indust. **1**(3) (2013). https://doi.org/10.24912/jitiuntar.v1i3.476
18. Dolnicar, S., Grün, B., Leisch, F., Rossiter, J.: Three good reasons NOT to use five- and seven-point Likert items. In: 21st CAUTHE National Conference (2011)
19. Aliev, R.A., Huseynov, O.H., Aliev, R.R., Alizadeh, A.V.: The Arithmetic of Z-Numbers: Theory and Applications. World Scientific, Singapore (2015)
20. Aliev, R., Huseynov, O., Serdaroglu, R.: Ranking of Z-numbers and its application in decision making. Int. J. Inf. Technol. Decis. Mak. **15**, 1–17 (2016)

21. Aliev, R.A., Gardashova, L.A.: Z-set based approach to control system design. In: Aliev, R.A., Kacprzyk, J., Pedrycz, W., Jamshidi, M.O., Babanli, M., Sadikoglu, F.M. (eds.) ICAFS 2020. AISC, vol. 1306, pp. 10–21. Springer, Cham (2021). https://doi.org/10.1007/978-3-030-64058-3_2
22. Aliev, R.A., Huseynov, O.H., Zulfugarova, R.X.: Z-distance based IF-THEN rules. Sci. World J. **2016**, 1673537 (2016). https://doi.org/10.1155/2016/1673537

Predicting Stock Prices Using Random Forest and Logistic Regression Algorithms

Aysel Aliyeva(✉)

Odlar Yurdu University, Baku, NJ AZ1072, Azerbaijan

Abstract. One of the most widely used ways of investing globally is stock markets. Especially nowadays, since the price barrier for making a trade is relatively low, earning capacity is promising, and accessing it is technically straightforward, the stock market has become a more attractive address for investment. The emergence of numerous speculative movements in the markets causes these stock markets to have exceptionally high ups and downs. Thus, developing swift and effective ways of predicting stock market prices in the current world is critical. As in many other industries, utilization of machine learning and deep learning algorithms [1] have extensively been adopted for forecasting trends of different financial tools, specifically for stock market prices.

In this study, data between 2015 and 2020 were used as the data set - opening and closing prices, high and low prices, and also the volume of the stocks of the Tesla company traded on the New York Stock Exchange (NYSE). In the light of machine learning methods and deep learning algorithms, it is aimed to estimate the closing prices of these stocks. Machine learning methods such as Random Forest (RF) and Logistic Regression (LR) have been used to construct a prediction model in this context. As a result of the investigation, it has been observed that Logistic Regression generates more consistent estimates than Random Forest. This conclusion is backed up by accuracy measurements, which determine how models are performed, and also various metrics are displayed for models.

Keywords: Forecasting · Stock prices · Machine Learning (ML) · Supervised learning · Random Forest (RF) · Logistic Regression (LR)

1 Introduction

Investors make educated guesses by analyzing data. They read the news, study company history, industry trends, and many more data points to make estimations on stock price changes. Based on the random walk theory [2], stock prices are random and unpredictable. Burton Malkiel noted, "A blindfolded monkey throwing darts at a newspaper's financial pages could select a portfolio that would do just as well as one carefully selected by experts" [3]. However, nowadays top companies such as UBS, JPMorgan Chase, Goldman Sachs, Morgan Stanley, Citigroup are hiring highly qualified quantitative analysts to build predictive models. Moreover, over 70% of all orders are now placed by software (algorithms) today on Wall Street. Is it possible to forecast stock prices by using machine learning?

R. A. Aliev et al. (Eds.): ICSCCW 2021, LNNS 362, pp. 95–101, 2022.
https://doi.org/10.1007/978-3-030-92127-9_16

These days, the number of publicly traded companies and investors funding them has tremendously increased. As different web services make financial markets more reachable and also extremely successful and famous companies such as Tesla, PayPal, Walmart, Apple, etc. go public, these trends result in enormous trading and turnover in financial markets. As an example, the New York Stock Exchange (NYSE) handles more than $50 billion worth of stock trades per day.

Thus, stocks and funds have become a very crucial and also at the same time mysterious topic for investors, researchers, and the scientific community. Thanks to the stock exchange, those who can correctly assess their investments can gain tremendous profits, while they can face the same big losses as a consequence of improper evaluations. Estimation trials, that were previously made with the help of mathematical and statistical models, provided better results, specifically with the improvement of IT technologies. Currently, the usage of machine learning and deep learning algorithms has developed, which possess the capability to generalize, can learn from samples, and are less inaccurate than statistical methods. Random Forest and Logistic Regression are used in the work.

1.1 Used Data and Descriptive Analysis

The data used in this study were obtained from Yahoo Finance website for Tesla company, the opening and closing, high and low daily prices, and daily volumes of stock between 10.07.2015 and 10.07.2020 are included. Data points of 1260 days were used for prediction. To have a better picture of the data set, we could review Fig. 1. Below:

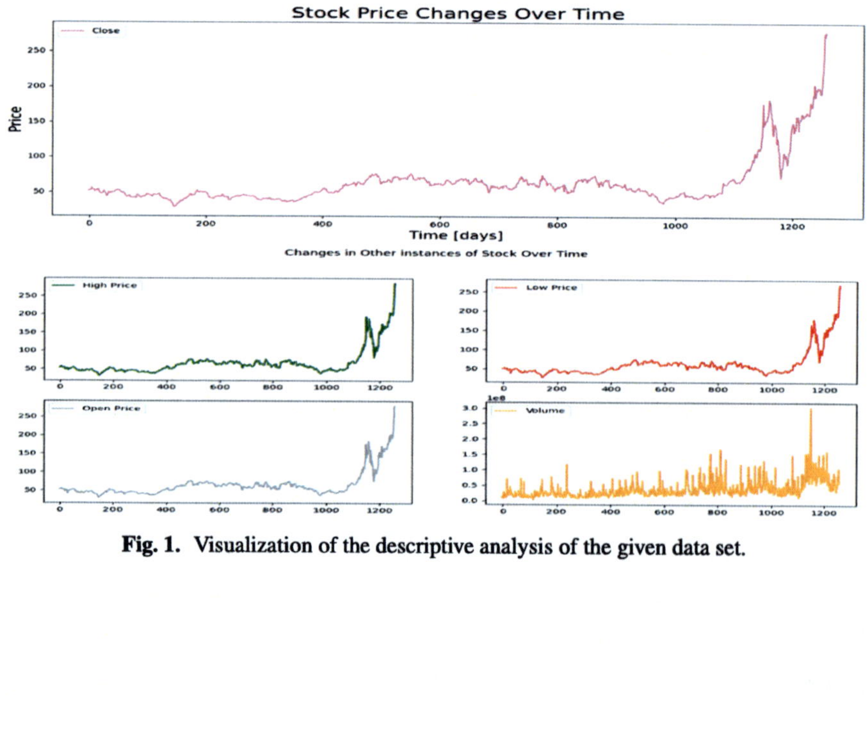

Fig. 1. Visualization of the descriptive analysis of the given data set.

1.2 Data Preprocessing

As the data obtained from Yahoo Finance website in comma-separated values (CSV) format, check for N/A values and trends for different features are visualized. Mean values with windows of two are calculated to fit into the models. And force indexes to illustrate how strong the actual buying or selling pressure is, are also calculated with the help of the Pandas Python library [4].

```
  data["DifCl"] = data.Close.diff()
  data["SMA2"] = data.Close.rolling(2).mean()
  data["ForceIndex"] = data["Close"] * data["Volume"]
  data["y"] = data["DifCl"].apply(lambda x: 1 if x > 0
else 0).shift(-1)

  data = data.drop(
     ["Open", "High", "Low", "Close", "Volume", "Diff",
"Adj Close"],
     axis=1,
  ).dropna()
```

As it might be visible from the aforementioned lines of code firstly, the difference of close prices has been calculated and stored into DifCl column, then mean of the close price with the size of two moving windows is computed and put into the SMA2 column, force index has been calculated by multiplying volume and close price. Lastly, values of "y" out of the DifCl column are fetched (as 0 and 1).

```
X = data.drop(["y"], axis=1).values
y = data["y"].values
```

After deleting N/A values and values that will not be used, the data frame is separated into X and y arrays, containing values of SMA2, Force Index, and y which are going to be used to train classification and regression models respectively.

2 General Overview of Methodology

The basic classifier proposed by me to solve this problem is the Logistic Regression method. It is good to use it for classification problems. Logistic regression is similar to linear regression in that it also requires finding coefficient values for input variables. The difference is that the output value is converted using a non-linear or logistic function. The logistic function looks like a large letter S and converts any value to a number between 0 and 1. As for comparison, I'm considering the classification model - Random Forest.

2.1 Random Forest

Random Forest is an ensemble of decision trees when each tree depends on a collection of defined features. A decision tree is the classification model that sequentially breaks objects into subsets based on the value of the features of the objects.

The random forest classification method is based on the creation of a large number of decision trees and the final classification of the object, based on the prediction of all the created decision trees. Thus, a random forest averages the predictions of decision trees. As a final prediction, the median of all predictions of the constructed decision trees is usually taken. Comparing to other classification techniques like support vector machines, neural networks, and discriminant analysis, RF performs very well [5]. We can conclude that RF uses the bagging method for classification.

The qualitative splitting of the subsets is going by a particular criterion. One way to get some information about the importance of features is to count the amount of impurity (disorder) of the subset that removes the partition by this attribute. A measure of uncertainty may be the Gini coefficient or entropy. And the splitting criterion used in my case is the Gini impurity index, which is a measure of "node impurity":

$$Gini = \sum_{y=1}^{y} p_y(1 - p_y) \tag{1}$$

Where p_y, is the proportion of data points in a leaf node assigned to class "y". As can be seen, this measure is zero for $p_y = 0$ and $p_y = 1$ and have a maximum at $p_y = 0.5$. This is a measure of how often a randomly chosen observation from the data set would be incorrectly classified if it was randomly classified according to the distribution of classes in the subset. Every split results in a decrease in Gini. A low Gini suggests that a known predictor feature plays a greater role in partitioning the data into the defined classes. There is a feature importance technique, which summing all decreases in the forest of trees [6].

According to Liaw and Wiene [5], random forests perform very well compared to other classifiers, like discriminant analysis, support vector machines, and neural networks.

Decision trees of great depth are usually overfitted on training data, that is, they have a small bias of their predictions relative to the correct predictions, but they also have a very large variance of their predictions. So as stated before, the purpose of a random forest is to reduce the variance of the prediction of individual decision trees.

Another important point is that when building the next decision tree and deciding on the choice of a sign for separation sets of objects belonging to the next node, into two subsets, a random subset of features is used. This approach allows twenty further reduce the correlation between pairs of trees in the forest, as it reduces the number of occurrences of very important traits that affect the total prediction to the tops of decisive trees. RF persistently trying to strive and therefore robust to any kind of overfitting [6].

As stated earlier, RF uses decision trees as base learners and the formula for prediction at the new point of RF looks like this:

$$f(x) = \sum_{k=1}^{K} I(h_k(x) = y) \tag{2}$$

Where, K = number of leaves of the tree, k = number of trees, *I* is the prediction function of the response feature at *x* using the *k-th* tree. The random forest method refers to nonlinear classification methods, which complicates the interpretation of the results of this method.

Also, interpretation of the model is important in the analysis of prediction errors since it allows you to find signs whose change influenced the transition of the prediction from right to wrong from object to object. Because the Random Forest model is often considered as a black-box model with a lot of deep trees, sometimes it makes the process of researching each tree not feasible. Moreover, even a single tree can lead to some difficult interpretations only with a small depth and a small number of signs. Even a decision tree with a depth of 10 may have thousands of nodes, which automatically means the impossibility of any clear explanation and decision of this model.

2.2 Logistic Regression

For our research problem, the first classification method to use will be Logistic Regression. It is quite a popular method for binary classification problems. It is a linear model for classification and allows predicting the probabilities of classes with the help of the sigmoid function, called the logit curve as in Fig. 2. In the basic definition of logistic regression, it is known that the values of the class feature are inside of {0, 1} and are binary, although it is also possible to model non-binary class features [7].

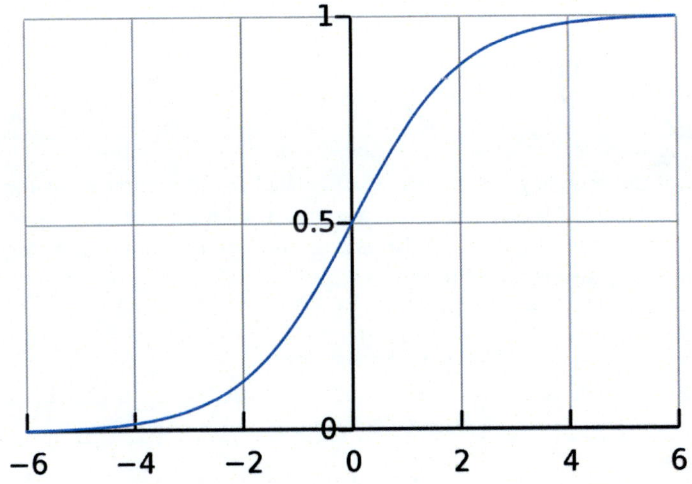

Fig. 2. Logistic sigmoid function, Source: Wikipedia

Logistic regression algorithm implementing the estimation of the probability of belonging to the positive class (1 instead of 0) as follows:

$$P(X) = \frac{1}{1 + e^{\beta_0 + \beta_1 X_1 + \beta_2 X_2 + \cdots + \beta_k X_k}} \tag{3}$$

Where Y = 1 represented a given loan instance would be good, X*i* is the feature for *i*-th numerical data, *i* is the coefficients for i-th feature, and p is the number of features. The dependent feature is modeled as a Bernoulli random feature cause of the binary classification problem. For estimation of parameters or coefficients, the maximum likelihood estimation is used [8]. It maximizes the probability of observing a given data set. We can define logistic regression as:

$$f(x) = P(Y * (x) = Y) \tag{4}$$

Where to the object (x) a class with the highest probability is assigned, which is calculated by the sigmoid function.

The advantages of logistic regression regard in its simplicity in implementation and the ease of interpretation, as each coefficient describes the magnitude and direction each feature has in determining the classification of each observation. Its simple nature meant the model could be trained and tuned very efficiently [7].

One of the main advantages of the logistic regression classifier is that it is a non-parametric method since no assumptions are made regarding the probability of feature distribution. Among the disadvantages are the sensitivity to outliers, the presence of a sufficiently large sample, and the necessity for data pre-processing. Because of the way the model is trained, logistic regression predictions can be used to display the probability that a sample belongs to class 0 or 1. This is useful in cases where you need to have more justification in the classification. The logistic regression model is fast to learn and is well suited for binary classification tasks.

3 Results and Conclusion

3.1 Results

The results of the accuracy measurements calculated for the estimates of the models are shown in Table 1. Metrics packages from scikit-learn are used to generate them. Accuracy measures are calculated and also as Logistic Regression is a regression problem R squared metric is also included to compare given models.

Table 1. Accuracy metrics.

Measures	Logistic regression	Random forest
Accuracy score	0.463	0.457
Mean squared error	0.537	0.542
Mean absolute error	0.537	0.542
R squared	−1.16	−1.18

Pang, Zhou, Wang, Lin, and Chang [1] proposed a long short-term deep neural network (LSMN) with an embedded layer that vectorizes data to predict the stock market. The experimental results show that the long short-term deep neural network with embedded layer is the most advanced technology with an accuracy rate of 57.2% for Shanghai A-stocks and 52.4% for individual stocks.

4 Conclusion

This work, it is tried to predict the price of Tesla company stocks with the help of machine learning algorithms. Logistic Regression (LR) and Random Forest (RF) models were established for this purpose. The analysis involves 5 years of daily stock prices and volume data between 10.07.2015 and 10.07.2020.

The Logistic Regression (LR) model, which is a kind of linear classification method, has been applied in many areas and it has been seen that successful results have been obtained in prediction studies. The reason why this model is preferred is that it gives easy, fast, flexible, and consistent results. In this thesis, Random Forest (RF), which is used as another model, has recently been successfully applied in classification, regression, and time-series estimation applications.

The results show that the LR and RF models in predicting stock prices yielded slightly diverse performance scores.

From a very abstract point of view, the answer to the question that will those models be able to predict stock price correctly, at any given time, is no. However, they are analytical tools that are helping us to make more educated guesses about the direction of the market. It is better than doing it haphazardly.

In general, we can say that both the LR and the RF models are useful for predicting stock prices in time series.

References

1. Pang, X., Zhou, Y., Wang, P., Lin, W., Chang, V.: Stock market prediction based on deep long short term memory neural network. In: 3rd International Conference on Complexity, Future Information Systems and Risk, pp. 102–108 (2018). https://doi.org/10.5220/0006749901020108
2. Fama, E.F.: Random walks in stock market prices. Financ. Anal. J. **21**(5), 55–59 (1965). https://doi.org/10.2469/faj.v21.n5.55
3. Malkiel, B.G.: A Random Walk Down Wall Street: Including a Life-Cycle Guide to Personal Investing. Norton, New York (1999)
4. Joshi, P.: Artificial Intelligence with Python. Packt Publishing Ltd., UK (2017)
5. Liaw, A., Wiener, M.: Classification and regression by random forest. R. News **2**(3), 18–22 (2002). http://CRAN.Rproject.org/doc/Rnews/
6. Breiman, L.: Random forests. Mach. Learn. **45**, 5–32 (2001). https://doi.org/10.1023/A:1010933404324
7. Aggarwal, C.C.: Data Mining, the Textbook. Springer, New York (2015). https://doi.org/10.1007/978-3-319-14142-8
8. Myung, I.J.: Tutorial on maximum likelihood estimation. J. Math. Psychol. **47**, 90–100 (2003). https://doi.org/10.1016/S0022-2496(02)00028-7

Explainable Artificial Intelligence: Rules Extraction from Neural Networks

Alexey Averkin[1,2] and Sergey Yarushev[3(✉)]

[1] Federal Research Centre of Informatics and Computer Science of RAS, Vavilova, 42, Moscow, Russia

[2] Educational and Scientific Laboratory of Artificial Intelligence, Neuro-Technologies and Business Analytics, Plekhanov Russian University of Economics, Moscow, Russia

[3] Departments of Informatics, Plekhanov Russian University of Economics, Moscow, Russia

Abstract. Researchers are currently working on the development of "Explainable Artificial Intelligence" or "Explainable Artificial Intelligence (XAI)". Such systems are designed to help the user understand the decisions made by the neural network, which will increase confidence in such systems, will allow making more effective decisions based on the results of the system operation. The first results of applying this approach allowed developers and users to study the factors that are used by the neural network to solve a specific problem and what parameters of the neural network need to be changed to improve the accuracy of its work. In addition, studying how neural networks extract, store, and transform knowledge may be useful for the future development of machine learning methods. To overcome this disadvantage of neural networks, it is proposed to consider methods for extracting rules from neural networks, which can become a link between symbolic and connectionistic models of knowledge representation in artificial intelligence. In this paper, we propose a neuro-fuzzy approach to rule extraction using time series forecasting and text recognition as examples.

Keywords: Artificial Intelligence · XAI · Explainable Artificial Intelligence · Rules extraction · Neural networks

1 Introduction

Recent advances in artificial intelligence make it an integral part of today's realities, whether it be recognizing emotions, images, or making strategic decisions. But the use of traditional models of artificial neural networks is limited by the impossibility of explaining the results obtained. For the solutions obtained with the help of artificial neural networks to have legal force, they must be able to prove the correctness of the solution obtained, to justify the decision chain. The development of methods for extracting knowledge from artificial intelligence can be a step towards the development of explainable artificial intelligence and expand the scope of its application [1].

R. A. Aliev et al. (Eds.): ICSCCW 2021, LNNS 362, pp. 102–109, 2022.
https://doi.org/10.1007/978-3-030-92127-9_17

2 Decision Trees for Extracting Rules from a Neural Network

Artificial neural networks are models in the form of a "black box", where we feed data to the input of the model and get the result at the output. But we do not have any information about the structure of the model and how the decision is made and what criteria are used in this case. Of course, the rules are implicitly contained in the weighting coefficients of the neural network, but the structure of these coefficients is extremely complex, and the number is enormous. Therefore, understanding them, obtaining with their help any knowledge about the initial data set, explaining and justifying the solutions obtained is not possible. The inability to substantiate the decision made, today, is considered a key disadvantage of artificial neural networks, in almost any architecture, therefore, the possibility of extracting decision rules from a trained neural network has long been of interest to researchers in this area. Many approaches have been developed to extract knowledge from network weights in a human-readable form. However, the vast majority of them, such as the NeuroRule algorithm described in [2], impose very strict restrictions on the network architecture, the type of threshold functions, the connectivity of neurons, and the set of input and output values of the neural network.

Among all approaches to extracting rules from artificial neural networks, the TREPAN algorithm developed in 1996 stands out [2], which builds a decision tree based on a trained neural network. This approach is interesting, first of all, because it does not impose absolutely any requirements on the architecture of the network, on the sets of its input and output values, the number of inputs and outputs, the presence of recurrent connections, the teaching method, etc. It is enough that the network is just a "black box" or, in the terminology of the authors of the algorithm, an "oracle", which is presented with a certain input vector and the output is the answer. Moreover, the "oracle" can be not only a neural network, but also an arbitrary classifier (since the generated decision tree is a classification tree).

In this work, an algorithm based on the TREPAN algorithm is implemented. The TREPAN algorithm itself was not taken in its pure form, for several reasons. First, it is very difficult to find a clear description of it. Even the most complete descriptions found in the literature, given in [2], is clearly insufficient to implement the algorithm. Second, the algorithm uses the accuracy estimate at each node of the tree as a local criterion for stopping learning, which is inapplicable in the attribution problem due to large amounts of data and the computational complexity of the algorithm. And thirdly, the TREPAN algorithm generates a tree that is slightly different in structure from classical decision trees, and in this work, we wanted to get a structure similar to ordinary decision trees in order to compare and unify the results and algorithms.

The maximum allowable depth of the generated tree is used as a global criterion for the completion of the algorithm. The local stopping criterion is fulfilled when in a given tree node there are examples belonging to only one class, or there are no more signs that allow further separation, or the test for the consistency of this separation has not been passed.

After reading the description of the algorithm, the question may arise: why, in fact, use the "oracle"? After all, you can simply build a decision tree based on the existing training sample, which already contains an exact indication of which class each example belongs to. Moreover, such an "oracle" as a neural network, no matter how well trained it is, will still make mistakes. Here the point is in the generalizing ability of artificial neural networks [2], which allows obtaining simpler decision trees. In addition, the use of the "oracle" makes it possible to compensate for the lack of data observed when constructing trees and decisions at the lower levels.

Thus, it is possible to extract structured knowledge not only from extremely simplified neural networks.

2.1 Rule Extraction Algorithm

The joint use of artificial neural networks and decision trees can turn the "black box" into a structure that is more understandable for a person by interpreting the result of training a neural network in the form of a hierarchically sequential structure of "ifto" rules. The task is to classify the input data set based on the perceptron and further analyze the neural network to extract the classifying rules specific to each class.

Let us consider this problem using the example of an artificial neural network of feedforward propagation with five input neurons and one output. The structure of the network is shown in Fig. 1. This neural network can be interpreted by a finite number of "IF-TO" rules [3] due to the finite number of possible input vectors.

The weights of the neural network take on the following values:

$$w_1 = 5, w_2 = 3, w_3 = 3, w_4 = 0, w_5 = 3$$

The activation threshold is 9. In this case, the following set of rules is extracted from the neural network.

Based on the obtained rules, the decision-making procedure consists in predicting the value y = true if the activation of the output neuron = 1, and y = false when the activation is 0.

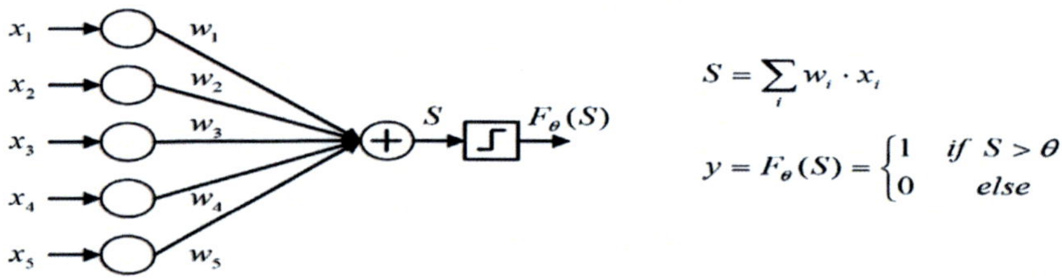

Fig. 1. The structure of the network

If for the output neuron, instead of the threshold activation function, a sigmoidal one is used, then the output value y will be true if the value activation function will exceed the threshold. When solving the classification problem, it is possible to use separate outputs for each classification class. In this case, the decision is made in favor of the maximum activation function. Thus, the resulting rule should characterize a set of input parameters for which an artificial neural network that has already passed the learning process predicts a certain class at the output.

One of the algorithms for extracting rules from neural networks trained to solve the classification problem is the NeuroRule method. This algorithm is based on local rule extraction and includes three main steps [4].

Stage 1. Training the neural network, when the two-layer perceptron is trained until a sufficient classification accuracy is obtained. At the initial moment, a large number of intermediate neurons are selected, and after training, unnecessary neurons and connections are discarded.

Stage 2. Thinning of the neural network. The trained neural network contains all possible connections between input neurons and neurons of the hidden layer, as well as between hidden and output neurons. The total number of these links is usually so large that it is impossible to extract classification rules that are understandable by the user from the analysis of their values. Thinning is the removal of redundant connections and neurons, which does not lead to an increase in the classification error by the network. The resulting network usually contains few neurons and connections between them, and the functioning of such a network is amenable to research.

Stage 3. Extracting rules. At this stage, rules are extracted from the thinned neural network in the form of "if - then". For this, preparation for rule extraction is carried out, which consists in coding continuous values, both at the input and within the network. The coding of the attributes of the classified objects is carried out if they are continuous values. To represent them, you can use binary neurons and the coding principle of the "thermometer" type.

The values that the neurons of the hidden layer takc are clustered and replaced with values that determine the centers of the clusters. The number of such clusters is not large. After such discretization of the activities of intermediate neurons, the accuracy of the classification of objects by the network is checked. If it remains acceptable, then preparation for rule extraction ends. Further, rules are extracted, while movement along the network occurs from the classifying output neurons to the network inputs. It is assumed that these rules are fairly straightforward to validate and can be easily applied to large databases.

However, this algorithm sets rather strict restrictions on the neural network architecture, the number of elements, connections, and the type of activation functions. So for intermediate neurons, a hyperbolic tangent is used and their states change in the interval [1], and for output neurons, the Fermi function is used with an interval of states [0,1].

The disadvantages of most rule extraction algorithms include the lack of universality and scalability. In this regard, the most interesting is the TREPAN algorithm [5] and its modifications, which are devoid of these drawbacks and do not impose any requirements on the network architecture, input and output values, learning algorithm, etc. This approach builds a decision tree based on the knowledge embedded in the trained neural network, and it is enough that the network is a kind of "black box" or "Oracle", to which you can ask questions and receive answers from it. Moreover, the algorithm is quite versatile and can be applied to a wide range of other trained classifiers. It also scales well and is not sensitive to the dimension of the input feature space and the size of the network.

2.2 Application of Decision Trees to Extract Rules from Neural Networks Using the Example of Attribution of Text

When conducting experiments, to assess the effectiveness of the described algorithms, in the task of classifying literary works according to the individual author's style, a specially prepared set of frequency features was used. The selected features were the number of paragraphs and the number of sentences in a fragment consisting of a certain number of words (from 1 to 39, the 40th feature is the number of sentences more than 39 words long). The experiment involved data on 10 writers and text fragments 200, 300, 400, ..., 1800 sentences long. Also, to estimate some parameters, a reduced set of data for four writers (Dostoevsky F.M., Kuprin A.I., Belyaev A.R., Dovlatov S.) was used, compiled on the basis of a fragment of 500 sentences long. To assess the effectiveness was applied method "k-subsets", for k = 10.

When conducting experiments with artificial neural networks, the data described above for ten writers were used. A neural network, unlike decision trees, has a large number of external parameters, such as the number of network layers, the number of neurons in intermediate layers, the type of the threshold function, the learning rate, and the number of learning iterations. Optimization of the listed parameters is beyond the scope of this work and is a topic for a separate large study. In this work, we used empirically selected parameters. It was decided to use an artificial neural network consisting of two layers of neurons. The number of inputs to the first layer was determined by the number of analyzed features (in our case, 41), the number of neurons in the first layer was taken equal to 10, and the number of neurons in the second output layer was equal to the number of writers. The encoding of the input and output values, as well as the normalization of the data. When training the network, we used a backpropagation algorithm with a decreasing step. The number of training iterations was taken to be 500.

Figure 2 shows a graph of the dependence of the efficiency of an artificial neural network on the length of a text fragment. You can see that on the same data an artificial neural network behaves somewhat differently than decision trees. First, the classification accuracy is noticeably higher. So, when using a decision tree, the maximum achieved efficiency was 59% with a fragment length of 1700 sentences. Whereas the neural network showed the maximum prediction efficiency equal to 71% with a fragment length of 1000 sentences. A similar picture is observed in relation to the minimum efficiency, which in both cases is achieved with a fragment length of 200 sentences and is equal to 39% for a decision tree and 59% for a neural network. At the same time, a neural network, in contrast to decision trees, has a significant decrease in efficiency with increasing fragment length. This may be due to the fact that with an increase in the length of the fragment, the influence of data irrelevant to the individual author's style increases, and the effect of overfitting begins to manifest itself. The decision tree uses a special tool, thinning, to avoid this effect, while the neural network does not have such a tool. A sharp decline in efficiency is observed when the length of the fragment becomes less than 700 sentences or more than 1300 sentences.

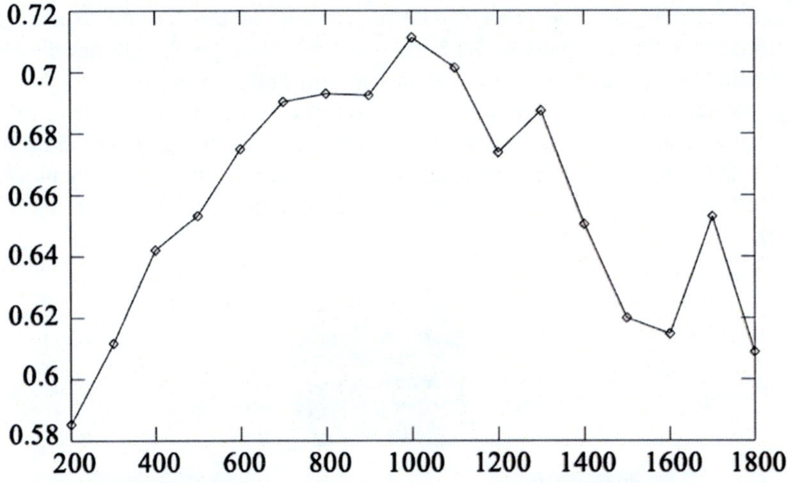

Fig. 2. A graph of the dependence of the efficiency of an artificial neural network

Decision trees were extracted from an artificial neural network, trained similarly to the one described in the previous paragraph. At the same time, an additional parameter of the algorithm appears as the minimum number of training examples used in each node of the tree. The graph of the dependence of the prediction accuracy on the minimum number of training examples is shown in Fig. 3.

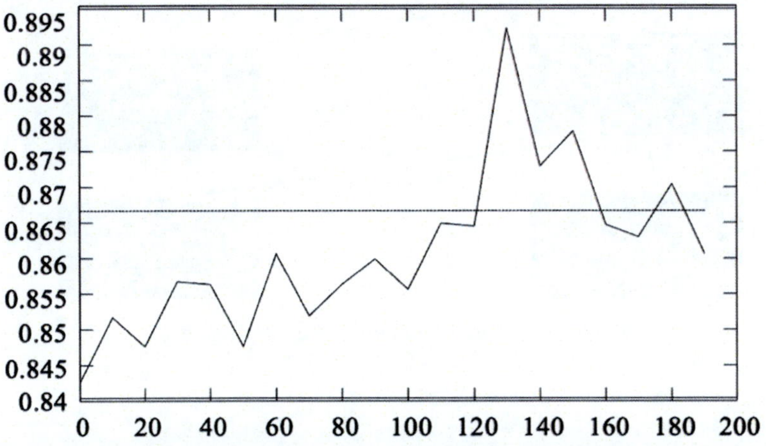

Fig. 3. The graph of the dependence of the prediction accuracy

The graph shows that with an increase in the number of additional examples, there is a noticeable increase in prediction accuracy, reaching an accuracy of 89% with 130 additional examples. Such accuracy cannot be achieved on this dataset using classical decision trees. For comparison, the horizontal line on the graph indicates the level corresponding to the accuracy of classical decision trees on the same dataset.

Figure 4 shows an illustration of how the extracted decision tree is able to reflect the internal structure of the trained neural network [6]. For this example, the neural network was trained in 10 steps, instead of 500. You can see that such a network is able to distinguish only two of the four writers, the differences between which are the strongest.

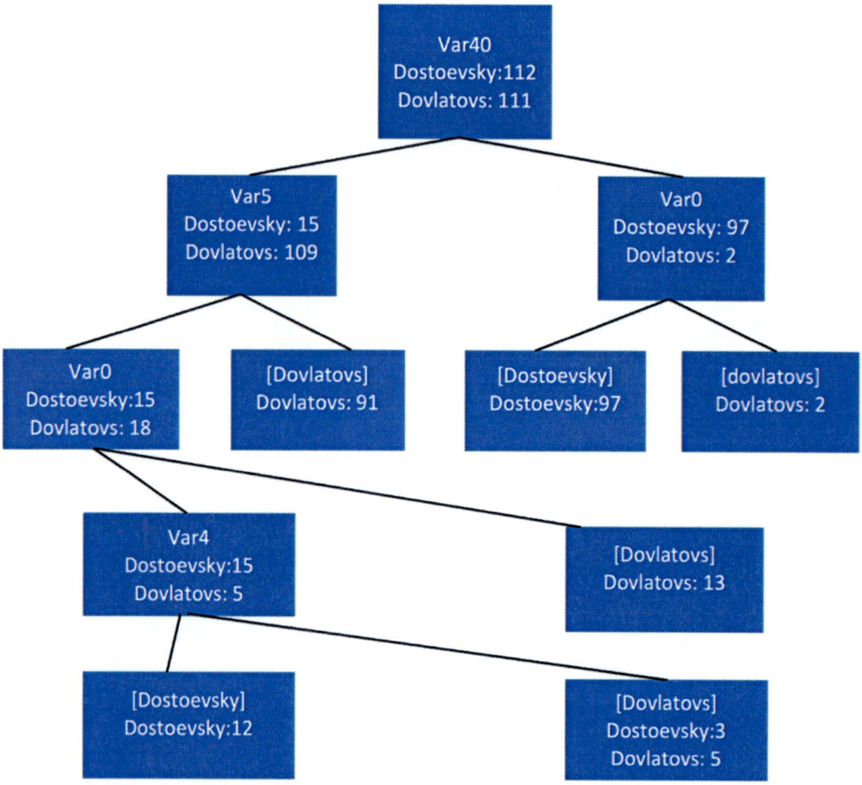

Fig. 4. An illustration of the extracted decision tree

3 Discussion

In this paper, we have presented a concise study of explainable artificial intelligence and its applications. Studies show that it is precisely the impossibility of human interpretation and substantiation of the results obtained on the basis of artificial neural networks [7] that imposes serious restrictions on the use of this powerful tool. We have demonstrated possible steps to extract knowledge from artificial neural networks based on decision trees and hybridization of neural networks with fuzzy logic. Including demonstrated examples in which a similar technique of extracting rules from artificial neural networks can be successfully applied. At the moment, we are actively working on the development of methods for hybridization of deep learning networks and fuzzy logic, which in turn can help make artificial intelligence more understandable for humans.

Acknowledgements. This research was performed in the framework of the state task in the field of scientific activity of the Ministry of Science and Higher Education of the Russian Federation, project "Development of the methodology and a software platform for the construction of digital twins, intellectual analysis and forecast of complex economic systems", grant no. FSSW-2020-0008.

References

1. Siau, K., Wang, W.: Building trust in artificial intelligence, machine learning, and robotics. Cut. IT J. **31**(2), 47–53 (2018)
2. Yezhov, A.A., Shumski, S.A.: Neurocomputing and its applications in economics and business. Mephi **224**, 10–14 (1998)
3. Aliev, R.A., et al.: Type-2 fuzzy neural networks with fuzzy clustering and differential evolution optimization. Inf. Sci. **181**(9), 1591–1608 (2011). https://doi.org/10.1016/j.ins.2010.12.014
4. Craven, M., Shavlik, J.: Extracting tree-structured representations of trained networks. Adv. Neural Inf. Process. Syst. **8**, 24–30 (1995)
5. Gridin, V.N., Solodovnikov, V.I., Evdokimov, I.A., Filippkov, S.V.: Building decision trees and extracting rules from trained neural networks. Artif. Intell. Decis. Mak. **4**, 26–33 (2013)
6. Shevelev, O.G., Petrakov, A.V.: Classification of texts using decision trees and feedforward neural networks. Bull. Tomsk State Univ. **290** (2006)
7. Yarushev, S.A., Averkin, A.N.: Time series analysis based on modular architectures of neural networks. Procedia Comput. Sci. **123**, 562–567 (2018). https://doi.org/10.1016/j.procs.2018.01.085

Application of Emotional Neural Network in Modeling Evaporation

Fahreddin Sadikoglu[1], Vahid Nourani[2](✉), Nardin Jabbarian Paknezhad[2], and Soheil Emamalipour[2]

[1] Faculty of Engineering, Department of Electrical and Electronic Engineering, Near East University, POBOX 99138, Nicosia, TRNC, Mersin 10, Turkey
fahreddin.sadikoglu@neu.edu.tr

[2] Faculty of Civil Engineering, Department of Water Resources Engineering, University of Tabriz, Tabriz, Iran
{nourani,n.jabbarian}@tabrizu.ac.ir

Abstract. Evaporation is dominant component of the hydrological cycle and it has vital impact on soil moisture, surface water resources and ground water resources. So accurately evaporation modeling could be useful in management of water resources. In this study the classic Artificial Neural Network (ANN) and its novel version, Emotional ANN (EANN) were applied for modeling evaporation in monthly scale in multiple stations with different climatological conditions. The stations were selected from Iran, Iraq, Turkey and Libya. Mean temperature, relative humidity and solar radiation were applied as inputs of modeling via the ANN and EANN. The obtained results of this study indicated that ANN and EANN showed different performance in various type of climates. EANN could properly be applied in modeling evaporation in semi-arid climates but ANN led to suitable performance in arid climates. In addition, results showed that EANN could perform better in modeling non-linear processes due to concluding the emotional units.

Keywords: Evaporation · Black box modeling · Artificial neural network · Emotional neural network

1 Introduction

Evaporation is one of main processes of the hydrologic cycle. Thermal energy and vapor pressure gradient affect the evaporation and its monitoring from soils, surface water and also its effect on groundwater is important issue in hydrology [1]. So accurately estimating evaporation is so vital, especially in areas with rare water resources [2] because it helps to manage and to develop the water resources.

Black box models, which attempt to model different processes independent of any information about physics of the phenomenon, considered as appropriate tools for hydrological modeling. Artificial Neural Networks (ANNs) are one of the black box models frequently applied in evaporation modeling [3, 4].

R. A. Aliev et al. (Eds.): ICSCCW 2021, LNNS 362, pp. 110–115, 2022.
https://doi.org/10.1007/978-3-030-92127-9_18

Despite wide application of ANN, one of its defeats is the ignorance of the biological basis of processes. In this way emotional ANN (EANN) is introduced, in which biological factors could be explained. The major superiority of EANN is due to its ability to model through hormonal glands [5]. EANN is more similar to the human emotional process so could led to better performance in training network. There are some studies used EANN in hydrological modeling [6–8]. In this study the classic ANN and new versions of it (EANN) were applied in modeling evaporation to assess the performance of different models in different climates.

2 Methods and Materials

2.1 Case Study

In this study different stations from different countries with distinct climates were considered in modeling evaporation. Data from some stations of Iran, Iraq, turkey Libya were used (see Fig. 1). The information about latitude, longitude, altitude and type of climate of each region are tabulated in Table 1.

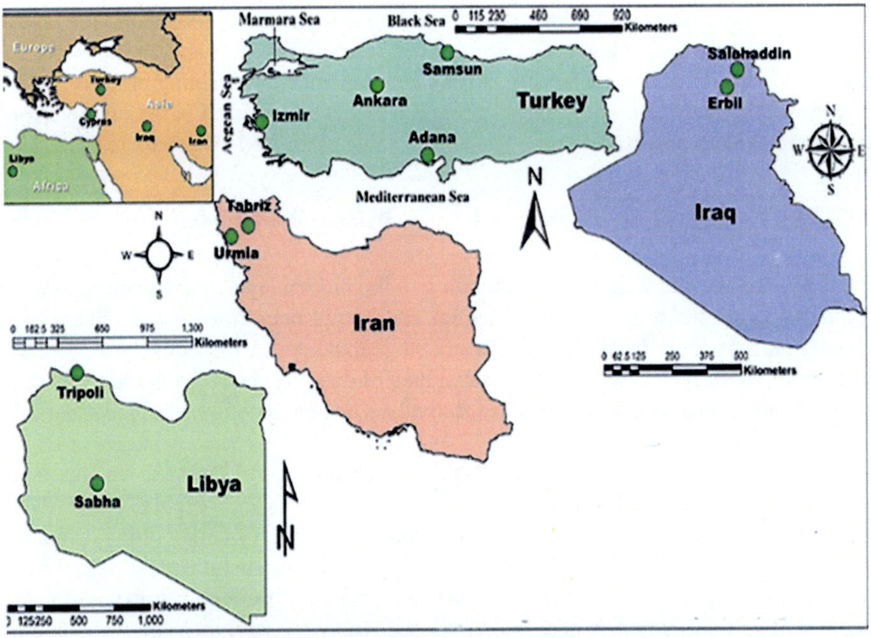

Fig. 1. Location of stations in this study.

Table 1. Latitude, longitude, altitude and type of the climate of each station.

Country	Station	Latitude	Longitude	Altitude	Climate
Turkey	Adana	36°99′N	35°33′E	27 m	Humid and semi-humid
	Ankara	39°93′N	32°86′E	938 m	Semi-arid steppe
	Izmir	38°42′N	27°14′E	30 m	Humid and semi-humid
	Samsun	41°28′N	36°34′E	4 m	Humid and semi-humid
Iraq	Erbil	35°55′N	43°57′E	390 m	Semi-arid and continental
	Salahaddin	36°15′N	44°07′E	1084 m	Semi-arid and continental
Iran	Tabriz	38°8′N	46°15′E	1,351 m	Semi-arid
	Urmia	37°34′N	44°58′E	1,332 m	Semi-arid
Libya	Tripoli	32°88′N	13°19′E	10 m	Dry arid
	Sabha	27°03′N	14°42′E	128 m	Dry arid

2.2 Methodology

ANNs consist of neurons and three layers, the first is input layer, second hidden layer and the third one is output layer. The neurons are linked by weights and their number is determined via trial and error process. ANNs are constructed in training process based on relationships between inputs and target and weights are updated based on the error between target and observed values. The output value of an ANN is calculated as [9]:

$$\hat{y}_k = f_0\left[\sum_{j=1}^{P} W_{kj} \times f_h\left(\sum_{i=1}^{L} W_{ji}X_i + W_{jo}\right) + W_{ko}\right] \tag{1}$$

where W_{ji} and W_{kj} are weights corresponding to the hidden layer and the output layer; W_{jo} and W_{ko} are the bias for hidden neuron and output neuron and f_h and f_0 are the activation functions of the hidden neuron and the output neuron, respectively. X_i, $\hat{y}_k$ and y are the input variable, output and actual values of data, respectively. L and P are the number of input and hidden neurons, respectively. The associated weights are assigned in training phase.

EANN is the new version of the ANNs and it consists of hormonal weights that monitor the performance of the neurons according to inputs and outputs. The EANN consists of four main learning blocks named thalamus, sensory cortex, amygdala and orbitofrontal cortex (OFC). The Thalamus is affected via the input pattern. Thalamus leads the input particularities directly to the amygdala and through sensory cortex to the OFC. The final output is determined via the Amygdala and then the weights are specified. In the EANN, the data is sent and received recurrently between input and output components via the nodes, which produce the dynamical hormones of H_a, H_b and H_c. y. The ith output of the EANN model with three hormones H_c,H_b and H_c is given as [6]:

$$Y_i = \underbrace{(\lambda_i + \sum_h \sigma_{i,h} H_h)}_{1} \times f(\textstyle\sum_j [\underbrace{(\beta_i + \sum_h \xi_{i,h} H_h)}_{2} \times \underbrace{(\theta_{i,j} + \Phi_{i,j,k} H_h) X_{i,j})}_{3} + \underbrace{(\alpha_i + \sum_h \chi_{i,h} H_h)}_{4}] \tag{2}$$

where *i*, *h*, and *j* represent the neurons of the input, hidden and output layers and $f()$ symbolizes an activation function. The artificial hormones are calculated as [6]:

$$H_h = \sum_i H_{i,h} (h = a,b,c) \tag{3}$$

In Eq. (2), term (1) is the applied weight to the activation function ($f()$) of weight of λ_i and $\sum_h \sigma_{i,h} H_h$. Term (2) is the weight of the net function, term (3) is the weight of the input of $X_{i,j}$ from *j*th neuron of previous layer and term (4) is the bias of net function, consists of α_i and $\sum_h \chi_{i,h} H_h$.

3 Results and Discussion

In this study, solar radiation, mean temperature and relative humidity were the inputs of modeling via the ANN and EANN, as these parameters are considered as most effective variables in evaporation modeling. Modeling was in monthly scale. The data set was split in two sets; 75% were considered as training and 25% as verification. The time periods of used time series for different stations are tabulated in Table 2. The time series are normalized in the range of -1 to 1, as active function of ANN and EANN is sensitive to this range. For modeling via ANN in order to obtain optimum structure of network, up to 12 hidden neurons and 300 epochs were checked in trial and error process. For modeling via EANN up to 12 emotional hormones were tested.

In order to assess the modeling implementation metrics of Root Mean Square Error (RMSE) and Determination of Coefficient (DC) were used. Lower values of RMSE indicated lower error of modeling. Range of the DC is between $(-\infty, 1]$. The values of DC, which are closer to 1 show higher adaptation of models output and observed values. The evaluation metrics of modeling via the ANN and EANN are tabulated in Table 3.

As shown in Table 3, the values of DC were higher than 0.73 which presented the proper implementation of modeling. As evaporation has a stable nature so both models led to acceptable results. Investigation of the obtained results indicated that ANN and EANN showed different performance in various regions with different climates. As shown in Table 3, DC values for modeling via the EANN is higher than those for ANN in Tabriz and Urmia stations which have semi-arid climate, but ANN outperforms EANN for modeling evaporation of Tripoli and Sabha stations with arid climate. Analysis of modeling performance in Turkey stations showed than EANN led to higher DC values for Ankara station which has semi-arid climate. However, EANN led to lower DC in Izmir, Adana Samsun stations with humid and semi humid climates. Therefore, results showed

Table 2. Training and verification periods of each station in modeling.

Station	Training periods	Verification periods
Tabriz	1992–2011	2012–2018
Urmia	1992–2009	2010–2018
Ankara	1998–2009	2010–2016
Izmir	1987–2009	2010–2017
Adana	1995–2006	2007–2010
Samsun	1987–2009	2010–2017
Erbil	1987–2009	2010–2017
Salahaddin	1992–2006	2007–2011
Tripoli	1995–2006	2007–2010
Sabha	1995–2006	2007–2010

Table 3. Modeling evaluation metrics.

	ANN				EANN			
Station	RMSE train	RMSE verify	DC Train	DC verify	RMSE train	RMSE verify	DC Train	DC verify
Tabriz	0.19	0.18	0.92	0.85	0.05	0.05	0.96	0.94
Urmia	0.12	0.14	0.91	0.85	0.08	0.09	0.91	0.87
Ankara	0.10	0.14	0.88	0.84	0.09	0.08	0.89	0.86
Izmir	0.51	0.52	0.95	0.88	0.07	0.06	0.92	0.94
Adana	0.48	0.48	0.90	0.85	0.08	0.07	0.9	0.89
Samsun	0.09	0.18	0.86	0.85	0.10	0.11	0.74	0.73
Erbil	0.10	0.14	0.97	0.91	0.07	0.1	0.93	0.91
Salahaddin	0.12	0.28	0.86	0.81	0.15	0.19	0.84	0.78
Trpoli	0.13	0.23	0.83	0.80	0.09	0.12	0.82	0.81
Sabha	0.53	0.49	0.87	0.84	0.1	0.11	0.88	0.78

that EANN could lead to more accurate results for modeling in semi-arid stations. This may be due to hormonal (emotional) parameters which handle the emotional conditions (e.g., peak states) of the model by acting as dynamic weights. Moreover, it can be concluded that the performance of models in stations with the same climatological condition was similar.

4 Conclusion

Black box models like artificial intelligence-based models are proper tools in modeling hydrological process like evaporation. Therefore, in this study the performance of the ANN and EANN was assessed for modeling evaporation in different regions with various climates. Solar radiation, mean temperature and relative humidity were the inputs of modeling. Results demonstrated that each of models may have different implementations according to the climate of the station. Comparison of the results showed that EANN may lead to better performance in semi-arid regions but ANN showed more accurate results in arid stations.

References

1. Shabani, S., et al.: Modeling daily pan evaporation in humid climates using Gaussian process regression. arXiv:1908.04267 (2019). https://doi.org/10.20944/preprints201907.0351.v1
2. Kişi, Ö.: Evolutionary neural networks for monthly pan evaporation modeling. J. Hydrol. **498**, 36–45 (2013). https://doi.org/10.1016/j.jhydrol.2013.06.011
3. Guven, A., Kisi, O.: Monthly pan evaporation modeling using linear genetic programming. J. Hydrol. **503**, 178–185 (2013). https://doi.org/10.1016/j.jhydrol.2013.08.043
4. Terzi, Ö., Erol Keskin, M., Dilek Taylan, E.: Estimating evaporation using ANFIS. J. Irrig. Drain. Eng. **132**(5), 503–507 (2006). https://doi.org/10.1061/(ASCE)0733-9437(2006)132:5(503)
5. Roshni, T., Jha, M.K., Drisya, J.: Neural network modeling for groundwater-level forecasting in coastal aquifers. Neural Comput. Appl. **32**(16), 12737–12754 (2020). https://doi.org/10.1007/s00521-020-04722-z
6. Nourani, V.: An emotional ANN (EANN) approach to modeling rainfall-runoff process. J. Hydrol. **544**, 267–277 (2017). https://doi.org/10.1016/j.jhydrol.2016.11.033
7. Sharghi, E., Paknezhad, N.J., Najafi, H.: Assessing the effect of emotional unit of emotional ANN (EANN) in estimation of the prediction intervals of suspended sediment load modeling. Earth Sci. Inf. **14**(1), 201–213 (2021). https://doi.org/10.1007/s12145-020-00567-1
8. Sharghi, E., Nourani, V., Najafi, H., Molajou, A.: Emotional ANN (EANN) and Wavelet-ANN (WANN) approaches for Markovian and seasonal based modeling of rainfall-runoff process. Water Resour. Manage **32**(10), 3441–3456 (2018). https://doi.org/10.1007/s11269-018-2000-y
9. Nourani, V., Alami, M.T., Vousoughi, F.D.: Wavelet-entropy data pre-processing approach for ANN-based groundwater level modeling. J. Hydrol. **524**, 255–269 (2015). https://doi.org/10.1016/j.jhydrol.2015.02.048

Malaria Detection Using Convolutional Neural Network

Khaled Almezhghwi(✉)

Electrical and Electronics Engineering, College of Electronics Technology, Tripoli, Libya

Abstract. Malaria detection and classification is still time and money costly. Identification of malaria cells can be done through some costly techniques. Those techniques are good, but they require time and high cost. Hence, there is a need of discovering alternative techniques to identify blood cells, that saves both time and reduce cost. In addition to time and cost, those new techniques should also be accurate and effective. Thus, in this work, we propose a transfer learning based GoogleNet approach for the classification of Malaria cells. The depth and inception of GoogleNet made it a very robust deep network that can classify accurately if trained and fine-tuned on enough amount of data. Thus, in this study, 27558 of the 2 types of cells are used for fine-tuning and testing the pre-trained network GoogleNet. Experimentally, the employed GoogleNet fine-tuned to classify Malaria, showed a great capability in generalizing accurate and correct diagnosis of images that were not seen during training, in which it achieved a testing accuracy of 95% with a relatively short time and small number of epochs.

Keywords: Malaria · Transfer learning · GoogleNet · Deep network · Classification

1 Introduction

Malaria is a possibly deadly parasitic disease of both human and creatures. Half of the total populace is at risk of this dangerous irresistible disease. Intestinal sickness is of extraordinary threat to pregnant lady and kids, particularly those under five [1]. As a rule, Malaria infection could be analyzed by tiny examination of blood films. Hence, to give a solid conclusion, important preparing and concentrated human asset are required. Patients experiencing Malaria disease ought to be analyzed first and ought to be given a compelling and reasonable treatment inside 24 h [2]. Shockingly, most infections happen in rustic regions, where assets are a long way from being sufficient. Additionally, inability to analyze on time may prompt off base medicines. This disturbing circumstance has incited scientists to create telemedicine answers for quick and precise recognizable proof of intestinal sickness infection. To give a thought of the red platelets engaged with this examination, we demonstrate a few examples in Fig. 1.

R. A. Aliev et al. (Eds.): ICSCCW 2021, LNNS 362, pp. 116–123, 2022.
https://doi.org/10.1007/978-3-030-92127-9_19

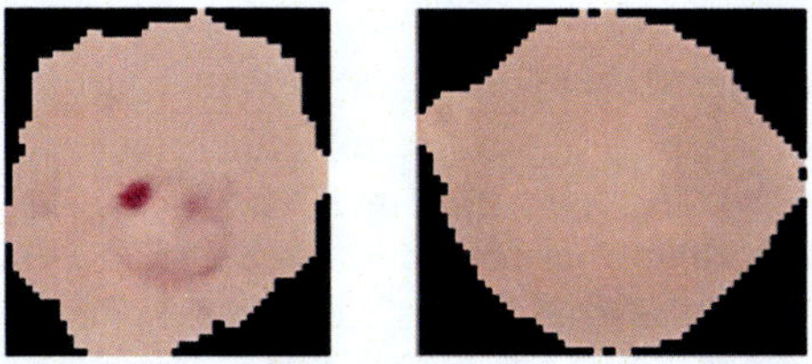

Fig. 1. Samples of parasitized and uninfected Malaria images.

Deep learning (DL) has been lately applied to medical field and it showed a great efficiency and accuracy [3, 4]. Deep networks showed a great capability in image classification in particularly, when transfer learning is used to transfer the knowledge extracted from one well-trained convolutional neural network (CNN) that is trained on millions of images to train on another task such medical image classification. Those convolutional neural networks (AlexNet, GoogleNet, VGG16, ResNet etc.) have been applied to many medical applications [5] and it was found that those networks can perform accurately without the need of large databases as they are already trained on millions of data which provided them with the good features extraction power due to their trained convolutional filters and feature maps [6]. In this thesis, a transfer learning based automatic classification system that is applied for the Classification of Malaria Infected Cells is presented. In order to achieve fully automated diagnosis without any manual feature extraction, we chose deep convolutional neural network (CNN) as the classifier. CNN can extract hierarchical representations of the input data. In this work, Inception network [7] was used to learn the inherent features of malaria infected and non-infected cells. Inception network is one of the best-known CNN architectures. In this work, we aim to apply transfer learning using Inception network that will be trained on large number of malaria images and evaluate their performances in classifying them into parasitic and uninfected. The network will be trained on a dataset of 27558 images collected from the Malaria Cell Images Dataset [7].

Infected and non-infected cells of malaria are being classified in this work. The pre-trained model employed in this work is the GoogleNet [8] which is due to its efficacy in images classification. The Network is fine-tuned on 2 malaria classes in order to learn the different level features that can classify each type and ends up with a good accuracy and minimum error. The images used in fine-tuning the pre-trained model GoogleNet are obtained from a public database in which microscopic images of the different classes cells are available. Those microscopic images are then used to train and test the GoogleNet and evaluate its performance in this medical classification task. Moreover, the network performance is compared to similar and related research that used other networks and methods in order to classify blood cells.

2 Methodology

In this paper, transfer learning is applied to solve the problem of white blood cells classification into 4 types. GoogleNet is fine-tuned in this work in order to add a new classification task to its functions, as shown in Figs. 2 and 3. This network contains 3 classification layers named as Loss1, Loss2, and Loss3. Those are fully connected layers, and they are retrained during fine-tuning of the network to its new classification task which is Malaria cells classification.

Figure 2 shows the fine-tuning process of GoogleNet to be trained to classify Malaria cells. As seen in the figure, all layers are fixed except the last three layers which are classification layers. Those three layers are retrained as they represent the fully connected layer, which is a traditional feedforward neural network. Those three layers are retrained using backpropagation learning algorithm in which error is reduced and weights are updated until the network reaches a minimum square error with a high classification rate.

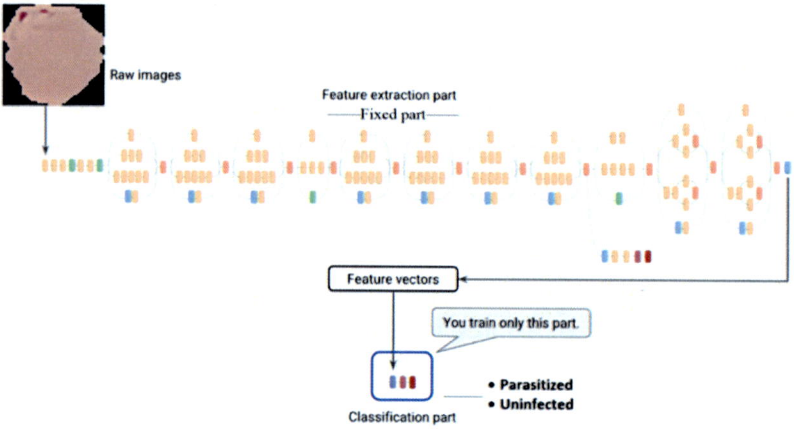

Fig. 2. Fine-tuning of GoogleNet to classify Malaria cells.

Figure 2 shows a block diagram of the whole process of fine-tuning GoogleNet to classify malaria into infected and uninfected images. As seen, images are first used for fine-tuning the network, but first pre-trained weights and parameters should be frozen as they are already well-trained using millions of images, which is the main aim of transfer learning. As seen in the block diagram, network is first trained for its new task, which is Malaria classification and then it is tested using new unseen images to measure its capability of generalizing accurate diagnosis of new images. Figure 3 shows a sample of the dataset images of the 2 different normal and Malaria blood cells.

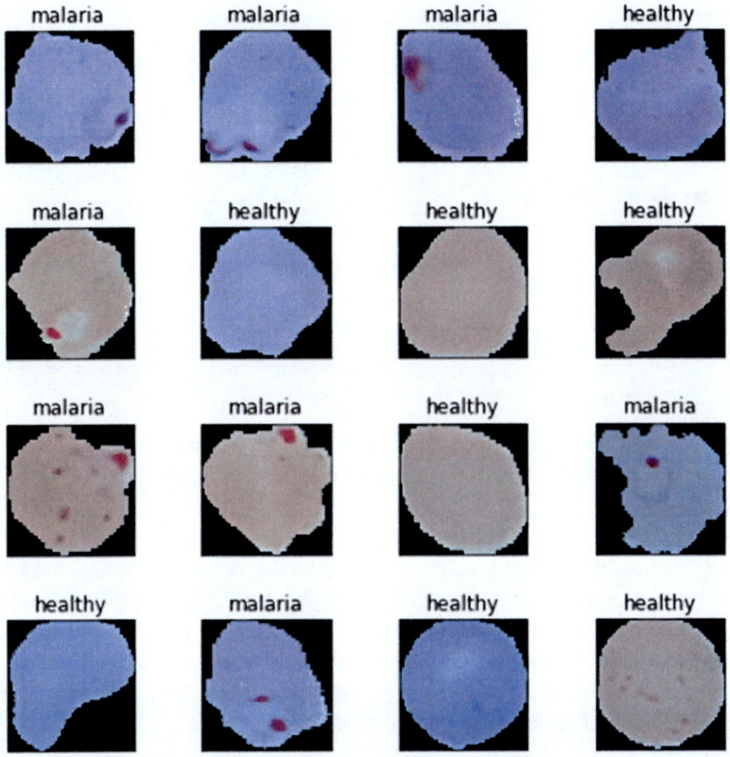

Fig. 3. Sample of dataset images of the two different Malaria cells

3 Experimental Results

The pre-trained model used in this research were trained and tested on a ratio of 60% of the available data. The performance of the networks was then evaluated using a held-out test set of the remaining 40% of the data. Note that images were all resized in order to fit the GoogleNet input which requires input images to be of size 224*224 pixels. Note that the publicly available weights of the network trained against the ILSVRC14 are used in this transfer learning-based research. As we are using a pre-trained model, its final fully connected layer Loss3 was removed, and a new layer was added, and it has 4 output neurons corresponding to the 2 Malaria cells categories. Note that the weights of this layer are initialized at random. On the other hand, the other layers are remained in the network, but their weights were frozen to act as a feature extractor. For training, a batch size of 200 images for each iteration is used via stochastic gradient descent. Also, the learning rates for the fully connected Loss1, Loss2, and Loss3 layers were fixed at 0.001, 0.001, and 0.01, respectively during training. Consequently, this allows the network to learn faster for the final fully connected layer. Moreover, the network is fine-tuned using 60% of the available data, and 100 epochs are set to train the network.

Table 1. GoogleNet learning parameters

GoogleNet	
Learning parameters	Values
Training ratio	60%
Learning rates (Loss1, Loss2, Loss3)	0.001, 0.001, 0.01
Number of epochs	20
Training accuracy	97.5%
Training time	1.5 h
Achieved mean square error (MSE)	0.017

Table 1 shows the learning parameters of the GoogleNet model. It can be seen that the network has achieved a training accuracy of 98.5% in approximately 40 min and 2 epochs.

Figure 4 shows the variations of accuracy in terms of epochs increase. It is seen that the network' accuracy was improving with the increase of epochs during training and testing until a minimum square error and accuracies of 97.5% and 95% are achieved, respectively. Once trained, GoogleNet was tested on 40% of the available data which includes the 2 types of Malaria cells. Figure 4 shows the testing accuracy variations in terms of epochs increasing. It can be seen that the network performed well in testing in which it was capable of reaching an accuracy of 95%.

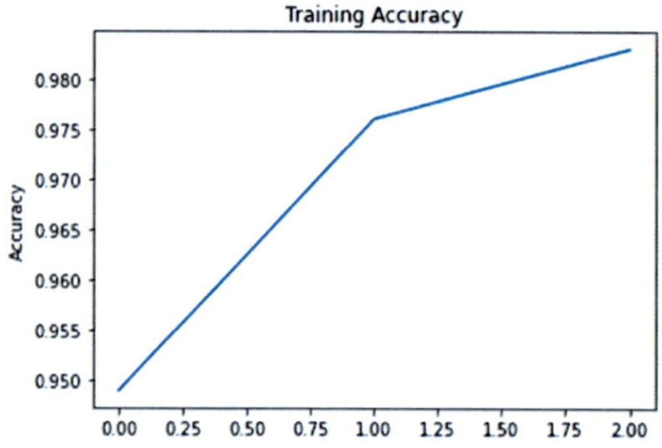

Fig. 4. Accuracy variations with the change of Epochs

Upon training, the employed pre-trained model is tested on 40% of the available data. Table 2 shows the performances of the model during training and testing. As seen, GoogleNet achieved 94% training accuracy; however, it was not capable of achieving such accuracy during testing, where it scored a lower recognition rate of 95%. On the other hand, this testing accuracy is also satisfactory as this classification task is tedious since there is a similarity between all Malaria cells. This might have made the network fall into local minima. Moreover, GoogleNet required 20 epochs to achieve such accuracy, which is relatively good to achieve such accuracy and a minimum square error of 0.017. In contrast, to achieve this accuracy and to reach that small error the network required a long training time of 1.5 h in order to converge and fine-tune. This time is obviously due to the depth of network as it contains many hidden layers. Also, it is because of the number of images which can be considered large number.

Figure 4 shows the learning curve of the fine-tuned GoogleNet. This figure shows the accuracy variations with respect to the Epochs increasing during training and testing of network. It can be seen network was trained well; however, the increase of depth of GoogleNet makes it more difficult to train, i.e., it required longer time and more epochs to reach the minimum square error (MSE) and converge. Furthermore, it is important to mention that this difference in time and epochs number of GoogleNet ended up with a low MSE and good accuracy.

Table 2. Performances of the model during training and testing

Number of epochs	Error reached (mse)	Training time	Training accuracy	Testing accuracy
2	0.017	40 min	98.5%	95%

For more understanding of the network learning performance and to have insight into the different level's features learned by the employed models, we sought to visualize the learned kernels or features in the convolutional layers.

Fig. 5. Learned filters of GoogleNet

Figure 5 shows the visualizations of extracted features of Malaria images by the GoogleNet pre-trained model. It is seen that different levels of abstractions are extracted at each layer which helps the network in learning the exact and appropriate features that distinguish two different classes of infected and uninfected Malaria cells.

4 Conclusion

A comparison of the developed networks employed in this work with some earlier works is shown in Table 3. Firstly, it is seen that the employed pre-trained CNNs achieved high recognition rates compared to other deep networks, which is obviously due to their powerful efficiency in extracting the important features from input images.

Table 3. Results comparison

	Classification objectives	Classifier used	Accuracy
Das et al. (2011)	2 types (Infected and uninfected)	Backpropagation neural network	8.78%
Das et al. (2013)	2 types (Infected and uninfected)	Machine learning and Light microscopic images	84%
Kaewkamnerd et al. (2012)	2 types (Infected and uninfected)	SVM CNN based device	90%
Sorgedrager, (2012)	2 types (Infected and uninfected)	CNN	92%
Our method	2 types (Infected and uninfected)	GoogleNet	95%

Finally, we conclude that GoogleNet, a well-designed and very deep architecture of sufficient complexity, was able to achieve significantly higher classification accuracy when distinguishing between normal and abnormal Malaria cells images, as compared to that of other methods such as CNN created from scratch, SVM, and decision tree. Furthermore, GoogleNet network learned features visualization demonstrates that mid and high-level features are learned effectively by the model.

Overall, it can be stated that the transfer of knowledge from a well-trained convolutional network to learn a new task can work accurately with a small margin of error, even when trained on a relatively small dataset. It is important to state that the depth of GoogleNet network contributed to a better understanding of the input images by extracting different levels of abstractions, which contributed to achieving higher recognition rates; however, required a long training time due to its depth.

References

1. Razzak, M.I.: Malarial parasite classification using recurrent neural network. Int. J. Image Process **9**(2), 27–32 (2015)
2. World Health Organization. Disease burden of malaria. http://www.who.int/mediacentre/factsheets/fs094/en/
3. Almezhghwi, K., Serte, S., Al-Turjman, F.: Convolutional neural networks for the classification of chest X-rays in the IoT era. Multim. Tool. Appl. **80**(19), 29051–29065 (2021). https://doi.org/10.1007/s11042-021-10907-y

4. Kaymak, S., Almezhghwi, K., Shelag, A.A.S.: Classification of diseases on chest X-rays using deep learning. In: Aliev, R.A., Kacprzyk, J., Pedrycz, W., Jamshidi, M.O., Sadikoglu, F.M. (eds.) ICAFS 2018. AISC, vol. 896, pp. 516–523. Springer, Cham (2019). https://doi.org/10.1007/978-3-030-04164-9_69
5. Abiyev, R.H., Maaitah, M.K.S.: Deep convolutional neural networks for chest diseases detection. J. Healthc. Eng. **2018**, 4168538 (2018). https://doi.org/10.1155/2018/4168538
6. Nguyen, D., et al.: A comparison of Monte Carlo dropout and bootstrap aggregation on the performance and uncertainty estimation in radiation therapy dose prediction with deep learning neural networks. Phys. Med. Biol. **66**(5), 054002 (2021)
7. Malaria Cell Images Dataset. https://www.kaggle.com/iarunava/cell-images-for-detecting-malaria
8. Szegedy, C., et al.: Going deeper with convolutions. In: IEEE Conference on Computer Vision and Pattern Recognition (CVPR), p. 12 (2015). https://doi.org/10.1109/CVPR.2015.7298594

Selection of Information - Measuring Components on the Basis of Layout Diagram of Flexible Manufacturing Cell

I. R. Aliyev[1], J. F. Mammadov[2(✉)], and Sh. R. Rakhimov[2]

[1] Azerbaijan State Oil Industrial University, Azadliq Prospekti, 34, Baku, Azerbaijan
[2] Sumgait State University, 43th Quartal, 1st Korpus, Sumgait, Azerbaijan

Abstract. The research of technological operations of manufacturing modules and their standard active components of the flexible manufacturing cell (FMC) for annealing and machine processing of billets (AMPB) carried out a mathematical analysis based on the use of N-oriented hypergraphs, their controlling actions and executive functions. Taking into account the functional connections of the active components in the layout diagram of the FMC, the data connections among the active components, the incidence matrix is determined. In accordance with the layout of technical units in a 2-dimensional reference frame, the general properties of layout diagram of the FMC for AMPB were determined for a 3-dimensional reference frame and a hyper-graph scheme was constructed.

Based on the unloading layout diagram for the conveyor of the annealing furnace of aluminum slags and loading them onto the conveyor of the manufacturing module of grinding machines, a step-by-step selection of sensor types for the servicing industrial robot is carried out, depending on its technological operations. An algorithm to select information and measurement components of ACS for manufacturing modules of annealing and machine processing of billets with the determination of their coordinate positions and planning of information flows within the automated control of manufacturing modules of the FMC for AMPB is proposed. The operating modes of the thermocouple are determined to provide flexibility in the manufacturing process of annealing billets, depending on its geometric parameters and material.

Keywords: Automated control system · Information and measurement components · Flexible manufacturing cell · Industrial robot · Sensor

1 Introduction

The research of the design stages of the automated control system (ACS) and its information-measuring level of the FMC determined that the productivity and reliability of its further functioning largely depends on the creation of information support, i.e., the formation of database and knowledge in the design of ACS, the implementation of the process of searching and selecting its information and measurement components of ACS at the initial stages [1–3]. The analysis of the existing ACS and its information-measuring

R. A. Aliev et al. (Eds.): ICSCCW 2021, LNNS 362, pp. 124–131, 2022.
https://doi.org/10.1007/978-3-030-92127-9_20

level of the FMC for machine processing in various industry fields of developed countries [4–6] showed that the effectiveness of the functioning of such large enterprises with hierarchical automated control levels largely depends on the correct selection of a sufficient number of components of the FMC's control system, evaluation of the productivity of the manufacturing process, effective use of modern innovative control tools with the further formation of an automated scheme of the FMC. The presence of a sufficiently large number of converters and a complex structure of the sensor system in each active component – industrial robots, manipulators, automatically transport systems, technological equipment and others elements of the FMC as a whole, identified from a comparative analysis of literary sources [7, 8], gives grounds to conclude that a number of issues need to be considered in this direction:

1. Analysis of the research object, i.e. the FMC for AMPB for the manufacture of evaporators of refrigeration units and justification of the selection of its layout diagram of the studied FMC.
2. Selection of information and measurement subsystems of the active components of the FMC for AMPB with the determination of their exact coordinate positions and planning of information flows within the automated control of its modules.

The purpose of the article is to select information and measurement components for the ACS of the FMC for AMPB depending on its layout diagram and functional research to ensure the efficiency of the control system design.

2 Process Flow Analysis of the Research Object

The investigation is concerned with the layout diagram of the FMC for AMPB with the further placement of the billet annealing furnace (BAF) (1), an industrial robot (IR1) (2) to service BAF and an automated conveyor (AC) (3), 2 grinding machines (GM1, GM2) (4), two IR2 and IR3 (5), serving GM1 and GM2, respectively ((4) Fig. 1).

Technological operations in the FMC for AMPB are carried out in accordance with the manufacturing requirements to a billet for evaporators. At the initial stage, cut according to the standard size of evaporators for household refrigerators, steel billets are annealed in BAF (1) at a temperature of 600–800 °C. Then IR1 (2), grabbing a metal billet from the automatic conveyor BAF (1), turning 90°, sets it on AC (3). The clamped billet on the AC (3) is moved to the IR2 position (4), where the positioned billet is caught by means of the IR2 (4) gripper and loaded into the GM1 (5). On GM1 (5), the surface of steel billets is ground by machine processing. The same way, the catching and grinding of the next billet is carried out in another manufacturing module of the FMC for AMPB.

3 Modeling the Layout Diagram of FMC for AMPB for Accurately Positioning of its Information-Measuring Elements

To formalize the problem and form the mathematical modeling constraint systems for the layout diagram of FMC for AMPB, we will use a generalized structure of the mathematical model based on the use of N-oriented hypergraphs with restrictions on the

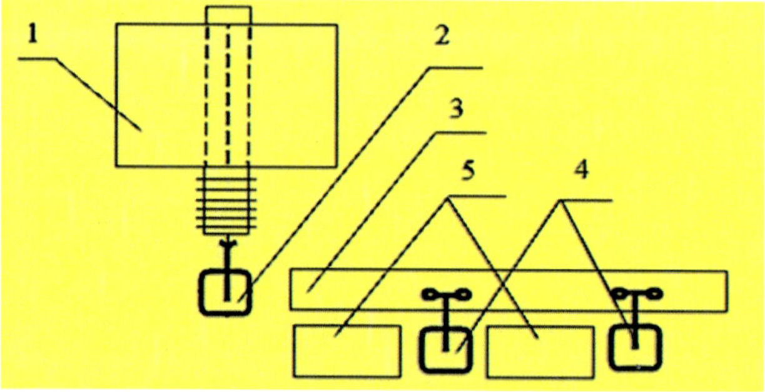

Fig. 1. Layout diagram of the FMC for AMPB

properties of vertices and edges [9, 10]:

$$G(P, U, UL) \tag{1}$$

where $P = \{p_i \mid i = \overline{1, N_0}\}$ – the set of vertices of the hypergraph; p_i – the i vertex; $U = \{U_m(P'_m) \mid m = \overline{1, M}\}$ is the set of hyper-edges of the hypergraph; $u_m(P'_m)$ is the m edge of the hypergraph; P'_m is the set of vertices incident to the m edge $P'_m \subseteq P$; $P'_m = \{p^V{}_k\}$, $\forall k \in K_m$, $K_m \subseteq \overline{1, I}$, $\overline{V}$- the vertex number in the edge of an oriented hypergraph has the form of a vector, $\overline{V} = \{V_n \mid n = \overline{1, N}\}$. The vertex number in the edge reflects a certain property of the vertex with specific values. $U^L = \{u_l^| = < p_{n1}; p_{n2} >_l \mid l = 1, N_l\}$- a set of edges defining a system of connections between objects of an oriented subgraph

$$G_i\left(P^L, U^L\right), \tag{2}$$

where $P^L \subset P -$ is a subset of connected vertices; N_l is the number of connections; $< p_{n1}; p_{n2} >$ is a pair of vertices incident to the l edge. The incidence matrix of a subgraph is written as $L = \{l_{p1,p2}\}$, where

$$l_{p1,p2} = \begin{cases} 0 & \text{if the vertex } p1 \text{ has no relation to edge } p2, \\ -1 & \text{if the vertex } p1 \text{ is the beginning of edge } p2, \\ 1 & \text{if the vertex } p1 \text{ is the end of edge } p2. \end{cases} \tag{3}$$

In the graphical representation, an N-oriented hypergraph is constructed taking into account the functional connections of the studied FMC for AMPB (Fig. 2). Taking into account the functional connections of the active components of the studied manufacturing process, shown in Fig. 2, and the expression (3), we write the incidence matrix as follows [11]:

$L_{Ni} =$	$P1$	$P2$	$P3$	$P41$	$P42$	$P51$	$P52$	
	0	1	0	0	0	0	0	$P1$
	−1	0	0	0	0	0	0	$P2$
	0	−1	0	1	1	0	0	$P3$
	0	0	−1	0	0	1	0	$P41$
	0	0	−1	0	0	0	1	$P42$
	0	0	0	−1	0	0	0	$P51$
	0	0	0	0	−1	0	0	$P52$

where p_1 is the vertex of the hypergraph corresponding to the position of BAF (1); p_2 – the vertex of the hypergraph corresponding to the position of IR1 (2); p_3 – the vertex of the hypergraph corresponding to the position of AC (3); p_{41} – the vertex of the hypergraph corresponding to the position of IR2 (4) of the first manufacturing module; p_{42} – the vertex of the hypergraph corresponding to the position of IR3 (4) of the second manufacturing module; p_{51} – the vertex of the hypergraph corresponding to the position of GM1 (5) of the first manufacturing module; p_{52} – the vertex of the hypergraph corresponding to the position of GM2 (5) of the second manufacturing module.

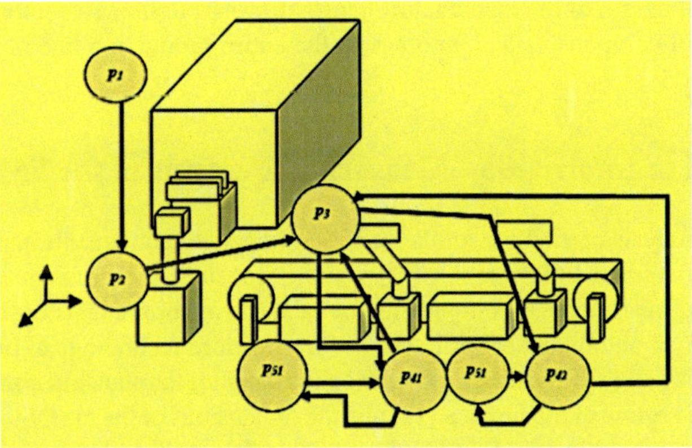

Fig. 2. 3-Dimensional representation of the FMC for AMPB together with the vertices and hyperedges of the hypergraph

Let's consider $S = \{s_j | j = \overline{1, J}\}$ - the set of all possible properties of vertices and edges, $SP_i = \{s_j | j \in J1_i \subset \overline{1, J}\} \subset S$ - the set of properties of the i vertex, $SU_m = \{s_j | j \in J2_m \subset \overline{1, J}\} \subset S$ - the set of properties of the m edge. Thus, the set of numbers of each vertex $V = \{v_n | n = \overline{1, N}\}$ is replaced by a set of properties SP_i, $i = \overline{1, I}$ likewise, each edge $u_m(P1_m)$ is aligned with a set of properties SU_m, $m = \overline{1, M}$. The edges of the subgraph $G_l(P^L, U^L)$, as well as the edges of the hypergraph $G(P, U, U^L)$, correspond to the set of properties $SU^L{}_l = \{s_j | j \in J3_l \subset 1, J\} \subset S$, $l = \overline{1, N_l}$.

In accordance with Fig. 1, we will determine the general properties of the layout of the FMC for AMPB for a 3-dimensional reference frame (Fig. 2) in a 2-dimensional reference frame. In this case, the initial coordinates and geometric dimensions of the manufacturing module, AC, IR and GM and the parameters of their location are set.

Permissible layout dimensions of the FMC for AMPB: the permissible area of the entire FMC for AMPB is determined based on the initial coordinates, the dimensions of 3 manufacturing modules. Thus, the permissible area of the FMC for AMPB is determined by the following formula:

$$S_{FMC} = \sum_{j=1}^{n} k_{fj} \sum_{fj\ i=1}^{n} (S_i + \Delta b_j), \quad i = \overline{1, n = 3} \tag{4}$$

where k_{fj} is the coefficient taking into account the working area to service BAF, AC, GM1 and GM2. For BAF $k_{f1} \approx 4{,}5 \div 5{,}0$; for AC $k_{f2} \approx 3{,}5 \div 4{,}5$; for GM1 and GM2 $k_{f1} \approx 3{,}0 \div 4{,}0$; S_i – areas of manufacturing modules ($i = \overline{1, 3}$); Δb_j – safety zones of the manufacturing modules.

Standard shapes and sizes of FMC for AMPB products: depending on the specifics of the manufacturing process of the FMC for AMPB, a square-shaped steel billet is required for annealing and grinding on the production line with dimensions $a \times a \times h$, where $h = \min$ *(5* mm*)*, $a = 300 \div 500$ mm and a flat shape.

Thus, the analysis of the technological process of the FMC for AMPB with the justification of the selection of active components in the manufacturing modules and the general layout diagram of the manufacturing cell allows us to implement the process of selecting control components and automating the entire production line of the studied FMC.

4 Selection of Information-Measuring Elements of the FMC

Based on the analysis carried out on the selection of active components and the layout diagram of the studied FMC for AMPB, we will consider the problem of selecting information and measurement components for each active components of the FMC for AMPB to form an automated control system for the entire technological process. The flexibility of this cell is provided by automatically changing the standard sizes of billets, the temperature regime of the furnace [12] and the dimensions of the machine processing of grinding machines using the SCADA-based cable-free ACS technology.

The process of selecting the information - measuring subsystem components starts with a step-by-step technological study of manufacturing modules and their active components. At the first stage, the manufacturing module, where the billets are annealed for further machine processing on grinding machines, is the target of research.

The operational environment characterized by physical properties F_{s_i}, geometric parameters G_{p_i} and parameters of handling objects O_{m_i} has a significant impact on the selection of one or another type of information and measurement component of the ACS of the FMC for AMPB, as well as on the reliability of data transfer channels, power consumption and constructive implementation of standard components [13].

The selection of sensors for controlling actions on the standard active components of the FMC for AMPB, depending on the initial parameters of the operational environment, I^{ec_i}, the initial geometric dimensions (the layout of the standard components of the studied FMC (based on the machine processing of billets) with information and measurement components (sensors) of ACS (Fig. 1)), I^{ho_i}, the initial parameters of the handling object (technical data of the manufactured product according to the manufacturing modules of the FMC for AMPB), I^{ho_I}, if the following conditions have been fulfilled [14]:

- normative data of the external environment of the production area: the atmospheric temperature $t_a \approx 18 \div 28$ °C; the temperature of the surface, $t_s \approx 12 \div 29$ °C;
- relative humidity $\varphi = (P_n/P_s) * 100\%$, where P_n is the airborne vapor pressure;
- P_s is the pressure of saturated vapor depending on temperature and air pressure (kg/m^3) – $\varphi \approx 15 \div 75\%$;
- the air velocity – *min (0,1), max (0,5)* (v, m/s);
- the thermal exposure intensity (J, W/m^2):

$$short\text{—}wavelength - J_{sw} \approx 0,76 \div 1,5\ \mu R;\ \ long\text{—}wavelength\ 1,5 \leq J_{lw}\mu R.$$

To select the temperature sensor (thermocouple) accurately, it is necessary to take into account the layout data of the billet annealing furnace for the billets in the shape of a hollow parallelepiped, the dimensions of which are from *300* × *300* × *5* to *500* × *500* × *5* (mm), and made of aluminum. The annealing temperature for billets reaches *600–800* °C. It is required to determine the furnace heating mode from *600 to 700* °C to ensure flexibility in this manufacturing process, depending on the parameters of the billet from *300* × *300* × *5 to 500* × *500* × *5 (*mm*)*. Consequently, the linguistic terms for these parameters will appear as follows:

Dimensions of Billets:

D_{b_1} - *the size of the billet is* ***minimum*** *(300* × *300* × *5* (mm)*);*
D_{b_2} - *the size of the billet is* ***medium*** *(400* × *400* × *5* (mm)*);*
D_{b_3} - *the size of the billet is* ***maximum*** *(500* × *500* × *5* (mm)*).*

Annealing Temperature of Billets:

T_{ann_1} - *the annealing temperature is* ***lowest*** *(600, 615, 630* (°C)*);*
T_{ann_2} - *the annealing temperature* ***is medium*** *(620, 655, 670* (°C)*);*
T_{ann_3} - *the annealing temperature is* ***high*** *(655, 675, 700* (°C)*);*
T_{ann_4} - *the annealing temperature is* ***highest*** *(720, 760, 800* (°C)*).*

Let's determine the operating conditions of the temperature sensor (furnace bottom thermocouple-*TI/GL/07-A/B*) in the annealing furnace for billets with a size from 300 × 300 × 5 (mm) to 500 × 500 × 5 (mm):

Rule 1. *If the billet size is* ***minimum,*** *then the annealing temperature is* ***lowest;***
Rule 2. *If the billet size is* ***medium***, *then the annealing temperature is* ***high;***
Rule 3. *If the billet size is* ***maximum***, *then the annealing temperature is* ***highest;***

The control component (the furnace bottom thermocouple-*TI/GL/07-A/B*) cannot process this controversial information directly, so the logical derivation process outcome from the rules above must be converted into clear numerical values. In this regard, a clear number to be determined (the annealing temperature of the billets in the furnace and the standard sizes of the billets) should provide a representation of the information contained in the controversial set. A centroid in the form of a "center" of the modes of controversial temperature indicators in the annealing furnace cavity is used to obtain a clear result. To find the central indicator among the controversial temperature data for each standard size of the billet, the following expression is used [15]:

$$\text{for minimum standard sizes of billet} \quad v_{\min} = \frac{\sum_{i=1}^{n} t_{ann\_i} a_{\min}}{m a_{\min}} \tag{5}$$

$$\text{for medium standard sizes of billets} \quad v_{med} = \frac{\sum_{i=1}^{n} t_{ann\_i} a_{med}}{m a_{med}} \tag{6}$$

$$\text{for maximum standard sizes of billets} \quad v_{med} = \frac{\sum_{i=1}^{n} t_{ann\_i} a_{med}}{m a_{med}} \tag{7}$$

where v – clear manipulation output (abscissa value of the "gravity center" of the temperature range in the annealing furnace (°C)); t_{ann_i}, where $i = \overline{1, n = 4}$ the value of controversial temperature data set for the operation mode of the annealing furnace's thermocouple depending on the sizes of billets (a_{min}, a_{med}, a_{max}); $m = \overline{1, 3}$ is the number of controversial parameters at each temperature scenario t_{ann_i} of the annealing furnace.

Based on (5), (6), (7), the abscissa values of the "gravity center" of the temperature ranges are determined.

5 Conclusion

1. For the technological analysis of the studied FMC for AMPB, the problem is formalized and the mathematical modeling constraint systems for its layout diagram based on the use of *N*-oriented hypergraphs are formed.
2. An algorithm has been developed to select information – measuring components of the annealing manufacturing module and an industrial robot, depending on the layout of the active components of the FMC for AMPB by means of the fuzzy linguistic terms of providing them of control regimes.
3. It is defined the abscissa values of the "gravity center" of the temperature ranges.

References

1. Astrom, K.J., Kumar, P.R.: Control: a perspective. Automat. J. IFAC Int. Federat. Automat. Control **50**(1), 3–43 (2014)

2. Albertos, P., Sala, A.: Multivariable Control Systems: An Engineering Approach. Springer-Verlag, London. 357 p. (2004). https://doi.org/10.1007/b97506
3. Ovechkin, M.V., Proskurin, D.A., Sergeev, A.I., Galina, L.V.: Automation of selecting the in-line equipment at machine-building enterprises. In: Theoretical and Practical View on the Current State of Science (TPVCSS): Materials of the International Scientific and Practical Conference, 20–30 Sept 2015, Kemerovo/West Siberian Scientific Center. - Electronic data, pp. 92–95. KuzSTU, Kemerovo (2015) (in Russian)
4. Titov, V.S., Besedin, A.V., Bobyr, M.V.: Analysis of the control automation methods for high accuracy of technological processes. Indust. ACSs Control. **7**, 29–32 (2003). (in Russian)
5. Tugov, V.V.: Designing of Automated Control Systems: Textbook. Tugov, V.V., Sergeev, A.I., Sharov, N.S. (eds.) Lan, St. Petersburg, 172 p. (2019) (in Russian)
6. Yan, L., Ruijie, K.: Design of Control System for Flexible Automatic Processing Line 10th International Conference on Computational Intelligence and Communication Networks (CICN) (2018). https://doi.org/10.1109/CICN.2018.8864961
7. Vorotnikov, S.A.: Information devices of robotic systems. In: Publishing House of the Bauman Moscow State Technical University (PH BMSTU), 384 p. (2005) (in Russian)
8. Pinto, R.: Smart sensing components in advanced manufacturing systems. Int. J. Adv. Intell. Syst. **9**, 181–198 (2016)
9. Reff, N., Rusnak, L.J.: An oriented hyper graphic approach to algebraic graph theory. Linear Algebra **437**(9), 2262–2270 (2012). https://doi.org/10.1016/j.laa.2012.06.011
10. Taha, Z., Rostam, S.: A hybrid fuzzy AHP-PROMETHEE decision support system for machine tool selection in flexible manufacturing cell. J. Intell. Manuf. **23**(6), 2137–2149 (2012). https://doi.org/10.1007/s10845-011-0560-2
11. Kulik, B.: Logical analysis of intelligence systems by algebraic method. In: Kulik, B., Fridman, A., Zuenko, A. (eds.) Cybernetics and Systems 2010: Proceedings of Twentieth European Meeting on Cybernetics and Systems Research (EMCSR 2010) Vienna, Austria, pp. 198–203 (2010) (in Russian)
12. Yevtushenko, A., Kuciej, M., Och, E.: Effect of the thermal sensitivity in modeling of the frictional heating during braking. Adv. Mech. Eng. **8**(12), 168781401668174 (2016). https://doi.org/10.1177/1687814016681744
13. El-Khalil, R., Darwish, Z.: Flexible manufacturing systems performance in U.S. automotive manufacturing plants: a case study. Prod. Plan. Control **30**(1), 48–59 (2019). https://doi.org/10.1080/09537287.2018.1520318
14. Serdyuk, A.I., Sergeev, A.I., Kornipayev, M.A., Proskurin, D.A.: Automated environment for pre-project research of flexible production systems FMS Concept. In: Automation in industry (AI), November, pp. 50–53 (2016) (in Russian)
15. Kaufmann, A., Gupta, M.M.: Introduction to Fuzzy Arithmetic. Thomson Computer Press (TCP) (1991). ISBN-10:0442008996, ISBN-13:978-0442008994

Minimizing Handover Process and Wireless Propagation Lose by Using Multilayer Perception Neural Network

Jamal Fathi(✉)

Near East University, North Nicosia, Turkish Republic of Northern Cyprus, Turkey

Abstract. The widely spread of using smart phone as a device to download and upload important documents in daily use, so that most users are complaining of the disconnection process even for very a short time in seconds. This disconnection yields to the loss of all processes in download or upload and then start from the beginning. In this paper incrementing an Artificial Neural Network (ANN) in order to have an appropriate point inside the building for each floor is arranged, and modified so that there will be almost zero disconnection process and minimum handover to obtain continuous connection. Moreover, this paper is arranged to maintain stability in IP addresses in the buildings at Near East University (NEU) in the Northern part of Cyprus (TRNC). Finally, we obtained accurate results of our proposed model as a solution of minimizing the handover and disconnection processes inside the buildings.

Keywords: Wireless Local Area Network · Routers · Artificial Neural Network · Handover · Download and upload

1 Introduction

In this paper, the handover problem inside the buildings and the disconnection process is involved and studied using a combination of the Artificial Neural Network for both stabilizing any IP addresses to be stable and decreasing the handover process in order to minimize the disconnection problems in each floor in the building. So that in each floor a detailed study is done and assigned the accurate places for the routers to be fixed in order to minimize the disconnection problems. Then applied to the ANN to stabilize the IP addresses and minimize the handover process.

A conventional plan as in [1] to be a threshing of Wireless Local Area Network (WLAN) finding base stations. In any case, cell phone clients might need to utilize it as indoor or open air, that makes solid walker dead retribution troublesome or unimaginable.

The used equation in [1, 2, 3, 4, 5, and 9] to describe the log-space and the sent sign to appraise the Radio inclusion is:

$$Y = 24.5 + 33.8\log(d) + 4.0K_{floor} - 16.6S_{win} - 9.8G_{G/1} - 0.25A_{elv} \quad (1)$$

Where, d is the distance among transmitter and collector, K_{floor} is the quantity of floors.

R. A. Aliev et al. (Eds.): ICSCCW 2021, LNNS 362, pp. 132–139, 2022.
https://doi.org/10.1007/978-3-030-92127-9_21

Abu Hasna [1], investigated the example of spread catastrophe inside and outside the plans utilizing frequencies ran between 900–1800 MHz as much as, applied straight supposition channel to the got signal. Zyad et al. [2], surveyed initiative troubles in multi cases as inside and outside the buildings to allocate the strength in the signal, then improved the quality of the transmitted signal. Fathi, [3], gave an idea to change the spots of the base stations to secure huger level and all the more consistent flexible signs. Fathi et al. [4], arranged a report discussing the misfortune inside lifts of the NEU structures. In addition, contrasted every single existing framework, and demonstrated a framework that can be substantial for discussing the main point in lifts inside the structures. Fathi et al. [5], organized and augmented an ANN to anticipate the progressions in the IP addresses as a significant highlight keep a steady IP transmission in the multi-floors working with lift. At last, demonstrated the precision of our model as an ideal answer for remote cause lose inside the lifts of structures. Obeidat et al. [6], arranged a path of misfortune forecast system so that contrasted it where indoor expectation systems by the utilization of reproduction information and ongoing estimations. Perez et al. [7], reviewed a system used in the indoor correspondence with 1.8 GHz, the model used in the model as a subjective variable is managed. Majed et al. [8], cultivated an estimation to exhibit the exactness of the expected structure for distant consideration. Zyad et al. [9], improved another numerical framework to limit the handover time delay. As much as, processed the point of the accessible base stations mulling over the distance between the base stations. Jadhav et al. [10], arranged an association of NN to get generally 2% to the blend of PDR and GPS analyzed structure to get a phase length assumption arrangement. Zreikat et al. [11], arranged a product program to deal with the handover interaction. Besides, broad examinations were done to contrast the got results and the pre-arranged programming. The outcomes got by the product demonstrated the exactness of the proposed programming. Bakinde et al. [12], improved nother way adversity a dependent system that is the Adaptive Neuro Fuzzy Interference System (ANFIS) is included and several transmitters so that to be considerable for VHF gatherings. Bose et al. [13], used Friis and Okumura models to set out an expected game plan of UHF. The delayed consequences of the used structure are differentiated and different systems and got another structure sensible for the city.

This paper is partitioned into four sections: Introduction, proposed system, conclusion of the proposed system, and then the references.

2 Proposed Model

As shown in Fig. 1, the received signal from the base station tower by the outdoor Antenna to be connected to YATE UCN in order to be able to connect YATE-BTS over GTP and SIP, then applying ANN to stabilize fixed IP address, then to indoor Omni Ceiling Antenna. Finally, to mobile signal booster.

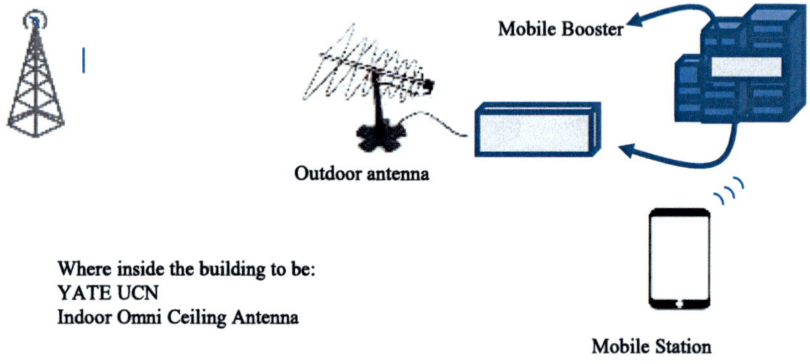

Fig. 1. Proposed model.

2.1 ANN Prediction Process

Out of disposal of using ANN by Matlab.R2013a, looking ahead to the all-round moving customers as a special case so that they have no exact position and the need to send connection request to more than one IP address as the case of stable position. The use of the ANN in this situation is to maintain same IP address as fixed and no need to be switched to another IP address in order to prevent the disconnection process while moving object.

2.1.1 Used Algorithm

1. Stabilized to have a goal coordinated to have fixed IP address,
2. Inputs are taken from the nearest mobile,
3. Target is consigned as d, y as addresses, and n tends to data,

$$E_j(n) = d_j(n) - y_i(n) \tag{2}$$

4. Adjustment interaction for the loads by

$$\varepsilon(n) = 0.5 \sum e_j^2(n) \tag{3}$$

5. Process of acclimation to recognize the progressions in loads as

$$\Delta w_{ij}(n) = -\eta \left(\frac{\partial_{\varepsilon}(n)}{\partial v_j(n)} \right) y_i(n) \tag{4}$$

   where η displays rate, y_i as passed case.
6. Arrangement interaction finished by

$$-\frac{(\partial \varepsilon(n))}{(\partial v_j(n))} = \varphi'(v_j(n)) \sum \left(-\frac{(\partial \varepsilon(n))}{\partial v_k(n)} \right) w_{kj}(n) \tag{5}$$

Table 1. ANN outcome

Used items	Value
Input	9
Output	1
Iterations	13
Time	00.00.06
Performance	18.8 3.26e + 04
Gradient	0.213 4.05e + 04
Mu	0.00100

In the above calculation, dealing with the dependability of the IP tends to communicated through the passage in each floor of the structure that incorporates lifts with no detachments. The ANN preparing results are arranged in Table 1.

As indicated by the acquired outcomes appeared in Table 1 versus cases, so Error equivalents to Target less Outputs, with the goal that precision of the framework came to 19.7214 at epoch 7 as shown in Fig. 2 and Fig. 3.

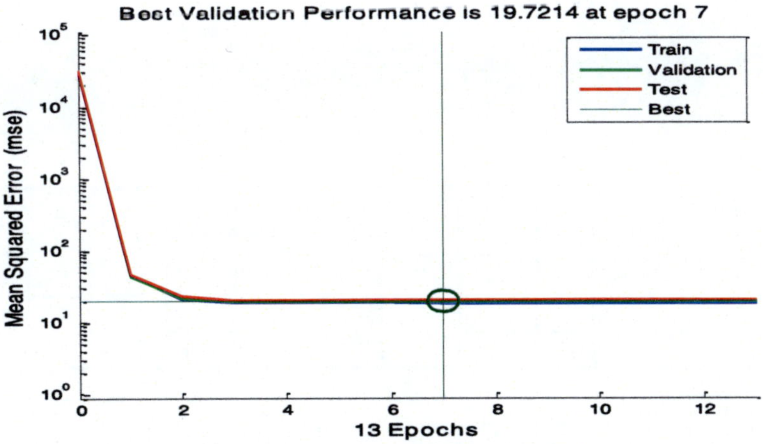

Fig. 2. Best outcome [MATLAB 2013b]

Utilizing the got results by the ANN to overview the way where episode utilizing the proposed condition, this outcome a precise evaluation guide of the sign strength inside multiwall structures. As for 2D and 3D LOS framework, which is utilized as an intensifier for the sign inside a multi-divider' structures for both transmitter and recipient. The proposed condition is a blend of all past standard models [1, 2, 3, 4, 5]. This model is genuine for any multi-dividers by the use of picture arranging progressions to see the dividers inside the development. Utilizing Hough-Transform (HT) to see the multi-dividers in the development which demonstrated in Fig. 4.

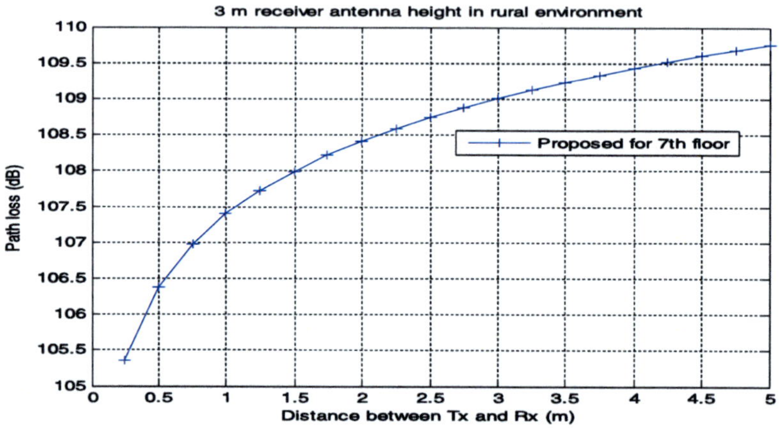

Fig. 3. Proposed model's results [MATLAB 2013b]

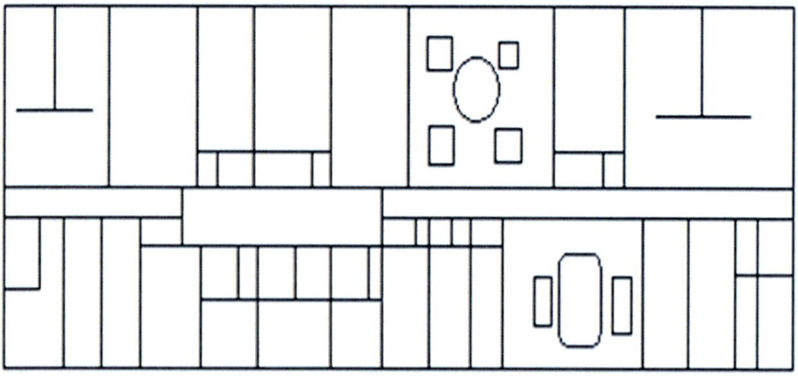

Fig. 4. Faculty of health sciences

The wave advancement case in the suggested structure is considered and all potential cases are thought about. As much as, the occasion of adding extra splitters to the base stations, which was another response for address the union region for the flexible customers to do their alluded to works in download measure. In this paper, the course toward investigating the specific space of the best way to deal with be presented and pointed so getting most perceptible idea with no lose in the sign. Inside the open base stations, the most sensible one is picked by its basic sign strength, typical weight and ordinary handover time.

The gathered information from the closest base station and used to perceive the specific area is arranged in Table 2.

Table 2. Obtained results

Signal strength	Distances	Handover time	Iteration
$-5.484353e+01$	2159.2668	$1.184947e-06$	93.0000
$-5.342145e+01$	490.3838	$4.147175e+00$	93.0000
$-5.192663e+01$	2943.7105	$1.219050e-06$	136.0000
$-4.990614e+01$	1348.8897	$1.187930e-06$	136.0000
$-4.749562e+01$	928.5197	$1.193734e-06$	149.0000
$-4.389168e+01$	3328.9528	$1.182256e-06$	152.0000
$-3.795956e+01$	5206.0784	$1.255120e-06$	151.0000

Regarding picked base station, the typical sign strength versus load are as exhibited in Fig. 5.

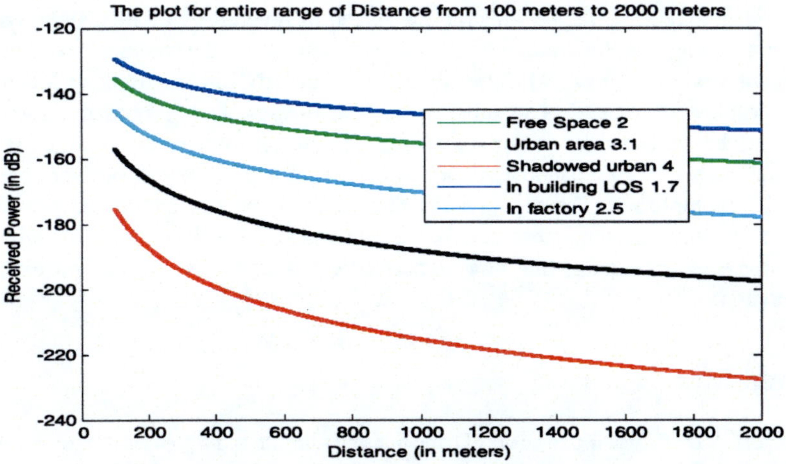

Fig. 5. In Building LOS [MATLAB 2013b]

Using the actually alluded data to track down the specific circumstance of the path as Tx, to keep the strength of the sign and the stability of the IP address. Presenting by straight-lines, picking any floor, and two brilliant lights of the picked floor. Additionally, picking circumstance of the way, and applying Eq. (1). The outcome by ANN as a stable IP address and determining the most probably location of the Router as a transmitter inside the building is showed up in Fig. 6.

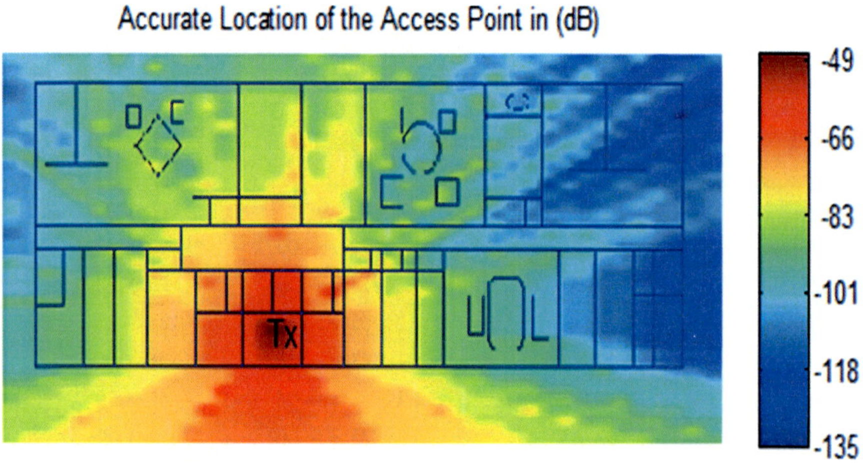

Fig. 6. Accurate location of the Tx inside the building [MATLAB 2013b]

3 Conclusion

In this paper, to take out the way misfortune spread in structures with lifts, and staying away from disengagement issues which prompts misfortune in the communicated information bundles is examined. Also, the utilization of the ANN in this framework enjoys the greatest benefit in the dependability of the IP addresses. As to proposed numerical model and the multi-divider working with lift, the exact situation of the passageway so that steady sign with least handover issues and wiping out in the way misfortune engendering is accomplished. These got results showed the impact of adding passageways inside structures to keep up steadiness in the sent sign, so that download cycle will be with ceaseless transmission and lower handover with a decline in the way misfortune proliferation.

References

1. Jamal, F.A.B.: Estimating coverage of radio transmission into and within buildings for line of sight visibility between two points in terrain by linear prediction filter. In: Third Mosharaka International Conference on Communications, Signals and Coding. MIC-CSC, pp. 1–5 (2009)

2. Zyad, N., Julius, D., Jamal, F.: A new algorithm to enhance radio wave propagation strength in dead spots for cellular mobile WiFi downloads. IEEE (2014). 978-1-4577-1343-9/12
3. Jamal, F.A.B., Firudin, K.: Direction prediction assisted handover using the multilayer perception neural network to reduce the handover time delays in LTE networks. Procedia Comput. Sci. **120**, 719–727 (2017)
4. Fathi, J.: Improvement in strength of radio wave propagation outside the coverage area of the mobile towers for cellular mobile WiFi. In: Aliev, R.A., Kacprzyk, J., Pedrycz, W., Jamshidi, Mo., Sadikoglu, F.M. (eds.) ICAFS 2018. AISC, vol. 896, pp. 464–471. Springer, Cham (2019). https://doi.org/10.1007/978-3-030-04164-9_62
5. Sadikoglu, F., Fathi, J.: ANN based solution to maintain connection continuity problems of mobile users in the buildings elevators. In: Aliev, R.A., Kacprzyk, J., Pedrycz, W., Jamshidi, Mo., Babanli, M., Sadikoglu, F.M. (eds.) ICAFS 2020. AISC, vol. 1306, pp. 328–336. Springer, Cham (2021). https://doi.org/10.1007/978-3-030-64058-3_41
6. Obeidat, H.A., Asif, R., Ali, N.T., Dama, Y.A., Obeidat, O.A., Jones, S.M.R.: An indoor path loss prediction model using wall correction factors for wireless local area network and 5G indoor networks. Radio Sci. **53**, 544–564 (2018). https://doi.org/10.1002/2018RS006536
7. Perez, C.V., Garcia, J.L.G., Higuera, J.M.: A simple and efficient model for indoor path-loss prediction. Meas. Sci. Technol. **8**, 1166–1173 (1997). https://doi.org/10.1088/0957-0233/8/10/020
8. Majed, M.B., Rahman, T.A., Aziz, O.A., Hindia, M.N., Hanafi, E.: Channel characterization and path loss modeling in indoor environment at 4.5, 28, and 38 GHz for 5G cellular networks. Int. J. Antennas Propag. **2018**, 1–14 (2018). https://doi.org/10.1155/2018/9142367
9. Zyad, N., Juliusm, D., Jamal, F.: A new selected points to enhance radio wave propagation strength outside the coverage area of the mobile towers in the dead spots of cellular mobile wifi downloads. In: Long Island Section Systems, Applications and Technology Conference. IEEE (2015). https://doi.org/10.1109/LISAT.2015.7160194
10. Jadhav, A.N., Kale, S.S.: Suburban area path loss propagation prediction and optimization using Hata model at 2375 MHz. Int. J. Adv. Res. Comput. Commun. Eng. **3**(1), 5004–5008 (2014)
11. Zreikat, A., Djordjevic, M.: Performance analysis of path loss prediction models in wireless mobile networks in different propagation environments. In: Proceedings of the 3rd World Congress on Electrical Engineering and Computer Systems and Science (EECSS 2017), Rome, Italy, Paper No. VMW 103 (2017). https://doi.org/10.11159/vmw17.103
12. Bakinde, N.T.S., et al.: Path loss predictions for multi-transmitter radio propagation in VHF bands using Adaptive Neuro-Fuzzy Inference System. Eng. Sci. Technol. Int. J. **21**, 679–691 (2018)
13. Bose, A., Chuan, H.F.: A practical path loss model for indoor WiFi positioning enhancement, 1-4244-0983-7/07. IEEE (2007)

Automatic Control of Reactive Power in the Load Node of the Power Supply System Based on Fuzzy Logic

Z. I. Farkhadov(✉) and R. Z. Azizov

Sumgait State University, 43th Quartal, 1st Korpus, Sumgait, Azerbaijan

Abstract. One of the directions of energy saving at an industrial enterprise is to increase the efficiency of electricity distribution in its power supply system. It provides for the solution of a number of technical problems, including the development of an automatic control system that provides rational management of reactive power in the power supply system. In this work, synchronous motors and batteries of static capacitors are used as reactive power regulators in the node of the electrical network of the power supply system. With an increase in the consumption of reactive power by the load, most of it is gradually compensated by static capacitors, and a smaller part is compensated by synchronous motors operating in the over excitation mode. Considering that in real conditions in the load node of the enterprise's power supply system, the load change is random, this may limit the operation of current regulators and compensating devices. In this paper, we propose the structure of a fuzzy reactive power regulator synthesized by a model based on fuzzy logic. A scheme and a fuzzy algorithm for regulating the reactive power have been developed. When studying the operation of the fuzzy controller for controlling reactive power compensation in the load node of the power supply system based on the Mamdani algorithm, it was found that the fuzzy algorithm better supports the specified value and direction of reactive power at the point where the load is connected to the buses of the electric network.

Keywords: Power supply systems · Fuzzy logic · Reactive power · Induction furnace

1 Introduction

The workload of the power supply systems of industrial enterprises and substations with full capacity, especially its reactive component, leads to electricity losses. Reactive power (RP) compensation is an important component of increasing the efficiency of electricity use at an industrial enterprse [1–3]. To solve the problem of optimal RP compensation, it is necessary to consistently solve a number of problems that should be formulated when developing its target integrated energy saving program [4, 5]. The initial material for their concretization is the results of the energy survey of the enterprise.

In general, these tasks include: analysis of the power supply system of the enterprise (PSSE); analysis of the modes of consumption of active and reactive power of the

R. A. Aliev et al. (Eds.): ICSCCW 2021, LNNS 362, pp. 140–147, 2022.
https://doi.org/10.1007/978-3-030-92127-9_22

enterprise as a whole and for the nodes of the electrical loads of the PSSE; assessment of the reserve of reactive power on the high and low side; study of the effect of changes in loads in the PSSE on the magnitude and direction of RP flows; studies of the effect of voltage changes in the nodes of the PSSE loads on RP flows; determination of the value of additional RP to ensure optimal compensation and its distribution among the nodes of electric loads in the PSSE.

2 The Power Supply System of the Enterprise Analysis and the Problem Definition

The final stage is the creation of an automatic control system (ACS) of the RP in the PSSE. For its comprehensive implementation at the initial stage, it is necessary to provide the first level of the RP ACS, namely, in the load nodes of the PSSE. At a voltage above *1000V (6, 10 kV)*, the following types of electric power receivers are usually included: asynchronous motors (AM) and synchronous motors (SM) and power transformers (PT), which supply various electrical installations with a voltage of up to *1000 V*. This is mainly a mixed active-inductive load-motors, transformers, induction furnaces, lighting devices. Also it has a significant share of nonlinear load, which includes semiconductor converters, welding installations, arc furnaces. Static capacitors (SC) are also connected to compensate for RP, since the reserve of the reactive power of the SM is not enough. Considering that in real conditions in the load node of the enterprise's power supply system, the load change is random, this may limit the operation of current regulators and compensating devices. In this paper, we propose the structure of a fuzzy reactive power regulator synthesized by a model based on fuzzy logic.

The structure of the fuzzy controller. The reactive power in the PSSE is determined from the expression:

$$Q_y = Q_H - Q_{CK} \pm Q_{CD},$$

where Q_H - the reactive power consumed by the load, while excluding SD; Q_{ck} - the reactive power generated by static capacitors ($C_1 \div C_N$) in the UES SEP; Q_{CD} is the reactive power at the connection point of synchronous motors in the UES SEP. In this case, synchronous motors operate in under excitation, over excitation modes and at *sos* $\varphi = 0$.

The scheme implementing this control algorithm is shown in Fig. 1, where the following designations are adopted: 1 - an electrical network node (ENN); 2 - a reactive power sensor (RPS); 3 - a load (H); 4-a reactive power sensor; 5 - a controlled capacitor bank; 6 - a synchronous motor; 7 - a static exciter; 8 - a static exciter control unit; 9 - a fuzzy controller. The fuzzy controller includes: a fuzzier 10 designed to transform clear signals into fuzzy sets; a table of linguistic rules (TLR) 11, i.e. a set of fuzzy rules describing the relationship between the input and output parameters of the controller; defuzzifier 12, where the resulting fuzzy value after defuzzification in the form of a clear control action enters the input of the control unit with a multi-stage contactless key; 13. The outputs of sensors 2 and 4 are connected to the adder 14, the output of which is connected to the first input of the fuzzy controller. The second output of the reactive power sensor 2 is connected to the second input of the fuzzy controller and to the block 15 of the load derivative of the reactive power, the output of which is connected to the third input of the fuzzy controller.

The scheme works as follows:

Loads 3 are powered from buses 1, which consume reactive power. Depending on the value of the reactive power measured by the sensor 2, and the value of the calculator of the derivative of the reactive power of the load (dQ_N/dt) of the block 15, one or another section of the capacitor bank 5 is switched on. Since the capacitor bank is regulated in a stepwise manner, it cannot accurately compensate for the reactive power consumed by loads 3. The difference between the reactive power of the load and the power generated by the capacitor bank is determined by the adder 14, which is compensated by the synchronous motor 6 by changing the excitation current of the synchronous motor.

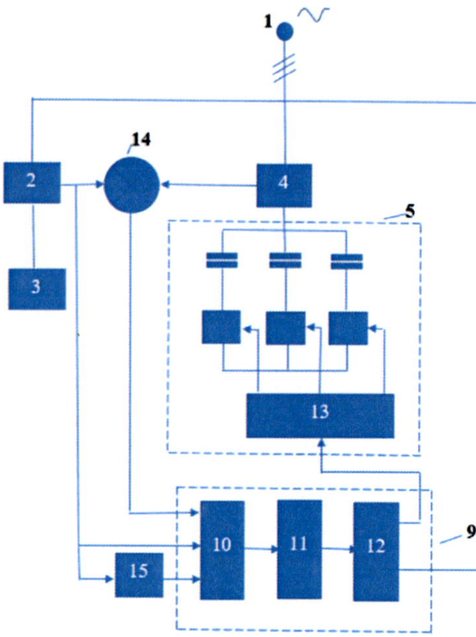

Fig. 1. The scheme of automatic control of reactive power in the load node of the power supply system

Thus, the main part of the reactive power of the load 3 is compensated by the capacitor bank 5 and only a small part of it is compensated by the synchronous motor. The fuzzy controller, depending on the value of the calculator of the derivative of the reactive power of the load (dQ_N/dt) of block 15, determines the speed and direction of the reactive power at the beginning of its change. At the same time, the total operating time with a deviation of the reactive power from the permissible value for a fuzzy controller is less than for a traditional one. The system itself turns out to be of higher quality and energy losses are reduced.

3 The Algorithm of Functioning of the Fuzzy Controller

To control the reactive power compensation in the load node of the power supply system, the Mamdani algorithm was chosen [6–8]. The construction of an intelligent control system for reactive power compensation in the load node of the enterprise's power supply system is carried out in the following sequence:

1) determination of the inputs and outputs of the reactive power compensation control system;
2) setting the membership function for each of the input and output variables;
3) development of linguistic rules;
4) selection and implementation of the fuzzy inference algorithm;
5) analysis of the control process of reactive power compensation in the load node of the power supply system.

The general logical output is performed according to the scheme shown in Fig. 2.

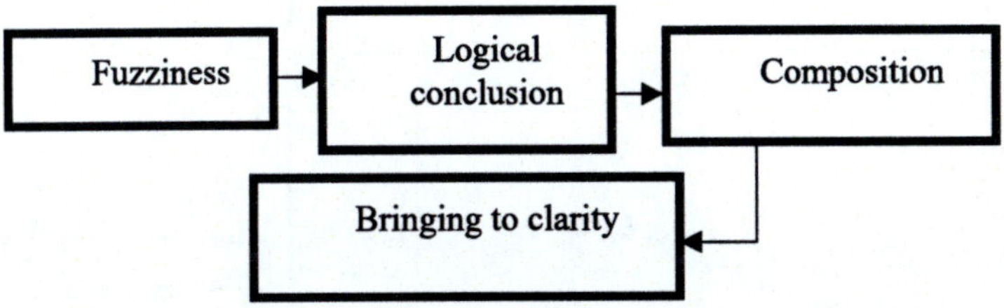

Fig. 2. The scheme of logical output

The membership functions used in this article were mainly of four types: triangular, trapezoidal, S - and Z-shaped. The parameters and formulas of the Membership function are shown in Table 1.

When the controller was operating according to the Mamdani algorithm, in order to maintain the deviation of the reactive power within the normalized limits, the input of the fuzzy controller was fed: the calculated value of the reactive power, the derivative of the reactive power dQ/dt and the difference of the reactive powers removed from the load and from the static capacitors (ΔQ). From the output of the controller, control actions were removed for switching the switched capacitance either with an increase or with a decrease in the total value, as well as a control signal proportional to the corresponding value of the excitation current of the SM. The RP SM regulation is smooth.

Table 1. Parameters and formulas of the membership function

The membership functions	Parameters	Formulas
S-shaped	*a*, *b*	$\mu(x) = \begin{cases} 0, x < a \\ \frac{x-a}{b-a}, a < x < b \\ 1, b < x \end{cases}$
Z-shaped	*b*, *c*	$\mu(x) = \begin{cases} 1, x < b \\ \frac{c-x}{c-b}, b < x < c \\ 0, c < x \end{cases}$
Triangular	*a*, *b*, *c*	$\mu(x) = \begin{cases} 0, x < a \\ \frac{x-a}{b-x}, a < x < b \\ \frac{c-x}{c-b}, b < x < c \\ 0, c < x \end{cases}$
Trapezoidal	a, b_1, b_2, c	$\mu(x) = \begin{cases} 0, x < a \\ \frac{x-a}{b_1-a}, a < x < b_1 \\ 1, b_1 < x < b_2 \\ \frac{c-x}{c-b_2}, b_2 < x < c \\ 0, c < x \end{cases}$

All the input values coming from the reactive power sensors are accurate values, since they are taken from real measuring devices that give accurate values of the measured parameters. Further, in the fuzzy controller itself, these values are already converted to fuzzy values. After the implementation of the fuzzy inference algorithm, the resulting output variables are again converted to the exact (normal) form. The following linguistic variables were used for the operation of the fuzzy controller:

1. The input of the fuzzy controller (input variables) receives:

   – Reactive power (*Q*) at the place where the load is connected to the buses of the electrical network; - The dynamics of changes in the reactive power (the derivative of the reactive power *dQ/dt*), the residual reactive power (ΔQ) compensated by the SM. The fuzzy values, membership functions, and parameters of the input linguistic variables are shown in Table 2.

2. The values of the following linguistic variables (output variables) were removed from the controller output:

Table 2. Fuzzy values, membership functions and input linguistic variables.

Variable values	The membership functions	Parameters
Reactive power		
Very small	Z-shaped	(0 0,4)
Small	Triangular	(0.01 0.045 0.9)
Average	Triangular	(0.46 0.905 1.36)
Big	Triangular	(0.91 1.35 1.8)
Very big	S-shaped	(1.4 1.76)
Dynamics		
Very negative	Z-shaped	(−0.8 −0.5)
Negative	Triangular	(−1 −0.5 0)
Zero	Triangular	(−0,5 0 0.5)
Positive	Triangular	(0 0.5 1)
Very positive	S-shaped	(0.5 0.8)
Residual reactive power		
Negative	Z-shaped	(−0.8 −0.5)
Zero	Triangular	(−0,5 0 0.5)
Positive	S-shaped	(0.5 0.8)

- Direction – ***Direction.*** Control actions for switching the switched capacity either with an increase or with a decrease in the total value. In the absence of the corresponding command, the capacity value did not change.
- Arousal - ***Excitation***. Control actions, proportional to the corresponding value of the excitation current of the SM. The fuzzy values, membership functions, and parameters of the output linguistic variables are shown in Table 3.

In the process of work, various options for the operation of the controller were tested. According to the above variables, 32 fuzzy inference rules were compiled for this system.

4 The Results of Computer Modeling

Modeling of operating modes in the PSSE load node taking into account the fuzzy regulation of the reactive power consumed, is performed on the example of the induction furnace (IF)-2.5/1000 of industrial frequency, with a nominal capacity of 2.5 t, a nominal voltage of 6 kV, a nominal voltage of the contour circuit of 495V and with a melting and overheating rate of 0.33 t/h. The total power of the induction furnace is 990kVA, and the power of the transformer of the industrial frequency induction furnace regulated under load is 1000kVA. The heating and melting time is 35 min [9].

The computational experiments were carried out in the Matlab environment using the Fuzzy Loqic Toolbox software package [10, 11].

Table 3. Fuzzy values, FP, and parameters of output linguistic variables.

Variable values	The membership functions	Parameters
Direction		
Up	Trapezoidal	(0.5 0.75 1.25 1.5)
Down	Trapezoidal	(-1.5-1.25-0.75-0.5)
Stop	Trapezoidal	(-0.5-0.25 0.25 0.5)
Excitation		
Very low	Z- shaped	(0.69 0.75)
Low	Trapezoidal	(0.85 0.9 0.95 1)
High	Trapezoidal	(1.01 1.05 1.1 1.15)
Very high	S- shaped	(1.051 1.101)

Figure 3 shows a graph of changes in reactive power at the load, as well as curves corresponding to the usual (traditional "clear") regulation of reactive power (blue curve) and fuzzy regulation (red curve). When using fuzzy (fuzzy) and classic, clear (crisp) regulators, it is clear that when the reactive power goes beyond the dead zone (200 kvar), the regulators are triggered. The fuzzy controller, depending on the value of the computer of the derivative of the reactive power of the load (dQ_N/dt), determines the speed and direction of the reactive power at the beginning of its change. Thanks to this, the fuzzy controller responds to changes in reactive power faster. At the same time, the total operating time with a deviation of the reactive power from the permissible value for a fuzzy controller is less than for a traditional one. During operation, the regulator "adjusts" depending on the current position of the sections of the capacitor bank.

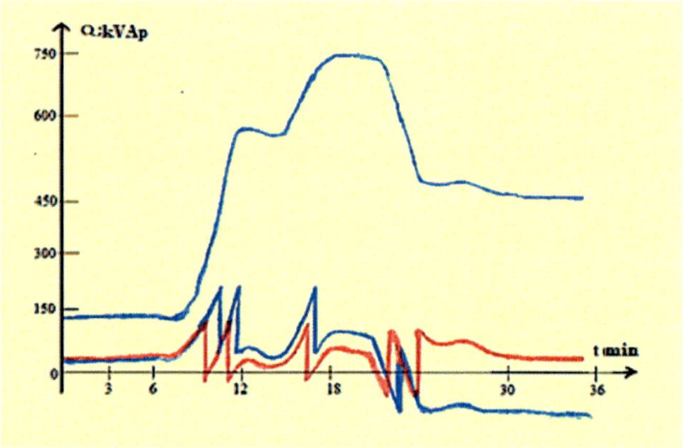

Fig. 3. Curves of changes in reactive power before and after control

It can be argued that devices based on fuzzy logic are more preferable for regulating the reactive power in the load node of the power supply system than devices based on conventional logic. The regulation itself turns out to be of higher quality and energy losses are reduced.

5 Conclusions

1. A scheme and a fuzzy algorithm for regulating the reactive power in the load node of the power supply system based on fuzzy logic have been developed.
2. When studying the operation of the fuzzy controller for controlling the reactive power compensation of an electric network node based on the Mamdani algorithm, it was found that the fuzzy algorithm better supports the specified value and direction of the reactive power at the point of connecting the load to the buses of the electric network.
3. A fuzzy controller responds faster to changes in reactive power. At the same time, the total operating time with a deviation of the reactive power from the 4permissible value for a fuzzy controller is less than for a classic one.
4. The regulation itself turns out to be of higher quality and energy losses are reduced.

References

1. Basalygin, M.Ya., Kapyrin, V.S.: Directory of electric power industry of non-ferrous metallurgy enterprises/Edited by M.: Metallurgy, 407 p. (1991). (in Russian)
2. Ilyashov, V.P.: Automatic power control of condenser installations. M. Energiya-2nd edition, 105 p. (1979). (in Russian)
3. Bakuta, V.P., Kopyrin, V.S.: Improving the efficiency of electricity distribution in the enterprise's power supply system. Industrial Power Engineering (İPE), no. 9, pp. 37–39 (1999). (in Russian)
4. Juice, N.I.: A device for regulating the reactive power of an electrical network node. Discoveries. Inventions, no. 35, pp. 34–42 (1990). https://doi.org/10.1109/ESS50319.2020.9160072
5. Guliyev, H.B., Farhadov, Z.I., Mammadov, J.F.: System of automatic control of reactive power using fuzzy logic. Reliability: Theory and Applications (RTA), vol. 10, no. 2(37). USA. San Diego, pp. 50–58 (2015). https://doi.org/10.21122/1029-7448-2016-59-4
6. Rakhmanov, N.R., Guliyev, G.B., Farkhadov, Z.I.: Identification of the structure of a fuzzy reactive power regulator. Electro. Electrical Engineering, Electric Power Industry, Electrical Industry (EEEEPİ), no. 4, pp. 28–31 (2015). (in Russian)
7. Zadeh, L.: Presentation of a new approach to the analysis of complex systems and decision-making processes. IEEE Trans. Syst. People Cybern. **SMC-3**, 28–44 (1973)
8. Aliev, R.A., Aliev, R.R.: Theory of intelligent systems and its application. Baku, Chashyoglu, 720 p. (2001). (in Russian)
9. Kopyrin, V., Tkachuk, A.: Control controller for reactive power compensation of an industrial frequency induction furnace. Power Electronics (PE), no. 1, pp. 96–99 (2005). (in Russian)
10. Shtoba, S.D.: Design of fuzzy systems by means of Matlab. Hotline-Telecom. M., 288 p. (2007)
11. A set of fuzzy logic tools. User's guide. Version 2. Math-Works. Lnc. (1999)

Fuzzy Aided Generalized Probability Distribution Function for Wireless Wearable Medical Sensors

Bülent Bilgehan and Fahreddin Sadikoglu(✉)

Near East University, Nicosia, Cyprus
{bulent.bilgehan,fahreddin.sadikoglu}@neu.edu.tr

Abstract. The accuracy of the wireless channel characterization is an important issue. There are several models applied in the planning of wireless communication systems. However, all statistical models rely on the local characteristics of the environments. This research article introduces a combination of a fuzzy model named Neuro-fuzzy (NF) to extract the random noise and model with a generalized probability function. The introduced model identifies and extracts the unknown noise in the signal. The NF system has a non-linear structure for noise prediction before the sampling process. This paper takes into consideration an Electrocardiogram signal generated by a wireless medical sensor. The setting parameters (Membership Function, training epochs) are processed using the available data. The second contribution in this research paper is that the probability function is generalized to be free of sensor location and distance. The model is validated with simulated experimental results and compared with most applicable probability distributions such as Rayleigh, Nakagami, Weibull, Normal, lognormal. The performance analysis of the NF aided generalized probability produced the best values in all respects. The newly introduced model produced the best matching criteria using Kozmogorov theorem analysis. The results verify to use of the introduced model for better representation of the wireless channel characteristics.

Keywords: Wireless channel · Probability distribution function · Channel characteristics · Wireless sensors · Fuzzy model

1 Introduction

An accurate representation of the signal noise is very significant in the design procedure of the communication systems [1]. The signal strength estimation at the receiver relies upon the channel characteristic between the transmitter and the receiver. The equipment involved in the wireless network usually demands to have a signal level above the pre-set threshold value [2]. This reason amongst many indicates the need for an accurate representation of the signal. The wearable wireless medical sensors in applications require knowing the source location. The survey paper [3] summarizes and lists the type of probability distribution needed at different locations.

The sensors in the application usually require pre-knowledge of the type and its location before any analysis. The Electrocardiogram (ECG) recorded data from sensors

R. A. Aliev et al. (Eds.): ICSCCW 2021, LNNS 362, pp. 148–155, 2022.
https://doi.org/10.1007/978-3-030-92127-9_23

usually includes noise that is very difficult to remove by any hardware system [4]. Such noise may mislead the medical staff on the ECG interpretation. The problem is that a linear filter cannot clean the non-linear ECG signal effectively [5].

The problem of filtering is solved by introducing a fuzzy aided nonlinear adaptive filtering technique. We will initially use this method to predict noise and introduce a generalized probability distribution function to represent the signal noise without the need to identify the location of the sensor. The fuzzy logic helps to formulate the mapping from a given input to an output. The aim is to discover the membership value. There are two well-known types namely the Mamdani and Sugeno type [6]. This work adopts the Sugeno method for the reason that the output is a representation of the weighted average method and does not need an additional method of defuzzification [7]. The contribution of the fuzzy terminates after the extraction process of the signal noise. The extracted data follows the next stage to determine the parameters of the generalized probability distribution function.

The main challenge is to model the noise level in transmission due to atmospheric variations, reflection, and refractions [8]. Such noise cannot be strictly formulated and require statistical distribution analysis [9]. Studies on such analysis highlight to use of probability distribution functions (PDF) such as the Weibull, lognormal, Gaussian (normal), Rayleigh, and Nakagami-m for such applications [10]. The research in this field remains to be interested and further analysis is needed [11]. The typical medical network system is given in Fig. 1.

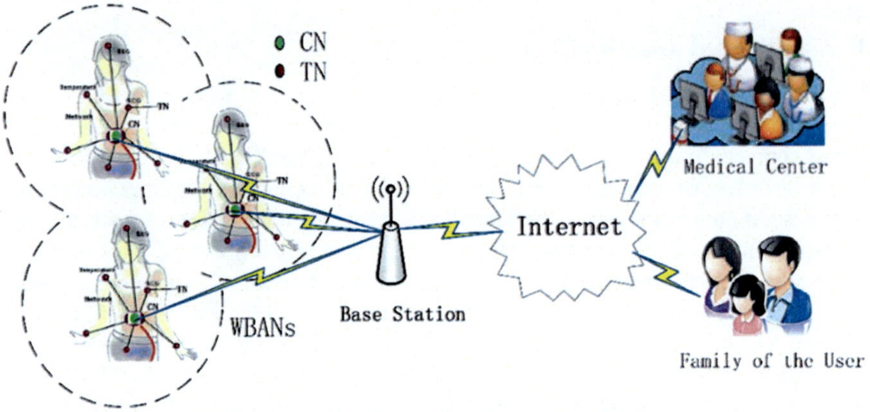

Fig. 1. Wireless medical sensors in the network.

The performance of the newly introduced model compared with the widely used classical distributions such as Normal, Weibull, Nakagami, lognormal, and Rayleigh. The measure of the performance was based on the Kolmogorov theorem [12].

2 Method

This paper uses Anfis based adaptive filter processing to identify noise in the signal and generate a probability density function for the randomly varying noise. The block diagram of the complete process is shown in Fig. 2.

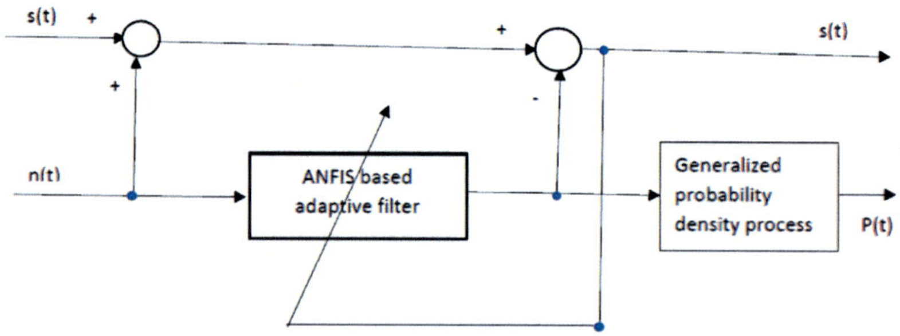

Fig. 2. System block diagram

The variables s(t) denotes the signal, n(t) noise, and p(t) probability density function.

There are two main parts under this section. The first section explains the fuzzy logic and the second section describes the application of the generalized probability function to the wireless medical sensors.

2.1 ANFIS Based Adaptive Filter

The primary signal s(t) is the one received at the receiver device and needs filtration. The second input n(t) is the reference input where it receives noise from internal and external sources. In this work, the filter is the ANFIS system itself. The ANFIS filter works as an adaptive filter because it can adjust and fix its impulse response according to the noise. It does not require prior knowledge about the noise and its characteristics.

ANFIS aims to identify the best features of fuzzy and neural networks. This analysis has the advantage of reducing the optimization research space. The gradient descent algorithm allows fine-tuning on the parameters that define the membership functions (MF) and ANFIS. The overall learning process includes both the least-squares method and the gradient descent algorithm. The learning and adoption process in ANFIS uses the Sugeno model. The ANFIS construction is based on two fuzzy if-then rules using the Sugeno model constructed as:

Rule 1:

If x is A_1 and y is B_1 then

$$f_1 = p_1 x + q_1 y + r_1 \tag{1}$$

Rule 2:

If x is A_2 and y is B_2 then

$$f_2 = p_2 x + q_2 y + r_2 \tag{2}$$

The layers in the architecture can be defined as:

Layer 1:
The node is an adaptable type and can be defined as;

$$L_j^1 = \mu A_i(x) i = 1, 2 \tag{3}$$

The variable x denotes the input to the *ith* node, A_i is the adaptable node having a membership function of $\mu A_i(x)$. This membership function can be taken as;

$$\mu A_i(x) = \frac{1}{1 + \left[\left(\frac{x - f_i}{d_i}\right)^2\right]^{e_i}} \tag{4}$$

Equation (4) is the standard bell membership function (MF) which gives the best result compare with the other membership functions. The variables $\{d_i, e_i, f_i\}$ are the previous values.

Layer2:
This layer holds fixed nodes only. The function of the fixed node is to solve the instantaneous weights of w_1 and w_2. The inputs are multiplied at layer 2 as:

$$L_i^2 = w_i = \mu A_i(x) X \mu B_i(y), i = 1, 2 \tag{5}$$

Layer 3:
The nodes in this layer are constants. The outputs of the nodes are determined as:

$$L_i^3 = w_I = \frac{w_i}{\sum w_i}, i = 1, 2 \tag{6}$$

Layer 4:
This layer includes the adaptable nodes. The outputs of the nodes are determined as:

$$L_i^4 = w_i f_i = w_i(p_i x + q_i y + r_i), i = 1, 2 \tag{7}$$

The variable values of *p, q, and r* are determined by the least-squares method.

Layer 5:
This layer sums all incoming data and the output is:

$$L_i^5 = \sum\nolimits_{i=1}^{2} w_I f_i = \frac{\sum w_i f_i}{\sum w_i} \tag{8}$$

The constructed fuzzy model has 5 layers as shown in Fig. 3. The next stage is to adopt these two rules in the ANFIS architecture as shown in Fig. 3.

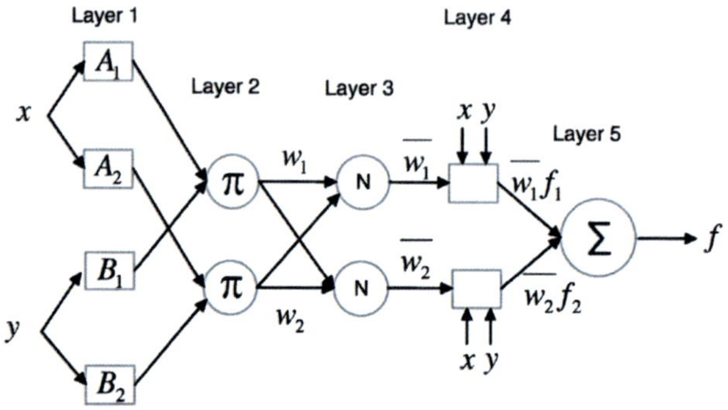

Fig. 3. Artificial intelligence and fuzzy structure.

The circular nodes represent the fixed nodes and the rectangular shapes represent the adaptive nodes.

2.2 Generalized Probability Density Function

The generalized probability function aims to efficiently represent randomly varying noise without the need to identify the location for the wireless sensor. The vital point is to disclose the uncertainty in both qualitative and quantitative ways. The model for such representation is important since human life is a concern. The introduced method uses exponential bases that capture the main issues in a mathematical model.

Let X represent a generalized exponentially distributed random variable such that X ~ GE (α, β, q). The probability density function (pdf) can be defined as:

$$f(x;\alpha,\beta,q)=\frac{1}{\Gamma(q+1)}(2\alpha x+\beta)(\alpha x^2+\beta x)^q.e^{-(\alpha x^2+\beta x)} \tag{9}$$

where $x > 0, \alpha, \beta \geq 0, \mathrm{Re}(q) > -1$.

Equation (9) is an exponential-based generalized probability density function. The introduction of the parameters enables Eq. (9) to process broad-based signals. The variables should satisfy the condition such that α, β ≥ 0, Re (q) > −1. The accurate derivation of the variables provides a closed representation of the probability distribution. The values of α, β ≥ 0 can be derived from the data using probability expectation, and Re (q) > −1 can be determined using the Kolmogorov-Smirnov theorem.

The random variable X exponentially distributed as $X \sim GE(\alpha,\beta,q)$ has some special cases that help to cover most classical distributions as:

1. If $X \sim GE\left(\frac{1}{2\alpha^2}, 0, -\frac{1}{2}\right)$, the random variable X follows a half-normal distribution
2. If $X \sim GE\left(0, \frac{1}{\beta}, 0\right)$, the random variable X follows an exponential distribution.
3. If $X \sim GE\left(\frac{1}{2\alpha^2}, 0, 0\right)$, the random variable X follows the Rayleigh distribution.
4. If $X \sim GE\left(\frac{1}{\alpha}, 0, \frac{q}{2} - 1\right)$, the random variable X follows a generalized gamma distribution.

5. If $X \sim GE(\alpha^2, 0, 0)$, the random variable X follows the Weibull distribution.

The values for $\alpha, \beta \geq 0, Re(q) > -1$ in Eq. (9) can be derived using the maximum likelihood estimation method. The derived values of $\alpha, \beta,$ *and* q are substituted in Eq. (9) to represent the probability density function of random data X.

The goodness fit of the $\alpha, \beta,$ *and* q values can be tested using the Kolmogorov–Smirnov test.

3 Results

The recorded data includes the original signal with noise. The noise removal stage processed using a fuzzy algorithm as described in the methodology section. The generalized probability density function used the separated noise data to identify the matching probability density function. This helps to identify the channel characteristics in wireless transmission. The histogram of the noise signal is plotted in Fig. 4.

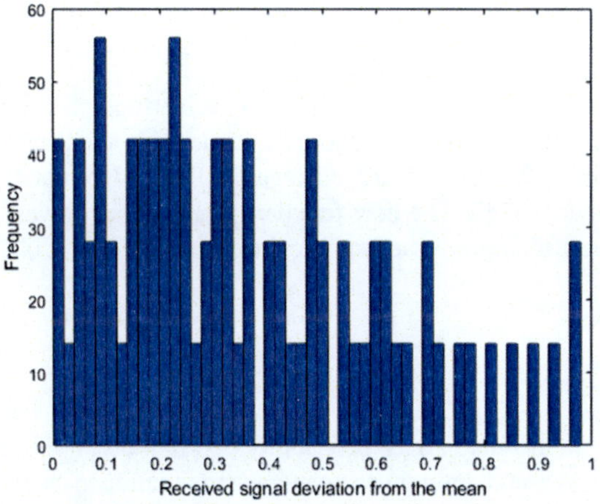

Fig. 4. Histogram for the wireless channel losses

The histogram normally helps to identify the type of probability distribution function to apply for the processing. There are several distributions (Rayleigh, Nakagami, Lognormal, Normal, Weibull) suitable for such application but they all need further examination to identify before the application. This paper tests the newly introduced generalized probability density function against the classical models. The randomly noisy signal tested and the results are plotted in Fig. 5.

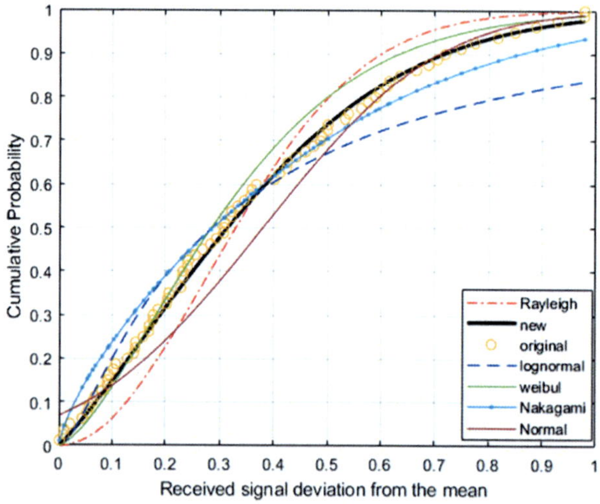

Fig. 5. Cumulative distribution plots for the wireless channel

The parametric values are determined as;

$\alpha = 2.3198$, $\beta = 1.7930$, and $q = 0.087$. The scientific validity of the generated function was carried out with the Kolmogorov-Smirnov test and produced results as;

Lognormal: 0.1627, Weibull: 0.0870, Nakagami: 0.1275, Normal: 0.0295, Rayleigh: 0.1159, New method: 0.0295. The new function produces the lowest value in the fit. Therefore, using the Kolmogorov-Smirnov test we can state that it is the best fitting to the recorded data.

4 Conclusion

This work is based on the generalized probability distribution of biomedical sensors in communication. The research introduces a generalized function to overcome the problem of the multi-distribution function requirement. The introduced function tested with the real data and produced successful results. This reduces the computational cost as well as overcoming the problem of prior knowledge of the source. The location and characteristics of different sensors in application demonstrated not to be a major concern using the generalized function. The generalized probability distribution function was tested, compared with classical distribution methods, and verified using the Kolmogorov–Smirnov test. The newly introduced function gives the best representation amongst the classical methods. Based on the achievements it can be stated that the generalized probability distribution function can be applied for such applications.

References

1. Latré, B., Braem, B., Moerman, I., Blondia, C.: Demeester, P.: A survey on wireless body area networks. Wirel. Netw. **17**(1), 1–18 (2011). https://doi.org/10.1007/s11276-010-0252-4

2. Filipe, L., Fdez-Riverola, F., Costa, N., Pereira, A.: Wireless body area networks for healthcare applications: protocol stack review. Int. J. Distrib. Sens. Netw. **2**, 1–23 (2015). https://doi.org/10.1155/2015/21370
3. Kaur, S., Malhotra, J.: Survey on empirical channel models for WBAN. Int. J. Future Gener. Commun. Netw. **8**, 399–410 (2015). https://doi.org/10.14257/ijfgcn.2015.8.2.34
4. Limaye, H., Deshmukh, V.: ECG noise sources and various noise removal techniques: a survey. Int. J. Appl. Innov. Eng. Manag. **5**(2), 86–92 (2016)
5. Chatterjee, S., Thakur, R.S., Yadav, R.N., Gupta, L., Raghuvanshi, D.K.: Review of noise removal techniques in ECG signals. IET Signal Proc. **14**(9), 569–590 (2020). https://doi.org/10.1049/iet-spr.2020.0104
6. Amri, S., Khelifi, F., Bradai, A., Rachedi, A., Kaddachi, M.L., Atri, M.: A new fuzzy logic-based node localization mechanism for wireless sensor networks. Future Gener. Comput. Syst. **93**, 799–813 (2019). https://doi.org/10.1016/j.future.2017.10.023
7. Medjahed, H., Istrate, D., Boudy, J., Baldinger, J. L., Dorizzi, B.: A pervasive multi-sensor data fusion for smart home healthcare monitoring. In: IEEE International Conference on Fuzzy Systems, Taipei, Taiwan, pp. 1466–1473. IEEE (2011). https://doi.org/10.1109/FUZZY.2011.6007636
8. Yang, G., Wu, X.-W., Li, Y., Ye, Q.: Energy efficient protocol for routing and scheduling in wireless body area networks. Wireless Netw. **26**(2), 1265–1273 (2019). https://doi.org/10.1007/s11276-019-02150-z
9. El Bouny, L., Khalil, M., Adib, A.: ECG signal filtering based on CEEMDAN with hybrid interval thresholding and higher-order statistics to select relevant modes. Multimed. Tools. Appl. **78**(10), 13067–13089 (2019)
10. Moungla, H., Haddadi, K., Boudjit, S.: Distributed interference management in medical wireless sensor networks. In: 13th IEEE Annual Consumer Communications & Networking Conference, Las Vegas, NV, USA, pp. 151–155. IEEE (2016). https://doi.org/10.1109/CCNC.2016.7444748
11. Cotton, S.L., Scanlon, W.G.: A statistical analysis of indoor multipath fading for a narrowband wireless body area network. In: 17th International Symposium on Personal, Indoor and Mobile Radio Communications, Helsinki, Finland, pp. 1–5. IEEE Press (2006). https://doi.org/10.1109/PIMRC.2006.254266
12. Giorgilli, A., Locatelli, U.: Kolmogorov theorem and classical perturbation theory. Z. Angew. Math. Phys. **48**(2), 220–261 (1997). https://doi.org/10.1007/PL00001475

Use of Decision Tree and Fuzzy Logic Methods to Predict Academic Achievement of University Freshmen

M. A. Salahli[1(✉)], T. Gasimzade[2], V. Salahli[3], F. Alasgarova[4], and A. Guliyev[4]

[1] Educational Technologies Application and Research Center, Çanakkale Onsekiz Mart University, Çanakkale, Turkey
msalahli@comu.edu.tr

[2] Department of Instrument Making Engineering, Azerbaijan State Oil and Industry University, 20 Azadlig Avenue, AZ 1010 Baku, Azerbaijan

[3] Department of Computer Engineering, Odlar Yurdu University, Koroglu Rehimov Street, 13, AZ 1072 Baku, Azerbaijan

[4] Department of Information Technology and Natural Sciences, Azerbaijan Tourism and Management University, Koroglu Rehimov Street, 822/23, AZ 1072 Baku, Azerbaijan

Abstract. It is not easy for students who have just entered the university to adapt to the new education and living environment. In this process, social-economic, educational, and personal factors have a great impact on the academic achievements of first-year students. In this study, the aim is to determine the factors affecting the academic success of university freshmen and to predict the academic success of students by using decision tree and fuzzy logic methods together, considering the most important factors. Students' achievement was predicted by the decision tree method considering the students' university entrance scores, satisfaction with the educational environment, accommodation, attendance to the course, and preference of the department as input of the prediction model. Fuzzy inference rules were created based on the decision tree obtained from implementation of the J48 method, which provides a better understanding of the dependence between students' academic achievement and impact factors.

Keywords: Academic achievement · Decision tree · Predictive model · Fuzzy inference

1 Introduction

Prediction of academic performance of students is potentially important for educational institutions to assist the students in improving their academic performance, and deliver high quality education. Developing an accurate student's performance prediction model is challenging task [1]. This issue is of particular importance for first-year students. Thus, it is not easy for students who have just entered the university to adapt the new education and living environment. In this process, many factors can have a great influence on the academic success of freshmen. These factors can be social-economic, educational,

R. A. Aliev et al. (Eds.): ICSCCW 2021, LNNS 362, pp. 156–164, 2022.
https://doi.org/10.1007/978-3-030-92127-9_24

and personal factors. It is very important to examine the influence of these factors on students to ensure that students are successful in their university life. Some studies have been conducted on the examination of various factors that affect the academic achievement of university students, especially first-year students. In the study [2] the authors explore the four factors that affect students' academic performance. These factors are students' communication skills, learning facilities, proper guidance, and family stress. Some researchers focus on university entrance exam scores as an important factor affecting academic success [3]. The study [4] is a prospective investigation of the academic, psychosocial, cognitive, and demographic predictors of academic performance of first year Australian university students. Previous academic performance was identified as the most significant predictor of university performance. Integration into university, self-efficacy, and employment responsibilities were also predictive of university grades. Identifying the factors that influence academic performance can improve the targeting of interventions and support services for students at risk of academic problems. The study [5] examines potential psychosocial predictors of freshman academic achievement and retention. College students were assessed on various dimensions, (i.e., demographics, prior academic record, smoking, drinking, health-related quality of life, social support) during the first week of their freshman, and at the beginning of the next academic year. It is thought that students' satisfaction with the department significantly affects their academic success [6, 7]. In the [8] GPAs of the students were taken as the basic parameter in the prediction model. [3, 9] use the factor of dropping out of university as the input parameter of the prediction model. The influence of accommodation on the academic performance of university students was investigated in [10]. In the study [11] the relationship between students' attendance and academic achievements was investigated.

2 Research Problem

The aim of this study is to investigate the dependence between social, economic, and educational factors and achievements of first year students studying in the computer education and Instructional Technology (CEIT) department by using predictive methods and fuzzy logic. Academic success factors have been investigated in many studies. However, the weight of these factors on student achievement varies from country to country, from university to university, and even between university departments.

CEIT departments prepare IT teachers; The curriculum of CEIT departments includes both computer technologies and teacher training courses. Department preference, interest in courses, attendance, satisfaction with the department and other factors affecting the academic success of students are related to this feature of CEIT departments. Both the advantages of the teaching profession and the word "computer" in the name of the department have an important place in the preference of CEIT departments. In this sense, the reasons for choosing the department have a significant effect on the academic achievement of CEIT students. [12] examined the reasons why 178 CEIT students preferred this department. It has been determined that students prefer CEIT departments primarily because their university exam scores hold this department, and also because students want to be teachers and they are interested in this department. [13] determined that more than half of the participants were reluctant to choose the department of CEIT, and the fact that the name of CEIT includes "computer" plays a role in choosing the department. It was found that only 39.3% of the students participating in the research did not regret that they chose the CEIT department and 29.5% of them regretted it. In his master thesis study, Uysal Hakan analyses the factors affecting the academic performance of CEIT students and created a prediction model to determine the success of the students [14].

3 Data Collection and Modeling Process

The steps of estimating the academic achievement process is shown in Fig. 1.

The first step in the process is to determine the input data for the model - the parameters that affect the success As mentioned above, there have been many studies to identify factors affecting the academic performance.

We can generally express these factors in the following groups:

1. data on students' education level of knowledge.
2. social-ekonomic aspects of the student's educational life
3. educational environment
4. personality traits of the student

The first group data includes the type of school the student graduated from, high school graduation grade, university entrance score, general grade point average at the university, semester grades, midterm exam grades.

The second group includes data on the socio-economic status of the students' families, scholarships, accommodation, social life of the city where the university is located, transportation, employment opportunities after graduation and economic living conditions in the city.

The third group includes factors such as the educational opportunities of the university and the department, laboratory, library, and free study opportunities, the difficulty of the curriculum, student-instructor relations, and measurement-evaluation methods used to evaluate academic achievements.

Finally, data such as student satisfaction with school, interest in the department, attendance to classes can be assigned to the last group.

Many of these factors are interrelated and can be included in several groups. For example, the prestige of universities or departments affects entrance scores. The satisfaction level of the student studying at a very prestigious university is always high. On the contrary, the educational environment of the university will not be satisfactory for the student who thinks he/she deserves a more prestigious university. Whether students choose the department willingly or unwillingly, affects their level of satisfaction with the school and their attitudes towards the school. In other words, students' satisfaction with the university depends on subjective values, not objective values, for many students. It has been determined that some factors are felt for a small number of students. For example, it has been observed that the opinion of the family is not dominant in the choice of profession.

The data in the study were collected from 3 sources:

- the student database where the academic and identity information of the students are saved.
- the results of the survey conducted with the students.
- course attendance documents, midterm, exam, and homework evaluation results

According to the results of the analysis of the collected data, it was determined that 5 factors were more dominant in the academic success of CEIT first-year students.

These are:

- the student's entrance score to the university (UEES),
- the level of satisfaction with the university life of students (SAT)
- the accommodation type (ACC)
- course attendance (ATT)
- order of preference of the department (PREF)

It is necessary to pass the entrance exams to study at Turkish Universities. Students should present the list of programs (departments) they want to study in order of preference. Students are admitted to universities according to the scores they get in the entrance exams (we expressed this score as UESS) and the choice they make. The admission of students to universities is determined by the points they collcct in the university admission and the preference list. The order of the student's department in the preference list is expressed as PREF. The factor SAT expresses students' satisfaction with the university education environment and the social-cultural life at the university. But, as stated above, students' subjective views are effective in the evaluation of satisfaction. For example, the satisfaction level of students who must study in the department they do not want is low. ATT- refers to the accommodation type and can take 3 categorical values (dormitory, flat for rent, living with family). As the value of the ATT parameter, the attendance rate of the students in the first 10 weeks from the beginning of the first semester was used.

Within the scope of the research, the data of 164 students who received undergraduate education in the CEIT department were examined. Since the survey data of 30 students included inconsistent data, they were not considered in the subsequent evaluations. Thus, the data of 134 students were used for the prediction model.

In terms of accommodation, most students (73 out of 134 participants, or 54,47%) participants live at dormitories. 53 students (37,8%) live on rent. Only 8 (5, 96%) students live with their families. 60 students declared that their satisfaction with the university was sufficient. 29 students (21.64% of all students) who participated in the survey stated that they were not satisfied with the university environment.

Data for the prediction model were collected for 2 consecutive years. The data of the previous year were used as training data for the model. The university entrance scores of the students were obtained from the University Student Information System. Data such as the accommodation type and the level of satisfaction with the educational environment of the students were obtained through the questionnaires made with the students. The attendance of the students to the course in 2 months from the beginning of the semester was taken into consideration. The semester grade point averages of the students were accepted as success criteria. With the 100-point system, students with a grade below 60 formed the class of unsuccessful students (risky students). The learning model and the data of new students are used to predict their success.

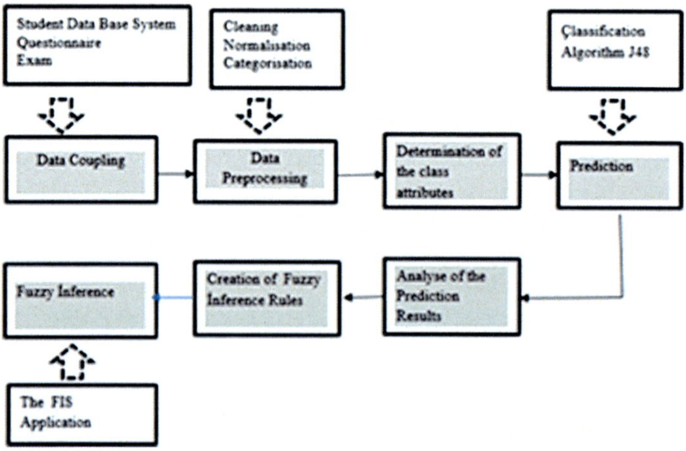

Fig. 1. The process of predicting first-year students' academic performance

The university entrance scores of the students were normalized in the range of 1–10, as they were close to each other and received continuous value. The values obtained because of normalization were categorized as *low, medium, and high.* In the questionnaires, students were asked to evaluate their satisfaction in the range of {1,10}. These categorical values were used in the decision tree model. Satisfaction levels were mapped into *low, medium, and high* linguistic values for use in the fuzzy inference rules. The places where the students stay are evaluated as {*dorm, rent, family*}. Linguistic values *low, medium,* and *high* were used for expressing attendance rates *of less than 30%, 30–80%,* and *more than 80%.*

In the Tables 1, 2, 3, 4 and 5 nominal, linguistic and fuzzy value ranges for the parameters UEES, SAT, ATT, ACC, and PREF are shown.

Table 1. Numerical and linguistic values of the parameter UEES

Nominal values range	Normalised values range	Categoric values	Linguistic value	Fuzzy value ranges
{238,427}	{1,10}	[1,2,3,4,5,6,7,8,9,10}	Low	{1,1,3,4}
			Medium	{3,4,5,6}
			Good	{5,6,10,10}

Table 2. Numerical and linguistic values of the parameter SAT

Nominal values range	Categoric value	Linguistic value	Fuzzy value ranges
{1,10}	[1,2,3,4,5,6,7,8,9,10}	Low	{1,1,3,4}
		Mod	{3,4,5,6}
		High	{6,7,10,10}

Table 3. Numerical and linguistic values of the parameter ATT

Nominal values tange	Categoric value	Linguistic value	Fuzzy value ranges
{0%,100%}	{0,40}	Low	{0.0.25,40}
	{41,75}	Medium	{25,40,60,75}
	{76,100}	High	{60,75,100,100}

Table 4. Numerical and linguistic values of the parameter ACC

Nominal value	Categoric value	Linguistic value
Family	Fam	Fam
Dorm	Dorm	Dorm
Rent	Rent	Rent

Table 5. Numerical and linguistic values of the parameter PREF

Nominal values range	Categoric values	Linguistic value
{1,10}	{1,3}	High
	{4,6}	Medium
	{7,10}	Low

According to the values of the input parameters the model predicts academic performance of students. If a student's performance value is below a certain value (in our applications, we take this value to 2.5), this student will be classified as risky student in the academic sense. Otherwise, the student is classified as not risky.

The results of the application of the decision tree method revealed that the students' attendance to the course was the most important factor. Students with high attendance need to be carefully monitored in cases where their university entrance scores are low, and their satisfaction levels are low. In other cases, where the student's attendance is high, the probability of the student falling into the risk group is low.

In cases where the student's attendance to the course is low, whether he falls into the risk group or not will be predicted largely depending on the university entrance score and the level of satisfaction with the university environment.

Students staying in dormitories and with their families are slightly less likely to fail than students staying in rented accommodation.

134 instants were used in the prediction model performed on the WEKA application [15]. The learning set used to learn the model contains 100 records. The data of 34 students were used to test the model. The number of correctly classified instants was 105 (81,34%).

In the Fig. 2 the results of the prediction model are described as decision tree.

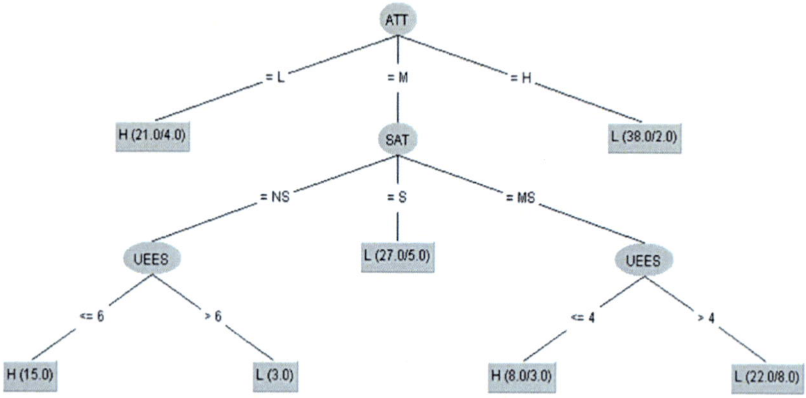

Fig. 2. Decision tree taken with J48 classification method (print screen obtained from WEKA application)

As it can be seen from the decision tree, the important factor in classification is the student's attendance to the lesson. It has been predicted that students with low attendance will mostly fail. On the contrary, students with high persistence constitute the majority among successful students. If the university entrance grades of the students who do not express satisfaction with the education and training environment are low, it is predicted that these students will be unsuccessful. It has been estimated that students living in dormitories and living with their families are less likely to fail than students living in rented accommodation. The reason why university entrance scores are not very effective can be related to the fact that the entrance scores of the students who have chosen the same department are usually very close and the courses are suitable for the average statistical student's knowledge level.

4 Fuzzy Inference Rules

In the last stage of our research, inference rules for fuzzy values were created based on production rules formed on categorical values with the decision tree method. For this purpose, the FisPro [16] application was used.

The left side of the inference rules consists of the level of satisfaction of the student with the educational environment, the level of attendance of the student to the course and the score of the university entrance exam. The right side of the rules is the determinant of the academic success of the student. The fuzzy variables are expressed in trapezoidal form, the extreme values of which are given in Table 1, Table 2 and Table 3. Figure 3 shows the trapezoidal expression of the fuzzy values (low, medium, high) of fuzzy variable ATT as an example.

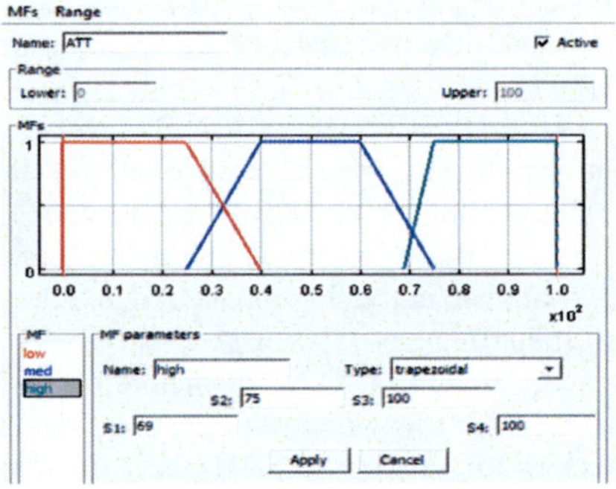

Fig. 3. Trapezoidal expression of the fuzzy variable ATT (attendance)

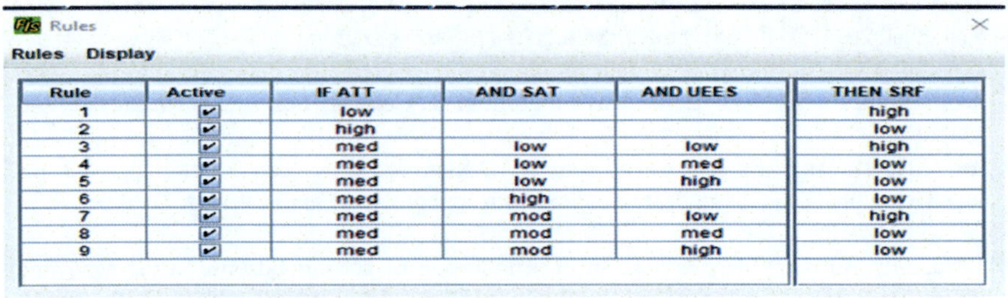

Rules

Rules Display

Rule	Active	IF ATT	AND SAT	AND UEES	THEN SRF
1	✓	low			high
2	✓	high			low
3	✓	med	low	low	high
4	✓	med	low	med	low
5	✓	med	low	high	low
6	✓	med	high		low
7	✓	med	mod	low	high
8	✓	med	mod	med	low
9	✓	med	mod	high	low

Fig. 4. Fuzzy inference rules (The screenshot)

In the given values of the input parameters of the model, the value of the output parameter (the value of the membership function of the student's belonging to the successful class) and the rules acquiring according to these values are shown in Fig. 4.

This research was carried out within the specified limits. The findings of the study should be evaluated with these limitations in mind.

5 Conclusion

The aim of this study is to predict the academic achievement of university freshmen in terms of various parameters. Students' attendance, satisfaction with the educational environment, university entrance scores, type of residence, department preference were taken as the input parameters of the prediction model. The estimation model was implemented using the J48 classification algorithm in the WEKA application. Based on the results of the prediction model, fuzzy inference rules were created and fuzzy classification of students as successful/unsuccessful was carried out in the FisPro application. The results of the study showed that the students' attendance to the course was the most important factor on their academic achievements.

References

1. Altaher, A., Barukab, O.: Prediction of student's academic performance based on adaptive neuro-fuzzy inference. Int. J. Comput. Sci. Netw. **17**, 165–169 (2018)
2. Mushtaq, I., Khan, S.N.: Factors affecting students' academic performance. Glob. J. Manag. Bus. Res. **12**(9), 17–22 (2012)
3. Alkhasawneh, R., Hargraves, R.H.: Developing a hybrid model to predict student first year retention in STEM disciplines using machine learning techniques. Int. J. STEM Educ. Innov. Res. **15**(3), 35–42 (2014). https://doi.org/10.1080/07924360120043621
4. McKenzie, K., Schweitzer, R.: Who succeeds at university? Factors predicting academic performance in first year Australian university students. High. Educ. Res. Dev. **20**(1), 21–33 (2010). https://doi.org/10.1080/07924360120043621
5. DeBerard, M.S., Spielmans, G.I., Julka, D.L.: Predictors of academic achievement and retention among college freshmen: a longitudinal study. Coll. Stud. J. **38**(1), 66–80 (2004)
6. Kuh, G.D., Kinzie, J., Buckley J.A., Bridges, B.K., Hayek, J.C.: What matters to student success: a review of the literature. Commissioned Report for the National Symposium on Postsecondary Student Success: Spearheading a Dialog on Student Success. National Postsecondary Education Cooperative (2006)
7. York, T.T., Gibson, C.W., Rankin, S.: Defining and measuring academic success. Pract. Assess. Res. Eval. **20**, 1–20 (2015). https://doi.org/10.7275/hz5x-tx03
8. Ogunde, A.O., Ajibade, D.A.: A data mining system for predicting university students' graduation grades using ID3 decision tree algorithm. J. Comput. Sci. Inf. Tech. **2**(1), 21–46 (2014)
9. Yu, C.H., DiGangi, S., Jannasch- Pennell, A., Kaprolet, C.: A data mining approach for identifying predictors of student retention from sophomore to junior year. J. Data Sci. **8**(2), 307–325 (2010)
10. Iqbal, M., Morrissey, H., Patrick, B.: The influence of accommodation on the academic performance of university students. Int. J. Current Res. **12**(04), 10900–10907 (2020). https://doi.org/10.24941/ijcr.38334.04.2020
11. Caviglia-Harris, J.L.: Attendance Rates and Academic Achievement: Do Attendance Policies and Class Size Effects Impact Student Performance? (2004). https://doi.org/10.2139/ssrn.605462
12. Gezgin, D.M.: Evaluation of computer education and instructional technologies students' selection causes of their department: the case of Trakya University. Trakya Univ. J. Soc. Sci. **17**(2), 297–311 (2015)
13. Sevim, N., Kaplan, O., Kaplan, G.: Perceptions of pre-service ICT teachers about their departments: the case of CEIT department at METU. Ahi Evran Univ. J. Kirsehir Education Faculty (KEFAD) **17**(1), 439–458 (2016)
14. Uysal, H.: Determining the factors that affect academic achievement of CEIT students and developing the academic achievement predictive model. Unpublished master's thesis, Canakkale Onsekiz Mart University, Institute of Education Sciences (2015)
15. WEKA. https://www.cs.waikato.ac.nz/ml/weka
16. FisPro: An open source portable software for fuzzy inference systems FisPro. https://www.fispro.org/documentation

Monthly Prediction of Reference Evapotranspiration in Northcentral Nigeria Using Artificial Intelligence Tools: A Comparative Study

Jazuli Abdullahi[1(✉)] and Gozen Elkiran[2]

[1] Department of Civil Engineering, Faculty of Engineering, Baze University, Abuja, Nigeria
jazuli.abdullahi@bazeuniversity.edu.ng
[2] Department of Civil Engineering, Faculty of Civil and Environmental Engineering, Near East University, Nicosia, Mersin 10, Turkey
gozen.elkiran@neu.edu.tr

Abstract. The aim of this study was to assess the potential of two artificial intelligence models including artificial neural network (ANN) and support vector regression (SVR) for the prediction of reference evapotranspiration (ET_0) in the northcentral stations of Ilorin and Minna, Nigeria. A conventional multiple linear regression (MLR) model was also developed for comparison. For this, several meteorological variables spanned for 34 years (1983–2016) duration including maximum, minimum and mean temperatures (T_{max}, T_{min} and T_{mean}), relative humidity (R_H) and wind speed (U_2) were obtained and used as input variables. Different 7 input combinations were utilized to come up with the best ET_0 prediction. The Nash-Sutcliffe efficiency (NSE) and root mean square error (RMSE) were applied as performance criteria of the developed models. The results showed that for AI based models, input combination that include U_2 such as M6 (T_{max}, R_H, U_2) mostly provided the best performance, which signifies the importance of U_2 on ET_0 while temperature related inputs (T_{max}, T_{min} and T_{mean}) were more efficient for MLR models. The general results depicted that all the applied models can be employed to achieve successful result but ANN provided the best performance with NSE values of 0.974 and 0.975 for Ilorin and Minna stations, respectively.

Keywords: Reference evapotranspiration · Nigeria · Relative humidity · Performance criteria · Temperature · Artificial neural network · Ilorin · Minna

1 Introduction

As through evapotranspiration (ET), more than 60% of global precipitation is returned to the atmosphere, ET is regarded as the main component of the hydrologic water cycle [1]. For direct ET measurement, the most accurate method is by the use of lysimeters, but the method is time consuming and costly. However, in view of the difficulty faced

R. A. Aliev et al. (Eds.): ICSCCW 2021, LNNS 362, pp. 165–172, 2022.
https://doi.org/10.1007/978-3-030-92127-9_25

in predicting ET for individual plants, generally, reference evapotranspiration (ET_0) is estimated initially and then using a crop coefficient the desired ET is obtained.

FAO 56 Penman Monteith given as FAO-56-PM is widely accepted and used as the reference method of estimating ET0 as recommended by the Food and Agricultural Organizations of the United Nations (FAO) [2]. Nonlinear techniques such as AI were developed by researchers to take care of the nonlinear and complex ET_0 phenomenon. The authors of [3] employed ANN, SVR, MLR and other AI and empirical models for ET_0 modeling in several climatic regions. The application of AI techniques in ET_0 modeling is limited in Nigeria especially the northcentral Nigeria that includes Kwara and Niger states. The global acceptability of AI techniques and their successful applications in ET_0 modeling make it necessary to employ such approaches for effective ET_0 modeling due to the climate change impact on hydro climatological variables. To fill this gap, two AI based models including ANN and SVR were employed in this study to estimate ET_0 and compare with linear conventional model (MLR) in northcentral stations of Ilorin and Minna, Nigeria.

2 Materials and Methods

Study Area. Ilorin is the capital of Niger state which has an area of 765 km^2, 25 °C average temperature, 79% humidity and 14 km/h wind. Minna is the capital city of Niger state which is the largest state in Nigeria in terms of land area of 76,363 km^2. The climate of the station include temperature average of 30 °C, 61% humidity and wind blows at 11 km/h. Figure 1 shows the study locations.

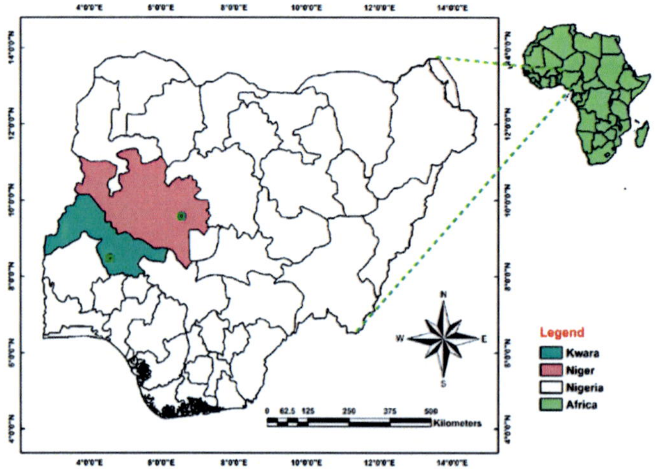

Fig. 1. Location of Nigeria and the study area

Study Data. In this study, a total of 408 number of monthly climate data samples including minimum and maximum temperatures (T_{min}, T_{max}) (°C), relative humidity (R_H) (%), and wind speed (U_2) (m/s) from 1983 to 2016 (34 years) were used in this study for the ET_0 modeling. The data used in this study were obtained from https://power.larc.nasa.gov/.

Performance Evaluation of the Models. To determine the performance of the models, Nash-Sutcliffe efficiency criteria (NSE) or determination coefficient and root mean square error (RMSE) were used. Details can be found from [4] study.

Proposed Methodology. The proposed methodology in this study is presented in Fig. 2.

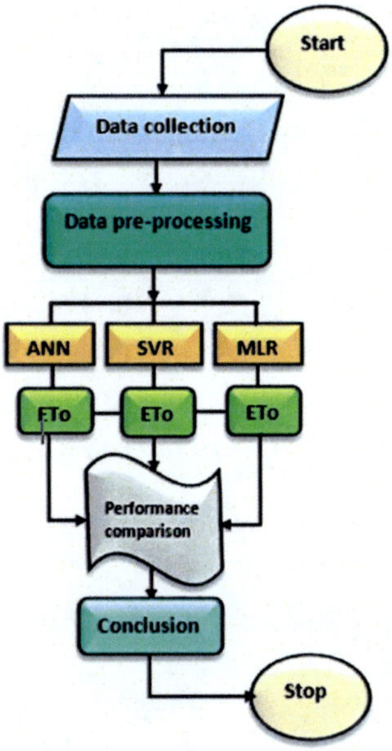

Fig. 2. Schematic representation of the proposed methodology

Artificial Neural Network constitutes a number of simple processing elements that are interconnected by nodes or neurons with fascination characteristics of information processing including parallelism, nonlinearity, generalization, capability, learning and noise tolerance. Detail information regarding the application of ANN can be found in [5] study.

Support vector regression developed on the basis of SVM concept is used for non-linear regression issues. [6] gave detail information about SVR and its application.

Multiple-Linear Regression. Details regarding MLR can be found in [7] study.

FAO-56-PM Method. The concept of ET_0 is fully discussed by [2] study. Details on how to compute the ET_0 by FAO-56-PM method can be found in [2] study.

3 Results

Results of the Applied Models. The results of the 3 applied models are presented in Tables 1, 2 and 3.

Table 1. Results of ANN models

Station	Model	Combination	Epoch	Training		Testing	
				R^2	RMSE	R^2	RMSE
Ilorin	M1	T_{min}, T_{max}, R_H	97	0.976	0.037	0.942	0.042
	M2	T_{min}, T_{max}, U_2	140	0.957	0.036	0.941	0.057
	M3	T_{min}, T_{max}, T_{mean}	98	0.972	0.047	0.916	0.058
	M4	T_{min}, R_H, U_2	100	**0.988**	**0.03**	0.967	**0.036**
	M5	T_{min}, R_H, T_{mean}	94	0.972	0.043	0.952	0.047
	M6	T_{max}, R_H, U_2	109	0.974	0.032	**0.974**	0.046
	M7	T_{max}, R_H, T_{mean}	126	0.985	0.034	0.959	0.04
Minna	M1	T_{min}, T_{max}, R_H	111	0.967	0.043	0.962	0.045
	M2	T_{min}, T_{max}, U_2	110	0.95	0.053	0.974	0.043
	M3	T_{min}, T_{max}, T_{mean}	125	0.948	0.063	0.886	0.081
	M4	T_{min}, R_H, U_2	124	0.978	0.036	0.973	0.046
	M5	T_{min}, R_H, T_{mean}	113	0.969	0.042	0.959	0.056
	M6	T_{max}, R_H, U_2	91	**0.978**	**0.035**	**0.975**	**0.037**
	M7	T_{max}, R_H, T_{mean}	168	0.976	0.043	0.967	0.044

Table 2. Results of SVR models

Station	Model	Combination	Training		Testing	
			R2	RMSE	R2	RMSE
Ilorin	M1	T_{min}, T_{max}, R_H	0.944	0.056	0.921	0.066
	M2	T_{min}, T_{max}, U_2	0.948	0.045	0.836	0.114
	M3	T_{min}, T_{max}, T_{mean}	**0.951**	0.062	0.894	0.065
	M4	T_{min}, R_H, U_2	0.927	0.054	0.923	**0.077**
	M5	T_{min}, R_H, T_{mean}	0.927	0.054	**0.924**	0.078
	M6	T_{max}, R_H, U_2	0.955	**0.042**	0.888	0.094
	M7	T_{max}, R_H, T_{mean}	0.917	0.057	0.917	0.081
Minna	M1	T_{min}, T_{max}, R_H	**0.956**	0.058	0.916	0.069
	M2	T_{min}, T_{max}, U_2	0.952	0.053	0.86	0.104
	M3	T_{min}, T_{max}, T_{mean}	0.873	0.092	0.851	0.099
	M4	T_{min}, R_H, U_2	0.947	0.064	**0.941**	**0.058**
	M5	T_{min}, R_H, T_{mean}	0.954	0.06	0.916	0.069
	M6	T_{max}, R_H, U_2	**0.956**	**0.05**	0.896	0.089
	M7	T_{max}, R_H, T_{mean}	0.953	0.06	0.911	0.071

Table 3. Results of MLR models

Station	Model	Combination	Training		Testing	
			R^2	RMSE	R^2	RMSE
Ilorin	M1	T_{min}, T_{max}, R_H	0.953	**0.052**	0.886	0.058
	M2	T_{min}, T_{max}, U_2	0.940	0.058	0.921	0.048
	M3	T_{min}, T_{max}, T_{mean}	0.946	0.065	0.888	0.066
	M4	T_{min}, R_H, U_2	0.936	0.071	0.913	0.059
	M5	T_{min}, R_H, T_{mean}	0.950	0.063	0.91	0.059
	M6	T_{max}, R_H, U_2	0.952	0.061	**0.934**	**0.051**
	M7	T_{max}, R_H, T_{mean}	**0.954**	0.060	0.909	0.060
Minna	M1	T_{min}, T_{max}, R_H	0.954	**0.051**	0.904	0.071
	M2	T_{min}, T_{max}, U_2	0.944	0.056	**0.931**	**0.060**
	M3	T_{min}, T_{max}, T_{mean}	0.913	0.082	0.842	0.095
	M4	T_{min}, R_H, U_2	0.952	0.061	0.918	0.069
	M5	T_{min}, R_H, T_{mean}	0.960	0.055	0.909	0.072
	M6	T_{max}, R_H, U_2	0.950	0.062	**0.931**	0.063
	M7	T_{max}, R_H, T_{mean}	**0.961**	0.054	0.907	0.073

3.1 Discussion

As seen in Table 1, the 7 models developed by ANN approach all produced satisfactory performance in both Ilorin and Minna stations with NSE ranges from 0.957 to 0.988 in the training phase and 0.916 to 0.974 in the validations phase of Ilorin station. For Minna station, the NSE ranges between 0.948 to 0.978 and 0.886 to 0.975 in the training and testing phases, respectively. However, as seen in the obtained results of Table 1, different epoch number used yielded different ANN performance. This is an indication that epoch number plays an important role in the development of ANN models, and higher or lower number of epoch does not signify better performance, trial and error process should be used to determine the best suited epoch number. For example, in the ANN results, the best model in the training phase is the one with 100 epoch (M4, NSE 0.988), whereas the least performance model is the one with 140 epoch (M2, NSE 0.957) for Ilorin station. These results demonstrate that despite the uncertainty and stochastic nature of ET_0, application of ANN could be employed successfully in the prediction of ET_0 in northcentral Nigeria.

The results of SVR models are presented in Table 2. Similar to ANN, the models developed by SVR approach depicted high performance mostly above NSE 0.9 in both Ilorin and Minna stations. This demonstrated the capability of AI techniques in dealing with uncertain phenomenon of ET_0. It can be seen from the result that different inputs combination led to different performance. For Ilorin station in the training phase in terms of NSE, M3 which has T_{min}, T_{max} and T_{mean} produced the best performance while M4 which has combination of T_{min}, R_H and U_2 provided the best result for NSE and RMSE performance criteria in the testing phase. For Minna station, a joint performance of 0.956 can be observed for M1 and M6 in the training phase while, M4 is found to be more skillful in comparison to others. By visual observation of these results, it is obvious that models which include U_2 as input mostly led to better performance. This may be because U_2 is more effective in the prediction of ET_0 than other models. Also, the models developed showed better accuracy in Minna station than Ilorin station which is as a result of higher U_2 Minna has than Ilorin. This suggestion is in agreement with conclusion drawn by [3] in which they stated that U_2 as a sole input may not have dominant performance in the prediction of ET_0 but will significantly lead to high performance when used in combination with other meteorological variables.

The results of MLR models is presented in Table 3. The results show that ET_0 as any other process, contains both linear and nonlinear aspects. MLR models with their inability to deal with nonlinear ET_0 phenomenon produced good results despite achieving less compared to AI techniques. It is observed that models that include the combination of either T_{min}, T_{max} or T_{mean} mostly produced better MLR results than other input variables and the results for Minna station is superior to the results obtained for Ilorin station. This is because T_{min}, T_{max} or T_{mean} are linearly more dominant variables to ET_0 modeling and have higher values in Minna than Ilorin. In other words, the temperature variables are directly proportional to ET_0, as the temperature on the earth surface increases, the ET_0 rate increases. Due to this, it can be seen in the testing phase of Minna station that M7 and M6 have the best performance in terms of NSE in both training and testing phases, respectively.

From the results obtained in Tables 1, 2 and 3, it can be deduced that, ANN, SVR and MLR with their unique capabilities, can be employed to predict ET_0 successfully in Ilorin and Minna stations of northcentral Nigeria. To compare the performance of one model over another and come out with the most suitable technique for the ET_0 prediction, Fig. 3 is presented for the best model from each technique with respect to NSE in both Ilorin and Minna stations in the testing phase.

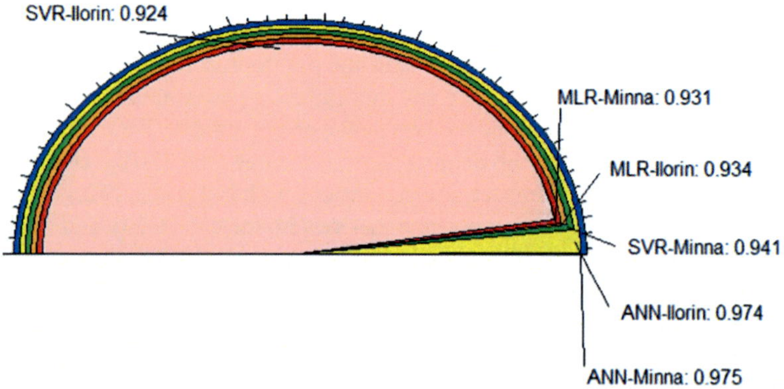

Fig. 3. Performance comparison of best models based on NSE in the testing phase

As seen in Fig. 3, each color is representing a given model and every color starts from left to right. The larger the portion occupied by the color, the better performance of the model. Also the higher the NSE the lower the position of the model from right. From the Fig. 3, it can be seen that ANN models for both Ilorin and Minna stations produced superior performance whereas all the applied models have NSE values above 0.9 which signifies a satisfactory ET_0 modeling. It can be understood that MLR model slightly performed better than SVR model for Ilorin station which could be because of the more linear behavior of the inputs owing to direct relationship they have with the ET_0. Apart from NSE, several other criteria could be used to determine the discrepancies of the approaches applied in terms of their performances, including correlation coefficient (R), standard deviation (SD) and RMSE. All these criteria are composed in Taylor diagram. Figures 4 shows a comparison of best models for the ET_0 prediction by Taylor diagrams.

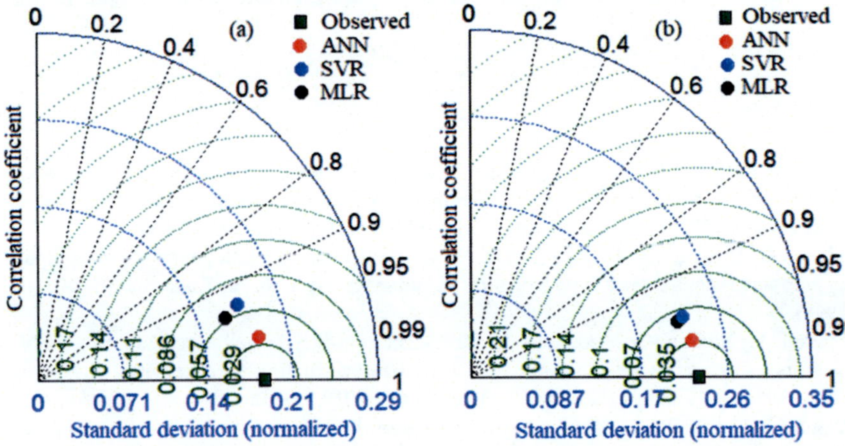

Fig. 4. Comparison of the best models for ET0 prediction in (a) Ilorin station and (b) Minna station

The Taylor diagrams (Figs. 4) displayed the performances of the best models from each of the applied techniques. Taylor diagram determines the performances of models based on 3 criteria, SD, R and RMSE. The SD is located in the x-axis of the graph with model value more close to the SD value of the observed ET_0 as the best performance, the curve extending from vertical (x) axis to horizontal (y) axis is R, the higher the R value, the better performance and the concave semi-circle (internal curve) represents RMSE, the best model is that which has less/lower RMSE. Based on the above conditions, it can be clarified that among all the applied models, ANN values are more close to the observed value, have higher R and are the values with lowest RMSE in both Ilorin and Minna stations and therefore, ANN is the most efficient technique in the prediction of ET_0 in northcentral Nigeria.

4 Conclusion

In this study, the potential of AI based approaches including ANN, SVR and a conventional MLR technique were assessed for the prediction of ET_0 in 2 northcentral stations of Ilorin and Minna located in Kwara and Niger states of Nigeria. To determine the best prediction results, 7 different models were developed from 7 distinct input combinations using each of the applied prediction techniques. The results revealed that the proposed input combinations can lead to successful prediction of ET_0 in the study stations but due to location, linear and nonlinear aspects of ET_0, some input combinations provided better results for a particular station and technique than others. The general results demonstrated that all the applied models can provide efficient ET_0 prediction but ANN model, performed better than SVR and MLR models in both Ilorin and Minna stations of northcentral Nigeria. It can be suggested that further studies should consider other AI models such as genetic programming (GP) and compare with the results of this study.

References

1. Mohammadi, B., Mehdizadeh, S.: Modeling daily reference evapotranspiration via a novel approach based on support vector regression coupled with whale optimization algorithm. Agric. Water Manag. **237**, 106145 (2020). https://doi.org/10.1016/j.agwat.2020.106145
2. Allen, R.G., Pereira, L.S., Raes, D., Smith, M.: Crop evapotranspiration-guidelines for computing crop water requirements-FAO Irrigation and drainage paper 56. Fao Rome **300**(9), D05109 (1998)
3. Nourani, V., Elkiran, G., Abdullahi, J.: Multi-station artificial intelligence based ensemble modeling of reference evapotranspiration using pan evaporation measurements. J. Hydrol. **577**, 123958 (2019). https://doi.org/10.1016/j.jhydrol.2019.123958
4. Abdullahi, J., Elkiran, G., Nourani, V.: Application of artificial neural network to predict reference evapotranspiration in Famagusta, North Cyprus. In: 11th International Scientific Conference on Production Engineering Development and Modernization of Production (2017)
5. Nourani, V., Elkiran, G., Abdullahi, J., Tahsin, A.: Multi-region modeling of daily global solar radiation with artificial intelligence ensemble. Nat. Resour. Res. **28**(4), 1217–1238 (2019). https://doi.org/10.1007/s11053-018-09450-9
6. Nourani, V., Elkiran, G., Abdullahi, J.: Multi-step ahead modeling of reference evapotranspiration using a multi-model approach. J. Hydrol. **581**, 124434 (2020). https://doi.org/10.1016/j.jhydrol.2019.124434
7. Abdullahi, J., Elkiran, G., Nourani, V.: Artificial intelligence based and linear conventional techniques for reference evapotranspiration modeling. In: Aliev, R.A., Kacprzyk, J., Pedrycz, W., Jamshidi, Mo., Babanli, M.B., Sadikoglu, F.M. (eds.) ICSCCW 2019. AISC, vol. 1095, pp. 197–204. Springer, Cham (2020). https://doi.org/10.1007/978-3-030-35249-3_25

Estimation of Benchmarking Influence in Buyer's Decision-Making Process by Using Fuzzy AHP

Khatira J. Dovlatova(✉)

Joint MBA Program, Azerbaijan State Oil and Industry University, 20 Azadlig Avenue, Baku AZ1010, Azerbaijan
khatira.dovlatova@asoiu.edu.az

Abstract. In most cases benchmarking approaches give the contrast between financial and operating metrics to those of the industry leader. Analyzing marketplace financial and organizational situations may not be sustainable. The aim of this paper is to define benchmarking influence in buyer's decision-making process by using fuzzy AHP. Benchmarking would provide which businesses are constantly refining and advancing their standards to satisfy the demands of advanced and competitive consumers. Benchmarking is the process which does not provide a stable path and strategies for learning from consumers when initiating value-added suggestions that meet their expectations. At the same time, it also helps to maintain a company's success and productivity. In the benchmarking situation the analysis of the given suggestions which meet consumers' expectations will support the long-term well-being and productivity of a business.

In the given problem for benchmarking process, factors which influence consumer buying behavior are analyzed. Decision making problem is characterized by using fuzzy AHP method for analyzing the best alternative which influences consumer buying behavior.

Keywords: Fuzzy AHP · Decision making · Eigenvalue · Eigenvector · Consumer behavior · Benchmarking

1 Introduction

Benchmarking is a method which is focused on consumer values in the center of process as well as the company's approach. Benchmarking main topics are success evaluation, opposition, clarifying best practices and ways. This paper is focused on psychological, social, cultural, economic and personal variables which have an influence on consumers. Different variables, particulars and features form an individual's identity as the customer's decision-making process, purchasing intention, shopping experience and other items influence the benchmarking process. Consumer behavior is impacted by different types of factors and they are classified as 5 major factors which influence consumer behavior [1]. Psychological factors are a major variable of consumer behavior, and these factors are difficult to identify and at the same time have powerful influence on buying behavior [2]. Social factors, individuals are initiated by other humans and want to

R. A. Aliev et al. (Eds.): ICSCCW 2021, LNNS 362, pp. 173–182, 2022.
https://doi.org/10.1007/978-3-030-92127-9_26

accept in social environment. Social factors give the shape to humans because they are social beings. Cultural factors influencing consumer behavior are formed on the basis of social classes, subcultures and the behavior of a particular society. Personal factors are special items, situations which influence consumer buying behavior. Personal factors differ from person to person and create different perceptions and consumer behavior. Economic factors, the consumer purchasing intention and decisions mostly depend on the business cycle of economy and influence market processes [3].

In some researches the benchmarking of the given products is analyzed according to the factors which influence consumer buying behavior, but in this paper, we research factors which affect consumer buying behavior in fuzzy environment by using AHP method. The fuzzy AHP was offered by [4]. Buckley was the first researcher who analyzed fuzzy eigenvalues [5, 6] for input –output of systems. Massa et al. obtained both eigenvalues problem was transformed to a common one to analyze of fuzzy data.

2 Preliminaries

Definition 1. The fuzzy The fuzzy pairwise comparison judgments [7]:

$$\tilde{c}_{ij} = (c_{ij,l}, c_{ij,m}, c_{ij,u}), \qquad \tilde{c}_{ij} = (1,1,1)\ \tilde{c}_{ij,l} > 0 \quad i,j = 1,2.......,k \quad i \neq j. \tag{1}$$

That comprises a fuzzy comparison matrix $\tilde{C}$ with the following elements,

$$\tilde{c}_{ij} = \frac{1}{\tilde{c}_{ij}} = \left(\frac{1}{c_{ij,u}}, \frac{1}{c_{ij,m}}, \frac{1}{c_{ij,l}}\right), \qquad i,j = 1,2,...k, \quad i \neq j. \tag{2}$$

This matrix is consistent if and only if

$$\tilde{c}_{ip} \approx \tilde{c}_{ip} \otimes \tilde{c}_{ip}, \qquad i,j = 1,2,...k, \tag{3}$$

Definition 2. Determination of fuzzy weights [7]:

$$\tilde{C} \otimes \tilde{w} = \tilde{\lambda}_l \otimes \tilde{w}, \tag{4}$$

where $\tilde{C}$, fuzzy reciprocal comparison matrix of type [k × k]. Elements of the fuzzy matrix $\tilde{C}$, fuzzy vector $\tilde{w}$ and fuzzy eigenvalue $\tilde{\lambda}$ are supposed as triangular fuzzy numbers that may be indicated as stated by $\tilde{w} = (w_l, w_m, w_u)$, $\tilde{\lambda} = (\lambda_l, \lambda_m, \lambda_u)$.

Definition 3. Consistency index and consistency ratio [8]:
Solve the fuzzy eigenvalue problem $\tilde{C} \otimes \tilde{w} = \tilde{\lambda}_l \otimes \tilde{w}$, as depicted in the previous part and define the principal fuzzy eigenvalue $\tilde{\lambda} = (\lambda_l, \lambda_m, \lambda_u)$ and corresponding fuzzy eigenvectors $\tilde{w} = (w_l, w_m, w_u)$. For the matrix C_l,C_m,C_u compute the CI and CR according to [9–13].

3 Statement of the Problem

The factors influencing consumer buying behavior are analyzed as criteria which play main role in benchmarking process. In fuzzy AHP process the pair-wise comparison matrix of both criteria and alternatives are exhibited by linguistics factors that are represented by triangular fuzzy numbers [14, 15]. MCDM problem involves 5 criteria C_1, C_2, C_3, C_4, C_5 and 5 alternatives A_1, A_2, A_3, A_4, A_5. C_1- Psychological Factors; C_2- Social Factors; C_3- Cultural factors; C_4- Personal Factors; C_5- Economic Factors. Table 1. cause to know comparison of all judgments under efficiency into triangular fuzzy numbers and final process to calculate consistency ratio of this shown matrix.

Table 1. Fuzzy matrix of the criteria

Criteria	C_1	C_2	C_3	C_4	C_5
C_1	(1, 1, 1)	(1/3, 1/2, 1/1)	(1/0.4, 1/0.3, 1/0.2)	(1, 1, 1)	(1/4, 1/3, 1/2)
C_2	(1, 2, 3)	(1, 1, 1)	(0.55, 0.56, 0.57)	(1, 2, 3)	(0.5, 0.6, 0.7)
C_3	(0.2, 0.3, 0.4)	(0.1, 0.2, 0.3)	(1, 1, 1)	(1, 2, 3)	(0.2, 0.3, 0.4)
C_4	(1, 1, 1)	(1/3, 1/2, 1/1)	(1/3, 1/2, 1/1)	(1, 1, 1)	(1/1, 1/0.9, 1/0.8)
C_5	(2, 3, 4)	(1/0.7, 1/0.6, 1/0.5)	(1/0.4, 1/0.3, 1/0.2)	(0.8, 0.9, 1)	(1, 1, 1)

4 Solution of the Problem

In this paper we investigate the best alternative to take into consideration the given criteria. Fuzzy pairwise comparison matrix of the criteria is shown in Table 1.

Step 1. First step formulate the pair-wise comparison reciprocal fuzzy matrix $\tilde{C}$ for the criteria $C_1, C_2, \ldots, C_n$ by evaluating the priority values as fuzzy numbers $\tilde{c}_{ij} = (c_{ij,l}, c_{ij,m}, c_{ij,u}), (i, j = 1, 2, \ldots, n)$. Shows the fuzzy matrix $\tilde{C}$ divided by three crisp matrices $C_l,\ C_m,\ C_u$ are obtained according to [7] as follows:

$$C_l = \begin{bmatrix} 1 & c_{12,l} & \cdots & c_{1k,l} \\ & & \cdots & \\ 1/c_{12,u} & 1 & \cdots & c_{2k,l} \\ 1/c_{1k,u} & 1/c_{2k,u} & \cdots & 1 \end{bmatrix} C_m = \begin{bmatrix} 1 & c_{12,m} & \cdots & c_{1k,m} \\ & & \cdots & \\ 1/c_{12,m} & 1 & \cdots & c_{2k,m} \\ 1/c_{1k,m} & 1/c_{2k,m} & \cdots & 1 \end{bmatrix} C_u = \begin{bmatrix} 1 & c_{12,u} & \cdots & c_{1k,u} \\ 1/c_{12,l} & 1 & \cdots & c_{2k,u} \\ & & \cdots & \\ 1/c_{1k,l} & 1/c_{2k,l} & \cdots & 1 \end{bmatrix} \quad (5)$$

Step 2. Obtained matrices are used for calculation of a system of fuzzy linear homogeneous equations.

$$\overline{C_l}w_l + \overline{C_m}w_m + \overline{C_u}w_u - \overline{\lambda_l}w_l - \overline{\lambda_m}w_m - \overline{\lambda_u}w_u = 0 \quad (6)$$

Where are:

$$\overline{C_l} = 2C_l + C_m = 3 \quad \overline{C_m} = C_l + 4C_m + C_u = 6 \quad \overline{C_u} = C_m + 2C_u = 3 \quad (7)$$

By using (5) we obtained three crisp matrices (Tables 2, 3 and 4):

Table 2. C_l and $\overline{C_l}$.

C_l				
1	0.33	2.5	1	0.25
1	1	0.55	1	0.5
0.2	0.1	1	1	0.2
1	0.33	0.33	1	1
2	1.43	2.5	0.8	1

$\overline{C_l}$				
3	1.17	8.33	3	0.83
4	3	1.66	4	1.6
0.7	0.4	3	4	0.7
3	1.17	1.17	3	3.11
7	4.52	8.33	2.5	3

Table 3. C_l and $\overline{C_l}$.

C_m				
1	0.5	3.33	1	0.33
2	1	0.56	2	0.6
0.3	0.2	1	2	0.3
1	0.5	0.5	1	1.11
3	1.67	3.33	0.9	1

$\overline{C_m}$				
6	3.33	20.83	6	2.08
12	6	3.36	12	3.6
1.8	1.2	6	12	1.8
6	3.33	3.33	6	6.69
18	10.1	25.83	5.4	6

Table 4. C_u and $\overline{C_u}$.

C_u				
1	1	5	1	0.5
3	1	0.57	3	0.7
0.4	0.3	1	3	0.4
1	1	1	1	1.25
4	2	5	1	1

$\overline{C_u}$				
3	2.5	13.33	3	1.33
8	3	1.7	8	2
1.1	0.8	3	8	1.1
3	2.5	2.5	3	3.61
11	5.67	23.33	2.9	3

Step 3. Eigenvalues of $\overline{C_l}$, $\overline{C_m}$ and $\overline{C_u}$ are obtained with MATLAB software program: $\overline{\lambda_l} = 14.57$; $\overline{\lambda_m} = 33.08$; $\overline{\lambda_u} = 20.18$.

$\lambda_l, \lambda_m, \lambda_u$ are founded as by using following process:

$$\begin{cases} \overline{\lambda_l} = 2\lambda_l + \lambda_m \\ \overline{\lambda_m} = \lambda_l + 4\lambda_m \\ \overline{\lambda_u} = \lambda_m + 2\lambda_u \end{cases} \Leftrightarrow \begin{cases} 14.57 = 2\lambda_l + \lambda_m \\ 33.08 = \lambda_l + 4\lambda_m + \lambda_u \Leftrightarrow \times(-2) \\ 20.18 = \lambda_m + 2\lambda_u \end{cases} \Leftrightarrow \begin{cases} -66.16 = -2\lambda_l - 8\lambda_m - 2\lambda_u \\ 20.18 = \lambda_m + 2\lambda_u \end{cases} \tag{8}$$

$$\Leftrightarrow \begin{cases} -45.98 = -2\lambda_l - 7\lambda_m \\ 14.57 = 2\lambda_l + \lambda_m \end{cases} \Leftrightarrow \lambda_l = 4.68,\ \ \lambda_m = 5.2,\ \ \lambda_u = 7.49.$$

An urgent consideration in terms of the quality of the eventual decision relates to the consistency of opinions that the decision maker displayed during the series of pairwise comparison.

Step 4. Eigenvectors also, is obtained by homogenous fuzzy linear Eq. (4):

Eigenvectors of $\overline{C_l}$, $\overline{C_m}$ and $\overline{C_u}$ are obtained with MATLAB software program:

$$\overline{w_l} = \frac{w_l \lambda_l}{s_l \lambda_m},\ \overline{w_m} = \frac{w_m}{s_m},\ \overline{w_u} = \frac{w_u \lambda_u}{s_u \lambda_m};\ where\ s_l = \sum_{i=1}^{n} w_{i,l},\ s_m = \sum_{i=1}^{n} w_{i,m},\ s_u = \sum_{i=1}^{n} w_{i,u}. \quad (9)$$

$$w_l = [0.37, 0.41, 0.23, 0.46, 0.65]\ \ S_l = 2.13$$

$$w_m = [-0.37, -0.44, -0.24, -0.34, -0.71]\ \ S_m = -2.09\ for\ example,\quad w_{l1} = \frac{0.37 * 4.68}{2.13 * 5.2} = 0.06$$

$$w_u = [-0.36, -0.42, -0.24, -0.31, -0.74]\ \ S_u = -2.06 \quad (10)$$

Applying the mentioned software program, the following eigenvectors are obtained: $\overline{w_l} = [0.16, 0.17, 0.1, 0.19, 0.28]$, $\overline{w_m} = [0.18, 0.21, 0.12, 0.16, 0.34]$, $\overline{w_u} = [0.25, 0.29, 0.16, 0.22, 0.52]$

Step 5. In accordance to [7] consistency index and consistency ratio for $\tilde{C}$ matrix are obtained by using following formulas [7]:

$$CI = \frac{5.2 - 5}{5 - 1} = 0.05,\ CR = \frac{0.05}{1.12} = 0.045,\quad CR = 0.045 \leq 0.10$$

Applying the proposed calculation, the eigenvalues, CI and CR ratio are defined. CR is 0.045, the suggested matrix $\tilde{C}$ is consistent. Calculation process is regularly repeated for each alternative to take into consideration the given criteria (Tables 5–9).

Table 5. Fuzzy matrix of the C_1 criteria

C1	A_1	A_2	A_3	A_4	A_5
A_1	(1, 1, 1)	(1/1.4, 1/1.3, 1/1.2)	(1/2.3, 1/2.2, 1/2.1)	(1, 1, 1)	(1/3, 1/2, 1/1)
A_2	(1.2, 1.3, 1.4)	(1, 1, 1)	(0.55, 0.56, 0.57)	(1, 2, 3)	(1.1, 1.2, 1.3)
A_3	(2.1, 2.2, 2.3)	(0.1, 0.2, 0.3)	(1, 1, 1)	(3.1, 3.2, 3.3)	(2.5, 2.6, 2.7)
A_4	(1, 1, 1)	(1/3, 1/2, 1/1)	(1/3.3, 1/3.2, 1/3.1)	(1, 1, 1)	(1/4, 1/3, 1/2)
A_5	(1, 2, 3)	(1/1.3, 1/1.2, 1/1.1)	(1/2.7, 1/2.6, 1/2.5)	(2, 3, 4)	(1, 1, 1)

By using (5) we obtained three crisp matrices and the matrices are used for calculation of a system fuzzy linear homogenous equations. Eigenvalues of $\overline{C_l}$, $\overline{C_m}$ and $\overline{C_u}$ are obtained $\lambda_l, \lambda_m, \lambda_u$ are founded as $\Leftrightarrow \lambda_l = 4.33,\ \lambda_m = 4.93,\ \lambda_u = 5.87$.

Applying the MATLAB software program, the following eigenvectors and CR are obtained: $\overline{w_l} = [0.15, 0.17, 0.1, 0.19, 0.27]$, $\overline{w_m} = [0.18, 0.21, 0.12, 0.16, 0.34]$, $\overline{w_u} = [0.21, 0.24, 0.14, 0.18, 0.43]$

$$CI = -0,017\ CR = -0.015$$

Eigenvalues are founded as $\Leftrightarrow \lambda_l = 4.9,\ \lambda_m = 5.49,\ \lambda_u = 6.64$.

Table 6. Fuzzy matrix of the C_2 criteria

C2	A_1	A_2	A_3	A_4	A_5
A_1	(1, 1, 1)	(1/0.5, 1/0.4, 1/0.3)	(1/0.4, 1/0.3, 1/0.2)	(1, 1, 1)	(1/3.6, 1/3.5, 1/3.4)
A_2	(0.3, 0.4, 0.5)	(1, 1, 1)	(0.55, 0.56, 0.57)	(1, 2, 3)	(1.5, 1.6, 1.7)
A_3	(0.2, 0.3, 0.4)	(0.1, 0.2, 0.3)	(1, 1, 1)	(1.2, 1.3, 1.4)	(0.2, 0.3, 0.4)
A_4	(1, 1, 1)	(1/3, 1/2, 1/1)	(1/1.4, 1/1.3, 1/1.2)	(1, 1, 1)	(1/1, 1/0.9, 1/0.8)
A_5	(3.4, 3.5, 3.6)	(1/1.7, 1/1.6, 1/1.5)	(1/0.4, 1/0.3, 1/0.2)	(0.8, 0.9, 1)	(1, 1, 1)

Applying the mentioned software program, the following eigenvectors and CR are obtained: $\overline{w}_l = [0.2, 0.18, 0.08, 0.15, 0.28]$, $\overline{w}_m = [0.24, 0.21, 0.09, 0.16, 0.3]$, $\overline{w}_u = [0.3, 0.25, 0.11, 0.2, 0.36]$

$$CI = 0.12, \quad CR = 0.10$$

Table 7. Fuzzy matrix of the C_3 criteria

C3	A_1	A_2	A_3	A_4	A_5
A_1	(1, 1, 1)	(1/0.8, 1/0.7, 1/0.6)	(1/3, 1/2, 1/1)	(1, 1, 1)	(1/4.3, 1/4.2, 1/4.1)
A_2	(0.6, 0.7, 0.8)	(1, 1, 1)	(0.3, 0.4, 0.5)	(1, 2, 3)	(1.1, 1.2, 1.3)
A_3	(1, 2, 3)	(0.1, 0.2, 0.3)	(1, 1, 1)	(1.2, 1.3, 1.4)	(0.2, 0.3, 0.4)
A_4	(1, 1, 1)	(1/3, 1/2, 1/1)	(1/1.4, 1/1.3, 1/1)	(1, 1, 1)	(1/1, 1/0.9, 1/0.8)
A_5	(4.1, 4.2, 4.3)	(1/1.3, 1/1.2, 1/1.1)	(1/0.4, 1/0.3, 1/0.2)	(0.8, 0.9, 1)	(1, 1, 1)

Eigenvalues are founded as $\Leftrightarrow \lambda_l = 4.9, \ \lambda_m = 5.49, \ \lambda_u = 6.64.$

Applying the mentioned software program, the following eigenvectors and CR are obtained: $\overline{w}_l = [0.2, 0.18, 0.08, 0.15, 0.28]$, $\overline{w}_m = [0.24, 0.21, 0.09, 0.16, 0.3]$, $\overline{w}_u = [0.3, 0.25, 0.11, 0.2, 0.36]$

$$CI = 0.123, \quad CR = 0.10$$

Table 8. Fuzzy matrix of the C_4 criteria

C4	A_1	A_2	A_3	A_4	A_5
A_1	(1, 1, 1)	(1/5.3, 1/5.2, 1/5.1)	(1/3, 1/2, 1/1)	(1, 1, 1)	(1/0.5, 1/0.4, 1/0.3)
A_2	(5.1, 5.2, 5.3)	(1, 1, 1)	(0.3, 0.4, 0.5)	(1, 2, 3)	(1.1, 1.2, 1.3)
A_3	(1, 2, 3)	(0.1, 0.2, 0.3)	(1, 1, 1)	(1.2, 1.3, 1.4)	(0.2, 0.3, 0.4)
A_4	(1, 1, 1)	(1/3, 1/2, 1/1)	(1/1.4, 1/1.3, 1/1.2)	(1, 1, 1)	(1/1, 1/0.9, 1/0.8)
A_5	(0.3, 0.4, 0.5)	(1/1.3, 1/1.2, 1/1.1)	(1/0.4, 1/0.3, 1/0.2)	(0.8, 0.9, 1)	(1, 1, 1)

Eigenvalues are founded as $\Leftrightarrow \lambda_l = 4.3,\ \lambda_m = 5.4,\ \lambda_u = 6.8$. Applying the mentioned software program, the following eigenvectors and CR are obtained:

$$\overline{w_l} = [0.12, 0.16, 0.11, 0.14, 0.27],\ \overline{w_m} = [0.14, 0.21, 0.15, 0.17, 0.33],$$
$$\overline{w_u} = [0.18, 0.26, 0.2, 0.21, 0.41]$$

$$CI = 0.1, \quad CR = 0.089$$

Table 9. Fuzzy matrix of the C_5 criteria

C5	A_1	A_2	A_3	A_4	A_5
A_1	(1,1,1)	(1/1.3,1/1.2,1/1.1)	(1/3,1/2,1/1)	(3.1,3.2,3.3)	(1/0.5,1/0.4,1/0.3)
A_2	(1.1,1.2,1.3)	(1,1,1)	(0.3,0.4,0.5)	(1,2,3)	(1.1,1.2,1.3)
A_3	(1,2,3)	(0.1,0.2,0.3)	(1,1,1)	(1.2,1.3,1.4)	(0.2,0.3,0.4)
A_4	(1/3.3,1/3.2,1/3.1)	(1/3,1/2,1/1)	(1/1.4,1/1.3,1/1.1)	(1,1,1)	(1/1,1/0.9,1/0.8)
A_5	(0.3,0.4,0.5)	(1/1.3,1/1.2,1/1.1)	(1/0.4,1/0.3,1/0.2)	(0.8,0.9,1)	(1,1,1)

Eigenvalues are founded as $\Leftrightarrow \lambda_l = 4,\ \lambda_m = 5,\ \lambda_u = 6.87$. Applying the mentioned software program, the following eigenvectors and CR are obtained: $CI = 0,\ \ CR = 0$.

$$\overline{w_l} = [0.23, 0.17, 0.13, 0.11, 0.17],\ \overline{w_m} = [0.27, 0.2, 0.18, 0.13, 0.22],$$
$$\overline{w_u} = [0.36, 0.27, 0.26, 0.17, 0.32]$$

Step 6. Local priority fuzzy matrix $\overline{\overline{P}} = (\overline{P_l}, \overline{P_m}, \overline{P_u})$ is expressed by using following formula:

$$\overline{P_l} = [\overline{P}_l^{(1)}\ \overline{P}_l^{(2)} \ldots \overline{P}_l^{(n)}],\ \ \overline{P_m} = [\overline{P}_m^{(1)}\ \overline{P}_m^{(2)} \ldots \overline{P}_m^{(n)}],\ \ \overline{P_u} = [\overline{P}_u^{(1)}\ \overline{P}_u^{(2)} \ldots \overline{P}_u^{(n)}]. \tag{11}$$

Multiply obtained priority matrices from the right by the priority vectors of the criteria respectively, which are defined in the fourth step.

$$\overline{w_l^T} = [\overline{w}_{1,l}\ \overline{w}_{2,l}\ \overline{w}_{n,l}\ ]^T, \quad w = [0.16\ 0.17\ 0.1\ 0.19\ 0.28]$$
$$\overline{w_m^T} = [\overline{w}_{1,m}\ \overline{w}_{2,m}\ \overline{w}_{n,m}\ ]^T, \ w = [0.48\ 0.21\ 0.12\ 0.16\ 0.34]$$
$$\overline{w_u^T} = [\overline{w}_{1,u}\ \overline{w}_{2,u}\ \overline{w}_{n,u}\ ]^T. \quad w = [0.25\ 0.29\ 0.16\ 0.22\ 0.52]$$

Vectors of global priorities g_l, g_m, g_u are obtained [7, 9].

$$g_l = P_l\overline{w_l} = [g_{1,l}\ g_{2,l} \cdots g_{m,l}]^T \qquad g_l = [0.16, 0.15, 0.1, 0.13, 0.22]$$
$$g_m = P_m\overline{w_m} = [g_{1,m}\ g_{2,m} \cdots g_{m,m}]^T\ g_m = [0.21, 0.21, 0.2, 0.16, 0.29]$$
$$g_u = P_u\overline{w_u} = [g_{1,u}\ g_{2,u} \cdots g_{m,u}]^T \quad g_u = [0.39, 0.37, 0.28, 0.3, 0.53]$$

These vectors compose fuzzy matrix of global priorities $\widetilde{G} = (g_l, g_m, g_u)$ of alternatives A_1, A_2, …,A_m. For every alternative A_i (i = 1,2,…,m), elements of these vectors are indicated by the corresponding approximate triangular fuzzy numbers.

$$\widetilde{g} = (g_{i,l}, g_{i,m}, g_{i,u}), \qquad i = 1, 2, \ldots, m.$$

Alternatives A_i (i = 1,2,…,m) are rated in last step as stated by to their global priorities which are indicated by triangular fuzzy numbers $\widetilde{g}_i (i = 1, 2, \ldots, m)$. In this paper, alternatives are ranked according to the expected value and standard deviation. For the triangular probability distribution [16, 17] of the triangular fuzzy number as a fuzzy event, these values for the fuzzy number $\widetilde{g}_i$ are calculated by the following formulas [7] (Table 10).

–Expected value:

$$g_{i,e} = \frac{g_{i,l} + 2g_{i,m} + g_{i,u}}{4}, \qquad i = 1, 2, \ldots m \tag{12}$$

–Standard deviation:

$$\sigma_i = \left[\frac{1}{80}(3g_{i,l}^2 + 4g_{m,l}^2 + 3g_{u,l}^2 - 4g_{i,l}g_{i,m} - 2g_{i,l}g_{i,u} - 4g_{i,m}g_{i,u})\right]^{1/2} \quad i = 1, 2, \ldots, m. \tag{13}$$

According to [7] $CV_i = \frac{\sigma_i}{g_{i,e}}$ $i = 1, 2, \ldots m$ $CV_3 = \frac{0.028}{0.77} = 0.036$

A_3 the alternative with smaller coefficient of variation CV_i is ranked better and as the best product which influenced on different factors and the suitable after benchmarking for buying process.

Table 10. Vectors of global priorities, expected values and standard deviations

Alternative	Vector g_l	Vector g_m	Vector g_u	Expected Value.$g_{i,e}$	Standard deviation (%)
A_1	0.16	0.21	0.39	0.98	0.040
A_2	0.15	0.21	0.37	0.94	0.036
A_3	0.10	0.20	0.28	0.77	0.028
A_4	0.13	0.16	0.30	0.75	0.030
A_5	0.22	0.29	0.53	1.33	0.053

5 Conclusion

The AHP method characterizes hierarchy between criteria, sub-criteria and alternatives. This process offered more flexible, completed, and practical conclusion particularly for the decision criteria in different spheres. This method is differentiated from others with using the pairwise comparison matrices of the criteria and alternatives and the main preference of this research paper is using of the new method for fuzzy AHP, which is based on calculation of consistency ratio. Using MATLAB environments, the eigenvalues, and eigenvectors for the matrix $\tilde{C}$ are λ_l, λ_m, λ_u. For the given the $\tilde{C}$ matrix consistency ratio is small or equal 10%. From obtained results we can say that $\tilde{C}$ matrices for each criterion are consistent and may be received. The offered method can be applied for various problems which related to consumer behavior to solve large decision-making problems by using software programs.

References

1. Edvardsson, B., Johnson, M.D., Gustafsson, A., Strandvik, T.: The effects of satisfaction and loyalty on profits and growth: Products versus services. Total Qual. Manage. **11**(7), 917–927 (2000). https://doi.org/10.1080/09544120050135461
2. Friese, M., Wänke, M., Plessner, H.: Implicit consumer preferences and their influence on product choice. Psychol. Mark. **23**(9), 727–740 (2006). https://doi.org/10.1002/mar.20126
3. Hallowell, R.: The relationships of customer satisfaction, customer loyalty, and profitability: an empirical study. Int. J. Serv. Ind. Manage. **7**(4), 27–42 (1996). https://doi.org/10.1108/09564239610129931
4. Kardi Teknomo's Homepage. https://people.revoledu.com/kardi/tutorial/AHP/purchase.html
5. Ayhan, M.B.: A fuzzy ahp approach for supplier selection problem: a case study in a gearmotor company. Int. J. Managing Value Supply Chains **4**(3), 11–23 (2013). https://doi.org/10.5121/ijmvsc.2013.4302
6. Dovlatova, K.J.: Decision making in investment by application of the analytic hierarchy process (AHP) WCIS 226–228 (2018)
7. Prascevic, Z.., Prascevic, N.: Application of fuzzy AHP method based on eigenvalues for decision making in construction industry. Tehnicki Vjesn. Tech. Gazette **23**(1), 57–64 (2016). https://doi.org/10.17559/tv-20140212113942
8. Buckley, J.J., Feuring, T., Hayashi, Y.: Fuzzy hierarchical analysis revisited. Eur. J. Oper. Res. **129**(1), 48–64 (2001). https://doi.org/10.1016/s0377-2217(99)00405-1

9. Cabala, P.: Using the analytic hierarchy process in evaluating decision alternatives. Oper. Res. Decisions **1**(1), 1–23 (2010)
10. van Laarhoven, P.J.M., Pedrycz, W.: A fuzzy extension of Saaty's priority theory. Fuzzy Sets Syst. **11**(1–3), 229–241 (1983). https://doi.org/10.1016/S0165-0114(83)80082-7
11. Bernasconi, M., Choirat, C., Seri, R.: The analytic hierarchy process and the theory of measurement. PsycEXTRA Dataset (2008). https://doi.org/10.1037/e636742011-022
12. Saaty, T.L.: The analytic hierarchy process. Open J. Soc. Sci. **4**(12), 9–29 (1980)
13. Saaty, T.L.: Theory and Applications of the Analytic Network Process: Decision Making with Benefits, Opportunities, Costs, and Risks, 3rd edn. RWS Publications, Pittsburgh (2005)
14. Vargas, V.R.: Using the analytic hierarchy process (AHP) to select and prioritize projects in a portfolio. In: PMI global Congress, North America Washington, DC, EUA, pp. 1–22 (2010). https://www.pmi.org/learning/library/analytic-hierarchy-process-prioritize-projects-6608
15. Real-Time Customer Experience Analytics for Enterprises. https://clootrack.com/knowledge_base/major-factors-influencing-consumer-behavior/
16. Aliev, R.A., Gardashova, L.A.: Z-set based approach to control system design. In: Aliev, R.A., Kacprzyk, J., Pedrycz, W., Jamshidi, M., Babanli, M., Sadikoglu, F.M. (eds.) ICAFS 2020. AISC, vol. 1306, pp. 10–21. Springer, Cham (2021). https://doi.org/10.1007/978-3-030-64058-3_2
17. Zadeh, L.A.: Fuzzy logic and approximate reasoning. Synthese **30**, 407–428 (1975)

Attitudes of Families of Children with Special Needs towards Technology

Rojda Kılınçaslan Akdağ[1], Şahin Akdağ[2], Çağla Kılınç[1], Yücehan Yücesoy[3], and Başak Bağlama[1](✉)

[1] Department of Special Education, Atatürk Education Faculty, Near East University, 10 Mersin, North Cyprus, Turkey
basak.baglama@neu.edu.tr
[2] Faculty of Sports Sciences, Near East University, 10 Mersin, North Cyprus, Turkey
sahin.akdag@neu.edu.tr
[3] Department of Classroom Teaching, Atatürk Education Faculty, Near East University, 10 Mersin, North Cyprus, Turkey
yucehan.yucesoy@neu.edu.tr

Abstract. The use of technology in special education facilitates learning and permanence. Technology infuses classrooms with digital learning tools, such as computers and hand held devices, expands course offerings, experiences and learning materials and supports learning. The role of families of children with special needs in integrating technology into students' lives is crucial as teachers since they spent much time with their children. This study aims to determine technology attitudes of parents with special needs children and using technology in courses. This study used quantitative research and a scale was used as data collection tool. Results of the study revealed that attitudes of parents of children with special needs towards technology and using technology in courses are positive at a moderate level. In addition to this, attitudes of parents showed significant difference based on parents' educational status and thoughts on the usefulness of technology in education. Suggestions for further research and practices were presented with relevant literature.

Keywords: Special education · Special needs · Families of children with special needs · Technology · Attitude

1 Introduction

In today's contemporary world, human life is deeply affected by current and rapid developments in science and technology. In addition to keeping up with these developments, individuals need to turn to education more than ever to contribute to science and technology since it is possible with providing appropriate education for individuals to be able to adapt to new situations in line with today's circumstances [1, 2]. Technology holds a very important place in human life. Technology has entered into educational life and pioneered further development. Educational technology is a set of information obtained

R. A. Aliev et al. (Eds.): ICSCCW 2021, LNNS 362, pp. 183–189, 2022.
https://doi.org/10.1007/978-3-030-92127-9_27

by applying the science of teaching/learning to the real-world conditions through the classroom environment [3].

Based on the importance of mainstreaming education, inclusion and universal design, the preparation of an environment that facilitates the access of all individuals, especially individuals with special needs, is gaining importance in recent times [4, 5]. In regard to the introduction of technology into educational environments, teaching processes in classroom environments have begun to be supported by technology. Technology-supported educational environments provide a learning experience that increases interest and motivation in line with the characteristics of the learners. In today's world, technology supported learning environments can be created for teaching various different skills, behaviors and concepts for individuals with special needs. One of the technologies required for preparing individuals with special needs for independent living is the assistive technologies. Assistive technologies support visual, hearing, reading, writing, social communication skills of individuals with special needs and facilitate their independent life skills. This allows individuals to learn the targeted information more easily, permanently and quickly [6, 7].

The role of parents, cognitive environment at home and the attitude of the parents towards the child's cognitive functions are effective for the cognitive development of the child. Research findings indicate that the quality of behavior and attitudes of parents towards their children creates changes in the development of children of developmental age. Positive attitudes of parents also positively affect children [8]. In this context, knowing the attitudes of the parents of the individuals with special needs to the technology and the use of technology in the lessons is of great importance in terms of revealing what can be done for improving educational practices for individuals with special needs. Therefore, this study aims to determine technology attitudes of parents with special needs children and using technology in courses. In line with this general aim, answers to the following questions were also sought in this study:

1. How are the technology attitudes of parents with special needs children and using technology in courses?
2. Do the technology attitudes of parents with special needs children and using technology in courses show a significant difference according to age, being mother or father, educational status, age of the child, technological device used frequently by the child and usefulness of technology in education variables?

2 Method

2.1 Research Model

Descriptive survey method was used in this study in order to reveal the attitudes of families of children with special needs towards technology and using technology in courses. The descriptive survey model is a research model that tries to describe and examine the events, assets, groups, current situation and characteristics under their own conditions [9].

2.2 Participants

A total number of 51 parents of children with special needs were included in this study. Demographic characteristics of the participants are provided in Table 1.

Table 1. Demographic characteristics of the participants

Demographic characteristics		f	%
Age	20–30	11	21.5
	31–40	24	47.1
	41–50	14	27.5
	51 and above	2	3.9
	Total	51	100
Gender	Female	39	76.5
	Male	12	23.5
	Total	51	100
Educational Status	Primary school graduate	7	13.7
	Secondary school graduate	14	27.5
	High school graduate	16	31.4
	University graduate	13	25.5
	Postgraduate	1	2.0
	Total	51	100
Age of the child	1–5	19	37.3
	6–10	26	51.0
	11–15	6	11.8
	Total	51	100
Technological device used frequently by the child	Tablet	20	39.2
	Mobile phone	17	33.3
	Computer	14	27.5
	Total	51	100
Usefulness of technology in education	Yes	36	70.6
	No	5	29.4
	Total	51	100

According to Table 1, 11 of the participants (21.5%) were between the ages of 20 and 30; 24 of them (47.1%) were between 31 and 40; 14 of them (27.5%) were 41 and 50 and 2 of them (3.9%) were 51 and above. 12 of the participants (23.5%) were male and 39 of them (76.5%) were female. When educational status of the participants are examined, it is seen that 7 of the participants were primary school graduate (13.7%); 14 of them were secondary school graduate (27.5%); 16 of them were high school graduate (31.4%); 13 of them were university graduate (25.5%) and 1 of them has postgraduate degree (2.0%). Ages of the child were changing between 6 and 15 in which 26 of the children were between 6 and 10 (51.0%) and 6 of them were between 11 and 15 (11.8%). When technological devices used frequently by the child are examined, it is seen that tablet (f = 20; 39.2%), mobile phone (f = 17; 33.3%) and computer (f = 14; 27.5%) were the most frequently used technological devices by the children of the participants. Results demonstrated that majority of the parents of children with special needs think that technology is usefulness in education (f = 36; 70.6%).

2.3 Data Collection Tools

Demographic Information Form and two different tools and parents' attitudes towards technology and using technology in courses scale was used for data collection. age in the demographic information form, being mother or father, educational status, age of the child, technological device used frequently by the child and usefulness of technology in education.

The scale on parents' attitudes towards technology and using technology in courses was originally developed by [10]. The scale included 25 statements and all expressions item is scored on a 5-point scale ranging from "Strongly Agree" (5) to "Strongly Disagree" (1). The scale consists of 14 positive and 11 negative statements.

2.4 Procedure

Data of the present study were collected in two different special education centers in Hatay, Turkey. Firstly, all of the families were informed about the aims and confidentiality of the study and parents who accept were invited to participate in the study when they are available.

2.5 Data Analysis

Data were analyzed with SPSS 20 program. Significance level was considered as $p < .05$ in statistical analysis. Percentage, frequency, t-test, one-way ANOVA were used to analyze data. The score ranges used in commenting the responses to scale are shown in Table 2:

Table 2. Score intervals

Option	Load	Interval
Strongly Disagree	1	1.00–1.79
Disagree	2	1.80–2.59
Neutral	3	2.60–3.39
Agree	4	3.40–4.19
Strongly Agree	5	4.20–5.00

3 Results

3.1 Descriptive Statistics on the Attitudes of Families of Children with Special Needs towards Technology and Using Technology in Courses

Table 3. Descriptive statistics on attitudes of families of children with special needs towards technology and using technology in course scale

Attitude	N	Number of statements	Minimum	Maximum	Mean	SD
Total Scores on attitudes	51	25	2.04	4.76	3.67	.74

Table 3 shows results of the participants on attitudes of families of children with special needs towards technology and using technology in course scale. It was shown that mean of total scores is between "*Agree*" score range. Based on this result, it can be inferred that attitudes of parents of children with special needs towards technology and using technology in courses are positive at a moderate level.

3.2 Independent Samples t-test Results for the Technology Attitudes of Parents with Special needs Children and Using Technology in Courses based on Gender and Thoughts on the Usefulness of Technology in Education

Table 4. Independent samples t-test results for the technology attitudes of parents with special needs children and using technology in course based on gender and thoughts on the usefulness of technology in education

Attitudes towards technology and using technology in course scale		N	Mean	SD	t	p
Being mother or father	Father Mother	12 39	96.50 90.43	18.90 17.67	− .98	.329
Usefulness of Technology in Education	Yes No	36 14	99.47 75.00	14.09 14.90	3.02	.000*

* $p < .05$

Independent samples t-test was used to figure out if attitudes of participants show significant difference based on gender and thoughts on the usefulness of technology in education. In Table 4, results for the technology attitudes of parents with special needs children and using technology in course based on gender and thoughts on the usefulness of technology in education are given. As it can be seen from the table, no significant difference was obtained between the technology attitudes of parents with special needs children and using technology in course and being mother and father. In other words, being mother and father did not affect attitudes towards technology and using technology in courses. In addition, it was revealed that attitudes of parents of children with special needs towards technology and using technology in courses show significant difference based on their thoughts on usefulness of technology in education. It is seen that participants who think that technology is useful in education have more positive attitudes towards technology and using technology in courses.

3.3 One-Way ANOVA Results on the Attitudes of Families of Children with Special Needs towards Technology and Using Technology in Courses based on Parents' Educational Status

Table 5. One-way ANOVA results of attitudes towards technology and using technology in courses based on parents' educational status

Variable		N	Mean	SD	Sum of Squares	df	Mean Square	F	p
Educational Status	Primary school	7	89.29	23.70	5242.83	4	1310.70	4.98	.002*
	Secondary school	14	77.29	12.33	12097.2	46	262.98		
	High school	16	96.00	17.16					
	University degree	13	104.08	13.96					
	Postgraduate	1	89.00	18.62					

One-way ANOVA was used to figure out whether the technology attitudes of parents with special needs children and using technology in courses differ significantly based on age, educational status, age of the child, technological device used frequently by the child. According to the results, only educational status of the parents showed significant difference on parents' attitudes towards technology and using technology in courses. Therefore, only educational status variable is provided in Table 5 in detail. One-way ANOVA results showed that parents who have university degree scored higher on the attitudes towards technology and using technology in courses.

4 Conclusion and Suggestions

In conclusion, it was determined that the technology attitudes of parents with special needs children and using technology in courses are positive at a moderate level. In addition to this, attitudes of parents showed significant difference based on parents' educational status and thoughts on the usefulness of technology in education. The following suggestions are provided based on the study:

- Awareness on advantages and potential risks of using technology in special education might be increased among families of children with special needs. Therefore, families

would be able to support using technology for the benefit of their children with special needs.
- This study should also be conducted using different research methods such as qualitative, experimental and phenomenological research design.

References

1. Courduff, J., Szapkiw, A., Wendt, J.L.: Grounded in what works: exemplary practice in special education teachers' technology integration. J. Spec. Educ. Tech. **31**(1), 26–38 (2016)
2. Martinez, W.: How science and technology developments impact employment and education. PNAS **115**(50), 12624–12629 (2018). https://doi.org/10.1073/pnas.1803216115
3. Mangal, S.K.: Essentials of Educational Technology. PHI Learning Pvt. Ltd, India (2019)
4. Armstrong, F., Armstrong, D., Barton, L.: Inclusive Education: Policy, Contexts and Comparative Perspectives. Routledge, US (2016)
5. Slee, R.: Inclusive education: from policy to school implementation. In: Towards Inclusive Schools? pp. 30–41. Routledge (2018)
6. Ahmad, F.K.: Use of assistive technology in inclusive education: making room for diverse learning needs. Transcience **6**(2), 62–77 (2015)
7. Carver, J., Ganus, A., Ivey, J.M., Plummer, T., Eubank, A.: The impact of mobility assistive technology devices on participation for individuals with disabilities. Disabil. Rehabil. **11**(6), 468–477 (2016)
8. Muller, C.: Parent involvement and academic achievement: an analysis of family resources available to the child. In: Parents, Their Children, and Schools, pp. 77–114. Routledge (2018)
9. Karasar, N.: Scientific Research Methods, 15th edn. Nobel Publishing, Ankara (2005).(in Turkish)
10. Kenar, I.: Development of parent attitude scale towards technology and technology use in lessons and tablet pc application. Ebad. **2**(2), 124–139 (2012). (in Turkish)

Some Aspects of Applying Fuzzy Logic Theory in Steel Metallurgy

R. I. Karimov(✉)

Baku Steel Company LLC, Baku, Azerbaijan
Ramin.Karimov@bakusteel.com

Abstract. In this paper, financial losses and limitations of product quality in melting of steel caused by human factor have been discussed. It is noted that loading of charge to furnace, thermic processing of melting and producing finished products are carried out based an experimental way all over the world. Such an approach does not always allow to choose an optimal variant.

Nowadays, application of fuzzy logic may be considered as an innovative approach in metallurgy. This approach would serve for both reducing of cost of steel products and improving of their quality. Thus, realization of artificial intelligence's (AI) capabilities with development of special programs on the basis of fuzzy logic theory in the entire technological chain has been substantiated in the article. An example of an application of adaptive neuro-fuzzy inference system (ANFIS) to modeling one process within steel production is provided to illustrate perspectives of using AI in metallurgy.

Keywords: Metallurgy · Fuzzy logic · Neural network · Melting · Thermic processing · Tabulation · Stratification

1 Introduction

One of innovative ways of increasing steel products' quality and decreasing the cost is the application of artificial intelligence (AI) in production based on improvement of their metallurgy technology [1].

In recent years electrometallurgy takes a particular place in high-quality steel manufacturing. A number of new melting aggregates formed off-furnace processing, continuous casting devices of steel and new thermic processing technologies.

Melting technologies of steel in marten furnace and converter already remained in furnace. Nowadays the most productive and qualitative process is significantly characterized by management of melting, diffusion, purification processes of steel on [2].

Production of qualitative electric steel product meeting all requirements of modern projects starts from melting process. At the same time, melting process in electric arc furnace is the most energy-saving process. In that case provision of energy-saving technology is not based on the full utilization of AI capabilities even in modern furnace aggregates. Fuzzy logic theory can be considered as one of the basics of innovative technologies in administration of technological chain from loading of charge (shikhta)

R. A. Aliev et al. (Eds.): ICSCCW 2021, LNNS 362, pp. 190–195, 2022.
https://doi.org/10.1007/978-3-030-92127-9_28

materials onto furnace to release of steel from furnace. At the same time, AI could play an important role in thermic processing, last operation after melting, casting and rolling procedures in attaining required quality of steel products, as well as tubes used for oil-gas industry. AI-based optimal choice of thermic processing modes (particularly, tabulation and stratification) and further choice of production mode depending on brand of steel, creates capabilities for full provision of obligations of producer before consumer [3, 4].

2 An Overview of Steel Melting Process

We consider an issue of application of fuzzy logic theory to process management in steel melting at "Baku steel Company" LLC. For example, currently melting process in Baku Steel Company LLC (BSC) is conducted in 60 ton. electric arc furnace (EAF) as follows: metal shikhta filled in to a special basin is loaded to AEF using bridge crane. On average, 4 basins are loaded into an alloy depending on quality and thickness of metal shikhta. Generally, for an alloy melted (release temperature-1620 °C) metal smelting is implemented under the impact of 55 min' electric arc.

The data on structure of buckets over a smelting process is shown in Table 1.

Table 1. Melting duration and metal output of buckets

№	Heat number	I bucket		II bucket		III bucket		IV bucket		Total time, min	Total weight, ton
		Time, min	Weight, ton	Time, min	Weight, ton	Time, min	Weight, ton	Time, min	Weight, ton		
1	105379	16	21.0	14	17.0	9	17.8	20	12.6	59	68.4
2	105380	13	20.6	9	18.0	13	15.4	23	11	58	65.0
3	105384	14	20.5	11	17.2	9	12.8	21	10	55	60.5
4	105386	13	22.0	10	17.4	11	14.2	22	11.7	56	65.3
5	105387	14	23.4	12	20.0	9	14.8	24	10	59	68.2
6	105388	13	22.6	12	20.1	9	14.7	22	9.9	56	67.3
7	105389	13	22.4	12	20.3	10	14.0	25	9.8	60	66.5
8	105391	14	20.4	14	18.6	11	16.4	19	13	58	68.4
9	105394	15	20.4	13	18.0	12	16.3	22	9	62	63.7
10	105397	13	21.7	10	18.5	9	14.2	28	11	60	65.4
11	105398	13	21.5	10	18.2	10	14.6	20	11.2	53	65.5
12	105399	12	25.2	10	17.4	10	12.5	19	10.2	51	65.3
13	105400	14	24.4	12	16.8	14	13.7	26	12.1	66	67.0
14	105401	14	25.1	13	17.0	14	13.8	23	12.5	64	68.4
15	105402	15	25.3	14	17.2	12	14.0	26	12.7	67	69.2
16	105403	13	24.8	16	17.2	9	14.3	28	11	66	67.3
17	105405	15	25.8	12	17.3	8	14.9	28	10.3	63	68.3

(continued)

Table 1. *(continued)*

№	Heat number	I bucket		II bucket		III bucket		IV bucket		Total time, min	Total weight, ton
		Time, min	Weight, ton	Time, min	Weight, ton	Time, min	Weight, ton	Time, min	Weight, ton		
18	105408	17	24.6	14	12.8	11	20.0	20	11.2	62	68.6
19	105410	14	24.4	12	20.1	13	15.0	14	9	53	68.5
20	105411	15	24.2	12	18.0	8	15.7	25	10.3	60	68.2
21	105415	14	23.4	11	18.8	13	15.7	24	10.6	62	68.5
22	105416	14	22.4	11	20.6	13	14.0	25	11.2	63	68.2
23	105417	14	23.0	13	20.7	12	14.2	25	10.6	64	68.5
24	105418	12	22.6	12	20.5	11	14.0	25	10.2	60	67.3
25	105420	15	22.3	12	18.6	11	14.7	21	10	59	65.6
26	105421	15	21.8	14	19.4	9	14.0	22	10	60	65.2
27	105422	14	22.0	11	19.6	8	15.0	24	9.6	57	66.2
28	105423	13	21.8	13	20.0	9	14.8	24	10.5	59	67.1
29	105424	11	22.4	11	19.8	9	15.0	23	11	54	68.2
30	105425	12	22.0	13	19.6	11	14.8	25	12	61	68.4

After complete melting of each basin, the process of transferring to the furnace is carried out on the basis of the decision of the steelmelter. This decision is made empirically by steelmelter, which mostly leads to extra time and energy losses in the melt. Thus, the analysis of a large number of alloy passports in the BSC confirms the overheating of the furnace under 4 min for each alloy.

Loading of the first basin - in this case the basin is transported to the furnace by crane and loading is carried out. The melting process is based on the experience of steel melting. He then turns to the steel crane driver to transport the second basin to the furnace. The crane operator, by his decision, transports the second basin and loads it into the furnace without saving time.

If the loading and melting processes are optimized, the melting duration can be up to 9 min [5]. In a 60-ton electric arc furnace, the electric power consumed for melting per minute is at least 550 kWt × hr. Then it is possible to calculate only the losses in the consumption of electricity in a smelting carried out by loading four baths with a simple empirical report (taking into account that we lose four minutes in the total solution):

$$4 \times 550\,\text{kWT} = 2200\,\text{kWT}$$

For 22 alloys this electricity loss makes up 2200 kWt × 22 = 48400 kWtt/day.
If factory operates 29 days per month then the loss of electricity will constitute:
48400 kWt × 29 working day (a month) = 1.403.600 kWt/hour.
If we express this loss by manats, then we have:

$$1403600\,\text{kWT} \cdot \text{h} \times 0.09\,\text{AZN} = 126324\,\text{AZN}$$

Therefore, annual losses on the melting process conducted through an ordinary habit experience in factory will be:

$$1234\,\text{AZN} \times 12\,\text{months} = 1515\,\text{mln. AZN}$$

With the purpose of prevention of these extreme losses it is a more innovative to carry out decision-making in melting process via AI.

It is also possible to conduct analogical calculations for the other operations of technological chain of metallurgy.

Baku Steel Company LLC produces a wide range of pipes for the oil and gas industry of our country and the Russian Federation. The required strength class of these pipes is provided not only by high-precision melting process, but also by heat treatment process. The heat treatment process is a complex technological process that involves heating the finished pipe in a furnace and then cooling it at different speeds [6]. Whether technological chain consists of tabulation and stratification, then these operations are repeated several times. Another important point is steel brand on which the pipes are made. Each steel brand requires the selection of modes according to the strength class in heat treatment. Adherence to these modes is mainly based on the empirical choice of the worker thermistor, which in often leads to loss of time and energy and reduced quality.

3 Toward an Application of Fuzzy Logic in Modeling of Steel Melting Process

Now let's look at the possibility of applying the theory of fuzzy logic to the technological chain in the example of the same operations.

For example, the exact time of smelting can be determined empirically by performing several experimental smelting operations in steel production. Based on this information, it is possible to program the entire technological chain (shikhta, transportation and loading of basins, melting, unloading, etc.). This means that the whole melting process may be guided by AI. In any case, using the capabilities of fuzzy logic theory, it is possible to ensure the optimal sequence of the technological chain in metallurgy-smelting technology. We believe that such experience will be useful for factories operating in different countries.

The same ideas mentioned above can be used as an example for heat treatment technology. It is possible to remove the tensions in the cast-in-workpiece and give them the required strength class only by heat treatment operations. These operations are both long-term and highly transitory, so losses are inevitable.

Now addressing these issues from the perspective of fuzzy logic theory. We can empirically take a different view by optimizing the work of the entire technology chain of thermal processing in working condition and then programming it as an AI-algorithm. The complexity here goes beyond creating a program for each steel brand. Of course, if there is such a program, then the worker can provide information about the brand of steel only on the computer. In result, there will be no extreme costs and the full quality of the product will be ensured.

Let us consider using adaptive neuro-fuzzy inference system (ANFIS) for modeling dependence between processing time and weight of buckets (the data are given in Table 1). For Bucket I, ANFIS training RMSE is 1.6 and testing RMSE is 1.12. The obtained fuzzy rules are shown in Fig. 1:

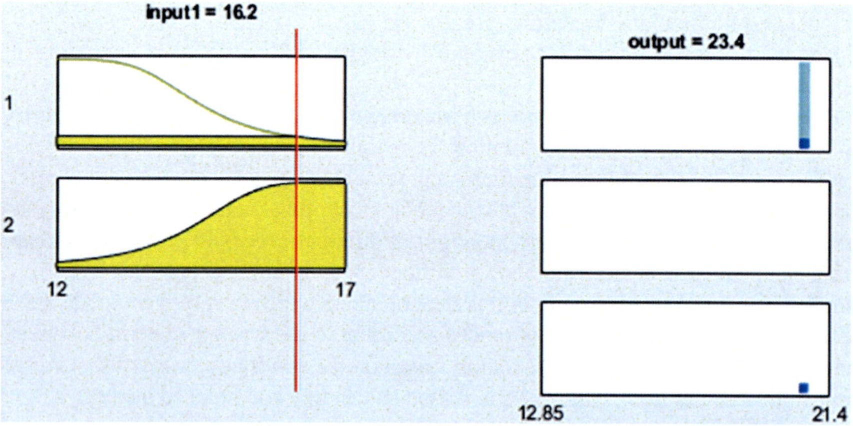

Fig. 1. Fuzzy rules for modeling dependence between time and weight (bucket 1).

The results for the other buckets are shown in Table 2:

Table 2. ANFIS-based modeling results

Bucket	No. of rules	Training RMSE	Testing RMSE
II	2	1.48	1.44
III	2	1.39	0.39
IV	2	0.98	0.88

Thus, the constructed ANFIS-based models for each bucket are of a good accuracy.

4 Conclusion

The use of traditional methods of control in the technological chain of metallurgy causes to financial loses and limiting capabilities on production of high-quality products. Fuzzy logic may allow to reduce financial losses and improve the product quality. This can be done by means of an optimal choice of smelting and heat processing modes in the entire technological chain of metallurgy. In this paper we considered a problem of modeling dependence between time of steel processing and weight of metallurgy buckets by means of neuro-fuzzy approach. Four ANFIS-based models of good accuracy were constructed for different buckets.

References

1. Karimov, R.T.: Thermodynamic features of increasing efficiency of metallurgical processes turing electric steel melting. Rensta San Gregorio Ecuader **2**, 845–864 (2020)
2. Karimov, R.T., Shakhov, S.I.: Using of metallized material in electric steel-making production. Black Metall. Moscow **1**, 70–78 (2019). (in Russian)
3. Karimov, R.T.: Development of technical strengthening technology of clutch couplings of protective tubes. J. Baku Eng. Univ. Mech. Ind. Eng. **2**(2), 98–104 2018. (In Azeri)
4. Smoyakov, A.S., Shakhov, S.I., Karimov, R.I., Rafiev, O.: Reconstruction of the operating section over continuous casting of the Baku steel company LLC from round billets for the needs of the pipe industry. Black Metall. **1**(1405), 43–46 (2017). (in Russian)
5. Karimov, R.T.: Improving steel melting intensity in the process of electro-smelting from waste and pellets (HBI). East. Eur. J. Enterp. Technol. **99**, 35–42 (2019). https://doi.org/10.15587/1729-4061.2019.168352
6. Karimov R.T.: Analysis of defects in protective drill pipe steels. J. Baku Eng. Univ. Mech. Ind. Eng. **2**(1), 47–57 (2018). (in Azeri)

Imperfect Knowledge Base Self-organization in Robotic Intelligent Cognitive Control: Quantum Supremacy

S. V. Ulyanov[1,2](✉), A. V. Shevchenko[2], A. A. Shevchenko[2], and A. G. Reshetnikov[1,2]

[1] Laboratory of Information Technology, Joint Institute for Nuclear Research, 6 Joliot-Curie St., Dubna 141980, Russian Federation

[2] Institute of System Analysis and Control, Dubna State University, 19 Universitetskaya St., Dubna 141982, Russian Federation

Abstract. The smart control design with secured achievement of information-thermodynamic trade-off interrelations is main goal for quantum self-organization algorithm of imperfect KB. Quantum genetic algorithm applied on line for the quantum correlation's type searching between unknown solutions in quantum superposition of imperfect knowledge bases of intelligent controllers designed on soft computing. Disturbance conditions of analytical information-thermodynamic trade-off interrelations between main control quality measures (as new design laws) discussed. Sophisticated synergetic quantum information effect in autonomous robot in unpredicted control situations) and swarm robots with imperfect KB exchanging between "master – slaves" introduced. A new robust smart controller on line designed from responses on unpredicted control situations of any imperfect KB applying quantum hidden information extracted from quantum correlation discussed. Within the toolkit of classical intelligent control, the achievement of the similar synergetic information effect is impossible.

Keywords: Quantum information theory knowledge base · Fuzzy controller · PID controller

1 Introduction

According to the definition of modern control problems, the achievement of required robustness property for a complex ill-defined control object models is possible with applying the computational intelligence toolkit. The goal of this article is the description of the applied aspects of developed intelligent design technology of robust knowledge bases (KB) [1–7] using the information synergy effects of quantum knowledge self-organization [2] in unpredicted and risky control conditions [5].

Four statements from quantum information theory and quantum thermodynamics are applied in this developed approach: 1) minimum entropy production rate principle of the system "control object + intelligent controller" that quadrantes the achievement of control goal with minimum of work waste in control object and in intelligent controller;

R. A. Aliev et al. (Eds.): ICSCCW 2021, LNNS 362, pp. 196–204, 2022.
https://doi.org/10.1007/978-3-030-92127-9_29

2) minimum information entropy principle for design intelligent cognitive controller that required minimum of initial information for intelligent controller action; 3) the amount of the work wasted on the extraction of hidden quantum information is less than the amount of work done on the received extracted quantum hidden information; and 4) the solution problem search of maximum extractable value work identical to a search of the minimum wasted entropy done on this work extraction [8, 9]. The article task is the description of the IT-design process a robust sophisticated KB of intelligent cognitive controller that produce control force that satisfied to these requirements.

2 Imperfect KB Quantum Self-organization Process

The role of specific quantum hidden information effects for smart control design described in [2]. The amount of hidden quantum information [2–4, 6] extracted from control classical states considered as the additional information-thermodynamic control force source.

In systems inspired by nature, robustness is determined by the natural process of self-organization [1, 2]. The process of quantum self-organization of KBs, in which the robustness property is achieved, is shown in Fig. 1. Natural evolution consists of the following stages: 1) creating a template; 2) self-assembling; 3) self-organization (see level 2 in Fig. 1).

As is known from the theory of quantum computing, each quantum algorithm contains such unitary quantum operators as interference, superposition, entanglement (quantum oracle) and measurement classical operator (irreversible and used for measurement of quantum computations). The quantum fuzzy inference (QFI) model is based on the corresponding quantum operators and accumulates the principles of self-organization. A quantum self-organization control algorithm, based on QFI, is shown in Fig. 1.

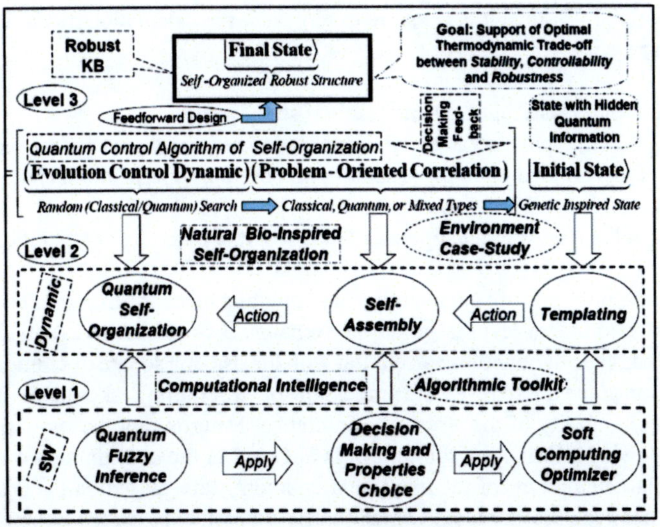

Fig. 1. Quantum search hierarchical structure of self-organization robust KB's design system.

2.1 Quantum Self-organization of Imperfect KBs

To design robust KBs of a fuzzy controller (FC), the developed self-organization quantum algorithm (QA) was used in unexpected control situations. Fulfillment of the requirements in information-thermodynamic criteria (see Fig. 1, level 3) guarantees the achievement of the goal of intelligent control with the required level of stability in operational invariant control. The design process is shown in [2, 4] as follows:

$$\begin{aligned}&\underbrace{|\text{Designed self-organized robust state}\rangle}=\\&=\left[\underbrace{\begin{pmatrix}\text{Evolution of self-}\\\text{organization process}\end{pmatrix}}_{\text{Quantum random search}}\bullet\begin{pmatrix}\textit{Quantum}\\\textit{computing}\end{pmatrix}\bullet\overbrace{\underbrace{(\text{Correlation type and form})}_{\text{Classical, quantum, mixed}}}^{\text{Problem orientation of control object}}\right]\\&\begin{Bmatrix}|\text{Initial state}\rangle\\\text{"Building"}\\\text{block}\end{Bmatrix}\bullet\begin{pmatrix}\text{Bio-inspired}\\\text{states reproduced}\\\text{by toolkit}\end{pmatrix},\end{aligned}\tag{1}$$

Figure 2 shows the structure of a robust QFI-based intelligent control system (ICS). Soft computing optimizer toolkit (SCOptKB™, the box "SCO" on Fig. 2) is a design KBs of fuzzy controllers (see boxes "FC1" and "FC2" on Fig. 2) on learning situations that can be non-robust (imperfect) in other control situations. The input of the quantum fuzzy inference (see box "QFI" on Fig. 2) with quantum genetic algorithm (QGA) extract a hidden quantum information from imperfect KB responses of two fuzzy controllers (FC1 and FC2) on unpredicted control errors realized quantum fuzzy controller in on line without changing production rules numbers of KBs in FC1 and FC2 [1].

In quantum computing, the design of a universal quantum simulator is possible with classical models [1–7], thus it is possible to represent the calculation model of the QA in the five stages:

1. Preparation of a classical or quantum initial state $|\psi_{out}\rangle$
2. Preparing the superposition state: applying Hadamard transforms for the initial state.
3. Applying an entangled operator (quantum oracle) to a superposition state.
4. Applying of the interference operator.
5. Applying of the measurement operator to the result of quantum computations $|\psi_{fin}\rangle$.

Based on the principle of superposition, a templating mechanism is implemented, as well as micro- and macro-level information exchange between active agents. Based on the choice of the type of quantum correlation with the use of a source of communication and information at the micro level, the process of self-assembling takes place, determining the level of stability of the KB fuzzy inference. Determining the most significant parts of the flow of information for control is engaged in the coordination of quantum oracle, making calculations of the intelligent quantum state. Based on the principle of interference, the result is extracted, which makes it possible to design a robust KBs in online.

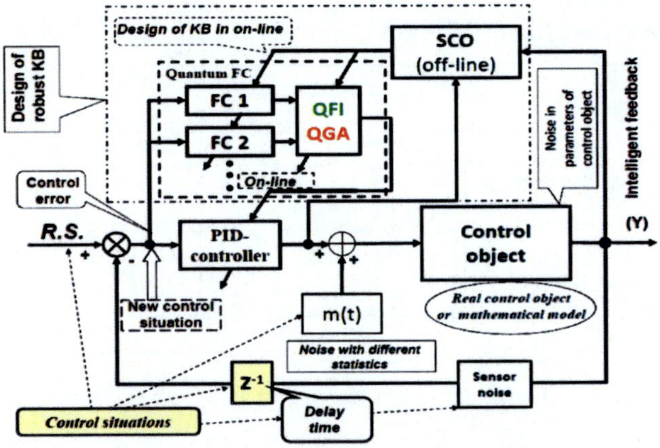

Fig. 2. Structure of a robust QFI-based ICS.

Accordingly, for the global optimization of the structures of basic ICS using quantum computing, quantum deep machine learning algorithms and quantum genetic search, it is possible to use the quantum gate approach [5, 7]. When implementing the quantum knowledge algorithm for QFI, the result of a fuzzy processed independently for each FC, extracts useful information hidden in individual KB. The online control signal is implemented in all knowledge sets of a fuzzy controller, and the QFI output signal is a set of PID controller effort coefficients. Thus, the principle of self-organization is realized.

Figure 3 demonstrates quantum algorithmic gate (QAG) for QFI based on QGA. The entangled state design of operator U_F correspond to design of the quantum search algorithm type. On the basis of such principles of quantum information theory as efficient compression of information (encoding the computational basis $\{|0\rangle, |1\rangle\}$ and the formation of quantum correlation between different computational bases), splitting information into classical and quantum components (by Hadamard transformations) and extracting hidden information (using the process of measuring classical correlation in a quantum state), it is possible to construct an optimal process for extracting valuable information from several KBs based on soft computing [6].

Summarizing what has been said, the QFI algorithm can be described in the following steps:

1. *Coding*: the construction of all normalized states is created $|0\rangle$ and $|1\rangle$ for the current input values of the gains $\{k_P^i(t), k_D^i(t), k_I^i(t)\}$ of fuzzy controllers, where the index is the NR number (or its knowledge base), the probability amplitudes are calculated $|0\rangle$ and $|1\rangle$ from the corresponding histograms.
2. *Choosing* the quantum correlation type to construct new values of output control signals. For example, consider the following type of quantum correlation: $\{\{k_P^1(t), k_P^2(t), k_D^1(t), k_D^2(t)\} \to k_P^{new}(t),$, where 1 and 2 are KB indices (or corresponding FC). Quantum state $|a_1a_2a_3a_4\rangle = |k_P^1(t)k_P^2(t)k_D^1(t)\ k_D^2(t)\rangle$ is an entangled state.
3. *Construction* of a superposition of entangled states.

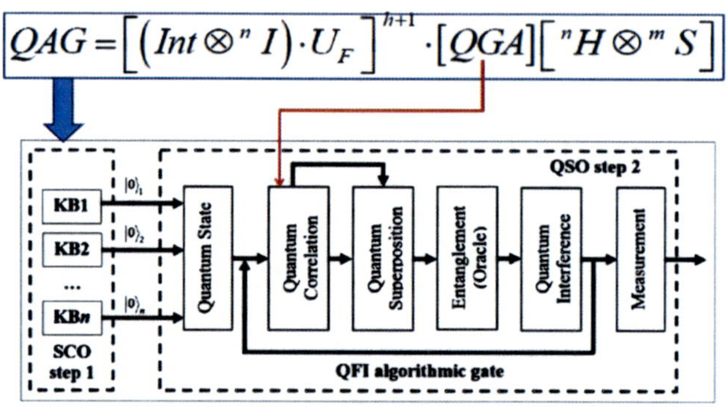

Fig. 3. QAG structure of QFI with QGA.

4. *Measurement* of an intelligent quantum state: the choice of the state $\overrightarrow{S} = |a_1a_2a_3a_4\rangle$ with maximum probability amplitude $|\alpha_k|^2$.
5. *Decoding*: calculating the output value of the corresponding gain as the norm of the vector $\overrightarrow{S}$ in the visual state: $k_P^{new}(t_i) = \frac{1}{\sqrt{2^n}}\sqrt{\langle a_1 \ldots a_n | a_1 \ldots a_n \rangle} = \frac{1}{\sqrt{2^n}}\sqrt{\sum_{i=1}^{n}(a_i)^2}$.
6. *Calculation* of the denormalized output value of the gain $k_P^{output} = k_P^{new}(t_i) \cdot maxK_p, k_D^{output} = k_D^{new}(t_i) \cdot maxK_D, k_I^{output} = k_I^{new}(t_i) \cdot maxK_I$.
7. (*offline*). Searching for robust scaling factors $\{gain_P, gain_D, gain_I\}$ using GA and the selected fitness function.
7. (*online*). Calculation of the actual output value of the corresponding gain, taking into account the scaling factors found at step 7 (*offline*): $k_P^{output} = k_P^{new}(t_i) \cdot gain_P, k_D^{output} = k_D^{new}(t_i) \cdot gain_D, k_I^{output} = k_I^{new}(t_i) \cdot gain_I$.

Thus, this quantum control algorithm uses the basic principles of quantum information theory, which are useful information resources of QFI. Based on these principles, it is possible to extract additional hidden information and eliminate the redundancy of information for the formation of a control action, thereby increasing the stability of the ICS, providing robustness and accuracy of control in conditions of uncertainty or poorly formalized description of the external environment. Consequently, the inclusion of stability in the architecture of the structure of an intelligent control system contributes to an increase in its efficiency.

3 Benchmark's Simulation of Smart Control with QFI

In Fig. 4 shows the results of an experiment of control in unexpected situations for an object "cart-*double* pole" and a 7 degrees of freedom *redundant* manipulator. The experiment compares the different controllers: PID controller, two fuzzy controllers (FC1, FC2) and three QFI controllers based on different types of correlations: Quantum-Time (Q-T), Quantum-Space (Q-S), Quantum-Space Time (Q-ST).

In the simulation and experiment, the structure of a robust ICS based on QFI (see Fig. 2) and QAG (see Fig. 3) was used. Based on the training signal taken directly from the control object, using the QCOptKB™ software toolkit, a KB of FC was designed. An abnormal situation was simulated by a threefold delay in the feedback sensor signal.

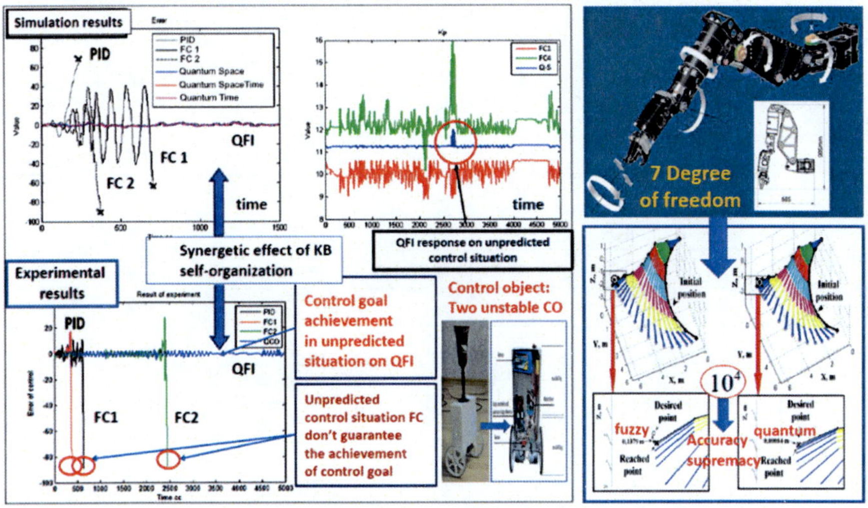

Fig. 4. The experiment of control in unexpected situations for an object "cart-double pole" and a 7 degrees of freedom redundant manipulator.

The experimental results show that the accuracy of a quantum controller is more than 10,000 (see Fig. 4, right side) times higher than that of a controller based on soft computing. Under conditions of uncertainty, the controller based on soft computing dramatically increases the control error, thereby failing to achieve the control goal (see Table 1). Comparison of controllers shows the presence of a synergistic effect of self-organization in the design of robust KBs based on imperfect KBs of FCs. The control coefficients of the PID controller are based on the feedback of imperfect KB (see the "QFI block" in Fig. 2), forming a control action in online. This is achieved by extracting an additional information resource using QFI in the form of quantum information hidden in the classical states of the control action as a new control error of the output signal of an imperfect KB [1, 2].

Remark. In [9–11], a reduced quantum genetic algorithm (RQGA) was proposed, which is an implementation of a genetic algorithm on a quantum computer. The search procedure for the desired solution is performed in one operation. Structurally, the algorithm consists of the following steps:

1. Initialization of the superposition of all possible chromosomes.
2. Assessment of the fitness function by operator F.
3. Applying Grover's algorithm.
4. Using a quantum oracle.
5. Using Grover's diffusion operator.
6. Evaluation of the solution.

Table 1. Comparison of the different regulators.

Time, sec	Cart motion, cm					
	PID	FC1	FC2	QFI (Q-S)	QFI (Q-ST)	QFI (Q-T)
1	−1	−1	−1	1	−1	−1
2	5	3	5	5	3	4
3	−35	−4	−26	−4	−2	−3
4	60	5	36	6	4	5
5	–	−5	−60	−5	−4	−7
6	–	10	–	5	8	6
7	–	−14	–	−4	−6	−9
8	–	23	–	4	5	7
9	–	−32	–	−6	−8	−3
10	–	50	–	9	6	4
11	–	–	–	−9	−4	−7

As can be seen from Fig. 5, after 1000 generations about 70% of spatio-temporal correlations have the best probability. After 5000 generations, the probability remains unchanged.

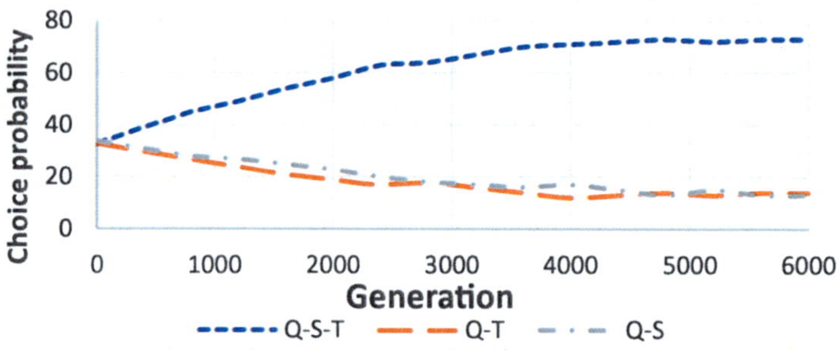

Fig. 5. The result of the QGA

However, after 200 generations the probability of spatio-temporal correlations decreases to 60% (see Fig. 6).

The described method is differed from others results described in [7, 8].

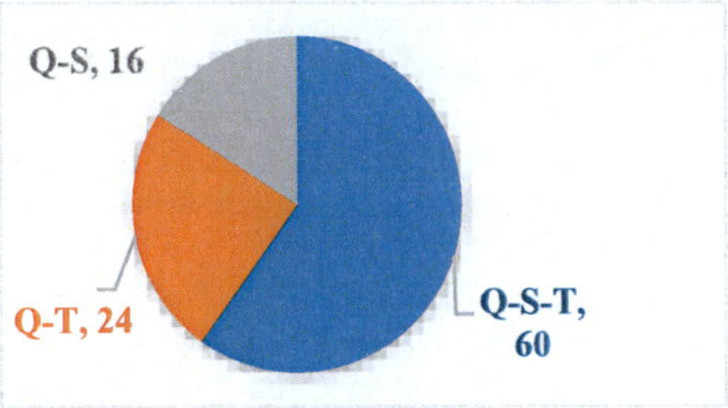

Fig. 6. The probability of spatio-temporal correlations after 200 generations

4 Conclusion

This paper describes a method for designing intelligent control systems using a quantum algorithmic gate for quantum fuzzy inference based on a quantum genetic algorithm. This method in online allows to achieve global sustainability in the face of unforeseen management situations. Building on the computing power of classical computers, new types of quantum computational intelligence tools such as quantum and soft computing are used. The presented QFI model implements a new type of quantum search algorithm with the introduction of a quantum genetic algorithm, which makes it possible to design a robust ICS with classical nonlinear objects (such objects can be considered as a standard for testing the effectiveness of an ICS) control in conditions of global instability [4, 6–8, 12]. Such a synergistic effect is achieved using hidden quantum information (as an additional resource), which obeys only the laws of quantum physics and has no analogues in classical physics.

References

1. Litvintseva, L.V., Ul'yanov, I.S., Ul'yanov, S.V., et al.: Quantum fuzzy inference for knowledge base design in robust intelligent controllers. J. Comput. Syst. Sci. Int. **46**, 908–961 (2007). https://doi.org/10.1134/S1064230707060081
2. Litvintseva, L.V., Ul'yanov, S.V.: Intelligent control systems. I. Quantum computing and self-organization algorithm. J. Comput. Syst. Sci. Int. **48**, 946 (2009). https://doi.org/10.1134/S1064230709060112
3. Barchatova, I.A., Ulyanov, S.V., Albu, V.A.: Intelligent robust control system based on quantum kb-self-organization: quantum soft computing and kansei/affective engineering technologies. In: Filev, D., et al. (eds.) Intelligent Systems'2014. AISC, vol. 323, pp. 37–48. Springer, Cham (2015). https://doi.org/10.1007/978-3-319-11310-4_4
4. Ulyanov, S.V.: Intelligent self-organized robust control design based on quantum/soft computing technologies and kansei engineering. Comput. Sci. J. Moldova **21**, 242–279 (2013)
5. Ulyanov, S.V.: Self-organizing quantum robust control methods and systems for situations with uncertainty and risk. Patent US 8788450 B2 (2014)
6. Ulyanov, S.V.: Self-organized Robust Intelligent Control. LAP Lambert Academic Publishing, Chisinau (2015)
7. Ulyanov, S.V.: Kvantovaya Relyativistskaya Informatika. LAP Lambert Academic Publishing, Chisinau (2014)

8. Ulyanov, S.V.: Quantum algorithm of imperfect kb self-organization pt i: smart control-information-thermodynamic bounds. Artif. Intell. Adv. **3**, 18–42 (2021). https://doi.org/10.30564/aia.v3i2.3171
9. Ahmadi, B., Salimi, S., Khorashad, A.S., et al.: The quantum thermodynamic force responsible for quantum state transformation and the flow and backflow of information. Sci. Rep. **9**, 8746 (2019). https://doi.org/10.1038/s41598-019-45176-1
10. Ulyanov, S.V., Litvintseva, L.V., Panfilov, S.A.: Design of self-organized intelligent control systems based on quantum fuzzy inference: Intelligent system of systems engineering approach. In: IEEE International Conference on Systems, Man and Cybernetics, vol. 4, pp. 3835-3840 (2005). https://doi.org/10.1109/ICSMC.2005.1571744
11. Ulyanov, S.V.: Quantum fast algorithm computational intelligence PT I SW HW smart toolkit. Artif. Intell. Adv. **1**, 18–43 (2019). https://doi.org/10.30564/aia.v1i1.619
12. Korenkov, V.V., Ulyanov, S.V., Shevchenko, A.A., Shevchenko, A.V.: Intelligent cognitive robotics. Pt. I: Quantum cognitive computing technologies. Kurs, Moscow (2021)

Solution of IT Investment Problem Using Fuzzy Logic-Based Approach

J. M. Babanli(✉)

Azerbaijan State Oil and Industry University, Azadlig Avenue, 20, Baku AZ1010, Azerbaijan
jale.babanli@asoiu.edu.az

Abstract. Information Technology (IT) has widely applied in organization to do services for users. It is reality, that IT investment should measure before the implementation and show the maximum result. There are some methods for determining the usefulness of an Information Technology investment prior to the application, however they are not effective. Existing methods partially operate uncertainty based information, when number of the criteria and alternatives are high. In this work we use a method for measuring usefulness of IT investment based on fuzzy type uncertainty. The kernel of this method is Zadeh's concept for aggregation of features and method proposed by author. For given study a new effective fuzzy decision making approach is used. A case of İT investment problem is considered for showing usefulness and sustainability of the given technique.

Keywords: Fuzzy number · IT investment · Aggregation

1 Introduction

Nowadays all enterprise makes investments in IT. Selection an IT investment which corresponds whole obligatory decision-making criteria that needs deep and careful investigation. In spite of high importance of information technology (IT) investments today very scarce effective methods for selection IT investment. In [1] authors analyze techniques to utilize for making Information Technology investment agreements, the way for selecting the technology and giving the reasons for its selection and by which mean the decision will affect the organization. Using fuzzy logics [2] confirm that the proposed models are more flexible and computationally efficient compared to the traditional crisp engineering economics.

In [3] given model determine recognizing profit creation for Information Technology investment in governmental organizations. The profit is counted related to decision makers in the organization.

A literature review for the Evaluation of Information Technology investments in E-Government is discussed in [4].

In [5] authors discuss express relation between market uncertainty and its affect on the return of latest IT investment.

[6] is devoted literature analysis about Multi-Criteria Decision-Making Methods in IT Investment. Authors was comparing 4 methods, such as Priority Order Technique

R. A. Aliev et al. (Eds.): ICSCCW 2021, LNNS 362, pp. 205–213, 2022.
https://doi.org/10.1007/978-3-030-92127-9_30

based on Ideal with Ideal Solutions (TOPSIS), Analytic Network Process (ANP), Data Envelopment Analysis (DEA) and Analytic Hierarchy Process (AHP). They found that from given four techniques, the AHP technique was more productive in IT investment. IT investment problem is related to a series of criteria and they are characterized by incomplete, imprecise and uncertain information, high complexity. Under such conditions, classical logic-based point of view is often not satisfactory, approximate reasoning method-based approach is more effective [7–9].

In this paper we discuss an IT investment problem.

The structure of the paper consists of 5 sections. Basic definitions used in the research are described in Sect. 2. Section 3 gives information about the statement of problem. When it comes to Sect. 4 it is related to the suggested approach of fuzzy decision making approach. An application of the methodology to Investment problem is discussed in Sect. 5.

2 Preliminaries

The considered issue is addressed through the application of the Group Decision Making Method. The method suggested by J.Babanli [10] consists of the following steps:

4.1 Establish an expert team for criteria evaluation of alternatives.

4.2 Each expert evaluates each a_i alternative on based criteria vector $F = (f_1, \ldots, f_n)$, so forms a vector $\overline{f}_i^k = (f_{i1}^k, \ldots, f_{ij}^k, \ldots, f_{in}^k)^T$.

4.3 For each a_i alternative vectors of values given by all experts $\overline{f}_i^k$ $k = 1, \ldots, K$ are aggregated to a vector as:

$$\overline{f}_i = \frac{\sum_{k=1}^{K} f_i^k}{K} \tag{1}$$

Here $\overline{f}_i$ - a_i is found based on overall opinion of the expert group. Each component of $\overline{f}_i = (f_{i1}, \ldots, f_{ij}, \ldots, f_{in})^T$. f_{ij} is calculated as:

$$\overline{f}_{ij} = \frac{\sum_{k=1}^{K} f_{ij}^k}{K} \tag{2}$$

4.4 It is calculated the arithmetic mean for every importance group of criteria for each alternative: $\varphi_l(a_i), l \in \{1, \ldots, L\}$, here L-is the number of subgroups.

4.5 Weighted average of values $\varphi_l(a_i)$ is a general aggregated index of the alternative:

$$\mathrm{Agg}(a_i) = \mathrm{IG}_1\varphi_1(a_i) + \ldots. + \mathrm{IG}_l\varphi_l(a_i) + \ldots + \mathrm{IG}_L\varphi_L(a_i). \tag{3}$$

Here, IG_l, $i = 1, \ldots L$ is the coefficient reflecting the importance of groups.

4.6 Alternatives a_i are ranked on basis of their indices, $i = 1, ..., n$. For that purpose, $Agg(a_i)$ is compared on basis of the distance to the fuzzy number Q, which represents the highest linguistic term of the scale of estimation.

$a_1 \succ a_2$ if $d(Agg(a_1), Q) < d(Agg(a_2), Q)$. Here d- is the distance among fuzzy numbers. Thus, an alternative which has a close distance to the fuzzy number Q is considered as superior. The distance d is defined as follows:

$$d(Agg, Q) = \left| \frac{agg_1 + 4 * agg_2 + agg_3}{6} - \frac{q_1 + 4 * q_2 + q_3}{6} \right| \tag{4}$$

Definition 1. Aggregation of Fuzzy Numbers. Let $C_1, \ldots, C_m$ be fuzzy numbers. An arithmetic mean-based aggregation of fuzzy numbers, C is defined as follows: $C = \frac{\sum_{i=1}^{m} C_i}{m}$,

Definition2. Distance Between Triangular Fuzzy Numbers.
Let $C_1 = (c_{11}, c_{12}, c_{13}), C_2 = (c_{21}, c_{22}, c_{23})$ be two triangular fuzzy numbers. A distance between C_1 and C_2 is defined as

$$d(C_1, C_2) = |P(C_1) - P(C_2)|,$$

were

$$P(C_1) = \frac{c_{11} + 4c_{12} + c_{13}}{6}, P(C_2) = \frac{c_{21} + 4c_{22} + c_{23}}{6}.$$

3 Statement of the Problem

Solution of IT investment problem using Fuzzy logic- based approach is the basic goal of the paper. In this work, $A = \{a_1, \ldots, a_m\}$, are alternatives and $F = (f_1, \ldots, f_n)$ are criteria vector.

Value of alternatives is determined by expert group. It shows that, every member of the expert group evaluates each alternative concerning the criteria provided (Table 1).

Table 1. Preference values of alternatives

	f_1	...	f_j	...	f_n
a_1	f_{11}^k	...	f_{1j}^k	...	f_{1n}^k
a_i	f_{1i}^k	...	f_{ij}^k	...	f_{in}^k
...	...	...	...	...	...
a_m	f_{m1}^k	...	f_{mj}^k	...	f_{mn}^k

Here, f_{ij}^{k} - is the evaluation of k-th expert for i-th alternative and j-th criterion. Criteria are grouped by importance degree. The considering issue of ranking leads to the search for an best alternative. So, a* is the best or optimal alternative is found as follow

$$\mathrm{Agg}(\mathrm{a}^*) = \max_{\mathrm{a} \in \mathrm{A}} \mathrm{Agg}(\mathrm{a}), \tag{5}$$

Here $\mathrm{Agg}(\mathrm{a}^*)$ is an aggregated index of the alternatives.

4 Application

Let's discuss the ranking problem of the IT investment. Other possible options are: a1 and a2. Criteria:

f_1 - Reducing cost; f_2 - Increasing productivity; f_3 - Accelerating process; f_4 - Reducing risk; f_5 - Increasing revenue; f_6 - Increasing accuracy; f_7 - Accelerating cash-in; f_8 - Increasing external services; f_9 - Increasing image; f_{10} - Increasing quality; f_{11} - Increasing internal services; f_{12} - Increasing competitive advantage; f_{13} - Avoiding cost.

Discussed problem was solved on the base of methodology above. A team consists of 5 experts. Each alternative is evaluated by all experts. The questionnaire was given to the a1 and a2 applications by expert. For a1 and a2 results of evaluation are represented in Tables 2 and 3.

Table 2. Values of A1 alternative

N	Criteria	Expert1	Expert2	Expert3	Expert4	Expert5
1	Reducing cost	VG	VG	G	VG	G
2	Increasing productivity	M	M	M	M	G
3	Accelerating process	VG	VG	G	VG	VG
4	Reducing risk	VG	VG	VG	VG	VG
5	Increasing revenue	M	M	M	L	L
6	Increasing accuracy	M	M	G	G	G
7	Accelerating cash-in	M	G	M	M	G
8	Increasing external services	M	M	L	L	L
9	Increasing image	M	L	L	L	M
10	Increasing quality	M	M	L	M	M
11	Increasing external services	M	L	M	L	L
12	Increasing competitive advantages	G	M	M	L	M
13	Avoiding cost	L	L	M	L	L

Table 3. Values of A2 alternative

N	Criteria	Expert1	Expert2	Expert3	Expert4	Expert5
1	Reducing cost	G	VG	G	VG	G
2	Increasing productivity	M	G	M	M	G
3	Accelerating process	VG	G	G	VG	VG
4	Reducing risk	VG	VG	G	VG	VG
5	Increasing revenue	M	M	M	VL	L
6	Increasing accuracy	M	M	G	M	G
7	Accelerating cash-in	M	G	M	G	G
8	Increasing external services	M	M	L	VL	L
9	Increasing image	M	VL	L	L	M
10	Increasing quality	M	M	L	G	M
11	Increasing external services	M	L	VG	L	L
12	Increasing competitive advantages	G	M	M	VL	M
13	Avoiding cost	L	L	G	L	L

Criteria vector $\overline{f}_i$ of expert group for each ai alternative are calculated using (2). The estimation of the criteria are represented by fuzzy numbers according to the codebook shown in Fig. 1.

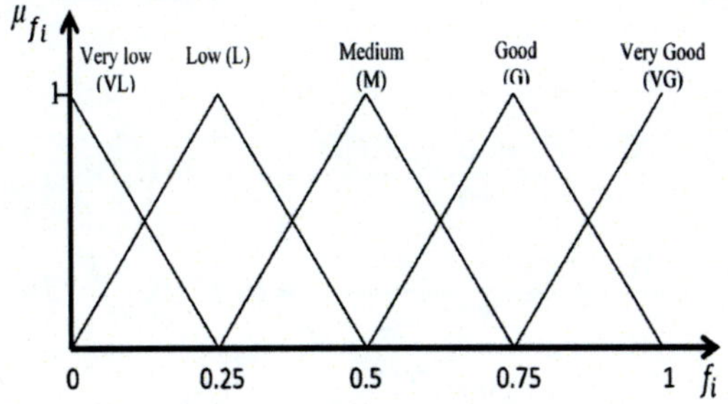

Fig. 1. Codebook of the linguistic terms

Using math expression (2), the vector $\overline{f}_i$ is determined. For instance, the determined values for a_1 are:

$$f_1 = \frac{VG + VG + G + VG + G}{5} = (0.65;\ 0.9;\ 1)$$

$$f_2 = \frac{M + M + M + M + G}{5} = (0.3; 0.55; 0.8)$$

$$f_3 = \frac{VG + VG + G + VG + VG}{5} = (0.7; 0.95; 1)$$

$$f_4 = \frac{VG + VG + VG + VG + VG}{5} = (0.75; 1; 1)$$

$$f_5 = \frac{M + M + M + L + L}{5} = (0.15; 0.4; 0.65)$$

$$f_6 = \frac{M + M + G + G + G}{5} = (0.4; 0.65; 0.9)$$

$$f_7 = \frac{M + G + M + M + G}{5} = (0.35; 0.6; 0.85)$$

$$f_8 = \frac{M + M + L + L + L}{5} = (0.1; 0.35; 0.6)$$

$$f_9 = \frac{M + L + L + L + M}{5} = (0.1; 0.35; 0.6)$$

$$f_{10} = \frac{M + M + L + M + M}{5} = (0.2; 0.45; 0.7)$$

$$f_{11} = \frac{M + L + M + L + L}{5} = (0.1; 0.35; 0.6)$$

$$f_{12} = \frac{G + M + M + L + M}{5} = (0.25; 0.5; 0.75)$$

$$f_{13} = \frac{L + L + M + L + L}{5} = (0.05; 0.3; 0.55)$$

Results of the a2 as follows:

$$f_1 = \frac{G + VG + G + VG + G}{5} = (0.6; 0.85; 1)$$

$$f_2 = \frac{M + G + M + M + G}{5} = (0.35; 0.6; 0.85)$$

$$f_3 = \frac{VG + G + G + VG + VG}{5} = (0.65; 0.91; 1)$$

$$f_4 = \frac{VG + VG + G + VG + VG}{5} = (0.7; 0.95; 1)$$

$$f_5 = \frac{M + M + M + L + L}{5} = (0.15; 0.35; 0.6)$$

$$f_6 = \frac{M + M + G + M + G}{5} = (0.35; 0.6; 0.85)$$

$$f_7 = \frac{M + G + M + G + G}{5} = (0.4; 0.65; 0.9)$$

$$f_8 = \frac{M + M + L + VL + L}{5} = (0.1; 0.3; 0.55)$$

$$f_9 = \frac{M + VL + L + L + M}{5} = (0.1; 0.3; 0.55)$$

$$f_{10} = \frac{M + M + L + G + M}{5} = (0.25; 0.5; 0.75)$$

$$f_{11} = \frac{M + L + VG + L + L}{5} = (0.2; 0.45; 0.65)$$

$$f_{12} = \frac{G + M + M + VL + M}{5} = (0.25; 0.45; 0.7)$$

$$f_{13} = \frac{L + L + G + L + L}{5} = (0.1; 0.35; 0.6)$$

For each alternative $\varphi_1(a_i)$, $\varphi_2(a_i)$, $\varphi_3(a_i)$ values are calculated. The subgroups of the importance criteria are described in Table 4.

Table 4. Importance rate of criteria

Importance rates	Criteria
IR 1 (High) (0.5 0.55 0.75)	f1 f3 f4
IR 2 (Medium) (0.2 0.35 0.45)	f2 f6 f7 f12
IR 3 (Low) (0 0.1 0 .25)	f8 f9 f10 f11 f13 f15

So,

$$\varphi_1(a_i) = \frac{f_1 + f_3 + f_4}{3}$$

$$\varphi_2(a_i) = \frac{f_2 + f_6 + f_7 + f_{12}}{4}$$

$$\varphi_2(a_i) = \frac{f_5 + f_8 + f_9 + f_{10} + f_{11} + f_{13}}{6}$$

Values for a1 as follows:

$$\varphi_1(a_1) = (0.7; 0.95; 1)$$

$$\varphi_2(a_1) = (0.325; 0.575; 0.825)$$

$$\varphi_3(a_1) = (0.11667; 0.36667; 0.61667)$$

Obtained values for a2:

$$\varphi_1(a_2) = (0.65; 0.9; 1)$$

$$\varphi_2(a_2) = (0.3375; 0.575; 0.825)$$

$$\varphi_3(a_2) = (0.15; 0.375; 0.61667)$$

The final aggregate grades for first alternative is calculated using (4): For a1:

$Agg(a_1) = (0.5; 0.55; 0.75)(0.7; 0.95; 1) + (0.2; 0.35; 0.45)*$
$*(0.325; 0.575; 0.825) + (0; 0.1; 0.25)(0.11667; 0.36667; 0.61667) = (0.415; 0.76042; 1)$

For the second alternative by the same rule, the final aggregated values were calculated. The values for a2 are shown:

$Agg(a_2) = (0.5; 0.55; 0.75)(0.65; 0.9; 1) + (0.2; 0.35; 0.45)(0.3375; 0.575; 0.825)+$
$(0; 0.1; 0.25)(0.15; 0.375; 0.61667) = (0.415; 0.76042; 1)$

Distance between aggregation value of the alternatives and very good are determined as follows according to (5):

$$d(Agg(a_1), \textit{very good}) = |0.78868 - 0.95833| = 0.16965$$

$$d(Agg(a_2), \textit{very good}) = |0.76715 - 0.95833| = 0.19118$$

The results obtained are given in Table 5:

Table 5. The rank of alternatives

Alternative	$d(\mathrm{Agg}(\mathrm{a}_1)$, very $good)$
(a_1)	0.16965
(a_2)	0.19118

The smallest distance has a higher rank. So, $a_1 \succ a_2$.

5 Conclusion

Fuzzy information based method was applied for solving investment problem. Proposed approach is based on Zadeh's idea. Selection IT investment problem consists of 13 criteria that helps to show the relevance of the method. Fuzziness is used for evaluation criteria and criteria importance. The determined results demonstrate efficiency of the proposed method.

References

1. Marc, J., Schniederjans, J.L., Hamaker, A.M.: Schniederjans: Information Technology Investment: Decision-Making Methodology. World Scientific Publishing Company, Jurong East (2010)
2. Sheen, J.N.: Information security investment decision by fuzzy economics. In: 3rd International Conference on Information Sciences and Interaction Sciences, pp. 648–653 (2010). https://doi.org/10.1109/ICICIS.2010.5534680
3. Haryanta, P., Azhari, A., Mustofa, K.: Model for measuring benefit of government IT investment using fuzzy AHP. Int. J. Adv. Comput. Sci. Appl. **11(1),** 188–192 (2020). https://doi.org/10.14569/IJACSA.2020.0110124
4. Pałka, W., Gülümser, Y., Jurisch, M.C., Wolf, P., Krcmar, H.: Basic approaches for the evaluation of IT-investments in e-government: a literature review. In: Electronic Government and Electronic Participation - Joint Proceedings of Ongoing Research of IFIP EGOV and IFIP ePart 2014, Gesellschaft für Informatik e.V., Bonn, pp. 27–34 (2014)
5. Jinlan, Ni., Deepak, K.: Information technology investment decisions under asymmetric information: a modified rational expectation model. https://digitalcommonsunomaha.edu/cgi/viewcontent.cgi?article=1022&context=isqafacpub
6. Anggraeni, W.P., Apol, P.S.: Literature review – using multi-criteria decisionmaking methods in information technology (IT) investment. In: ICBLP 2019, 13–14 February, Indonesia Copyright, Sidoarjo (2019). https://doi.org/10.4108/eai.13-2-2019.2286076
7. Zadeh, L.A.: A very simple formula for aggregation and multicriteria optimization. Int. J. Uncertainty Fuzziness Knowl. Based Syst. **24**(6), 961–962 (2016). https://doi.org/10.1142/S0218488516500446
8. Aliev, R.A., Aliev, R.R.: Soft Computing and its Application. World Scientific, Hong Kong (2001). https://doi.org/10.1142/4766
9. Gardashova, L.A.: Z-number based TOPSIS method in multi-criteria decision making. In: Aliev, R.A., Kacprzyk, J., Pedrycz, W., Jamshidi, M., Sadikoglu, F.M. (eds.) ICAFS 2018. AISC, vol. 896, pp. 42–50. Springer, Cham (2019). https://doi.org/10.1007/978-3-030-04164-9_10
10. Babanli, J.M.: Fuzzy approach for evaluation of student's performance. In: 14th International Conference on Theory and Application of Fuzzy Systems and Soft Computing, ICAFS-2020. Advances in Intelligent Systems and Computing, vol. 1306, pp. 140–147 (2020). https://doi.org/10.1007/978-3-030-64058-3_18

Application of ELECTRE Method to Decision Making Under Z-number-valued Information

K. I. Jabbarova[1(✉)] and Aynur I. Jabbarova[2]

[1] Department of Computer Engineering, Azerbaijan State Oil and Industry University, 20 Azadlig Avenue, AZ1010 Baku, Azerbaijan
konul.jabbarova@mail.ru
[2] Azerbaijan State Economic University, Baku, Azerbaijan
stat_aynur@mail.ru

Abstract. In this paper, ELECTRE method is applied to supplier selection problem under Z-valued information. All values of criteria and importance weights are described by Z-numbers. The obtained results show effectiveness of the approach.

Keywords: Multi-criteria decision making · ELECTRE method · Concordance matrix · Discordance matrix · Z-number

1 Introduction

Nowadays, supplier selection process is essential for competitive advantage of manufacturing organizations [1]. Supplier selection is a multi-criteria decision making (MCDM) problem. In scientific literature, a number of models and techniques of supplier selection exist. In [1] a case study of an application of ELECTRE method to supplier selection is considered. In [2] an investment group's local acquisition opportunities are estimated by using ELECTRE method. Six factors that affect local acquisition alternatives selection in Turkey are considered: financial performance, technical infrastructure, staff expertise and others. For estimating the criteria, AHP method is used. The ELECTRE method is used for ranking the alternatives. An application of fuzzy ELECTRE method to project selection is considered in [3]. The goal of work [4] is to compare crisp and fuzzy ELECTRE methods for supplier selection (for a case of manufacturing company in Turkey).

ELECTRE is one of the methods extensively used to solve MCDM problems. Real-world problems are usually characterized by both fuzziness and partial reliability of information. Therefore, Prof. Zadeh introduced the Z-number concept. A Z-number consists of a pair of fuzzy numbers Z = (A,B), where A is a soft constraint on a value of a variable of interest, and B is a soft constraint on a value of a probability measure of A, playing a role of reliability of A. In this article, we apply ELECTRE method to problem of supplier selection under Z-number-valued information.

The paper is structured as follows. In Sect. 2 we give a prerequisite material used in the sequel. In Sect. 3, we apply ELECTRE method under fuzziness and partial reliability of information. Section 4 concludes.

R. A. Aliev et al. (Eds.): ICSCCW 2021, LNNS 362, pp. 214–220, 2022.
https://doi.org/10.1007/978-3-030-92127-9_31

2 Preliminaries

Definition 1 A discrete Z-number [5–8]. A discrete Z-number is an ordered pair $Z = (A, B)$ where A is a discrete fuzzy number a fuzzy constraint role on values of a random variable X: X *is* A. Is a discrete fuzzy number with a membership function $\mu_B : \{b_1, ..., b_n\} \rightarrow [0, 1], \{b_1, ..., b_n\} \subset [0, 1]$, playing a fuzzy constraint role on the probability measure of A: $P(A) = \sum_{i=1}^{n} \mu_A(x_i)p(x_i)$ *is* B.

Definition 2 Operations over Discrete Z-numbers [5–8]. Assume X_1 and X_2 are discrete Z-numbers and explain information on values of X_1 and X_2. Contemplate referring to $Z_{12} = Z_1 * Z_2,\ * \in \{+, -, \cdot, /\}$. This is a computation of $A_{12} = A_1 * A_2$ as the initial stage.

The second stage includes construction of B_{12}. We put into practice that in Z-numbers Z_1 and Z_2, the 'true' probability distributions p_1, p_2 are not known precisely. Contrary, fuzzy restrictions outlined in terms of the membership functions are available:

$$\mu_{p_1}(p_1) = \mu_{B_1}\left(\sum_{k=1}^{n_1} \mu_{A_1}(x_{1k})p_1(x_{1k})\right), \mu_{p_2}(p_2) = \mu_{B_2}\left(\sum_{k=1}^{n_2} \mu_{A_2}(x_{2k})p_2(x_{2k})\right)$$

Probability distributions $p_{jl}(x_{jk}), k = 1, .., n$ infer probabilistic uncertainty over $X_{12} = X_1 + X_2$. Given any likely pair p_1, p_2, the convolution $p_{12} = p_1 \circ p_2$ is calculated as

$$p_{12}(x) = \sum_{x_1+x_2=x} p_1(x_1)p_2(x_2),\ \forall x \in X_{12}; x_1 \in X_1, x_2 \in X_2.$$

Given p_{12s}, the value of probability measure of A_{12} is calculated:

$$P(A_{12}) = \sum_{k=1}^{n} \mu_{A_{12}}(x_{12k})p_{12}(x_{12k}).$$

While p_1 and p_2 are described by fuzzy restrictions which infer fuzzy set of convolutions:

$$\mu_{p_{12}}(p_{12}) = \max_{\{p_1,p_2:p_{12}=p_1 \circ p_2\}} \min\{\mu_{p_1}(p_1), \mu_{p_2}(p_2)\}$$

Uncertainty of information on p_{12} infers fuzziness of A_{12} as a discrete fuzzy number B_{12}. The membership function $\mu_{B_{12}}$ is interpreted as

$$\mu_{B_{12}}(b_{12}) = \max \mu_{p_{12}}(p_{12})$$

$$\text{s.t. } b_{12} = \sum_{i=1}^{n} \mu_{A_{12}}(x_i)p_{12}(x_i)$$

As a result, $Z_{12} = Z_1 * Z_2$ is obtained as $Z_{12} = (A_{12}, B_{12})$.

Definition 3 Fuzzy Pareto optimality (FPO) principle-based comparison of Z-numbers [7]. Fuzzy Pareto optimality (FPO) principle allows to determine degrees of

Pareto Optimality of multiattribute alternatives. This principle may be used to compare Z-numbers as multiattribute alternatives – one attribute measures value of a variable, the other one measures the associated reliability. According to this approach, by directly comparing Z-numbers $Z_1 = (A_1, B_1)$ and $Z_2 = (A_2, B_2)$ one arrives at total degrees of optimality of Z-numbers: $do(Z_1)$ and $do(Z_2)$. These degrees are determined on the basis of a number of components (the minimum is 0, the maximum is 2) with respect to which one Z-numbers dominates another one. Z_1 is considered higher than Z_2 if $do(Z_1) > do(Z_2)$.

3 Statement of the Problem and a Solution

Let us consider MCDM problem under Z-number-valued information. The textile company placed in the Bursa textile industrial area produces shirts [9]. For the company, customer satisfaction, quality and pricing issues are important. Thus, choosing the supplier is strategically significant for the company [9]. Such decision problems are characterized by uncertain information. In view of this, we consider application of ELECTRE method to problem of supplier selection under Z-information. There are three supplier alternatives f_1, f_2, f_3 and five criteria $c_1,\ c_2,\ c_3,\ c_4,\ c_5$.

Let us solve supplier selection problem by using the algorithm of the ELECTRE method under Z-number-valued information.

At the *first step*, we construct Z-number-based decision matrix. The criteria values and importance weights are described by Z-numbers $f_{ij} = (A_{ij}, B_{ij})$ (Tables 1 and 2).

Table 1. Z-number-valued decision matrix

c_1		c_2
f_1	$\{{}^{0}/_{0.82}\ {}^{1}/_{0.96}\ {}^{0}/_{0.99}\}\{{}^{0}/_{0.7}\ {}^{1}/_{0.8}\ {}^{0}/_{0.9}\}$	$\{{}^{0}/_{0,76}\ {}^{1}/_{0.92}\ {}^{0}/_{0.99}\}\{{}^{0}/_{0.8}\ {}^{1}/_{0.9}\ {}^{0}/_{1}\}$
f_2	$\{{}^{0}/_{0.56}\ {}^{1}/_{0.76}\ {}^{0}/_{0.92}\}\{{}^{0}/_{0.7}\ {}^{1}/_{0.8}\ {}^{0}/_{0.9}\}$	$\{{}^{0}/_{0.36}\ {}^{1}/_{0.56}\ {}^{0}/_{0.76}\}\{{}^{0}/_{0.8}\ {}^{1}/_{0.9}\ {}^{0}/_{1}\}$
f_3	$\{{}^{0}/_{0.71}\ {}^{1}/_{0.88}\ {}^{0}/_{0.98}\}\{{}^{0}/_{0.7}\ {}^{1}/_{0.8}\ {}^{0}/_{0.9}\}$	$\{{}^{0}/_{0.44}\ {}^{1}/_{0.64}\ {}^{0}/_{0.84}\}\{{}^{0}/_{0.8}\ {}^{1}/_{0.9}\ {}^{0}/_{1}\}$
c_3		c_4
f_1	$\{{}^{0}/_{0.76}\ {}^{1}/_{0.92}\ {}^{0}/_{0.99}\}\{{}^{0}/_{0.7}\ {}^{1}/_{0.8}\ {}^{0}/_{0.9}\}$	$\{{}^{0}/_{0.69}\ {}^{1}/_{0.89}\ {}^{0}/_{0.99}\}\{{}^{0}/_{0.8}\ {}^{1}/_{0.9}\ {}^{0}/_{1}\}$
f_2	$\{{}^{0}/_{0.36}\ {}^{1}/_{0.56}\ {}^{0}/_{0.76}\}\{{}^{0}/_{0.7}\ {}^{1}/_{0.8}\ {}^{0}/_{0.9}\}$	$\{{}^{0}/_{0.36}\ {}^{1}/_{0.56}\ {}^{0}/_{0.76}\}\{{}^{0}/_{0.8}\ {}^{1}/_{0.9}\ {}^{0}/_{1}\}$
f_3	$\{{}^{0}/_{0.44}\ {}^{1}/_{0.64}\ {}^{0}/_{0.84}\}\{{}^{0}/_{0.7}\ {}^{1}/_{0.8}\ {}^{0}/_{0.9}\}$	$\{{}^{0}/_{0.58}\ {}^{1}/_{0.78}\ {}^{0}/_{0.94}\}\{{}^{0}/_{0.8}\ {}^{1}/_{0.9}\ {}^{0}/_{1}\}$
c_5		
f_1	$\{{}^{0}/_{0.69}\ {}^{1}/_{0.89}\ {}^{0}/_{0.99}\}\{{}^{0}/_{0.7}\ {}^{1}/_{0.8}\ {}^{0}/_{0.9}\}$	
f_2	$\{{}^{0}/_{0.42}\ {}^{1}/_{0.62}\ {}^{0}/_{0.82}\}\{{}^{0}/_{0.7}\ {}^{1}/_{0.8}\ {}^{0}/_{0.9}\}$	
f_3	$\{{}^{0}/_{0.18}\ {}^{1}/_{0.38}\ {}^{0}/_{0.58}\}\{{}^{0}/_{0.7}\ {}^{1}/_{0.8}\ {}^{0}/_{0.9}\}$	

Table 2. Importance weights of criteria

Criteria		Weights
C_1	Quality	$\{0/0.2\ 1/0.23\ 0/0.24\}\{0/0.8\ 1/0.9\ 0/1\}$
C_2	Pricing	$\{0/0.15\ 1/0.2\ 0/0.23\}\{0/0.8\ 1/0.9\ 0/1\}$
C_3	Delivery time	$\{0/0.18\ 1/0.22\ 0/0.24\}\{0/0.8\ 1/0.9\ 0/1\}$
C_4	Technology	$\{0/0.14\ 1/0.18\ 0/0.22\}\{0/0.8\ 1/0.9\ 0/1\}$
C_5	Flexibility	$\{0/0.12\ 1/0.17\ 0/0.21\}\{0/0.8\ 1/0.9\ 0/1\}$

At the second step, we calculate weighted normalized decision matrix as follows [10]:

$$Y_{ij} = \begin{bmatrix} w_1f_{11} & w_2f_{12} & \ldots & w_nf_{1n} \\ w_1f_{21} & w_2f_{22} & \ldots & w_nf_{2n} \\ \ldots & \ldots & \ldots & \ldots \\ w_1f_{m1} & w_2f_{m2} & \cdots & w_nf_{mn} \end{bmatrix} \text{ and } \sum_{j=1}^{n} w_{ij} = 1. \tag{1}$$

The obtained results are given in Table 3.

Table 3. Z-valued weighted normalized decision matrix

	C_1	C_2
y_1	$\{0/0.16\ 1/0.22\ 0/0.24\}\{0/0.59\ 1/0.73\ 0/0.84\}$	$\{0/,11\ 1/0.18\ 0/0.23\}\{0/0.66\ 1/0.81\ 0/0.94\}$
y_2	$\{0/0.11\ 1/0.17\ 0/0.22\}\{0/0.59\ 1/0.73\ 0/0.84\}$	$\{0/0.05\ 1/0.11\ 0/0.17\}\{0/0.66\ 1/0.81\ 0/0.94\}$
y_3	$\{0/0.14\ 1/0.2\ 0/0.24\}\{0/0.59\ 1/0.73\ 0/0.84\}$	$\{0/0.07\ 1/0.13\ 0/0.19\}\{0/0.66\ 1/0.81\ 0/0.94\}$
	C_3	C_4
y_1	$\{0/0.14\ 1/0.2\ 0/0.24\}\{0/0.59\ 1/0.73\ 0/0.86\}$	$\{0/0.1\ 1/0.16\ 0/0.22\}\{0/0.66\ 1/0.82\ 0/0.93\}$
y_2	$\{0/0.06\ 1/0.12\ 0/0.18\}\{0/0.59\ 1/0.73\ 0/0.86\}$	$\{0/0.05\ 1/0.1\ 0/0.17\}\{0/0.66\ 1/0.82\ 0/0.94\}$
y_3	$\{0/0.08\ 1/0.14\ 0/0.2\}\{0/0.59\ 1/0.73\ 0/0.86\}$	$\{0/0.08\ 1/0.14\ 0/0.21\}\{0/0.66\ 1/0.82\ 0/0.93\}$
	C_5	
y_1	$\{0/0.08\ 1/0.15\ 0/0.21\}\{0/0.58\ 1/0.73\ 0/0.9\}$	
y_2	$\{0/0.05\ 1/0.1\ 0/0.17\}\{0/0.59\ 1/0.73\ 0/0.9\}$	
y_3	$\{0/0.02\ 1/0.06\ 0/0.12\}\{0/0.58\ 1/0.73\ 0/0.9\}$	

At the third step, we construct concordance and discordance sets as follows [10]:

$$C_{ab} = \{j|y_{aj} \geq y_{bj}\};\ D_{ab} = \{j|y_{aj} < y_{bj}\}\ \ a, b \in \{f_i : i = 1, ..., M\}.$$

The obtained results are:

$C_{12} = \{1, 2, 3, 4, 5\}$, $C_{13} = \{1, 2, 3, 4, 5\}$, $C_{21} = \{\varnothing\}$, $C_{23} = \{5\}$, $C_{31} = \{\varnothing\}$, $C_{32} = \{1, 2, 3, 4\}$;

$D_{12} = \{\varnothing\}$, $D_{13} = \{\varnothing\}$, $D_{21} = \{1, 2, 3, 4, 5\}$, $D_{23} = \{1, 2, 3, 4\}$, $D_{31} = \{1, 2, 3, 4, 5\}$, $D_{32} = \{5\}$.

In the next step, we calculate of the concordance and discordance matrices as follows:

$$C_{ab} = \sum_{j \in C_{ab}} w_j,\ C = \begin{vmatrix} - & c_{12} & \dots & c_{1m} \\ c_{21} & - & \dots & c_{2m} \\ \vdots & & \vdots & \\ c_{m1} & c_{m2} & \dots & - \end{vmatrix};$$

$$d_{ab} = \frac{\max\limits_{j \in D_{a,b}} |v_{aj} - v_{bj}|}{\max\limits_{j} |v_{aj} - v_{bj}|},\quad D = \begin{vmatrix} - & d_{12} & \dots & d_{1m} \\ d_{21} & - & \dots & d_{2m} \\ \vdots & & \vdots & \\ d_{m1} & d_{m2} & \dots & - \end{vmatrix}.$$

The obtained results:

$$C = \begin{vmatrix} - & (0.79\ 1\ 1.14)(0.43\ 0.64\ 0.85) & (0.79\ 1\ 1.14)(0.43\ 0.64\ 0.85) \\ - & - & (0.12\ 0.17\ 0.21)(0.8\ 0.9\ 1) \\ - & (0.58\ 0.83\ 1.02)(0.37\ 0.59\ 0.85) & - \end{vmatrix}$$

$$D = \begin{vmatrix} - & - & - \\ (0.04\ 1\ 22.5)(0.33\ 0.52\ 0.78) & - & (0.06\ 1\ 16)(0.35\ 0.53\ 0.76) \\ (0.06\ 1\ 15.8)(0.3\ 0.42\ 0.7) & (0.03\ 1\ 21.4)(0.25\ 0.44\ 0.73) & - \end{vmatrix}$$

At the fifth step we determine dominant concordance matrix:

$$\overline{c} = \frac{1}{M(M-1)} \sum_{a \in \{f_i\}, a \neq b} \ \sum_{b \in \{f_i\}, a \neq b} c_{ab},$$

$$\begin{aligned} \overline{c} = \frac{1}{3(3-1)} \cdot ((0.79\ 1\ 1.14)(0.43\ 0.64\ 0.85) + (0.79\ 1\ 1.14)(0.43\ 0.64\ 0.85) \\ +(0.12\ 0.17\ 0.21)(0.8\ 0.9\ 1) + (0.58\ 0.83\ 1.02)(0.37\ 0.59\ 0.85)) \\ = (0.38\ 0.5\ 0.585)(0.13\ 0.3\ 0.67) \end{aligned}$$

Here $\overline{c}$ is the critical value, which can be determined by average dominance index. Thus, a Boolean matrix (Q) is given by:

$$\begin{cases} q_{ab} = 1 & if \quad c_{ab} \geq \overline{c} \\ q_{ab} = 0 & if \quad c_{ab} < \overline{c} \end{cases},\quad Q_{a,b} = \begin{pmatrix} 0 & 1 & 1 \\ 0 & 0 & 1 \\ 0 & 1 & 0 \end{pmatrix}.$$

In the sixth step we determine dominant discordance matrix. The preference of dissaticfaction can be measured by discordance index:

$$\overline{d} = \sum_{a=1}^{m}\sum_{b}^{m} d(a,b)/m(m-1),$$

$$\overline{d} = \frac{1}{3(3-1)} \cdot ((0.04\ 1\ 22.5)(0.33\ 0.52\ 0.78) + (0.06\ 1\ 16)(0.35\ 0.53\ 0.76)$$
$$+(0.06\ 1\ 15.8)(0.3\ 0.42\ 0.7) + (0.03\ 1\ 21.4)(0.25\ 0.44\ 0.73))$$
$$= (0.03\ 0.67\ 12.62)(0.02\ 0.1\ 0.38)$$

Based on the discordance index, the discordance index matrix (G) is as follows:

$$\begin{cases} g(a,b) = 1 & if \quad d(a,b) \leq \overline{d} \\ g(a,b) = 0 & if \quad d(a,b) > \overline{d} \end{cases} \qquad G(a,b) = \begin{pmatrix} 1 & 1 & 1 \\ 0 & 1 & 0 \\ 0 & 0 & 1 \end{pmatrix}.$$

In the next step, we calculate the aggregate dominance index matrix:

$$e(a,b) = \begin{pmatrix} 0 & 1 & 1 \\ 0 & 0 & 1 \\ 0 & 1 & 0 \end{pmatrix} \times \begin{pmatrix} 1 & 1 & 1 \\ 0 & 1 & 0 \\ 0 & 0 & 1 \end{pmatrix} = \begin{pmatrix} 0 & 1 & 1 \\ 0 & 0 & 0 \\ 0 & 0 & 0 \end{pmatrix}.$$

Matrix E provides the order of choice, that is, if $e(a,b) = 1$ then alternative a is a better choice than b. So, the row of E matrix which has least amount of $e = 1$ will be eliminated. Thus, the best alternative is f_1.

4 Conclusion

In this paper we apply ELECTRE method to supplier selection problem under Z-valued information. The calculation of concordance and discordance matrices is described. The obtained results demonstrate applicability of this approach.

References

1. Birgün, S., Cihan, E.: Supplier selection process using ELECTRE method. In: IEEE International Conference on Intelligent Systems and Knowledge Engineering, pp.634–639. IEEE, Hangzhou, China (2010). https://doi.org/10.1109/ISKE.2010.5680767
2. Yücel, M.G., Görener, A.: Decision making for company acquisition by ELECTRE Method. Int. J. Supply Chain Manag. **5(1),** 75–83 (2015)
3. Rouyendegh, B.D., Erol, S.: Selecting the best project using the fuzzy ELECTRE method. Math. Probl. Eng. **2012** (2012). (1024-123X), https://doi.org/10.1155/2012/790142
4. Sevkli, M.: An application of the fuzzy ELECTRE method for supplier selection. Int. J. Prod. Res. **48**(12), 3393–3405 (2010). https://doi.org/10.1080/00207540902814355
5. Aliev, R.A., Alizadeh, A.V., Huseynov, O.H.: The arithmetic of discrete Z-numbers. Inf. Sci. **290**, 134–155 (2015). https://doi.org/10.1016/j.ins.2014.08.024

6. Aliev, R.A, Alizadeh, A.V., Huseynov, O.H., Jabbarova, K.I.: Z-number based linear programming. Int. J. Intell. Syst. **30**(5), 563–589 (2015)
7. Aliev, R.A., Huseynov, O.H.: Decision theory with Imperfect Information. World Scientific, Singapore (2014)
8. Aliev, R.A., Huseynov, O.H., Aliyev, R.R., Alizadeh, A.V.: The Arithmetic of Z-numbers. World Scientific Theory and Applications, Singapore (2015)
9. Arikan Kargi, V.S.: Supplier selection for a textile company using the fuzzy TOPSIS method. Yönetim ve Ekonomi **23**(3), 789–803 (2016)
10. Triantaphyllou, E., Shu, B., Sanchez, S. N., Ray, T.: Multi-criteria Decision Making: An Operations Research Approach. Encyclopedia of Electrical and Electronics Engineering, vol. 15, pp. 175–186. Wiley, New York (1998). (Webster, J.G. (ed.))

Fuzzy Logic Evaluation of Teacher Competencies Through Students Perceptions

Saide Sadikoglu[1](✉) and Konul Memmedova[2]

[1] Faculty of Tourism and Hotel Management, Near East University, North Cyprus, Mersin 10, Mersin, Turkey
saide.sadikoglu@neu.edu.tr

[2] Department of Psychological Counselling and Guidance, Near East University, North Cyprus, Mersin 10, Mersin, Turkey
konul.memmedova@neu.edu.tr

Abstract. Today, the importance of foreign language teaching and learning has increased, countries have participated in the race to teach their language as a foreign language. In this study is evaluated teacher competencies through students perceptions of teaching Russian as a foreign language. A 5-point Likert style questionnaire having 21 items was applied to measure the opinion of the international students taking Russian as a foreign language. Results of measurement are evaluated using the fuzzy-logic approach. This approach reduces a subjective error introduced by survey participants in the process of measurement and to grades the competency of each teacher individually.

Keywords: Teaching Russian · Foreign language · Teacher competencies · Fuzzy logic evaluation

1 Introduction

The 21st century has witnessed many rapid developments and changes in many areas. As the needs of societies change, it needs to be restructured. It has become imperative for societies to keep up with these needs. Foreign language education, a long-term process, imposes some duties and responsibilities on language trainers and learners [1]. Care should be taken to select and use materials following the criteria of teaching methods and theories to develop basic language skills. Many activities, including cognitive processes, assessment and evaluation stages, play an essential role in developing basic language skills. "The specific goals of people (and groups) who want to learn a language may differ from each other, but the general goals in teaching a foreign language- to be able to understand, read and speak the foreign language taught correctly; furthermore, to develop positive attitudes towards the counterculture, literature, art and science and become able to understand the elements of this culture; verbal and written communication in the target language [2].

R. A. Aliev et al. (Eds.): ICSCCW 2021, LNNS 362, pp. 221–228, 2022.
https://doi.org/10.1007/978-3-030-92127-9_32

The teachers' professional competence is understood as a complex integrative education, including a set of professionally significant knowledge, skills and abilities, personal qualities, specific ways of thinking and an understanding of responsibility.

The teacher's professional competence who teaches Russian as a foreign language (RFL) is the possession of professional significant knowledge, skills, and abilities that ensure the effectiveness of language education at university [3]. In this context, it is crucial using of the efficiency of the cognitive and brain-based approaches in teaching RFL [4].

In the structure of the professional competence of an RFL teacher, the following competencies can be distinguished: Socio-linguistic, Communicative, literature, Sociocultural, Social, Methodological, Psychological, Pedagogical, Personal and Technological competencies.

The professional competence of a faculty member is one of the important factors that determines the quality of education. Many studies [5, 6] have been shown that "Students are the recipients of educational services and should therefore measure the quality of the output". The measurement of the quality of education through students perception has been applied in many countries. Therefore, it is important to measure student perception for evaluation the teacher competencies. teacher, who teaches Russian for foreigner.

In [7] is developed a highly reliable scale for measuring teaching effectiveness through student perception. In [8], the authors considered the impact of the teaching-learning process on student perception in teaching Turkish to foreigners.

Literature analysis of all studies mentioned above shows that published papers use statistical approaches to find the relationship between teacher competence and student perception.

Li in [9] developed a fuzzy Likert scale with high precision measurements compared to the traditional Likert scale.

In [10] developed fuzzy the Likert scale with high precision by applying fuzzy logic, based on fuzzification of survey participants opinion, validity and discrimination of each question. In [11] is considered the application of fuzzy logic in educational and psychological researches. The efficiency of the Likert scale is improved by using the Z-number approach in [12].

2 Set of Problem

This study aims to the fuzzy-logic evaluation of the teachers' competencies teaching Russian to foreigners according to student perceptions. In the study, teacher competencies were investigated for the teaching-learning process (teaching profession). The authors adapted the questionnaire used to measure the students' perception.

The conceptual model establishing a relationship between the teaching-learning process and students' perception based on the fuzzy logic approach is given in Fig. 1. This model shows a link between the teacher competence defined by the quality of the learning-teaching process and student perceptions.

Data, which are collected via questionnaires, have been evaluated using the fuzzy logic approach.

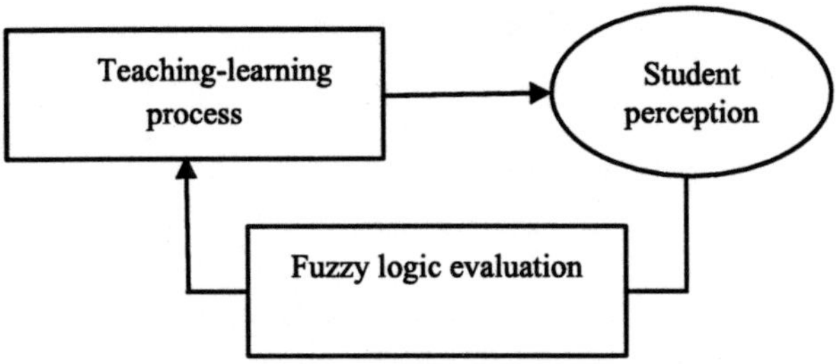

Fig. 1. Conceptual model of the study

3 Methodology

In the research it is used a survey technique (see Table 1) involving 21 items (questions) as a data collection tool. The participants were 50 overseas students in 3rd and 4th years from Turkey, the Middle East and African countries studying at the faculty of Tourism and Hotel Management of the Near East University of North Cyprus. The data were measured using a 5-point Likert-type scale (1-*Strongly Disagree, 2—Agree, 3-Neither agree nor disagree, 4-Disagree, 5-Strongly Agree*).

Table 1. Questionnaire-competencies for the learning-teaching process/tcaching Russian

	Questions	1	2	3	4	5
1	Knows and applies for the Russian teaching method as a foreign language					
2	Competences about the comprehension and expression of language skills					
3	Using the Russian as the predominant means of the classroom communication					
4	Encouraging students to speak Russian beginning on the first-day of class					
5	Using activities and assignments that draw learners' attention to specific grammatical features					
6	Competence in communicative language					
7	Proficient tat working with small and large group					
8	Providing opportunities for students to speak in the lesson					
9	Using technology effectively					
10	Motivate students in the learning process					
11	Know at least one foreign language as an instrument of language					
12	Giving clear, understandable classroom instructions					
13	Using textbooks and teaching materials effectively					

(continued)

Table 1. (*continued*)

	Questions	1	2	3	4	5
14	Interacts satisfactorily with the students					
15	Using the teaching materials that support teaching apart from the textbook					
16	Using activities and assignments that draw students' attention					
17	Using recast (correct reformulations of students' speech) as a preferred method of corrective feedback					
18	Using drama and poem to strengthen teaching					
19	Using didactic and educational games to reinforce the teaching process in lessons					
20	Prepares and directs discussion environments to reinforce classroom teaching					
21	Competences related to the evaluation and assessment of the students					

The fuzzy-logic evaluation of the data is based on the structure shown in Fig. 2.

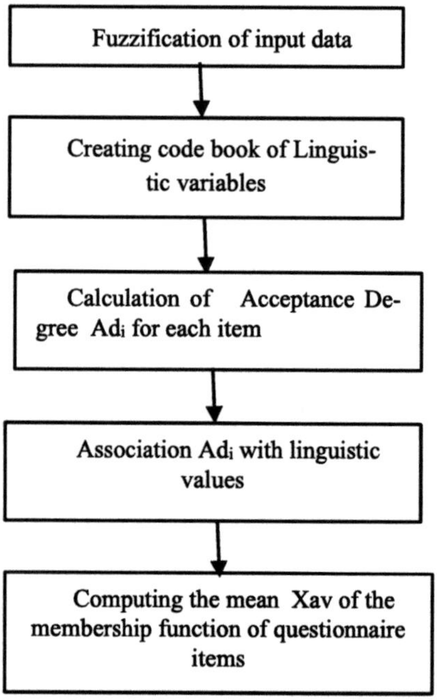

Fig. 2. Fuzzy logic evaluation steps

4 Results

Fuzzy evaluation of the teacher competency is based on the approach given in [13].

Step 1. Fuzzification of the input variable.

For fuzzification of input variables as used the trapezoidal membership functions with the degree of acceptance: ***"Strongly Agree" -1, "Agree" -2, "Neither agree nor disagree" -3, "Disagree" -4, "Strongly Disagree"*** is shown in Fig. 3.

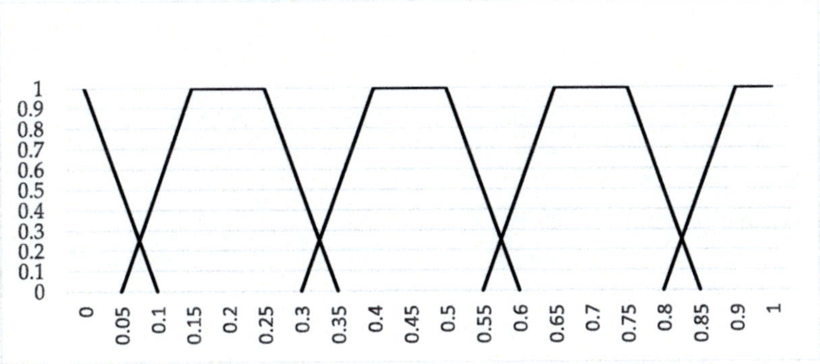

Fig. 3. Graphically representation of linguistic variable

Step 2. Creating the codebook for a linguistic variable given in Table 2.

Table 2. Codebook for linguistic variable

Scale	Level	Trapezoid MF value
1	Strongly Disagree	1/0, 0.5/0.05, 0/0.1
2	Disagree	0/0.05, 1/0.15, 1/0.25,0/0.35
3	Undecided	0/0.3, 1/0.4, 1/0.5, 0/0.6
4	Agree	0/0.55, 1/0.65, 1/0.75, 0/0.85
5	Strongly Agree	0/0.8 1/0.9, 1/1, 1/1

Step 3. Calculate the value of the Acceptance degree based on the survey results. For each question, the crisp value of Degree of Acceptance (Ad) is calculated according to the following rules:

$$\text{if } X_{i-yes} \neq X_{i-no} \text{ then } Ad_i = X_{i-yes} - X_{i-no}, \text{ otherwise } 0. \quad (1)$$

Then obtained values of Acceptance degree are normalized according to this formula

$$Ad_{i\_norm} = \frac{Ad_i}{|\max\{Ad_i, \ldots, Ad_n\} - \min\{Ad_i, \ldots, Ad_n\}|} \quad (2)$$

Table 3. The linguistic values of degree of acceptance

N	Questions	Ad_{i_norm}	Degree of acceptance
X_1	Knows and applies for the Russian teaching method as a foreign language	0,45	Neither agree nor disagree
X_2	Competences about the comprehension and expression of language skills	0, 72	Agree
X_3	Using the Russian as the predominant means of the classroom	1.00	Strongly agree
X_4	Encouraging students to speak in Russian beginning on the first day of class	0,42	Neither agree nor disagree
X_5	Using activities and assignments that draw learners' attention to specific grammatical features	0,06	Disagree
X_6	Competence in communicative language	0,56	Agree
X_7	Proficiency to work with a small and large group	0,65	Neither agree nor disagree
X_8	Providing opportunities for students to speak in the lesson	0,67	Agree
X_9	Using technology effectively	0,35	Disagree
X_{10}	Motivates students in learning process	1,00	Strongly agree
X_{11}	Knowing at least one foreign language as an instrument language	0,46	Neither agree nor disagree
X_{12}	Giving clear, understandable classroom instructions	0,67	Agree
X_{13}	Using textbooks and teaching materials effectively	0.72	Agree
X_{14}	Interacts satisfactorily with the students	0,68	Agree
X_{15}	Using the teaching materials that support teaching apart from the textbook	0,31	Disagree
X_{16}	Using activities and assignments that draw students attention	0.45	Neither agree nor disagree
X_{17}	Using recast (correct reformulations of students speech) as a preferred method of corrective feedback	0,87	Disagree
X_{18}	Uses drama and poem to strengthen teaching	1.00	Strongly agree
X_{19}	It makes use of didactic and educational games to reinforce the teaching process in lessons	0.52	Neither agree nor disagree
X_{20}	Prepares and directs discussion environments that reinforce classroom teaching	0,48	Neither agree nor disagree

(*continued*)

Table 3. (*continued*)

N	Questions	Ad_{i_norm}	Degree of acceptance
X_{21}	He/she is competent related to the evaluation and assessment of the students	0.41	Agree

where n = 1,20, and for scale [0,1]

$$Ad_{i\_final} = Ad_{i\_norm} + \left|\min\left\{Ad_{1_{norm}}, \ldots, Ad_{n_{norm}}\right\}\right|$$

Step 4. Then, based on the results obtained in Step 2 (Ad_{final}), the linguistic values of the Degree of Acceptance are outlined for every question, and therefore the following table is created:

It is plain that some values of Ad_{i_final} may be assigned concurrently to two overlapping fuzzy sets. For example, Ad_{3_final} can belong each- to "Undecided" and to "Disagree", Ad_{8_final} - each to "Undecided" and to "Disagree", Ad_{16_final} - each "Undecided" and to "Disagree", Ad_{20_final} - each to "Undecided" and to "Disagree". Ultimately, associating to a fuzzy set is decided through by the higher value of the membership function.

Step 5. The overall Degree of Acceptance is calculated as the mean value of of the linguistic variables [13].

For n trapezoidal fuzzy numbers (a_i,b_i,c_i,d_i), $i = 1,\ldots,n$ the average value is:

$$X_{avg} = \left(\frac{1}{n}\sum\nolimits_1^n a_i, \frac{1}{n}\sum\nolimits_1^n b_i, \frac{1}{n}\sum\nolimits_1^n c_i, \frac{1}{n}\sum\nolimits_1^n d_i\right) \quad (3)$$

With this formula we get the $X_{avg} = (0.5, 0.59, 0.68, 0.75)$, which corresponds to linguistic value "Agree".

For arithmetic mean, $Ad_i = 0.61$, which also relates to the linguistic meaning of "Agree".

5 Conclusions

The quality of the teaching-learning process is one of the important dimension related to teacher competencies. In the study is developed a fuzzy logic model for evaluating teacher competencies measured through student perception.

Applying the fuzzy logic evaluation in the present study enables handling subjective measurement errors of the survey participants. Moreover, it offers the possibility to evaluate the teaching-learning process individually. Results of this research can be used for individual scoring of teachers competencies.

This model postulates that if teacher competency is higher, the greater its impact on students perceptions.

In the paper, students perception is analyzed using one important dimension (learning-teaching process). In real cases, teachers competencies can be defined in the areas such as subject area/subject area knowledge, program and content knowledge and tracing and assessing skills. Results obtained in this study can be extended for defining students perceptions in the multidimensional cases.

References

1. Pennycook, A.: The concept of method, interested knowledge, and the politics of language teaching. TESOL Q. **23**, 589–618 (1989). https://doi.org/10.2307/3587534
2. Ceyhan, E.: Foreign Language Teaching Technology. Morpa, Istanbul (2007)
3. Shibko, N.L.: Communicative competence as a component of the professional competence of an RFL teacher. Bullet. BOOPRYAI. Sci. Method. J. **05**(1), 62 – 66 (2016). https://doi.org/10.2307/3587534
4. Demirel, Ö.: Foreign Language Teaching. Pegem, Ankara (2012)
5. Malechwanzi, J.M., Hongde, L., Lu, W.: Students perceptions and faculty measured competencies in higher education. Int. J. Higher Educ. **5**(3), 55–69 (2016). https://doi.org/10.5430/ijhe.v5n3p56
6. Sakthivel, P.B., Rajendran, G., Raju, R.: TQM implementation and students' satisfaction of academic performance. TQM Mag. **17**(6), 573–589 (2005). https://doi.org/10.1108/09544780510627660
7. Catanova, V.M., Steve, H.: Student perception of teaching effectiveness: development and validation of the evaluation of teaching competencies scale (ETCS). Assess. Eval. High. Educ. **36**(6), 701–717 (2011)
8. Yıldız, Ü., Tepeli, Y.: a study on teacher qualification in teaching Turkish as a foreign language. Int. J. Lang. Acad. **2**(4), 564–578 (2014)
9. Qing, L.: A novel likert scale based on fuzzy sets theory. Expert Syst. Appl. **40**(5), 1609–1618 (2013)
10. Paothai, V.: Application of fuzzy logic to improve the Likert scale to measure the latent variable. Kasetsart J. Soc. Sci. **38**(3), 337–344 (2017)
11. Sadikoglu, F., Huseynov, O., Memmedova, K.: Z-Regression analysis in psychological and educational researches. Proc. Comput. Sci. **102**, 385–389 (2016)
12. Aliev, R., Memmedova, K.: Application of-Z-number based modelling in psychological research. Comput. Intell. Neurosci. **1**(1), 1–7 (2017). https://doi.org/10.1155/2015/760403
13. Škoda, M., Brožova, H.: Weighted fuzzy group decision-making in recruitment. RELAIS **2**(3), 1–15 (2019). https://repositorio.lasalle.mx/handle/lasalle/1771

Graph-Based Load Balancing Model for Exascale Computing Systems

Araz R. Aliev and Nigar T. Ismayilova(✉)

Azerbaijan State Oil and Industry University, Baku, Azerbaijan
nigar.ismailova@asoiu.edu.az

Abstract. This study investigates the possibility of load balancing handling in an Exascale environment by application of matching algorithms, rather than a Hungarian algorithm to the definition of similarities between computing system structure after and before dynamic and interactive event occurrence, which is the most important benchmark of Exascale environment. To illustrate the relations between the processes and also resources at two consecutive time moments was used bipartite graph. These graphs can be extended to Bayesian belief networks and by using statistical inference methods it is possible to optimize load balancing. Was found a way for designating the optimal machine for execution of the given task, guiding by optimal load balancing solution at the previous time. A suitable bipartite graph maximum matching algorithm was applied for the definition of an optimal mapping between requests and machines after calculating the distances – similarities between resources and requests vectors in two consecutive periods. Considering the uncertain nature of the process was proposed to use fuzzy graphs for getting better results.

Keywords: Exascale systems · Load balancing · Dynamic and interactive event · Fuzzy graphs

1 Introduction

Research on optimization of load balancing in high performance computing systems by application of different algorithms has met great success. Representation of nodes and tasks in distribution systems as vertices of graph models has been widely adopted for the definition of the optimal assignment between processes and machines [1–8].

Proposed methods for task scheduling in distributed systems for both static and dynamic load balancing have been used for many years in the field of distributed systems. A challenging problem that emerges in this situation is uncertainty in the behavior of the computing system related to modification of resource attributes and process requirements after dynamic and interactive (D&I) event occurrence. Divergence of hardware requirements and also heterogeneity of tasks make it difficult to handle load balancing in the Exascale environment [9–20].

One of the simplest ways for managing load balancing in the Exascale environment is representing computing system structure by graph models. There is previous research

R. A. Aliev et al. (Eds.): ICSCCW 2021, LNNS 362, pp. 229–236, 2022.
https://doi.org/10.1007/978-3-030-92127-9_33

using a bipartite graph modeling approach for handling static load balancing problems by characterizing vertex sets as resource attributes and process requirements of the distributed system [21, 22]. However, the solution of the optimization problem for tasks assignment to machines in an Exascale environment by utilization of maximum matching algorithms remains an open problem in this field.

Hence, we have analyzed the opportunity of application of bipartite graph model to dynamic load balancing in Exascale environment and investigated whether information about assignment between the nodes of the computing system at the initial time can be useful for handling load balancing at the current time moment. Our research aims to find the maximum matching between process requirements (resource attributes) at two consecutive time moments by calculating similarities between them. Ways for the definition of optimal assignment between the nodes of the computing system and tasks were investigated based on the solution of optimized load balancing at the previous time.

Possibility of dissimilarity or partial similarity between assignments of the requests to resources on two consecutive time moments causes to use an alternative approach. In this work, we suggest applying fuzzy graphs for measuring similarities between matchings in two consecutive time moments for the distributed computing system.

2 Literature Review

There have been numerous studies to investigate optimizing load balancing in distributed systems such as cloud computing, grid computing, peer-to-peer computing systems which have been successfully applied in practice. A series of recent studies have indicated that load balancing in cloud computing can be solved by using algorithms for network optimization, task scheduling, optimal strategies from games theory, genetic algorithms, etc. [1–3].

A large number of existing studies in the broader literature have been examined grid load balancing systems, where computational nodes are characterized by heterogeneity, dynamicity, and so on [4–8]. Several methods reported in literature addressed to optimize the distribution of items to nodes in distributed hash tables in P2P systems [9–11].

The heterogeneous and dynamic nature of Exascale computing systems complicates the solution of system management, job scheduling, and load balancing problems, which requires dynamic and adaptive solution mechanisms by implementing load balancing in all levels of the computing platform [12, 13].

Several theories have been proposed to obtain an optimal load balancing scheme in heterogeneous distributed systems, some focusing on static load balancing others on job scheduling in the dynamic environment from different points of view [14–18]. Advantage of the method introduced by Catalyurek et al. [18] for distribution of the data among processors based on hypergraph partitioning is that, this method decreases the repartitioning time, finds optimal load balancing for the new data, and reduces costs for data migration and node communication [19].

However, the existing research has many problems in finding the best assignment between processes and resources of the distributed system in the Exascale environment after the D&I event occurring. The previous studies propose different methods and algorithms for dynamic load balancing which are usually problematic to load balancing under the diversity of hardware capabilities of the computing system and dynamic complexion of executing processes [20, 21]. One of the tough challenges disturbing scientists in this field is establishing an efficient algorithm for dynamic load balancing related to both resource attributes and process requirements dynamicity and heterogeneity.

This was a productive solution to apply matching algorithms for weighted and unweighted bipartite graphs, where was defined the optimal matching between the tasks and computational machines in distributed systems [22]. Satisfactory results of solution of the problem using matching algorithms in bipartite graphs for system management in high performance computing systems demonstrate the efficiency of using bipartite graph models for optimization of load balancing in Exascale environment. This paper addresses the finding of optimal matching in bipartite graphs, where vertex sets contain process requirements (resource attributes) at the current time and at the initial time connected with edges weighted by similarities of these vectors.

3 Proposed Method

The main problem of load balancing in the Exascale environment is the modification of process requirements and resource attributes after the D&I event occurring. In this case, load balancing should regularize the system, assign processes to resources in an optimal way, in the condition that resource usage must be 100%.

For determining the optimal assignment for load balancing it is possible to use the assignment of processes to resources at the previous step. Let's assume that, at the time moment t^0 load balancing has assigned processes to resources in an optimal way. Process requirements at this moment are

$$PR_j^0 = < PR_{j1}^0;\ PR_{j2}^0; \ldots;\ PR_{jL}^0 >$$

where L – is the number of parameters of process requirements vector, $j = 1, \ldots, n$ and n is the number of processes and resource attributes is represented by the following expression:

$$RA_i^0 = < RA_{i1}^0;\ PR_{i2}^0; \ldots;\ PR_{iK}^0 >$$

where K – is the number of parameters of resource attributes vector, $i = 1, \ldots, m$ and m is the number of resources. After the D&I event occurrence new process requirements and resource attributes vectors appear PR_q^1 and RA_h^1, where $q = 1, \ldots, Q$, Q is the number of processes after the D&I event occurred and $h = 1, \ldots, H$, H – is the number of corresponding resources:

$$PR_q^1 = < PR_{q1}^0;\ PR_{q2}^0; \ldots;\ PR_{qL}^0 >$$
$$RA_h^1 = < RA_{h1}^0;\ PR_{h2}^0; \ldots;\ PR_{hK}^0 > .$$

For assignment of the processes to resources after the D&I event occurred it is possible to use the mapping between the domain and the range of load balancing at the previous step. From this perspective, we can define relations between the process requirement vector at the current time and the previous moment PR_j^0 and PR_p^1, identically between the vectors RA_i^0 and RA_h^1. It is attainable to represent this process by bipartite graphs $G_1(U_1, V_1, E_1)$ and $G_2(U_2, V_2, E_2)$, where the U_1 vertices set represents process requirements at the time moment t^0 and V_1 the set represents process requirements at the time moment t^1, correspondingly PR_j^0 and PR_p^1. Similarly, U_2 vertices set represents resource attributes at the time moment t^0 and V_2 the set represents resource attributes at the time moment t^1, correspondingly RA_i^0 and RA_h^1. In each graph edge sets, E_1 and E_2 are regarded as connections between the vertices sets. Let's consider that, given graphs are weighted graphs. As weights, it is credible to use the similarities between mentioned vectors, PR_j^0 and PR_p^1. Similarities can be defined as Euclidean distance between these vectors:

$$D^{PR}(PR_j^0; PR_p^1) = \frac{1}{\sqrt{n}}\sqrt{\sum\nolimits_{l=1}^{L}\left(PR_{jl}^0 - PR_{pl}^1\right)^2}, \tag{1}$$

analogously

$$D^{RA}(RA_i^0; RA_h^1) = \frac{1}{\sqrt{m}}\sqrt{\sum\nolimits_{k=1}^{K}\left(RA_{ik}^0 - PR_{hk}^1\right)^2} \tag{2}$$

Calculation of these distances, more clearly the similarities between process requirements at the initial time and at the time after D&I occurred, also between resource attributes at the initial time and current time provide information about weights of edges in the bipartite graph (Fig. 1).

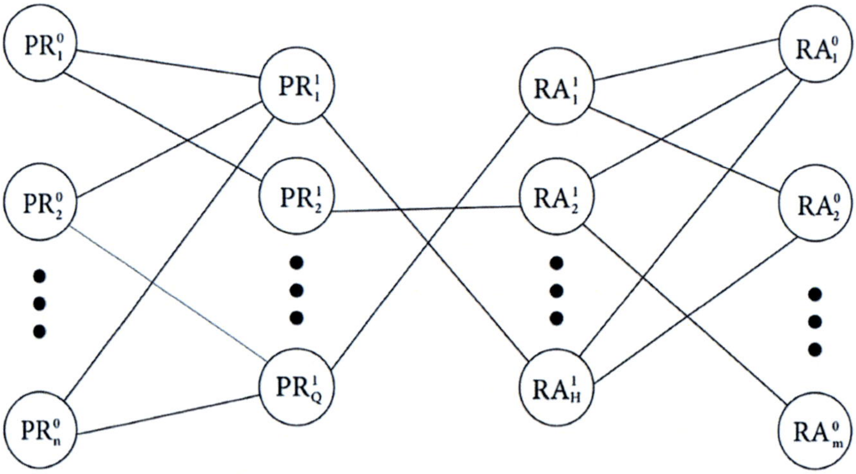

Fig. 1. Matching between resource attributes and process requirements at two consecutive moments

Let's assume that, number of processes after D&I occurring is greater than the number of processes at the previous time moment, i.e. $n \geq P$. Similarly, the quantity of processes after the D&I event occurring is larger than the number of processes at the previous time moment, i.e. $m \geq H$. Matrices contained distances calculated by Eq. 1 and Eq. 2 are the adjacency matrices of mentioned bipartite graph.

Application of Hungarian algorithm for finding optimal matching between vertex sets of a mentioned bipartite graph, more accurately definition of the best assignment between processes and resources in Exascale computing environment is the most appropriate method; hence adjacency matrix needs to be square, with the same number of rows and columns. Addition *n-P* number of zero rows at the end converts the matrix to symmetric:

$$D^{PR} = \begin{array}{l} P\left\{\begin{array}{l} d_{11}\ d_{12}\ \ldots\ d_{1n} \\ d_{21}\ d_{22}\ \ldots\ d_{2n} \\ \quad\ldots \\ d_{Q1}\ d_{Q2}\ \ldots\ d_{Qn} \end{array}\right. \\ n-Q\left\{\begin{array}{l} 0\,0\ldots0 \\ \quad\ldots \\ 0\,0\ldots0 \end{array}\right. \end{array} \quad (3)$$

here D^{PR} is the adjacency matrix of the bipartite graph. In this graph vertex sets correspond to process requirements at the initial time and process requirements after D&I occurrence. Elements of the matrix represent the weights of edges in the graph, more accurately similarities between mentioned vectors computed by the Eq. 1.

$$D^{RA} = \begin{array}{l} H\left\{\begin{array}{l} \widetilde{d_{11}}\ \widetilde{d_{12}}\ \ldots\ \widetilde{d_{1m}} \\ \widetilde{d_{21}}\ \widetilde{d_{22}}\ \ldots\ \widetilde{d_{2m}} \\ \quad\ldots \\ \widetilde{d_{H1}}\ \widetilde{d_{H2}}\ \ldots\ \widetilde{d_{Hm}} \end{array}\right. \\ m-H\left\{\begin{array}{l} 0\,0\ldots0 \\ \quad\ldots \\ 0\,0\ldots0 \end{array}\right. \end{array} \quad (4)$$

D^{RA} – is an adjacency matrix of weights related to edges between the vertex sets of bipartite graph which represents resource attributes vectors at the initial time and current time. Elements of the matrix are regarded as similarities between process attribute vectors at two consecutive time moments. Following these operations application of the Hungarian algorithm is used for finding optimal matching between process requirements and resource attributes.

After the definition of sequences of optimal elements in each row and each column of the matrices D^{PR} and D^{RA} which corresponds to the maximum sum of weights in the adjacency matrix (maximum sum of similarities between process requirements vectors and resource attributes vectors at two consecutive time moments) via Hungarian algorithm application, optimal assignment for the domain, and range of load balancing becomes manageable. Determined element in the first row of the matrix D^{PR} is d_{1i^*} indicates that the first process after the D&I occurring has the optimal similarity with the i^*th process for the moment before D&I occurring. On condition that i^*th process for the moment before D&I occurring was executed on j^*th resource, found the element of the j^*th column in the matrix D^{RA} after application of Hungarian algorithm gives the number of the resource at the current time related to the best assignment for the 1st process after D&I occurring, i.e. if the searched element in the j^*th column of the matrix D^{RA} is $\widetilde{d_{h^*j^*}}$, the then satisfied resource is the h^*th resource.

Let's consider that, during analyzing of matchings between assignments of requests to resources in two consecutive time moments besides the similarities of two vertices of the graph we can measure non-membership of the one vertex to the given relation. In this case, the best way for optimization of the load balancing in the Exascale computing system is to use fuzzy graphs and apply corresponding algorithms for matching in fuzzy graphs [23, 24].

4 Conclusion

The mentioned aspect of the research suggested that analyzing the optimization structure of load balancing at the previous step makes it possible to find an appropriate machine for the task extracted after the D&I event is found. This allows the application of the Hungarian algorithm for finding maximum matching after representing vertex sets of the bipartite graph by process requirements (and resource attributes) at the current time and the initial time and using similarities between them as weights of the edges connected them. Finding the most similar process for arising task after D&I occurrence among the processes executed at the previous time moment allows assignment of the given task to the resource allotted for the similar process. Due to the uncertain nature of the process, it is possible to define that given assignments are close to each other with the calculated value of the distance, at the same time they are not similar by another measurement. For elimination of this problem and realization of the optimal load balancing construction of the fuzzy graphs based on the investigated process and application of the suitable algorithms was suggested for the improvement of the given approach.

The reported approach provides a good starting point for handling load balancing in Exascale systems by application of graph models and utilization of matching algorithms. The possibility of implementation of the investigated method guarantees further research.

References

1. Kaur, R., Luthra, P.: Load balancing in cloud computing. In: International Conference on Recent Trends in Information, Telecommunication and Computing, ITC (2012)
2. Fang, Y., Wang, F., Ge, J.: A task scheduling algorithm based on load balancing in cloud computing. In: Wang, F.L., Gong, Z., Luo, X., Lei, J. (eds.) WISM 2010. LNCS, vol. 6318, pp. 271–277. Springer, Heidelberg (2010). https://doi.org/10.1007/978-3-642-16515-3_34
3. Al Nuaimi, K., Mohamed, N., Al Nuaimi, M., & Al-Jaroodi, J.: A survey of load balancing in cloud computing: challenges and algorithms. In: Second Symposium Network Cloud Computing and Applications (NCCA), pp. 137–142. IEEE (2012). https://doi.org/10.1109/NCCA.2012.29
4. Krishna, P.V.: Honey bee behavior inspired load balancing of tasks in cloud computing environments. Appl. Soft Comput. **13**(5), 2292–2303 (2013). https://doi.org/10.1016/j.asoc.2013.01.025
5. Yagoubi, B., Slimani, Y.: Dynamic load balancing strategy for grid computing. Trans. Eng. Comput. Technol. **13**, 260–265 (2006)

6. Li, Y., Lan, Z.: A Survey of load balancing in grid computing. In: Zhang, J., He, J.-H., Fu, Y. (eds.) CIS 2004. LNCS, vol. 3314, pp. 280–285. Springer, Heidelberg (2004). https://doi.org/10.1007/978-3-540-30497-5_44
7. Li, Y., Yang, Y., Ma, M., Zhou, L.: A hybrid load balancing strategy of sequential tasks for grid computing environments. Future Gener. Comp. Sy. **25**(8), 819–828 (2009). https://doi.org/10.1016/j.future.2009.02.001
8. Nandagopal, M., Uthariaraj, V.R.: Decentralized dynamic load balancing for multi cluster grid environment. In: Meghanathan, N., Kaushik, B.K., Nagamalai, D. (eds.) CCSIT 2011. CCIS, vol. 133, pp. 149–160. Springer, Heidelberg (2011). https://doi.org/10.1007/978-3-642-17881-8_15
9. Lua, E.K., Crowcroft, J., Pias, M., Sharma, R., Lim, S.: A survey and comparison of peer-to-peer overlay network schemes. IEEE Commun. Surv. Tut. **7**(2), 72–93 (2005)
10. Karger, D. R., Ruhl, M.: Simple efficient load balancing algorithms for peer-to-peer systems. In: the sixteenth annual ACM symposium on Parallelism in algorithms and architectures, pp. 36–43. ACM (2004). https://doi.org/10.1145/1007912.1007919
11. Saroiu, S., Gummadi, P. K., Gribble, S. D.: Measurement study of peer-to-peer file sharing systems. In: International Society for Optics and Photonics Multimedia Computing and Networking, vol. 4673, pp. 156–171 (2001)
12. Reed, D.A., Dongarra, J.: Exascale computing and big data. Commun. ACM **58**(7), 56–68 (2015). https://doi.org/10.1145/2699414
13. Khaneghah, E.M., ShowkatAbad, A.R., et al.: A ExaMig matrix: process migration based on matrix definition of selecting destination in distributed exascale environments. Azerb. J. HPC **1**(1), 20–41 (2018). https://doi.org/10.32010/26166127.2018.1.1.20.41
14. Gupta, A., Sarood, O., Kale, L. V., Milojicic, D.: Improving hpc application performance in cloud through dynamic load balancing. In: 13th IEEE/ACM International Symposium on Cluster, Cloud and Grid Computing (CCGrid), pp. 402–409. IEEE (2013). https://doi.org/10.1109/CCGrid.2013.65
15. Dong, B., Li, X., Wu, Q., Xiao, L., Ruan, L.: A dynamic and adaptive load balancing strategy for parallel file system with large-scale I/O servers. J. Parallel Distr. Com. **72**(10), 1254–1268 (2012). https://doi.org/10.1016/j.jpdc.2012.05.006
16. Cao, X., Gao, S., Chen, L.: Gossip-based load balance strategy in big data systems with hierarchical processors. Wireless Pers. Commun. **98**(1), 157–172 (2017). https://doi.org/10.1007/s11277-017-4861-4
17. Geist, A., Lucas, R.: Major computer science challenges at exascale. Int. J. High Perform. C. **23**(4), 427–436 (2009). https://doi.org/10.1177/1094342009347445
18. Grosu, D., Chronopoulos, A.T.: Noncooperative load balancing in distributed systems. J. Parall. Distr. Comp. **65**(9), 1022–1034 (2005). https://doi.org/10.1016/j.jpdc.2005.05.001
19. Catalyurek, U. V., Boman, E. G., Devine, K. D., Bozdag, D., Heaphy, R., Riesen, L. A. Hypergraph-based dynamic load balancing for adaptive scientific computations. In: International Parallel and Distributed Processing Symposium, pp. 1–11. IEEE (2007). https://doi.org/10.1109/IPDPS.2007.370258
20. Hendrickson, B., Devine, K.: Dynamic load balancing in computational mechanics. Comput. Meth. Appl. M. **184**(2–4), 485–500 (2000). https://doi.org/10.1016/S0045-7825(99)00241-8
21. Khayyat, Z., Awara, K., Alonazi, A., Jamjoom, H., Williams, D., Kalnis, P.: Mizan: a system for dynamic load balancing in large-scale graph processing. In: the 8th ACM European Conference on Computer Systems, pp. 169–182. ACM (2013). https://doi.org/10.1145/2465351.2465369
22. Harvey, N.J.A., Ladner, R.E., Lovász, L., Tamir, T.: Semi-matchings for bipartite graphs and load balancing. In: Dehne, F., Sack, J.-R., Smid, M. (eds.) WADS 2003. LNCS, vol. 2748, pp. 294–306. Springer, Heidelberg (2003). https://doi.org/10.1007/978-3-540-45078-8_26

23. Low, C.P.: An approximation algorithm for the load-balanced semi-matching problem in weighted bipartite graphs. Inform. Process. Lett. **100**(4), 154–161 (2006). https://doi.org/10.1016/j.ipl.2006.06.004
24. Khalili, M., Borzooei, R.A., Deldar, M.: Matching numbers in fuzzy graphs. J. Appl. Math. Comput. **67**(1–2), 1–22 (2021). https://doi.org/10.1007/s12190-020-01463-z

The Impact of Store Environment on Purchase Intention in Supermarkets

Gunay Sadikoglu[1], Nigar E. Adilova[2(✉)], and Promise Isioma Anene[1]

[1] Department of Marketing, Near East University, Lefkosa, Mersin 10, Mersin, Turkey
gunay.sadikoglu@neu.edu.tr, promisetemianene@gmail.com
[2] Joint MBA Program, Azerbaijan State Oil and Industry University, 20 Azadlig Ave., AZ1010 Baku, Azerbaijan
nigar.adilova@asoiu.edu.az

Abstract. This study aims to investigate the most sensitive factor of the in-store environment on purchase intention in supermarkets. Primary data were collected through a questionnaire-based survey from 300 customers in the Turkish Republic of Northern Cyprus. The paper's main purpose is to define how much environmental factors impact customers' purchase intention. The realization of the fuzzy-rule interpolation method allows identifying one of the most sensitive factors that affect the customer's intention. The results showed that the highest impact factor on customers' purchase intention within considering the store's environment is defined with only one of the main factors, such as ambient, design and social factors.

Keywords: Purchase intention · Store environment · Sensitivity · Fuzzy rule interpolation · Possibility measure

1 Introduction

In the past, consumers primarily relied on the functions or qualities of the product to choose a place to shop. However, today's consumers require additional valuable elements to pick out retail outlets for their purchases, like supermarkets. Deciding on a specific store to enter, the amount of time to spend inside, whether to make a purchase or not, is significantly triggered by the store's environmental condition. It is important to note that today's consumers do not only buy the product or service itself; they pay money for the experience they have. One of the ways to create a great customer experience is to have an exciting store design and atmosphere. It is, therefore, of great importance to understand the effects of the shopping environment on consumers' purchase intention.

With the developments towards the advancement of innovation and growth in retail stores, managers and retail store owners are now curious to know the experience of the shoppers who visit their stores and how they feel when they shop. As it was mentioned earlier, customers do not just respond to the quality of the product or service itself; they also acknowledge and pay special attention to the store environment. Thus, it can be said that a store environment has become one of the most critical determinants in the purchasing process [1–3]. When consumers feel satisfied with the store's atmosphere,

R. A. Aliev et al. (Eds.): ICSCCW 2021, LNNS 362, pp. 237–245, 2022.
https://doi.org/10.1007/978-3-030-92127-9_34

they are more likely to spend more time and money than they planned. This is because pleasurable environmental conditions have the power to affect customer emotions and create a positive customer experience. Many studies have shown that a particular environment can trigger specific emotional responses in a person, affecting their behaviours in terms of approach or avoidance.

It is also essential to state that paying more attention to the store environment helps retailers communicate with customers about the brand's value and increase buyers' purchase intention. To achieve long-term objectives, supermarkets must understand all the parameters that have an impact on customers' purchase intention. Customers' purchase intention is estimated as the extent to which a supermarket, its services, or products meets or exceeds the customers' expectations. The convenience of store location, salesmen's expertise, and the product's qualities have been noticed in [4]. Some attributes such as merchandise, service, physical facilities, convenience, promotion, store atmosphere were identified in [5]. In [6], essential factors like parking, lighting, ambient factors, design factors, and sales personnel characteristics have been advocated. Authors in [7] focused on examining and identifying the positions of select stores and causal relationships among store attributes, product attributes, and intention to purchase. The authors [8] concentrate on the moderating effect of the shopping environment and the time pressure on travellers' shopping motivation and buying behaviour. The application of Fuzzy Logic (FL) allows us to handle both uncertainty and the imprecision inherent in input data and develop a universal model to find the input-output relationship. Although many research studies have been conducted in the area of the store environment, no research has been found that considers affecting factors with respect to fuzzy logic in supermarkets. Therefore, the paper is dedicated to specifying the in-store environment's impact on purchasing intention using FL.

The goals of this paper is structured as follows: Sect. 1 presents a brief introduction to this sphere. Section 2 is contained with preliminary information about the analyzed problem. The statement of the problem and conceptual model are determined in Sect. 3. The solution of the problem is analyzed in Sect. 4 with the usage of fuzzy rule interpolation techniques. Finally, Sect. 5 concludes with obtained results from the investigation.

2 Preliminaries

Definition 1. ***Fuzzy number*** [9]: A fuzzy number is a set A on R which has the following properties: a) A is a normal fuzzy set; b) A is a convex fuzzy set; c) $\alpha-$ cut of A, A^{α} is a closed interval for every $\alpha \in (0, 1]$; d) the support of A, A^{+0} is bounded.

Definition 2. ***Fuzzy If-Then rules*** [10]: Fuzzy if-then rule statements are commonly used to define the conditional statements that possess fuzzy logic. A single fuzzy if-then rule is shown in this form:

$$\text{If x is } A \text{ then y is } B$$

Definition 3. ***Possibility measure*** [11]: If a and b are triangular fuzzy numbers, The possibility measure is described in Fig. 1, and calculated by the following formula.

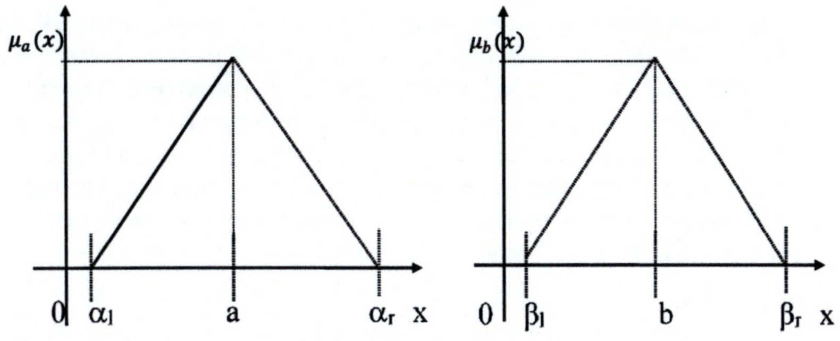

Fig. 1. Fuzzy numbers

$$Poss(a/b) = \begin{cases} 1 - \frac{a-b}{\alpha_l+\beta_r}, & if 0 < a - b < \alpha_l + \beta_r \\ 1 - \frac{b-a}{\beta_l+\alpha_r}, & if 0 < b - a < \beta_l + \alpha_r \\ 0, & otherwise \end{cases} \quad (1)$$

Definition 4. ***Fuzzy rule interpolation*** [12, 13]**:** Fuzzy Rule Interpolation (FRI) methods can release the need for the complete rule-base by replacing the rule matching reasoning concept with fuzzy interpolating function. The Fuzzy Rule Interpolation (FRI) methods were produced to handle the case of sparse rule-base. FRI methods are suitable to produce a conclusion even if some observations are not covered directly by the fuzzy rules.

3 Statement of the Problem

A customer's purchase intention within the retail store may depend on a complex set of factors, such as ambient, design, and social factors. Figure 2 shows the conceptual model of the research. The research model is extensive and includes interdependent relations between three independent variables, which represents stimulus (ambient factors, design factors and social factors) and one dependent variable, which represents response (customer purchase intention). This study aims to find the most sensitive factor on purchase intention among the three store environmental aspects.

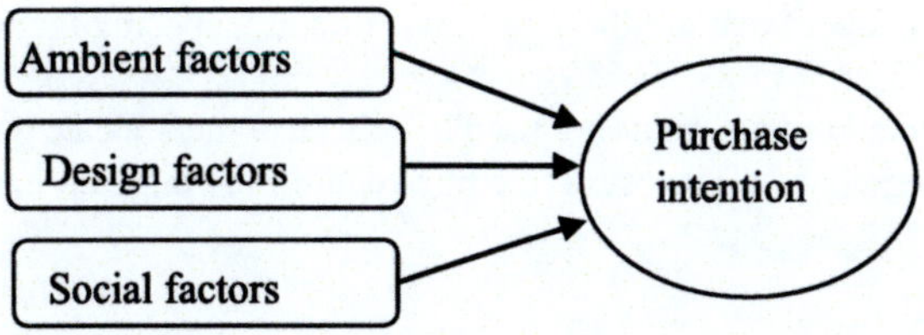

Fig. 2. The influential store environmental factors of purchase intention.

Enhancing and facilitating the supermarket's current attributes to attract and retain customers is vital. The fact that supermarkets face competition from local and international players, making it more important to retain their existing customers. The value of store attributes was widely discussed when selecting a supermarket. When choosing a supermarket to shop in, most customers use the store environment as a criterion. So, it can be said that having an appealing store environment can help supermarkets to differentiate themselves from their competitors. [14]. The main store environmental factors influencing consumers' purchase intention are widely represented as follows:

Ambient Factor (AF)

Ambient factors are considered non-visual background conditions that exist below our current customer awareness level. Characteristics of the environment may include music, light, smell, neatness, loud sounds and relative humidity such as temperature.

*Music:*Music can be viewed as a pleasurable melody influencing the decision making of a consumer either consciously or unconsciously. In a retail outlets music played has a significant impact on customer's intention to buy.

Lighting: Lighting is used in highlighting products to make them more visible and attractive. It creates excitement and impacts consumer buying behavior positively.

Scent: The existence or lack of scent at the store environment has a visible effect on the consumer's intention to make a purchase. Scent does have a significant effect on how consumers evaluate products for sale.

*Temperature:*Measuring temperature in stores have a significant impact on the consumer's intention to purchase. Customers are dissatisfied, and consequently, spend limited time in stores and produce bad reviews when extreme temperatures.

Design Factors (DF)

Design factors including floor and wall coverings, color schemes, orderliness, aisles, design, signages, etc., which are at the forefront of our awareness. In contrast, the ambient factors are described as the non-visual elements.

Cleanliness: Cleanliness as the state or quality of being clean is regarded as very essential by frequent and occasional shoppers in regular stores or supermarket selection.

*Color:*To create a particular atmosphere for customers, retailers use different colors in their store. In aspects of participants' perceived feelings, several colors have been discussed in many research in terms of agitated-relaxing, uninteresting-stimulating, awful-good etc.

Display: To attract customers, products should be displayed in the retail chain outlets in such a way that they are pleasing and appealing to them.

*Layout:*Layout can direct customers in a specific direction that impact purchase decisions. Customers buy more unintended items if exposed to too many shelves and aisles because of a layout that aid them to move around, exposing them to many other products.

Signs: In the retail outlets, signs are used to help customers locate goods by leading them to specific segments, aisles or service lines while showing their ease of access.

Social Factors (SF)

Social factors include other people at the store such as customers and staff. Social factors focus on the social conditions and crowding as represented by the attitudes and performance of the interacting customers and staff in the supermarket.

Crowding: Customers in a store recognize crowding across several ways, for instance, the allocation of space, the existence and number of other customers in the aisle waiting, the confined feelings of the customers and any form perceived restriction in movement.

Personnel in Store: The perceptions of service quality and customers sensations of excitement may be related to personnel. For example, stores with friendlier floor staff have a relatively high quality of service intrusion and arouse the feelings of less friendly staff.

Purchase Intention (PI)

Purchasing behavior begins with components in the store environment which arouse the behavior of the customers. So, the perception of the store impacts the information and affective state of the customer, which inevitably affects the purchasing behavior. Purchasing behavior comprises of several different facets, such as the total time spent, the sum of money spent, the number of transactions and the intentions to repurchase.

The problem is to determine the most sensitive environmental factor that impacts customers' purchase intention with the help of the fuzzy rule interpolation method in supermarkets.

4 Solution of the Problem

The methodology was carried out by online survey method using well-structured questionnaires. The participants were 300 students from different countries studying in the Turkish Republic of Northern Cyprus. The result of questionnaires is structured in the

Table 1. IF-Then rules

Rule	IF			Then
	AF	DF	SF	PI
1	VH	VH	VH	VH
2	L	H	L	N
3	H	H	H	H
4	VH	VH	VH	VH
5	N	N	N	N
6	H	H	H	H
7	H	N	H	H
8	N	VH	N	H
9	H	VH	VH	VH
10	L	N	L	L
11	L	L	L	L
12	VL	VL	VL	VL
13	N	L	N	L

shape of fuzzy If-Then rules regarding to the opinion of the experts and results of pilot studies (Table 1).

VL, L, N, H, VH are linguistic terms for *Very low, Low, Neutral, High, Very high* respectively. The codebook for linguistic terms is described in Table 2.

Table 2. Codebook

Very low	(0, 0.2, 0.35)
Low	(0.2, 0.35, 0.5)
Neutral	(0.35, 0.5, 0.75)
High	(0.5, 0.75, 1)
Very high	(0.75, 1, 1)

In accordance with the mentioned 3 antecedent factors (AF, DF, SF,) and 1 consequent fuzzy rule-based system's variables and their membership functions are showed in Fig. 3.

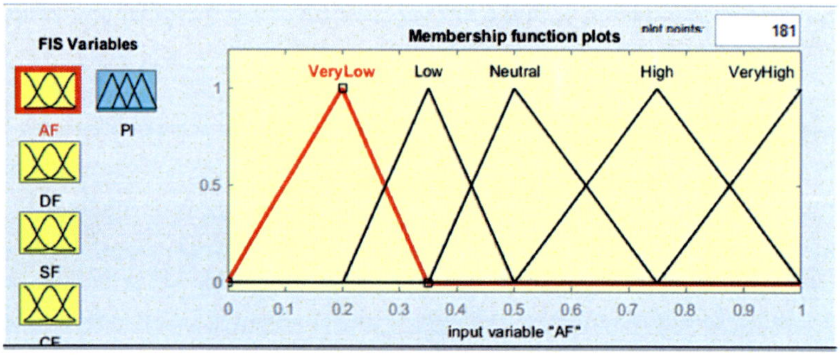

Fig. 3. Membership functions for AF factor

The solution of the problem under the classical fuzzy reasoning techniques can be caused sparse fuzzy rule base. Therefore, fuzzy rule-based interpolation (FRI) techniques were introduced to generate inference for sparse fuzzy rule base. Basically, FRI techniques perform interpolative approximate reasoning by taking into consideration the existing fuzzy rules for cases where there is no matching fuzzy rule [12] (Fig. 3).

Initially, the general case based on fuzzy rules is represented with fuzzy inference systems in the Matlab program, and then to find the most sensible factor we used the fuzzy rule interpolation method – KH method [15].This method is referred to determine the conclusion by its α-cuts in such a way that the ratio of distances between the conclusion and the consequents should be identical with the ones between the observation and the antecedents for all important α-cuts. From the neutral case, we increased the level with high and decreased the level with low per each antecedent one-by-one and investigated their individual sensitivity for the updated statement. Figure 4 demonstrates observation data of AF factor. The realization of input and output universes with the help of the type of KH technique was presented with a bold line.

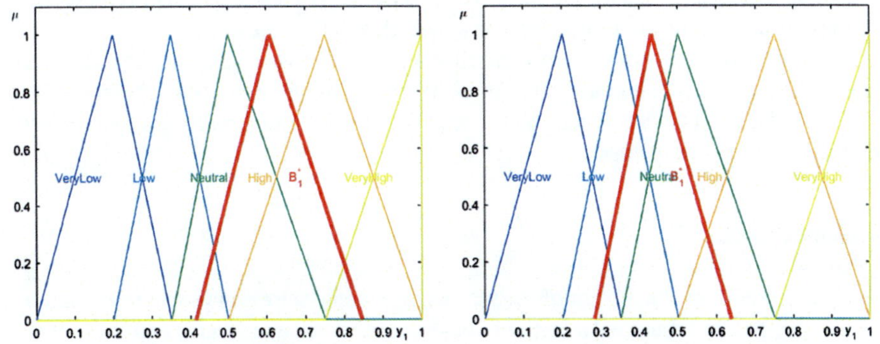

a) Result for high observation data b) Result for low observation data

Fig. 4. Consequent universe for a factor with stabilized KH method

After continuously doing the same actions to all antecedents, to find the most sensitive one among them we use (1) formula [11]. Possibility measure between updated consequent universe and initial data is measured (Table 3).

Table 3. Possibility measures

	High/Very low	High/Low	High/Neutral	High/High	High/Very high
AF	0.206224663	0.55504537	0.77934232	0.92436559	0.922514224
DF	0.208223857	0.558055312	0.782380296	0.927225057	0.92635
SF	0.204853862	0.553251855	0.77760631	0.922761932	0.923814684

The most sensitive factor for in-store environment on purchase intention among 4 factors is obtained from the closest value to *High/Very high* level. It demonstrates when the level is increased from neutral case to high case, only one parameter reflects the highest value that means its closeness to *Very high* case. It reveals DF (Design factor) which is the closest to the top position is the most sensitive for changes than other antecedents. The second influential parameter is SF (Social factor), and the third one is AF (Ambient factor). The results show that the design factor is more important in determining consumer behavioural responses than social and ambient factors. It can be said that the design factor has a significant positive impact on customers' experiences, which can increase purchase intention in most cases. For instance, shoppers like shopping in supermarkets with a well-structured layout, which helps them move around easily. Also, a creative and systematic arrangement of products, a clean environment, attractive and well-decorated shops and supermarkets with a good display of in-store information gives an excellent shopping experience. Authors in [16] also found in their research that design factors have a significant direct effect on consumer behaviour at shopping centres. The study [17], findings suggested that customer sensory valuations such as perceived service quality and perceived product value and their positive emotional state while in-store can significantly be improved when design factors at a retail outlet are also improved. And this, in turn, produces approached behaviours.

According to the low observation data, all antecedents reflect approximately the same degrees. It is explained with the quality of the rules. So, to obtain the best inference, the quality of rules must be developed and covered with not only the positive cases but also the positive and negative ones.

5 Conclusion

The paper introduced the impact of in-store environment factors on purchase intention in supermarkets. We examined a method to find the most sensitive factor with respect to three influencing parameters and a related parameter on purchase intention. Applying the fuzzy rule interpolation method and calculating possibility measures supported us to identify the most sensitive parameter. Results show that all environmental factors (ambient, design and social) positively impact customers' purchase intentions. Therefore, it is vital that supermarket managers improve the shopping experience for the customer by putting in place all necessary environmental factors and attributes to enable customers to have an enjoyable shopping experience, which will, in turn, influence their purchase intention. In addition, the results show that the design factors have the most effect on customers' purchase intention. If retailers wish to promote purchase intention, they have to understand the impact of each element that is a part of these factors. So, it is recommended that future research focus on each category's elements to know which elements are more significant for retail store managers to enhance.

References

1. Kotler, P.: Atmospherics as a marketing tool. J. Retail. **49**(4), 48–65 (1973)
2. Juhl, H.J., Kristensen, K., Ostergaard, P.: Customer satisfaction in European food retailing. J. Retail. Consum. Serv. **9**(6), 327–334 (2002). https://doi.org/10.1016/s0969-6989(02)00014-0
3. Liljander, V., Polsa, P., Riel, V.A.: Modelling consumer responses to an apparel store brand: Store image as a risk reducer. J. Retail. Consum. Serv. **16**, 281–290 (2009). https://doi.org/10.1016/j.jretconser.2009.02.005
4. Schiffman, L.G., Dash, J.F., Dillon, W.R.: Contribution of store-image characteristics to store-type choice. J. Retail **53**(2), 3–14 (1977). https://doi.org/10.1177/0972262917750236
5. Lindquist, J.D.: Meaning of image: survey of empirical and hypothetical evidence. J. Retail. **50**(4), 29–38 (1974)
6. Heitmeyer, J., Kind, K.: Retailing in my back yard: Consumer perceptions of retail establishments located within new urbanist communities. J. Shopping Center Res. **11**(1), 33–53 (2004)
7. Hasan, A.: Impact of store and product attributes on purchase intentions: an analytical study of apparel shoppers in indian organized retail stores. J. Bus. Perspec. **22**(1), 1–18 (2018). https://doi.org/10.1177/0972262917750236

8. Gunay, S.: Modeling of the travelers' shopping motivation and their buying behavior using fuzzy logic. Proc. Comput. Sci. **120**(2017), 805–811 (2017). https://doi.org/10.1016/j.procs.2017.11.311
9. Aliev, R.A.: Fundamentals of the Fuzzy Logic-Based Generalized Theory of Decisions. Springer-Verlag Berlin Heidelberg (2013). https://doi.org/10.1007/978-3-642-34895-2
10. Aliev, R., Tserkovny, A.: Systemic approach to fuzzy logic formalization for approximate reasoning. Inform. Sci. **181**(6), 1045–1059 (2011). https://doi.org/10.1016/j.ins.2010.11.021
11. Aliev R.A., Aliev R.R.: Soft computing and its applications. World Sci., 1–460 (2001)
12. Johanyák, Z. C., Tikk, D., Kovács, S., Wong, K. W.: Fuzzy rule interpolation Matlab toolbox – FRI toolbox. In: 2006 IEEE International Conference on Fuzzy Systems Sheraton Vancouver Wall Centre Hotel, Vancouver, BC, Canada, pp. 351–357 (2006). https://doi.org/10.1109/FUZZY.2006.1681736.
13. Alzubi, M., Johanyak, Z. C., Kovács, S.: Fuzzy rule interpolation methods and FRI toolbox. J. Theoretical App. Inf. Tech. **96**(21), 7227–7244 (2018)
14. Lockett, A., Moon, J., Visser, W.: Corporate Social Responsibility in Management Research: Focus, Nature, Salience and Sources of Influence (2006). https://doi.org/10.1111/j.1467-6486.2006.00585.x
15. Kóczy, L. T., Hirota, K.: Rule interpolation by α-level sets in fuzzy approximate reasoning. In: J. Busefal, Automne, URA-CNRS., Toulouse, France, vol. 46, 115–123, (1991)
16. Odeh, M.R., Abu-Rumman, A.: The impact of Jordanian shopping malls' physical surrounding on consumer buying behaviour: field research. Int. J. Market. Stud. **6**(3), 135–141 (2014)
17. Chen, H., Hsieh, T.: The effect of atmosphere on customer perceptions and customer behaviour responses in chain store supermarkets. Afr. J. Bus. Manage. **5**(24), 10054–10066 (2011)

Simulation of Electrodynamic Processes in the Cylindrical-Rectangular Microwave Waveguide Systems Transmitting Information

Islam J. Islamov[1(✉)], Mehman H. Hasanov[1], and Mikael H. Abbasov[2]

[1] Department of Radioengineering and Telecommunication, Azerbaijan Technical University, H. Javid ave. 25, AZ 1073 Baku, Azerbaijan

[2] Radio Television Broadcasting and Satellite Communication Production Union, The Ministry of Transport, Communications and High Technologies of Azerbaijan Republic, A. Abbaszade str. 2, AZ 1004 Baku, Azerbaijan

Abstract. A rigorous three-dimensional electrodynamic model of hybrid electromagnetic oscillations in a "cylindrical to rectangular waveguide" waveguide branch filled with a dielectric is constructed. The eigen types of oscillations are classified into branching eigen resonances at transcendental modes and resonances of the waveguide-dielectric type. For resonances of the first type, electromagnetic field in the communication region of the waveguides is described with the sum of the fields of the damped waves of the partial waveguides, and for resonances of the second type with the sum fields of damped and propagating waves. Oscillations of the first type exist both when the waveguide branching is filled with a dielectric and in empty branching. The second type of resonance exists only in structures with a dielectric constant greater than unity. The structure under study can be used to measure the electrical parameters of dielectric samples. Since, the spectral branching characteristics are mainly determined by the size of the central communication area waveguides and electrical parameters of parts of the dielectric that is located there, the measurements are local in nature.

Keywords: Electromagnetic field · Dielectric samples · Numerical analysis · Electrodynamic processes · Hertz vectors

1 Introduction

Electromagnetic oscillations in cruciform and T-shaped waveguide branches dedicated a significant number of works have been published [1–5]. We investigated free two-dimensional [1] and three-dimensional [2] vibrations in the branching of rectangular waveguides, axially symmetric [3] and unbalanced metric [4–6] oscillations in the branching of cylindrical waveguides. The spectrum of quasi-proper vibrations in such structures was studied [7]. The interest in such structures is explained by their widespread used in the microwave technology: as components of passive and active devices [8], measuring devices for determining the electrical parameters of dielectrics without their destruction [9].

R. A. Aliev et al. (Eds.): ICSCCW 2021, LNNS 362, pp. 246–253, 2022.
https://doi.org/10.1007/978-3-030-92127-9_35

Problems on vibrations in branches with different cross-sectional shapes of waveguides belong to the category of the most complex centuries-torus boundary value problems. This is also connected with the representation of the total field in the form of a superposition of the fields H and E of the types of waves in all waveguides and especially with the fact that these representations are written in different coordinate systems. Work [8, 10–12] are devoted to algorithms for calculating S-matrices of an unfilled tee connection of circular and rectangular waveguides. The aim of this work is to study the electrodynamic characteristics of the "cylindrical - rectangular waveguides with a dielectric" waveguide branching, since the structure under study allows us to measure the electrical parameters of dielectric samples both cylindrical and rectangular cross section. Strict partial method areas with a common communication area waveguide and represents the field in it in the form of superposition of the fields of partial waveguides with filling the circular waveguide and the central field of dielectric cylinder applied to solve a vector problem.

2 Development of Mathematical Models in the Microwave System "Cylindrical - Rectangular" Waveguide

The studied structure is presented in Fig. 1. We distinguish three regions in the structure. Consider hybrid vibrations of $HE\ (EH)_{nmg}$ - types (n, m, g is the number of fields of semi-variations along the axes (φ, r, x).

We carry out the solution by the method of partial regions presenting the fields in the regions II and III through the electric and magnetic Hertz vectors $\vec{K}^e$ and $\vec{K}^h$ in the form of expansion into its own decayed functions (modes) of the cylindrical and rectangular areas (waveguides).

In the region II ($|x| \geq b, r \leq a, 0 \leq \varphi \leq 2\pi$)– there is a cylindrical waveguide of radius a (filled with a dielectric, with a permittivity ε_2),

$$\vec{K}^{h(2)} = \vec{x}_0 \sum_{m,n} A_{mn} J_n(p_{nm} r) e^{\pm \gamma_{nm}^{(2)} (x-b)} e^{in\varphi}, \tag{1}$$

$$\vec{K}^{e(2)} = \vec{x}_0 \sum_{m,n} C_{mn} J_n(q_{nm} r) e^{\pm \gamma_{nm}^{(2)} (x-b)} e^{in\varphi}, \tag{2}$$

Where, the upper signs correspond to a wave propagating in the positive direction of the Ox axis, the lower signs correspond to a wave propagating in the opposite direction;

$$(\gamma_{nm}^{(2)})^2 = p_{nm}^2 - k_0^2 \varepsilon_2, (\gamma_{nm}^{\prime(2)})^2 = q_{nm}^2 - k_0^2 \varepsilon_2 - \tag{3}$$

constant propagation (attenuation);

$$m = 1, 2, ...; n = \pm 1, \pm 2, ...; p_{nm} = \frac{\mu_{nm}}{a}, \mu_{nm} - \tag{4}$$

roots of the equation

$$J_n'(\mu_{nm})^2 = 0; q_{nm} = \frac{\nu_{nm}}{a}, \nu_{nm} - \tag{5}$$

roots of the equation

$$J_n(\nu_{nm})^2 = 0; J_n(p_{nm}r), J_n'(p_{nm}r)- \tag{6}$$

n-th order Bessel function of the first kind and its derivative;

$$i = \sqrt{-1}; A_{nm}, C_{nm}- \tag{7}$$

decomposition coefficients (amplitudes of H and E waves).

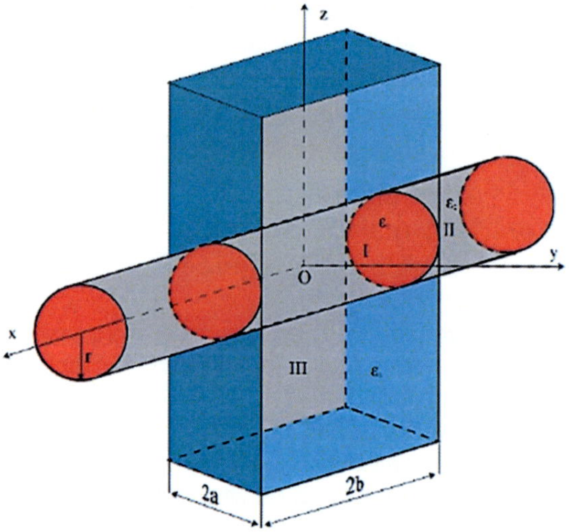

Fig. 1. Branching "cylindrical – rectangular waveguides".

In the region III ($|x| \leq b$, $|y| \leq a$, $|z|) \geq a-$ the straight line angular waveguide (unfilled $\varepsilon_3 = 0$) with transverse dimensions 2b × 2a,

$$\vec{K}^{h(3)} = \vec{z}_0 \sum_{m.g} B_{mg} \cos(p_{my}(y-a)) \cos(p_{gx}(x-b)) e^{\pm\gamma_{mg}^{(3)}(z-a)}, \tag{8}$$

$$\vec{K}^{e(3)} = \vec{z}_0 \sum_{m.g} D_{mg} \sin(p_{my}(y-a)) \sin(p_{gx}(x-b)) e^{\pm\gamma_{mg}^{(3)}(z-a)}, \tag{9}$$

Where,

$$(\gamma_{mg}^{(3)})^2 = p_{my}^2 + p_{gx}^2 - k_0^2 (m, g - 0, 1, 2, ..., m \neq g = 0) \tag{10}$$

and

$$(\gamma_{mg}^{\prime(3)})^2 = p_{my}^2 + p_{gx}^2 - k_0^2 (m, g - 0, 1, 2, ..., m \neq g = 0) \tag{11}$$

constant distribution (attenuation);

$$p_{my} = \frac{m\pi}{2a}; p_{gx} = \frac{g\pi}{2b}; B_{mg}, D_{mg}- \tag{12}$$

decomposition coefficients.

Region I ($|x| \leq b, |y| \leq a, 0 \leq r \leq a$)– is the region of intersection of the waveguides (filled with a dielectric, with permittivity ε_1). For matching projection bases cylindrical and rectangular waveguides in the selected area I, we represent the fields in the form of superposition of the fields of a cylindrical and radial (formed by metal planes $x = -b, x = b$) waveguides, since the modes radial waveguide connect to the modes of the projection bases of a cylindrical and rectangular waveguide:

$$\begin{aligned}\vec{K}^{h(1)} &= \vec{x}_0 \sum_{m.g} \left\lfloor A_{1mn} e^{-\gamma_{nm}^{(1)} x} + A_{2mn} e^{\gamma_{nm}^{(1)x}} \right\rfloor \times J_n(p_{nm} r) e^{in\varphi} + \\ &+ \vec{x}_0 \sum_{n.g} F_{ng} I_n(\gamma_g^{(1)} r) \sin(p_{gx}(x-b)) e^{in\varphi},\end{aligned} \tag{13}$$

$$\begin{aligned}\vec{K}^{e(1)} &= \vec{x}_0 \sum_{m.g} \left\lfloor C_{1mn} e^{-\gamma_{nm}^{(1)} x} + C_{2mn} e^{\gamma_{nm}^{(1)x}} \right\rfloor \times J_n(q_{nm} r) e^{in\varphi} + \\ &+ \vec{x}_0 \sum_{n.g} G_{ng} I_n(\gamma_g^{(1)} r) \cos(p_{gx}(x-b)) e^{in\varphi}.\end{aligned} \tag{14}$$

Where,

$$(\gamma_{nm}^{1})^2 = p_{nm}^2 - k_0^2 \varepsilon_1, (\gamma_{nm}^{\prime(1)})^2 = q_{nm}^2 - k_0^2 \varepsilon_1; (\gamma_n^{(1)})^2 = p_{gx}^2 - k_0^2 \varepsilon_1 -$$

constant distribution (attenuation) in the region I; $I_n(\gamma_g^{(1)} r)$– n-th order modified by Bessel function of the first kind; F_{ng}, G_{ng}– the amplitudes H and E of the waves of the radial waveguide.

To satisfy the boundary conditions in the recording fields of a rectangular waveguide goes to a cylindrical coordinate system. The further solution of the problem is reduced to satisfy the boundary conditions for the tangent components of the electric and magnetic fields:

$$\left[E_\varphi^{(1)} - E_\varphi^{(2)}\right]_{x=\pm b} = 0, \left[E_r^{(1)} - E_r^{(2)}\right]_{x=\pm b} = 0, \tag{15}$$

$$\left[E_\varphi^{(1)} - E_\varphi^{(3)}\right]_{r=a} = 0, \left[E_x^{(1)} - E_x^{(3)}\right]_{r=a} = 0, \tag{16}$$

$$\left[H_\varphi^{(1)} - H_\varphi^{(2)}\right]_{x=\pm b} = 0, \left[H_r^{(1)} - H_r^{(2)}\right]_{x=\pm b} = 0, \tag{17}$$

$$\left[H_\varphi^{(1)} - H_\varphi^{(3)}\right]_{r=a} = 0, \left[H_x^{(1)} - H_x^{(3)}\right]_{r=a} = 0, \tag{18}$$

and obtaining a system of functional equations, which we will design for a system of basic functions of a cylindrical and radial waveguide. As a result, we get an infinite systems of linear algebraic equations (SLAE) of the second kind:

$$A_{sm}^{-} + C_{sm}^{+} \frac{R_{3sm} R_{4sm}}{R_{1sm} R_{2sm}} + \sum_{g} \left[1 + (-1)^g\right] \times (k_0 \varepsilon_1 G_{sg} + p_{gx} F_{sg}) \frac{T_{1g}(p) + T_{1g}(q)}{R_{1sm} R_{2sm}}, \tag{19}$$

$$A_{sm}^{-} - C_{sm}^{+}\frac{R_{6sm}R_{4sm}}{R_{5sm}R_{2sm}} + \sum_{g}\left[1+(-1)^{g}\right]\times(k_0\varepsilon_1 G_{sg} + p_{gx}F_{sg})\frac{T_{1g}(p)+T_{1g}(q)}{R_{5sm}R_{2sm}}, \tag{20}$$

$$G_{sg} + \sum_{m} B_{sm} i k_0 \frac{R_{8mg}}{R_{7mg}}(I_{2sm} + I_{2sm}^{*}) - D_{sm}\frac{\gamma_{mg}^{\prime(3)} R_{9mg}}{R_{7sg}} p_{gx}(I_{2sm} + I_{2sm}^{*}) = 0, \tag{21}$$

$$G_{sg} + F_{sg}\frac{T_{3sg}}{T_{2sg}}\sum_{m} B_{sm} k_0 \frac{k_0 R_{8mg} p_{gx}}{T_{2sg}}(I_{1sm} + I_{1sm}^{*}) - D_{sm}\frac{i p_{my}\gamma_{mg}^{\prime(3)} R_{9mg}}{T_{2sg}}(I_{1sm}^{\prime} + I_{1sm}^{\prime\prime *}) = 0, \tag{22}$$

$$\begin{aligned} &G_{sg} + F_{sg}\frac{T_{2sg}}{\varepsilon_1 T_{3sg}} + \frac{1}{\varepsilon_1 T_{3sg}}\sum_{m}\Big[A_{sm}^{+}[1-(-1)^{g}]ch(\gamma_{sm}^{(1)}b)\Big] \\ &-A_{sm}^{-}\Big[1+(-1)^{g} sh(\gamma_{sm}^{(1)}b)\Big]R_{10sm} + +\Big[C_{sm}^{+}[1+(-1)^{g}]sh(\gamma_{sg}^{\prime(1)}b) - C_{sm}^{-}[1-(-1)^{g}]\Big] \\ &\times ch(\gamma_{sm}^{\prime(1)}b)R_{11sm} - iB_{sm}RI_{1sm} + D_{sm}RI_{2sm} = 0, \end{aligned} \tag{23}$$

$$\begin{aligned} &F_{sg} - \sum_{m}\Big[A_{sm}^{+}[1+(-1)^{g}]sh(\gamma_{sm}^{(1)}b)\Big] - A_{sm}^{-}\Big[1-(-1)^{g} ch(\gamma_{sm}^{(1)}b)\Big]\frac{R_{12sm}}{R_{7mg}} \\ &+B_{sm}RI_{3sm} + iD_{sm}RI_{4sm} = 0. \end{aligned} \tag{24}$$

The following notation is introduced here:

$$A_{sm}^{\pm} = A_{sm}\left(1 \pm e^{-2\gamma_{sm}^{(2)}b}\right)\begin{cases}(2sh(\gamma_{sm}^{(1)}b))^{-1},\\ (-2sh(\gamma_{sm}^{(1)}b))^{-1},\end{cases} \tag{25}$$

$$R_{1sm} = p_{sm}(\gamma_{sm}^{(1)} ch(\gamma_{sm}^{(1)}b) + \gamma_{sm}^{(2)} sh(\gamma_{sm}^{(1)}b)), \tag{26}$$

the upper signs in the expression for $A_{sm}^{\pm}$ correspond to the upper function, the lower signs to the lower one;

$$R_{2sm} = \frac{a^2}{2}\left[J_{s-1}^2(p_{sm}a) - J_s^2(p_{sm}a) - J_{s-2}^2(p_{sm})a\right], \tag{27}$$

$$R_{3sm} = k_0 q_{sm}(\varepsilon_1 ch(\gamma_{sm}^{\prime(1)}b) - \varepsilon_2 sh(\gamma_{sm}^{\prime(1)}b)), \tag{28}$$

$$R_{4sm} = \frac{a^2}{2}J_{s-1}^2(q_{sm}a). \tag{29}$$

3 Numerical Analysis

Matrix SLAE operators of the second kind, similar to SLAE (25), have been studied in detail [13, 14]. For the fields, the conditions on the edge and the conditions of finiteness of energy are satisfied in any limited region of space, as well as the discreteness and finiteness of the spectrum of frequencies of free vibrations, and the SLAE itself is solvable by the reduction method. For the numerical analysis [15, 16] of the dispersion of the equation obtained from the condition that the determinant from SLAE (25), an algorithm based on the Mueller method (the quadratic interpolation procedure together with the half division procedure) was built, and programs for a personal computer were developed on its basis. With their help, a number of graphs were obtained. The calculations took into account two waves in each of the waveguides: in a rectangular waveguide - one main wave of H- and E-type, in a cylindrical and radial - one wave H-type with left ($s = +1$) and right ($s = -1$) rotation. Such an approximation is justified for calculating the natural oscillations. The graph in Fig. 2 characterizes the dependence of the reduced of the natural frequency $2a/\lambda$ of natural oscillation HE $(EH)_{111}$ - type of the filling dielectric constant (a dielectric fills a cylindrical waveguide and a coupling region, $\varepsilon_1 = \varepsilon_2 = \varepsilon_3$) for $a/b = const$. The graph in Fig. 3 reflects the dependence of the reduced resonant frequency $2a/\lambda$ oscillations of HE $(EH)_{111}$ on the geometric dimensions of the structure ab for fixed values of permittivity ε.

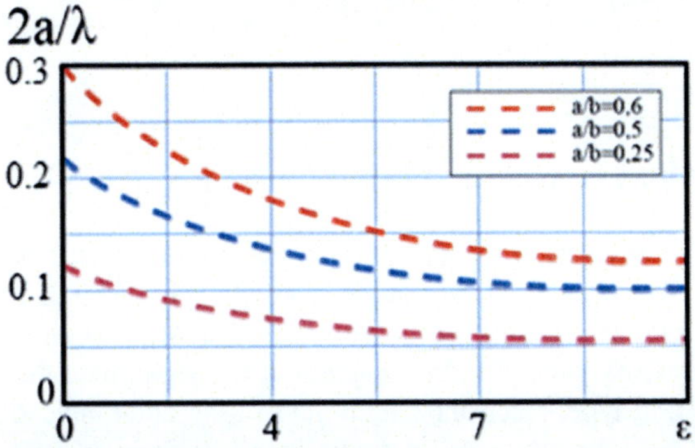

Fig. 2. The dependence of the reduced resonant frequency 2a/λ oscillations of HE_{111} from the dielectric constant of the filling ε for several values a/b.

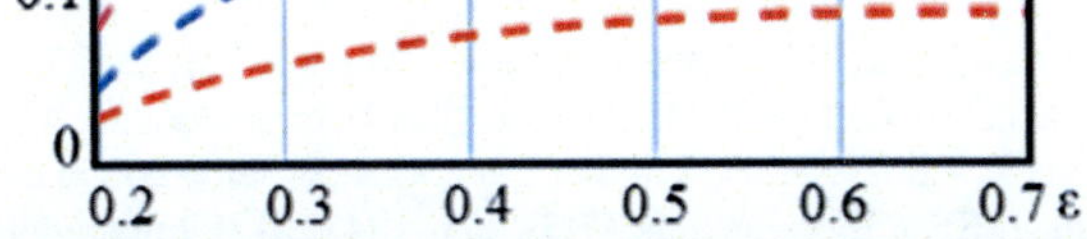

Fig. 3. The dependence of the reduced resonant frequency 2a/λ oscillations of HE_{111} from the dielectric constant of the filling ε for several values *ab*.

The maximum difference between the experimentally measured f_{exp} and the theoretically computed f_{cal} calculation of the resonance frequencies of the HE_{111} vibration did not exceed 5% (see Table 1, f_{cr1} is the critical frequency of a cylindrical waveguide with dielectric, f_{cr2} - critical frequency of an empty rectangular waveguide). Since all waveguides are transcendent at the resonant frequency, then the resonator is mainly the central region and the volume of the dielectric, which located in this area. Due to this, moving the sample in one of the waveguides relative to the central region, it is possible to carry out local measurements of material parameters.

Table 1. Measurement results.

ε	f_{exp}, MHz	f_{cal}, MHz	f_{cr1}, MHz	f_{cr2}, MHz
1	12950	13180	17570	14990
2,20	10990	10750	11985	14990
4,75	7395	7700	8150	14990

4 Conclusion and Recommendation

A rigorous three-dimensional electrodynamic model of hybrid electromagnetic oscillations in a "cylindrical - rectangular waveguide" waveguide branch filled with a dielectric is constructed. The eigen types of oscillations are classified into branching eigen resonances at transcendental modes and resonances of the waveguide-dielectric type. For resonances of the first type, electromagnetic field in the communication of region of the waveguides is described the sum of the fields of the damped waves of the partial waveguides, for resonances of the second type, the sum fields of damped and propagating waves. Oscillations of the first type exist both when the waveguide branching is filled with a dielectric and in empty branching. The second type of resonance exists only in structures with a dielectric constant greater than unity. The structure under study can be used to measure the electrical parameters of dielectric samples. Since the spectral branching characteristics are mainly determined by the size of the central communication area of waveguides and electrical parameters of parts of the dielectric that is located there, the measurements are local in nature.

References

1. Kumar, V., Dwivedi, S., Jain, P.K.: Circular sectoral waveguide TM01 to TE11 mode converter. Microwave Opt. Technol. Lett. **61**(7), 1697–1701 (2019). https://doi.org/10.1002/mop.31789
2. Yeap, K.H., Wong, E.V.S., Nisar, H., Lai, K.C., Ng, C.A.: Attenuation in circular and rectangular waveguides. IEEE J. Electromagn. **37**(3), 171–184 (2017). https://doi.org/10.1080/02726343.2017.1301198
3. Ceccuzzi, S., Ponti, C., Ravera, G.L., Schettini, G.: Physical mechanisms and design principles in mode filters for oversized rectangular waveguides. IEEE Trans. Microwave Theory Tech. **65**(8), 2726–2733 (2017). https://doi.org/10.1109/TMTT.2017.2684119
4. Islamov, I.J., Ismibayli, E.G., Gaziyev, Y.G., Ahmadova, S.R., Abdullayev, R.: Modeling of the electromagnetic feld of a rectangular waveguide with side holes. Prog. Electromagn. Res. **81**, 127–132 (2019). https://doi.org/10.2528/PIERL19011102

5. Islamov, I.J., Shukurov, N.M., Abdullayev, R.S., Hashimov, K.K., Khalilov, A.I.: Diffraction of electromagnetic waves of rectangular waveguides with a longitudinal. In: IEEE Conference 2020 Wave Electronics and its Application in Information and Telecommunication Systems (WECONF) (2020). https://doi.org/10.1109/WECONF48837.2020.9131457
6. Khalilov, A.I., Islamov, I.J., Hunbataliyev, E.Z., Shukurov, N.M., Abdullayev, R.S.: Modeling microwave signals transmitted through a rectangular waveguide. In: IEEE Conference 2020 Wave Electronics and its Application in Information and Telecommunication Systems (WECONF) (2020). https://doi.org/10.1109/WECONF48837.2020.9131525
7. Rahimzadeh, R.M., Lamecki, A., Sypek, P., Mrozowski, M.: Residue-pole methods for variability analysis of s-parameters of microwave devices with 3d fem and mesh deformation. J. Radio. Eng. **29**(1), 10–20 (2020). https://doi.org/10.13164/re.2020.0010
8. Sun, D.Q., Xu, J.P.: Real time rotatable waveguide twist using contactless stacked air-gapped waveguides. IEEE Microwave Wireless Compon. Lett. **27**(3), 215–217 (2017). https://doi.org/10.1109/LMWC.2017.2661881
9. Li, F., et al.: Design and microwave measurement of a Ka-band HE11 mode corrugated horn for the Faraday rotator. IET Microw. Antennas Propag. **11**(1), 75–80 (2017). https://doi.org/10.1049/iet-map.2016.0288
10. Liu, Z.Q., Sun, D.Q.: Transition from rectangular waveguide to empty substrate integrated gap waveguide. Int. J. Electron. Lett. **55**(11), 654–655 (2019). https://doi.org/10.1049/el.2019.1042
11. Yousefian, M., Hosseini, S.J., Dahmardeh, M.: Compact broadband coaxial to rectangular waveguide transition. J. Electromagn. Waves Appl. **33**(9), 1239–1247 (2019). https://doi.org/10.1080/09205071.2019.1606737
12. Menachem, Z.: A new technique for the analysis of the physical discontinuity in a hollow rectangular waveguide with dielectric inserts of varying profiles. J. Electromagn. Waves Appl. **33**(9), 1145–1162 (2019). https://doi.org/10.1080/09205071.2019.1601035
13. Singh, R.R., Priye, V.: Numerical analysis of film-loaded silicon nanowire optical rectangular waveguide: an effective optical sensing. Micro Nano Lett. **13**(9), 1291–1295 (2018). https://doi.org/10.1049/mnl.2018.0140
14. Taghizadeh, H., Ghobadi, C., Azarm, B., Majidzadeh, M.: Grounded coplanar waveguide-fed compact MIMO antenna for wireless portable applications. Radio. Eng. **28**(3), 528–534 (2019). https://doi.org/10.13164/re.2019.0528
15. Aliev, R.A., Gardashova, L.A.: Z-set based approach to control system design. In: Aliev, R.A., Kacprzyk, J., Pedrycz, W., Jamshidi, Mo., Babanli, M., Sadikoglu, F.M. (eds.) ICAFS 2020. AISC, vol. 1306, pp. 10–21. Springer, Cham (2021). https://doi.org/10.1007/978-3-030-64058-3_2
16. Dadasheva, A.N.: Analysis of consistency of pairwise comparison matrix with fuzzy type-2 elements. In: 11th International Conference on Theory and Application of Soft Computing, Computing with Words, Perception and Artificial intelligence ICSCCW 2021 (2021). In press

Ranking Integration Factors Using Fuzzy TOPSIS Method

Tural Suleymanli(✉)

Department of Economic Relations, Odlar Yurdu University, AZ1072 Baku, Azerbaijan
suleymanli.tural@gmail.com

Abstract. The improvement of integration processes as the main aspect of economic activity plays a significant role in development of the country. It is closely connected with the factors, which have an effect on integration processes. Existing models on the direction for structuring integration processes with respect to the factors are usually dedicated to theoretical research and fundamentally, there is no practical approach to the matter. In this paper, author suggests a method for ranking factors, which provide and intensify integration processes considering uncertainty in real-life situation.

Keywords: Decision making · Fuzzy TOPSIS · Integration processes · Economics

1 Introduction

In contemporary times, regional economic development of the country and improvement of foreign trade relations are closely linked with integration process. It is also considered as an economic cooperation among enterprises. Thus, integration process is the process to combine the activities of economic structures and formation of a new unity to perform socio-economic functions. During the economic integration, all the combined systems act together in the economic sphere that ensures the goal of integration subjects. The union of manufacturing enterprises with other enterprises under integration process is determined as a perspective factor to increase production efficiency. Therefore, scientists and researchers strictly control to improve the efficiency of integration process for the sake of economic development. Unfortunately, they have not paid enough attention to consider imprecise, uncertain real-life conditions in this field. The motivation of the research is devoted to the application of fuzzy logic which is the best method to solve decision making problems under uncertainty. For this aim the author will group main factors affecting to the integration process and rank all the factors by using fuzzy Ideal solution method corresponding to the survey's results.

Ideal solution technique – TOPSIS plays a leading role in multi factor decision making method. This advantage resulted to aplly TOPSIS method in many fields, such as technology, medicine, economics etc. Historically, TOPSIS was successfully initiated by Hwang, and Yoon to represent fuzzy set approach to MADM [1], and well developed in the comparative analysis of VIKOR and TOPSIS methods [2, 3]. These studies caused to

R. A. Aliev et al. (Eds.): ICSCCW 2021, LNNS 362, pp. 254–260, 2022.
https://doi.org/10.1007/978-3-030-92127-9_36

take a close interest of researchers in the application of fuzzy ranking technique. Inspired by mentioned works, the ranking method was offered in some fields, including economics [4–6]. Some researchers investigated the combination of AHP and TOPSIS methods, or other applied one of them to experimental works. For instance, authors in [5] used from the combination of AHP and TOPSIS techniques for the assessment of sustainable development in emerging economy. In [6] authors well-practiced fuzzy AHP method to solve the ranking problem according to multiple factors and obtained remarkable comparative results. Later on, authors in [7] were chosen fuzzy TOPSIS method as a practical way to illustrate material analysis for economic production. Nowadays a series of works still exist devoted to fuzzy TOPSIS method [8–11]. In this research, the author applies fuzzy TOPSIS method to rank affecting factors which provides and intensify integration processes in the country.

The rest of the paper is arranged as follows: Sect. 2 contains preliminary information that contributes the solution of the matter. Section 3 summarizes the statement of the problem. Experimental analysis is illustrated in Sect. 4. Eventually, Sect. 5 points out some conclusions.

2 Preliminaries

Definition 1. Fuzzy Fuzzy sets [12]: Let U is a universal set $U = \{u_1, u_2, \ldots, u_n\}$, then the fuzzy set A of U-universal set will be defined:

$$A = \mu_A(u_1)/u_1 + \mu_A(u_2)/u_2 + \ldots \mu_A(u_n)/u_n$$

Definition 2. Normalized decision matrix [2]: All the elements given in x_{ij} matrix are normalized by the following formula r_{ij}:

$$r_{ij} = \frac{x_{ij}}{\sqrt{\sum_{i=1}^{M} x_{ij}^2}} \tag{1}$$

Definition 3. Weighted normalized decision matrix [2]: Assume that normalized r_{ij} matrix and weights $W = (w_1, w_2, w_3, \ldots, w_n)$ are given. The final weighted normalized decision matrix R_{norm} will be as follows:

$$R_{norm} = \begin{bmatrix} w_1 r_{11} & w_2 r_{12} & w_3 r_{13} & \ldots & w_N r_{1N} \\ w_1 r_{21} & w_2 r_{22} & w_3 r_{23} & \ldots & w_N r_{2N} \\ . & & & & . \\ . & & & & . \\ . & & & & . \\ w_1 r_{M1} & w_2 r_{M2} & w_3 r_{M3} & \ldots & w_N r_{MN} \end{bmatrix} \tag{2}$$

Definition 4. Euclidean distance [2]: The Euclidean distance is considered as a distance of each alternative to the positive ideal solution and negative-ideal solution.

$$S_i^+ = \sqrt{\sum_{j=1}^{m} (v_{ij} - v_j^+)^2} \tag{3}$$

$$S_i^- = \sqrt{\sum_{j=1}^{m} (v_{ij} - v_j^-)^2}, \quad j = 1, 2, 3, \ldots, m. \tag{4}$$

Definition 5. Nearness to the ideal solution [2]: The relative nearness to positive and negative ideal solutions is calculated as follows:

$$C_{i*} = S_{i-}/(S_{i+} + S_{i-}), \quad 0 \leq C_{i*} \leq 1, \quad i = 1, 2, 3, \ldots, M. \tag{5}$$

3 Statement of the Problem

It is well known that integration processes of manufacturing enterprises are realized in different ways. In general, all the integration processes are linked with some key factors to improve their efficiency. Factors that determine the effectiveness of the integration process can be structured as follows (Fig. 1).

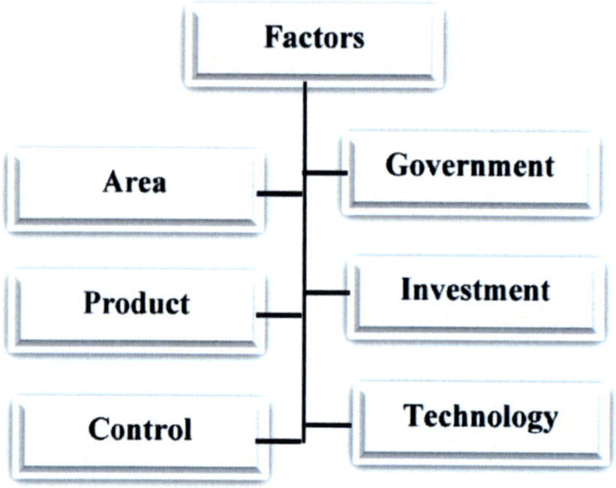

Fig. 1. Factors to determine the effectiveness of integration process

As seen in Fig. 1 the efficiency of integration processes depend on the 6 main factors:

- Area: Proximity of economic enterprises for the area;
- Product: Manufacturing the product that is in high demand in the regional market;
- Control: The implementation of control among integrated enterprises;
- Government: Protection of government-business partnership;
- Investment: The amount of investment;
- Technology: The usage of modern technological means.

The problem is to find the most affecting factor to integration process, after receiving initial data from the survey, which realized among 50 respondents [13]. Fuzzy TOPSIS method is one of the most comprehensive ranking method to solve this problem.

4 Experimental Analysis

Experiments are on the result of online survey method which carried out well-structured questions about effecting factors to integration processes. Respondents answered the questions by estimation the factors regarding to the linguistic variables. The codebook for the linguistic variables is given below (Table 1):

Table 1. Linguistic variables for estimating 6 factors

Linguistic variables	Triangular fuzzy numbers
Strongly agree – *SA*	SA = {0.75; 1; 1}
Agree – *A*	A = {0.5; 0.75; 1}
Neutral – *N*	N = {0.25; 0.5; 0.75}
Disagree – *D*	D = {0; 0.25; 0.5}
Strongly disagree - *SD*	SD = {0; 0; 0.25}

The analysis focus on the solution of ranking problem for determined 6 factors by using fuzzy TOPSIS method. The genralized survey result is represented in Table 2.

Table 2. Final result of the survey [9]

Factors	*SA*	*A*	*N*	*D*	*SD*
Area	8	24	13	4	0
Product	24	23	3	0	0
Control	11	28	10	0	0
Government	11	23	15	0	0
Investment	16	23	11	0	0
Technology	21	20	7	2	0

First, we should normalize survey result given in Table 2 by using (1) formula, and consider linguistic variables as weight vectors. For the second step, with the assistance of (2) formula fuzzy weighted normalized decision matrix is obtained (Table 3).

Table 3. Normalized fuzzy decision matrix with weights

Factors	Normalized degrees				
Area	{0.15; 0.2; 0.2}	{0.21; 0.31; 0.31}	{0.13; 0.25; 0.38}	{0; 0.22; 0.45}	{0; 0; 0}
Product	{0.45; 0.6; 0.6}	{0.2; 0.3; 0.3}	{0.03; 0.06; 0.09}	{0; 0; 0}	{0; 0; 0}
Control	{0.21; 0.28; 0.28}	{0.24; 0.36; 0.36}	{0.1; 0.19; 0.29}	{0; 0; 0}	{0; 0; 0}
Government	{0.21; 0.28; 0.28}	{0.2; 0.3; 0.3}	{0.14; 0.29; 0.43}	{0; 0; 0}	{0; 0; 0}
Investment	{0.3; 0.4; 0.4}	{0.2; 0.3; 0.3}	{0.11; 0.21; 0.32}	{0; 0; 0}	{0; 0; 0}
Technology	{0.4; 0.53; 0.53}	{0.17; 0.26; 0.26}	{0.07; 0.13; 0.2}	{0; 0.11; 0.22}	{0; 0; 0}

The next step of experimental analysis is devoted to find positive and negative ideal solutions. For the purpose of getting ideal solutions, object functions are characterized as below using (6) formula.

$$\begin{aligned} &SA \rightarrow max; \\ &A \rightarrow max; \\ &N \rightarrow max; \\ &D \rightarrow min; \\ &SD \rightarrow min; \end{aligned} \tag{6}$$

Considering goal functions, positive and negative ideal solutions will be obtained as below (Table 4):

Table 4. Ideal solutions

Ideal solutions	*SA*	*A*	*N*	*D*	*SD*
Positive ideal	{0.45; 0.6; 0.6}	{0.24; 0.36; 0.36}	{0.14; 0.29; 0.43}	{0; 0; 0}	{0; 0; 0}
Negative ideal	{0.15; 0.2; 0.2}	{0.17; 0.26; 0.26}	{0.03; 0.06; 0.09}	{0; 0.22; 0.45}	{0; 0; 0}

Subsequently, the Euclidean distance will be calculated using the formula (3)–(4) to reveal the distance of each alternative to the value of positive ideal solution and the value of negative solution. Result is described in Table 5.

Table 5. Euclidian distance for ideal solutions

	Area	Product	Control	Government	Investment	Technology
S + (SA)	0.64	0	0.51	0.51	0.32	0.11
S − (SA)	0	0.64	0.13	0.13	0.32	0.53
S + (A)	0.08	0.09	0	0.09	0.09	0.16
S − (A)	0.08	0.06	0.16	0.06	0.06	0
S + (N)	0.06	0.42	0.18	0	0.14	0.29
S − (N)	0.36	0	0.25	0.42	0.29	0.14
S + (D)	0.5	0	0	0	0	0.25
S − (D)	0	0.5	0.5	0.5	0.5	0.25
S + (SD)	0	0	0	0	0	0
S − (SD)	0	0	0	0	0	0

Eventually, the greatest and least factors which affect on integration processes will be obtained by calculating the nearness to the ideal solutions in accordance with (5) formula. Table 6 describes summarized result.

Table 6. Selection of the best alternative by degrees of nearness to the ideal solutions.

Area	Product	Control	Government	Investment	Technology
0.45	0.8	0.93	0.87	0.86	0.55

In consequence, the most effecting factor to integration processes will be *Control*, the least effecting factor will be *Area*. In a generalized way, ranking result of the given 6 factors is as below:

$$\text{Control} \succsim \text{Government} \succsim \text{Investment} \succsim \text{Product} \succsim \text{Technology} \succsim \text{Area}$$

5 Conclusion

In this paper, the author applied fuzzy TOPSIS method to one of the most important economic activities, which is well known as an integration process. The method was examined by solving the ranking problem within six impressive factors to the development of integration processes. The result of the investigation showed that the best impressive factor is *Control* factor, which determines the management of integrated enterprises, and the least impressive factor is *Area*, which defines proximity of economic enterprises.

References

1. Hwang, C.L., Yoon, K.: Multiple Attributes Decision Making Methods and Applications, vol. 186. Springer, Heidelberg (1981). https://doi.org/10.1007/978-3-642-48318-9
2. Chen, C.-T.: Extension of the TOPSIS for group decision-making under fuzzy environment. Fuzzy Set. Syst. **114**(1), 1–9 (2000). https://doi.org/10.1016/S0165-0114(97)00377-1
3. Opricovick, S., Tzeng, G.H.: Compromise solution by MCDM methods: a comparative analysis of VIKOR and TOPSIS. Eur. J. Oper. Res. **156**, 445–455 (2004). https://doi.org/10.1016/S0377-2217(03)00020-1
4. Olcer, A.I., Odabasi, A.Y.: A new fuzzy multiple attributive group decision making methodology and its application to population/maneuvering system selection problem. Eur. J. Oper. Res. **166**(1), 93–114 (2005). https://doi.org/10.1016/j.ejor.2004.02.010
5. Dang, V.T., Wang, J., Dang, W.V.-T.: An integrated fuzzy AHP and fuzzy TOPSIS approach to assess sustainable urban development in an emerging economy. Int. J. Environ. Res. Public. Health **16**(16), 2902 (2019). https://doi.org/10.3390/ijerph16162902
6. Aliyev, R.R., Temizkan, H., Aliyev, R.R.: Fuzzy analytic hierarchy process-based multi-criteria decision making for universities ranking. Symmetry **12**(8), 1351 (2020). https://doi.org/10.3390/sym12081351
7. Emovon, I., Aibuedefe, W.O.: Fuzzy TOPSIS application in materials analysis for economic production of cashew juice extractor. Adv. Soft Comp. **12**(1), 1–18 (2020). https://doi.org/10.1080/16168658.2020.1775332
8. Huseynov, O.H., Adilova, N.E.: Multi-criterial optimization problem for fuzzy if-then rules. In: Aliev, R.A., Kacprzyk, J., Pedrycz, W., Jamshidi, Mo., Babanli, M., Sadikoglu, F.M. (eds.) ICAFS 2020. AISC, vol. 1306, pp. 80–88. Springer, Cham (2021). https://doi.org/10.1007/978-3-030-64058-3_10
9. Gardashova, L.A.: Z-number based TOPSIS method in multi-criteria decision making. In: Aliev, R.A., Kacprzyk, J., Pedrycz, W., Jamshidi, Mo., Sadikoglu, F.M. (eds.) ICAFS 2018. AISC, vol. 896, pp. 42–50. Springer, Cham (2019). https://doi.org/10.1007/978-3-030-04164-9_10
10. Aliyeva, K.: Multifactor personnel selection by the fuzzy TOPSIS method. In: Aliev, R.A., Kacprzyk, J., Pedrycz, W., Jamshidi, Mo., Sadikoglu, F.M. (eds.) ICAFS 2018. AISC, vol. 896, pp. 478–483. Springer, Cham (2019). https://doi.org/10.1007/978-3-030-04164-9_64
11. Huseynova, F.: Exploring the factors affecting the happy marriages by using fuzzy TOPSIS. In: Aliev, R.A., Kacprzyk, J., Pedrycz, W., Jamshidi, Mo., Babanli, M.B., Sadikoglu, F.M. (eds.) ICSCCW 2019. AISC, vol. 1095, pp. 914–922. Springer, Cham (2020). https://doi.org/10.1007/978-3-030-35249-3_122
12. Aliev, R.A.: Fundamentals of the Fuzzy Logic-Based Generalized Theory of Decisions, vol. 293. Springer, Heidelberg (2013). https://doi.org/10.1007/978-3-642-34895-2
13. Survey link. https://docs.google.com/forms/d/e/1FAIpQLSdgI8FWQLhlLwVHzQeXFQfudtzcMdBVS3D6TCeG-sXpvenv-Q/viewform

Evaluation of HCV Infection Laboratory Test Results Using Machine Learning Methods

Hamit Altıparmak[1(✉)], Şerife Kaba[2], and Meliz Yuvalı[3]

[1] Department of Computer Engineering, Near East University, Near East Boulevard, P.O. Box: 99138, Nicosia, TRNC, Mersin 10, Turkey
hamit.altiparmak@neu.edu.tr

[2] Department of Biomedical Engineering, Near East University, Near East Boulevard, P.O. Box: 99138, Nicosia, TRNC, Mersin 10, Turkey
serife.kaba@neu.edu.tr

[3] Faculty of Medicine, Department of Biostatistics, Near East University, Near East Boulevard, P.O. Box: 99138, Nicosia, TRNC, Mersin 10, Turkey
meliz.yuvali@neu.edu.tr

Abstract. The Hepatitis C virus (HCV) infection is a serious health problem associated with a significant risk of death in different parts of the world. Hepatitis C disease, which can cause liver inflammation and, in severe cases, liver damage, is defined as a viral infection transmitted by contaminated blood. If hepatitis C infection continues for many years, severe diseases such as scarring of the liver (cirrhosis), liver failure and liver cancer may occur. According to data from the World Health Organization (WHO), there are about 34 million new infections each year. The reason is that hepatitis C virus is the etiology of chronic hepatitis and is often capable of developing cirrhosis and hepatocellular carcinoma (HCC). In this study, HCV laboratory test results of 615 patients are included. The aim of this study is to compare HCV-related diseases and test results in the HCV dataset. In this study, 9 different machine learning methods were applied to obtain the highest accuracy. In this study, 9 different machine learning methods were applied to obtain the highest accuracy. Machine learning methods used in this study were implemented using Python Programming Language. Among the methods used in this study, the highest accuracy rate of 95.77% was obtained with the logistic regression machine learning method.

Keywords: Hepatitis C · Cirrhosis · Fibrosis · Liver function test · Artificial neural network

1 Introduction

Viral hepatitis describes an infection caused by a virus that affects the liver. This is a major public health problem all over the world especially in the United States. Viral hepatitis not only carries a high rate of morbidity, but it also stresses health resources and causes serious economic crisis. Most cases of viral hepatitis have a chance of being prevented. The most widely known types of hepatitis viruses include viruses A, B, C,

R. A. Aliev et al. (Eds.): ICSCCW 2021, LNNS 362, pp. 261–269, 2022.
https://doi.org/10.1007/978-3-030-92127-9_37

and D [1]. Hepatitis (refers to liver inflammation) affected humanity in the fifth century BC. This disease was mentioned in early biblical literature, and it was explained that it appeared in epidemics, especially during wartime. The contagious nature of hepatitis II was found after World War II [1].

The Hepatitis C virus (HCV) is the main cause of the worldwide known diseases of hepatitis (acute and chronic) and a chronic degenerative condition called cirrhosis, in which healthy liver cells are replaced by damaged scar tissue [2]. HCV is mainly transmitted via parenteral exposure to infectious blood or body fluids containing blood, most commonly with injection drug use [3]. Before a blood donor is transfused into another patient, a blood test is done to determine if they have HCV disease. One of the blood samples taken from the donor is checked by performing screening tests. Thus, antibodies against HCV infection are detected; Nucleic Acid Amplification Test (NAT) is applied to detect viral genetic material. If a donor receives repeated reactive (positive) results after antibody screening or NAT tests are performed, this indicates that a donor is infected with HCV and the blood is not healthy, so all the blood is destroyed and cannot be used for transfusion. Thus, blood donation is permanently delayed [4]. Diseases commonly caused by HCV infection include liver diseases such as cirrhosis, hepatocellular carcinoma (HCC), and fibrosis. Liver fibrosis is defined as the result of the wound healing reaction to tissue injury due to chronic HCV infection [5].

The author in [6] used the HCV dataset containing 73 individuals. There are 52 male and 21 female in this dataset from UCI artificial intelligence. It applied 4 artificial intelligence methods to classify this dataset, these are, KNN, Naive Bayes, Neural Network (NN) and Random Forest (RF). As a result of this study, NN gave the highest percentage of accuracy with a value of 95.12%. In another study [7], 3 methods were applied on a similar data set. These, SVM, Logistic Regression and CART. Looking at the results, CART (100%) and SVM (98.7%) methods gave the best results. The study in [8] suggested that decision tree, genetic algorithm, particle swarm optimization and multiple linear regression models were developed for fibrosis risk estimation. The analyzed machine learning algorithms could predict advanced fibrosis in patients with AUROC ranging from 73% to 76%, with an accuracy between 66.3% and 84.4%. This study [9] used a dataset of patients from maize and aims to know the performance comparisons of the dataset between multiple and binary class labels. The highest accuracy is indicated by KNN (51.06%, R) and random forest (54.56%, Python). In a study conducted in [10], clinical risk prediction models in chronic hepatitis c virus were studied. They developed and compared two ML algorithms to predict cirrhosis development in the cohort. They found superior predictive performance for the longitudinal Cox model applied in the study compared to the CS Cox model (conformity 0.764 vs 0.746) as well as the longitudinal augmented survival tree mod-el compared to the linear Cox model (conformity 0.774 vs 0.764).

Although important studies have been done, new methods are being tested for the accuracy of machine learning for HCV. In our study, the purpose is to use machine learning to come up with the method which would provide the most accurate result.

2 Methodology

2.1 Database

In this study, a data set containing HCV laboratory results of 615 patients from UCI (California Irvine University) was used [11]. Dataset contains 12 variables and result categories, also this data has some missing values [11]. The test results used in the database used in this study are available in Table 2. A liver function test is a blood test that diagnoses and monitors liver disease or damage. These tests allow it to measure the levels of certain proteins and enzymes in the blood. Among the frequently performed liver function tests are Alanine transaminase (ALT), Aspartate transaminase (AST), Alkaline phosphatase (ALP), Albumin (ALB) and total protein, Bilirubin (BIL), Gamma-glutamyl transferase (GGT) tests [12]. Cholinesterase (CHE) is a blood test to check the levels of two substances that help the nervous system function properly. The creatinine (CREA) blood test is used to evaluate kidney function. Cholesterol test (CHOL) is done to determine the risk of forming fat deposits that can lead to narrowing or blockage of the arteries in your body [13]. The test to measure the total amount of protein (PROT) in the blood is called the Total serum protein test, and it can also check and measure the amount of albumin and globulin, the two main protein groups in the blood [14].

In the tables below, the distribution of the categories in the data by gender and the minimum and maximum values of the laboratory tests given were determined by using the IBM SPSS (Statistics Package for Social Sciences, version 21) package program. The category values in Table 1 are as follows: 0: Blood Donor, 0s: Suspect Blood Donor, 1: Hepatitis, 2: Fibrosis, 3: Cirrhosis.

Table 1. Output for HCV Data set 1.

Category	Female	Male
0	215	318
0s	1	6
1	4	20
2	8	13
3	10	20
Total	238	377

Table 2. Input for HCV Data set 2.

Lab tests	Min-Max
ALB	15–82
ALP	11–417
ALT	1–325
AST	10.6–324
BIL	8–254
CHE	1.42–16.41
CHOL	1–10
CREA	8–1079.1
GGT	4.5–650.9
PROT	45–90
AGE	19–77
SEX	1 or 0

The abbreviations in Table 2 are as follows: ALB: Albumin, ALP: Alkaline Phosphatase, ALT: Alanine amino-transferase, AST: Aspartate aminotransferase, BIL: Bilirubin, CHE: Choline esterase, CHOL: Cholesterol, CREA: Creatinine, GGT: Gamma-Glutamyl Transpeptidase, PROT: Protein.

3 Classification

Unlike other artificial intelligence applications, machine learning does not need manually entered rules while imitating human intelligence. In machine learning, artificial intelligence improves itself by assimilating the data presented to it. Machine learning makes predictions by inferring from mathematical and statistical data. Processing of mathematical and statistical data can be done with many machine learning methods. In this study, HCV data set was studied with 9 different machine learning methods. The created 9 methods have different parameters and mathematical operations. 9 different machine learning methods used in this study are listed below.

3.1 Logistic Regression

It is often used to describe data and explain the relationship between dependent and independent data. This algorithm is often used to analyze numerical data [15]. In this study, our logistic regression algorithm obtained the best accuracy for the HCV dataset with the parameters Solver = "liblinear" and Tol = "0.0001".

3.2 Gaussian Naive Bayes

This algorithm is a machine learning algorithm for classifying data. It tries to predict the class of new data entered the system with high accuracy by calculations made according to probability conditions [16]. The naive bayes algorithm can adapt itself as the data changes in the system. This algorithm re-scans the entire dataset for each new classification operation. That's why it's a little slow.

3.3 KNN (K-nearest Neighbors)

The working logic of this algorithm makes clustering process according to the proximity relations between the objects. It works in the coordinate plane with the linear decomposition method [17]. In this study, it has been seen that the parameters Leaf. Size = "30", Metric = "minkowski" and N-neighbors = "5" for the KNN algorithm are the best parameters for this data set.

3.4 Support Vector Clustering (SVC)

This algorithm is one of the supervised learning algorithms [18]. It is used to predict an optimal hyperplane in an N-dimensional space. In this study, the parameters Degree = "3", Kernel = "linear", Tol = "0.001" and C = "1" gave the highest accuracy for the SVC method.

3.5 Radial Basis Function (RBF)

It is a radial-based network that uses basic radial functions as activation functions. It is used for time series analysis, system controls and classification of categorical data [19]. In this algorithm, the outputs are obtained by applying the sum function. Training and learning times are short. In the RBF SVC method applied in the study, C = "10" and Gamma = "0.0001" parameters were the best parameters.

3.6 Artificial Neural Network (ANN)

Artificial neural networks are algorithms based on mathematical modeling of the human brain. ANN mimics the human learning process. Just like a human, he makes himself more successful by repeating the examples he encounters [17]. In this method, information enters the input layer, is processed in the hidden layer, and leaves the output layer. It has been observed that the best parameters for the ANN algorithm used in this study are Activation = "logistic", Alpha = "0.0001" and Hidden layer = "(5,3)".

3.7 Classification and Regression Trees (CART)

This algorithm is used to construct a decision tree. It has a binary tree structure and branches out from the parent node into two different nodes. This algorithm is frequently used on numerical values [20]. It can show high success in noisy data and is a fast algorithm. The highest accuracy for the CART algorithm used in this study was obtained with the parameters Max depth = "5" and Min.samples. split = "19".

3.8 Random Forest

Random Forest algorithm is a supervised learning algorithm. With this algorithm, a random forest is created. There are decision trees that are trained in this forest. It is often used in classification and regression problems [21]. This algorithm is a flexible and easy to use machine learning algorithm. It has been observed that the best parameters for this algorithm used in this study are Max.depth = "10", Min.samples. split = "10" and N.estimators = "1000".

3.9 Gradient Boosting Machines (GBM)

The GBM algorithm is a machine learning algorithm used in regression and classification problems that generates prediction models in batches [22]. A model is created by combining decision trees. In this study, the highest accuracy for the GBM algorithm was obtained by using the learning_rate = "0.01", max_depth = "3", min_samples_split = "5", and n_estimators = "500" parameters.

3.10 Success Evaluation Criterion

In this study, accuracy was used as success evaluation criterion. Accuracy is a widely used method among success evaluation methods. To calculate the accuracy, firstly a confusion matrix is created and TP (True Positive), TN (True Negative), FP (False Positive) and FN (False Negative) values are obtained. Accuracy can be briefly explained as the ratio of all correct answers (TP and TN) to all answers (TP, TN, FP, FN). The model is considered successful if the Accuracy score is between 0 and 1, with scores approaching 1. The equation of accuracy is shown in formula (1).

$$Accuracy = \frac{TP + TN}{TP + TN + FP + FN} \tag{1}$$

4 Experimental Results

In this study, the HCV data set, whose details are given in Table 1 and Table 2, is presented to 9 different machine learning algorithms. The result obtained with all these algorithms can be reached in Table 3. In this study, anaconda program and Jupiter lab module were used as working environment. Through this module, 9 different machine learning algorithms were created with the python programming language and training processes were realized with the data set. The parameters that offer the highest accuracy for all machine learning algorithms created with the Python programming language have been determined and all parameters are presented in the classification part of our study. Trainings for all algorithms took place at the same time on average. All algorithms presented close accuracy on average. The HCV dataset, which is available as open source, is directly submitted to the system. There is no need for any normalization process beforehand from this established system.

Table 3. Accuracy table.

Method	Accuracy
Logistic Regression	0.9577
KNN	0.9395
RBF	0.9243
ANN	0.9027
SVC	0.9027
Random Forest	0.9027
Gradient Boosting Machines	0.8972
Gaussian Naive Bayes	0.8918
CART	0.8702

5 Conclusion

Hepatitis C virus (HCV) infection is one of the major causes of death and morbidity in the world. Examples of symptoms in HCV disease include yellow discoloration of the eyes and skin, itchy skin, fatigue and weakness, weight loss, loss of appetite, and possibly similar symptoms. People infected with HCV often develop chronic infections A long-term (chronic) hepatitis C Infection may show no symptoms and is very difficult to detect, so it is described as a "silent" infection that is usually present in the blood for many years, until it damages the liver. There is currently no vaccine available for Hepatitis C. Early detection of hepatitis C disease is very important for treatment. In the case of early detection of the HCV virus, it has been observed that nearly 60% of the patients recovered with treatment. While early detection of HCV virus is easy in developed countries, it may be impossible in some rural areas due to the lack of doctors in poor countries. The aim of this study is to help reliable early diagnosis of HCV virus according to laboratory blood test results in countries where the number of specialist doctors is not sufficient by using the system we proposed in our study. Thus, if the virus is detected early, the spread of the virus to the body can be prevented during the treatment process. 70% of the data set was used for the training of all machine learning algorithms presented in this study. 30% test data was separated from the data set to determine accuracy of this study. In this study, when 9 different machine learning algorithms applied to the HCV data set were compared, the highest accuracy was obtained with the logistic regression algorithm with 95.77%. The next best accuracy was obtained with the KNN algorithm with 93.95%. All machine learning algorithms used in this study and the obtained accuracies are presented. In the present study, the lowest accuracy in the HCV data set was obtained with the CART algorithm. With the CART algorithm, 87.02% accuracy was obtained.

References

1. Johns Hopkins Medicine. https://www.hopkinsmedicine.org/gastroenterology_hepatology/_pdfs/liver/viral_hepatitis_b.pdf
2. Salas-Villalobos, T.B., Lozano-Sepúlveda, S.A., Rincón-Sánchez, A.R., Govea-Salas, M., Rivas-Estilla, A.M.: Mechanisms involved in liver damage resolution after hepatitis C virus clearance. Med. Univ. **19**(75), 100–107 (2017). https://doi.org/10.1016/j.rmu.2017.05.005
3. Centers for Disease Control and Prevention. https://www.cdc.gov/mmwr/volumes/69/rr/rr6902a1.htm
4. Vitalant Hospital. https://hospitals.vitalant.org/getattachment/Products-Services/Donor-Test-Information/BS_910.pdf.aspx?lang=en-US
5. Khatun, M., Ray, R.B.: Mechanisms underlying hepatitis C virus-associated hepatic fibrosis. Cells **8**(10), 1249 (2019). https://doi.org/10.3390/cells8101249
6. Syafa'ah, L., Zulfatman, Z., Pakaya, I., Lestandy, M.: Comparison of machine learning classification methods in hepatitis C virus. J. Online Informatika **6**(1), 73 (2021). https://doi.org/10.15575/join. v6i1.719
7. Suwardika, G.: Pengelompokan dan klasifikasi pada data hepatitis dengan menggunakan support vector machine (SVM), classification and regression tree (cart) dan regresi logistik biner. J. Educ. Res. Eval. **1**(3), 183 (2017). https://doi.org/10.23887/jere.v1i3.12016
8. Hashem, S., Esmat, G., Elakel, W., Habashy, S., Raouf, S., Elhefnawi, M., et al.: Comparison of machine learning approaches for prediction of advanced liver fibrosis in chronic hepatitis C patients. IEEE/ACM Trans. Comput. Biol. Bioinf. **15**(3), 861–868 (2018). https://doi.org/10.1109/tcbb.2017.2690848
9. Nandipati, S.C.R., Xin Ying, C., Wah, K.K.: Hepatities C virus (HCV) prediction by machine learning techniques. Appl. Model. Simul. **4**, 89–100 (2020)
10. Konerman, M., Beste, L., Van, T., Liu, B., Zhang, X., Zhu, J., et al.: Machine learning models to predict disease progression among veterans with hepatitis C virus. PLoS ONE **14**(1), e0208141 (2019). https://doi.org/10.1371/journal.pone.0208141
11. UCI (California Irvine University). https://archive.ics.uci.edu/ml/datasets/HCV+data
12. Mayo Clinic. https://www.mayoclinic.org/tests-procedures/liver-function-tests/about/pac-20394595
13. LAB TESTS ONLINE. https://labtestsonline.org/tests/cholinesterase-tests
14. University of Michigan Health System (UMHS). https://www.uofmhealth.org/health-library/hw43614
15. Ali, L., et al.: A feature-driven decision support system for heart failure prediction based on statistical model and Gaussian naive bayes. Comput. Math. Methods Med. (2019). https://doi.org/10.1155/2019/6314328
16. Kambi Beli, I.L., Guo, C.: Enhancing face identification using local binary patterns and k-nearest neighbors. J. Imaging **3**(3), 37 (2017). https://doi.org/10.3390/jimaging3030037
17. Altıparmak, H., Salama, R., Gökçekuş, H., Uzun Ozsahin, D.: Predict future climate change using artificial neural networks. In: Uzun Ozsahin, D., Gökçekuş, H., Uzun, B., LaMoreaux, J. (eds.) Application of Multi-Criteria Decision Analysis in Environmental and Civil Engineering. PPES, pp. 57–63. Springer, Cham (2021). https://doi.org/10.1007/978-3-030-64765-0_9
18. Hao, Z., Ge, H., Gu, T.: Automatic image annotation based on particle swarm optimization and support vector clustering. Math. Probl. Eng. (2017). https://doi.org/10.1155/2017/8493267
19. Helwan, A., Idoko, J.B., Abiyev, R.H.: Machine learning techniques for classification of breast tissue. Procedia Comput. Sci. **120**, 402–410 (2017). https://doi.org/10.1016/j.procs.2017.11.256

20. Zhang, B., Wei, Z., Ren, J., Cheng, Y., Zheng, Z.: An empirical study on predicting blood pressure using classification and regression trees. IEEE Access **6**, 21758–21768 (2018). https://doi.org/10.1109/ACCESS.2017.2787980
21. Iwendi, C., et al.: COVID-19 patient health prediction using boosted random forest algorithm. Front. Public Health **8**, 357 (2020). https://doi.org/10.3389/fpubh.2020.00357
22. Zvarevashe, K., Olugbara, O.O.: Gender voice recognition using random forest recursive feature elimination with gradient boosting machines. In: International Conference on Advances in Big Data, Computing and Data Communication Systems (icABCD), pp. 1–6. IEEE (2018). https://doi.org/10.1109/ICABCD.2018.8465466

Fuzzy Logic Approach to the Amount of Medication Taken During Breathing with an Inhalator

Aynur J. Jabiyeva(✉)

Azerbaijan State Oil and Industry University, 20 Azadlig Ave., Baku AZ1010, Azerbaijan

Abstract. The fuzzy logic is form of logic that helps professionals solve various issues in the field of health care by grading right and incorrect in classical reasoning. Experts decide whether to change the inhalation time medication dose using the inhaler depends on the patient's age, weight, height, and co-morbidities. In the study, the doses of more regularly prescribed inhaler medications were calculated by applying fuzzy logic. The study uses fuzzy logic to compute the medication dose that should be given to the patient set on the patient's age, human body's mass index, and patient class at the outset. The research was based on calculating the medicine dosage without use of any other medications and assuming no premedication. The medication dosage levels discovered at the conclusion of the trial were quite close to the values recommended by specialists. As a result, fuzzy logic estimator utilized in this work may be thought of as a source of assistance for inhaler specialists and users.

Keywords: Fuzzy logic · Drug dose · Inhalator · Expert systems · Inhalation process · Fuzzy systems · Mamdani technique: medication dosage

1 Introduction

Nowadays with the advancement of technology and engineering applications, it is easy to find answers to the majority of issues encountered in the fields of health and pharmacology. In addition, solutions for processes such as pharmacological modeling, medication dosage, concentration modification, patient follow-up, illness diagnosis systems are designed to help experts by utilizing intuitive information supplied by experts [1–3]. As uncertain variable expressions, heuristic rules and expert knowledge are required in modeling complex systems, fuzzy logic is one of the most helpful techniques. In classical logic, fuzzy logic determines right and incorrect through grading [4–7]. Fuzzy logic employs linguistic variables from everyday life. The inference rules of the fuzzy logic rule base are imperfect yet approximate. Individual and nonlinear differences in pharmacology, such as age, sex, weight, illness state, genetic polymorphisms, concurrent medications, challenge modeling to predict results. As a result, in a variety of applications, fuzzy modeling has been utilized to regulate medication administration mechanically. Closed-loop systems are capable of making independent decisions and attempting to accomplish a predefined objective. In closed-circuit systems that can adjust dosage or

R. A. Aliev et al. (Eds.): ICSCCW 2021, LNNS 362, pp. 270–277, 2022.
https://doi.org/10.1007/978-3-030-92127-9_38

drug concentration, control approaches like fuzzy logic and artificial neural networks are extremely prevalent [7]. The medication dose is calculated by specialists using variables that different in each patient based on the patient's specific features. The specialists take into account the patient's age, BMI and co-morbidities as the most relevant considerations. With this research, we believe that the fuzzy logic modeling technique might be useful in assisting specialists in determining medication dose.

2 Materials and Methods

Fuzzy logic is a form of logic that holds everything is a question of degree. It encompasses classical logic by selecting to rank the truth as well as the dual and clear thinking features of classical logic. When assessing a thought, it represents the uncertainty in thinking by incorporating the values of true or false (0, 1) in classical logic and ranks it in the range of [0 1] how true and false it is. Fuzzy logic provides a solution when classical logic is insufficient, i.e. when the accuracy criteria cannot be stated precisely. The majority of physical processes are based on erroneous human thinking [3, 8]. This is a sort of knowledge that may not be completely accurate, yet it may still be extremely valuable to individuals. The capacity to incorporate reasoning in unintelligible and complicated issues is the criterion for assessing fuzzy logic's efficacy.

In the classical set notion, an element is either a member of the set or not a member of the set (0, 1). The membership degree notion, on the other hand, defines whether or not an element belongs to a set-in fuzzy set. The degree of membership specifies how closely the element is related to the set. The degree of membership specifies how closely an element is related to the set. The degree of membership is represented by the symbol "μ" and has a value in the range [0 1]. The membership degree is determined by several membership functions [9–11] such as bell, pi, triangle and trapezoid.

Systems that use fuzzy logic to model them. Figure 1 depicts the overall structure of fuzzy systems.

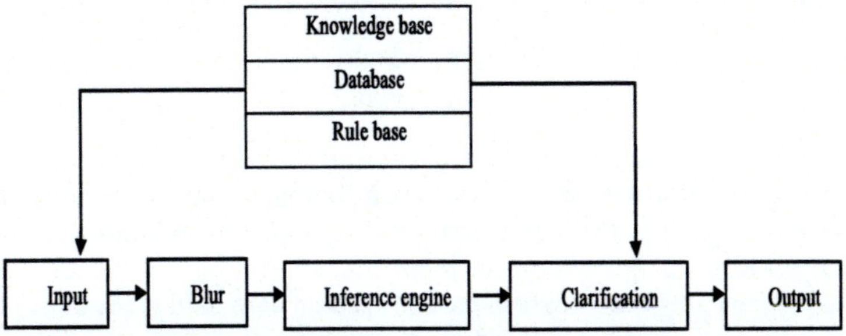

Fig. 1. General structure of fuzzy systems

The database is the structure in which the fuzzy system's membership functions are stored. Inputs to the fuzzy system are represented as net values in the form of sharp set elements. Net values are transformed to fuzzy values utilizing membership functions throughout the fuzzification process. The rule base contains the fuzzy rules that will be used by the system to make conclusions. With the inference engine, the system uses fuzzy rules to produce the output matching to the fuzzy inputs as fuzzy values. Defuzzification is the process of transforming fuzzy outputs into net values. As a result of going through these processes, a fuzzy system can provide crisp outputs. Mamdani is a fuzzy logic algorithm that is commonly employed in fuzzy systems. The inputs and outputs of fuzzy systems are in the form of fuzzy sets [4]. Fuzzy subsets, particularly membership functions, are defined by linguistic variables for each input and output universal fuzzy sets. The degree of membership defines how much of the net input or output value is part of these subgroups. Fuzzy outputs matching to fuzzy inputs are generated in a Mamdani type inference engine by employing the max-min operator from fuzzy set operations. Using fuzzy rules, the inference engine determines which inputs and outputs to infer. Inference rules in fuzzy systems are unclear yet approximate in their validity. Uncertainty and linguistic variables are present in fuzzy logic truth tables and inference methods. For example, in a Mamdani-type rule, the patient is described as a "child" with a "weak" mass index and the "DD1" class as a "large" beginning dose of the medication. The rule base is made up of heuristic rules based on expert knowledge. Experts create membership functions and rule bases in fuzzy systems; as a result, it is critical that the expert possess knowledge and expertise with the topic.

3 Statement of the Problem

Defuzzification refers to the process of transforming fuzzy outputs to net values in fuzzy systems. The most widely used defuzzification methods are center of gravity, area bisector, middle of the biggest, largest of the largest and smallest of the largest. The center of gravity technique as described in Eq. (1) is the most widely used method; the net output value is the center of gravity of the resultant solution surface.

$$\mathbf{Z} = \frac{\int \mu_A(x)x\,dx}{\int \mu_A(x)\,dx} \tag{1}$$

In general, professionals consider the patient's age, weight, height and co-morbidities when estimating the first dose of the medication. Based on these criteria, the expert calculates how many mg per kg of the initial dosage should be administered based on his familiarity with the drug's therapeutic properties.

This application uses fuzzy logic to predict the drug dose used in the most popular inhaler device in mg/kg, without premedication or the usage of extra medicines.

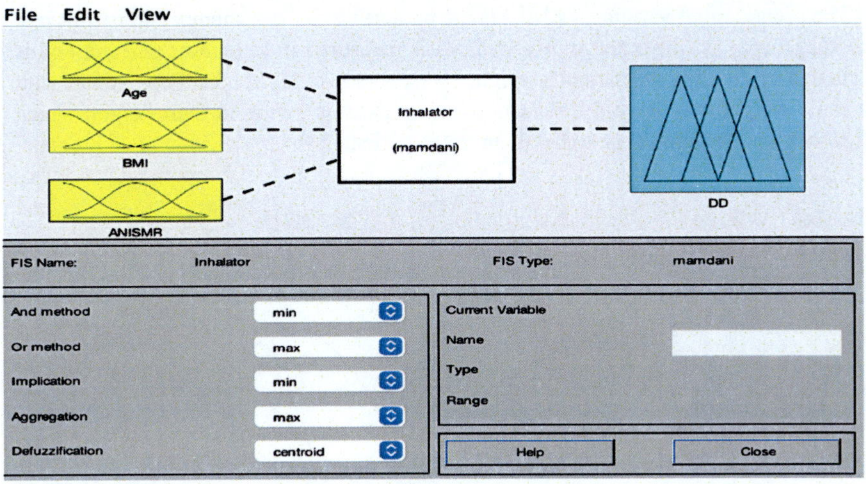

Fig. 2. Application on FIS file

The medication dosage is calculated using the categorization entries "AGE," "BMI," and "ANISMR." As illustrated in Fig. 2, the Mamdani technique was used in the implementation, which was done with the MATLAB Fuzzy logic toolbox.

As shown in Fig. 3, the first input variable "AGE" is split into four subgroups. Because the inhaler device is not utilized in children under the age of five, the age range (5–100) is used.

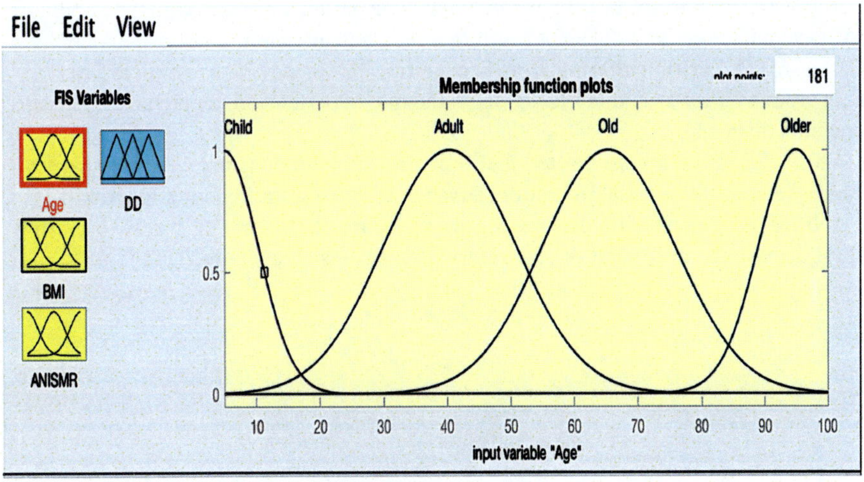

Fig. 3. Input of "AGE"

The second input variable, "BMI," is the patient's body mass index and is split into four subgroups, as illustrated in Fig. 4. BMI is measured in kilograms per unit and is calculated by dividing the patient's weight by the square of his height. Two patients with the same weight may have different body mass indexes owing to their height, which influences the medicine dose to be given to the patients.

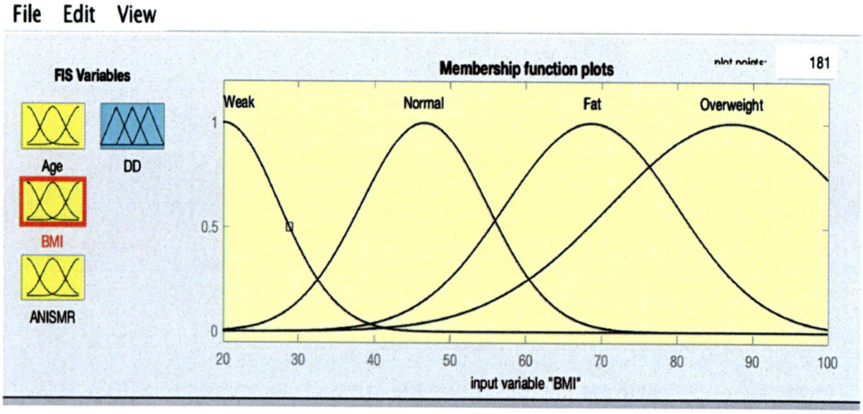

Fig. 4. Input of "BMI"

The patient's "ANISMR" categorization is the third input variable. The Azerbaijan National Institute of Sports Medicine and Rehabilitation (ANISMR) categorization system categorizes patients treated based on their physical health state:

ANISMR1: A normal, healthy individual with no systemic issues, with the exception of surgical pathology that does not result in a systemic disease.

ANISMR2: Person suffering from a condition that necessitates surgical intervention or from a minor systemic illness (mild anemia, hypertension, emphysema, chronic bronchitis, diabetes).

ANISMR3: Person with disease that limits activity but does not debilitate (hypovolemia, previous myocardial infarction, advanced diabetes, limited lung function).

ANISMR4: Person with serious illness that poses a continuing threat to his life (shock, decompensated, heart or respiratory disease, unstable angina) (Fig. 5).

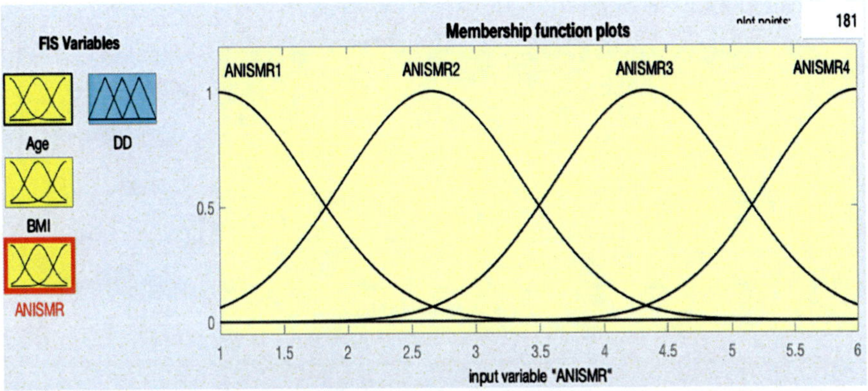

Fig. 5. Input of "ANISMR"

As shown in Fig. 6, the medication dosage was taken in mg/kg as the output variable. The dosage range was [0.5 5].

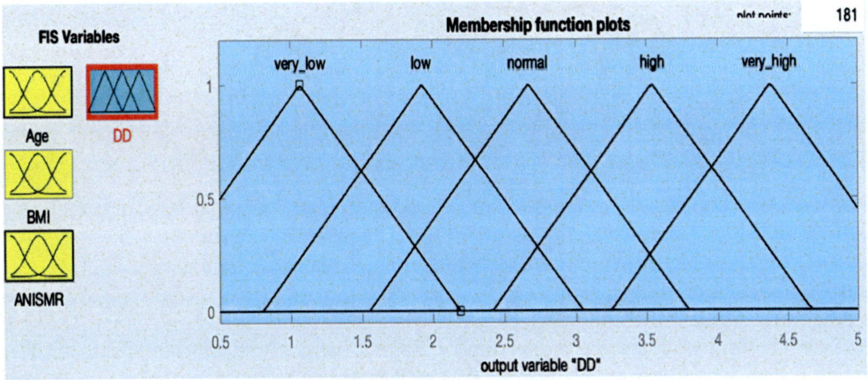

Fig. 6. Output of drug dosage

The figure below calculates the average drug dose based on the parameters of a 50-year-old patient weighing 60 kg depending on the data from Azerbaijan National Institute of Sports Medicine and Rehabilitation (Fig. 8).

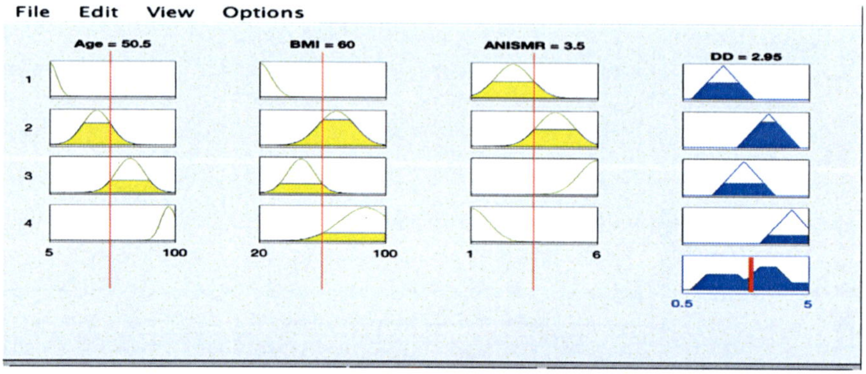

Fig. 7. Example of datas with diagrams

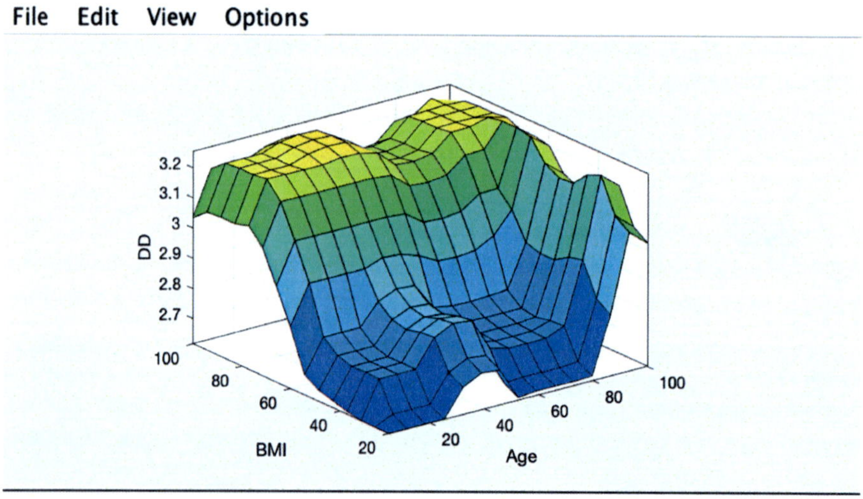

Fig. 8. Surface viewer

4 Conclusion

It was intended to create a system that assists specialists by calculating medicine doses using fuzzy logic and the system dose levels were fairly similar to the specialist physicians' values. Thanks to the data, the average starting dose for a specific patient type was computed using fuzzy logic in Fig. 7. For example, for a patient weighing 60 kg, standing 1.55 m tall, being 50 years old and having an ANISMR of 3.5, specialists received an average medication dose of 3 mg/kg, while the study received 2.95 mg/kg. As a result, the fuzzy logic estimator utilized in this work may be regarded as a resource for specialists in the field of inhaler devices.

References

1. Zadeh, L.A., Aliyev, R.A.: Fuzzy Logic Theory and Applications: Part 1 and Part 2. 2018, p. 612 (2018). https://doi.org/10.1142/10936
2. Ross, T.J.: Fuzzy Logic With Engineering (2010). ISBN: 978-0-470-74376-8
3. Sproule, B.A., Naranjo, C.A., Türksen, I.B.: Fuzzy pharmacology: theory and applications. Trends Pharmacol. Sci. **23**(9), 412–417 (2002). https://doi.org/10.1016/s0165-6147(02)02055-2
4. Sable, P., Khanvikar, V.V., Sable, P., Khanvilkar, V.V.: Pharmaceutical applications of artificial intelligence. Int. J. Pharma Res. Heal. Sci. **6**(2), 2342–2345 (2018)
5. Allahverdi, N.: Applied mathematics, electronics and computers design of fuzzy expert systems and its applications in some medical areas. Int. J. Appl. Math. Electron. Comput. **2**(1), 1–8 (2014)
6. Bazoon, M., Shuhuan, K.I.: Fuzzy logic pharmacokinetic modeling: application to lithium concentration prediction. Clin. Pharmacol. Ther. **62**, 29–40 (2019). https://doi.org/10.1016/S0009-9236(97)90149-1
7. Saritas, I., Ozkan, A., Allahverdi, N.: Determination of the drug dose by fuzzy expert system. Determination of the drug dose by fuzzy expert system in treatment of chronic intestine inflammation 2009, June 2015
8. Levy, G.: Predicting effective drug concentrations for individual patients. Clin. Pharmacokinet. **34**(4), 323–333 (1998). https://doi.org/10.2165/00003088-199834040-00005
9. Aliev, R.A., Aliev, R.R.: Soft Computing and its Application. World Scientific, New Jersey, London, Singapore, Hong Kong (2001). https://doi.org/10.1142/4766
10. Gardashova, L.A.: Z-Number based TOPSIS method in multi-criteria decision making. In: Aliev, R., Kacprzyk, J., Pedrycz, W., Jamshidi, M., Sadikoglu, F. (eds) 13th International Conference on Theory and Application of Fuzzy Systems and Soft Computing — ICAFS-2018. Advances in Intelligent Systems and Computing. vol. 896, pp. 42–50. Springer, Cham (2018). https://doi.org/10.1007/978-3-030-04164-9_10
11. Babanli, J.M.: Fuzzy approach for evaluation of student's performance. In: 14th International Conference on Theory and Application of Fuzzy Systems and Soft Computing, ICAFS-2020. Advances in Intelligent Systems and Computing, vol. 1306, pp. 140–147 (2020). https://doi.org/10.1007/978-3-030-64058-3_18

Fuzzy Modeling of the Relationship Between Tax Administration Efficiency and Tax Obligations Fulfillment

A. F. Musayev[1,2,3] and M. Kh. Gazanfarli[1](✉)

[1] ANAS Institute of Economy, H. Javid Avenue, 115, AZ1143 Baku, Azerbaijan
akif.musayev@gmail.com, m.qezenferli@gmail.com, q.miraa@gmail.com
[2] Baku Business University, H. Zardabi Street, 88a, AZ1122 Baku, Azerbaijan
[3] The Azerbaijan University, J. Hajibeyli Street, 71, AZ1007 Baku, Azerbaijan

Abstract. One of the essential factors that can indicate the effective functioning mechanism of tax administration is the realization level of tax compliance and liabilities by taxpayers. This paper focuses on developing a model to evaluate fulfillment level of tax obligations by using fuzzy inference system, as well as to define interrelationship between tax obligations and tax administration. Proposed approach has been used to estimate realization level of tax obligations in Azerbaijan based on the accurate data in 2020. Unfortunately, the result is unsatisfactory (0.51), however high fulfillment level can be achieved by strengthening of tax administration activity (increasing the both quantity and quality of administrative measures).

Keywords: Tax administration · Fulfillment of tax obligations · Fuzzy inference system · Efficiency level

1 Introduction

Large scale investigation about taxation proves that most of researches focus on eliminating or reducing tax evasion, detecting illegal activities, ways to achieve effective administration, etc. As taking into consideration more than 30% of the world economy is an underground activity, it can be mentioned that not only Azerbaijan, but also developed countries' tax systems have not achieved effective organization of administration yet [1]. For this reason, the necessity to define indicators on system of tax administration activity, to develop new methodologies and models to measure its efficiency have remained relevant for many years. There are approaches proposed by both other scientists and us to measure the effectiveness of tax administration. Let's look through some of them.

In our previous research, a new system of indicators characterizing the activities of tax administration and expressed in 3 groups - internal, external micro and external macro environment was proposed [2]. As a follow-up to this study, a new model - tax administration efficiency index was developed to measure the efficiency of tax administration using this system of indicators.

R. A. Aliev et al. (Eds.): ICSCCW 2021, LNNS 362, pp. 278–285, 2022.
https://doi.org/10.1007/978-3-030-92127-9_39

$$I = \frac{\sum_{i=1}^{3} I_i}{r_t}$$

$$\begin{cases} I_1 = \sqrt[n_1']{\prod_{j'=1}^{n_1'} (s_{1j'}^{\dot{q}y})*} + \sqrt[n_1'']{\prod_{j''=1}^{n_1''} (v_{1j''}^{\ddot{q}f'})} \\ I_2 = \sqrt[n_2']{\prod_{j'=1}^{n_2'} (s_{2j'}^{\dot{q}y})*} + \sqrt[n_2'']{\prod_{j''=1}^{n_2''} (v_{2j''}^{\ddot{q}f'})} \\ I_3 = \sqrt[n_3']{\prod_{j'=1}^{n_3'} (s_{3j'}^{\dot{q}y})*} + \sqrt[n_3'']{\prod_{j''=1}^{n_3''} (v_{3j''}^{\ddot{q}f'})} \end{cases}$$

Since the main duty of tax administration is to regulate and manage tax relations, it is obvious that the system of indicators that characterizes it is expressed not only in quantitatively but also in qualitatively. The necessary information base for the assessment was formed by the State Tax Service under the Ministry of Economy of the Republic of Azerbaijan, the State Statistical Committee of the Republic of Azerbaijan and a survey of 27 taxpayers, and the efficiency of tax administration in Azerbaijan was calculated by the proposed index [3]. By using efficiency index of tax administration new methodology was proposed to define detection level of probability level of tax evasion [4].

A method of unified assessment with a hierarchical system was proposed by Ilona Skackauskiene as considering of tax administration to be one of the three indicator groups (main, partially and comprehensively integrated) of the tax system. Using the importance levels of the selected indicators in this study, fairness, efficiency and complexity of tax administration were calculated using the formulas as below.

$$T = \sum_{i=1}^{n} \omega_i t_i, \quad E = \sum_{i=1}^{n} \omega_i e_i, \quad S = \sum_{i=1}^{n} \omega_i s_i$$

As a result, the tax system was evaluated by the equation $M = \omega_1' T + \omega_2' E + \omega_3' S$ [5].

A new method was proposed in 2016 by several Indian researchers to measure the effectiveness of tax administration. The study is based on a simple conceptual equation proposed by these authors in 1995 that links the effectiveness of tax administration to tax revenue:

$$R_j = t_j B_j C_j; \quad B_j = b_0 Y_j^{b_1} N_j^{b_2}; \quad C_j = c_0 E_j^{c_1} T_j^{c_2}$$

R_j – tax revenue for a certain period (j^{th} subject), t_j – effective tax rate, B_j – the potential tax base in j, C_j – the part of the potential tax base that is actually taxed in j, Y_j – the national income, N_j – the number of taxpayers (during that period), E_j – the attitude of the taxpayer, T_j – the effectiveness of tax administration. b_0, b_1, b_2 *and* c_0, c_1, c_2 – positive constants. The model was used to assess quantitatively the impact of tax administration on tax revenues by making it empirically through external audits of VAT administrators by Indian government agencies [6]. In another approach, the efficiency of tax administration was measured by the amount of tax revenue per employee of a tax authority [7]. In case of problems related to the taxpayers' economic activities, various concessions are applied within the legislation to compensate for the damage. A new methodology was proposed using multi-criteria decision-making methods to provide these allowances to the most deserving taxpayers (the most affected) [8]. Global analysis show that tax burden is one of the most crucial indicators used to measure the efficiency of tax administration. There are different approaches to assess the tax burden, one of which is the methodology for assessing the tax burden, taking into account the shadow economy under uncertainty [9].

2 Statement of Problem

The level of tax obligations[1] fulfillment is one of the factors that allows to measure the how tax administration mechanism operate effectively. In our previous study [3], the individual components of tax obligation were taken into account in the assessment of the tax administration efficiency index. However, the relationship between the general level of fulfillment of tax obligations and tax administration has not been analyzed.

Therefore, this paper investigates the relationship between the level of tax obligations of taxpayers (businesses) and the activities of tax administration. For this purpose:

Initially, the process of tax obligations fulfillment and the role of tax administration in this process, will be analyzed;

Secondly, this process will be modeled by using a fuzzy inference system, as this process being an issue of decision-making depending on the various components;

Finally, the level of tax administration effectiveness and fulfillment level of tax obligations will be defined with appropriate assessments.

3 Methodology for Modeling the Relationship Between Fulfilment of Tax Liabilities and the Tax Administration Effectiveness

The fulfillment of the tax obligations initiates with the registration of the taxpayers and the acceptance of their duties. However, it does not happen effortlessly. Thus, some of engaged citizens with entrepreneurial activities try to hide their activities either ignorantly or deliberately. The formation of a tax culture among the population is one of the prerequisites for solving serious problems in the country, such as reducing the level of the shadow economy, increasing budget revenues and consequently improving the living standards of the population, ensuring socio-economic welfare.

Seminars, various trainings, regular publications on taxation, especially various competitions for the formation of knowledge about taxes in children and youth in secondary schools, the organization of competitions play an invaluable role in enlightenment and awareness policy. Undoubtedly, all of this is possible in a tax system that possesses a strong, efficient administration. As a result, the official declaration of business activities (registration as a taxpayer) reduces the level of the shadow economy by preventing tax evasion.

Other components of the tax obligations are the tax liability (amount of tax depending on their activity) and tax compliance (timely, accurately filling of tax return). Informing taxpayers, prompt and complete response to inquiries to eliminate difficulties related to

[1] **Note:** Tax obligations cover both tax compliance and tax liability.

tax liabilities, sending warning letters, etc. helps to complete this stage completely and legally. However, in some cases, taxpayers try to evade taxes by committing various illegalities (showing less than their income, etc.). Although these attempts are often detected through various inspections and other administrative measures, in some cases they cannot be detected. Such cases occur as a result of poor organization of administration.

The relationship between tax obligations and tax administration can be expressed in Fig. 1:

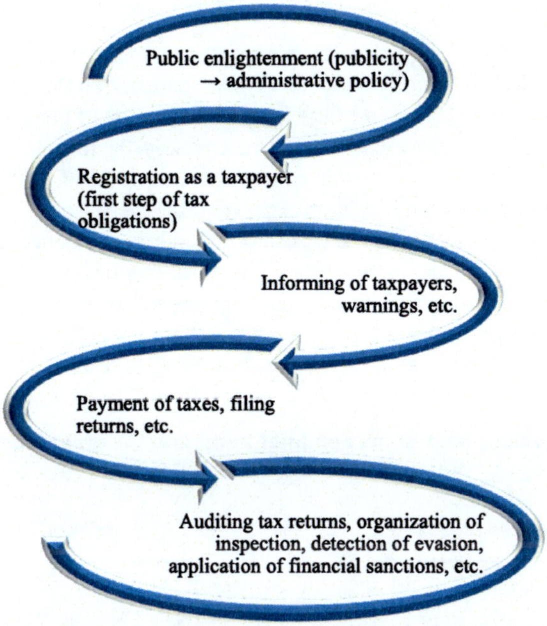

Fig. 1. Interrelationship between tax obligations and tax administration

All above demonstrate that effective tax administration leads to a high fulfillment on level of tax obligations as well as reduction of the level shadow economy, the application of financial sanctions and fines by preventing tax evasion in the country. That means,

$$If\ x_1\ is\ vh(h)\ and\ x_2\ is\ vh(h)\ then\ y_1\ is\ vl(l)\ and\ y_2\ is\ vl(l)$$

Since the process under consideration is a matter of decision-making depending on its various components, it is more expedient to express it through a fuzzy inference system:

Herein, x_1, x_2 – input variables, y_1, y_2 – output variables.

x_1 – volume of registration as a taxpayer;

x_2 – number of accurate (or accepted) tax returns;

y_1 – level of shadow economy;

y_2 – amount of income from fines.

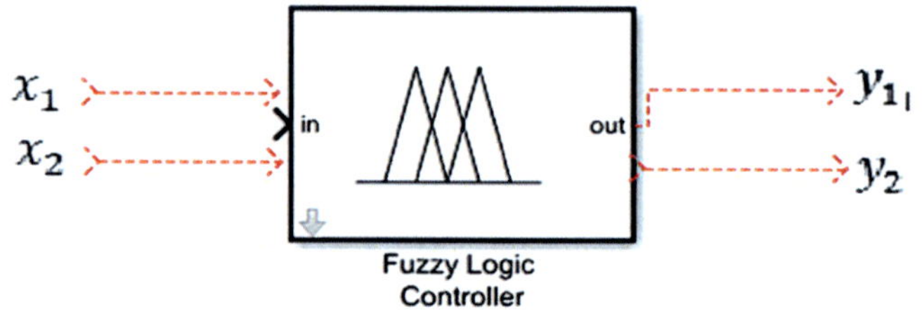

Fig. 2. Components and consequences of process of tax obligations fulfillment

The system (see Fig. 2) expresses the entire process of tax obligations fulfillment. Undoubtedly, the effective functioning of this system depends on the effective organization of tax administration. That is, the high level of realization of the tax obligations, its sufficient and legitimate implementation and consequently, the few numbers of financial sanctions imposed an amount of income from fines. Low level of shadow economy in the country shows how tax administration organize effectively.

Therefore, taking into account the necessity to analyze the components of the tax obligations and the relationship of the results of its fulfillment within tax administration, a dual inference system is built for modeling this process that both of them consist of 2 inputs and 1 output variables.

The schematic configuration of the process will be as follows.

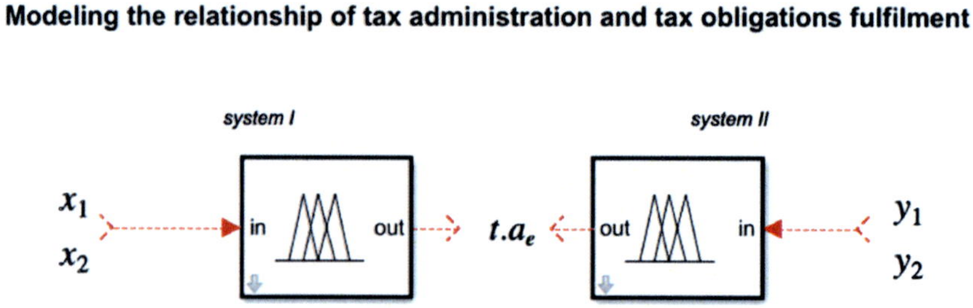

Fig. 3. Fuzzy relationship model of tax administration and tax obligations fulfillment

Herein,

x_1, x_2 – input variables of system I;

y_1, y_2 – input variables of system II;

$t.a_e$ – output variable of both fuzzy inference systems, the level of efficiency of tax administration, and it is assessed in the range of [0.1].

Figure 3 was built to demonstrate obviously the role of tax administration in the process of tax obligations fulfillment, by authors.

Initially the input and output variables were fuzzified. Membership functions of each variables were built by expressing of the input variables with 5 linguistic variables {very low, low, average, high, very high} as well as the output variable with 5 linguistic variables {non-effective, less effective, average, more effective, the most effective}. In the next stage, a fuzzy rule set was developed. The result was obtained by aggregating the input variables according to the selected inference system type on the basis of the generated knowledge base and defuzzification of it. Tax administration can be effective in different combinations of input variables, but the highest level is possible for both inference systems under the following conditions.

$$\textit{If } x_1 \textit{ is vh and } x_2 \textit{ is vh then } t.a_e \textit{ is most effective} \rightarrow I$$

$$\textit{If } y_1 \textit{ is vl and } y_2 \textit{ is vl then } t.a_e \textit{ is most effective} \rightarrow II$$

4 Measuring of Tax Administration Effectiveness Depending on Tax Obligations Results, in the Sample of Azerbaijan

Using the proposed model, we have analyzed the level of fulfillment of tax obligations depending on the tax administration in the Republic of Azerbaijan. Since the general volume of entrepreneurial activity (registered and non-registered), the quality of filled returns by taxpayers and the degree of acceptance are unknown, we can determine the level of tax obligation fulfillment by the degree of efficiency of administration. Therefore, first of all, we have to assess the efficiency level of tax administration through system II. To realize this evaluation, we have used Mamdani fuzzy inference system [10].

Rule base that is applied to implement this process consists of 13 rules as follows (see Table 1):

Table 1. Applied rule base

Rules	İnputs		Output - $t.a_e$
	y_1	y_2	
R_1	vl	vl	most effective
R_2	l	vl	most effective
R_3	vl	l	most effective
R_4	l	l	more effective
R_5	l	avrg	more effective
R_6	avrg	l	more effective
R_7	avrg	avrg	avrg
R_8	avrg	h	avrg
R_9	h	avrg	avrg
R_{10}	h	h	less effective
R_{11}	vh	h	less effective
R_{12}	h	vh	less effective
R_{13}	vh	vh	non-effective

It has to be noted that the data used for the input variables of the system II are obtained from the State Tax Service under the Ministry of Economy of the Republic of Azerbaijan and Prof. Friedrich Schneider's study [11] about shadow economy.

The assessment covers the year 2020 and used data (input variables) is as below:

$$y_1 - \text{level of shadow economy} = 44,6(\%)$$

$$y_2 - \text{amount of income from fines} = 5.62e + 08(m)$$

Therefore, depending on the data in 2020, the output of fuzzy system (result), that means the level of efficiency of tax administration ($t.a_e$) is equal to 0.51 or 51%.

5 Conclusion

This research investigated the interrelationship between tax obligations and organization of tax administration, in addition, proposed new approach to evaluate the realization level of tax obligations and practically assessed it for the Republic of Azerbaijan. The result shows that the tax administration in Azerbaijan is on average at 0.51 points. Taking into consideration the relationship between tax obligations and administration, we can mention that the tax obligations which must be fulfilled by businesses (general), as well as registered legal entities and individuals engaged in entrepreneurship in our country is unsatisfactory and requires further administrative measures.

References

1. Musayev, A.F.: Modeling the process of emergence of the shadow economy through latent profitability. Soc. Econ. **8**, 43–60 (2020). https://doi.org/10.31857/S020736760010800-9
2. Gazanfarli, M.Kh.: Tax administration and organization of its effective activity. News of ANAS: Economy Series, vol. 1, pp. 113–120 (2020). http://www.economics.com.az/images/fotos/xeberler_pdf/2020_1/16.Mirvari.pdf
3. Gazanfarli, M.Kh.: The effectıveness model of tax admınıstratıon process. Sci. Rev. UNEC **8**(8), 92–102 (2020). http://unec.edu.az/application/uploads/2015/03/UNEC-EX-1-2020.pdf
4. Musayev, A.F., Gazanfarli, M.: Modeling the probability of the detection process of tax evasion taking into account quality and quantity indicators. Asian J. Econ. Bus. Account. **18**(4), 28–37 (2020). https://doi.org/10.9734/ajeba/2020/v18i430291
5. Skackauskiene, I.: Tax system evaluation model. In: 6th International Scientific Conference, Business and Management, Selected Papers, Vilnius, Lithuania, pp. 719–727 (2010). https://doi.org/10.3846/bm.2010.095
6. Das-Gupta, A., Estrada, G.B., Donghyun, P.: Measuring tax administration effectiveness and its ımpact on tax revenue. ERIA Discussion Paper Series (2016). https://www.eria.org/ERIA-DP-2016-17.pdf
7. Musayev, A.F., Aliyev, A.G., Musayeva, A.A., Gazanfarli, M.K.: Evaluation of tax administration effectiveness using brown-gibson model. In: Aliev, R.A., Kacprzyk, J., Pedrycz, W., Jamshidi, Mo., Babanli, M.B., Sadikoglu, F.M. (eds.) ICSCCW 2019. AISC, vol. 1095, pp. 221–228. Springer, Cham (2020). https://doi.org/10.1007/978-3-030-35249-3_28

8. Musayeva, A.A., Musayev, T.M., Gazanfarli, M.K.: Evaluation of the impact of the changing the term of tax liability performance on tax receipts by minmax composition method. In: Aliev, R.A., Kacprzyk, J., Pedrycz, W., Jamshidi, Mo., Sadikoglu, F.M. (eds.) ICAFS 2018. AISC, vol. 896, pp. 378–385. Springer, Cham (2019). https://doi.org/10.1007/978-3-030-04164-9_50
9. Musayev, A.F., Musayeva, A.A.: A Study of the ımpact of underground economy on ıntegral tax burden in the proportional growth model under uncertainty. Adv. Fuzzy Syst. **1**, 1–5 (2018). https://doi.org/10.1155/2018/6309787
10. Mamdani, E.H., Assilian, S.: An experiment in linguistic synthesis with a fuzzy logic controller. Int. J. Man Mach. Stud. **7**(1), 1–13 (1975). https://doi.org/10.1016/S0020-7373(75)80002-2
11. https://audit.gov.az/Upload/Files/2019/Friedrich_Schneider_Shadow_EcScanned_around_the_world_September_Azerbaijan_final-%D0%BA%D0%BE%D0%BD%D0%B2%D0%B5%D1%80%D1%82%D0%B8%D1%80%D0%BE%D0%B2%D0%B0%D0%BD.pdf

Project Selection Under U-Number-Valued Information

K. I. Jabbarova(✉)

Department of Computer Engineering, Azerbaijan State Oil and Industry University, 20 Azadlig Avenue, AZ1010 Baku, Azerbaijan
konul.jabbarova@mail.ru

Abstract. In this paper PROMETHEE method is applied to project selection problem under U-number-valued information. All criteria values and importance weights are described by U-numbers. The obtained results show validity and effectiveness of the performed analysis.

Keywords: PROMETHEE method · U-number · Preference function · Project selection · Decision matrix

1 Introduction

A selection of an optimal project is very important problem for decision makers [1]. For a decision maker (DM), project selection can be considered as a difficult multi-criteria decision making (MCDM) problem, which needs consideration of a number of conflicting criteria. In addition, for expressing criteria values, DMs tend to use linguistic terms due to uncertain and incomplete relevant information. In [1], a problem of project selection is solved by combining fuzzy AHP and VIKOR methods.

The basic goal of [2] is project selection in a Big-Four company located in Turkey. An MCDM problem is considered under fuzziness of selection criteria. Paper [3] focuses on developing a model based on Shannon Entropy and fuzzy TOPSIS technique for solving of a problem of project portfolio selection.

Often, real-world decision problems occur in 'usual' situations. Prof. Zadeh introduced the concept of usuality for formalization of information relevant to such situations [4–6]. From the other side, decision relevant information in real-world problems is characterized by partial reliability. In order to deal with this issue, Prof. Zadeh introduced the concept of a Z-number, as a pair of two fuzzy numbers Z = (A, B) [7]. A is a fuzzy constraint on values of a random variable X. B is a fuzzy reliability of A, and is considered as a value of probability measure of A. One can see that we deal with a combination of fuzziness and probabilistic uncertainty. In [8, 9] a general and effective approach to computation with discrete Z-numbers is proposed.

Aliev introduced a concept of U-number as a special case of a Z-number to form a new bridge to formalization of usuality. In U-number, the reliability part ranges within the class of "rarely", "often", "usually" and other usuality terms.

R. A. Aliev et al. (Eds.): ICSCCW 2021, LNNS 362, pp. 286–293, 2022.
https://doi.org/10.1007/978-3-030-92127-9_40

In turn, no works exist on application of U-number concept as a new view on usuality paradigm. In this paper we use the concept of U-number in a project selection problem. Criteria values and importance weights are described by U-numbers.

The paper is organized as follows. In Sect. 2 we give a prerequisite material used in the sequel. In Sect. 3, we use PROMETHEE method to solve the considered problem U-number. Conclusion is presented in Sect. 4.

2 Preliminaries

Definition 1. U-numbers [10, 11]. Let X be a random variable and A be a fuzzy number playing a role of fuzzy constraint on values that the random variable may take: X is A. The definition of a usual value of X may be expressed in terms of the probability distribution of X as follows [5]. If $p(x_i)$ is the probability of X taking x_i as its value, then

$$usually(X\ is\ A) = \mu_{usually}\left(\sum_i p(x_i)\mu_A(x_i)\right) \tag{1}$$

or

$$usually(X\ is\ A) = \mu_{most}\left(\sum_i p(x_i)\mu_A(x_i)\right) \tag{2}$$

A usual number describing, "usually, professor's income is medium" is shown Fig. 1.

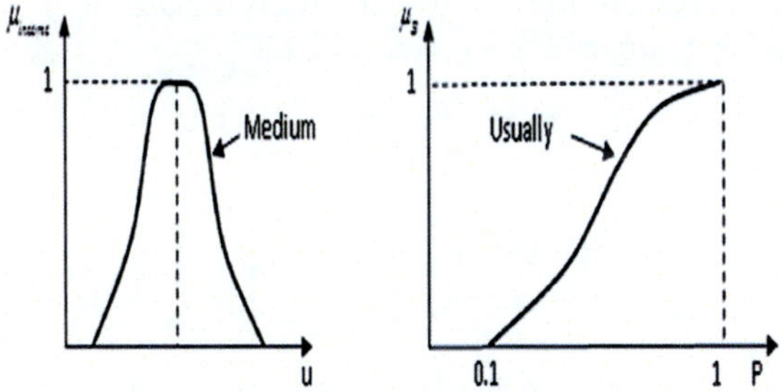

Fig. 1. An example of U-number

Formula (2) indicates that the probability that the event A occurs as the value for the variable X, is "*most*". As it was mentioned above, Zadeh provided an outline for the theory of usuality, however this topic requires further investigation. It is needed a more general approach for other usuality quantifiers. In this paper "usuality" will be a composite term characterized by fuzzy quantities as *always, usually, frequently/often, occasionally, seldom, almost never/rarely, never*. The codebook for "usuality" is shown in Fig. 2.

Definition 2. Operation on U-numbers. [10, 11]. Let $U_1 = (A_1, B_1)$ and $U_2 = (A_2, B_2)$ be U-numbers (B_1 and B_2 are fuzzy terms of the usuality codebook) describing values of random variables X_1 and X_2. Assume that it is needed to compute the result $U_{12} = (A_{12}, B_{12})$ of a two-place operation $* \in \{+, -, \cdot, /\}$: $U_{12} = U_1 * U_2$.

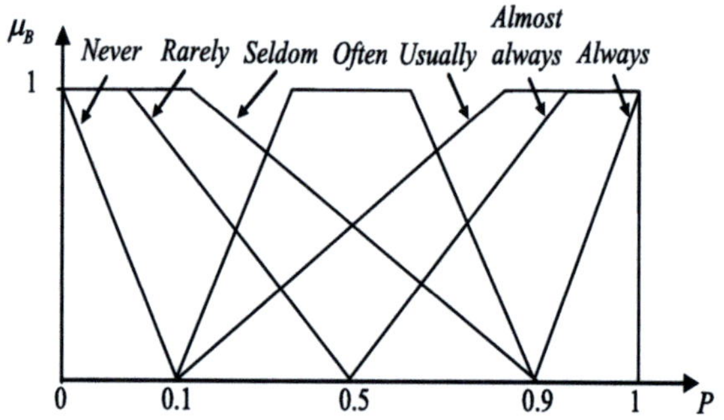

Fig. 2. The codebook of the fuzzy quantifiers of usuality

The first stage is the computation of two-place operations $*$ of fuzzy numbers A_1 and A_2 on the basis of fuzzy arithmetic.

The second stage involves step-by-step construction of B_{12} and is related to propagation of probabilistic restrictions. We realize that in U-numbers $U_1 = (A_1, B_1)$ and $U_2 = (A_2, B_2)$, the 'true' probability distributions p_1 and p_2 are not exactly known. In contrast, the information available is represented by the fuzzy restrictions:

$$\mu_{p_1}(p_1) = \mu_{B_1}\left(\sum_{k=1}^{n_1} \mu_{A_1}(x_{1k})p_1(x_{1k})\right), \mu_{p_2}(p_2) = \mu_{B_2}\left(\sum_{k=1}^{n_2} \mu_{A_2}(x_{2k})p_2(x_{2k})\right).$$

Given these fuzzy restrictions, extract probability distributions $p_j, j = 1, 2$ by solving the following goal linear programming problem [7, 8]:

$$c_1 v_1^l + c_2 v_2^l + \ldots + c_n v_n^l \to b_{jl} \tag{3}$$

subject to

$$\left.\begin{array}{l} v_1^l + v_2^l + \ldots + v_n^l = 1 \\ v_1^l, v_2^l, \ldots, v_n^l \geq 0 \end{array}\right\}. \tag{4}$$

where $c_k = \mu_{A_j}(x_{jk})$ and $v_k = p_j(x_{jk}),\ k = 1, .., n_j, k = 1, .., n_j$.

Distributions $p_{jl}(x_{jk}), k = 1, .., n_j$ naturally induce probabilistic uncertainty over the result $X_{12} = X_1 * X_2$. This implies that given a pair p_{1l_1}, p_{2l_2}, the convolution $p_{12s} = p_{1l_1} \circ p_{2l_2}, s = 1, \ldots, m^2$ is to be computed.

Given p_{12s}, the value of probability measure of A_{12} can be computed: $P(A_{12}) = \sum_{k=1}^{n} \mu_{A_{12}}(x_{12k})p_{12}(x_{12k})$. However, the 'true' p_{12s} is not exactly known as the 'true' p_{1l_1}, p_{2l_2} are described by fuzzy restrictions. These fuzzy restrictions induce the fuzzy set of convolutions $p_{12s}, s = 1, \ldots, m^2$ with the membership function defined as [6–8]

$$\mu_{p_{12}}(p_{12s}) = \max_{p_{12s}=p_{1l_1} \circ p_{2l_2}}[\mu_{p_1}(p_{1l_1}) \wedge \mu_{p_2}(p_{2l_2})] \tag{5}$$

subject to

$$\mu_{p_j}(p_{jl_j}) = \mu_{B_j}\left(\sum_{k=1}^{n_j} \mu_{A_j}(x_{jk})p_{jl_j}(x_{jk})\right), j = 1, 2 \tag{6}$$

where $\wedge$ is *min* operation.

The membership function of $Z_1^+ = (A_1, R_1)$ is defined as [6–8]

$$\mu_{B_{12}}(b_{12_s}) = \max(\mu_{p_{12}}(p_{12_s})) \tag{7}$$

subject to

$$b_{12s} = \sum_k \mu_{A_{12}}(x_k) p_{12s}(x_k) \tag{8}$$

As a result, $U_{12} = U_1 * U_2$ is obtained as $U_{12} = (A_{12}, B_{12})$.

Definition 3. Fuzzy Pareto optimality (FPO) principle based comparison of Z-numbers [7]. Fuzzy Pareto optimality (FPO) principle allows to determine degrees of Pareto Optimality of multiattribute alternatives. We apply this principle to compare Z-numbers as multiattribute alternatives – one attribute measures value of a variable, the other one measures the associated reliability. According to this approach, by directly comparing Z-numbers $Z_1 = (A_1, B_1)$ and $Z_2 = (A_2, B_2)$ one arrives at total degrees of optimality of Z-numbers: $do(Z_1)$ and $do(Z_2)$. These degrees are determined on the basis of a number of components (the minimum is 0, the maximum is 2) with respect to which one Z-numbers dominates another one. Z_1 is considered higher than Z_2 if $do(Z_1) > do(Z_2)$.

Let us consider a MCDM problem under Z-valued information.

Definition 4 [8]. **A distance between Z-numbers.** The distance between Z-numbers $Z_1 = (A_1, B_1)$ and $Z_2 = (A_2, B_2)$ is defined as

$$D(Z_1, Z_2) = \frac{1}{n+1} \sum_{k=1}^{n} \left\{ \left| a^L_{1\alpha_k} - a^L_{2\alpha_k} \right| + \left| a^R_{1\alpha_k} - a^R_{2\alpha_k} \right| \right\} + \frac{1}{m+1} \sum_{k=1}^{m} \left\{ \left| b^L_{1\alpha_k} - b^L_{2\alpha_k} \right| + \left| b^R_{1\alpha_k} - b^R_{2\alpha_k} \right| \right\}$$

where $a^L_\alpha = \min A^\alpha$, $a^R_\alpha = \max A^\alpha$, $b^L_\alpha = \min B^\alpha$, $b^R_\alpha = \max B^\alpha$.

3 Statement of the Problem and a Solution Method

Let us consider MCDM problem with U-number-valued information. Three projects P1, P2, P3 are evaluated by four criteria: net present value (C1), quality (C2), contractor's technology (C3) and contractor's economic status (C4) [12]. The criteria values and importance weights are described by U-numbers $U = (A, B)$ (see Tables 1 and 2).

Let us solve this problem by using the algorithm of the PROMETHEE method under U-number-valued information [13].

At the *first stage*, decision matrix is weighted by multiplying the importance weights of evaluation criteria and the criteria values. The weighted normalized U-numbers are as follows (Table 3):

Table 1. U-number-valued criteria evaluations

	C_1	C_2
f_1	$\{0/0.55\ 1/0.61\ 0/0.67\}\{0/0.5\ 1/0.7\ 0/0.9\}$	$\{0/0.44\ 1/0.48\ 0/0.53\}\{0/0.3\ 1/0.5\ 0/0.7\}$
f_2	$\{0/0.39\ 1/0.43\ 0/0.48\}\{0/0.5\ 1/0.7\ 0/0.9\}$	$\{0/0.53\ 1/0.59\ 0/0.65\}\{0/0.3\ 1/0.5\ 0/0.7\}$
f_3	$\{0/0.6\ 1/0.66\ 0/0.73\}\{0/0.5\ 1/0.7\ 0/0.9\}$	$\{0/0.58\ 1/0.65\ 0/0.71\}\{0/0.3\ 1/0.5\ 0/0.7\}$
	C_3	C_4
f_1	$\{0/0.52\ 1/0.57\ 0/0.63\}\{0/0.5\ 1/0.7\ 0/0.9\}$	$\{0/0.50\ 1/0.56\ 0/0.61\}\{0/0.5\ 1/0.7\ 0/0.9\}$
f_2	$\{0/0.44\ 1/0.49\ 0/0.54\}\{0/0.5\ 1/0.7\ 0/0.9\}$	$\{0/0.46\ 1/0.51\ 0/0.56\}\{0/0.5\ 1/0.7\ 0/0.9\}$
f_3	$\{0/0.6\ 1/0.66\ 0/0.72\}\{0/0.5\ 1/0.7\ 0/0.9\}$	$\{0/0.59\ 1/0.65\ 0/0.72\}\{0/0.5\ 1/0.7\ 0/0.9\}$

Table 2. Importance weights of the criteria

N	Weights
w_1	$\{0/0.16\ 1/0.25\ 0/0.38\}\{0/0.5\ 1/0.7\ 0/0.9\}$
w_2	$\{0/0.16\ 1/0.24\ 0/0.36\}\{0/0.5\ 1/0.7\ 0/0.9\}$
w_3	$\{0/0.15\ 1/0.23\ 0/0.35\}\{0/0.5\ 1/0.7\ 0/0.9\}$
w_4	$\{0/0.18\ 1/0.28\ 0/0.42\}\{0/0.5\ 1/0.7\ 0/0.9\}$

Table 3. The weighted Z-number-valued decision matrix

	C_1	C_2
f_1	$\{0/0.09\ 1/0.15\ 0/0.25\}\{0/0.3\ 1/0.52\ 0/0.82\}$	$\{0/0.07\ 1/0.12\ 0/0.19\}\{0/0.19\ 1/0.39\ 0/0.64\}$
f_2	$\{0/0.06\ 1/0.11\ 0/0.18\}\{0/0.3\ 1/0.53\ 0/0.82\}$	$\{0/0.08\ 1/0.14\ 0/0.23\}\{0/0.19\ 1/0.39\ 0/0.64\}$
f_3	$\{0/0.1\ 1/0.17\ 0/0.23\}\{0/0.3\ 1/0.53\ 0/0.82\}$	$\{0/0.09\ 1/0.16\ 0/0.26\}\{0/0.19\ 1/0.39\ 0/0.64\}$
	C_3	C_4
f_1	$\{0/0.08\ 1/0.13\ 0/0.22\}\{0/0.3\ 1/0.53\ 0/0.82\}$	$\{0/0.09\ 1/0.16\ 0/0.26\}\{0/0.3\ 1/0.53\ 0/0.82\}$
f_2	$\{0/0.07\ 1/0.11\ 0/0.19\}\{0/0.3\ 1/0.53\ 0/0.82\}$	$\{0/0.08\ 1/0.14\ 0/0.24\}\{0/0.3\ 1/0.53\ 0/0.82\}$
f_3	$\{0/0.09\ 1/0.15\ 0/0.25\}\{0/0.3\ 1/0.53\ 0/0.82\}$	$\{0/0.11\ 1/0.18\ 0/0.3\}\{0/0.3\ 1/0.53\ 0/0.82\}$

At the *second stage*, for comparing two alternatives g and f on each criterion, distances values $D(U_{gj}(A, B), (1, 1))$, $D(U_{fj}(A, B), (1, 1))$, $D(U_{gj}(A, B), U_{fj}(A, B))$ are computed. We consider that $U_{gj}(A,B) \geq U_{fj}(A,B)$ iff $D(U_{gj}(A,B),(1,1)) \leq D(U_{fj}(A,B),(1,1))$ [13]. Then preference function is formalized as:

$$P(U_{gj}(A,B),U_{fj}(A,B)) = \begin{cases} 0, U_{gj}(A,B) \leq U_{fj}(A,B) \\ D(U_{gj}(A,B),U_{fj}(A,B)), U_{gj}(A,B) > U_{fj}(A,B) \end{cases}.$$

The obtained results are given in Table 4.

Table 4. The preference function $P_j(U_g(A,B),U_f(A,B))$

	C_1	C_2	C_3	C_4
$P(1,2)$	0.08	0	0.03	0.03
$P(1,3)$	0	0	0	0
$P(2,1)$	0	0.04	0	0
$P(2,3)$	0	0	0	0
$P(3,1)$	0.03	0.07	0.03	0.04
$P(3,2)$	0.09	0.03	0.07	0.07

At the *third stage,* Z-number valued preference index is computed to define the value of the outranking relation ($j = 1, 2, \ldots, n$):

$$\pi(g,f) = \sum_{j=1}^{n}\left[w_j P_j(g,f)\right].$$

The obtained results are shown in Table 5.

Table 5. Z-number valued preference index

	f_1	f_2	f_3
f_1	–	(0.023,0.035,0.054) (0.22,0.43,0.76)	–
f_2	(0.006, 0.01, 0.014), (0.5, 0.7, 0.9)	–	–
f_3	(0.0282, 0.043, 0.0634), (0.15, 0.35, 0.71)	(0.042,0.066,0.099), (0.18,0.37,71)	–

At the *fourth stage,* the leaving and entering flows are computed for ranking of alternatives:

$$\phi^{+}(g) = \frac{1}{n-1}\sum_{\substack{f \neq g \\ f=1}}^{m}\pi(g,f), \phi^{-}(g) = \frac{1}{n-1}\sum_{\substack{f \neq g \\ f=1}}^{m}\pi(f,g), (f \neq g) \quad (9)$$

where n is the number of alternatives.

The obtained results are shown in Table 6.

Table 6. Z-number valued flows for each alternative

	Leaving flows	Entering flows
f_1	(0.0115, 0.0175, 0.027) (0.22, 0.43, 0.76)	(0.017, 0.0265, 0.0385), (0.15, 0.35, 0.73)
f_2	(0.003, 0.005, 0.007), (0.5, 0.7, 0.9)	(0.0325, 0.0505, 0.0765), (0.18, 0.37, 0.73)
f_3	(0.035, 0.055, 0.08), (0.15, 0.35, 0.73)	–

At the *fifth stage,* net flow is computed as follows:

$$\phi(g) = \phi^{+}(g) - \phi^{-}(g)$$

The obtained results are shown in Table 7.

Table 7. Z-number valued net flows for the alternatives

	$\phi(g)$
f_1	(−0.006, −0.009, −0.012)(0.19,0.35,0.62)
f_2	(−0.03, −0.05, −0.07)(0.21, 0.4, 0.62)
f_3	(0.04,0.06,0.08),(0.15,0.35,0.73)

At *sixth stage*, the alternatives are ranked by using Definition 3 (see Table 8).

Table 8. Preference rank of each alternatives

Ranking
f_3
f_2
f_1

Thus, alternative f_3 is the best one.

4 Conclusion

In this paper PROMETHEE is applied to project selection problem. The values and importance weights of criteria are described by U-numbers. For ranking of alternatives, fuzzy Pareto optimality principle-based comparison of Z-numbers is used.

References

1. Salehi, K.: A hybrid fuzzy MCDM approach for project selection problem. Decis. Sci. Lett. **4**, 109–116 (2015). https://doi.org/10.5267/j.dsl.2014.8.003
2. Beldek, T., Konyalıoğlu, A.K., Camgöz-Akdağ, H.: A fuzzy MCDM approach for project selection criteria prioritization in a big-four company: evidences from Turkey consultancy sector. In: Kahraman, C., Cebi, S., Cevik Onar, S., Oztaysi, B., Tolga, A.C., Sari, I.U. (eds.) INFUS 2019. AISC, vol. 1029, pp. 796–803. Springer, Cham (2020). https://doi.org/10.1007/978-3-030-23756-1_95
3. Haddadha, A.K., Namazian, A.,Yakhchali, S.H.: Project selection problem by combination of Shannon entropy and MCDM techniques. In: International Conference on Literature, History, Humanities and Social Sciences, Dubai, pp. 32–35 (2017). https://doi.org/10.15242/icehm.ed0117032
4. Zadeh, L.A.: Fuzzy sets and commonsense reasoning. Institute of Cognitive Studies report 21, University of California, Berkeley (1984)
5. Zadeh, L.A.: Fuzzy sets as a basis for the management of uncertainty in expert systems. Fuzzy Sets Syst. **11**, 199–227 (1983)
6. Zadeh, L.A.: A computational theory of dispositions. In: International Conference on Computation Linguistics, pp. 312–318. Stanford (1984)
7. Zadeh, L.A.: A note on Z-numbers. Inf. Sci. **181**, 2923–2932 (2011)
8. Zadeh, L.A.: Generalized theory of uncertainty (GTU) – principal concepts and ideas. Comput. Stat. Data An. **51**, 15–46 (2006)
9. Aliev, R.A., Alizadeh, A.V., Huseynov, O.H.: The arithmetic of discrete Z-numbers. Inf. Sci. **290**, 134–155 (2015). https://doi.org/10.1016/j.ins.2014.08.024
10. Aliev, R.A.: Approximate arithmetic operations of U-numbers. Procedia Comput. Sci. **102**, 378–384 (2016)
11. Aliev, R.A.: Introduction to U-number calculus. Intell. Autom. Soft Comput. **24**(1), 1–6 (2017). https://doi.org/10.1080/10798587.2017.1330311
12. Rouyendegh, B.D., Erol, S.: Selecting the best project using the fuzzy ELECTRE method. Math. Probl. Eng. **2012**, 1–12 (2012). https://doi.org/10.1155/2012/790142
13. Jabbarova, K.I., Jabbarova, A.I.: Z-information based PROMETHEE method. In: Aliev, R.A., Yusupbekov, N.R., Kacprzyk, J., Pedrycz, W., Sadikoglu, F.M. (eds.) WCIS 2020. AISC, vol. 1323, pp. 287–293. Springer, Cham (2021). https://doi.org/10.1007/978-3-030-68004-6_37

Downscaling Precipitation from GCM Parameters Using Recurrent Neural Networks

Vahid Nourani[1] (✉), Fahreddin Sadikoglu[2], Kasra Khodkar[1], and Arash Shahli[1]

[1] Center of Excellence in Hydroinformatics and Faculty of Civil Engineering, University of Tabriz, 5166616471 Tabriz, Iran
nourani@tabrizu.ac.ir, {kasrakhodkar98, a.shahli99}@ms.tabrizu.ac.ir

[2] Department of Electrical and Electronic Engineering, Faculty of Engineering, Near East University, POBOX: 99138 Mersin 10, Nicosia, Turkey
fahreddin.sadikoglu@neu.edu.tr

Abstract. In this study, the accuracy of two recurrent neural network (RNN) units, long short-term memory (LSTM) and gated recurrent unit (GRU) are compared for downscaling precipitation in Tabriz synoptic station from Can-esm2 general circulation model (GCM) parameters. Also feedforward neural network (FFNN) is also employed as a base model to compare. To avoid redundant calculations, Among GCM parameters, the most effective parameters for the proposed modeling are selected via k-means clustering and mutual information (MI) measure. Then with a split sampling approach, 80% of the data is used for training of the models and remaining 20% is utilized to compare the performance of the models. Results show that two employed RNNs, are performing similarly in terms of coefficient of determination (DC). Although GRU modeling demanded more hidden units, which results in similar trainable neurons with LSTM, LSTM modeling needed more epochs to achieve optimal results. Therefore, due to the obtained results, GRU is more time efficient and requires less computational cost. Predictions made by both RNNs resulted similarly and around 35% higher DCs from results obtained with (FFNN).

Keywords: General circulation models · Downscaling precipitation · Statistical downscaling · Long short-term memory · Gated recurrent unit

1 Introduction

General circulation models (GCMs) are indispensable tools for investigation and climate change modeling. However, practical applications of GCM data are limited due to the coarse spatial grading of the data. The purpose of Coupled Model Intercomparison Project (CMIP) is to provide better understanding of the past, present and future changes of the climate with respect to natural variability and radiative forcing. CMIP5 model includes concentration and emission scenarios and has finer spatial grading, which makes the model superior to previous CMIPs. Among important climatologic parameters, the

R. A. Aliev et al. (Eds.): ICSCCW 2021, LNNS 362, pp. 294–301, 2022.
https://doi.org/10.1007/978-3-030-92127-9_41

precipitation is more pivotal due to undisputed importance of water. Moreover, modeling precipitation is a challenging task in comparison to other parameters like temperature, due to complexity of natural processes affecting precipitation, which are lifting the uncertainty involved in modeling. Statistical downscaling GCM data for precipitation is a researching filed, which benefits from future scenarios for forecasting precipitation in a region.

Artificial Neural Network (ANN) modeling is a common approach for statistical downscaling of GCMs among studies. Shallow ANN structures such as feedforward neural networks (FFNN), have advantages like being non-linear and having generalization capacity, which makes them efficient for downscaling GCM parameters. However, disadvantages such as challenge of selecting number of inputs and outputs, overfitting and immutability of weights and biases after training phase, brought the interest of researches to use deep learning (DL) methods like Recurrent Neural Networks for modeling sequences. Although classic RNNs have been shown to be an effective tool for modeling hydrologic time series [1], RNNs are observed to be not effective for long sequences due to vanishing and exploding gradient problems [2]. Long Short-Term Memory (LSTM) is an advanced DL method, which has been designed to overcome exploding and vanishing gradient problems [3]. LSTM is proven to be effectively include long-term dependencies via the Cell State and gated structure while maintaining the inclusion of the short dependencies with the Hidden State. More recently, a new variant of RNNs, the Gated Recurrent Unit (GRU), has been designed with gated structure, similar to LSTM, which has comparable results with LSTM with less computational effort [4]. Although there are very limited studies for downscaling precipitation via LSTM [5], to the best of authors knowledge there is no study employing GRU for downscaling GCM parameters.

2 Materials and Methods

2.1 Study Area and Dataset

In this study, in order to investigate the proposed methods performance, data from station of Tabriz synoptic station were used as targets for modelling. Tabriz is the capital of East Azerbaijan province, which is among the industrial cities in Northwestern Iran. Its latitude, longitude and altitude are respectively 38.07 °N, 46.14 °E and 1364 m above sea level. Tabriz is surrounded by the Einali and Sahand mountains. As indicated by the De Martonne aridity index, the region is classified as a semi-arid area. The mean yearly precipitation of Tabriz is nearly 290 mm, which is very low compared with the world normal, which is 800 mm. This city is characterized by relatively hot summers (up to 42 °C) and very cold winters (down to −25 °C). Rainfall events tend to be intensive due to the proximity of mountains.

In this study, the observed data for downscaling precipitation and temperature were retrieved from the Iran meteorological office for the periods of 1951–2005 for Tabriz station. GCM data under CMIP5 were extracted from (http://cera-www.dkrz.de/). For downscaling and modeling past data, historical data from the years 1951 to 2005 for the Tabriz stations were applied.

Among different GCMs CMIP5, the Can-ESM2 was used in this study since some previous studies indicated that these GCMs could lead to acceptable performance over

Iran [6]. Data from four grid points around the stations were retrieved, since each grid point from different areas may have its individual impact on the considered parameters.

2.2 Dominant Input Selection and Data Pre-processing

Since GCMs include high number of parameters, selecting dominant inputs for down-scaling is an essential step for reducing calculation time and increase the efficiency of the model. Thus, k-means clustering is applied to determine dominant inputs and group similar parameters. Then, Mutual Information is applied as a nonlinear measure and select dominant input parameters from clusters [7].

Selected input data and target output data were normalized to the range of [−1, 1] which is the sensitive range of the activation functions used in the ANN structures.

A split data method is applied to evaluate the generalization capability of the models in which, 80% of the data are used for training and 20% of the data are used for validation.

2.3 Recurrent Neural Networks

Long Short-Term Memory networks have been developed to overcome the vanishing and exploding gradient problems of the classical RNNs with flow of the data in the hidden state through the gates and activation functions and in the cell state without going through any activation function. The LSTM cell is consisting of four gates, each with an activation function, which are considered as neurons and are trainable. The mentioned gates are input, forget, modulation and output gates. The process in a LSTM cell is described via Eqs. 1–4 [3] as:

$$f_i = \sigma_f\left(W^f_{data}x_t + W^f_{state}h_{t-1} + b^f\right)(forget\ gate) \tag{1}$$

$$i_i = \sigma_i\left(W^i_{data}x_t + W^i_{state}h_{t-1} + b^i\right)(input\ gate) \tag{2}$$

$$g_i = \sigma_g\left(W^g_{data}x_t + W^g_{state}h_{t-1} + b^g\right)(modulation\ gate) \tag{3}$$

$$o_i = \sigma_o\left(W^o_{data}x_t + W^o_{state}h_{t-1} + b^o\right)(output\ gate) \tag{4}$$

These gates make it possible for LSTM to decide to whether forget the information (via weights) or pass it to the next cell as:

$$c_t = f_t \odot c_{t-1} + i_t \odot g_t \tag{5}$$

The cell state c_t gets updated with Eq. 9. $\odot$ indicates the element-wise multiplication and:

$$h_t = o_t \odot \sigma_h(c_t) \tag{6}$$

where σ_t is the non-linear activation function and the hidden state h_t is divided from output gate and passes to next cell too.

The hidden state is the short-term memory unit which carries information from the immediately previous events and overwrites at every time step, which is common in RNN structures. On the other hand, the cell state is the long-term memory unit that saves and loads the information from the previous time steps (not necessarily the immediately previous time step).

Gated Recurrent Unit networks are developed to reduce the calculations and speed up the modeling process by simpler structure. In GRU structure, h_t and c_t are merged into into one h_t and number of gates is reduced to two control gates, the update and the reset gates. The update gate controls the amount of h_{t-1} to brought in current timestep (t) and the reset gate controls the amount of information from h_{t-1} to written into the current candidate set. The internal process of GRU is described as:

$$r_t = \sigma_r\left(W^r_{data}x_t + W^r_{state}h_{t-1} + b^r\right)(reset\ gate) \tag{7}$$

$$z_t = \sigma_i\left(W^z_{data}x_t + W^z_{state}h_{t-1} + b^z\right)(update\ gate) \tag{8}$$

$$\hat{h}_t = tanh\left(W^c_{data}x_t + W^c_{state}(r_t \odot h_{t-1}) + b^c\right) \quad (new\ candidate\ state) \tag{9}$$

$$c_t = (1 - z_t) \odot c_{t-1} + z_t \odot \hat{c}_t \quad (output\ gate) \tag{10}$$

2.4 Evaluation Criteria

For evaluation of accuracy of employed methods in this study, two evaluation criteria of root mean square error (RMSE) and coefficient of determination (DC or Nash-Sutcliffe) were used as:

$$RMSE = \sqrt{\frac{\sum_{i=1}^{N}(R_i - Z_i)^2}{N}} \tag{11}$$

$$DC = 1 - \frac{\sum_{i=1}^{N}(Z_i - R_i)^2}{\sum_{i=1}^{N}(Z_i - \overline{Z})^2} \tag{12}$$

where R_i is the estimated value, Z_i is the target, $\overline{Z}$ is the mean value of the targets and N is the sample size. RMSE is sharing the dimension with observations while DC is dimensionless and the range of DC is $(-\infty, 1]$, in which the more value of DC gets close to 1, the more regression is accurate.

3 Results

Dominant input selection has resulted in selecting huss (near-surface specific humidity) from grid point located in 37.67308963 and 45 latitude and longitudes respectively and huss, evspsbl (water evaporation flux) and hur (relative humidity) from grid point located in 40.46364818 and 45 latitude and longitudes respectively.

For evaluation of the employed RNN models, results are compared with modeling via classic FFNN. Number of hidden neurons and epochs are defined by trial and error process and best achieved results are reported in this study. The training data set consists of time steps for 1951 to 1995 and validation data set is chosen from 1995 to 2005. Results, which are tabulated in Table 1, show the superiority of RNNs over classic FFNN for downscaling precipitation using GCM parameters.

Table 1. Results of modeling with FFNN, GRU and LSTM networks

Model	DC train	DC validation	RMSE* train	RMSE* validation	Elapsed time	No. of hidden neurons	No. of epochs
FFNN	0.3829	0.3106	0.301	0.2586	6.79	4	20
GRU	0.285	0.4292	0.3242	0.2353	80.93	9	40
LSTM	0.3543	0.4197	0.3081	0.2373	113.84	6	80

* Denotes normalized RMSE values.

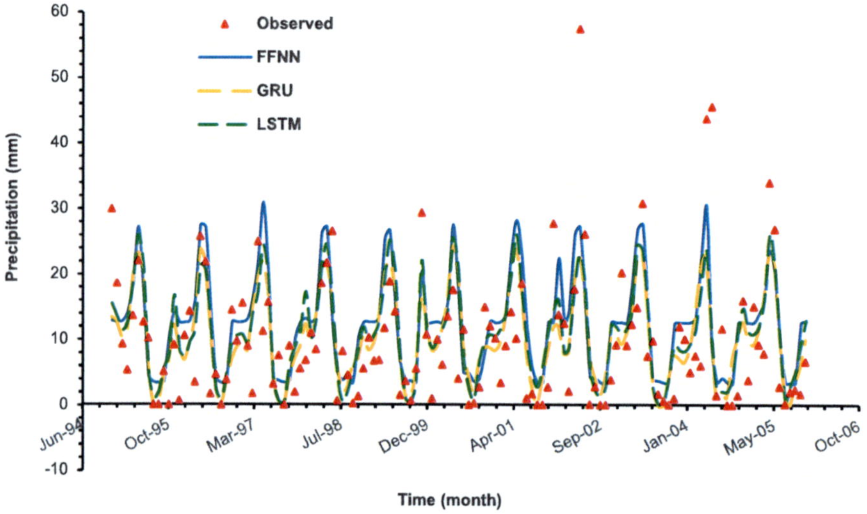

Fig. 1. Comparison of observed precipitation values in Tabriz station with downscaled values via FFNN, GRU and LSTM

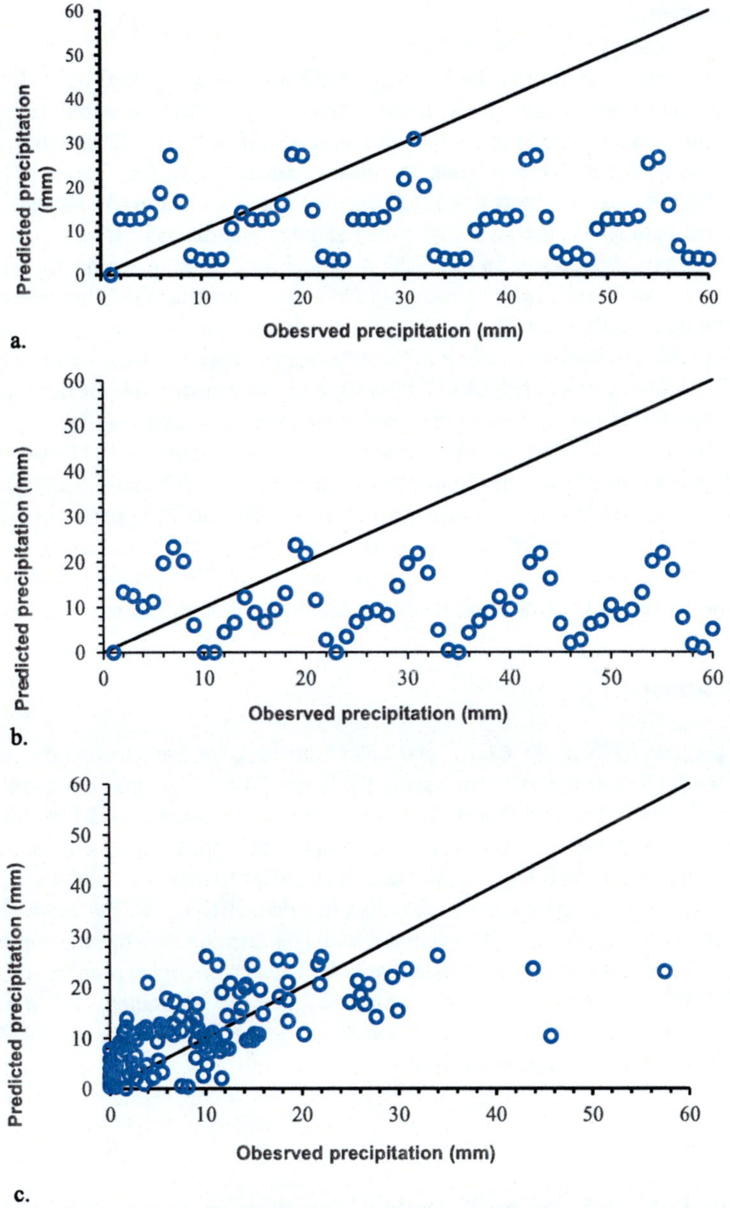

Fig. 2. Scatter plot for validation dataset of a) FFNN, b) GRU and c) LSTM models outputs

4 Discussion

To obtain the best possible structure, 1 to 15 hidden neurons are tried with 10 to 100 epochs. As it is shown in Table 1, the more complicated structure of neural network is, the more epochs demanded to reach the optimal result. However, GRU structure which has less gates (as a result less weights and biases), needed a greater number of hidden units than LSTM, which is more complex (has more gates and more parameters). Thus, it is clear that both RNNs need a certain range of gates (internal activation functions) to reach the best possible results. In turn, LSTM model needed more epochs for reaching the best results which resulted in consuming 40% more time than GRU modeling.

The obtained results show that employed RNNs in this study can improve the robustness of downscaling precipitation by 35%, in validation dataset, when compared with classic FFNN. One possible reason of RNNs superior performance in downscaling precipitation with GCM parameters is the capability to include past events of precipitation, which model generated in previous time steps, in modeling. However, FFNN resulted in higher DC in training dataset than both RNNs, which means RNNs have better generalization capability than FFNNs. On the other hand, LSTM and GRU performed similar to each other with very low difference is accuracy (see Fig. 1 and 2). It is also concluded in other hydrological modeling studies [5] that LSTM and GRU are subjected to equal performance, with GRU demanding less computational labor and time.

5 Conclusion

In this paper, two RNN models, GRU and LSTM, are used for statistically downscaling precipitation for Tabriz city synoptic station from the CMIP5. The most dominant input candidates for modeling are chosen from parameters of can-esm2 GCM by k-means clustering and MI measure. Then with a split sampling approach, 80% of data were utilized as training dataset for models and remaining 20% were used as validation dataset for evaluating the accuracy of models. Results show that GRU and LSTM are predicting the precipitation values with similar accuracy, while performing 35% more accurate than the classical FFNN. Although GRU model resulted better with higher number of hidden units than LSTM, which means similar number of gates to be trained, LSTM needed more epochs to reach similar accuracy. Thus, GRU can perform similar to LSTM with less calculations and less time consumed.

References

1. Coulibaly, P., Anctil, F., Aravena, R., Bobée, B.: Artificial neural network modeling of water table depth fluctuations. Water Resour. Res. **37**(4), 885–896 (2001). https://doi.org/10.1029/2000WR900368
2. Bengio, Y., Simard, P., Frasconi, P.: Learning long-term dependencies with gradient descent is difficult. IEEE Trans. Neural Netw. **5**(2), 157–166 (1994). https://doi.org/10.1109/72.279181
3. Hochreiter, S., Schmidhuber, J.: Long short-term memory. Neural Comput. **9**, 1735–1780 (1997). https://doi.org/10.1162/neco.1997.9.8.1735

4. Cho, K., van Merriënboer, B., Gulcehre, C., Bougares, F., Schwenk, H., Bengio, Y.: Learning phrase representations using RNN encoder-decoder for statistical machine translation. In: Proceedings of the 2014 Conference on (EMNLP), Stroudsburg, PA, USA, pp. 1724–1734. Association for Computational Linguistics (2014). https://doi.org/10.3115/v1/D14-1179
5. Tran Anh, D., Pham, S., Dang, T., Hoang, L.: Downscaling rainfall using deep learning long short-term memory and feedforward neural network. Int. J. Climatol. **39**, 4170–4188 (2019). https://doi.org/10.1002/joc.6066
6. Nourani, V., Baghanam, A.H., Gokcekus, H.: Data-driven ensemble model to statistically downscale rainfall using nonlinear predictor screening approach. J. Hydrol. **565**, 538–551 (2018). https://doi.org/10.1016/j.jhydrol.2018.08.049
7. Nourani, V., Jabbarian Paknezhad, N., Sharghi, E., Khosravi, A.: Estimation of prediction interval in ANN-based multi-GCMs downscaling of hydro-climatologic parameters. J. Hydrol. **579**, 124226 (2019). https://doi.org/10.1016/j.jhydrol.2019.124226

Using Z-Number-Based Information in Personnel Selection Problem

Latafat A. Gardashova(✉) and Salman Salmanov

Azerbaijan State Oil and Industry University, Azadlig 35, Nasimi, Baku, Azerbaijan
{l.qardashova,s.salmanov}@asoiu.edu.az

Abstract. Personnel choice problem is one of the main problems in management. This problem has been an extensively discussed topic during the last decade. If considered not as a pure mathematical task but rather as a complex real-world problem, choosing sufficiently qualified personnel while having multiple alternatives requires to process perception based information about multiple persons (specialists) to get hired. The perception based type of information can be mathematically processed by applying Zadeh's and Aliev's theories: the application of fuzzy logic theory and its extension – Z number theory is shown in the corresponding scientific literature, where the outlined theories are described as quite effective approaches to process the perception based type of information. Z-number has an ability to represent restraints and reliability of human knowledge. In this research, for solving personnel selection problem a new method based on Z-numbers is proposed.

Keywords: Z-number · Personnel selection · State of nature · Decision matrix · Alternatives

1 Introduction

Personnel selection problem is a complicated problem of the real life. Determining values of several personnel-selection criteria is compulsory for Personnel selection task. Each criterion is characterized by the different level of importance.

In the scientific literature, there have been several methods proposed and applied for solving the personnel selection task, that is able to be split into three groups: deterministic, stochastic and fuzzy methods. Each method has its own features, advantages and limitations; and TOPSIS, AHP (analytic hierarchy process), Expert Systems, ANP (analytic network process), fuzzy set theory are among the frequently used ones.

There are a lot of works in the scientific researches reviewing the state of personnel selection problem [1–5], and the main drawback, emerging from these reviews, it is possible to show shortly as follows. Significant number of studies treat different personnel-selection criteria as independent from each other, whereas this is often not the case in the real-world tasks: the criteria can have some level of cognation, which means that they represent to-some-extent overlapping qualities of a person. Determining the relations among criteria should provide to the objectiveness and improve the quality of the corresponding decision making process. Currently, there are a number of reported studies

R. A. Aliev et al. (Eds.): ICSCCW 2021, LNNS 362, pp. 302–307, 2022.
https://doi.org/10.1007/978-3-030-92127-9_42

on the theme of computer-based rational decision making. However, they are focused on either fuzzy or probability based information and cannot represent both restraints and validity. So, a Z-number based theory, in such case, can be considered as a more adequate concept for representation of information [7–13]. A Z - number consist of an ordered pair as $Z = (A, B)$ of fuzzy numbers representing a random variable X, where $\tilde{A}$ is restriction and $\tilde{B}$ is an estimation of the reliability of $\tilde{A}$. Application of Z- number is effective tool for describing information [6], thus Z-number based computations are applied in this study to solve personnel selection problem.

The structure of the paper is as follows. In Sect. 2, we describe preliminaries. In Sect. 3, a statement of the problem and the proposed approach are presented. A numerical example to demonstrate the suggested approach is shown in Sect. 4. The paper ends with some conclusions.

2 Preliminaries

Definition 1. Discrete Z-number [7]. A discrete Z-number is an ordered pair $Z = (A, B)$ of discrete fuzzy numbers A and B. A plays a role of the fuzzy constraint on the values that a random variable X may take. B is a discrete fuzzy number with a membership function $\mu_B : \{b_1, ..., b_n\} \to [0, 1]$, $\{b_1, ..., b_n\} \subset [0, 1][0, 1]$, playing a role of a fuzzy constraint on the probability measure of A, $P(A)$.

Definition 2. Operations Over Discrete Z-numbers [8–12]. Let X_1 and X_2 be discrete Z-numbers describing information about values of X_1 and X_2. Consider computation of $Z_{12} = Z_1 * Z_2$, $* \in \{+, -, \cdot, /\}$ and, square root, power. The first state is the computation of $A_{12} = A_1 * A_2$.

The second state involves construction of B_{12}. We realize that, in Z-numbers Z_1 and Z_2, the 'true' probability distributions p_1 and p_2 are not exactly known. In contrast, the fuzzy restrictions represented in terms of the membership functions are available:

$$\mu_{p_1}(p_1) = \mu_{B_1}\left(\sum_{k=1}^{n_1} \mu_{A_1}(x_{1k})p_1(x_{1k})\right), \mu_{p_2}(p_2) = \mu_{B_2}\left(\sum_{k=1}^{n_2} \mu_{A_2}(x_{2k})p_2(x_{2k})\right),$$

Probability distributions $p_{jl}(x_{jk})$, $k = 1, ..., n$ induce probabilistic uncertainty over $X_{12} = X_1 + X_2$. Given any possible pair p_1 and p_2, the convolution $p_{12} = p_1 \circ p_2$ is computed as follows:

$$p_1(x) = \sum_{x_1+x_2=x} p_1(x_1)p_2(x_2),\ \forall \mathrm{x} \in \mathrm{X}_{12};\ x_1 \in X_1, x_2 \in X_2$$

Given p_{12s}, the value of probability measure of A_{12} is computed as follows:

$$P(A_{12}) = \sum_{k=1}^{n} \mu_{A_{12}}(x_{12k})p_{12}(x_{12k}),$$

However, p_1 and p_2 are described by fuzzy restrictions which induce fuzzy set of convolutions: $\mu_{p_{12}}(p_{12}) = \max_{\{p_1,p_2:p_{12}=p_1 \circ p_2\}} \min\{\mu_{p_1}(p_1), \mu_{p_2}(p_2)\}$.

The fuzziness of information on p_{12} induces fuzziness of $P(A_{12})$ as a discrete fuzzy number B_{12}. The membership function $\mu_{B_{12}}$ is defined as follows:

$$\mu_{B_{12}}(b_{12}) = \max \mu_{p_{12}}(p_{12})$$

$$\text{subject to } b_{12} = \sum_{i=1}^{n} \mu_{A_{12}}(x_i) p_{12}(x_i)$$

As a result, $Z_{12} = Z_1 * Z_2$ is obtained as $Z_{12} = (A_{12}, B_{12})$.

3 Statement of the Problem

The statement of the problem is described by using states of nature, alternatives and utilities.

States of nature are described with the personal attributes: Extraversion, Conscientiousness, Emotional Stability. During the personnel evaluation, it is very effective to identify the real state of the personal attributes. Personal attributes are not sharply defined, they overlap with each other. In this case, it seems rather effective to represent the attributes by using Z-number: $S = \{S_1, S_2, S_3\} = \{(S_1, Z_{P_1}), (S_2, Z_{P_2}), (S_3, Z_{P_3}\}$, $Z_{P_1} = (\tilde{A}_{p_1}, \tilde{B}_{p_1}), Z_{P_2} = (\tilde{A}_{p_2}, \tilde{B}_{p_2}), Z_{P_3} = (\tilde{A}_{p_3}, \tilde{B}_{p_3})$,

$$S_1 - \text{Emotional Stability}, S_2 - \text{Extraversion}, S_3 - \text{Conscientiousness}.$$

Thus, in this paper, linguistic information on likelihood $Z_{P_i}, i = 1, 2, 3$ of the states of nature is described by Z-value based probabilities $Z_{P_i} = (\tilde{A}_{p_i}, \tilde{B}_{p_i}), i = 1, 2, 3$ of the states S_i.

In this problem alternatives are given by the staff of IT and control faculty: Dean, Head of department, Head of IT center. Alternatives, which described Z-number is as follows

$$A_f = \{f_1, f_2, f_3\}$$

Where f_1-dean, f_2 -head of department, f_3- head of IT center.

Usefulness of an alternative $f_j, j = 1, 2, 3$ on a state $s_i, i = 1, 2, 3$ is examined as a productiveness of an existing staff employed at the existing personal attributes. Evaluation of a considered person at the considered attributes can be adequately described by a decision maker. In this case, usefulness or utility of an alternative f_j chosen at the s_i will be taken into account as a $Z_{U(F)}$. In the given situation $Z_{U(f)}$ is determined as

$$Z_{U(f)} = Z_{X_1} Z_{P_{s1}} + Z_{X_2} Z_{P_{S2}} + Z_{X_3} Z_{P_{S3}}$$

where multiplication and addition operation are determined according to definition given in Sect. 2.

Table 1. Decision matrix

	S_1(unimportant, high)	S_2(very unimportant, high)	S_3(important, high)
f_1	(very unimportant, about 0.8)	(very important, high)	(more important, very high)
f_2	(more important, about 0.8)	(fair, about 0.5)	(unimportant, very high)
f_3	(very important, about 0.7)	(fair, about 0.5)	(unimportant, high)

4 Numerical Example

Decision maker determine the staff at various state the personal attributes by using z-information. Decision matrix is given in Table 1. Linguistic evaluation of linguistic variables and their fuzzy restrictions and reliability are given below:

The notations: S_1 - Emotional Stability, S_2 - Extraversion, S_3 - Conscientiousness. The Z-numbers used for S_i, $i = 1, ..., 3$:

S_1(unimportant, high) = {(0.2, 0.28,0.3,0.4),(0.8,0.88,0.9,0.95)};
S_2(very unimportant, high) = {(0.1,0.18,0.2,0.3), (0.8,0.88,0.9,0.95)};
S_3(important, high) = {(0.4,0.48,0.5,0.6), (0.8,0.88,0.9,0.95)}.
1-st alternative:
(very unimportant, about 0.8) = {(0.1, 0.18,0.2,0.3), (0.7,0.78,0.8,0.9)};
(very important, high) = {(0.5,0.58,0.6,0.7), (0.8,0.88,0.9,0.95)};
(more important, very high) = {(0.8,0.88,0.9,0.95), (0.95,0.99,1,1)}.
2-nd alternative:
(more important, about 0.8) = {(0.8, 0.88,0.9,1), (0.7,0.78,0.8,0.9)};
(fair, about 0.5) = {(0.3,0.38,0.4,0.5), (0.4,0.48,0.5,0.6)};
(unimportant, very high) = {(0.2,0.28,0.3,0.4), (0.95,0.99,1,1)}.
3-rd alternative:
(very important, about 0.7) = {(0.5, 0.58,0.6,0.7), (0.6,0.68,0.7,0.8)};
(fair, about 0.5) = {(0.3,0.38,0.4,0.5), (0.4,0.48,0.5,0.6)};
(unimportant, high) = {(0.2,0.28,0.3,0.4), (0.9,0.94,0.95,1)}.

We will formulate the personnel selection by using optimality degree. Goal to determine $f^* \in A_f$, when $Z_{U(f^*)} \geq Z_{U(F)}$, $\forall f \in A_f$.

Using the above given data we calculate the value of expected utility over alternatives f_1, f_2, f_3 and obtain the results as follows:

$$Z_{U(f1)} = ((0.38\ 0.57\ 0.631\ \ 0.9), (0.492\ 0.64\ \ 0.67\ \ 0.79))$$

$$Z_{U(f2)} = ((0.27\ 0.45\ 0.5\ \ 0.79), (0.37\ 0.47\ \ 0.50\ \ 0.64))$$

$$Z_{U(f3)} = ((0.21\ 0.37\ 0.41\ \ 0.66), (0.34\ 0.44\ \ 0.47\ \ 0.6))$$

We need to determine the best alternatives using obtained utilities that described by Z-number. The method proposed by Aliyev and et al. is used for defining degree of optimality over alternatives. Using corresponding method, degrees of optimality of the alternatives are obtained as:

Degree of optimality over alternative 1: do = 1;
Degree of optimality over alternative 2: do = 0.001;
Degree of optimality over alternative 3: do = 2.204e−16

For this reason, f_1 is selected as best alternative among alternatives- f_1 -dean, f_2 -head of department, f_3 - head of IT center.

5 Conclusion

To sum up, a Z-number based approach for personnel selection problem is analyzed. The goal of this study is the application of a method for Z-information based decision making in education enterprise. The priority of the method is the possibility for implementing both the uncertainty and reliability features of data. Utilities for the alternatives are determined by the calculation over Z-numbers. For finding best alternative ranking of Z-numbers is used.

References

1. Chien, C.-F., Chen, L.-F.: Data mining to improve personnel selection and enhance human capital: a case study in high-technology industry. Expert Syst. Appl. **34**(1), 280–290 (2008). https://doi.org/10.1016/jeswa.2006.09.003
2. Wei, J., Li, X., Bi, R.: Using ANP to do the information personnel evaluation. Key Eng. Mater. **439–440**, 749–753 (2010). https://doi.org/10.4028/www.scientific.net/KEM.439-440.749
3. Oostrom, J.K., Van der Linden, D., Born, M.P., Van der Molen, H.T.: New technology in personnel selection: how recruiter characteristics affect the adoption of new selection technology. Comput. Hum. Behav. **29**, 2404–2415 (2013). https://doi.org/10.1016/j.chb.2013.05.025
4. Dağdeviren, M.: A hybrid multi-criteria decision-making model for personnel selection in manufacturing systems. J. Intell. Manuf. **21**(4), 451–460 (2010). https://doi.org/10.1007/s10845-008-0200-7
5. Alguliyev, R.M., Aliguliyev, R.M., Mahmudova, R.S.: Multicriteria personnel selection by the modified fuzzy VIKOR method. Recent Trend. Techniq. Comput. Inf. Intell. **2015**, 1–16 (2015). https://doi.org/10.1155/2015/612767
6. Aliev, R.R., Derar, A.T.M., Huseynov, O.H.: Expected utility based decision making under information and its application. J. Comput. Intell. Neurosci. (2015). https://doi.org/10.1155/2015/364512
7. Aliev, R., Alizadeh, A., Huseynov, O.: The arithmetic of discrete Z-numbers. Inf. Sci. **290**, 134–155 (2015). https://doi.org/10.1016/j.ins.2014.08.024
8. Zadeh, L.: A note on Z-numbers. Inf. Sci. **181**(14), 2923–2932 (2011). https://doi.org/10.1016/j.ins.2011.02.022
9. Aliev, R., Gardashova, L., Huseynov, O.: Z-numbers-based expected utility. In: Proceedings of Tenth International Conference on Application of Fuzzy Systems and Soft Computing, pp. 49–61. Lisbon (2012)
10. Yager, R.: On Z-valuations using Zadeh's Z-numbers. Int. J. Intell. Syst. **27**, 259–278 (2012)
11. Aliev, R.A., Huseynov, O.H., Aliyev, R.R., Alizadeh, A.V.: The Arithmetic o-Numbers: Theory and Applications. World Scientific, Singapore (2015)

12. Aliyev, R.R.: Construction of consistent Z-preferences in decision making for a foreign market selection. In: Aliev, R.A., Kacprzyk, J., Pedrycz, W., Jamshidi, M.O., Babanli, M., Sadikoglu, F.M. (eds.) ICAFS 2020. AISC, vol. 1306, pp. 30–37. Springer, Cham (2021). https://doi.org/10.1007/978-3-030-64058-3_4
13. Aliyev, R.R., Adilova, N.E.: Solution of Zadeh's "Fast Way" problem under Z-information. In: 10th International Conference on Theory and Application of Soft Computing, Computing with Words and Perceptions - ICSCCW-2019. Advances in Intelligent Systems and Computing, vol. 1095, pp. 86–91 (2020)

Portfolio Selection Model Using Z-Numbers Theory

Leyla Hasanova(✉)

Azerbaijan State Oil and Industry University, 20 Azadlig Avenue, AZ1010 Baku, Azerbaijan
hasanova.leyla@asoiu.edu.az

Abstract. Business environment includes many uncertainties, such as the insufficient data that presents difficulties for decision makers in portfolio selection process. The problem of allocating wealth among several assets, arises due to unpredictable behavior of financial markets.

The problem of portfolio selection presented by Markowitz, is a common issue encountered in the realm of finance. It is concerned with determining an optimal strategy for designing acceptable portfolios. Many other modeling approaches have been developed because of Markowitz's key work to offer suitable investment strategies.

This paper examines a new method of combining the mean-variance model and the Z-numbers theory, with the aim of collecting of an investment portfolio. Modelling of a sample portfolio was conducted by using statistical data taken from the Yahoo finance website. The simulation of the Z-portfolio model was performed using the MATLAB Package. Obtained results justify the efficiency of the proposed method.

Keywords: Z-number · Investment · Portfolio selection · Mean-variance model · Utility function

1 Introduction

Portfolio selection process includes two main components: return and risk. Markowitz's method aims to maximize a portfolio's return while minimizing risks to the investment. This involves incorporating one of the two components as a constraint, while embedding the other an objective function.

Portfolio selection is a decision-making process related to allocation of wealth among multiple assets with the aim of making financial profit, while taking into consideration relative unpredictability of the financial market behavior. Such decision-making process requires information, and it is crucial for this information to be valid and relevant.

In this regard, Lotfi A. Zadeh introduced the Z-number theory which provides a foundation for calculating with numbers that aren't entirely reliable. A Z-number described by Zadeh, where Z = (A, B), consists of an order pair of fuzzy numbers. The first component (A) is a restriction on a real variable such as X, and the second component (B) is the reliability of the first component [1].

R. A. Aliev et al. (Eds.): ICSCCW 2021, LNNS 362, pp. 308–315, 2022.
https://doi.org/10.1007/978-3-030-92127-9_43

This concept was further improved upon in several papers by other authors. For instance, the authors of work [2] offered theoretical aspects of addition, subtraction, multiplication, division, square root of a Z-number, and other arithmetic operations over discrete Z-numbers.

Because Z number theory can describe a wide range of routines, it soon gained popularity in a variety of realms.

Contribution of R. Aliyev and L. Zeinalova [3] relates to decisions made under uncertain conditions, when decision-relevant information is represented by Z-numbers. They define the processes of decision making as probabilities of states of nature and outcomes of alternatives given by Z-numbers. There are two primary steps in the proposed decision analysis. First involves, conversion of Z-numbers to fuzzy numbers. Fuzzy utility function values for the alternatives are calculated in the second step, and the alternative with the greatest fuzzy utility value is chosen as the best. A fuzzy expected utility and a fuzzy Choquet expected utility are viewed as fuzzy utility models. The significant contribution of the proposed study is that it is the first paper on handling decision problems under uncertainty defined by Z-information. Primary drawback of this method is the loss of information that occurs during a conversion from Z numbers to fuzzy numbers.

Furthermore, in the paper [4] Z-numbers are used in linear programming problems (LP) and optimized by differential evolution (DE) algorithm.

The theory of Z-numbers used in portfolio selection process is described in the work written by Jirofi and Najafi [5].The utility function here, is used to examine the theory of Z-number in a portfolio selection problem. In the relevant research, the Z-number is usually reduced to a classical fuzzy number. Although this method decreases computing complexity, it also results in the loss of considerable information when converting Z-numbers to classical fuzzy numbers. For this reason, authors of this paper offer two models based on the Z-number method, which they optimize both with and without converting Z numbers to classical fuzzy numbers.

Indeed, the Z-number theory can be widely employed in the financial markets, notably in optimal portfolio construction, due to the absence of a precise vision for the future tendencies of financial markets.

Nevertheless, there are only a handful of papers focused on constructing portfolios based on Z- numbers theory.

The aim of this paper is to study portfolio selection model using Z-numbers. The proposed approach consists of following steps:

- Define constraints on risk and return for efficient portfolios by solving linear programming problem;
- Generate alternative investment portfolios justifying these constraints;
- Compute utility function values with probabilities represented as Z+ discrete numbers for the alternative portfolios;
- Rank obtained utility function values of portfolios through defuzzification of triangular fuzzy numbers.

As result, alternative with the greatest value will be chosen as the best.

The developed model will enable investors to make the best possible investment decisions.

The rest of sections are organized as follows. Section 2 provides a prerequisite material on mean-variance model and Z-numbers. Section 3 presents statement of the problem and solution method based on real datasets of assets. Finally, Sect. 4 concludes the paper.

2 Preliminaries

Definition 1. Mean–variance approach. The MV approach [6] estimates the portfolio return and risk by mean and standard deviation of return distribution. Let n be the number of assets for investment and $R_t = (r_{1t}, r_{2t}, r_{3t}, \ldots, r_{nt})$ be their returns at the tth outcome, t = 1, …, T.

A portfolio x = $(x_1, \ldots, x_n) \in X \subseteq R_n$, refers to the vector of proportions of the inceptive budget to be invested in each asset, assume X is a set of reasonable portfolios, given by, X = $\{(x_1,\ldots,x_n) \mid \sum_{i=1}^{n} x_i = 1, x_i \geq 0$, i = 1, …, n}.

Here, $\sum_{i=1}^{n} x_i = 1$ is a normalized budget restriction. Let [0, T] be a timespan of investment, and is divided into t possible outcomes or time points. To calculate the return of a portfolio x at t, t = 1, …, T, by $R_t(x) = \sum_{i=1}^{n} r_{it} x_i$.

Let p_t be the probability of tth outcome. The portfolio return is a random variable R(x), finitely distributed over $\{R_1(x), \ldots, R_T(x),\}$ with respective probabilities $\{p_1, \ldots, p_T\}$. Thus, the mean return of portfolio is:

$$E(x) = \sum_{t=1}^{T} (\sum_{i=1}^{n} r_{it} x_i) p_t \tag{1}$$

Let $\bar{r}_i = \sum_{t=1}^{T} r_{it} p_t$ be the mean return from i^{th} asset, then:

$$E(x) = \sum_{t=1}^{n} \bar{r}_i x_i \tag{2}$$

The variance of portfolio returns R(x) is given by:

$$\sigma^2 = \sum_{i=1}^{n} \sum_{k=1}^{n} \sigma_{ik} x_i x_k \tag{3}$$

where σ_{ik} is the covariance between returns of ith and k^{th} assets.

Definition 2. A discrete Z-number [2]. A discrete Z-number is an ordered pair Z = (A, B), where A is a discrete fuzzy number playing a role of a fuzzy constraint on values that a random variable X may take: X is A. B is a discrete fuzzy number with a membership functionμ_B: {b1, …, bn} → [0, 1], {b1, …, bn} ⊂ [0, 1], playing a role of a fuzzy restriction on the probability measure of A: P(A) = $\sum_{i=1}^{n} \mu_A(x_i) p(x_i)$ is B.

A concept of a discrete Z^+ number is closely related to the concept of a discrete Z-number. Given a discrete Z-number Z = (A, B), Z^+ number is a pair consisting of a fuzzy number, A and a random number R: Z^+ = (A, R), where A plays the same role as it does in a discrete Z-number Z = (A, B), and R plays the role of the probability distribution p: P(A) ∈ supp (B).

Definition 3. Addition of discrete Z-numbers. In this case, $A_1 * A_2$ is a sum $A_1 + A_2$ defined on the basis of Definitions given in [4] and $R_1 * R_2$ is a convolution defined as:

$$p_{12}(x) = \sum\nolimits_{x=x_1+x_2} p_{R_1}(x_1) p_{R_2}(x_2) \tag{4}$$

So, we will have $Z^+{}_{12}$ as:

$$Z^+{}_{12} = (A_1 + A_2, p_{12}) \tag{5}$$

Next, the resulting Z-number is found:

$$Z_{12} = (A_1 + A_2, B_{12}) \tag{6}$$

Scalar multiplication of Z-number. A scalar multiplication of a Z-number $Z = \lambda \cdot Z_1$, λ is a real number and is determined as $Z = (\lambda A_1, B_1)$.

Definition 4. Defuzzification of triangular fuzzy numbers [7]. The crisp real number S_a corresponding to the triangular fuzzy number a = (l, m, u) is obtained from the following relation:

$$S_a = \frac{l + m + u}{3} \tag{7}$$

3 Problem Definition and Solution Method

Suppose that nine assets $A_i (i = 1, 2, \ldots, 9)$ are considered by investor for the selection of the best investment portfolio, which will provide the optimal distribution of funds between these assets.

The success of an investment strategy is appraised on higher returns and lower risk. With an upper limit on risk and a lower limit on risk, the mean-variance portfolio optimization models strive to maximize returns while minimizing risk. The variations in these confines compile the efficient frontier. For the purpose of defining these confines for risk and return, investor has to solve the following optimization problems:

$$minC_1 = \sum\nolimits_{i=1}^{n} \sum\nolimits_{k=1}^{n} \sigma_{ik} x_i x_k \tag{8}$$

subject to $x \in X$

$$\max\ C_2 = \sum\nolimits_{t=1}^{n} \overline{r}_i x_i \tag{9}$$

Let $(x_1', \ldots, x_n')$be an optimal solution of Problem 1 with optimal C_1 and $(x_1'' \ldots, x_{n1}'')$ be an optimal solution of (5) with optimal value ofC_2. The mean and variance of portfolios range in the intervals $[\sum_{t=1}^{n} \overline{r}_i x_i'; \sum_{t=1}^{n} \overline{r}_i x_i'']$ and $[\sum_{i=1}^{n} \sum_{k=1}^{n} \sigma_{ik} x_i' x_k'; \sum_{i=1}^{n} \sum_{k=1}^{n} \sigma_{ik} x_i'' x_k'']$ respectively.

At the next step, five different investment portfolio alternatives are defined which satisfy these intervals. Further, approach that determines the Z-number-valued description of utility value for each investment alternative is suggested.

$$Z_V^+ = \sum_{t=1}^{T} Z_t^+ \left(\sum_{i=1}^{n} r_{it} x_i\right) \quad (10)$$

subject to

$$\sum_{i=1}^{n} \sum_{k=1}^{n} \sigma_{ik} x_i x_k \leq \sigma_0^2; \sum_{i=1}^{n} x_i = 1; x_i \geq 0; \mathrm{i} = 1, 2, \ldots, \mathrm{n}$$

where $\sigma_0^2 \in [\sum_{i=1}^{n} \sum_{k=1}^{n} \sigma_{ik} x_i' x_k'; \sum_{i=1}^{n} \sum_{k=1}^{n} \sigma_{ik} x_i'' x_k'']$, and x′ and x″ are the optimal solutions of (1) and (2), respectively. $Z^+{}_t$ -is the probability of tth scenario. For convenience, we consider $Z^+{}_t = (A, R)$ as $Z^+{}_t = (A, P(A))$, where $P(A)$ is as in Definition 2.

The analysis is executed on the base of dataset of assets for the period March 1, 2017, to April 1, 2018 is taken from Yahoo finance website. In this analysis, nine assets issued by different companies were randomly selected. The assets are denoted by *Ai*, i = 1, ..., 9. The data comprising of the monthly closing prices of above nine assets have been taken. The monthly return of the ith asset is computed by $r_{it} = \frac{c_{i(t+1)} - c_{it}}{c_{it}}$, t = 1, 2, ..., 12, where $c_{i(t+1)}$ and c_{it} are the closing prices of ith asset in months t + 1 and t, respectively. In order to facilitate calculation, the quarterly average returns for assets were defined. The Table 1 represents the quarterly average returns, and the Table 2 information on covariance of nine assets.

Table 1. Quarterly returns of the assets

Assets/Periods	Q1	Q2	Q3	Q4	$\bar{r}_i$	σ^2
A_1	−0.0058	0.0108	0.0355	0.0220	**0.0156**	**0.0002**
A_2	0.0435	0.0397	0.1412	0.0355	**0.0650**	**0.0019**
A_8	0.0023	0.0301	0.0395	−0.0364	**0.0089**	**0.0009**
A_9	0.0588	0.0243	0.0171	0.0113	**0.0279**	**0.0003**

Table 2. Covariance matrix

	A_1	A_2	A_3	A_4	A_5	A_6	A_7	A_8	A_9
A_1	0.0002	0.0005	0.0001	0.0001	0.0005	−0.0001	−0.0002	0.0001	−0.0002
A_2	0.0005	0.0019	0.0003	0.0009	−0.0001	−0.0001	−0.0003	0.0008	−0.0002
A_8	0.0001	0.0008	0.0002	0.0008	−0.0003	0.0005	−0.0001	0.0009	0.0000
A_9	−0.0002	−0.0002	−0.0001	0.0000	−0.0010	0.0001	0.0002	0.0000	0.0003

Using the input data from Tables 1 and 2, the following problems were solved:

$$minC_1 = \sum_{i=1}^{9} \sum_{k=1}^{9} \sigma_{ik} x_i x_k \quad (11)$$

subject to

$$\sum_{i=1}^{9} x_i = 1; x_i \geq 0; \mathrm{i} = 1, \ldots, 9.$$

The optimal value of $C_1 = 0.0000000000000016$ and the optimal interpretation is $x' = (0.2978; 0.0089; 0.0351; 0; 0.0705; 0; 0.3022; 0; 0.2855)$, with the appropriate expected return E(x) = 0.02093.

$$\max C_2 = \sum_{t=1}^{n} \bar{r}_i x_i \qquad (12)$$

The optimal value of $C_2 = 0.06497$ and the optimal strategy is investment in asset 2 (A_2), because $x_2'' = 1$ and all others $x_i'' = 0$; i = 1, …9; i ≠ 2. The relevant variance σ^2 = 0. 00195. Thus, the intervals of expected return E(X) and variance σ^2 are [0.02093; 0.06497] and [0.0000000000000016; 0.00195] respectively.

Next, by using linear programming the efficient portfolios which satisfy intervals mentioned above were defined. As result, five alternatives of investment portfolio with different combinations of risk-return characteristics were suggested. In Table 3 five investment alternatives are represented.

Table 3. Suggested investment portfolios

№										E(x)	σ^2
	x_1	x_2	x_3	x_4	x_5	x_6	x_7	x_8	x_9		
1	0	0.1157	0	0	0.1564	0	0.0399	0	0.3281	0.028	0.000001
2	0	0.8659	0	0	0.0373	0	0	0	0.0968	0.060	0.001416
3	0	0.7534	0	0	0.0639	0	0	0	0.1827	0.056	0.001035
4	0	0.1767	0	0	0.2003	0	0	0	0.6230	0.035	0.000019
5	0	0.4843	0	0	0.1276	0	0	0	0.3882	0.046	0.000366

In the context of finding the best investment alternative, we solve the objective function (6). Let X be a portfolio and $\sum_{i=1}^{n} r_{it} x_i$ be returns from it with probabilities p_t, t = 1, …T, over period divided into T points of observations. In this work quarterly returns of the portfolios are considered. Therefore, based on experts' opinion four probabilities respectively to quarters were determined and represented as Z^+-numbers:

$$Z_1^+ = ((0.09, 0.1, 0.11), 0.8); Z_2^+ = ((0.45, 0.5, 0.55), 0.8);$$
$$Z_3^+ = ((0.27, 0.3, 0.33), 0.8); Z_4^+ = ((0.01, 0.1, 0.19), 0.8)$$

For each investment alternative given in Table 3, we use relevant input data to solve proposed approach (6) as follows:

$$Z_V^+ = \sum_{t=1}^{4} Z_t^+ \left(\sum_{i=1}^{9} r_{it} x_i\right) \qquad (13)$$

subject to

$$\sum_{i=1}^{9}\sum_{k=1}^{9}\sigma_{ik}x_ix_k \le \sigma_0^2; \sum_{i=1}^{9} x_i = 1; x_i \ge 0; \mathrm{i} = 1, 2, \ldots, 9$$

where $\sigma^2{}_0 \in [0.00000000000000016; 0.00195]$.

As result, the function values for each investment portfolio alternative are represented in the shape of Z^+ numbers, $Z^+{}_{Vi} = ((a_{1i}, a_{2i}, a_{3i}), (\mu, \sigma))$ as follows:

$$Z^+_{V1} = ((0.0368,\ 0.0431,\ 0.0705),\ (0.0083,\ 0.0061))$$

$$Z^+_{V2} = ((0.0546,\ 0.0639,\ 0.0730),\ (0.0083,\ 0.0067))$$

$$Z^+_{V3} = ((0.0500,\ 0.0585,\ 0.0671),\ (0.0083, 0.0061))$$

$$Z^+_{V4} = ((0.0256,\ 0.0311,\ 0.0364),\ (0.0083, 0.0062))$$

$$Z^+_{V5} = ((0.0387,\ 0.0457,\ 0.0528),\ (0.0083, 0.0066))$$

The obtained results are also represented from the viewpoint of reliability in the shape of Z^+ numbers as $Z^+{}_{Vi} = ((a_{1i}, a_{2i}, a_{3i}), \mathrm{P(A)})$:

$$Z^+_{V1} = ((0.0368, 0.0431, 0.0705), 0.47)$$

$$Z^+_{V2} = ((0.0546, 0.0639, 0.0730), 0.50)$$

$$Z^+_{V3} = ((0.0500, 0.0585, 0.0671), 0.48)$$

$$Z^+_{V4} = ((0.0256, 0.0311, 0.0364), 0.48)$$

$$Z^+_{V5} = ((0.0387, 0.0457, 0.0528), 0.50)$$

Further, we convert Z^+ numbers into fuzzy numbers and carry out defuzzification of obtained triangular fuzzy numbers.

$$\mathrm{S_{Vi}} = \frac{\mathrm{R_i} * \mathrm{a_{1i}} + \mathrm{R_i} * \mathrm{a_{2i}} + \mathrm{R_i} * \mathrm{a_{3i}}}{3} \tag{14}$$

Results are shown below:

$$S_{V1} = 0.0236; S_{V2} = 0.0319; S_{V3} = 0.0281; S_{V4} = 0.0149; S_{V5} = 0.0229$$

Ranking of alternative investment portfolios:

Order	Investment alternative	Value
1	S_{V2}	0.0319
2	S_{V3}	0.0281
3	S_{V1}	0.0236
4	S_{V5}	0.0229
5	S_{V4}	0.0149

4 Conclusion

The concept of Z-portfolio selection problem was investigated in this paper. Initially, some concepts and definitions about Z-number were reviewed. The real data were used to illustrate the approach. Analysis of obtained results shows that investors may choose any of suggested alternatives as their risk preferences, because all portfolios are efficient. Nevertheless, based on ranking performed at last step, investment alternative #2 is considered as the best proposal. Despite of higher risk, this portfolio diversified enough due to negative covariance, and consequently, correlation between the assets.

To sum up, the results show that proposed approach is effective and allow us to obtain reliable information for further decision-making process.

References

1. Zadeh, L.A.: A note on Z-numbers. Inf. Sci. **181**(14), 2923–2932 (2011)
2. Aliev, R.A., Alizadeh, A.V., Huseynov, O.H.: The arithmetic of discrete Z-numbers. Inf. Sci. **290**, 134–155 (2015). https://doi.org/10.1016/j.ins.2014.08.024
3. Aliev, R.A., Zeinalova, L.M.: Decision making under Z-information. In: Guo, P., Pedrycz, W. (eds.) Human-Centric Decision Making Models for Social Sciences, Studies in Computational Intelligence, vol. 502, pp.233–252. Springer, Heidelberg (2014). https://doi.org/10.1007/978-3-642-39307-5-10
4. Aliev, R.A., Alizadeh, A.V., Huseynov, O.H., Jabbarova, K.I.: Z-number-based linear programming. Int. J. Intell. Syst. **30**(5), 563–589 (2015). https://doi.org/10.1002/int.21709
5. Jirofti, A., Najafi, A.A.: Portfolio selection using Z-number theory: two solution methodologies. Int. J. Fuzzy Syst. **20**(8), 2484–2496 (2018). https://doi.org/10.1007/s40815-018-0513-4
6. Markowitz, H.: Portfolio selection. J. Fin. **7**(1), 77–91 (1952). https://doi.org/10.2307/2975974
7. Rahmani, A., Hosseinzadeh, L.F., Rostamy-Malkhalifeh, M., Allahviranloo, T.: A new method for defuzzification and ranking of fuzzy numbers based on the statistical beta distribution. Advances in Fuzzy Systems, vol. 2016, pp.1–8 (2016). https://doi.org/10.1155/2016/6945184

Predicting the Mechanical Power of a New-Style Savonius Wind Turbine Using Machine Learning Techniques and Multiple Linear Regression: Comparative Study

Youssef Kassem[1,2](✉), Hüseyin Çamur[1], and Mohamed Almojtba Hamid Ali Abdalla[1]

[1] Faculty of Engineering, Mechanical Engineering Department, Near East University, Nicosia 99138, North Cyprus
{yousseuf.kassem,huseyin.camur}@neu.edu.tr

[2] Faculty of Civil and Environmental Engineering, Near East University, Nicosia 99138, North Cyprus

Abstract. The ability and accuracy of machine learning techniques have been investigated for static modeling of the Savonius wind turbine. The objective of this paper is to predict the mechanical power (MP) of the new-style Savonius rotor as a function of aspect ratio, overlap ratio, number of the blade, wind speed, and rotational speed. In this paper, the MP of the proposed rotors was evaluated through Multilayer Feed-Forward Neural Network (MFFNN), and Cascade Feed-forward Neural Network (CFFNN), and Elman neural network (ENN) based on experimental data. Additionally, the proposed models were compared with multiple linear regressions (MLR) to show the ability and accuracy of the proposed models. The results indicated that the ENN model has higher predictive accuracy compared to other models.

Keywords: Machine learning models · Mechanical power · Multiple linear regression · Savonius turbine · New-style

1 Introduction

Wind turbines are used to convert wind energy into mechanical energy then into electrical energy. They are categorized into two types, namely, horizontal axis wind turbine (HAWT) and vertical axis wind turbine (VAWT). The HAWTs are commonly used to generate electricity particularly in the regions that has high wind speed. The VAWTs such as Savonius wind Turbine have a simple structure and it is suitable to operate at low wind speed. These types of the turbine are independent of the wind direction and can be utilized for generating electricity for domestic applications [1]. In this study, a new Savonius style wind turbine was designed to generate electricity for domestic applications.

R. A. Aliev et al. (Eds.): ICSCCW 2021, LNNS 362, pp. 316–323, 2022.
https://doi.org/10.1007/978-3-030-92127-9_44

Moreover, the performance of the Savonius rotor has been investigated by several scientific researchers [2–4]. For instance, Roy and Saha [2] compared the performance of the new style of Savonius turbine with semi-circular, semi-elliptic, Benesh, and Bach types. They found that the power coefficient of the developed turbine was found to be 34% higher than the other types of Savonius turbines. Jeon et al. [4] studied the influence of the endplates with various sizes and shapes on the power coefficient (CP) of the helical Savonius rotor. The results showed that the CP increased by 36% when the upper and lower end-plate are used. Based on the above, the investigation of the effect of turbine geometries and the shape of the blade is important for evaluating the Savonius performance. In the literature, the behaviour of the performance of the Savonius turbine shows high nonlinearity. Several empirical models including machine learning models and mathematical have been used to predict the performance of the Savonius turbine including power coefficient, mechanical power, and torque. For example, Sargolzaei and Kianifar [5] used an adaptive neuro-fuzzy inference system, fuzzy inference system and radial basis function to predict the torque of the Savonius rotor.

As an ongoing study of authors investigation the performance of small-scale vertical axis wind turbine, this study's goal is to predict the mechanical power of new style Savonius wind turbine using three machine learning tools, namely, Multilayer Feed-Forward Neural Network (MFFNN), Cascade Feed-forward Neural Network (CFFNN), and Elman neural network (ENN). Also, the accuracy of models is compared with multiple linear regressions (MLR).

2 Experimental Data

Figure 1 shows the 2D and 3D views of the proposed rotors. In this study, the effect of blade number (NB), blade height (H), blade diameter (D), external gap (L′), and wind speed (WS) on the mechanical power (MP) of the proposed rotors are investigated as shown in Fig. 1. The blade and the shaft of the rotor are made from PV and stainless steel, respectively. Also, the desks are made from fiberglass. Table 1 lists the dimensions of the developed rotor. Details of the experiment setup and measurements were given in Ref. [6]. In this research, the experimental data of the new style Savonius wind turbine were collected to develop and validate the proposed models (MFFNN, CFFNN, and ENN) in comparison with the MLR model. In this study, aspect ratio, overlap ratio, wind speed, rotational speed, and the number of the blade are used as input variables. The mechanical power is the output variable.

3 Prediction Methods

Many models and techniques are such as machine learning models and mathematical models are used as alternative tools to descript a complex system. They are utilized in a wide variety of applications. In this study, four empirical models (Multilayer Feed-Forward Neural Network, Cascade Feed-forward Neural Network, and Elman neural network and multiple linear regression) are developed to estimate the mechanical power of the new-style Savonius rotor. In this study, TRAINLM is utilized as a training function. Also, Mean squared error (MSE) is estimated to find the best performance of the training algorithm. The descriptions of developed models in detail were given in Ref. [7–9]. MATLAB software was used to develop the proposed models.

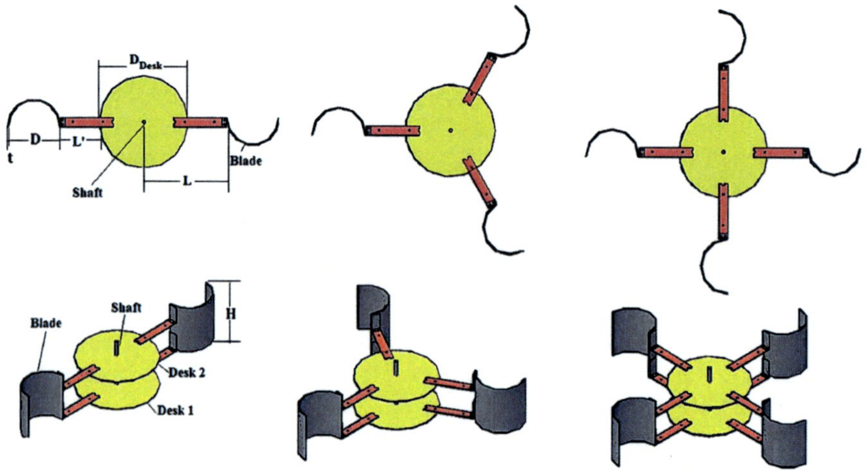

Fig. 1. The 2D and 3D view of the new-style Savonius rotors

Table 1. Fixed and variable parameters of the design

Category	Design parameter	Value
Physical features	1. Blade	Semi-cylindrical
	2. Number of blades (N)	N = 2, 3 and 4 blades
	3. Blade material	Light plastic (PVC)
	4. Desk material	Fiberglass
	5. Shaft material	Stainless steel
Dimensional	6. Blade diameter (d)	d = 200 mm
	7. Blade thickness (t)	t = 3 mm
	8. Desk diameter (D)	D = 700 mm
	9. Blade height (H)	H = 300, 600, 900 and 1200
	10. External Overlap (L′)	L′ = 0, 150, 300 and 400 mm
Operational	11. Rated wind speed (V)	V = 2, 3, 4, 6, 8, 10 and 12 m/s

4 Results and Discussions

4.1 Artificial Models

The descriptive statistics of the experimental data are presented in Table 2. In this study, the data are divided into training and testing groups and the results by the models are compared with each other. The optimum network architecture for all models was determined through the trial and error method. It should be noted that the optimum number of HLs and NNs in the MFFNN, CFFNN, and ENN models were estimated based on the minimum value of MSE.

It is found that the best transfer function for the hidden neurons is the tangent-sigmoid function. Based on the value of MSE, it is found that one hidden layer and 5 neurons are selected as the best for the MFFNN model (5:1:1) with an MSE value of 8.86×10^{-3}. While it found that 1 hidden layer and 7 neurons are chosen as an optimum number for the CFFNN model (5:1:1) with an MSE value of 2.7×10^{-3}. Also, it is observed that the ENN model (5:1:1) with 5 neurons has the minimum MSE with a value of 3.43×10^{-3}. For the training phase, the R-squared value was found to be 0.967, 0.919, and 0.951 for MFFNN, CFFNN, and ENN, respectively as shown in Fig. 2.

Table 2. Selected parameters used in this study

Parameter	Variable	Explanation	Standard deviation	Variation coefficient	Minimum	Maximum
Input 1	NB	Number of blades	0.8171	27.24	2	4
Input 2	H/D	Aspect ratio	1.3094	34.16	1.875	6.25
Input 3	L′/D	Overlap ratio	0.5876	81.75	0	1.875
Input 4	WS	Wind speed	3.187	44.46	3	12
Input 5	RPM	Rotational speed	38.02	59.66	11.8	173.6
Output	MP	Mechanical power	2.807	120.77	0.013	12.241

4.2 MLR Model

The development of MLR was implemented to predict the mechanical power of the proposed rotator. The data of blade number, aspect ratio, overlap ratio, wind speed, rotational speed, and mechanical power were used to generate a mathematical equation as given in Eq. (1). It is observed that R-squared value for training data was found to be 0.7697 as shown in Fig. 3.

$$MP = -5.543 + 1.0499 \cdot NB + 0.1544 \cdot (H/D) + 0.3988 \cdot (L/D) - 0.0066 \cdot WS + 0.060980 \cdot RPM \quad (1)$$

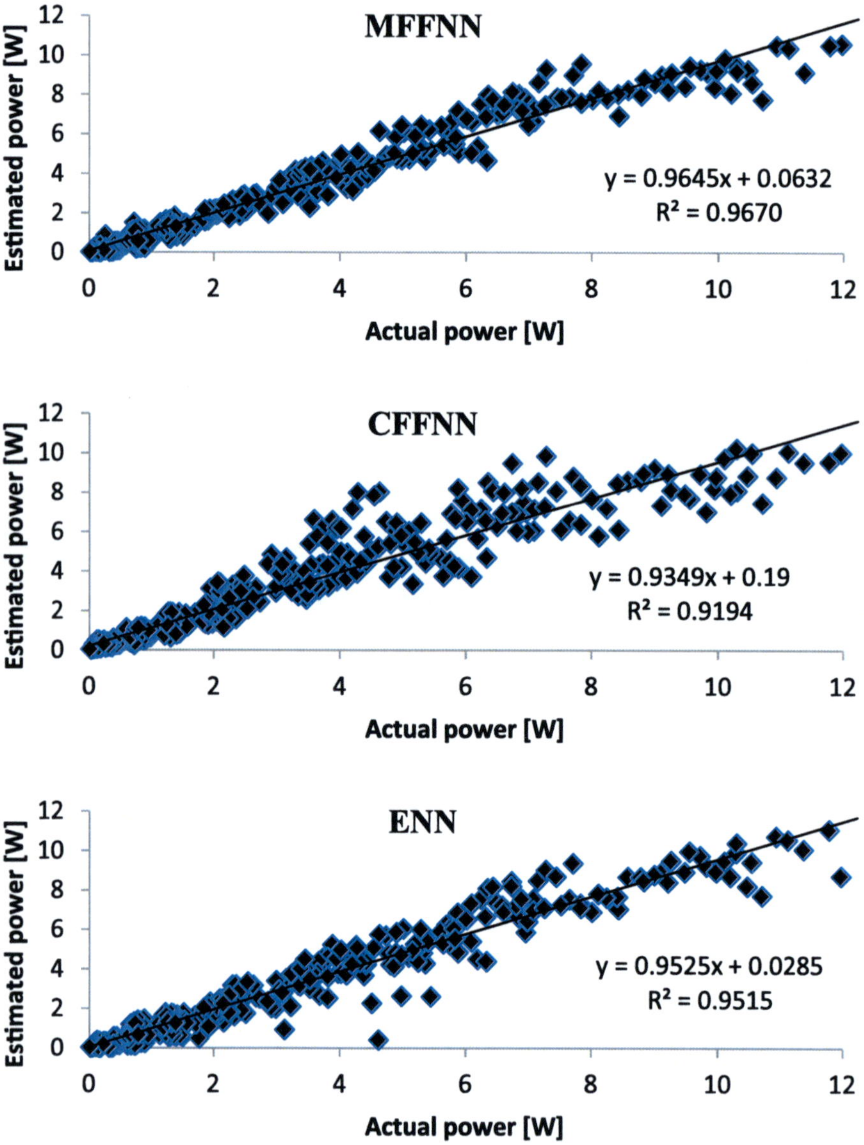

Fig. 2. Comparison of experimental data and the estimated values found by machine learning models

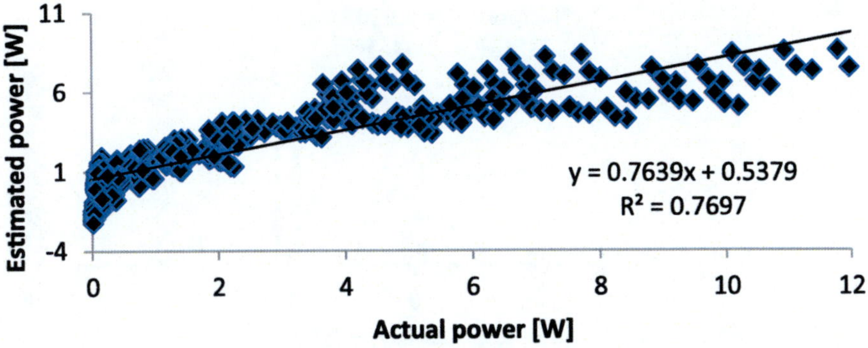

Fig. 3. Comparison of experimental data and the estimated values found by the MLR model

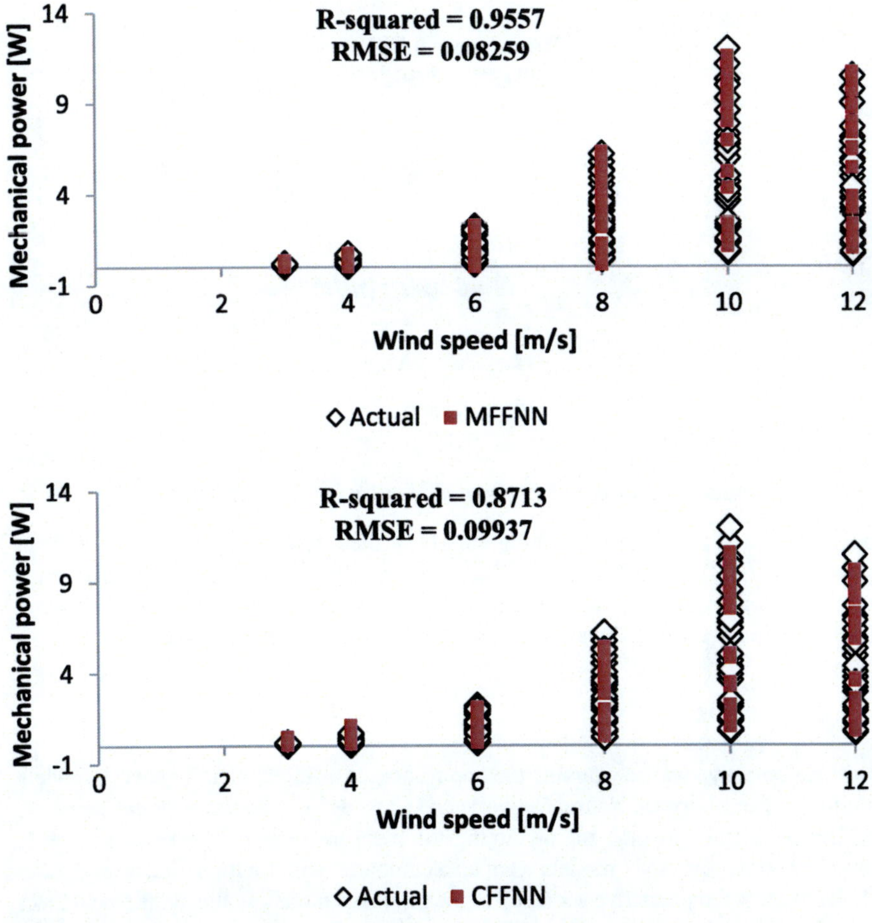

Fig. 4. Comparison of experimental data and the estimated values found by empirical models

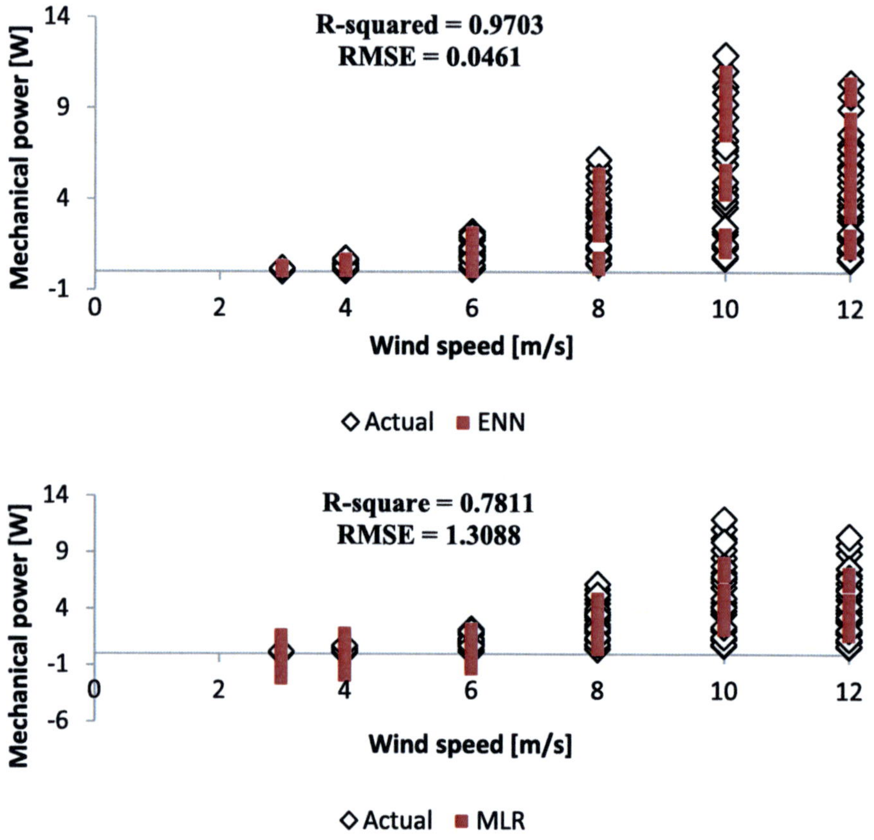

Fig. 4. continued

4.3 Performance Evaluation of Empirical Models for Testing Data

To select the best model for estimating the MP of the rotors, R-squared and root mean squared error (RMSE) are determined. The comparison of the predicted and actual values of the MP for all models is shown in Fig. 4. It is found that the highest R-squared value of 0.9703 and lowest RMSE value of 0.0461 are obtained from ENN model.

5 Conclusions

The main objective was to examine the application of artificial neural network models (Multilayer Feed-Forward Neural Network, and Cascade Feed-forward Neural Network, and Elman neural network) for predicting the mechanical power (MP) of new-style Savonius rotors. The ANN models were also compared with multiple linear regression (MLR) to show the predictive accuracy of the proposed model. In this work, the impact of the blade number, aspect ratio, overlap ratio, wind speed, rotational speed on the mechanical power was investigated and the experimental data were used to develop the proposed models. Moreover, the coefficient of determination (R^2) and root mean squared error (RMSE) were used to assess the best empirical model. It is found that the ENN model was found to be the best model for estimating the EP of the new-configuration Savonius rotor and more precise compared to CFFNN, MFFNN, MLPNN and RBFNN models.

References

1. Kassem, Y., Sefik, A., Çamur, H., Bahroun, A.A.: Experimental and numerical investigation of the influence of blade geometries and blade number on the performance of a newly developed Savonius-Style wind rotor. J. Eng. Appl. Sci. **14**, 9788–9805 (2019). https://doi.org/10.36478/jeasci.2019.9788.9805
2. Roy, S., Saha, U.K.: Wind tunnel experiments of a newly developed two-bladed Savonius-style wind turbine. Appl. Energy **137**, 117–125 (2015). https://doi.org/10.1016/j.apenergy.2014.10.022
3. Kassem, Y., Çamur, H., Hasan, H.: Power generation with different types of the vertical wind turbine for domestic use in northern cyprus: a case study. J. Eng. Appl. Sci. **14**, 9745–9754 (2019). https://doi.org/10.36478/jeasci.2019.9745.9754
4. Jeon, K.S., Jeong, J.I., Pan, J.-K., Ryu, K.-W.: Effects of end plates with various shapes and sizes on helical Savonius wind turbines. Renew. Energy **79**, 167–176 (2015). https://doi.org/10.1016/j.renene.2014.11.035
5. Sargolzaei, J., Kianifar, A.: Neuro–fuzzy modeling tools for estimation of torque in Savonius rotor wind turbine. Adv. Eng. Softw. **41**(4), 619–626 (2010). https://doi.org/10.1016/j.advengsoft.2009.12.002
6. Kassem, Y., Gökçekuş, H., Çamur, H.: Artificial neural networks for predicting the electrical power of a new configuration of Savonius rotor. In: Aliev, R.A., Kacprzyk, J., Pedrycz, W., Jamshidi, M.O., Babanli, M.B., Sadikoglu, F.M. (eds.) ICSCCW 2019. AISC, vol. 1095, pp. 872–879. Springer, Cham (2020). https://doi.org/10.1007/978-3-030-35249-3_116
7. Kassem, Y., Gokcekus, H.: Do quadratic and Poisson regression models help to predict monthly rainfall? Desalin. Water Treat. **215**, 288–318 (2021). https://doi.org/10.5004/dwt.2021.26397
8. Kassem, Y., Gokcekus, H., Camur, H., Esenel, E.: Application of artificial neural network, multiple linear regression, and response surface regression models in the estimation of monthly rainfall in Northern Cyprus. Desalin. Water Treat. **215**, 328–346 (2021). https://doi.org/10.5004/dwt.2021.26525
9. Li, X., Han, Z., Zhao, T., Zhang, J., Xue, D.: Modeling for indoor temperature prediction based on time-delay and Elman neural network in air conditioning system. J. Build. Eng. **33**, 101854 (2021). https://doi.org/10.1016/j.jobe.2020.101854

Analysis of Consistency of Pairwise Comparison Matrix with Fuzzy Type-2 Elements

Aygul Dadasheva(✉)

Azerbaijan State Oil and Industry University, BA Programs, 20 Azadlig Avenue, AZ1010 Baku, Azerbaijan
aygul.dadasheva@asoiu.edu.az

Abstract. The consistency is used in the estimation of preference knowledge quality. Preference densities' rationality of compared elements are kept by the consistency conditions set. In this paper, consistency analysis is represented for construction of consistent pairwise comparison matrices from inconsistent one and the consistent matrix with minimum distance to inconsistent matrix has been selected from the options. The elements of matrix are presented by fuzzy type-2 numbers. In pairwise matrix three criteria of Information Systems (IS) Project have been selected: contribution of profitability, availability of skilled personnel, information processing efficiency.

Keywords: Fuzzy type-2 number · Multicriteria group decision making · Pairwise comparison matrix · Consistency

1 Introduction

The pairwise comparison matrices are used in multicriteria group decision making for determining a ranking of criteria or alternatives and if consistency condition for the matrix satisfied, it allows to determine a weighted ranking. Information about the comparison degrees of criteria or alternatives is presented in pairwise matrices. If to assume that this matrix is $(\tilde{A}_{ij})$ and in this matrix $\tilde{A}_{ij}$ is fuzzy type-2 number entry which denotes preference degree of *i-th* criteria (alternative) to *j-th* one [1]. The main conditions for $(\tilde{A}_{ij})$ is [2]:

$$\tilde{A}_{ii} = \tilde{A}(1) \tag{1}$$

where $\tilde{A}(1)$ is fuzzy singleton.

The second condition is reciprocity:

$$A_{ji} = {}^{1}/_{A_{ij}}, \ \forall i, j = 1, \ldots, n. \tag{2}$$

The basis of consistency is multiplicative transitivity condition which is represented below [3]:

$$A_{ji}A_{jk} = A_{ik} \tag{3}$$

R. A. Aliev et al. (Eds.): ICSCCW 2021, LNNS 362, pp. 324–330, 2022.
https://doi.org/10.1007/978-3-030-92127-9_45

An inconsistency depends on the violation of the multiplicative transitivity condition. The consistency ratio term was suggested by Saaty [4]. Consistency indexes suggested by Cavallo and D'Apuzzo [5], Kou and Lin [6] are prominent ones.

In the following sections, multicriteria pairwise matrix [7] with type-2 fuzzy elements has been represented. In sequence, the new algorithm model has been shown for the fuzzy type-2 multicriteria group decision making problem solution. Section 2 includes preliminaries and construction of consistent pairwise comparison matrix from inconsistent matrix problem has been presented in Sect. 3. In Sect. 4 includes the problem solution.

2 Preliminaries

Definition 1. Fuzzy type-2 sets [8] $\widetilde{A}$ in the discourse universe X can be characterized by 2 type membership function.

$$\widetilde{A} = \{((x, u), \mu_{\widetilde{A}}(x, u)) | \forall x \in X, \forall u \in J_x \subseteq [0, 1], 0 \leq \mu_{\widetilde{A}}(x, u) \leq 1\} \quad (4)$$

where J_x denotes an interval [0,1].

Figure 1 presents trapezoidal interval type 2 fuzzy set sample.

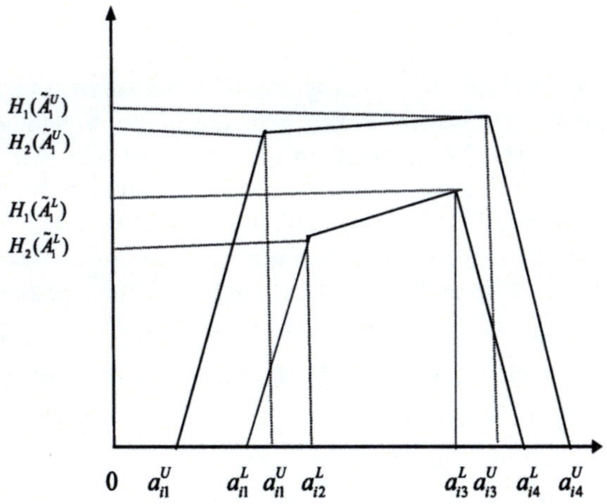

Fig. 1. Fuzzy type-2

Where: a_{i1}^U, a_{i2}^U, a_{i3}^U, a_{i4}^U, a_{i1}^L, a_{i2}^L, a_{i3}^L, a_{i4}^L are the references points of the type-2 fuzzy $\widetilde{A}$; $H_j(\tilde{A}_i^U)$ denotes the membership value of the element $a_{i(j+1)}^U$ in the upper trapezoidal membership function $\widetilde{A}_i^U$, $1 \leq j \leq 2$, $H_j(\tilde{A}_i^L)$ denotes the membership value of the element $a_{i(j+1)}^U$ in the upper trapezoidal membership function $\widetilde{A}_i^U$; and $1 \leq j \leq 2$, $H_j(\tilde{A}_i^U)$, $H_j(\tilde{A}_i^L) \in [0, 1]$, $1 \leq i \leq n$ [8].

Definition 2. Multicriteria group decision matrix of different decision makers is determined by ranking method by Wang, comparing each criterion for each decision maker and choosing the maximum valued option [9].

Let $\tilde{U} = (u_1, u_2, u_3, u_4; w_u)$ and $\tilde{V} = (v_1, v_2, v_3, v_4; w_v)$ be the fuzzy type-2 numbers, where w_u and w_v are membership functions, (x_{0u}, y_{0u}) and (x_{0v}, y_{0v}) be the centroid points of these numbers, then:

$$1.\ \text{If } x_{0u} \succ x_{0v}\ (x_{0u} \prec x_{0v}) \text{ then } \tilde{U} \succ \tilde{V}\ (\tilde{U} \prec \tilde{V}), \tag{5}$$

$$2.\ \text{If } x_{0u} = x_{0v} \text{ and } y_{0u} = y_{0v}, \text{ then } \tilde{U} \approx \tilde{V} \tag{6}$$

$$3.\ \text{If } x_{0u} = x_{0v} \text{ then, } \tilde{U} \succ \tilde{V}\ (\tilde{U} \prec \tilde{V}) \text{ is resulted from } y_{0u} \succ y_{0v}\ (y_{0u} \prec y_{0v}) \tag{7}$$

Definition 3. Assume that the $\tilde{A}_{ij}$ and $\tilde{A}'_{ij}$ are the 2-type fuzzy numbers which are characterized by membership function and M($\tilde{A}_{ij}$) and M ($\tilde{A}'_{ij}$) are the centroids of $\tilde{A}_{ij}$ and $\tilde{A}'_{ij}$. The distance d_M between C($\tilde{A}_{ij}$) and C ($\tilde{A}'_{ij}$) is represented by the following formula [10]:

$$d_M(\tilde{A}_{ij}, \tilde{A}'_{ij}) = \left|M_l(\tilde{A}_{ij}) - M_l(\tilde{A}'_{ij})\right| + \left|M_u(\tilde{A}_{ij}) - M_u(\tilde{A}'_{ij})\right| \tag{8}$$

3 Statement of the Problem

Fuzzy multicriteria group decision making approach has been represented for evaluation and selection of Information Systems Project Selection. Evaluation imprecision [11] is modeled by trapezoidal type-2 fuzzy numbers which are characterize linguistic terms.

Suppose that the multicriteria group decision making problem involves 3 criteria: C_1- Contribution of Profitability; C_2- Availability of Skilled Personnel; C_3- Information Processing Efficiency. We will construct consistent pairwise comparison matrix for criteria importance. In the given problem ($\tilde{A}_{ij}$) is given inconsistent pairwise comparison matrix ($\tilde{A}'_{ij}$) and ($\tilde{A}''_{ij}$) are the consistent matrices constructed from ($\tilde{A}_{ij}$). In ($\tilde{A}'_{ij}$) and ($\tilde{A}''_{ij}$) all consistency conditions are verified. Our main problem is to choose the matrix between ($\tilde{A}'_{ij}$) and ($\tilde{A}''_{ij}$) with minimum distance to inconsistent matrix. Assume that the following pairwise comparative matrix ($\tilde{A}_{ij}$) is given:

$$\tilde{A}_{ij} = \begin{bmatrix} 1 & \tilde{A}_{12} & \dots & \tilde{A}_{1n} \\ \tilde{A}_{21} & 1 & \dots & \tilde{A}_{2n} \\ \vdots & \vdots & \ddots & \vdots \\ \tilde{A}_{n1} & \tilde{A}_{n2} & \dots & 1 \end{bmatrix} \tag{9}$$

The components of consistent matrix have to meet the conditions in formulas (1), (2), (3).

For the updated pairwise comparison matrices the distance between inconsistent and consistent matrices will be calculated by the formula (8).

4 Solution of the Problem

The problem may be solved based on the algorithm shown below:

Step-1: Identification of the criteria for the information systems project selection.

Step-2: Construction of pairwise comparison matrices including all the criteria. Linguistic variables for the pairwise comparisons and the scales of these matrices in type-2 fuzzy numbers are shown in Table 1.

Table 1. Linguistic variables and 2-type fuzzy scales

Linguistic variables	Trapezoidal interval 2-type fuzzy numbers
Extremely Strong (ES)	(7,8,9,9; 1) (7.2,8.2,8.8,9; 0.8)
Very Strong (VS)	(5,6,8,8; 1) (5.2,6.7,8,8; 0.8)
Intermediate Strong (IntS)	(3,4,6,7; 1) (3.4,4.2,6,6.8; 0.8)
Moderately Strong (MS)	(1,2,4,5; 1) (1.4,2.2,3.8,4.8, 0.8)
Equally Strong (E)	(1,1,1,1; 1) (1,1,1,1; 1)

Assume that in our problem we have 3 decision makers (DM) and each of them suggests an idea about the comparison of criteria. Thc pairwise comparison matrices with fuzzy type-2 elements are represented in Tables 2, 3 and 4:

Table 2. The pairwise comparison matrix for the first DM

	C_1	C_2	C_3
C_1	E	1/MS	1/MS
C_2	MS	E	1/MS
C_3	MS	MS	E

Table 3. The pairwise comparison matrix for the second DM

	C_1	C_2	C_3
C_1	E	E	1/IS
C_2	E	E	1/IS
C_3	IS	IS	E

Table 4. The pairwise comparison matrix for the third DM

	C_1	C_2	C_3
C_1	E	1/MS	1/MS
C_2	MS	E	1/MS
C_3	MS	MS	E

Step-3: Determination of group decision making pairwise matrix by using formulas (5), (6), (7). This matrix has been presented in Table 5.

Table 5. The pairwise comparison matrix for the group decision making

	C_1	C_2	C_3
C_1	(1,1,1,1; 1) (1,1,1,1; 1)	(1,0.5,0.25, 0.2; 1) (0.7,0.45,0.26, 0.2; 0.8)	(1,0.5,0.25,0.2; 1) (0.7,0.45,0.26, 0.2; 0.8)
C_2	(1,2,4,5; 1) (1.4.2.2.3.8,4.8; 0.8)	(1,1,1,1; 1) (1,1,1,1; 1)	(1,0.5,0.25,0.2; 1) (0.7,0.45,0.26, 0.2; 0.8)
C_3	(3,4,6,7; 1) (3.4,4.2,3, 3.8, 0.8)	(3,4,6,7; 1) (3.4,4.2,3, 3.8; 0.8)	(1,1,1,1; 1) (1,1,1,1; 1)

Step-4: If to consider that the pairwise matrix is inconsistent, for construction of consistent matrix by using formulas (1), (2), (3) we need to change the matrix components and to create new consistent matrix. In this case we have 2 options where all constraints are appropriate. These matrices are represented in Tables 6 and 7:

Table 6. First option where group decision making matrix components are in compliance

	C_1	C_2	C_3
C_1	(1,1,1,1; 1) (1,1,1,1; 1)	(1,1,1,1; 1) (1,1,1,1; 1)	(1,0.5,0.25,0.2; 1) (0.7,0.45,0.26, 0.2; 0.8)
C_2	(1,1,1,1; 1) (1,1,1,1; 1)	(1,1,1,1; 1) (1,1,1,1; 1)	(1,0.5,0.25,0.2; 1) (0.7,0.45,0.26, 0.2; 0.8)
C_3	(1,2,4,5; 1) (1.4.2.2.3.8,4.8; 0.8)	(1,2,4,5; 1) (1.4.2.2.3.8,4.8; 0.8)	(1,1,1,1; 1) (1,1,1,1; 1)

Step-5: As we have 2 new consistent pairwise comparison matrices where all conditions (1), (2), (3) are met, for choosing the matrix with minimum distance we will calculate the distance by the formula (8). The results for the distance for both consistent pairwise matrices have been shown.

Table 7. Second option where group decision making matrix components are in compliance

	C1	C2	C3
C1	(1,1,1,1; 1) (1,1,1,1; 1)	(1,0.5,0.25,0.2; 1) (0.7,0.45,0.26, 0.2; 0.8)	(1,0.25,0.06,0.04; 1) (1,0.51,0.21,0.07, 0.04; 0.8)
C2	(1,2,4,5; 1) (1.4.2.2.3.8,4.8; 0.8)	(1,1,1,1; 1) (1,1,1,1; 1)	(1,0.5,0.25,0.2; 1) (0.7,0.45,0.26, 0.2; 0.8)
C3	(1,4,16,25; 1) (1.96,4.84,14.44, 23.04; 0.8)	(1,2,4,5; 1) (1.4.2.2.3.8,4.8; 0.8)	(1,1,1,1; 1) (1,1,1,1; 1)

For first constructed consistent matrix the obtained distance is as follows:

$$d_M(\tilde{A}_{ij}, \tilde{A}'_{ij}) = 13.25 \tag{10}$$

The distance for the second constructed consistent matrix has been represented below:

$$d_M(\tilde{A}_{ij}, \tilde{A}''_{ij}) = 14.35 \tag{11}$$

Thereby, $\tilde{A}'_{ij}$ has the minimum distance to the inconsistent matrix.

5 Conclusion

The main purpose of the paper is to transfer the inconsistent pairwise comparison matrix to consistent pairwise comparison matrix. Reliable preferences over choice criteria are presented by fuzzy type-2 numbers in this work. By using the proposed approach, the consistent matrix with minimum distance to inconsistent matrix is obtained. As the first consistent matrix is with minimum distance to inconsistent matrix, we have chosen the first option.

References

1. Brunelli, M.: A survey of inconsistency indices for pairwise comparisons. Int. J. Gen. Syst. **47**(8), 751–771 (2018). https://doi.org/10.1080/03081079.2018.1523156
2. Aliev, R.A., Guirimov, B.G., Huseynov, O.H., Aliyev, R.R.: A consistency-driven approach to construction of Z-number-valued pairwise comparison matrices. Iran J. Fuzzy Syst. **18**(4), 37–49 (2021)
3. Lin, L.Z., Yeh, H.R.: A perceptual measure of mobile advertising using fuzzy linguistic preference relation. Iran J. Fuzzy Syst. **10**(5), 25–46 (2013)
4. Saaty, T.L.: A scaling method for priorities in hierarchical structures. J. Math. Psychol. **15**(3), 234–281 (1977). https://doi.org/10.1016/0022-2496(77)90033-5
5. Cavallo, B., D'Apuzzo, L.: A general unified framework for pairwise comparison matrices in multicriterial methods. Int. J. Intell. Syst. **24**(4), 377–398 (2009). https://doi.org/10.1002/int.20329
6. Kou, G., Lin, C.: A cosine maximization method for the priority vector derivation in AHP. Eur. J. Oper. Res. **235**(1), 225–232 (2014). https://doi.org/10.1016/j.ejor.2013.10.019

7. Dovlatova, K.J.: Estimation of consumer buying behavior for brand choosing by using fuzzy IF-THEN rules. In: Aliev, R., Kacprzyk, J., Pedrycz, W., Jamshidi, M., Babanli, M., Sadikoglu, F. (eds.) 10th International Conference on Theory and Application of Soft Computing, Computing with Words and Perceptions - ICSCCW-2019. ICSCCW 2019. Advances in Intelligent Systems and Computing, vol. 1095, pp. 805–812. Springer, Cham (2019). https://doi.org/10.1007/978-3-030-35249-3_106
8. Zadeh, L.A.: The concept of a linguistic variable and its application to approximate reasoning—I. Inf. Sci. **8**(3), 199–249 (1975). https://doi.org/10.1016/0020-0255(75)90036-5
9. Wang, Y.-M., Yang, J.-B., Xu, D.-L., Chin, K.-S.: On the centroids of fuzzy numbers. Fuzzy Sets Syst. **157**(7), 919–926 (2006). https://doi.org/10.1016/j.fss.2005.11.006
10. Celemin, C.E., Melgarejo, M.A.: A proposal to speed up the computation of the centroid of an interval Type-2 fuzzy set. Adv. Fuzzy Syst. **2013**, 1–17 (2013). https://doi.org/10.1155/2013/158969
11. Aliev, R.A., Gardashova, L.A.: Z-set based Approach to control system design. In: 14th International Conference on Theory and Application of Fuzzy Systems and Soft Computing, ICAFS-2020. Advances in Intelligent Systems and Computing, vol. 1306, pp. 10–21 (2020). https://doi.org/10.1007/978-3-030-64058-3_2

Analysis and Assessment of Accuracy of the Oil Recovery Factor Calculations by Using the Fuzzy Clustering Algorithm

B. N. Koilybayev[1(✉)], M. K. Karazhanova[1], G. M. Efendiyev[2], A. S. Strekov[2], and L. B. Zhetekova[1]

[1] Yessenov University, 32 Microregion, Main Building, 130000 Aktau, Republic of Kazakhstan
nomad_bk@bk.ru

[2] Institute of Oil and Gas of Azerbaijan, National Academy of Sciences, 9, F. Amirov Street, Baku 1000, Azerbaijan

Abstract. Up today, various multivariate scholastic models have been developed for various oil-producing regions in order to calculate the oil recovery factor. They are based on the results of processing actual data on the considered fields, which are in long-term development and have similar geological and physical conditions. The methods used for the "direct" calculation of the oil recovery factor can be combined into three groups: extrapolational; hydro-dynamical; statistical.

The performed analysis shows a different degree of estimates using the considered statistical dependences when changing the values of the parameters included in them. For example, the influence of the values of permeability or net-to-gross ratio can be noted, which can become apparent both insignificantly and significantly.

As the result, a comparative analysis of the previously built statistical models for estimating the oil recovery factor was carried out, the errors in their application were calculated; linear and multiplicative models have been built for predictive estimation of the oil recovery factor, the need to determine the area of their application from the point of view of individual conditions has been substantiated due to the ambiguity of the calculation results; with the help of fuzzy cluster analysis according to the data of the Karazhanbas field, four classes are distinguished, characterized by certain values of the parameters corresponding to different geological and physical conditions; classes of congruence of geological and physical characteristics and errors for the two obtained models were established, which made it possible to define the conditions for applicability of these models.

Keywords: Uncertainty · Cluster analysis · Error · Oil recovery · Oil · Density · Viscosity · Permeability

1 Introduction

In control theory it is often to face problem of mathematical mode-building under uncertainty, described by outstanding complexity, nonlinearity and insufficient knowledge

R. A. Aliev et al. (Eds.): ICSCCW 2021, LNNS 362, pp. 331–339, 2022.
https://doi.org/10.1007/978-3-030-92127-9_46

of connections between numerous variables, presence of significant interferences and uncertainty of measurement. In this circumstances fuzzy models are showing high efficiency, containing rules like "if ..., then ..." and membership function, which specifies intervals of change in input and output variables in each of the rule. In field development normally it is performed assessment of oil recovery estimation. Oil recovery (oil recovery rate ORR) is usually assessed as a ratio of cumulative oil production to original-oil-in-place. Depending on numerous factors it's value normally varies in the range from 0,09 to 0,75. By it's physical meaning the oil recovery represents completeness of oil recovery from the flooded volume. The oil recovery factor is influenced by various factors, including geological-physical and technological ones.

Numerous scientific works provide a review and analysis of the results of calculations of reservoir recovery rate from field data [1–4]. The need for building of models, which allows reliably predict the oil recovery rate is obvious, due to the importance and role of this indicator in the analysis of field development and decision-making. In this regard, currently a large number of models have been accumulated, that describe the desired dependence in relation to the conditions of various fields. However, in most cases, statistical models do not accurately predict this indicator. In this paper, it is considered the possibility of building fuzzy models and the procedure for their identification based on fuzzy clustering.

2 Analysis of Statistical Models for Predicting Oil Recovery in Various Conditions

As it is highlighted by researchers, oil recovery rate «should be considered as a technical and economical indicator, since, from one side, it affects the profitability of investment in the development of the field, and on the other side, it characterizes the success of the proposed process and technical solutions for the field development. Up today, various multivariate scholastic models have been developed for various oil-producing regions in order to calculate the oil recovery factor. They are based on the results of processing actual data on the considered fields, which are in long-term development and have similar geological and physical conditions. As for the "direct" calculation of the oil recovery factor, various methods are used, which, according to [5], can be combined into three groups: extrapolational; hydro-dynamical; statistical.

The first group of the techniques is used to update the final oil recovery factor, they are also called as a "characteristics of desaturation" [5].

Extrapolation techniques are based on the processing of data on oil production at the considered field for the past period and extrapolation of production profile for the future. Poor accuracy is possible herein. This is the main disadvantage of these techniques – all of them require a long-term production period, reliable production profile data, maximum drilling-out of the field already in the first years of development. By applying these methods, in fact, it is possible to assess with a sufficient degree of reliability only the success of the development system proposed for the field on the basis of calculations of the "base production" of oil, and only for a short period.

Hydrodynamic techniques foresees the building of hydrodynamic models, which, along with geological 3D models, constitute the majority of prediction calculations.

It should be noted that statistical dependences are generally valid only under conditions similar to those on the basis of which they were obtained [6, 7].

Based on the best practice of long-term development of fields, statistical methods allows establishment of a statistical relationship between the oil recovery factor and a large number of factors that have a significant impact on the completeness of oil recovery. There are existing many statistical models [5, 6]. Using the multiple correlation equations, which are reflecting the influence of geological, physical and technological factors on the value of the oil recovery factor (ORR) it seems possible to make expected estimates for the conditions under consideration. The most accurate results of estimating oil recovery factor by oil pools can be obtained via applying these models in conditions close to those on which these models were built initially.

After analyzing the literature [8], the author of [6] selected several statistical models in such a way that "the initial parameters of the objects considered in the work had values that fit into the intervals suggested by the statistical models".

In order to check the possibility of calculating the oil recovery factor for the conditions of the Karazhanbas field under consideration, we carried out calculations and a comparative analysis of the calculation results for the conditions of the central part of the Karazhanbas field.

It should be noted that for the conditions we are considering, all of the previously built models provided poor accuracy, which is due to the difference in the conditions for building these models and the actual conditions we are considering.

3 Prediction of the Oil Recovery Factor Based on Statistical Analysis of Statistical Models for Predicting Oil Recovery in Various Conditions

Prediction of development indicators under the conditions noted above is very difficult, which is associated with the difficulties in selecting a more adequate model. In this case, it is necessary to use methods that allow, based on the analysis of errors, to estimate the conditions of applicability of the model, based on the smallest error.

Initially, similar to the models described in the above mentioned works, we have built a model based on the collection and statistical processing of data on the central section of the Karazhanbas field as a result of correlation analysis. For this purpose, data has been collected that characterize the object of the field. These data were put on the statistical processing, correlation analysis was performed, for which a standard multiple linear regression program was used.

As a result, a relevant model has been built, which expresses the dependence of the oil recovery factor on perviousness, average oil saturation, well grid density, net-to-gross ratio, oil-to-water viscosity ratio, porosity coefficient, oil-saturated thickness of the reservoir in the form of:

$$\begin{aligned}\eta = 0.074027 + 0.004991 * \lg\left(K_{per.}\right) - 0.0368 * S_{oil} + 0.0211 * S_{gen.} \\ - 0.034866 * K_{N/G} + 0.000013 * \mu_0 + 0.102827 * m - 0.000038 * h\end{aligned} \quad (1)$$

where η - oil recovery rate (ORR), *fraction units,* $K_{per.}$ - perviousness, *micron*^{2}, S_{oil} - oil saturation, *fraction units,* $S_{gen.}$ - well grid density, *ha/well,* $K_{N/G.}$ - net-to-gross ratio, *fraction units,* μ_0 - oil-to-water viscosity ratio, m - porosity, *fraction units,* h - net oil pay, *m.*

In the model, instead of *perviousness*, its logarithm is given, due to the fact that the distribution of perviousness values complies to a logarithmic-normal law. If we plot the dependence of the oil recovery factor on each factor used in the statistical analysis, then it can be seen that the data are scattered in the form of a "cloud" of an indefinite shape. This is due to the nonsufficiency, nonhomogenity and inaccuracy of field data, although many represent the average of several observations. In general, a strictly linear relationship is also not traced: a straight line closer to most of the points, in this case, is very difficult to draw. However, multiple regression analysis, on which all existing statistical methods for oil recovery estimation are based, emerges from the basic assumption that there is a certain linear functional relationship between the initial data and the result of calculations [5]. From the graph given in [5], the nonlinearity of the functional dependence of the oil recovery factor on the factors under consideration is obvious, therefore, the author of this work considers it unreasonable to apply multiple regression analysis with building of a linear model, although the linear model we have built gives better results for the conditions under consideration compared to others. Nevertheless, a low degree of convergence of the calculated and actual values of oil recovery factor is observed.

As the analysis shows, there is no linear relationship between the oil recovery factor and the geological and physical characteristics of the formations, therefore, for a more accurate calculation of oil recovery factor, it is necessary to take this fact into account in the mathematical model. To solve this problem, it is need to use a method that allows to identify an unknown nonlinear functional dependence with a large number of input variables, as in our case, (different parameters of oil reservoirs).

In this case, one of the possible ways is to build a multiplicative model. For building of such a model the data are first translated into their logarithms, according to the values of which and by implementing a linear regression program, a linear model is built, each variable of which is the logarithm of the input and output variables. After the exponentiating, a multiplicative model was obtained, which, after simple transformations, looks as follows:

$$\eta = 0.06551 \frac{K_{per.}^{0,03236} \cdot S_{gen.}^{0,45136} \cdot \mu_0^{0,05385} \cdot m^{0,28073} \cdot h^{0,0235}}{S_{oil}^{0,07413} \cdot K_{N/G}^{0,15162}} \tag{2}$$

Comparative analysis of actual and calculated data showed that the maximum and average relative errors in calculations using linear and multiplicative models are 55.002%; 27.16% & 49.16%; 24.9% respectively. As it was mentioned above, during build-up of these models for predictive estimation of oil recovery factor, it was utilized the experience of building of similar models accumulated by various specialists in the conditions of various fields. A comparative analysis of the initial information on the geological and physical parameters of the reservoir used in the building of statistical models and the results of calculations based on the built-up models it was demonstrated that among the factors considered there exists also those whose influence is insignificant. In this case, such factors are the porosity coefficient and the density of the well grid: the values of these factors within the considered set are homogeneous, and therefore their influence on the final result turned out to be insignificant. Thus, the calculated oil recovery factor according to the linear model is almost not affected by the porosity and density of the well grid, since in this case, in the considered range, they change insignificantly, despite the fact that this relationship in other models can be traced much more clearly. In our case, the porosity varies within 0.29–0.37, and the density of the wells grid is 1.95–2.87. In general, for these conditions, high values of oil recovery factor are not observed, the highest value is 0.21. Under mentioned conditions, in order to elaborate the model, we have made an attempt to rebuild it without involving insignificant factors. However, even in this case, it was not possible to avoid high errors, although they turned out to be less than in the previously built models, but not so much.

The performed analysis shows a different degree of estimates using the considered statistical dependences when changing the values of the parameters included in them. For example, the influence of the values of permeability or net-to-gross ratio can be noted, which can become apparent both insignificantly and significantly. Naturally, all of the considered dependences, including ours, cannot take into account the variety of factors, their degree of influence: the complexity of the geological structure, to one degree or another, registered in most of the fields. Thus, taking into account the results of the studies, methodological approaches to predicting oil recovery factors, as well as complicated geological and physical conditions of the field under consideration, it is not expected more accurate results from statistical models.

Analysis of the results of calculations based on the obtained models allows us to conclude that the main attention in studying the influence of geological and physical conditions on the indicators of field development needs to be directed to the development of methods that allow obtaining of results that could provide the possibility of predictive assessment of the oil recovery factor both quantitatively and qualitatively, taking into account also uncertainties, associated primarily with the lack of a aprior information about the properties of the reservoir and reservoir fluids, formulations during decision making process. In recent years, such methods have been widely used in solving problems of oilfield practice. These methods make it possible to build predictive models when it is impossible to obtain more accurate results by statistical methods that allow assessing the influence of factors and performing the estimates. In this regard, we have performed an analysis to establish the congruence of the geological and physical parameters and the oil recovery factor, by taking into account the uncertainty. At the same time, the possibility of establishing this congruence on the basis of fuzzy cluster analysis is considered.

4 Study of the Impact of Geological and Physical Factors on the Oil Recovery Factor Based on the Results of Fuzzy Cluster Analysis

During process of solving various problems related to declustering of various objects into homogeneous classes, establishing the conformance of the characteristics of these classes to any parameter, the value of which is usually expressed qualitatively, the utilization of cluster analysis plays one of the important roles.

The clustering matter is expressed by breakdown of the initial set of objects into subsets of classes, so that the elements of one class are as similar as possible to each other, i.e. are homogeneous within a class, and at the same time, the elements that

Table 1. Coherence of quantitative and qualitative values of reservoir characteristics and the results of calculation errors by linear and multiplicative models, respectively, within the iMdentified clusters.

	Perviousness, micron^2	Average oil saturation, fr. units	Net-to-gross ratio, fr. units	Oil viscosity/water viscosity	Net oil pay, m	Oil recovery rate (ORR), fr. units	Calculation of Oil recovery factor, fr. units		Error, %		Cluster
							l	m	l	m	
Average	0.878	0.678	0.384	446.985	5.018	0.119	0.129	0.126	11.840	11.307	**1**
Quant. value	High	High	Low	High	High	Low			Low	Low	
Average	0.212	0.521	0.411	492.783	18	0.136	0.126	0.120	30.069	29.104	**2**
Quant. value	Medum	Low	Medium	Medium	High	High			High	High	
Average	0.283	0.672	0.386		7.047	0.127	0.124	0.119	23.752	21.984	**3**
Quant. value	Medium	High	Medium	Low	Medium	Medium			Medium	Medium	
Average	0.137	0.648	0.585	482.481	15.947	0.111	0.114	0.109	24.003	21.738	**4**
Quant. Value	Low	High	High	Medium	High	Low			High	Medium	

l – linear model, m - multiplicative model.

make up different classes shall be significantly different. Traditional, or as they are called otherwise, "hard" methods of cluster analysis work with objects, the parameters of which are clearly specified. Currently, methods based on fuzzy clustering have been developed and continue to develop for clustering fuzzy objects. The contribution of famous scientists are dedicated to research in this area: J.C. Bezdek, Y.V. Pedrycz, L.A. Zadeh, I.B. Turksen, A.H. Averkin, R.A. Aliev, I.Z. Batyrshin et al. [9–11].

When conducting a research on the classification of hard-to-recover reserves, we have also previously performed clustering using the fuzzy cluster analysis algorithm [12]. In this case, in order to identify objects of the Karazhanbas field that are homogeneous in terms of geological and physical characteristics, we used the noted program of

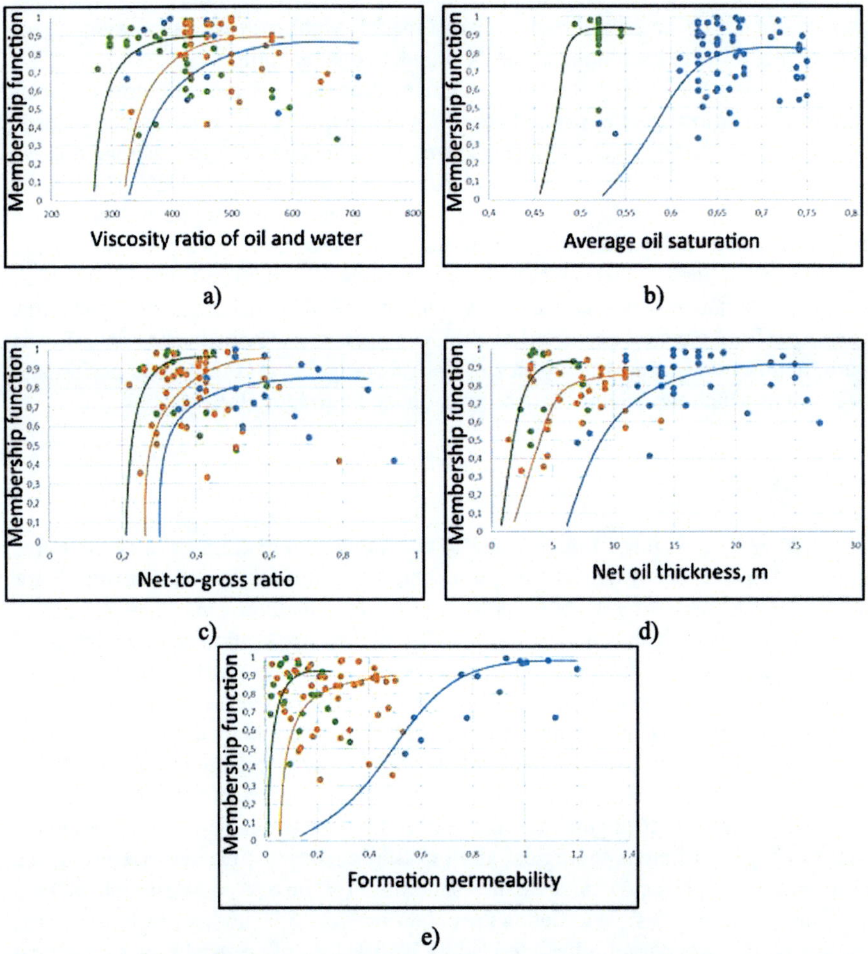

Fig. 1. Term-sets of different values: a) the ratio of the viscosities of oil and water; b) average oil saturation fraction units; c) net-to-gross ratio; d) formation permeability; e) oil-saturated formation thickness.

fuzzy cluster analysis. The conducted fuzzy cluster analysis revealed 4 groups of objects with similar estimates of the following geological characteristics: permeability, average oil saturation, net-to-gross ratio, ratio of oil viscosity to water viscosity, oil-saturated reservoir thickness (Table 1).

According to the data obtained, it can be concluded that the first cluster includes objects with high values of permeability, average oil saturation and viscosity ratio of oil and water, low values of the net-to-gross ratio and oil-saturated thickness of the reservoir, etc. Figure 1 (a–e) shows term sets of different levels of input variables.

From the above, we can conclude the following.

During the research, the results of which are presented in this work, a statistical analysis was carried out of the complex of geological and physical parameters, the results of laboratory analyses, information on the values of oil recovery for the considered field, as a result of which linear and multiplicative models were built for calculating the oil recovery factor in the conditions of the field under consideration.

Built models for the quantitative predictive assessment of the oil recovery factor depending on various geological and physical characteristics of the developed objects, as a result of comparative calculations, showed that the maximum and average relative errors in calculations using linear and multiplicative models is 55.002%; 27.16% & 49.16%; 24.9% respectively. As it can be seen, in some cases, high error values are obtained. In order to determine the conditions for the possible use of a particular model and at least a qualitative assessment of the influence of the considered characteristics on the oil recovery factor, it is necessary to establish classes that are homogeneous in terms of geological and physical characteristics of the objects of the Karazhanbas field. In this regard, in this work, using a fuzzy cluster analysis program, clustering was performed, as a result of which the original data array is divided into four classes.

5 Conclusion

The results obtained allowed us to come to the following conclusions: a comparative analysis of the previously built statistical models for estimating the oil recovery factor was carried out, the errors in their application for the conditions under consideration were calculated, which showed the impossibility of their application and confirmed the opinion about the necessity to build such models for the conditions under consideration; linear and multiplicative models have been built for predictive estimation of the oil recovery factor, the need to determine the area of their application from the point of view of individual conditions has been substantiated due to the ambiguity of the calculation results; with the help of fuzzy cluster analysis according to the data of the Karazhanbas field, four classes are distinguished, characterized by certain values of the parameters corresponding to different geological and physical conditions; classes of congruence of geological and physical characteristics and errors for the two obtained models were established. It made possible to define the conditions for applicability of these models, specifically: the first cluster, characterized by high values of permeability, average oil saturation and viscosity ratio of oil and water, low values of the net-to-gross ratio and oil-saturated thickness of the reservoir, the calculation errors for which are estimated as low; the multiplicative model can also be applied to the third and fourth clusters;

however, these models are inapplicable for the second cluster due to high error values; for the second cluster, the assessment can be made according to the qualitative agreement obtained as a result of fuzzy cluster analysis.

References

1. Davarpanah, A., Nassabeh, M.M.: Recommendations for optimizing the efficiency of polymer flooding techniques in production operation of an oilfield. Electronic J. Biol. **13**(3), 1–2 (2017)
2. White, J.U., Goddard, J.E., Philips, A.M.: Use of polymers to control water production in oil wells. J. Pet. Technol. **25**(02), 143–150 (1973). https://doi.org/10.2118/3672-PA
3. Abasov, M.T., Jafarova, N.M., Strekov, A.S., et al.: Prediction of the duration of limiting water inflows with polymer solutions in production wells. News Natl. Acad. Sci. Azerbaijan Earth Sci. **2**, 97–102 (2001). (in Russian)
4. Ghilayev, G.G.: Control of technological processes for the intensification of oil production. Oil Ind. **10**, 74–77 (2004). (in Russian)
5. Bulgakov, S.A.: A neural network technique for assessing the final oil recovery factor (on the example of fields in the Samara region). Electron. Sci. J. "Oil Gas Bus." **1**, 59–72 (2013). (in Russian)
6. Korolev, M.I.: Justification of the technology for extracting residual oil from heterogeneous terrigenous reservoirs using microemulsion compositions. Ph.D. thesis, p. 127 (2018). (in Russian)
7. Lyugay, D.V., Minakov, I.I., Burakova, S.V.: Results of the analysis of the effectiveness of the application of methods for calculating the coefficients of oil recovery from oil rims of oil and gas condensate deposits to the conditions of the Chayandinskoye field. Sci. Tech. Collect. "Vesti gazovoy nauki" **2(26)**, 10–18 (2016). (in Russian)
8. Abasov, M.T., Strekov, A.S., Efendiyev, G.M.: Improving the Efficiency of Limiting Water Inflows in Oil Wells. "Nafta-Press" Publishing House, Baku (2009). (in Russian)
9. Bezdek, J.C., Ehrlich, R., Full, F.: FCM: the fuzzy c-means clustering algorithm. Comput. Geosci. **10**(2–3), 191–203 (1984). https://doi.org/10.1016/0098-3004(84)90020-7
10. Aliev, R.A., Guirimov, B.G.: Type-2 Fuzzy Neural Networks and Their Applications. Springer, Switzerland (2014). https://doi.org/10.1007/978-3-319-09072-6
11. Turksen, I.B.: Full Type 2 to Type n Fuzzy System Models. In: Seventh International Conference on Soft Computing, Computing with Words and Perceptions in System Analysis, Decision and Control, Turkey, Izmir, 2013, p. 21 (2013)
12. Akhmetov, D.A., Efendiyev, G.M., Karazhanova, M.K., Koylibaev, B.N.: Classification of hard-to-recover hydrocarbon reserves of Kazakhstan with the use of fuzzy cluster-analysis. In: Aliev, R.A., Kacprzyk, J., Pedrycz, W., Jamshidi, M., Sadikoglu, F.M. (eds.) ICAFS 2018. AISC, vol. 896, pp. 865–872. Springer, Cham (2019). https://doi.org/10.1007/978-3-030-04164-9_114

Using Fuzzy Linear Regressions by the Estimation of Expected Changes in the Oil and Gas Saturation's Capacity at Great Depths

Adalat B. Hasanov(✉) and Mubarize S. Ibrahimli

Azerbaijan State Oil and Industry University, Azadlig Avenue, 20, Baku, Azerbaijan

Abstract. Identification of productive formations within oil and gas deposits is based mainly on the assessment of reservoir properties of rocks, the values of which usually determine the nature of the pore space and fluid saturation. Typically, reservoir property variability is predicted using 2D and 3D petrophysical models that reflects of well logging and laboratory core studies data. At the same time, it is known that the main oil reservoirs of Azerbaijani fields are located in terrigenous-sedimentary rocks, consisting of a structured matrix with a texture-organized pore space. In this reason classical petrophysical models based only on core analysis cannot be considered completely perfect.

Another serious problem of petrophysical modeling is the effect of scatter of experimental data values in the phase space of parameters, which is not explained only by calculation or measurement errors. The listed problems can be avoided by using fuzzy petrophysical models based on fractal concepts of the multimodal distribution of the main parameters. From the point of view of the presentation of experimental data, such models are the most optimal both in terms of polygonal scattering and in terms of phase space parameters.

Keywords: Petrophysical models · Fuzzy logic · Productive formations · Deposits · Properties of rocks

1 Introduction

One of the basic concepts, actively discussed in recent years in the scientific literature, is associated with the formation of oil and gas deposits due to the migration and accumulation of hydrocarbons in the upper layers of the earth's crust. According to this concept, the determining condition in this process is the effective volume and structure of the pore space within reservoirs. It is assumed that in general, the natural productivity potential of terrigenous reservoirs is largely determined by their intergranular porosity and the nature of grain packing. However, in addition to this, the shape of the grains, and most importantly, the ratio of the content and distribution of grains of different sizes in the volume of the rock also have a significant effect on the effective porosity of the reservoirs. In support of this, there exist several published attempts of simulations of the total effect of influencing factors on the multimodal distribution of intergranular porosity. The description of such attempts is well illustrated by a number of studies [1, 2],

R. A. Aliev et al. (Eds.): ICSCCW 2021, LNNS 362, pp. 340–346, 2022.
https://doi.org/10.1007/978-3-030-92127-9_47

whose authors also used parameters averaging the granulometric composition of rocks to establish the dependence of porosity on mechanical compaction. These parameters were used for different levels of rocks grain size (Md), as well as some coefficients reflecting the sorting of sediments (Ksort, Hr, plus the maximum content of the remaining fractions Mf). The clarification of pair correlation dependencies between fractions and parameters averaging the particle size distribution and reflecting the sorting of rocks is given in Table 1. As follows from these data, the average grain size, on the one hand, and on other hand, the sorting of rock grains are independent and not related with any functional dependencies. At the same time, there is a close correlation between some defined indicators (for example, Md).

From the above mentioned, it follows that forecasting the reservoir properties of rocks is a modern and very urgent task. This task, on the one hand, is characterized by the complexity of the relationship between individual properties of rocks, and, on the other hand, by the uncertainty of the relevant information.

At the same time, it is known that the main oil reservoirs of deposits in Azerbaijan are formed in terrigenous-sedimentary rocks, consisting of a structured matrix with a texture-organized pore space. Such a structured matrix is composed of grains of minerals of different sizes and looks like a chaotic system. Taking this into account, typical natural reservoirs are characterized by a set of parameters, such as tortuosity, shape of pore channels, thickness of the layer of bound water and other factors on the core analysis and cannot be considered completely perfect. In particular, changes in the content of individual fractions depending on the depth and the calculation of the pair correlation coefficient between the fractions and the parameters, average particle size distribution and sorting of rocks show that they are independent and not related functions. Moreover, the influence of individual fractions, and most importantly, their ratio on the value of intergranular porosity is uneven.

In this regard, a more effective solution to these problems was considered by many researchers from the standpoint of using methods of fuzzy logic and soft computing. So, in [3], the authors consider a multistage forecast of five geological parameters. However, the approach used in [3, 4] turned out to be ineffective for solving forecasting problems based on extrapolation. From the point of view of presenting experimental data in the form of polygonal scattering and phase space parameters, fuzzy petrophysical models were discussed in [3]. Here authors analyzed the shale porosity model in SCB. They compiled this model using the method of correlation and regression analysis. A possible forecast of petrophysical parameters based on actual logging data is mentioned in [4]. In doing so, the authors use the artificial intelligence method, which is devoted to predicting the PVT parameters of oil and gas reservoirs in the fields of the Middle East. The researchers used a combination of a 2nd type fuzzy logic system. The 2nd type fuzzy logic system's ability to handle uncertainty made the resulting model more robust. In [5], the application of fuzzy logic in petrophysics was discussed, in which the basic concepts underlying lithophiles are calculated using fuzzy logic methods. In this case, the uncertainty of the parameters is determined by the heterogeneity of the studied geological objects and is estimated in quantitative indicators of reliability [6]. Some authors propose to predict the reliability of graphic images based on the relationships between the parameters [7]. According to the proposed method, the forecast is performed

Table 1. Paired correlation coefficients between fractions and parameters that reflect average grain size distribution and sorting of rocks (title of the parameters, see the text).

Parameters		Grain size, мм						Indicators					
		F1	F2	F3	F4	F5	F6	Md	ВЗП	Ксорт	So	Hr	Mf
		1–0,5	0,5–0,25	0,25–0,1	0,1–0,05	0,05–0,01	< 0,01						
Indicators	Mf	−0,07	−0,17	−0,32	−0,67	−0,27	−0,85	−0,5	0,56	0,22	0,27	-0,93	
	Hr	0,15	0,27	0,47	0,71	0,03	0,86	0,58	0,7	−0,7	−0,24		
	So	0,07	0,06	−0,48	−0,46	0,09	0,61	−0,73	−0,54	0,97			
	Ксорт	0,08	0,07	−0,35	−0,38	0,007	0,5	−0,61	−0,41				
	ВЗП	0,24	0,3	0,91	0,51	−0,39	−0,81	0,93					
	Md	−0,007	0,08	0,9	0,53	−0,35	−0,83						
Grain size, мм	(F6) < 0,01	−0,03	−0,12	−0,62	−0,75	−0,05							
	(F5) 0,05–0,01	0,008	−0,04	−0,49	−0,22								
	(F4) 0,1–0,05	−0,04	0,02	0,25									
	(F3) 0,25–0,1	−0,009	0,09										
	(F2) 0,5–0,25	0,57											
	(F1) 1–0,5												

in the form of a fuzzy model of the distribution of possible values for each spatial point [4]. The disadvantage of this method is the need to store and use large amounts of information. However, it is balanced by a high degree of visibility and the possibility of achieving reliability for each parameter of the forecast up to the expected values [7, 8].

In this article, we also apply a fuzzy regression model [4] to predict reservoir rock properties. This type of model can be realized both on the basis of extrapolation and prediction based on interpolation. Differential evolution optimization was used to build this model. Main objectives: to study the modern scientific situation in vector time sequences. The study purpose is to choose appropriate forecasting vector's mathematical model for sediment's parameters of the South Caspian Basin based on a review of the results of well known published researches in this sphere. Conducted review showed that a qualitative forecasting of oil and gas deposits is impossible by using of models of artificial neural networks and fuzzy sets for vector time sequences. This task is better solved with method of data clustering (fuzzy Sugeno models) with a selection of affiliated functions (membership functions).

Within paper describes the results of the forecasts for hydrocarbon deposits in the South Caspian basin realized by the above mentioned method.

2 Statement of the Problem and Solution Method

The issue, considered as an example, is to predict the properties of rocks in oil and gas deposit [5]. For this direct purpose, a predictive model based on fuzzy linear regression is used to describe the interdependence of reservoirs properties and the natural uncertainty of information. The fuzzy regression model is considered as an equation:

$$y_k(x_1, x_2, ..., x_N) = \tilde{a}_{k0} + \sum\nolimits_{i=1}^{N} \tilde{a}_{ki} x_i \tag{1}$$

$x_i,\ i = \overline{1, N}$ - values of the parameters of the properties of rocks (input data), $y_j, j = \overline{1, N}$ - the predicted values of the parameters of the properties of rocks (the output of the regression model), and $\tilde{a}_{k0}$, $\tilde{a}_{ki}$ - coefficients of the regression model, described by triangular fuzzy numbers (TFNs), $k = \overline{1, K},\ i = \overline{0, N}$.

The construction of model (1) consists of determination of values of fuzzy coefficients $\tilde{a}_{k0}, \tilde{a}_{ki}$ which minimize the error:

$$RMSE = \sqrt{\frac{\sum_{h=1}^{n} (y_i(h) - y_i^*(h))^2}{n}} \to min \tag{2}$$

$$\tilde{a}_{k0}, \tilde{a}_{ki} \in A,\ \ k \in 1, ..., K.$$

here $y_i^*(h)$ - are real values of rock quality indicators at depth h; A - is the search space. The built-in model (1) allows estimating y_k values of rock quality indicators using the values of previous depth parameters.

Verification of the described method was carried out on real quality indicators of the reservoirs of the well-known group of deposits of the Baku Archipelago in Azerbaijan (see Table 2).

Table 2. Reservoirs' quality indicators

h (depth)	z_1 (relative clay)	z_2 (permeability)	z_3 (density)
4062,5	0,75	9,6	2,39
4599,5	0,93	23,3	2,65
4902,5	0,85	0,3	2,51
4902,5	0,69	53,3	2,56
4943	0,89	4,3	2,63
4963,5	0,68	61,5	2,48
5089,5	0,69	107,3	2,58
5089,5	0,7	107,28	2,4
5248,5	0,22	98,6	2,02
5256	0,84	8,4	2,3

The data given in Table 1 were preliminary interpolated in depth with the same step within interval 4300–5200 m, and then recalculated according to the models (1) described above:

$$\tilde{y}_1(x_1, x_2, ..., x_6) = \tilde{a}_{10} + \sum_{i=1}^{6} \tilde{a}_{1i}x_i,$$

$$\tilde{y}_2(x_1, x_2, ..., x_6) = \tilde{a}_{20} + \sum_{i=1}^{6} \tilde{a}_{2i}x_i,$$

$$\tilde{y}_3(x_1, x_2, ..., x_6) = \tilde{a}_{30} + \sum_{i=1}^{6} \tilde{a}_{3i}x_i.$$

Here $\tilde{y}_k(x_1, x_2, ..., x_6) = z_k(h), x_1 = z_1(h-1), x_2 = z_1(h-2), x_3 = z_2(h-1)$ $x_4 = z_2(h-2), x_5 = z_3(h-1), x_6 = z_3(h-2)$, $\tilde{a}_{k0}, \tilde{a}_{ki}$ the fuzzy coefficients in the form of triangles: h - are the depth, k = 1, 2, …, 3.

To calculate fuzzy coefficients, optimization algorithms of differential sets were used, with parameters: mutation rates $F = 0.8$; connection probabilities $CR = 0.7$ and population size $PN = 80$. Optimized RMSE is approximately equal 0.0006.

Particularly, values of fuzzy coefficients were calculated as:

$$\tilde{a}_{10} = (-11.34, -12.6, -13.86), \tilde{a}_{11} = (-2.07, -2.3, -2.53),$$
$$\tilde{a}_{12} = (0.495, 0.55, 0.605), \tilde{a}_{13} = (2.097, 2.33, 2.563),$$
$$\tilde{a}_{14} = (-0.045, -0.05, -0.055), \ \tilde{a}_{15} = (-0.441, -0.49, -0.539),$$
$$\tilde{a}_{16} = (-0.072, -0.08, -0.088);$$
$$\tilde{a}_{20} = (0.261, 0.29, 0.319), \ \tilde{a}_{21} = (0.936, 1.04, 1.144),$$
$$\tilde{a}_{22} = (1.062, 1.18, 1.298), \ \tilde{a}_{23} = (-0.495, -0.55, -0.605),$$
$$\tilde{a}_{24} = (-3.708, -4.12, -4.532), \ \tilde{a}_{25} = (-0.171, -0.19, -0.21),$$
$$\tilde{a}_{26} = (-0.093, -0.104, -0.11).$$

Using the established coefficients and based on the initial data, it is possible to predict the quality indicators of reservoirs (Table 3) which exist at great depths (from 5300 to 5500 m).

Table 3. Predicted reservoir quality indicators

h	$\tilde{z}_1$	$\tilde{z}_2$	$\tilde{z}_3$
5300	(0.164, 0.189, 0.214)	(100.18,100.21,100.23)	(2.29, 2.32, 2.34)
5400	(3.12, 3.46, 3.81)	(86.12, 95.69, 105.27)	(2.8, 3.27, 3.66)
5500	(0.147, 0.163, 0.18)	(76.21, 84.68, 93.16)	(0.353, 0.523, 0.635)

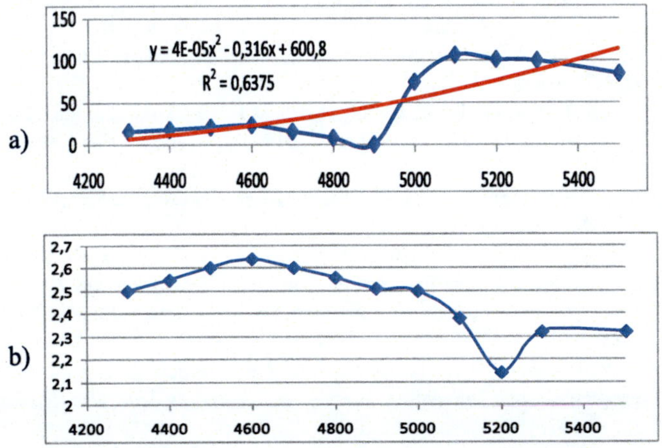

Fig. 1. Real and predicted quality indicators of reservoirs a) fluid permeability, b) matrix density

The consistency of predicted quality indicators of reservoirs with the initial ones is shown in Fig. 1 from the consideration of which it follows that in the cross section of the studied deposits at depths above 4900 m a decrease can be predicted in the relative clay content and density of reservoirs, while fluid permeability increases.

3 Conclusion

The paper describes the use of a fuzzy regression model to predict reservoir quality indicators using real data which belongs to a well-known group of oil deposits of the Baku archipelago in Azerbaijan. The composed model was first used when extrapolating the initial data, and then when calculating the predicted values of reservoir quality indicators at great depths.

References

1. Levchuk, M.A., Bukreeva, G.F.: On the sorting of terrigenous sediments and values averaging the particle size distribution. Geocyclicity – Novosibirsk, IGiG, pp. 17–25 (1976). (in Russian)
2. Romanovsky, S.I.: Application of information theory to solve some problems of lithology, pp. 75–92. Math. Methods Geol., VSEGEI (1968). (in Russian)
3. Aliyarov, R.Y., Ramazanov, R.A.: Prediction of multivariable properties of reservoir rocks by using fuzzy clustering. In: Proceedings of 12th International Conference on Application of Fuzzy Systems and Soft Computing, ICAFS 2016, Vienna, Austria, pp. 434–440 (2016). https://doi.org/10.1016/j.procs.2016.09.423
4. Aliyarov, R.Y., Hasanov, A.B., Ibrahimli, M.S., Ismayilova, Z.E., Jabiyeva, A.J.: Forecasting oil and gas reservoirs properties using of fuzzy-logic based methods. In: Aliev, R.A., Kacprzyk, J., Pedrycz, W., Jamshidi, M.O., Sadikoglu, F.M. (eds.) ICAFS 2018. AISC, vol. 896, pp. 769–773. Springer, Cham (2019). https://doi.org/10.1007/978-3-030-04164-9_101
5. Filev, D., et al. (eds.): Intelligent Systems 2014. AISC, vol. 323. Springer, Cham (2015). https://doi.org/10.1007/978-3-319-11310-4
6. Qu, Z., Lin, J. (eds.): Proceedings of the International Field Exploration and Development Conference 2017. SSGG, Springer, Singapore (2019). https://doi.org/10.1007/978-981-10-7560-5
7. Latyshova, M.G.: Practical guidance on the interpretation of geophysical research of wells (in Russian), p. 377. "Nedra", Moscow (1991)
8. Aliyarov, R.Y., Hasanov, A.B.: Forecasting of qualitative characteristics of oil reservoirs. In: Materials of the Republican Scientific-Practical Conference Devoted to the 95th Anniversary of H. Aliyev: Unity of Science, Education and Production at the Present Stage of Development, pp. 22–31 (2018)
9. Wendelstein, B.Y., Rezvanov, R.A.: Geophysical methods for determining the parameters of oil and gas reservoirs: when calculating reserves and designing field development, p. 145. "Nedra", Moscow (1978). (in Russian)

Fuzzy Control of Mechanical Ventilation System

Gorkem Say(✉), Nurullah Akkaya, Ersin Aytac, Sanan Abizada, Tolga Yirtici, Kemal Ruso, Irfan Gunsel, Murat Tuzunkan, and Rahib H. Abiyev

Applied Artificial Intelligence Research Centre, Robotics Lab, Near East University, Mersin-10, P.O. Box 670, Lefkosa, North Cyprus, Turkey
{gorkem.say,nurullah.akkaya,ersin.aytac,sanan.abizada, tolga.yirtici,igunsel,murat.tuzunkan,rahib.abiyev}@neu.edu.tr, kemal.ruso@robotics.neu.edu.tr

Abstract. COVID-19 virus affects the respiratory system of the patient and can obstruct the breathing process and lead to the death of the patient. The development of ventilation systems was in primary concern in order to help the patients who suffered from COVID-19. In this paper, the electronic and software design of the mechanical ventilation system, its control algorithm is considered. The electronic design of the system is implemented on STM32 Arm. The breathing system is based on AmbuBag compressed by the actuator. The fuzzy logic control system is proposed for the control of AmbuBag pressure. The stages of development of mechanical, electronic and software designs of the ventilation system are presented. Simulation results are presented for the verification of the developed control system. The obtained results indicate the suitability of using the designed automatic mechanical ventilation system (AMVS) in the help of COVID-19 patients to be able to breathe.

Keywords: Mechanical ventilation system · COVID-19 · Fuzzy control · Breathing

1 Introduction

COVID-19 were becoming a global world pandemic that grows exponentially. According to the World Health Organization [1], the number of people infected by COVID-19 has reached more than 155 billion. COVID-19 is a very fast spread virus in the world that in a couple of months spread worldwide [1, 2]. People having age of 60 and over carries a higher risk of developing the series illness. In most hospitals, the number of critical patients exceeds the number of ventilation systems. For this reason, the fast design of ventilation systems is acquired more importance and it is a vital task for many health centers.

The mechanical ventilation systems (MVS) control the airflow volume and pressure of oxygen given to the patient. The system allows to help ventilation and breathing process of the patients. The ventilation system provides a flow of oxygen through the ventilation system. The designed artificial MVS (AMVS) can support patients in a short

R. A. Aliev et al. (Eds.): ICSCCW 2021, LNNS 362, pp. 347–354, 2022.
https://doi.org/10.1007/978-3-030-92127-9_48

time and long time needs. The paper [3] designed MVS where AmbuBag is compressed by means of two paddles that are actuated electrical motors.

The fast design of AMVS took place in different countries, because of the vital necessity AMVS in hospitals during a pandemic. Infection of a huge number of patients with coronavirus, increasing deaths need to apply a more cordial solution of the health problem. As it was noted in [4] the patients' morality undergoing ventilation during critical diseases is 31–37%. Several factors need to be evaluated for using the AMVS. These are the evaluation of oxygenation, ventilation, lung mechanisms and condition of the patient. The evaluation of the current and previous conditions of the patients are also important. Based on these parameters the AMVS should be optimized for the patient. A set of research studies presented the design of AMVS. The reference [5] developed a microcontroller-based mechanical ventilator. The AmbuBag is pressed by the arm and can work in several modes as child, pediatric and adult modes. The reference [6] designed a low-cost open-source mechanical ventilator. [7, 8] developed simple ventilators with the portable AmbuBag compression system. The breathing rate of the ventilator changes in the range of 5–40 breaths/minute. In [9] the ventilator is controlled by pressure-controlled ventilation. The authors have designed an adaptive neuro-fuzzy control of a pneumatic ventilator. As it has been seen from previous researches the designing an automatic compressing system for AmbuBag is a key point in the design of a mechanical ventilation system. In this paper, we are designing the AMVS using two electrical motors. The motors will actuate the arms for pressing AmbuBag and providing air gas for breathing of patient lungs. For this purpose in the paper fuzzy logic inference is proposed for the design of the control system. The design of such architecture will provide more accurate control of the respiratory system of a human.

2 Mechanical and Hardware Design of Ventilator

The AMVS is developed and manufactured by the researchers of the Robotics lab of Applied Artificial Intelligence Centre, Near East University with the appearance of coronavirus. Figure 1 (a,b) shows the designed AMVS. The AmbuBag of the ventilator is compressed by the two arms, from the left and right sides. The arms are connected to the electrical motors that are controlled by the control circuit STM32 Arm microcontroller. There are DAC that transmits the output signals of control circuits to the electrical motors. The circuit contains two sensors for measuring pressure and airflow. The sensors data are displayed on the screen that is mounted set ventilation system (Fig. 1a). The operator doctor can set reference pressure signal and other parameters such as volume, airflow from the screen of AMVS.

The structure of the system is given in Fig. 2. The values of control parameters pressure and airflow are compared with the setpoint signals that are set by the doctor. According to this error signal, the control circuit determines the control signal and send it to step motors through drivers. The motors are moving the left and right arms that compress the Ambubag. By pressuring Ambubag the briefing of the patient is organized.

The electronic section consists of different subsections such as the mainboard, motor driver, sensors and Screen. The main working principle is to set the required specs of the Ambubag. RPM is related to how many beats the user wants to set and the

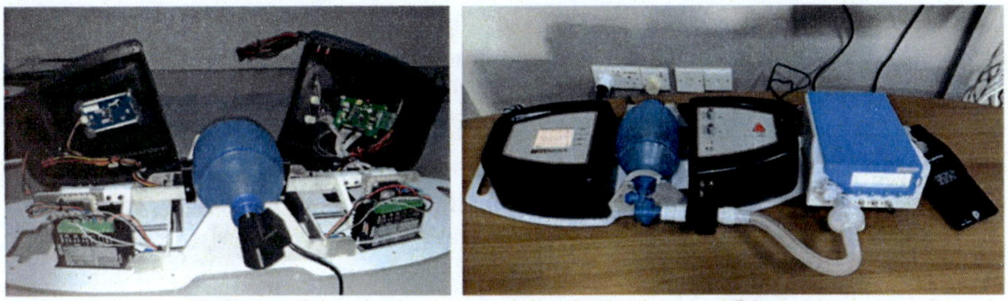

(a) (b)

Fig. 1. The designed AMVS. (a) electronic parts, (b) AMVS with artificial lung

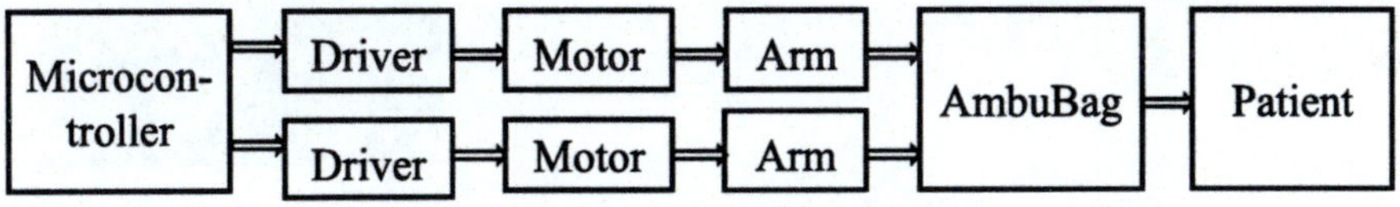

Fig. 2. Structure of AMVS.

pressure value is also set by the user depending on the patient. The pressure value is detected with the BMP180 pressure sensor and feedback to the MCU. As the main MCU, STM32F103CBT6 has been selected due to different reasons. The MCU has an ARM 32-bit Cortex M3 CPU core. PWM timers, I2C, SPI, USART, CAN interface and USB 2.0 interface communications protocols, analogue to digital converter channels are specified in [10, 11]. The bootloader processes are done by using Serial Wire Debug (SWD). As a bootloader, the generic boot20 pc13 file has been flashed to MCU by using ST-LINK V2 debugger/programmer. The bootloader allows the user to program a custom STM32F103 circuit board with Arduino IDE via USB. STMs AN2606 application note is the guide for system memory boot mode.

Lithium-Ion based battery packs are designed to supply 16.8 Vmax. Furthermore, BMS is used for reliable charging. The battery pack powers the mainboard and stepper motor driver directly. The 3.3V regulated voltage has powered the STM32F103CB, sensors and screen. LM2596 power converter is designed to convert 16.8 Vmax to 3.3 V. The output voltage can be regulated according to feedback resistors. LM2596 schematics can be seen in Fig. 3. Here $Vout = Vref(1 + R2R1)$.

The mainboard controls the stepper driver, reads sensor values and drives the LCD screen. TB6600 stepper motor drivers have been used to drive Nema 17 Stepper motors. There are three different control inputs. The EN enables the driver, and the DIR controls the direction of the motor due to HIGH or LOW situation of the pin. The PUL is the pulse pin that controls the speed of the motor by changing PWM. The stepper motor positions should be known because the stepper driver does not give any feedback position of the motor. So, the potentiometers are placed to the bottom of the stepper motors and driven by stepper motors with specific spur gear to give feedback of the stepper motor positions to the MCU. One of the other sensors is the MAX30100 which measures pulse oximeter and heart rate. Both sensor values are shown on the LCD screen. Sensors use the I2C communication protocol and both sensors have pulled to 3.3 V via 4.7 Kohms. The UART protocol has been applied for LCD screen communication and supplied with 3.3 V. In the mainboard, 100 nf, 10 uf and 1 uf capacitors are placed and used as decoupling capacitors which are placed between the ground and 3.3 V supply in the mainboard. Moreover, the ferrite bead is added to suppress noises on an analog supply voltage of MCU. Also, to specify each beat of the Ambu, the blink sound is given by the buzzer. The custom mainboard can be seen in Fig. 4.

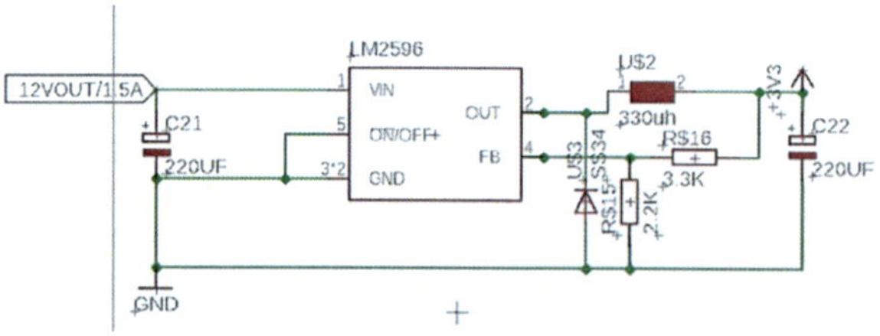

Fig. 3. LM2596 schematics.

Fig. 4. Hardware design.

3 Control System Design

The basic aim of the designed control system is the control of required pressure in the lungs of patients using AmbuBag compressed by the actuator. The control of pressure in AmbuBag is provided by gradual compressing of AmbuBag. After the very subtle release of AmbuBag, its pressure decreases rapidly. The pressure decreased more than two times. This lower value of pressure should also be controlled for the next phase of breathing when the actuator retracts and extends. Any oscillating course of pressure is not desirable. As it can be seen, the control includes two modes. In the first mode, the pressure should be increased to the value of reference pressure. After achieving the desired pressure, the current pressure should be maintained to the end of inspiration time. When pressure is decreased to the lower value it should be maintained to the end of expiration time, that is the next cycle of breathing. In the next mode, when the patient starts breathing according to the sensor measurements of positive flow, the ventilator provides tidal volume for the patient. In the next stage when a patient starts to exhale then according to the sensor measures of the negative flow ventilator is deactivating. During control, these two modes should be controlled by the controller.

The inspiration and expiration process is shown in Fig. 5. Here the blue line is the reference pressure signal, the green line is the current value of the pressure.

The control system of AMVS is designed using an STM32 Arm microcontroller. The values of control parameters pressure and airflow are compared with the set-point signals that are set by the doctor. In the paper, we presented a fuzzy control system to control the pressure of the ventilator. The fuzzy controller is designed using knowledge of expert-doctor that was formulated using if-then rules [12–14]. These rules are used for controlling the pressure of AmbuBag. According to the determined error signals $e(k) = G(k) - x(k)$ and change-in-error signal $de(k) = e(k) - e(k-1)$ the fuzzy controller, based on the If-Then rule base, determines the output signal of the controller. This signal is given to the input of the electrical step motor through drivers.

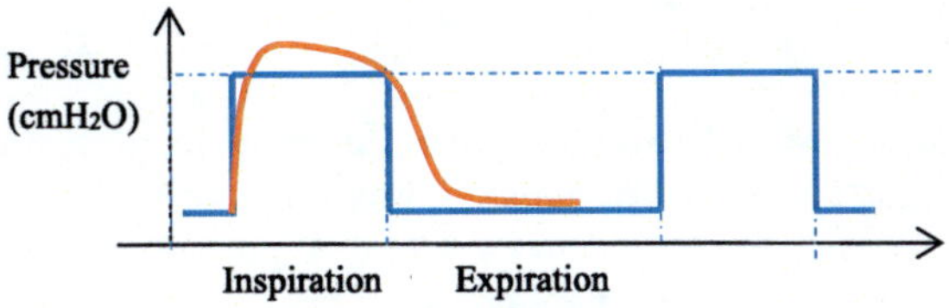

Fig. 5. Breathing cycle

Fuzzy rule-based systems are designed using error and change-in-error input signals. The membership functions of linguistic values of the input and output variables used for the construction of rule bases are presented in triangular form. Each input and output variables have seven linguistic values. Using these linguistic terms the If-Then rule base is constructed [15–20]. The triangle type membership functions are taken to represent linguistic values.

Table 1 describes the if-then rule base used for controlling the value of pressure. Error and change-in error signals have five linguistic terms. These signals are used to determine the control signal used to control the pressure in AmbuBag. Here NL – negative large, NM-negative medium, NS – negative small, Z – zero, PS – positive small, PM-positive medium, PL – positive large. Using these linguistic values the construction of a fuzzy rule base is performed. Based on designed if-then rules and input signals, a fuzzy inference engine determines the output signal. A compositional rule of inference is used for calculating the output signal of the controller.

$$\tilde{e}, \tilde{e}' \to \tilde{u}, \ \mu(u) = \max_{X1,X2} \min_{R}\{\mu_{X1}(e), \mu_{X2}(e'), \mu_R(e, e', u)\} \quad (1)$$

Using the center of average the defuzzification is performed in order to find the output signal and control the breathing system.

Table 1. The rule base of the controller

Pressure u		Change-in-error e'						
		NL	NS	NS	Z	PS	PM	PL
Error e	NL	PL	PL	PL	PL	PM	PS	Z
	NM	PL	PL	PL	PM	PS	Z	NS
	NS	PL	PL	PM	PS	Z	NS	NM
	Z	PL	PM	PS	Z	NS	NM	NL
	PS	PM	PS	Z	NS	NM	NL	NL
	PM	PS	Z	NS	NM	NL	NL	NL
	PL	Z	NS	NM	NL	NL	NL	NL

4 Simulations

Simulations of the designed AMVS system were carried out in our research laboratory and tested using laboratory tools. The problem was the control of air pressure in the lungs of the patients. The system structure for control of pressure is given in Fig. 6. We used a fuzzy controller for control of the step motor. By controlling the step motor we can control AmbuBag pressure that controls the pressure in the lungs.

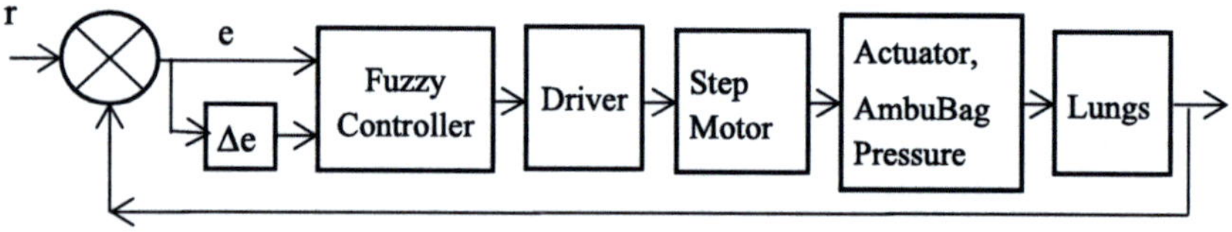

Fig. 6. Control system structure.

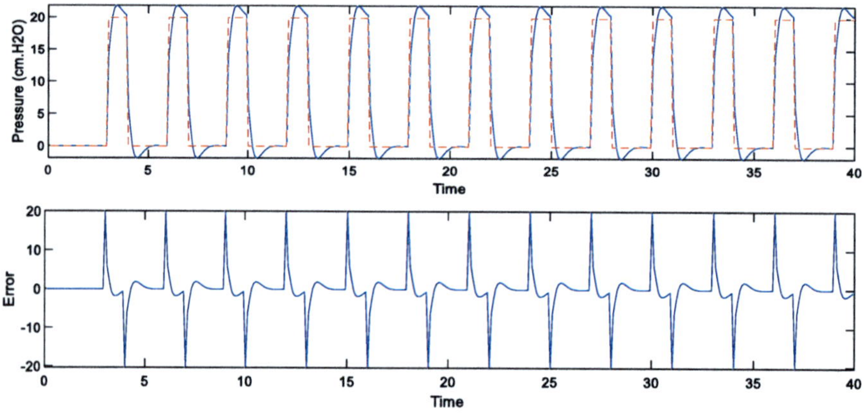

Fig. 7. Simulation results with reference pressure 20 cm.H_2O: Plot of pressure and Error plot

We used the fuzzy controller presented in the previous section. The laboratory design of this control system is presented in Fig. 1(b). In the figure, AMVS is connected to the laboratory tools imitating the functions of physical lungs which are shown on the right side of the figure. When the system is working, the desired signals, the high value of pressure are set using buttons shown on a ventilator. According to set-point pressure, the value of the control signal is generated by a fuzzy controller. This signal controls the pressure of the AmbuBag. The AmbuBag through the pipe is connected to the artificial lung. The increase of pressure in Ambubag causes the increase of the value of pressure in the lungs of the patient. For feedback connection and measuring the current value of pressure sensors are used. Through sensors, the current value of pressure in AmbuBag and lungs are measured and sent to the microcontroller of AMVS. The microcontroller

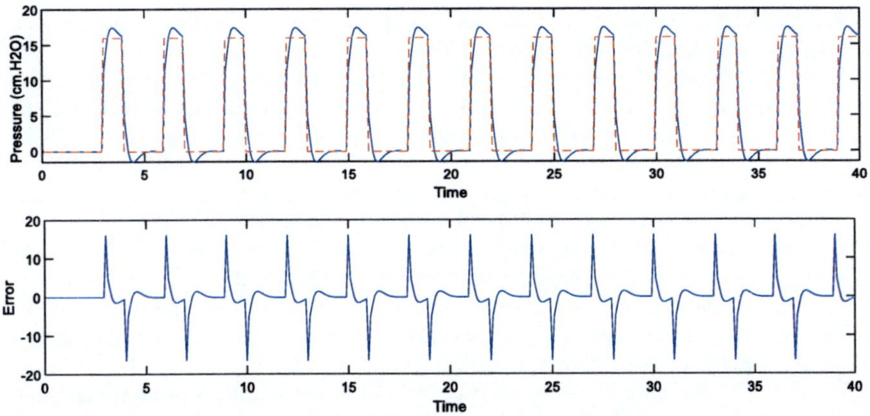

Fig. 8. Simulation results with reference pressure 16 cm.H_2O: Plot of pressure and Error plot

controls the whole breathing process of patients. All reference and current values of pressure signal are displayed on the window of AMVS (left part) and on the window of the lung oscillator.

The fuzzy controller is designed for pressure control in the lungs. We have obtained response characteristics of the fuzzy control system of the lung. Figure 7 depicts the plot of the current and error signals obtained for pressure when the reference signal is equal to 20 cm.H_2O. In the next simulation, the reference signal is set equal to 16 cm.H_2O. Figure 8 depicts the plot of current (a) and error (b) signals. The obtained simulation results demonstrate the suitability of using a fuzzy control system to control the pressure in the human lung.

5 Conclusions

The paper presents the design of an automatic mechanical ventilation system for the for patients suffered from COVID-19. The mechanical, electronics and software design of the AMVS has been implemented. The control system has been developed using a fuzzy set theory. The design of electronic system is carried out using STM32 Arm. The design of a fuzzy controller for controlling the pressure of AmbuBag has been presented. The fuzzy rule bases are developed for controllers. The inference in fuzzy controllers is carried out using the compositional rule of inference. The obtained results demonstrate the efficiency of using the fuzzy system in controlling the pressure in the Ambubag.

References

1. WHO Coronavirus Disease (COVID-19) Dashboard. https://covid19.who.int/
2. Li, Q., et al.: Early transmission dynamics in Wuhan, China, of novel coronavirus-infected pneumonia. Engl. J. Med. **3821**, 1199–1207 (2020). https://doi.org/10.1056/NEJMoa2001316
3. Gattinoni, L., et al.: Lung recruitment in patients with the acute respiratory distress syndrome. N. Engl. J. Med. **354**, 1775–1786 (2006)

4. Esteban, A., et al.: Evolution of mechanical ventilation in response to clinical research. Am. J. Respir. Crit. Care Med. **177**, 170–177 (2008)
5. Islam, R., Ahmad, M., Hossain, S., Islam, M.M., Ahmed, S.F.U.: Designing an electro-mechanical ventilator based on double CAM integration mechanism. In: Proceedings of the 2019 1st International Conference on Advances in Science, Dhaka, Bangladesh, 3–5 May 2019, pp. 1–6 (2019)
6. Acho, L., Vargas, A.N., Pujol-V´azquez, G.: Low-cost, open-source mechanical ventilator with pulmonary monitoring for COVID-19 patients. Actuators **9**(3), 84 (2020)
7. Castro-Camus, E., et al.: Simple ventilators for emergency use based on bag-valve pressing systems: lessons learned and future steps. Appl. Sci. **10**, 7229 (2020)
8. Petsiuk, A., Tanikella, N.G., Dertinger, S., Pringle, A., Oberloier, S., Pearce, J.M.: Partially RepRapable automated open source bag valve mask-based ventilator. HardwareX **8**, e00131 (2020). https://doi.org/10.1016/j.ohx.2020.e00131
9. Živcák, J., et al.: An adaptive neuro-fuzzy control of pneumatic mechanical ventilator. Actuators **10**(3), 51 (2021). https://doi.org/10.3390/act10030051
10. STM32F103. https://www.st.com/en/microcontrollers-microprocessors/stm32f103cb.html
11. STM32 microcontroller system memory boot mode. https://www.st.com/resource/en/application_note/cd00167594-stm32-microcontroller-system-memory-boot-mode-stmicroelectronics.pdf. Accessed 5 May 2021
12. Abiyev, R.H., Günsel, I., Akkaya, N., Aytac, E., Çağman, A., Abizada, S.: Fuzzy control of omnidirectional robot. In: Proceedings of 9th International Conference on Theory and Application of Soft Computing, Computing with Words and Perception, ICSCCW 2017, Book Series: Procedia Computer Science, vol. 120, pp. 608–616 (2017)
13. Abiyev, R.H., Günsel, I., Akkaya, N., Aytac, E., Çağman, A., Abizada, S.: Robot soccer control using behaviour trees and fuzzy logic. In: Proceedings of ICAFS 2016, Book Series: Procedia Computer Science, vol. 102, pp. 477–484 (2016). https://doi.org/10.1016/j.procs.2016.09.430
14. Abiyev, R.H., Akkaya, N., Gunsel, I.: Control of omnidirectional robot using z-number-based fuzzy system. IEEE Trans. Syst. Man Cybern. Syst. **49**(1), 238–252 (2019). https://doi.org/10.1109/TSMC.2018.2834728
15. Abiyev, R.H., Günsel, I., Akkaya, N.: Z-number based fuzzy system for control of omnidirectional robot. In: Aliev, R.A., Kacprzyk, J., Pedrycz, W., Jamshidi, M.O., Babanli, M.B., Sadikoglu, F.M. (eds.) ICSCCW 2019. AISC, vol. 1095, pp. 470–478. Springer, Cham (2020). https://doi.org/10.1007/978-3-030-35249-3_60
16. Abiyev, R.H.: A type-2 fuzzy wavelet neural network for time series prediction. In: García-Pedrajas, N., Herrera, F., Fyfe, C., Benítez, J.M., Ali, M. (eds.) IEA/AIE 2010. LNCS (LNAI), vol. 6098, pp. 518–527. Springer, Heidelberg (2010). https://doi.org/10.1007/978-3-642-13033-5_53
17. Abiyev, R., Abizada, S.: Type-2 fuzzy wavelet neural network for estimation energy performance of residential buildings. Soft. Comput. **25**(16), 11175–11190 (2021). https://doi.org/10.1007/s00500-021-05873-4
18. Abiyev, R., Arslan, M., Günsel, I., Çagman, A.: Robot pathfinding using vision based obstacle detection. In: 3rd IEEE International Conference on Cybernetics (CYBCONF), Book Series: IEEE International Conference on Cybernetics-CYBCONF, pp. 29–34, Exeter, 21–23 June 2017
19. Abiyev, R.H.: Credit rating using type-2 fuzzy neural networks. Math. Probl. Eng. **2014**, 460916 (2014). https://doi.org/10.1155/2014/460916
20. Abiyev, R.H., Akkaya, N., Aytac, E., Ibrahim, D.: Behaviour tree based control for effi-cient navigation of holonomic robots. Int. J. Robot. Autom. **29**(1), 44–57 (2014). https://doi.org/10.2316/Journal.206.2014.1.206-3788

Modeling of Relationship Between the Quality Indicators of Plastic Details

Djahid A. Kerimov(✉)

Azerbaijan State Oil and Industry University, Azadlig 34, Baku, Azerbaijan

Abstract. In this work, it is considered an analytical relationship between quality indicators and shrinking deformation, which is a function of the difference in temperature voltages. The connection between volumetric deformation and strength was determined by controlling the cooling process of polymer details in production conditions, knowing the change in thermic voltage difference during cooling. Let us note that quality indicators are characterized by imprecise information. In view of this, we discuss issues of fuzzy modeling of relationship between quality indicators and its influential factors.

Keywords: Plastic details · Quality indicators · Strength · Accuracy · Shrinking deformation · Temperature voltages · Fuzzy function · Fuzzy integral

1 Introduction

The main indicators of quality are strength and accuracy. But on the other hand, accuracy and strength depend on the amount of shrinking deformation of the plastic details. Thus, it is necessary to establish an analytical relationship directly between the quality index of strength and shrinking deformation, which is a function of the difference in temperature voltages. Currently, the difference in temperature voltages (residual voltage) is not controlled in the manufacturing conditions of plastic details due to the lack of reliable methods and their determination and control. However, analytically, it is possible to determine the amount of volumetric or linear deformation of finished details by governing and controlling them with other quality indicators [1–7].

The determination of the relationship between the volume strain and the strength is possible by controlling the process of cooling the polymer details under production conditions, knowing the change in the difference in thermal stress during cooling. Thus, volumetric deformation is a linear thermal shrinkage during cooling in three directions x, y, z and is a controllable quality criterion.

On the other hand, for the cooling process of plastic details, due to the low thermal conductivity of plastics, a gradient of the cooling rate over the cross section is characteristic. This leads to the appearance of thermo elastic voltages. The greater the temperatures difference between the central and surface layers, the greater the voltages to be expected. For each infinitesimal layer, the difference in the rates of free thermal and general shrinkage will be compensated by the elastic deformation of the material.

R. A. Aliev et al. (Eds.): ICSCCW 2021, LNNS 362, pp. 355–364, 2022.
https://doi.org/10.1007/978-3-030-92127-9_49

At the first moment, the cooling rate of the outer layers is more than that of the inner layers. Farther cooling results in a state where the cooling rate of the inner regions exceeds the cooling rate of the outer layer. As a result, a more intensive reduction in the volume of internal layers of details begins. The outer layers prevent this reduction. The resistance of some layers of plastic details to shrinkage of others causes the appearance of temperature voltages, which at high temperatures can cause the destruction of details if the magnitude of the resulting voltages exceeds the limit of strength of the material [2].

After cooling compressive tangential voltages are fixed in the outer layers of the plastic details and stretching voltages are fixed in the inner. The residual voltages will be the greater; the greater the temperature difference over the section, the higher the cooling rate, and the smaller part of the voltages will have time to relax [1]. Therefore, we will consider in more detail the regularities of the distribution of the temperature stress difference in cylindrical plastic details during their cooling down. One of the main issues of the mentioned modeling problem is to account for uncertainty. In this paper we consider issues of fuzzy modeling of the mentioned relationships. This involves consideration of fuzzy functions, fuzzy derivatives, fuzzy integrals, fuzzy differential equations (FDEs) [8–20]. Particularly, a fuzzy heat equation is considered in [8].

2 Study of the Temperature Distribution of Plastic Details in Production Conditions

In this case, consider the temperature distribution of plastic parts in the production conditions under possibilistic uncertainty. Suppose that the axis of a uniformly cooled cylinder coincides with the axis $0z$ and the temperature distribution does not depend on the angle Q. Then the temperature difference ΔT of the isotropic cylindrical body (plastic part) depends on r(cylinder height and radius), z and t. Due to fuzzy relevant information intrinsic to heat processes, we consider ΔT and r as variables whose values are convex and compact fuzzy sets [12]. Then, dependence of ΔT on r will be determined by solving the fuzzy thermal conductivity equation in cylindrical coordinates (Fig. 1):

$$\frac{\partial \Delta T}{\partial t} = a\left(\frac{1}{2}\frac{\partial \Delta T}{\partial r} + \frac{\partial^2 \Delta T}{\partial r^2} + \frac{\partial^2 \Delta T}{\partial z^2}\right) = a\nabla \Delta T, \quad (1)$$

with initial conditions

$$\Delta T(r, z, 0) = T(r, z, 0) - T_0 = T_n - T_0 = \Delta T_{in}, \quad (2)$$

and with the following boundary conditions

$$\begin{cases} \frac{\partial \Delta T(R_1, z, t)}{\partial r} - \frac{\alpha_1}{\lambda}\Delta T(R_1, z, t) = 0, \\ \frac{\partial \Delta T(R_2, z, t)}{\partial r} + \frac{\alpha_2}{\lambda}\Delta T(R_2, z, t) = 0, \\ \frac{\partial \Delta T(z, 0, t)}{\partial z} = 0, \\ \frac{\partial \Delta T(r, h, t)}{\partial z} + \frac{\alpha_3}{\lambda}\Delta T(r, h, t) = 0, \end{cases} \quad (3)$$

where a – thermal conductivity of the material; z and r – variables (cylinder height and radius); t – time; ∇ – Laplace operator; R_1 and R_2 – radii of the inner and outer surfaces of the hollow cylinder; α_1, α_2 and α_3 – heat transfer coefficients between the cooled body and the cooling medium in the inner cavity, on the outer surface and at the ends of the cylinder, respectively; λ – thermal conductivity coefficient; $2h = AB$ – cylinder height; $T(r, z, t)$ – variable temperature; T_0 – cooling medium temperature; $\Delta T(r, z, t) = T(r, z, t) - T_0$ – temperature drop; T_{in} – the initial temperature of the cooled body. As information on the values of the mentioned variables is often imprecise, we consider that values of R_1, R_2, α_1, α_2 and α_3, h are described by fuzzy numbers. Indeed, while R_1, R_2, h may be more or less precisely measured, the imprecision of α_1, α_2, α_3 and λ is conditioned by the fact that thermal conductivity and heat transfer depends significantly on composition of material (plastic details). In turn, information on composition is, as a rule, imprecise. Then, fuzzy form of Eqs. (1)–(3) may be constructed on the basis of theory of FDEs.

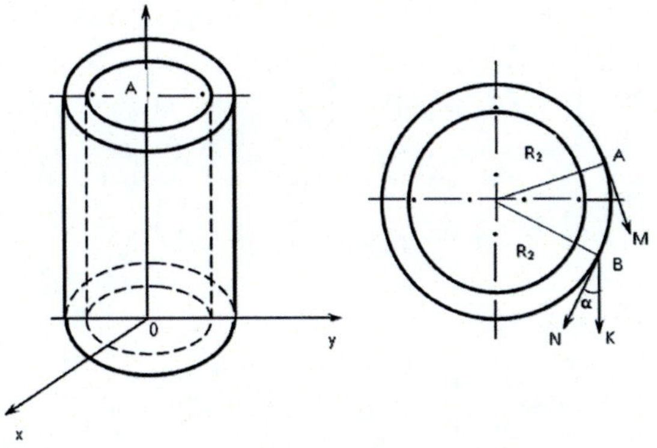

Fig. 1. Cylindrical coordinates used to define the temperature difference ΔT of the isotropic cylindrical body (plastic detail).

To determine the solution to problem (1)–(3), we apply the Hankel integral transform [3, 4]:

$$\{\Delta T(r, z, t)\}_H = \int_{R_1}^{R_2} \Delta T(r, z, t) r U_0\left(\mu_1 \frac{r}{R_1}\right) dr, \quad (4)$$

where

$$U_0\left(\mu_n \frac{r}{R_1}\right) = \left[J_0(\mu_n) + \frac{\mu_n}{B_{i_1}} J_1(\mu_n)\right] \cdot J_0\left(\mu_n \frac{r}{R_1}\right)$$
$$-\left[J_0(\mu_n) + \frac{\mu_n}{B_{i_2}} J_1(\mu_n)\right] \cdot J_0\left(\mu_n \frac{r}{R_1}\right), B_{i_1} = \frac{\alpha_1 R_1}{\lambda}, \quad B_{i_2} = \frac{\alpha_2 R_2}{\lambda}, \tag{5}$$

μ_n – root of the equation. The fuzzy case of (4) and (5) can be considered on the basis of the theory of fuzzy integrals[9–11, 13]. Then we have:

$$\frac{U_0(k\mu_n)}{U_1(k\mu_n)} = \frac{k\mu_n}{B_{i_2}}, \quad n = 1, 2, ...,$$

$$U_1\left(\mu_n \frac{r}{R_1}\right) = \left[J_0(\mu_n) + \frac{\mu_n}{B_{i_1}} J_1(\mu_n)\right] \cdot J_1\left(\mu_n \frac{r}{R_1}\right)$$
$$-\left[J_0(\mu_n) + \frac{\mu_n}{B_{i_2}} J_1(\mu_n)\right] \cdot J_1\left(\mu_n \frac{r}{R_1}\right),$$

$$J_0(x) = \sum_{n=0}^{\infty} (-1)^n \frac{x^{2n}}{(n!)^2 \cdot 2^{2n}}, \quad J_n(x) = \sum_{\nu=0}^{\infty} \frac{(-1)^\nu}{\nu!\Gamma(n+\nu+1)} \left(\frac{x}{2}\right)^{n+2\nu},$$

$\Gamma(P)$ – gamma Euler function, at that $\Gamma(n+1) = n!$. Therefore

$$J_1(x) = \sum_{\nu=0}^{\infty} \frac{(-1)^\nu}{(\nu!)^2(\nu+1)} \cdot \left(\frac{x}{2}\right)^{2\nu+1},$$

$J_0(x)$ – Bessel functions of the first kind of n-order; $J_n(x)$ – Bessel functions of the second kind of n-order. One has:

$$J_0(x) = \frac{2}{\pi}\left[c + \ell n \frac{x}{2}\right] J_0(x) - \frac{1}{\pi} \sum_{k=0}^{\infty} \frac{(-1)^k}{(\gamma!)^2} \cdot \left(\frac{x}{2}\right)^{2k} \cdot 2\Phi(k),$$

$$J_1(x) = \frac{2}{\pi}\left[c + \ell n \frac{x}{2}\right] J_1(x) - \frac{1}{\pi} \sum_{k=0}^{\infty} \frac{(-1)^k}{(k+1)\cdot(k!)^2} \cdot \left(\frac{x}{2}\right)^{2k+1} \cdot [\Phi(k+1) + \Phi(k)],$$

c – Euler's constant, $c \approx 0, 5772,$. At the same time,

$$\Phi(k) = \sum_{s=1}^{k} \frac{1}{s}, \quad \Phi(k+1) = \Phi(k) + \frac{1}{k+1}.$$

Then

$$\Phi(k+1) + \Phi(k) = 2\Phi(k)\frac{1}{k+1},$$

therefore

$$J_1(x) = \frac{2}{\pi}\left[c + \ell n \frac{x}{2}\right] J_1(x) - \frac{1}{\pi} \sum_{k=0}^{\infty} \frac{(-1)^k}{(k+1)\cdot(k!)^2} \cdot \left(\frac{x}{2}\right)^{2k+1} \cdot \left[2\Phi(k) + \frac{1}{k+1}\right],$$

it can easily be shown that the circulation formula for the transformation [4] for a given task will have the form:

$$\Delta T(r,z,t)=\sum_{n=1}^{\infty}2\left\{\Delta T(r,z,t)\right\}_{+1}\cdot U_0\left(\mu_n\frac{r}{R_1}\right)\cdot R_1^{-2}$$
$$\times\left\{k^2U_0^2(k\mu_n)\cdot\left[1+\left(\frac{B_{i_2}}{k\mu_n}\right)^2\right]-\frac{4}{\pi^2B_{i_1}^2}\left[1+\left(\frac{B_{i_1}}{k\mu_n}\right)^2\right]\right\}^{-1}.$$

From here, after applying the cosine- transformation, we obtain the general solution of problem (4) – (5) in the form:

$$\Delta T(r,z,t)=\sum_{n=1}^{\infty}\sum_{m=1}^{\infty}4\left(B_{i_{pl}}^2+\gamma_m^2\right)\int_{R_1}^{2}r\Delta T_{in}\cdot U_0\left(\mu_n\frac{r}{R_1}\right)\cdot\cos\gamma_m\frac{z}{h}drdz\times\exp\left[-\left(\mu_m^2+b^2\mu_n^2\right)F_{O1}\right]\times$$
$$\left[R_1^2b\left(B_{i_{pl}}^2+B_{i_{pl}}+\gamma_m^2\right)\cdot\left\{k^2U_0^2(k\mu_n)\cdot\left[1+\left(\frac{B_{i_1}}{k\mu_n}\right)^2\right]-\frac{4}{\pi^2B_{i_1}^2}\left[1+\left(\frac{B_{i_1}}{\mu_n}\right)^2\right]\right\}\right]^{-1} \tag{6}$$

where $k=\frac{R_2}{R_1}$; $b=\frac{\pi R_2}{h}$; $F_{O_1}=\frac{at}{R_1^2}$; $B_{n\pi}=\frac{ah}{\lambda}$.

If the initial temperature of the body is constant $T_n=const$, then the temperature drop is determined in the form:

$$\Delta T(r,z,t)=\Delta T_{in}\left\{1-\sum_{n=1}^{\infty}\sum_{m=1}^{\infty}M_nB_mU_0\left(\mu_n\frac{r}{R_1}\right)\cdot\cos\mu_m\frac{z}{h}\cdot\exp\left[-\left(\mu_m^2+b^2\mu_n^2\right)F_{O_1}\right]\right\}, \tag{7}$$

where

$$M_n=\frac{2\left[B_{i_2}U_0(k\mu_m)+\frac{2}{\pi}\right]}{k^2U_0^2(k\mu_m)\cdot\left[\mu_m^2+\left(\frac{B_{i_2}}{k}\right)^2\right]-\frac{4}{\pi^2B_{i_1}^2}\left[\mu_m^2+B_{i_1}^2\right]}, \tag{8}$$

$$B_m=(-1)^{m+1}\frac{2B_i\sqrt{B_i^2+\mu_n^2}}{\mu_n^2\left(B_i^2+B_i+\mu_n^2\right)},\quad B_i=\frac{ah}{\lambda}. \tag{9}$$

With the same values of the heat transfer coefficient on the outer and inner cylindrical surfaces ($\alpha_1=\alpha_2=\alpha$), the decomposition coefficient M_n takes the form:

$$M_n=\frac{2B_{i_1}}{\left(\mu_n^2+B_{i_1}^2\right)\cdot\left[kU_0(k\mu_n)-\frac{2}{\pi B_{i_1}}\right]}, \tag{10}$$

The rows representing solutions to non-stationary thermal conductivity problems quickly converge and one or two members of a series are enough for practical calculations. Thus, at $n=1$, $m=1$, the approximate solution of the problem has the form:

$$\Delta T(r,z,t) = \Delta T_{in}\left\{1 - \frac{2B_{i1}U0\left(\mu_1 \frac{r}{R_1}\right)}{\left(\mu_n^2 + B_{i_1}^2\right).\left[kU_0(k\mu_1) - \frac{2}{\pi B_{i_1}}\right]} \cdot \frac{2B_{i_1}\sqrt{B_i^2 + \mu_1^2}}{\mu_1^2(B_i^2 + B_i + \mu_1^2)} \cdot \cos\gamma_1 \frac{z}{n} \times \exp[-(\mu_1^2 + B_i^2\mu_1^2)F_{O_1}]\right\} =$$
$$\Delta T_{in} \cdot 1 - \left\{\frac{2B_{i_1}U_0\left(\mu_1 \frac{r}{R_1}\right)}{\mu_1^2 + B_{i_1}^2 \cdot \left[\frac{R_2}{R_1}U\left(\frac{R_2}{R_1}\mu_1\right) - \frac{2}{\pi B_{i_1}}\right]} \times \frac{2B_{i_1}}{\mu_1^2(B_i^2 + B_i + \mu_1^2)} \cdot \cos\mu_1 \frac{z}{h} \cdot e - \mu_1^2\left(1 + \frac{\pi R_2}{n}\right)^2 \frac{at}{R_1^2}\right\}. \tag{11}$$

Here, to determine temperature voltages, let's consider the process of cooling plastic parts of the bushing type (Fig. 1).

According to Duhamel's hypothesis [5, 6], if the inertial effects not be taking the regard (not take into account), then the temperature voltages distribution is quasi-static.In this case, the temperature distribution satisfies the thermal conductivity equation [1]. Then the thermoelastic potential of displacements for the quasi-static case is determined by the following formula:

$$\nabla\Phi = \frac{1+\mu}{1-\mu}\alpha\nabla\Delta T, \tag{12}$$

from here

$$\Phi(t,r,z) = \frac{1+\mu}{1-\mu}a \cdot \alpha\Delta T = \beta_0\Delta T, \tag{13}$$

where

$$\beta_0 = \frac{1+\mu}{1-\mu}\alpha \cdot aT_0.$$

For the quasi-static case, temperature stresses using the thermoelastic potential of displacements are determined by the formula:

$$\begin{cases} \overline{\sigma}_{rr} = \dfrac{\sigma_{rr}}{2G} = \dfrac{\partial^2\Phi}{\partial r^2} - \Delta\Phi, \\ \overline{\sigma} = \dfrac{\sigma_{\theta\theta}}{2G} = \dfrac{1}{r} \cdot \dfrac{\partial\Phi}{\partial r} - \Delta\Phi, \\ \overline{\sigma}_{zz} = \dfrac{\sigma_{zz}}{2G} = \dfrac{\partial^2\Phi}{\partial z^2} - \Delta\Phi. \end{cases} \tag{14}$$

Considering, that

$$\Delta T(r,z,t) = \Delta T_{in}\left\{1 - \sum_{n=1}^{\infty}\sum_{m=1}^{\infty} M_n B_m U_0\left(\mu_n \frac{r}{R_1}\right) \cdot \cos\mu_m \frac{z}{n} \cdot e^{-\frac{\mu_m^2 + b^2\mu_n^2}{R_1^2} \cdot at}\right\} \tag{15}$$

we receive

$$\Phi(t,r,z) = \frac{1+\mu}{1-\mu}a \cdot$$
$$\alpha\Delta T_{in}\left\{1 - \sum_{n=1}^{\infty}\sum_{m=1}^{\infty} M_n B_m U_0\left(\mu_n \frac{r}{R_1}\right) \cdot \cos\mu_m \frac{z}{n} \cdot e^{-\frac{\mu_m^2 + b^2\mu_n^2}{R_1^2} \cdot at}\right\}. \tag{16}$$

As it was noted, a row representing solutions to the tasks of non-stationary thermal conductivity quickly converges and one or two elements are sufficient for practical calculations. Therefore the temperature potential of displacements is determined by the formula [7]:

$$\Phi(t,r,z)=\beta_0\Delta T_{in}\left\{1-\frac{2B_{i_1}U_0\left(\mu_1\frac{r}{R_1}\right)}{\left(\mu_1^2+B_{i_1}^2\right)\cdot\frac{R^2}{R^1}U_0\left(\frac{R_2}{R_1}\mu_1\right)-\frac{2}{\pi B_{i_1}}}\cdot\frac{2B_iB_i^2+\mu_1^2}{\mu_1^2\left(B_i^2+B_i+\mu_1^2\right)}\times\cos\mu_1\frac{z}{n}\cdot e^{-\mu_1^2\left(1+\frac{\pi R_2}{n}\right)^2\frac{at}{R_1^2}}\right\}$$

$$=\beta_0\Delta T_{in}\left\{1-A\cos\mu_1\frac{z}{n}\cdot U_0\left(\mu_1\frac{r}{R_1}\right)\cdot e^{-A_0\cdot\frac{at}{R_1^2}}\right\} \tag{17}$$

where the symbol is entered

$$A=\frac{2B_{i_1}}{\left(\mu_1^2+B_{i_1}^2\right)\cdot\left[\frac{R_2}{R_1}U_0\left(\frac{R_2}{R_1}\mu_1\right)-\frac{2}{\pi B_{i_1}}\right]}\cdot\frac{2B_{i_1}\sqrt{B_i^2+B_1^2}}{\mu_1^2\left(B_i^2+B_i+\mu_1^2\right)};\ A_0=\mu_1^2\left[1+\left(\frac{\pi R_2}{n}\right)^2\right] \tag{18}$$

(value A is given in [1]).

Now we calculate the temperature voltages:

$$\overline{\sigma}_{rr}=\frac{\partial^2\Phi}{\partial r}-\Delta\Phi=-\left[\frac{1}{r}\frac{\partial\Phi}{\partial r}+\frac{\partial^2\Phi}{\partial z^2}\right]=\frac{1}{r}\beta_0\Delta T_{in}\cdot A\cdot\cos\mu_1\frac{z}{n}\cdot U_0'\left(\mu_1\frac{r}{R_1}\right)\times$$
$$\times e^{-A_0\cdot\frac{at}{R_1^2}}\cdot\frac{\mu_1}{R_1}-\beta_0\Delta T_{in}\cdot A\cdot\left(\frac{\mu_1}{n}\right)^2\cdot\cos\mu_1\frac{z}{n}\cdot U_0\left(\mu_1\frac{r}{R_1}\right)\cdot e^{-A_0\cdot\frac{at}{R_1^2}}=$$
$$=\beta_0\Delta T\cdot A\left[\frac{1}{r}\frac{\mu_1}{R_1}\cdot U_0'\left(\mu_1\frac{r}{R_1}\right)-\left(\frac{\mu_1}{n}\right)^2U_0\left(\mu_1\frac{r}{R_1}\right)\right]\cdot\cos\mu_1\frac{z}{n}\cdot U_0'\left(\mu_1\frac{r}{R_1}\right)\cdot e^{-A_0\cdot\frac{at}{R_1^2}}, \tag{19}$$

$$\overline{\sigma}_{\theta\theta}=\frac{1}{r}\frac{\partial\Phi}{\partial r}-\Delta\Phi=-\left[\frac{\partial^2\Phi}{\partial r^2}+\frac{\partial^2\Phi}{\partial z^2}\right]=\left(\frac{\mu_1}{R_1}\right)^2\cdot\beta_0\Delta T_{in}\cdot A\cdot\cos\mu_1\frac{z}{n}\cdot U_0''\left(\mu_1\frac{r}{R_1}\right)\times$$
$$\times e^{-A_0\cdot\frac{at}{R_1^2}}-\beta_0\Delta T_{in}\cdot A\cdot\left(\frac{\mu_1}{n}\right)^2\cdot\cos\mu_1\frac{z}{n}\cdot U_0\left(\mu_1\frac{r}{R_1}\right)\cdot e^{-A_0\cdot\frac{at}{R_1^2}}=\beta_0\Delta T_{in}\cdot A\cdot\mu_1^2\times$$
$$\times\left[\frac{1}{R_1^2}U_0''\left(\mu_1\frac{r}{R_1}\right)-\frac{1}{n^2}U_0\left(\mu_1\frac{r}{R_1}\right)\right]\cdot\cos\mu_1\frac{z}{n}\cdot e^{-A_0\cdot\frac{at}{R_1^2}}, \tag{20}$$

$$\overline{\sigma}_{zz}=\frac{\partial^2\Phi}{\partial z^2}-\Delta\Phi=-\frac{1}{r}\frac{\partial\Phi}{\partial r}+\frac{\partial^2\Phi}{\partial r^2}=\frac{1}{2}\beta_0\Delta T_{in}\cdot A\cdot\cos\mu_1\frac{z}{n}\cdot U_0'\left(\mu_1\frac{r}{R_1}\right)\times$$
$$\times e^{-A_0\cdot\frac{at}{R_1^2}}\cdot\frac{\mu_1}{R_1}+\left(\frac{\mu_1}{R_1}\right)^2\beta_0\Delta T_{in}\cdot A\cdot\cos\mu_1\frac{z}{n}\cdot U_0''\left(\mu_1\frac{r}{R_1}\right)\cdot e^{-A_0\cdot\frac{at}{R_1^2}}=$$
$$=\frac{\mu_1}{R_1}\beta_0\Delta T_{in}\cdot A\left[\frac{1}{r}U_0'\left(\mu_1\frac{r}{R_1}\right)+\frac{\mu_1}{R_1}U_0''\left(\mu_1\frac{r}{R_1}\right)\right]\cdot\cos\mu_1\frac{z}{n}\cdot e^{-A_0\cdot\frac{at}{R_1^2}}. \tag{21}$$

It is known that the sum of normal voltages is a function of volumetric deformation and is determined by the formula:

$$\theta = \overline{\sigma}_{rr} + \overline{\sigma}_{\theta\theta} + \overline{\sigma}_{zz} = \frac{\partial^2 \Phi}{\partial r^2} + \frac{1}{r}\frac{\partial \Phi}{\partial r} + \frac{\partial^2 \Phi}{\partial z^2} - 3\Delta\Phi = -2\Delta\Phi = 2\beta_0 \Delta T_{\mathcal{H}}$$
$$\times \mathrm{A}\left[\left(\frac{\mu_1}{R_1}\right)^2 \cdot U_0''\left(\mu_1 \frac{r}{R_1}\right) + \frac{1}{r} U_0'\left(\mu_1 \frac{r}{R_1}\right) - \left(\frac{\mu_1}{n}\right)^2 U_0\left(\mu_1 \frac{r}{R_1}\right)\right] \cdot \cos \mu_1 \frac{z}{n} \cdot e^{-\mathrm{A}_0 \cdot \frac{at}{R_1^2}}. \tag{22}$$

Numerous experiments show that the destruction of samples during their cooling, the development of plastic deformation occurs under the action of the maximum difference in stresses. Taking into account the above formula for the voltages difference, we obtain the following voltage:

$$\overline{\sigma}_{rr} - \overline{\sigma}_{\theta\theta} = \beta_0 \Delta T_{in} \cdot \mathrm{A}\left[\frac{1}{r}\frac{\mu_1}{R_1} \cdot U_0'\left(\mu_1 \frac{r}{R_1}\right) - \frac{\mu_1^2}{R_1^2} U_0''\left(\mu_1 \frac{r}{R_1}\right)\right] \cdot \cos \mu_1 \frac{z}{n} \cdot e^{-\mathrm{A}_0 \cdot \frac{at}{R_1^2}} \tag{23}$$

$$\overline{\sigma}_{rr} - \overline{\sigma}_{zz} = \frac{\partial^2 \Phi}{\partial r^2} - \frac{\partial^2 \Phi}{\partial z^2} = -\beta_0 \Delta T \cdot \mathrm{A}\left[\frac{\mu_1^2}{R_1^2} U_0''\left(\mu_1 \frac{r}{R_1}\right) + \frac{\mu_1^2}{n^2} U_0\left(\mu_1 \frac{r}{R_1}\right)\right] \times$$
$$\times \cos \mu_1 \frac{z}{n} \cdot e^{-\mathrm{A}_0 \cdot \frac{at}{R_1^2}}, \tag{24}$$

$$\overline{\sigma}_{zz} - \overline{\sigma}_{\theta\theta} = \frac{\partial^2 \Phi}{\partial z^2} - \frac{1}{r}\frac{\partial \Phi}{\partial r} = \beta_0 \Delta T_{in} \cdot$$
$$\mathrm{A}\left[\left(\frac{\mu_1}{n}\right)^2 U_0\left(\mu_1 \frac{r}{R_1}\right) + \frac{1}{r}\left(\frac{\mu_1}{R_1}\right) U_0'\left(\mu_1 \frac{r}{R_1}\right)\right] \times \cos \mu_1 \frac{z}{n} \cdot e^{-\mathrm{A}_0 \cdot \frac{at}{R_1^2}}. \tag{25}$$

Let us note, that a shift from classical description of the considered process to fuzzy representation requires delicate consideration of a lot of aspects. Namely, types of fuzzy derivatives [12, 14], approaches to numerical solutions of FDE and fuzzy equations[9, 20] etc. An important issue is reducing computational complexity when dealing with fuzzy models.

3 Discussion and Conclusions

Analysis of these expressions shows that the first voltage difference is the largest. Thus, the most dangerous, from the point of view of specimen cracking, is the maximum voltage difference $\overline{\sigma}_{rr} - \overline{\sigma}_{\theta\theta}$, which should be minimized. Due to the limited volume of the collection "Achievement in the field of intellectual and computing systems", this article discusses the general formulation of the problem and methods for their solution.

References

1. Kerimov, D.A.: Scientific Bases and Practical Methods of Optimization of Parameters of Quality of Plastic Details of The Oil-Field Equipment. Dissertation of Doctor of technical sciences., 298 p., Baku (1985) (in Russian)
2. Mustafaev, A.D., Kerimov, D.A., Kaziev, R.A.: Study the Stability of Size For Plastic Products. Collection of Reports of The Conference "Ways to Improve the Quality and Improvement of Equipment - Planning The Production of Oilfield Equipment", Baku (1969) (in Russian)
3. Trubin, A.N.: Nonlinear tasks of voltages concentration in machine parts. In: Publishing House "Mechanical Engineering", Leningrad, 157 p. (1972) (in Russian)
4. Krushevsky, A.E.: Variational Methods for Calculating Corpus Details of Machines. Science and Technology, Minsk, 224 p. (1967) (in Russian)
5. Aponson, A.Ya., Bugov, A.U., other: Calculation on the Strength of Hydraulic Turbine Details. Mashgaz (1965) (in Russian)
6. Kaganov, L.M.: Foundations of the Theory of Plasticity. Moscow, Publishing house "Science" (1969) (in Russian)
7. Kerimov, D.A., Gasanova, N.A.: Development of optimal regimes of cooling for plastic parts of thermosets. In: SAEQ Science 8 Applied Engineering Quarterly, Publ. in the UK, London, Jan-Feb-Mar, Issue 11, p. 17–19 (2017). ISSN: 2054–2763 11
8. Abbasbandy, S., Allahviranloo, T., Armand, A., Gouyandeh, Z.: A fuzzy solution of heat equation under generalized Hukuhara differentiability by fuzzy Fourier transform. Fuzzy Sets Syst. **309**, 81–97 (2017). https://doi.org/10.1016/j.fss.2016.04.010
9. Ahmady, N.: A fuzzy newton-cotes method for integration of fuzzy functions. J. Fuzzy Set Valued Anal. **2**, 171–178 (2015). https://doi.org/10.5899/2015/jfsva-00247
10. Bede, B., Gal, S.G.: Generalizations of the differentiability of fuzzy-number-valued functions with applications to fuzzy differential equations. Fuzzy Sets Syst. **151**, 581–599 (2005). https://doi.org/10.1016/j.fss.2004.08.001
11. Bede, B., Gal, S.G.: Quadrature rules for integrals of fuzzy-number-valued functions. Fuzzy Sets Syst. **145**, 359–380 (2004). https://doi.org/10.1155/2014/413570
12. Bede, B., Stefanini, L. Generalized differentiability of fuzzy-valued functions. Fuzzy Sets Syst. 1–23. https://doi.org/10.1016/j.fss.2012.10.003
13. Costa T.M.: Jensen's inequality type integral for fuzzy-interval-valued functions. Fuzzy Sets Syst. **327**. https://doi.org/10.1016/j.fss.2017.02.001
14. Gomes L.T., Barros L.C.: Fuzzy calculus via extension of the derivative and integral operators and fuzzy differential equations. In: Annual Meeting of the North American Fuzzy Information Processing Society (NAFIPS), 2012, 6–8 Aug. pp. 1–5 (2012). https://doi.org/10.1109/NAFIPS.2012.6290965
15. Gomes, L.T., Barros, L.C.: Fuzzy Differential Equations with Arithmetic and Derivative via Zadeh's Extension. Mathware Soft Comput. Mag. **20**, 70–75 (2013)
16. Mazandarani, M., Najariyan, M.: Differentiability of type-2 fuzzy number-valued functions. Commun Nonlinear Sci. Numer. Simulat. **19**, 710–725 (2014). https://doi.org/10.1016/j.cnsns.2013.07.002
17. Stefanini, L.: A generalization of Hukuhara difference and division for interval and fuzzy arithmetic. Fuzzy Sets Syst. **161**, 1564–1584 (2010). https://doi.org/10.1016/j.fss.2009.06.009
18. Stefanini L.: A generalization of Hukuhara difference. In: D. Dubois, Lubiano, M.A., Prade, H., Gil, M.A., Grzegorzewski, P., Hryniewicz, O., (eds.), Soft Methods for Handling Variability and Imprecision, Series on Advances in Soft Computing, vol. 48, Springer, 2008. https://doi.org/10.1007/978-3-540-85027-4_25

19. Stefanini, L., Bede, B.: Generalized Hukuhara differentiability of interval-valued functions and interval differential equations. Nonlinear Anal. **71**, 1311–1328 (2009). https://doi.org/10.1016/j.na.2008.12.005
20. Wu, X.: Optimal fuzzy quadrature formula for classes of fuzzy-number-valued functions of Lipschitz type. Fuzzy Sets Syst. 1-9. https://doi.org/10.1016/j.fss.2012.09.001

Fuzzy-Multiple Model of Gamified Mobile Application "Smart House" Based on Bartle Classification of Psychotypes of Players

S. Alexeev Anton[1], A. Batishcheva Galina[1(✉)], A. Kiyaschenko Tatyana[2], V. Lukyanova Galina[1], and A. Ovsyannikov Victor[1]

[1] Economics, B. Sadovaya Street, 69, 344002 Rostov-on-Don, Russia
[2] Rostov State Transport University, 2, Rostovskogo Strelkovogo Polka Narodnogo Opolcheniya Sq., Rostov-on-Don 344038, Russia

Abstract. A fuzzy-multiple model has been developed for adjusting the interface of the Smart Home mobile application to the psychotype of the user in accordance with the classification of gamers by R. Bartle. Achieveer, Socialist and Researcher were chosen as the three main psychotypes in relation to the gamification of a mobile application. It was suggested to implement the adjustment based on the tabs settings in the mobile application menu: 1) Achievement level tab (primarily focused on Achievements); 2) Communication tab (focused on the Socialist); 3) Tab System Information (focused on Researchers). For each tab, it is proposed to implement three levels: basic, expanded and psychotype-oriented. The choice of the level is carried out on the basis of the work of the system of productive rules, and the formation of fuzzy sets for the system is proposed to be formed on the basis of the statistics of user calls during the control period to each of the tabs.

Keywords: Fuzzy-multiple model · Productive rules · Gamification

1 Introduction

Smart home means a system that controls electrical loads, lighting, security, climate control, media devices. Currently, one of the most promising technologies is the "Smart Home" technology, as a system that ensures safety, resource conservation and comfort for all users of residential premises. The Smart Home system consists of many subsystems that provide work with climate control, energy saving, lighting, security, house maintenance, watering, control over the adjacent territory, etc. The problems of functioning and maintenance of the Smart Home system are widely covered at the present time in the specialized literature, including the mathematical modeling of the relevant systems.

Thus, in 2018, as part of the revised Directive on energy efficiency in buildings, the European Commission introduced an "intelligent readiness indicator" as a means to make energy efficiency in buildings more understandable and useful for users, owners and tenants of buildings [1]. By the end of 2016, households around the world were

R. A. Aliev et al. (Eds.): ICSCCW 2021, LNNS 362, pp. 365–370, 2022.
https://doi.org/10.1007/978-3-030-92127-9_50

using four thousand million connected devices, including smart meters, energy devices, and connected thermostats CTs. [2] The main purpose of the control system HVAC is to maintain temperature and air quality in a comfortable range with minimal energy consumption [3]. In [4], CTs is defined as an electrical device that connects smart homes to smart grids. CTs are designed to automatically learn tenants' schedules and turn heating and cooling on or off on behalf of users [3]. CTs are designed to consume little energy, increase consumer comfort and inform users about the energy status of thermostats via mobile device interfaces. However, CTs were not fully accepted by consumers because their interface did not meet consumer expectations [5].

The problem of creating applications that provide a comfortable user interaction with the system is one of the most frequently discussed in the specialized literature. This is due to the fact that the corresponding mobile application, created without taking into account the level, needs and psychological characteristics of the user, may turn out to be cumbersome and not always acceptable for a particular user. An interface that is comfortable and pleasant for one user may be annoying and redundant/insufficient for another. In addition, there is an acute issue of training the user to work with the application, familiarization with its functions and capabilities. Training, ideally, should be done by the application itself and be user-friendly. This is how the idea of using the elements of gamification in teaching arose,

So, to save energy, work [6] proposes a system of gamification and serious games, adapted by a fuzzy logic system to motivate connected thermostat consumers. Gamification is the use of game elements in non-game contexts to improve user experience and user engagement [7]. Gamification in energy systems has emerged as a tool to improve energy efficiency by stimulating customer engagement and changing their energy-related behavior by targeting a wide range of motives that customers may have, including economic, environmental and social incentives [8]. In [4], the authors propose a reward for consumers by sending incentives to change consumer behavior. Five mechanisms have been proposed for the gamification system [9].

Despite the proliferation of different applications with different interfaces, users may find it difficult to master the mobile application. Moreover, these difficulties are associated both with the achieved level of computer technology, and with psychological characteristics associated with the personal characteristics of users.

This study proposes a methodology for developing a mobile application based on fuzzy multiple modeling taking into account gamification (Bartle's typology) and assessing the user's level of application ownership.

2 Methods: User Psychotype in Terms of Gamification

At the initial stage of the application development, Bartle's typology was used. Richard Bartle's model consists of two intersecting segments that form the players' field, which he called the plane of interests: "action - interaction" and "player - system". Consider the four psychotypes from the point of view of Richard Bartle, their driving motives and potential leverage on them in the game.

The Accumulators (Achiever) or Careerists. It is important for them to accumulate power, money, cool artifacts - any game benefits and resources. They accumulate everything: bonus points, game currency, stickers for the sixth coffee as a gift. The main thing for them is to get an award in any way. Careerists love to receive all kinds of gaming benefits.

The Killer. For them, the main motivation is superiority over other players, dominance, and domination. They only crave victory. It is important for them to be the first: to be at the top of the leaderboard, collect as many prize points as possible, be the first to know about everything and win.

The Explorer. They are interested in studying the game world and discovering its secrets. They strive to get to know the game universe and would rather explore the location than try to complete the game. Researchers love gaming reviews. This class is focused on knowledge, on the absorption of game content, on the study of game mechanics and game capabilities: the larger the game world, the better.

The Socializer. Party people. Communication with other players, social interaction and mutual understanding are important for them. This type loves to be in the spotlight and communicate in online and offline channels. Social workers are happy to go to forums and conferences, happily book tickets for open lectures, and, speaking of business, most likely, they have already tried several products similar to yours.

Thus, Killers pride themselves on dominating others; The accumulators - by the fact that they have the most resources and benefits; Researchers because they are smarter than others, and Socialists because they are more popular. This classification is effective in the game, however, in relation to the mobile application, the classification is simplified. The Killer and Accumulator classes are characterized by the level of achievement in using the mobile application: the amount of information acquired, the level achieved, the position in the amount of accumulated benefits and bonuses, as well as superiority in these areas over other users of the application. The combined class is proposed to be named Achievers. So, we have identified three classes of users of the Smart Home mobile application: Achievements, Socialists and Researchers.

3 Implementation of Application Selection in Accordance with Bartle's Typology Based on a System of Productive Rules

In accordance with the introduced classification, it is proposed to introduce three tabs in the mobile application menu: 1) Achievement level tab (focused primarily on Achievements); 2) Communication tab (focused on Socialists); 3) Tab System Information (focused on Researchers). However, almost every user is a combination of several psychotypes, so the interface of each of the tabs should be expanded for each user, taking into account his needs. In accordance with this, it was proposed to develop three levels for each of the tabs: basic (1), expanded (2) and psychotype-oriented (3). Thus, we get three interfaces of the tabs:

A1 (Basic Achievements) - minimum information about the levels passed in mastering the application, accumulated points and bonuses; A2 (Achievements detailed) - detailed information about the levels passed in mastering the application, accumulated points and bonuses; elements of leaderboards and badges when assessing achievements, taking into account the wishes of the user; A3 (Achievements psycho-oriented) - in detail detailed information about the passed levels, accumulated points and bonuses; active use of leaderboards and badges in a playful way, comparison with others and users of the application;

S1 (Basic communication) - access to the forum, the ability to communicate and consult with other users; S2 (Expanded Communication) - access to the forum, expanded communication with other users, the ability to participate in collective games based on the application; S3 (Communication psycho-oriented) - extensive communication with other users on the forum and in social, participation in collective games based on the application;

E1 (Basic information) - basic information about the system and its capabilities; E2 (Information expanded) - expanded information about the system, access to the library, a system of suggestions for studying the system; E3 (Information psycho-oriented)- detailed information about the system, including in a game form, access to the library, a system of suggestions for studying the system, quests and levels showing the level of proficiency in the mobile application.

It is proposed on each tab to introduce an automatic scoring system upon contact. Moreover, the accumulation of points is carried out in such a way that different actions are assessed with different numbers of points. For example, 1 point is awarded for a simple access to the forum. For participation in the discussion of the problem 2 points. For inviting a friend to a group - 5 points, etc. After that, a correspondence is established between the interface levels and the number of points accumulated by the user for the assessed period, based on expert assessments. For the intervals established in this way, the centers of gravity are calculated and on their basis fuzzy sets are formed SX1 = "social activity is low", SX2 = "social activity is average", SX3 = "social activity is high".

After that, the rules of fuzzy production are formed to select the interface of the Communication tab:

RULE Soc1: IF X1 = SX1 ("Social activity is low"), THEN Y1 = SY1 ("interfaceBasic communication");
RULE Soc2: IF X1 = SX2 ("Social activity is average"), THEN Y1 = SY2 ("interfaceCommunication expanded");
RULE Soc3: IF X1 = SX3 ("Social activity is high"), THEN Y1 = SY3 ("interfaceCommunication psycho-oriented").

Similarly, based on the analysis of statistics of work with the Achievement Level tab, its interface is formed:

RULE Ach1: IF X2 = AX1 ("The level of interest in achievements is low"), THEN Y2 = AY1 ("interface Basic achievements");

RULE Ach2: IF X2 = AX2 ("Level of interest in achievements is average"), THEN Y2 = AY2 ("interface Achievements expanded");
RULE Ach3: IF X2 = AX3 ("Level of interest in achievements is high"), THEN Y3 = AY3 ("interface Achievements psycho-oriented").

Accordingly, based on the analysis of statistics of working with the Information tab, its interface is formed:

RULE Exp1: IF X2 = EX1 ("The level of interest in information is low"), THEN Y2 = EY1 ("interface Basic information");
RULE Exp2: IF X2 = EX2 ("The level of interest in information is average"), Then Y2 = EY2 ("interface Information expanded");
RULE Exp3: IF X2 = EX3 ("The level of interest in information is high"), THEN Y3 = EY3 ("interface Information psycho-oriented").

The result of the work of the production rules is the choice of the interface of the mobile application for the user for a certain regulated period of time. Figure 1 shows the results of defining the interface for the Achievements tab in Fuzzy Toolbox.

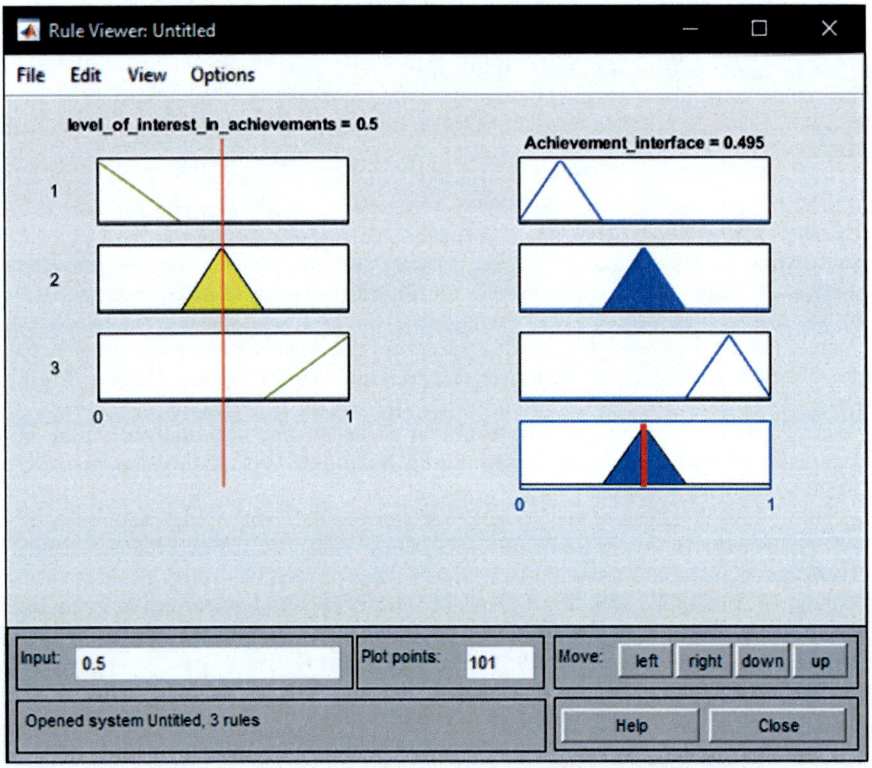

Fig. 1. Implementation of the fuzzy inference system in Fuzzy Toolbox.

It should be noted that the initial interface of the application (for a control period of time) can be selected in accordance with the results of user testing according to the Bartle method.

4 Conclusions

A fuzzy-multiple model has been developed for adjusting the interface of the Smart Home mobile application to the psychotype of the user in accordance with Bartle's classification of gamers. Achieve, Socialist and Researcher were chosen as the three main psychotypes in relation to the gamification of a mobile application. A correspondence was established between the characteristic features of psychotypes in relation to the game, as well as the main methods of their most effective involvement in the game (in this case, gamified training in working with the Smart Home mobile application). It was suggested to implement the adjustment based on the tabs settingsin the mobile application menu: 1) Achievement level tab (primarily focused on Achievements); 2) Communication tab (focused on the Socialist); 3) Tab System Information (focused on Researchers). For each tab, it is proposed to implement three levels: basic, expanded and psychotype-oriented. The choice of the level is carried out on the basis of the work of the system of productive rules, and the formation of fuzzy sets for the system is proposed to be formed based on the statistics of user calls during the control period to each of the tabs. The initial interface during the control period (evaluation) will be formed on the basis of the user's passing the classic Bartle test.

References

1. Gas 2018: Market Report Series: Gas (2018). https://doi.org/10.1787/gas_mar-2018-en
2. Gas 2017: Market Report Series: Gas (2017). https://doi.org/10.1787/gas_mar-2017-en
3. Soltanaghaei, E., Whitehouse, K.: Practical occupancy detection for programmable and smart thermostats. Appl. Energy **220**, 842–855 (2018). https://doi.org/10.1016/j.apenergy.2017.11.024
4. Ponce, P., Peffer, T., Molina, A.: Usability perceptions and beliefs about smart thermostats by chi-square test, signal detection theory, and fuzzy detection theory in regions of Mexico. Front. Energy **13**(3), 522–538 (2018). https://doi.org/10.1007/s11708-018-0562-2
5. Ponce, P., Peffer, T., Molina, A.: Framework for communicating with consumers using an expectation interface in smart thermostats. Energy Build. **145**, 44e56 (2017). https://doi.org/10.1016/j.enbuild.2017.03.065
6. Mendez, J.I., et al.: Empower saving energy into smart homes using a gamification structure by social products. In: 2020 IEEE (ICCE) (2020). https://doi.org/10.1109/icce45568.2020.90431174
7. Johnson, D., Horton, E., Mulcahy, R., Foth, M.: Gamification and serious games within the domain of domestic energy consumption: a systematic review. Renew. Sustain. Energy Rev. **73**, 249–264 (2017). https://doi.org/10.1016/j.rser.2017.01.134
8. AlSkaif, T., Lampropoulos, I., van den Broek, M., van Sark, W.: Gamification based framework for engagement of residential customers in energy applications. Energy Res. Soc. Sci. **44**, 187–195 (2018). https://doi.org/10.1016/j.erss.2018.04.043
9. Lucassen, G., Jansen, S.: Gamification in consumer marketing - future or fallacy? Procedia Soc. Behav. Sci. **148**, 194–202 (2014). https://doi.org/10.1016/j.sbspro.2014.07.034

Decision on Facility Location on Base of Interval Data

Kamala Aliyeva(✉)

Department of Instrument Engineering, Azerbaijan State Oil and Industry University, 20 Azadlig Avenue, AZ1010 Baku, Azerbaijan

Abstract. Choosing a place to facility is very important in the strategic decision making of manufacturing companies. Globalization transcends national boundaries, that's why the entire world becomes an area of site selection. This of the essence changes the nature of facility location issue. Change will be especially important when considering criteria that influence selection decisions. A lot of studies considered at the general dimensions of facility location and identified the social, economic, and political criteria that affect production operations. Complication of site selection problem joint with the appearance general criteria affecting place selection for production operations creates challenging research topics. In this paper, we consider the problem of a hierarchical multidimensional decision of the facility location under interval data.

Keywords: Facility location · Interval data · Interval arithmetic · Multiple criteria decisions making

1 Introduction

The choice of location is very important process for many companies in decreasing costs and increasing apply of resources. From a narrow perspective, the location of the facility is where companies continue their production. In extensive sense, an enterprise location is the most appropriate place where companies can implement basic economic targets, logistics, production, selling functions, distributions [1]. Facility location selection is very important part of organizational strategies. The decision contains organizations seeking to place, replace or broaden their operations. The decision-making process involves identifying, analyzing and evaluating and selecting alternatives. Therefore, facility location problem begins with the recognition of a need for additional capacity or change [2]. These problems have obtained more attention over the years and different aspects, both qualitative and quantitative, have been suggested. Selection place for facility has improved theoretic knowledge [3]. Commonly, research in this area is based on optimizing technology of facility location selection [4]. The consistency of the facility location depends on the criteria that are determined during the first evaluation step.

Some basic affects for choosing place are infrastructure, total cost, rents, labor costs, energy costs, taxes, transportation, closeness to raw production resources, costs

R. A. Aliev et al. (Eds.): ICSCCW 2021, LNNS 362, pp. 371–374, 2022.
https://doi.org/10.1007/978-3-030-92127-9_51

of resources, closeness to workforce, closeness to the customers, government barriers, environmental regulation, competitive advantage, host community, quality of labor, environmental opportunities, suppliers e.g. [5]. Since the choice of location of an object depends on different factors, solution of this problem needs a delicate decision-making process. In the process of making the best decision, managers must simultaneously consider and define different factors. Therefore, a lot of various methods have been proposed and experienced over time to solve this problem. Some of methodologies involve mathematical techniques, intuitive, financial methods, simulations and modern technologies based on hierarchy such AHP, Fuzzy AHP, TOPSIS, Fuzzy TOPSIS, Fuzzy Logic, ANP [6, 14, 16]. Baumol and Wolfe solved the facility location problem with nonlinear programming [3]. Another method utilized statistic functions [7].

Other methodologies that have been adopted are dynamic programming [8], multivariate statistics and heuristic and search procedures [9]. Some uses of the recent methods include Ballı and Korukoğlu [10], Liang and Wang's works [11], employ using multi factor decision methods with fuzzy computing. Chen's study which picks optimum solution for distribution center using decision maker's linguistic expressions is also another example of the mixed method foregoing [12]. Kaboli et al. and Tabari et al. uses the AHP with fuzzy computing to select facility location [13].

This paper is constituted as next. Section 2 introduces the basic definitions of interval arithmetic for the problem. In Sect. 3, example presents applying interval arithmetic for the facility location problem, Sect. 4 presents the main results developed in the paper.

2 Preliminaries

Definition 1. An interval is a set of data, determined as $[a\ b] = \{x | a \leq x \leq b\}$.

An interval $[a\ b]$ and $[c\ d]$ are said to be equal if $a = c$ and $b = d$.

Definition 2. Conjunction of two intervals

$$[a\ b] \cap [c\ d] = [\max(a, c), \min(b, d)] \tag{1}$$

Definition 3. Disjunction of two intervals

$$[a\ b] \cup [c\ d] = [\min(a, c), \max(b, d)] \tag{2}$$

Definition 4. Superiority of intervals

$$d(X, Y) = \begin{cases} \frac{\overline{X}-\overline{Y}}{(\overline{X}-\overline{Y})+(\underline{X}-\underline{Y})} & \overline{X} > \overline{Y},\ \underline{X} > \underline{Y} \\ 1 & \overline{X} = \overline{Y},\ \underline{X} > \underline{Y}\ or\ \overline{X} > \overline{Y},\ \underline{X} > \underline{Y}\ or\ \overline{X} = \overline{Y},\ \underline{X} = \underline{Y} \\ 1 - d(X, Y) & other \end{cases} \tag{3}$$

is used for comparison and evaluating superiority of $d(X, Y)$ intervals.

3 Statement and Solution of the Problem

The proposed method applied for the choice of the facility location for manufacturing. For the choice of the best location, three alternatives are selected: A_1, A_2 and A_3. Four basic criteria have been used: C_1 – Infrastructure, C_2 – Suppliers, C_3 – Business climate, C_4 – Labor quality. The comparative weights of the factors were defined in interval form as $w_1 = [0.20\ \ 0.22]$, $w_2 = [0.3\ \ 0.32]$, $w_3 = [0.18\ \ 0.20]$, $w_4 = [0.32\ \ 0.34]$ respectively. Assume that a linguistic performance rating and codebook are represented in next form.

Very low = [0 1], Low = [1 2], High = [2 3], Very high = [3 4] (Table 1).

Table 1. Linguistic performance rating.

Attribute	C_1 infrastructure	C_2 suppliers	C_3 business climate	C_4 labor quality
A_1	High	Very high	High	Low
A_2	Very high	Very low	Low	High
A_3	Low	High	Very high	High

$$Y_i = \sum_{j=1}^{4} y_j \cdot w_j$$

$$[a, b] \times [c, d] = [\min(ac, ad, bc, bd), \max(ac, ad, bc, bd)].$$

$$Y_{A_1} = [0.20\ \ 0.22] \times [2\ \ 3] + [0.3, 0.32] \times [3\ \ 4] + [0.18, 0.20] \times [2\ \ 3] \\ + [0.32, 0.34] \times [1\ \ 2] = [0.4\ \ 0.66] + [0.9\ \ 1.28] + [0.36\ \ 0.6] + [0.32\ \ 0.68] \\ = [1.98\ \ 3.22\ ].$$

$$Y_{A_2} = [0.20\ \ 0.22] \times [3\ \ 4] + [0.3, 0.32] \times [0\ \ 1] + [0.18, 0.20] \times [1\ \ 2] \\ + [0.32, 0.34] \times [2\ \ 3] = [0.6\ \ 0.88] + [0\ \ 0.20] + [0.18\ \ 0.4] + [0.64\ \ 1.02] \\ = [1.42\ \ 2.5\ ].$$

$$Y_{A_3} = [0.20\ \ 0.22] \times [1\ \ 2] + [0.3, 0.32] \times [2\ \ 3] + [0.18, 0.20] \times [3\ \ 4] \\ + [0.32, 0.34] \times [2\ \ 3] = [0.2\ \ 0.44] + [0.6\ \ 0.96] + [0.48\ \ 0.80] \\ + [0.64\ \ 1.02] = [1.92\ \ 2.84\ ].$$

The next step ranks the overall assessments for the three locations. For this purpose, we calculate $d(X, Y)$ for each alternative by using formula (3).

$d_{A_1}(X, Y) = 0.45$, $d_{A_2}(X, Y) = 0.49$, $d_{A_3}(X, Y) = 0.53$.

After these calculations we determine that best place for facility location for manufacturing is place A_3.

4 Conclusion

In this paper is represented applying of interval arithmetic to hierarchical multicriteria decision making on facility location problem. The results represent reality of the suggested method on an optimal facility location selection by using interval arithmetic.

References

1. Ko, J.: Solving a distribution facility location problem using an analytic hierarchy process approach. In: ISAHP Proceedings Honolulu Hawaii, pp. 1991–1996 (2005)
2. Rao, R.V.: Facility Location Selection. Decision Making in the Manufacturing Environment: Using Graph Theory and Fuzzy Multiple Attribute Decision Making Methods, pp. 305–314 (2007)
3. Baumol, W.J., Wolfe, P.: A warehouse-location problem. Oper. Res. **6**(2), 252–263 (1958). https://doi.org/10.1287/opre.6.2.252
4. Brown, P.A., Gibson, D.F.: A quantified model for facility site selection-application to a multiplant location problem. AIIE Trans. **4**(1), 1–10 (1972). https://doi.org/10.1080/05695557208974822
5. Rao, R.V.: Decision Making in the Manufacturing Environment, Facility Location Selection, (SpringerLink), p. 305 (2007)
6. MacCarthy, B.L., Atthirawong, W.: Factors affecting location decisions in international operations-a Delphi study. Int. J. Oper. Prod. Manag. **23**(7), 794–818 (2003). https://doi.org/10.1108/01443570310481568
7. Wesolowsky, G.O.: Probabilistic weights in the one-dimensional facility location problem. Manag. Sci. **24**(2), 224–229 (1977). https://doi.org/10.1287/mnsc.24.2.224
8. Geoffrion, A., Bride, R.M.: Lagrangean relaxation applied to capacitated facility location problems. AIIE Trans. **10**(1), 40–47 (1978). https://doi.org/10.1007/978-3-540-68279-09
9. Bowen, W.M.: A Thurstonian comparison of the analytic hierarchy process and probabilistic multidimensional scaling through application to the nuclear waste site selection decision. Socioecon. Plann. Sci. **29**(2), 151–163 (1995). https://doi.org/10.1016/0038-0121(95)00003-5
10. Ballı, S., Korukoğlu, S.: Development of a fuzzy decision support framework for complex multi-attribute decision problems: a case study for the selection of skilful basketball players. Expert. Syst. **31**(1), 56–69 (2014). https://doi.org/10.1111/exsy.12002
11. Liang, G.S., Wang, M.J.J.: A fuzzy multi-criteria decision-making method for facility site selection. Int. J. Prod. Res. **29**(11), 2313–2330 (1991). https://doi.org/10.1016/0278-6125(96)84211-7
12. Chen, C.T.: A fuzzy approach to select the location of the distribution center. Fuzzy Sets Syst. **118**(1), 65–73 (2001). https://doi.org/10.1016/S0165-0114(98)00459-X
13. Tabari, M., Kaboli, A., Aryanezhad, M., Shahanaghi, K., Siadat, A.: A new method for location selection: a hybrid analysis. Appl. Math. Comput. **206**(2), 598–606 (2008). https://doi.org/10.1016/j.amc.2008.05.111
14. Aliev, R., Huseynov, O., Serdaroglu, R.: Ranking of Z-numbers and its application in decision making. Int. J. Inf. Technol. Decis. Mak. **15**, 1–17 (2016)
15. Aliev, R.A., Gardashova, L.A.: Z-set based approach to control system design. In: Aliev, R.A., Kacprzyk, J., Pedrycz, W., Jamshidi, M., Babanli, M., Sadikoglu, F.M. (eds.) ICAFS 2020. AISC, vol. 1306, pp. 10–21. Springer, Cham (2021). https://doi.org/10.1007/978-3-030-64058-3_2
16. Aliev, R.A., Huseynov, O.H., Zulfugarova, R.X.: Z-distance based IF-THEN rules. Sci. World J. **2016**, 9 p. (2016). https://doi.org/10.1155/2016/1673537. Article ID 1673537

Fuzzy Gain Scheduling Controller for Quadrotor

Rahib H. Abiyev(✉), Nurullah Akkaya, Ersin Aytac, and Sanan Abizada

Applied Artificial Intelligence Research Centre, Near East University, Lefkosa, North Cyprus, Turkey

{rahib.abiyev,nurullah.akkaya,ersin.aytac, sanan.abizada}@neu.edu.tr

Abstract. Quadrotor is an unmanned aerial vehicle, characterized by high order nonlinear nature and high sensitivity to external disturbances. The control of these vehicles is very challenging. This paper proposes a fuzzy gain scheduling control system for a quadrotor. A brief description of the unmanned aerial vehicle (UAV) system, its kinematics and dynamic equations are given. Dynamic equations are derived using Newton low. The fuzzy rules were developed to determine the parameters of the fuzzy PID controller. The structure of the fuzzy gain scheduling control system is proposed. The control structure and presented gain scheduling fuzzy controller were used for control of attitude and altitude of the quadrotor. The design algorithms of the fuzzy controller were presented. The fuzzy control algorithm is applied for control altitude, pitch, roll, yaw angles of a quadrotor. The results obtained from the simulations demonstrate the efficiency of using the fuzzy gain scheduling controller in real-time control of the quadrotor.

Keywords: Quadrotor · Fuzzy controller · Gain scheduling · Fuzzy rules · Control system

1 Introduction

Quadrotor unmanned aerial vehicle (UAV) can also be called helicopter has four rotors. These rotors are designed in a cross configuration and joined to propellers to provide movement of the quadrotor. The roll, pitch, yaw angles and up-thrust action are employed for control of the quadrotor. The quadrotor is used in civil and military applications. In civil applications, it is used for aerial photography of mapping, traffic monitoring, an inspection of power lines, analysis of atmosphere, detection and control of fire, rescue operations, crop monitoring, monitoring of natural disasters.

The control of the quadrotor is carried out using variations in motor speeds. However, the control of the quadrotor is not easy because of the combined underactuated design configuration, combined dynamics and unstable nature [1, 2]. The quadrotor is very sensitive to external disturbances. The dynamics of the quadrotor are highly nonlinear and complex, it has encountered several uncertainties during mission [3, 4]. For this purpose, the design of the control system of a quadrotor is a challenging task.

R. A. Aliev et al. (Eds.): ICSCCW 2021, LNNS 362, pp. 375–383, 2022.
https://doi.org/10.1007/978-3-030-92127-9_52

Quadrotor control includes navigation, obstacle avoidance and optimal path tracking which makes it more challenging. Several algorithms were developed for the quadrotor's control. In [4, 5] the authors used a PID controller for control of attitude, also, for regulation of the position of quadrotor. Because of the high order nonlinearity, the application of PID controller to the control of the quadrotor limits its performances. The paper [6] considering the variation of the aerodynamical coefficients describes the Integral Backstepping control for full control of the quadrotor. In [7] LRQ path following controller was used for collision-free operations of the quadrotor. The paper [8] used backstepping control for quadrotor control.

Adaptive control algorithms were applied to improve the control system performance. [9] used an adaptive control with feedback linearization for quadrotors. In [10] robust and the L_1 - the optimal algorithm was used for tracking control of attitude and heading. The paper [11] compared the performances of feedback linearization and adaptive sliding mode control approaches. In [12] fuzzy controller is developed and used to control the orientation and position of the quadrotor. In [13] output feedback controller was developed using neural networks (NN) and used for control of UAV having six degrees of freedom. [14] used NN for control of quadrotor in the presence of sinusoidal disturbance. The paper [15] presented the development of a hybrid fuzzy control system using the backstepping and sliding mode approach for the quadrotor.

The analysis of previous researches demonstrates that the quadrotor has high order nonlinear dynamic structure and is characterized by uncertainty in terms of parameters and outside disturbances. One of the efficient approaches for these quadrotors UAV is the usage of a gain scheduling fuzzy system for their effective control. In this paper, a gain scheduling fuzzy system is proposed for the control of the quadrotor.

2 Problem Description

Quadrotor has four rotors driven by corresponding motors, as shown in Fig. 1(a), that are mounted in orthogonal directions. The motors by rotation propellers produce powers F_1, F_2, F_3, F_4. The front and the rear rotors that are numbered as 1 and 3 turn counterclockwise, but the left and right rotors - 2 and 4 turn clockwise (see Fig. 1(b)).

By controlling the pitch θ, yaw ψ and roll ϕ angles the quadrotor track the desired position. Each rotor rotates at equal angular velocities. By selecting the velocities of the forward and backward propellers the pitch angle, by selecting the velocities of left and right propellers the roll angle is controlled. The velocities differences between the two pairs of forward-backward and left-right propellers allow selecting positive or negative yaw angle values. In quadrotor control, the basic problem is the control of rotors' velocities that will control the position and speed of the controller.

Kinematics: Quadrotor kinematics is presented by two coordinate systems. These are the earth fixed (global) frame described by $E = \{x, y, z\}$ and the body-fixed (local) frame described by $B = \{x_b, y_b, z_b\}$. These are coordinates of the quadrotor in global and local coordinate systems. The coordinate systems are presented in Fig. 1(b). Here O is the center of the mass point. There are relations between the local and global coordinates of the quadrotor. The relation is presented by the following rotation matrix [4].

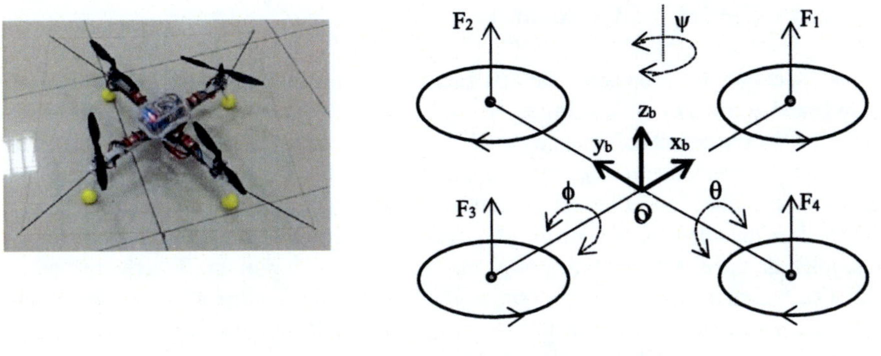

(a) (b)

Fig. 1. Configuration of quadrotor. (a) Quadrotor, (b) structure

$$R = \begin{vmatrix} \cos\psi\cos\theta & \cos\psi\sin\theta\sin\phi - \cos\phi\sin\psi & \cos\phi\cos\psi\sin\theta + \sin\phi\sin\psi \\ \cos\theta\sin\phi & \cos\phi\cos\psi + \sin\phi\sin\psi\sin\theta & \cos\phi\sin\psi\sin\theta - \cos\psi\sin\phi \\ -\sin\theta & \cos\theta\sin\phi & \cos\phi\cos\theta \end{vmatrix} \quad (1)$$

where θ, ψ and ϕ are pitch, yaw and roll angles respectively.

Dynamics: Quadrotor's dynamic model is obtained using the Newton-Euler approach. The values input forces will be formulated as:

$$\begin{aligned} &\ddot{x} = (\cos\psi\sin\theta\cos\phi + \sin\phi\sin\psi)\frac{U1}{m}; && \ddot{y} = (\sin\psi\sin\theta\cos\phi - \sin\phi\cos\psi)\frac{U1}{m} \\ &\ddot{z} = -g + (\cos\theta\cos\phi)\frac{U1}{m}; && \ddot{\phi} = \frac{(I_y - I_z)}{I_x}\dot{\theta}\dot{\psi} + \frac{J_r}{I_x}\dot{\theta}\Omega + \frac{l}{I_x}U_2 \\ &\ddot{\theta} = \frac{(I_z - I_x)}{I_y}\dot{\phi}\dot{\psi} - \frac{J_r}{I_y}\dot{\phi}\Omega + \frac{l}{I_y}U_3; && \ddot{\psi} = \frac{(I_x - I_y)}{I_z}\dot{\theta}\dot{\phi} + \frac{l}{I_z}U_4 \end{aligned} \quad (2)$$

where ϕ, θ, ψ are roll, pitch and yaw angles, J_r is the inertia of the rotor; Ω is the rotor angular velocity, I_x, I_y, I_z are the mass moment of inertia in the x, y and z axes respectively; l is the length of the rotor arm from the origin; U_1, U_2, U_3 and U_4 are the input signals of the four rotors defined as

$$U_1 = b\left(\Omega_1^2 + \Omega_2^2 + \Omega_3^2 + \Omega_4^2\right); \quad U_2 = b\left(-\Omega_2^2 + \Omega_4^2\right)$$

$$U_3 = b\left(\Omega_1^2 - \Omega_3^2\right); \quad U_4 = d\left(-\Omega_1^2 + \Omega_2^2 - \Omega_3^2 + \Omega_4^2\right) \quad (3)$$

where b and d are trust and drug coefficients respectively, Ω_i is the angular velocity of the *i*-th rotor (i = 1,.., 4) [4].

The quadrotor has four inputs and six outputs (x, y, z, ϕ, θ, ψ) as shown in a dynamic model of (2). More details can be found in [4].

3 Fuzzy Control of Quadrotor

In the paper, we are proposing a fuzzy gain scheduling PID controller for control of altitude and attitude of the quadrotor. As we know the PID controller is widely used in industries for the control of linear dynamic plants. The values of coefficients of the PID controller are selected according to the plant models. When the plant is characterized by nonlinear dynamics, uncertainties of external disturbances, then the used classical controllers does not adequately describe the considered problems. In a very large operation time interval, the effective control of plants become very important. For effective control of such dynamic plants, the determination of the corresponding values of controller coefficients become very important. In such cases, one of the approaches is manipulating the values of the controller's coefficients for producing an acceptable time response of the control system. Finding near-optimal values of the controller need to be considered [16–21]. The structure of the proposed fuzzy control system that uses a fuzzy ruler for producing the fuzzy values of the coefficients of the controller are given in Fig. 2. In this figure, according to the values of error and change of error the coefficients of the controller are determined. The association between the error and change-in-error and fuzzy controller's coefficients are given in tables. The associations between the error and change-in-error and proportional coefficient Kp, differential coefficient Kd and the integral coefficient Ki are presented in Table 1, Table 2 and Table 3 correspondingly. Here PB-positive big, PM-positive medium, PS-positive small, Z-zero, NS-negative small, NM-negative medium, NB-negative big, S-small, M-medium, B-big. These Associations are represented using If-Then rules. The linguistic values of input variables and controller coefficients are represented by the triangular membership functions.

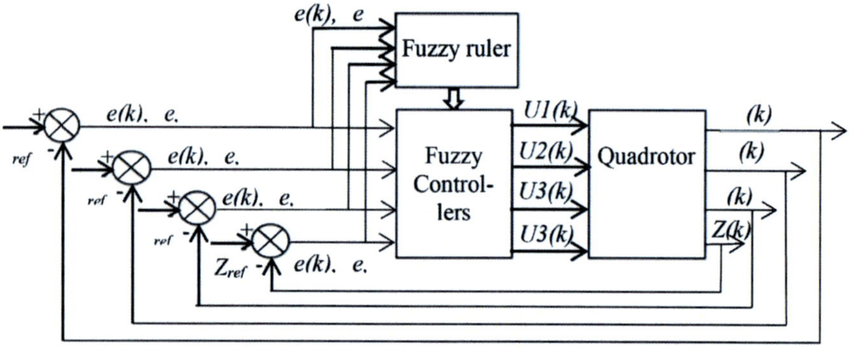

Fig. 2. Structure of fuzzy gain scheduling control system

In the paper, max-min composition is applied for finding the values of the Kp, Kd and Ki coefficients. The design of fuzzy controllers are carried out for control of altitude Z, for control of θ pitch angle, for - ϕ roll angle and for - ψ yaw angle. Online mode for each controlled variable the values of parameters of controllers are determined. After, all the determined values of K = {Kp, Kd, Ki} parameters of the fuzzy controller are

Table 1. Rules for proportional coefficients

Kp'		Change in error (e')						
		NB	NM	NS	Z	PS	PM	PB
Error (e)	NB	B	B	B	B	B	B	B
	NM	S	M	B	B	B	M	S
	NS	S	S	M	B	M	S	S
	Z	S	S	S	M	S	S	S
	PS	S	S	M	B	M	S	S
	PM	S	M	B	B	B	M	S
	PB	B	B	B	B	B	B	B

Table 2. Rules for differential coefficients

Kd'		Change in error (e')						
		NB	NM	NS	Z	PS	PM	PB
Error (e)	NB	S	S	S	S	S	S	S
	NM	B	M	S	S	S	M	B
	NS	B	B	M	S	M	B	B
	Z	B	B	B	M	B	B	B
	PS	B	B	M	S	M	B	B
	PM	B	M	S	S	S	M	B
	PB	S	S	S	S	S	S	S

scaled using maximum and minimum values of the parameters for the given plant as $K = Kmin + (Kmax - Kmin) * K$. The scaled values of parameters are used for control of the Z altitude, θ pitch angle, ϕ roll angle and ψ yaw angle.

Table 3. Rules for integral coefficients

Ki'		Change in error (e')						
		NB	NM	NS	Z	PS	PM	PB
Error (e)	NB	B	B	B	B	B	B	B
	NM	S	M	B	B	B	M	S
	NS	S	S	M	B	M	S	S
	Z	S	S	S	M	S	S	S
	PS	S	S	M	B	M	S	S
	PM	S	M	B	B	B	M	S
	PB	B	B	B	B	B	B	B

4 Simulations

The quadrotor and its control system are developed in our research laboratory. The constructed quadrotor is presented in Fig. 1(a). At first, before the implementation of the control system, we have tested the fuzzy controllers that are designed in Sect. 3 using a Matlab program. For the control purpose, we used dynamic equations of the quadrotor presented in Sect. 2. The simulation of the proposed control system is accomplished using the following parameters: Mass m = 0.38 kg, half-length l = 0.17 m, thrust coefficient b = $3.13 * 10^{-5}$, drag coefficient d = $1.1 * 10^{-6}$, moment of inertia of rotor Jtp = 0.00006 kg.m^2, moment of inertia on z-axis Iz = 0.0172 kg.m^2, moment of inertia on x-axis Ix = 0.0086 kg.m^2, moment of inertia on y-axis Iy = 0.0086 kg.m^2.

We have done simulations of altitude and attitude control of the quadrotor. The simulation results have been obtained for different values of reference signals G, where G = (Z, ϕ, θ, ψ). Starting points are taken as (0, 0, 0, 0), next we have changed reference signals as (3;0.3;0.3;0.3) then (5;0.5;0.5;0.5), (3;0.3;0.3;0.3), (5;0.5;0.5;0.5). In the simulation, the unit of altitude is meter, the unit of roll, yow and pitch angles are radian. By controlling these angles the position of a quadrotor is controlled. Their values are important for keeping a stable position. To measure the performance of the designed control system we used root mean square error (RMSE). Figure 3 presents the plot of the time response characteristic of altitude obtained using the fuzzy control system. Figure 4 presents error plots obtained for (a) altitude Z, (b) roll angle ϕ, (c) pitch angle θ and (d) yow angle ψ. The performance of the control system is evaluated using RMSE values of response characteristics. Table 4 depicts RMSE values of the control system obtained from the simulation. The performance of the gain scheduling fuzzy control system is compared with the performance of the PID control system. Obtained results indicate the advantage of a proposed control system.

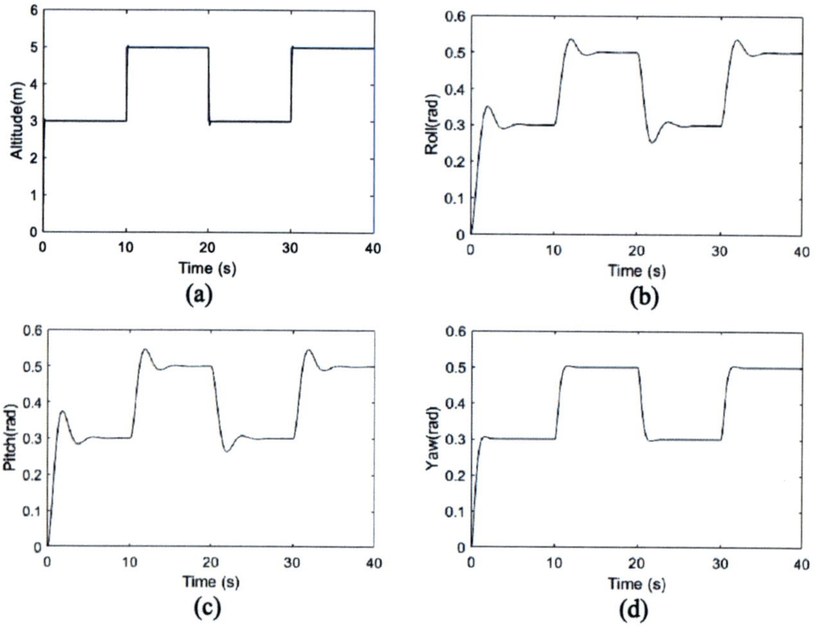

Fig. 3. Response characteristics: (a) altitude Z, (b) roll angle ϕ, (c) pitch angle θ and (d) yow angle ψ

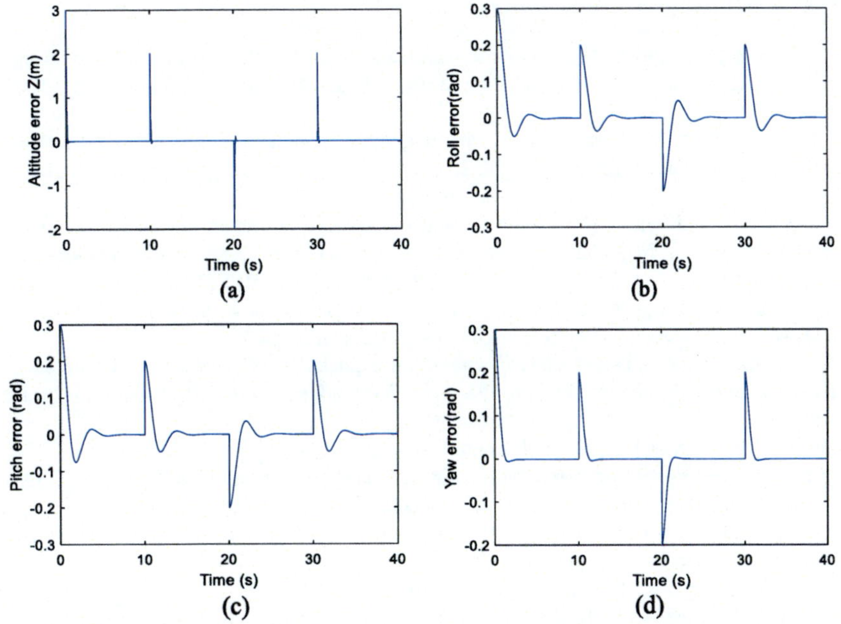

Fig. 4. Error plot for (a) altitude Z, (b) roll angle ϕ, (c) pitch angle θ and (d) yow angle ψ.

Table 4. Performance characteristics

Type of controller	Altitude (Z)		Roll angle ϕ,		Pitch angle θ		Yow angle ψ	
	SSE	RMSE	SSE	RMSE	SSE	RMSE	SSE	RMSE
PID control	64.384	0.1795	13.268	0.0815	14.08	0.08393	8.294	0.0644
Fuzzy control with gain scheduler	53.189	0.163	11.847	0.0769	11.69	0.0764	7.42	0.0609

5 Conclusions

Fuzzy control of quadrotor altitude and attitude is proposed using a fuzzy ruler of gains. The kinematics and dynamic models of the quadrotor are presented. The structure and design stages of the gain scheduling fuzzy control system are presented. Fuzzy rule bases were developed for determining the coefficients of the controller. The designed fuzzy controller is used for control of altitude, pitch, roll and yow angles. The performance characteristics of designed control systems have been obtained through simulations. Comparative results are provided in order to evaluate the efficiency of the proposed control system. The obtained results indicate the efficiency of constructed gain scheduling fuzzy controller. The designed fuzzy controllers have been implemented on real quadrotors and provided satisfactory results.

References

1. Huo, X., Huo, M., Karimi, H.R.: Attitude stabilization control of a quadrotor UAV by using backstepping approach. Math. Probl. Eng. **2014**, 1–9 (2014). https://doi.org/10.1155/2014/749803
2. Mahony, R., Kumar, V., Corke, P.: Multirotor aerial vehicles: modeling, estimation, and control of quadrotor. Robot. Automat. Mag. **19**, 20–32 (2012). https://doi.org/10.1109/MRA.2012.2206474
3. Lee, B.Y., Lee, H.I., Tahk, M.J.: Analysis of adaptive control using on-line neural networks for a quadrotor UAV. In: 13th International Conference on Control, Automation and Systems, ICCAS, pp. 1840–1844 (2013). https://doi.org/10.1109/ICCAS.2013.6704240
4. Bouabdallah, S.: Design and control of quadrotors with application to autonomous flying. Ph. D. Dissertation, Lausanne Polytechnic University, Lausanne (2007)
5. Li, J., Li, Y.: Dynamic analysis and PID control for a quadrotor. In: International Conference on Mechatronics and Automation, pp. 573–578 (2011). https://doi.org/10.1109/ICMA.2011.5985724
6. Bouabdallah, S., Siegwart, R.: Full control of a quadrotor. In: Proceedings of the 2007 IEEE/RSJ International Conference on Intelligent Robots and Systems, San Diego, CA, USA, pp. 153–158 (2007). https://doi.org/10.1109/IROS.2007.4399042
7. Cowling, I.D., Yakimenko, O.A., Whidborne, J.F., Cooke, A.K.: A prototype of an autonomous controller for a quadrotor UAV. In: European Control Conference, Kos, pp. 1–8 (2007). https://doi.org/10.23919/ECC.2007.7068316
8. Madani, T., Benallegue, A.: Backstepping control for a quadrotor helicopter. In: Proceedings of the 2006 IEEE/RSJ International Conference on Intelligent Robots and Systems, Beijing, pp. 3255–3260 (2006). https://doi.org/10.1109/IROS.2006.282433
9. Palunko, I., Fierro, R.: Adaptive control of a quadrotor with dynamic changes in the center of gravity. In: Proceedings of the 18th IFAC World Congress, Milan, pp. 2626–2631 (2011). https://doi.org/10.3182/20110828-6-IT-1002.02564
10. Satici, A., Poonawala, H., Spong, M.: Robust optimal control of quadrotor UAVs. IEEE Access **1**, 79–93 (2013). https://doi.org/10.1109/ACCESS.2013.2260794
11. Lee, D., Kim, H.J., Sastry, S.: Feedback linearization vs adaptive sliding mode control for a quadrotor helicopter. Int. J. Control Automat. Syst. **7**, 419–428 (2009). https://doi.org/10.1007/s12555-009-0311-8
12. Santos, M., Lopez, V., Morata, F.: Intelligent fuzzy controller of a quadrotor. In: Proceedings of the 2010 International Conference on Intelligent Systems and Knowledge Engineering (ISKE), Hangzhou, pp. 141–146 (2010). https://doi.org/10.1109/ISKE.2010.5680812
13. Dierks, T., Jagannathan, S.: Output feedback control of a quadrotor UAV using neural networks. IEEE Trans. Neural Network **21**, 50–66 (2010). https://doi.org/10.1109/TNN.2009.2034145
14. Boudjedir, H., Yacef, F., Bouhali, O., Rizoug, N.: Adaptive neural network for a quadrotor unmanned aerial vehicle. Int. J. Found. Comput. Sci. Technol. **2**, 1–13 (2012). https://doi.org/10.5121/ijfcst.2012.2401
15. Zeghlache, S., Saigaa, D., Kara, K., Harrag, A., Bouguerra, A.: Backstepping sliding mode controller improved with fuzzy logic: application to the quadrotor helicopter. Arch. Control Sci. **22**, 315–342 (2012). https://doi.org/10.2478/v10170-011-0027-x
16. Abiyev, R., Abizada, S.: Type-2 fuzzy wavelet neural network for estimation energy performance of residential buildings. Soft. Comput. **25**(16), 11175–11190 (2021). https://doi.org/10.1007/s00500-021-05873-4
17. Abiyev, R.H., Akkaya, N., Gunsel, I.: Control of omnidirectional robot using z-number-based fuzzy system. IEEE Trans. Syst. Man Cybern.-Syst. **49**(1), 238–252 (2019). https://doi.org/10.1109/TSMC.2018.2834728

18. Abiyev, R.: Credit rating using type-2 fuzzy neural networks. Math. Probl. Eng. **2014**, 1–8 (2014). https://doi.org/10.1155/2014/460916
19. Abiyev, R.H.: Fuzzy wavelet neural network for prediction of electricity consumption. AIEDAM: Artif. Int. Eng. Des. Anal. Manuf. **23**(2), 109–118 (2009). https://doi.org/10.1017/S0890060409000018
20. Abiyev, R.H.: Dynamic system control using z-number based fuzzy neural network. In: Kahraman, C., Cevik Onar, S., Oztaysi, B., Sari, I.U., Cebi, S., Tolga, A.C. (eds.) INFUS 2020. AISC, vol. 1197, pp. 964–972. Springer, Cham (2021). https://doi.org/10.1007/978-3-030-51156-2_112
21. Abiyev, R.H., Günsel, I., Akkaya, N., Aytac, E., Çağman, A., Abizada, S.: Robot soccer control using behaviour trees and fuzzy logic. Procedia Comput. Sci. **102**, 477–484 (2016). https://doi.org/10.1016/j.procs.2016.09.430

Application of Fuzzy Logic Model for Optimal Solution of Light Reflection Value in Lighting Calculations

Tarlan Abdullayev[1](✉), Rahib Imamguluyev[2](✉), and Niyar Umarova[3]

[1] Scientific Affairs, Odlar Yurdu University, Baku AZ1072, Azerbaijan
[2] Department of IT and Engineering, Odlar Yurdu University, Baku AZ1072, Azerbaijan
[3] "General Disciplines" Department of the Azerbaijan Academy of Labor and Social Relations, Baku, Azerbaijan

Abstract. The light reflectance value (LRV) of colors is one of the important factors in lighting calculations. Thus, in order to illuminate any space, architects, designers, and engineers first make lighting calculations in accordance with the requirements of that space. In light calculations, the light reflectance factor is taken from a standard table. In the standard tables, the light reflection coefficient of the colors is given between 0–100. Here 0-full black color 100-full white color, other color types must correspond to one of these ranges. Due to a large number of color types, it is not possible to obtain optimal results with the values given in the standard table in lighting calculations. For example, in any room, the colors of the ceiling, walls, and floor should be "very light", "very very light", "dark", "very dark", "very very dark", "a little light"., "a little dark", "really dark", "really light", "not too dark", "not too light" and so on can be noted. As can be seen, these language variables cannot be applied to classical mathematical calculations. Therefore, the fuzzy logic method is the best solution to such problems. Thus, the theory of fuzzy logic was applied to the optimal solution of the light reflection factor of the ceiling, wall, and floor in lighting issues, which allows using the knowledge and experience of designers as the main goal in this work.

Keywords: Fuzzy logic · Light reflectance value · LRV · Design · Matlab

1 Introduction

The role of light in everyday life can hardly be overestimated because without lighting the comfort of our home will be greatly reduced. Light has an impact not only on the safety of our movement around the apartment but also on health indicators. If the room is not illuminated by an insufficient number of lamps, then the following health problems may arise [1]:

R. A. Aliev et al. (Eds.): ICSCCW 2021, LNNS 362, pp. 384–391, 2022.
https://doi.org/10.1007/978-3-030-92127-9_53

- significant drop in visual acuity. In the worst case, you may need glasses and consultation with an ophthalmologist;
- decrease in the general health of households;
- the appearance of excessive irritability;
- a drop in immunity and an increase in the incidence of colds;

Note! Incorrect lighting of the room affects the health of children especially negatively.

- decrease in work productivity;
- sleep disturbance;
- decrease in the emotional background of household members.

As you can see, for each room it is necessary to calculate the required number of bulbs, with the help of which sufficient room lighting will be created.

As we found out, the lighting in the house plays a huge role. Lamps should give as much light as needed for a particular room [2].

In an apartment or a private house, each room has its own purpose and features of operation (kitchen, bedroom, living room, corridor, etc.). Particular attention in this matter must be paid to the children's room since for children even a slight deviation of the luminous flux from the norm can lead to a negative effect on the body. Each room should have its own indicator of the number of bulbs and fixtures [3].

Calculate the number of lamps that you need for a particular room should be done using special formulas. Ideally, illumination should be taken into account even at the design stage of buildings and rooms. With proper planning, lamps will provide enough light for a person to stay comfortably in a particular room. [3].

In the lighting calculations, the number of luminaires to be used, their power, and the total lighting power are revealed as a result of the lighting calculation made according to the minimum illuminance level. The following mathematical calculations are traditionally used in lighting calculations (Fig. 1) [4].

As can be seen from the formulas given in Fig. 1, several factors are used in lighting calculations. One of these factors is the light reflectance factor. The light reflectance factor of the colors is given in Table 1 as standard.

As can be seen from Table 1, the light reflectance coefficient of the colors is given between 0–100. Here, 0-full black, 100-white, and other shades of color must match one of these ranges. Due to a large number of types of one color, it is not possible to achieve optimal results with the values given in the table in lighting calculations. For example, "very light", "very very light", "dark", "very dark", "very very dark", "a little light", "a little dark", "really dark", "really light", "often" It is not possible to apply the linguistic variables mentioned in "classical" to classical mathematical calculations. Therefore, the fuzzy logic method is the optimal solution in solving such problems.

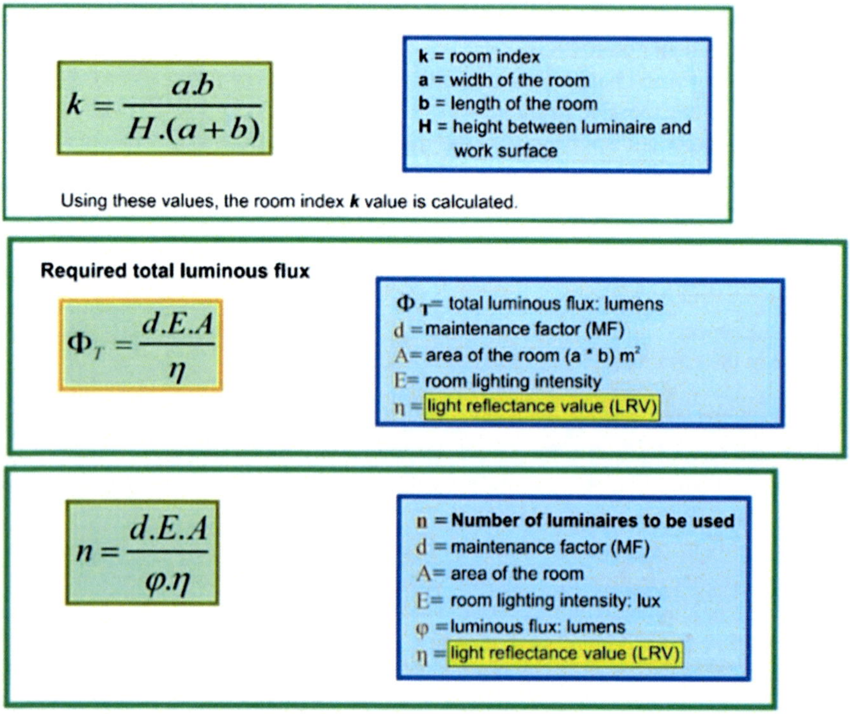

Fig. 1. Lighting account by traditional method [5].

Table 1. Light reflectance values of colors [6].

Colors	LRV
Dark colors	0–30%
Medium colors	30–70%
Light colors	70–100%

The following example illustrates the multiplicity of colors. Formally it can be written as follows:

$B = \{\text{set of dark colors}\}$

Given that the dark color starts at 0, the lower limit of this set must be 0. It is a bit difficult to set the upper limit. First, let's take the upper limit as 30. Thus, the boundaries of B are a smooth interval A = [0.30]. The question arises: if we increase 30 by 1, that is, if we make 31, how dark is it? Apparently, this problem persists no matter how much the upper limit increases [7].

The simplest way to create a set B is to weaken the critical boundary values of "dark" and "non-dark". Use more flexible judgments such as "Yes, it's dark enough" or "No, it's not too dark" without using the obvious "Yes, it's in the dark" or "No, it's not in the dark." Let's create a set B.

Let's define the phrase "not so dark" with the help of a fuzzy plural. In the example above, we rated all elements of the set as 0 or 1 in terms of affiliation. The simplest way to continue this idea is to evaluate affiliation with values between 0 and 1. More precisely, an infinite number of values between 0 or 1 can be used in the interval I = [0, 1] [8, 9].

In this way, the interpretation of all the elements of the set becomes more complicated. Thus, the value 1 corresponds to an element that is unambiguously included in set B, and 0 indicates that the element is not included in the set. Other values determine the degree of belonging to set B [10].

2 Practical Application of Fuzzy Logic

Fuzzy logic is one of the most flexible modeling methods studied in recent years. In classical logic, something is either all black or completely white. According to fuzzy logic, something can be black and part white. In his concept of fuzzy logic, Zadeh defined certain value ranges with verbal expressions that are closer to human thought, instead of Aristotle's classical plural logic, which combines two definite and distinct states that exist and does not exist, such as 1 or 0 [11–17].

Classical logic:

- Is the room light in color? NO/YES
- Is the room dark in color? NO/YES.

Fuzzy logic:

- To some extent Yes (Fig. 2).

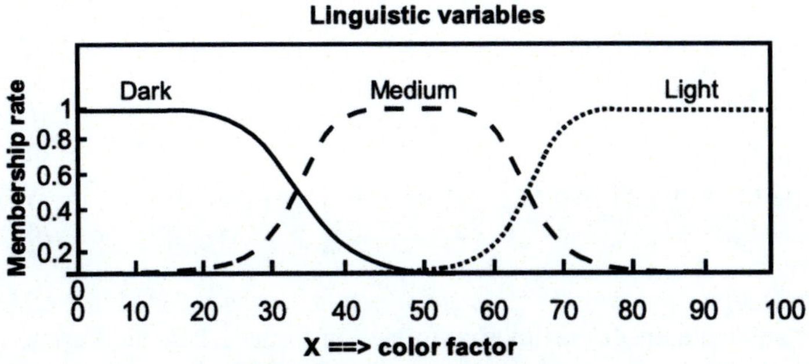

Fig. 2. Linguistic variables of color.

Now let's define linguistic variables in the following abbreviated form.

Dark-D, Light-L, Very Dark-VD, Very Very Dark-VVD, A Little Light-LL, A Little, Dark-LD, Not Very Light-NVL, Not Too Dark-NTD, Really Dark-RD, Really, Light-RL.

Definition of concepts.

$$\mu_D(x) = \left\{ \begin{array}{cc} 1, & x < 30 \\ \frac{70-x}{35}, & x \in [30, 70] \\ 0, & x > 60 \end{array} \right\}$$

$$\mu_L(x) = \left\{ \begin{array}{cc} 0, & x < 40 \\ \frac{x-40}{35}, & x \in [40, 75] \\ 1, & x > 75 \end{array} \right\}$$

$$\mu_{VD}(x) = CON\mu_D(x)^2 = \left\{ \begin{array}{cc} 1, & x < 30 \\ \left[\frac{70-x}{35}\right]^2, & x \in [30, 70] \\ 0, & x > 70 \end{array} \right\}$$

$$\mu_{VVD}(x) = CON\mu_D(x)^4 = \left\{ \begin{array}{cc} 1, & x < 30 \\ \left[\frac{70-x}{35}\right]^4, & x \in [30, 70] \\ 0, & x > 70 \end{array} \right\}$$

$$\mu_{LD}(x) = DIL\mu_D\sqrt{x} = \left\{ \begin{array}{cc} 1, & x < 30 \\ \sqrt{\left[\frac{70-x}{35}\right]}, & x \in [30, 70] \\ 0, & x > 70 \end{array} \right\} \quad (1)$$

$$\mu_{LL}(x) = DIL\mu_L\sqrt{x} = \left\{ \begin{array}{cc} 1, & x < 40 \\ \sqrt{\left[\frac{x-40}{35}\right]} & x \in [40, 75] \\ 0, & x > 75 \end{array} \right\}$$

$$\mu_{RD}(x) = \left\{ \begin{array}{c} 2\mu_D\left(\frac{70-x}{35}\right)^2, 0 \le \mu_D\left(\frac{70-x}{35}\right) \le 0.5 \\ 1 - 2\left[1 - \mu_D\left(\frac{70-x}{35}\right)\right]^2, 0.5 < \mu_D\left(\frac{70-x}{35}\right) \le 1 \end{array} \right.$$

$$\mu_{NVL}(x) = 1 - CON\mu_{NVL} = 1 - \mu_L(x)^2 = \left\{ \begin{array}{cc} 1, & x < 40 \\ 1 - \left[\frac{x-40}{35}\right]^2 & x \in [40, 75] \\ 0, & x > 75 \end{array} \right\}$$

$$\mu_{NVD}(x) = 1 - CON\mu_{NVD} = 1 - \mu_D(x)^2 = \left\{ \begin{array}{cc} 0, & x < 30 \\ 1 - \left[\frac{70-x}{35}\right]^2, & x \in [30, 70] \\ 1, & x > 70 \end{array} \right\}$$

342 production rules with 3 inputs and 1 output were developed in the MATLAB program to automate the calculations performed in formula 1 by the Sugeno method. Input parameter Color factor -» "Very Very Dark" -VVD, "Very Dark" -VD, Dark-D, "Average" -A, "Light" -L, "Very Light" -VL, Very Very Light-VVL, Output The parameter is unique: Value (Fig. 3, Fig. 4).

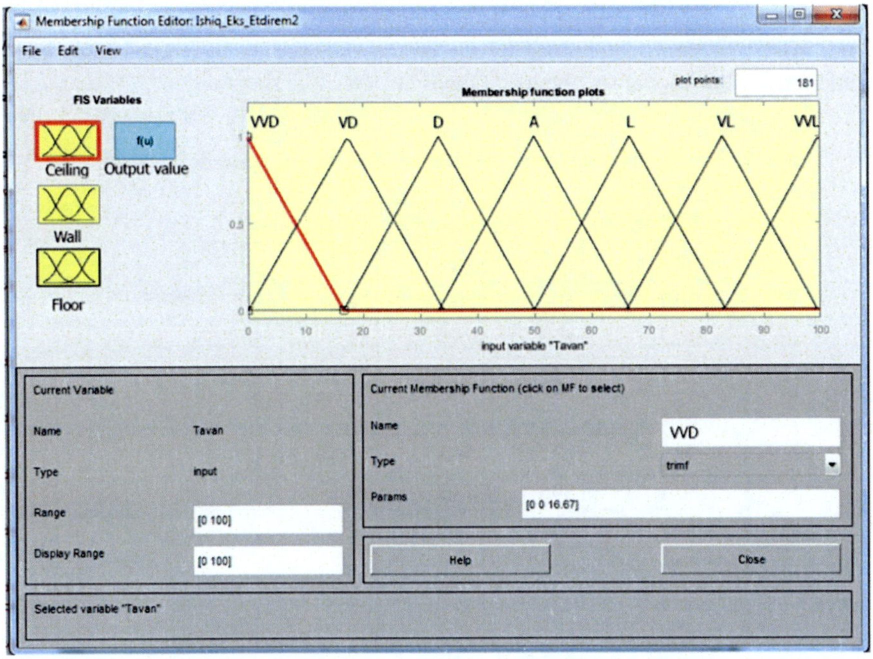

Fig. 3. Belonging functions of the ceiling linguistic variable.

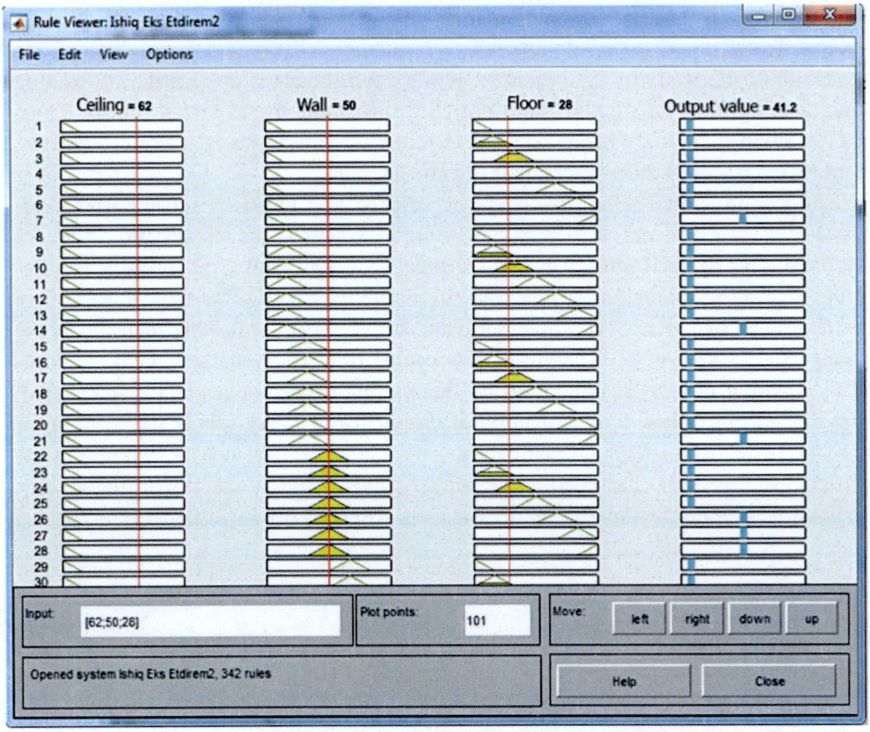

Fig. 4. Description of the rules of logical derivation.

Fluent values can be obtained by incorporating the light reflectance value of the ceiling, wall, and floor, determined by the designers based on their knowledge and experience, into the model we obtained (Table 2).

Table 2. Results obtained for the application of the logical derivation module.

Ceiling (%)	Wall (%)	Floor (%)	Output value (%)
62	50	28	41.2
30.7	50	50	18.6
36.7	69.3	78.9	57.1
78.9	75.3	35.5	55.2
84.9	64.5	28.3	50
91	82.5	54.8	65.4
80.1	86.1	78.9	90
24.7	65.7	23.5	11
53.6	47.6	18.7	16.6
66.9	42.8	45.2	39

3 Conclusion

As a result of the study of the literature, it was found that the application of existing formulas in lighting calculations does not allow obtaining the expected results. Thus, light reflection factors are based on predetermined accurate interval values. However, the transition between these intervals is not considered.

In the lighting process, the knowledge, experience, and intuition of specialists (architects, designers, engineers) are often widely used, as well as the wishes and desires of customers. This in itself often reduces the effectiveness of analytical methods in solving the lighting problem. Because analytical methods do not allow the use of human knowledge and experience. On the other hand, some of the parameters involved in the lighting process are verbal. Therefore, they cannot be used in mathematical methods, and the lighting process is influenced by factors that cannot be accurately measured. From this point of view, the application of the theory of fuzzy sets to the solution of lighting problems, which allows using the knowledge and experience of designers as the main goal in this work, has been implemented algorithmically and programmatically. The analysis of other factors in lighting issues is also relevant as future work.

References

1. Villalba, A., Monteoliva, J.M., Pattini, A.: Development of a simplified light reflectance value assessment tool for indoor surface coverings. Indoor Built Environ. **30**, 970–984 (2020). https://doi.org/10.1177/1420326X20925138

2. Shapiro, I.M.: Lighting reflectances. In: Energy Audits and Improvements for Commercial Buildings (April 2016). https://doi.org/10.1002/9781119174851.app8
3. Grzesiak, W., Solnica, D.: Intelligent lighting systems with a very high value of color rendering index. Przeglad Elektrotechniczny **94**(8), 9–12 (2018). https://doi.org/10.15199/48.2018.08.03
4. Yoo, S., Javanainen, J.: Light reflection and transmission in planar lattices of cold atoms. Opt. Express **28**(7), 9764–9776 (2020). https://doi.org/10.1364/OE.389570
5. http://epokulu.com/aydinlatma-hesabi/
6. https://thelandofcolor.com/lrv-light-reflectance-value-of-paint-colors/
7. Imamguluyev, R.A.: Determination of correct lighting based on fuzzy logic model to reduce electricity in the workplace. In: Conference: International Conference on Eurasian Economies, Baku, Azerbaijan (September 2020). https://doi.org/10.36880/C12.02456
8. Imamguluyev, R.A.: Optimal solution of indoor lighting system based on fuzzy logic. In: Conference: 2nd International Baku Conference on Scientific Research, Baku, Azerbaijan (May 2021)
9. Valiyev, A., Imamguluyev, R., Ilkin, G.: Application of fuzzy logic model for daylight evaluation in computer aided interior design areas. In: Aliev, R.A., Kacprzyk, J., Pedrycz, W., Jamshidi, M., Babanli, M., Sadikoglu, F.M. (eds.) ICAFS 2020. AISC, vol. 1306, pp. 715–722. Springer, Cham (2021). https://doi.org/10.1007/978-3-030-64058-3_89
10. Imamguluyev, R., Suleymanli, T., Umarova, N.: Evaluation of the effectiveness of integration processes in production enterprises based on the fuzzy logic model. In: Aliev, R.A., Kacprzyk, J., Pedrycz, W., Jamshidi, M., Babanli, M., Sadikoglu, F.M. (eds.) ICAFS 2020. AISC, vol. 1306, pp. 133–139. Springer, Cham (2021). https://doi.org/10.1007/978-3-030-64058-3_17
11. Imamguluyev, R.: Application of fuzzy logic model for correct lighting in computer aided interior design areas. In: Kahraman, C., Cevik Onar, S., Oztaysi, B., Sari, I.U., Cebi, S., Tolga, A.C. (eds.) INFUS 2020. AISC, vol. 1197, pp. 1644–1651. Springer, Cham (2021). https://doi.org/10.1007/978-3-030-51156-2_192
12. Aliev, R., Tserkovny, A.: Fuzzy logic for incidence geometry. In: Kosheleva, O., Shary, S.P., Xiang, G., Zapatrin, R. (eds.) Beyond Traditional Probabilistic Data Processing Techniques: Interval, Fuzzy etc. Methods and Their Applications. SCI, vol. 835, pp. 49–85. Springer, Cham (2020). https://doi.org/10.1007/978-3-030-31041-7_4
13. Zadeh, L.A., Aliev, R.: Fuzzy Logic Theory and Applications: Part I and Part II, p. 61 (December 2018). https://doi.org/10.1142/10936
14. Zadeh, L.A.: Toward a theory of fuzzy information granulation and its centrality in human reasoning and fuzzy logic. Fuzzy Sets Syst. **90**, 111–127 (1997)
15. Aliev, R., Gardashova, L., Huseynov, O.: Z-numbers-based expected utility. In: Proceedings of Tenth International Conference on Application of Fuzzy Systems and Soft Computing, Lisbon, pp. 49–61 (2012)
16. Aliev, R.A., Huseynov, O.H., Aliyev, R.R., Alizadeh, A.V.: The Arithmetic o-Numbers: Theory and Applications. World Scientific, Singapore (2015)
17. Aliyev, R.R.: Construction of consistent z-preferences in decision making for a foreign market selection. In: Aliev, R.A., Kacprzyk, J., Pedrycz, W., Jamshidi, M., Babanli, M., Sadikoglu, F.M. (eds.) ICAFS 2020. AISC, vol. 1306, pp. 30–37. Springer, Cham (2021). https://doi.org/10.1007/978-3-030-64058-3_4

Fuzzy-Multiple Model of Adjustment of the "Smart House" Mobile Application Under the User Level

Ansari Mustafa[1], Alexander N. Kuzminov[1](✉), Michael B. Stryukov[1], and Oleg A. Ternovsky[2]

[1] Rostov State University of Economics, Rostov-on-Don, Russian Federation
[2] Platov South-Russian State Polytechnic University (NPI), Novocherkassk, Russian Federation

Abstract. The methodological foundations for adapting the Smart Home application to the user level have been developed for its most efficient operation and optimization of learning to work with the system. We propose to use three levels of mastering a custom mobile application: Beginner, Basic, Classic. At each level, we use our own sequence of work with the system, permissible actions and opportunities for working with the Smart Home, as well as training moments in work. It is assumed that training in working with the system is organized in such a way that the user can eventually move from one level to another. To recognize the level of the user, a system of productive rules, fuzzy inferences, based on the inference of Mamdani, was used. The creation of productive rules is based on the statistics of the user's daily actions when working with the Smart Home system.

Keywords: Smart Home mobile application · Fuzzy sets · Productive rules · Fuzzy-multiple model · User level

1 Introduction

Currently, one of the most Smart Homes means a system that controls electrical loads, lighting, security, climate control, media devices. The main advantage of the "Smart Home" system is that it simplifies the daily life of a person and increases the comfort of living. One notable example is that lights throughout the house can be turned off thanks to the voice assistant, without touching the switch. Almost limitless options for possible scenarios will allow you to put the home on the alarm, thereby de-energize all the sockets, leaving only a couple of necessary ones, turn on the music, turn off the gas supply, close the curtains, regulate the temperature and leave it with a calm sense of security. The basic principles of forming the Smart Home system are considered in a wide range of articles. So, in [1], the main aspects of the reliability of the "smart home" system are considered: the presence in the building of a unified control system for all engineering systems; availability of computing power sufficient for decision-making; elimination of all maintenance personnel of the building and transfer of control functions, decision-making and execution of actions to the "intelligent center" and its subsystems; the ability to instantly turn off computer control of a smart home and transfer control

R. A. Aliev et al. (Eds.): ICSCCW 2021, LNNS 362, pp. 392–400, 2022.
https://doi.org/10.1007/978-3-030-92127-9_54

functions to a person; own automation, ensuring their safe operation when the central control is turned off; the system must be configured for expansion. In [2], an algorithm for optimizing information exchange processes in security and monitoring systems of an automated control system, based on the use of a closed exponential model of a queuing network and characterized by the ability to operate with fuzzy sets, is considered. Article [3] gives the concept of a smart home, which means a system that controls electrical loads, lighting, security, climate control, media devices. Almost unlimited options for possible scenarios will allow you to set the apartment on an alarm, turn off all the sockets, leaving only a couple of necessary ones, turn on music, etc. Smart home scenarios are pre-programmed behavior of the smart home system for a particular event. An event can be a specific keystroke, a command from a control panel or a computer, a timer triggering and a scheduled operation. However, the most promising and complex is the automatic way of managing scenarios for various events. In work [4], methods of constructing and managing an intelligent house are considered. It also describes the increase in comfort, thanks to the technology "Smart Home". The following functions are affected: optimization of lighting, heating, protection and control of the house remotely, using mobile devices and special key fobs.

The article [5] is devoted to the creation of an integrated application solution for smart home control based on Arduino. A list of tasks for automation, a conceptual architecture of such a solution for implementation on a mobile platform is proposed. The solution of several problems is shown on a practical example. At the same time, one of the main tasks of research work in this direction is the development of intelligent applications that optimize the user's work with the Smart Home system. In particular, one of the most important tasks is the creation of intelligent applications focused on a specific user, his level of knowledge of computer technology and preferences, as well as allowing training the user in the process. The author's approach to the integration of methods and technical systems (cenoses) is described in detail in [6].

Our article is aimed at developing a methodological basis for solving this problem.

2 Methods and Materials: The Proposed System of User Levels and Its Implementation by the "Smart Home" System

It is proposed to introduce the following types of users into consideration: 1) Beginner (for example: children); 2) Basic, classic (a person who understands technology and how to use it, but does not know the meanings); 3) Advanced, experienced (a person who understands equipment and is technically savvy). Description of user types is performed with examples. *Purpose:* The user needs to set up a smart split and automate it so that it works from 17:00 to 19:00 in order to cool the apartment by the arrival of the owner in the summer.

Note: The user just bought a split and never used it, and never entered the "smart home" application.

Elementary. Upon entering the application, the user is asked to pass a test that determines the level of his knowledge of the "Smart Home" system, as well as the level of his smartphone skills in general. Let's assume that the test determines that the user is of the "initial type". In addition, the user is invited to pass the Bartle test, designed to determine his psychotype as a gamer, which will be used further to select the most preferred interface, considering psychological characteristics, as well as learning based on gamification (the theory of serious games). After passing the proposed tests, the entry-level user sees the following interface, similar to the interfaces of ticket vending machines in banks or hospitals. The specified interface contains all possible actions that the user can perform at this stage. This is done intentionally so as not to overload the user with redundant information and give him the opportunity to figure it out on his own. In addition, a button for connecting "smart devices" to a "smart home" appears on the screen. Pressing this button leads to the appearance of new buttons in the interface; the arrows indicate the buttons that must be pressed in order, for example, to connect a split. After connecting the split, all possible functions and actions with this split will be displayed, and its main functionality will be displayed on the main screen (enable/disable). Next to the main tips, a hint "how to use the advanced settings of a smart split" will appear. The performance of the recommended actions is accompanied by the appearance of a message (achievement), notifying the user of his initial achievement. When you press the button in the form of an arrow, a full-fledged split settings menu is displayed, where it can be adjusted.

A similar picture is observed when working with other smart devices of a smart home. An n-number of devices can be connected to the Smart Home application, and they can be repeated. Having dealt with the on/off buttons and split temperature adjustment, the user wants to automate this, but does not yet know how to do it. The hint on the home screen will help him by demonstrating how to set the on/off timer for a specific time. Thanks to the implementation of the listed actions, the goal is realized - to show the initial user how the smart home system works and teach him the basics of working with the application. After that, the user can go to the knowledge section and continue his training in using the Smart Home application, which will eventually lead to the transition to the next level.

Base. Suppose the initial test identifies the user as a "baseline". After passing the test, he is greeted by an interface with standard application functionality and a search system. All buttons are already unlocked, and for questions or documentation, the user can refer to the "knowledge" section. To connect to the split, the user is invited to use an intuitive design and an icon for connecting smart devices. Using the knowledge and skills he already has in working with computer equipment, the user will connect the split using Wi-Fi or Bluetooth technologies (depending on the device) and configure it. In our example, a split appears, so the user will connect the split and figure out how to adjust the temperature on his own. However, he may wonder how to automate the split before his return home? He can find the answer in the "knowledge" section, or by using the search bar inside the application, where there are typical examples. After that, the user will have the opportunity to adjust the settings according to their needs. The user training does not end there: he has options for saving energy. That is, if at the appointed time, for example, at 19:00, the user does not come home from work, the split will stop working and the house will heat up again, and the user will have to turn on the split again. The user, at his discretion, can visit the knowledge base and learn new functions, as well as their meaning, in order to switch to the "advanced type". It is essential that the system periodically, with a given frequency, will remind the user what can be improved in his actions.

Advanced. The interface for the "Advanced user" does not differ in functionality from the basic one, but with one caveat. The user can already use the entire system and knows where the desired button is or will be. He is not afraid of the situation when he cannot find the necessary lever, since he knows where to look or learn this information. The advanced type of user does not require intensive training. But even in this case, training is present, since, for example, the user may know how to automate the split when entering the application for the first time. In addition, a human factor is required: for example, the user can forget that open windows and the enabled split contradict each other, and the system will remind him of this. Further, the main role includes the Achievements, Communication and Information system, which help the user not only master the management system, but also accumulate points and bonuses, participate in leaderboards, communicate (advise others and receive advice himself) in chats, as well as take part in contests and group games. It should be noted some more common features characteristic of the interface of different levels. The general level system shows the obvious reasons for the compatibility or incompatibility of the equipment (open windows and split on, closed doors and appliances in the house, morning 12:00 and the lights are on, etc.)

For advanced users (experienced), allows on/off system of prompts, describes unloading from statistics for a month/week/day. Suggests possible ways of saving.

For basic users (classic), the system cautiously talks about new features (so as not to get bored) and about the values obtained, trying not to scare off the user (it is disabled in the settings if the basic user does not want to develop to an advanced one).

For beginners, the system guides the user, showing and, if desired, telling how to use the application. At the same time, it hides all redundant functions, buttons, etc. and opens them as needed, in the process of learning from an intuitive interface.

User Level Test. All questions in this work are ordered in ascending order of answers, since it is easier for a person to perceive information, in the final version they will be scattered randomly so that there is no realization that the 4th answer is for an advanced user, and the first is for a beginner. Speaking about the test (see Appendix 1), it should be noted that it was not fully developed and this test does not have a final version, since many formulations are incorrect and will need to be corrected as testing progresses.

To move the user to the next level, a test will be given again to check his level of knowledge during the time that he was at his level, as well as there will be an interactive game where the user will be in the virtual "smart home" application, that is, from his clicks, change or nothing will happen, and to complete the transition to the next level, he will need to indicate where the desired button is located and click on it, which he learned during the moment he was at the level.

Achievement System. The system is intended for people who are happy that they have been awarded for this or that activity and also for those who want to explore the full functionality of the application, thereby being able to show their friends their achievements and how they figured out the application.

The achievement system is structured as follows. A person should be encouraged every time he has done something new for himself and the achievement should end by demonstrating all the possibilities of the application.

Examples. On/off light for the first time. Connect a smart vacuum cleaner. Open the directory. Share a report on the energy savings with your friends. Get 1000 views per thread. Save electricity more than last month.

Points are awarded for each action.

Family Access. Since in the modern world smartphones are an integral part of our life and not only an adult, but even children have it, a family access system was invented, where the child is given the opportunity to control certain smart devices, that is, to use only those devices that were provided to him adult. In addition, just as there can be not only one adult in a family, but also several, the main one can share the configured system with another user of the application. Having general access to all the functionality with his own profile and his own training system, an additional user can also control all devices.

Guest Mode: Provides functionality to allowed components, so that guests can control certain devices, for example, if they are staying overnight and need to adjust something.

Rating System. Included in the system of achievements, shows how many percent of users have achieved the same result. On the social tab there is functionality and a description of the user himself, including his achievements, on the same tab a leaderboard is displayed, which depends on the formula Ai * E/100, where A is the number of achievements, E is the electricity saved for the period.

The "socialization" tab includes some aspects, like a kind of forum on various topics, where users can create threads to demonstrate how you can make everyday life at home easier and share their results. There will be a section on the negative system so that you can update the search for various devices and connections in the house and the most cuddled ones never remain in the top, thereby the system guarantees an endless feed and support for new publications from the community. This system will work in the order of dislikes, where 0 dislikes will be shown at the very top, and the more negative likes, the lower the article, this will help make the feed universal. In order not to lose the best developments, there will be a section for the best in time (month/year/day), where the outdated system will be located, which will contain the best work. This social mechanism is designed to keep the audience inside the application and provide them with all sorts of options for communicating with each other on this topic.

3 Fuzzy-Multiple Model for Determining the Level of the User

We will introduce a system of designating user levels of ownership of the Smart Home mobile application: *U1 - Beginner; U2 - Basic (classic); U3 - Advanced (experienced).*

To assess the knowledge, skills and abilities available to the user, a fuzzy-multiple variable X is introduced – "LEVEL OF PROFESSION OF THE APPLICATION SMART HOUSE". It is proposed to introduce a system of fuzzy rules:

Rule 1: If X = X1 ("The level of mastery of the Smart Home system is low"), then Y = Y1 ("User's level is initial");

Rule 2: If X = X2 ("The level of mastery of the Smart Home system is average"), then Y = Y1 ("User's level is basic (classic");

Rule 3: If X = X3 ("The level of mastery of the Smart Home system is high"), then Y = Y3 ("The level of the user is advanced (experienced)".

The corresponding fuzzy sets are formed on the basis of statistics of user actions and reflect the number of points scored by him for the period under consideration.

A possible variant of the scale is: (1–25) - development of the initial user, where level 25 is the transition to the basic user; (26–75) - development of the basic; (76–100) - development of advanced (experienced).

Condition: If the user immediately chose the basic or advanced user mode for himself, the previous levels will not be passed and it will be necessary to also enter the training system to confirm the knowledgeable information. After choosing the estimation intervals corresponding to the user levels, the centers of the corresponding intervals are determined and fuzzy sets are formed. After that, the recognition of the user level is carried out based on the above-described system of productive rules.

4 Conclusion

This article presents a fuzzy logic-based mechanism for developing a smart home model based on the inference of Mamdani. An analysis was carried out to determine possible criteria for ranking users of elements of such an IoT system. The results showed that the use of a grading of users based on the results of test assessment of their competencies provides the best solution to the problem, since it has an important advantage of reducing possible risks during use. Fuzzy logic provides an assessment of dynamic changes in user competencies. The reference values were obtained based on empirical modeling of the composition of components typical for a large city, and a variant of ranking users was proposed. Further work will be related to the multi-criteria pairing of the tasks of deploying mobile applications for Smart Home systems.

Acknowledgements. The study was carried out with the financial support of the Russian Federal Property Fund in the framework of the scientific project No. 19-310-90046.

Appendix 1

- Let's start with the basics of a smartphone:

1. What does the home button do with one press?
a) Builds a house; b) Pressed; c) Out to the desktop; d) Exits, from applications or games and to navigate smartphone screens.

2. What is a notification?
a) Unknown; b) Document; c) Message with sound; d) Small pop-ups on the device screen.

3. When you press the "lock" button and the "home" or "quieter" button, what happens?
a) I don't know; b) The application will close; c) The phone is locked; d) Take a screenshot.

4. The front camera is:
a) The camera behind the phone; b) The camera used to shoot the video for TikTok; c) Selfie camera; d) Camera located on the front of the phone.

5. What is the "lock screen" for?
a) Block users; b) Watch movies; c) Watch the time; d) View notifications and do not accidentally tap the screen.

6. What is displayed in the "notification panel"?
a) I don't know; b) All documents; c) Messages from Whatsapp; d) Notifications from the entire smartphone system.

• Let's move on to the knowledge of the application interface.

7. What does the word "tap" mean?
a) I don't know; b) Crush cockroaches; c) Click on the screen; d) Single finger touch.

8. What is this word "swipe"?
a) I don't know; b) Electronic cigarette; c) Finger movement; d) Swiping your finger across the screen.

9. In order to postpone, what needs to be done?
a) I don't know; b) Turn the binoculars; c) Click on "-"; d) Bring two fingers together to one point.

10. What is Sidebar?
a) I don't know; b) Bar with lemon juice only; c) Menu; d) Left menu, which can be accessed by swiping from left to right.

11. What is the "floating button" for? for example: "+".
a) I don't know; b) To be caught; c) For additional settings; d) To perform certain actions.

12. What gesture is there to go to the tab on the right?
a) I don't know; b) By hands; c) Click on the desired tab; d) Swipe from right to left.

13. What's in the menu button or "3) x dot"?
a) I don't know; b) Add-ins; c) Application Editor; d) Application settings.

• The following questions are aimed at knowing the values in the smart home application:

14. What are degrees Fahrenheit?
a) I don't know; b) Weather; c) Temperature; d) Temperature unit.

15. What is Celsius degree?
a) I don't know; b) Real weather; c) Also temperature; d) Common temperature unit.

16. How many seconds are there in an hour?
a) 18; b) 100; c) 60; d) 3600.

17. How is household electricity measured?
a) I don't know; b) Volts; c) Watts; d) W / h.

18. How is the average determined?

a) I don't know; b) Calculate; c) Number of numbers divided by their group; d) Group of numbers divided by their number.

19. What is MB?

a) I don't know; b) Mercedes-Benz; c) Megabits; d) Megabyte.

20. How to calculate the percentage?

a) I don't know; b) At a discount; c) Divide by 100; d) Divide the number by 100 and multiply by the number of percent.

Keys to the Answers: Based on the test results, you can determine the level of user proficiency: 1) from 0 to 100 – Initial; 2) from 100 to 200 – Basic; 3) from 200 to 300 – Advanced.

		Answers			
		A	B	C	D
Questions	1	0	0	5	10
	2	0	0	5	10
	3	0	0	5	10
	4	0	0	5	10
	5	0	0	5	10
	6	0	0	5	10
	7	0	0	5	10
	8	0	0	5	10
	9	0	0	5	10
	10	0	0	5	10
	11	0	0	5	10
	12	0	0	5	10
	13	0	0	5	10
	14	0	0	5	10
	15	0	0	5	10
	16	0	0	5	10
	17	0	0	5	10
	18	0	0	5	10
	19	0	0	5	10
	20	0	0	5	10

References

1. Rychkova, V.A.: Reliability of the "Smart Home" system as the main criterion for the efficiency of the object's functioning. Bull. Sci. Educ. **4–2**(58) (2019). (in Russian)

2. Eremenko, V.T.: Optimization of information exchange processes in the security and monitoring system of the automated control system using the example of the Smart House technology. Inform. Comput. Eng. Manag. **7**, 146–151 (2017). https://doi.org/10.12737/article_5940f01a8d60b0.87715921
3. Firouzi, R., Rahmani, R., Kanter, T.: An autonomic IoT gateway for smart home using fuzzy logic reasoner. Procedia Comput. Sci. **177**, 102–111 (2020). https://doi.org/10.1016/j.procs.2020.10.017
4. Hakpyeong, K., Heeju, C., Hyuna, K., Jongbaek, A., Seungkeun, Y., Taehoon, H.: A systematic review of the smart energy conservation system: from smart homes to sustainable smart cities. Renew. Sust. Energy Rev. **140**, 110755 (2021). https://doi.org/10.1016/j.rser.2021.110755
5. Jasmine Gnana Malar, A., Agees Kumar, C., Gnana Saravanan, A.: IoT based sustainable wind green energy for smart cites using fuzzy logic based fractional order Darwinian particle swarm optimization. Measurement **166**, 108208 (2020). https://doi.org/10.1016/j.measurement.2020
6. Ansary, M., Kuzminov, A., Ternovsky, O.: Technocenosis paradigm sustainability management of a forest enterprise. In: IOP Conference Series: Earth and Environmental Science, Voronezh, p. 012008 (2020). https://doi.org/10.1088/1755-1315/595/1/012008

Generation of Fuzzy-Chaotic Behavior in Secure Communication System

K. M. Babanli(✉)

Istanbul University, Cerrahpasa 34320, Avcılar, Istanbul, Turkey

Abstract. In this paper, we propose a new approach to generation of chaotic masking signal which has a higher level of security compared to traditional masking methods. An originality of the approach is conditioned by its relatively low computational complexity due to the use of fuzzy rule base instead of intensive simulations of a fuzzy chaotic system. Due to efficiency and accuracy requirements, we use combination of fuzzy logic and chaos theories to model uncertainty and complexity of real complex secure communication system. Complexity of this hybrid fuzzy-chaotic approach assures an increased level of security. Trade-off between complexity and security may be achieved by generation of transmitted information using fuzzy modeling of specific desired chaotic behavior. Computer simulations are provided to verify feasibility and effectiveness of the proposed approach.

Keywords: Fuzzy sets · Fuzzy numbers · Chaotic behavior · Lorenz system · Secure communication · Fuzzy differential equation

1 Introduction

Communication systems are characterized by high level of uncertainty. In view of this, we intend to consider such systems as fuzzy chaotic systems. The latter can theoretically be described by using fuzzy differential equations (FDEs). However, achieving a desired chaotic behavior by simulating of FDEs system is computationally costly. We propose to build fuzzy rule base to describe influence of Lorenz system parameters and initial conditions to a level of chaotic behavior. This helps to generate a desired chaotic behavior under lower analytical and computational complexities. The latter contains information on parameters and initial conditions of Lorenz system and corresponding values of Lyapunov dimension [1]. Complexity of this hybrid fuzzy-chaotic approach assures an increased level of security.

Let us shortly overview existing works on modeling of circuits with chaotic behavior by using classical and fuzzy methods.

In [2, 3] they propose a new chaotic system based on the Euler equation of the motion of rigid body. To implement this system, an electronic circuit was designed. The dynamic behavior of the system is described by three ordinary differential equations with quadratic nonlinear terms.

R. A. Aliev et al. (Eds.): ICSCCW 2021, LNNS 362, pp. 401–411, 2022.
https://doi.org/10.1007/978-3-030-92127-9_55

In [4], the fractional-order Chen-Lee system implementation by using chain fractance and tree fractance circuits is considered. It is shown that such system can be used to create electronic circuits for secure communication purposes.

Control of chaos in the Chen-Lee system by multiple time delays is discussed in [5]. The time delay constants efficiency was discussed. Proposed system displays fixed points and chaotic motion.

In [6] they propose a methodology for application of fuzzy modeling to synthesis of discrete-time chaotic systems of Lure type. The advantages of the proposed cryptosystem: 1) it is systematic; 2) it is flexible in terms of chaotic signals selection for secure key generator; 3) it is flexible in terms of masking the ciphertext using state or output; 4) it is characterized by time-varying super increasing sequence.

In [7] authors use FDE for modeling a resistor–inductor circuit. By using an FDE, dynamic behavior of system, control and anti-control of chaos are analyzed. Also, the implementation of Takagi-Sugeno (TS) fuzzy-chaotic systems on electronic circuits is proposed.

In [8], they discuss construction of electronic circuit using the Chen-Lee system and consider stability of the given equilibrium points. It is shown that such system is useful for encryption purposes. In view of this, using FDEs to model Chen-Lee system under uncertainty is effective for electronic circuit analysis. The latter, in turn, can be applied for analysis of numerous uncertain factors in telecommunication area.

As it is shown in [9], chaotic behavior in nonlinear dynamics is difficult to detect and control. Particularly, theoretical results of chaos analysis are hard to use in practice, especially taking into account imprecision intrinsic to the results of numerical simulations of dynamical systems. For this purpose, fuzzy set theory can be used to model uncertainty of chaotic behavior.

In [10], a new chaotic secure communication scheme is proposed. The scheme is based on reduction of channel noise by using a Modified Gravitational Search Algorithm. The proposed method is verified in a unified chaotic map. It is shown a better performance of the method as compared to existing secure communication schemes in terms of lower bit error rate.

To the best of the authors' knowledge, works on the use of fuzzy chaotic masking of transmitted information to increase security level of communication system do not exist in the literature to date. This interesting property of fuzzy chaotic systems makes them very useful in secure communication systems.

The paper is structured as follows. An overview of existing works is given in Sect. 1. Some preliminaries used in the paper, including definitions of a fuzzy number, Lorenz system, Lyapunov dimension are provided in Sect. 2. Section 3 is devoted to a statement of problem, Lorenz system parameters and initial conditions on Lyapunov exponents (which are a measure of chaotic behavior). Section 4 is devoted to solution of a problem stated in Sect. 3.

2 Preliminaries

***Definition 1. Fuzzy Number** [11].* A fuzzy number is a fuzzy set A on real line R which possesses the following properties: a) A is a normal fuzzy set; b) A is a convex fuzzy set;

c) α-cut of A, A^{α} is a closed interval for every $\alpha \in (0, 1]$; d) the support of A, supp$A$ is bounded.

***Definition 2. A Generalized Hukuhara Difference** [12, 13].* A generalized Hukuhara difference between fuzzy numbers A, B is a fuzzy number C, if it exists, such that

$$A -_{gH} B = C \Leftrightarrow \begin{cases} 1)\ A = B + C \\ or\ 2)\ B = A + (-1)C \end{cases},$$

where addition and scalar multiplication of fuzzy numbers are defined as usual [14].

***Definition 3. Lorenz System** [15].* Lorenz system is described by the following differential equations:

$$\frac{dx_1}{dt} = a(x_2 - x_1),\ \frac{dx_2}{dt} = cx_1 - x_1x_3 - x_2,\ \frac{dx_3}{dt} = x_1x_2 - bx_3$$

where x_1, x_2, x_3 are state variables and a, b, c are parameters. The behavior of this system under some values of a, b and c is chaotic.

***Definition 4. Lyapunov Exponents** [11, 16].* Lyapunov exponents are a measure of a dynamical system's predictability and sensitivity to changes in its initial conditions (stability). They can be thought of as the average logarithmic rate of separation or convergence of two nearby points of two time series $x(k)$ and $y(k)$, $k = 0, ..., M, ...,$ separated by an initial distance $\Delta R_0 = \|x(0) - y(0)\|^2$:

$$\lambda = \lim_{M \to \infty} \frac{1}{M} \sum_{k=1}^{M} \ln\left|\frac{\Delta R_k}{\Delta R_0}\right|$$

***Definition 5. Lyapunov Dimension** [11].* Lyapunov dimension D_λ is calculated as:

$$D_\lambda = j + \sum_{i=1}^{j} \lambda_i / |\lambda_{j+1}|,$$

where Lyapunov exponents are ranked as $\lambda_1 \geq \lambda_2 \geq ... \geq \lambda_n$, and value of j is found from the requirements:

$$\sum_{i=1}^{j} \lambda_i \geq 0, \sum_{i=1}^{j+1} \lambda_i < 0.$$

The concept of Lyapunov dimension was introduced [17] as an estimation of fractal dimension of attractors.

3 Lorenz System with Fuzzy Parameters

Modern complex secure communication systems are characterized by high order uncertainty and, therefore, need an increased level of security. An effective way to solve this problem is to use fuzzy logic. Namely, generation of fuzzy-chaotic complex dynamic behaviors allows to significantly increase security level of communication system by fuzzy-chaotic masking the transmitted message.

In analyzing fuzzy-chaotic systems, we need to consider possible complex dynamical behaviors stemming from initial conditions. However, it is more important to account for influence of system parameters. The issue is that even small changes in parameters values may lead to significant changes in the system behavior modes, yielding chaotic behavior.

The most well-known nonlinear chaotic system is Lorenz system (Definition 3). Let us consider a fuzzy Lorenz system as a system with fuzzy number-valued initial conditions $x_1(0)$, $x_2(0)$, $x_3(0)$ and fuzzy parameters a, b, c. In this case, trajectories of state variables will be described by fuzzy functions $x_1(t)$,$x_2(t)$, $x_3(t)$ of time t, that satisfy:

$$\begin{aligned} \frac{dx_1(t)}{dt} &= a(x_2(t) - x_1(t)), \\ \frac{dx_2(t)}{dt} &= cx_1(t) - x_1(t)x_3(t) - x_2(t), \\ \frac{dx_3(t)}{dt} &= x_1(t)x_2(t) - bx_3(t), \end{aligned} \tag{1}$$

Thus, uncovering desired chaotic behavior of fuzzy Lorenz system requires solving of system (1) under various combinations of parameters a, b, c and initial conditions $x_1(0)$, $x_2(0)$, $x_3(0)$. Conducting such intensive simulations is computationally complex, and we will use fuzzy If-Then rules to describe influence of fuzzy initial conditions and fuzzy parameters to system dynamics. This may help to uncover desirable fuzzy-chaotic behavior of the Lorenz system. It is known that the chaotic behavior can be analyzed on the basis of Lyapunov dimension D_λ. The latter is computed on the basis of Lyapunov exponents λ_1, λ_2, λ_3.

4 Solution of Lorenz System With Fuzzy Parameters

At first, let us provide the necessary definitions related to fuzzy-number-valued functions and derivatives used in existing literature. For convenience of notation, we denote fuzzy sets by using small letters.

Denote by E^n the space of all normal, convex, and upper semicontinuous fuzzy sets $x \subset R^n$ with compact support supp x [22].

In order to model derivatives of fuzzy functions in the right-hand side of Eqs. (1), let us consider the concept of strongly generalized differentiability as a formulation of differentiability of fuzzy number-valued mappings [12]. In this concept, a generalized Hukuhara difference (Definition 2) is used implicitly for the first time. Namely, the two possible cases of this difference are used (each case is denoted "–" as a special case of

"$-_{gH}$"). These two cases form four types of differentiability. We will consider only two main types here.

Strongly Generalized Differentiability *[12].* Let $x : T \to E^n$, and $t_0 \in T \subset R$. We say that x is strongly generalized differentiable at t_0 if there exists an element $x'(t_0) \in E^n$, such that

(i) for all $h > 0$ sufficiently small, $\exists\, x(t_0 + h) - x(t_0), x(t_0) - x(t_0 - h)$ and the limits (in supremum metric [22])

$$\lim_{h \to 0+} \frac{x(t_0 + h) - x(t_0)}{h} = \lim_{h \to 0+} \frac{x(t_0) - x(t_0 - h)}{h} = x'(t_0), \text{ or}$$

(ii) for all $h > 0$ sufficiently small, $\exists\, x(t_0) - x(t_0 + h), x(t_0 - h) - x(t_0)$ and the limits

$$\lim_{h \to 0+} \frac{x(t_0) - x(t_0 + h)}{(-h)} = \lim_{h \to 0+} \frac{x(t_0 - h) - x(t_0)}{(-h)} = x'(t_0),$$

(h and $(-h)$ at denominators mean $1/h$ and $1/(-h)$ respectively).

The type (i) is not suitable for engineering purposes as the support of a fuzzy number $x(t)$ increases, and, for majority of cases increases unboundedly. In the type (ii) the support of $x(t)$ decreases. We will use the type (ii) in this study.

Let $x : T \to E^1$ be strongly generalized differentiable. Then $x_l^\alpha(t)$ and $x_r^\alpha(t)$ are differentiable and one has [13]:

$$\left[x'(t)\right]^\alpha = [\min([x_l^\alpha(t)]', [x_r^\alpha(t)]'), \max([x_l^\alpha(t)]', [x_r^\alpha(t)]')].$$

The fuzzy analog of Lorenz system, system (1) will be considered as a system of fuzzy differential equations in terms of (ii)-type of strongly generalized differentiability [12]:

$$x' = f(t, x) \tag{2}$$

where $f \in C^1\left[R_+ \times E^n, E^n\right]$, $x(t_0) = x_0 \in E^n, t \geq t_0, t_0 \in R_+$.

Now, we proceed to modeling of a fuzzy Lorenz system. Assume that a, b, c parameters in (1) are described by fuzzy numbers. Then by using strongly generalized differentiability, one can approximate solution to (1) by using difference equations:

$$x_1(k + 1) = x_1(k) + a(x_2(k) - x_1(k))\Delta t,$$

$$x_2(k + 1) = x_2(k) + (cx_1(k) - x_1(k)x_3(k) - x_2(k))\Delta t,$$

$$x_3(k + 1) = x_3(k) + (x_1(k)x_2(k) - bx_3(k))\Delta t,$$

$$x_1(0) = x_{10}, x_2(0) = x_{20}, x_3(0) = x_{30}, k = 1, \ldots K.$$

Following the existing tradition, we proceed with α-cuts-based representation of these equations. For convenience of notations, denote $a^\alpha = a(\alpha)$, $b^\alpha = b(\alpha)$, $c^\alpha = c(\alpha)$, $x_i^\alpha = x_i(k, \alpha)$, $i = 1, \ldots, 3$. We obtain:

$$x_1(k+1,\alpha) = x_1(k,\alpha) + a(\alpha)(x_2(k,\alpha) - x_1(k,\alpha))\Delta t,$$

$$x_2(k+1,\alpha) = x_2(k,\alpha) + (c(\alpha)x_1(k,\alpha) - x_1(k,\alpha)x_3(k,\alpha) - x_2(k,\alpha))\Delta t,$$

$$x_3(k+1,\alpha) = x_3(k,\alpha) + (x_1(k,\alpha)x_2(k,\alpha) - b(\alpha)x_3(k,\alpha))\Delta t,$$

$$x_1(0,\alpha) = x_{10}(\alpha),\ x_2(0,\alpha) = x_{20}(\alpha),\ x_3(0,\alpha) = x_{30}(\alpha).$$

For lower and upper bounds of the α-cuts, we have the following two systems:

$$x_{1l}(k+1,\alpha) = x_{1l}(k,\alpha) + a_l(\alpha)(x_{2l}(k,\alpha) - x_{1l}(k,\alpha))\Delta t,$$

$$x_{2l}(k+1,\alpha) = x_{2l}(k,\alpha) + (c_l(\alpha)x_{1l}(k,\alpha) - x_{1l}(k,\alpha)x_{3l}(k,\alpha) - x_{2l}(k,\alpha))\Delta t,$$

$$x_{3l}(k+1,\alpha) = x_{3l}(k,\alpha) + (x_{1l}(k,\alpha)x_{2l}(k,\alpha) - b_l(\alpha)x_{3l}(k,\alpha))\Delta t,$$

$$x_{1l}(0,\alpha) = x_{10l}(\alpha),\ x_{2l}(0,\alpha) = x_{20l}(\alpha),\ x_{3l}(0,\alpha) = x_{30l}(\alpha);$$

$$x_{1r}(k+1,\alpha) = x_{1r}(k,\alpha) + a_r(\alpha)(x_{2r}(k,\alpha) - x_{1r}(k,\alpha))\Delta t,$$

$$x_{2r}(k+1,\alpha) = x_{2r}(k,\alpha) + (c_r(\alpha)x_{1r}(k,\alpha) - x_{1r}(k,\alpha)x_{3r}(k,\alpha) - x_{2r}(k,\alpha))\Delta t,$$

$$x_{3r}(k+1,\alpha) = x_{3r}(k,\alpha) + (x_{1r}(k,\alpha)x_{2r}(k,\alpha) - b_r(\alpha)x_{3r}(k,\alpha))\Delta t,$$

$$x_{1r}(0,\alpha) = x_{10r}(\alpha),\ x_{2l}(0,\alpha) = x_{20r}(\alpha),\ x_{3r}(0,\alpha) = x_{30r}(\alpha).$$

For solving the obtained systems of difference equations, we will use Euler method. Consider the following TFN-based initial conditions and a, b, c parameters:

$$x_{10} = (0.4, 0.5, 0.6), x_{20} = (0.3, 0.4, 0.5), x_{30} = (0.5, 0.6, 0.7);$$

$$a = (10,\ 15,\ 20),\ \ b = (3,\ 4,\ 5),\ \ c = (40,\ 45,\ 50).$$

One has a chaotic behavior. The values of D_λ computed by fuzzy equations are $D_\lambda = 2.005$.

The value of D_λ computed by using fuzzy rule base describes this behavior closely.

Thus, a huge number of experiments shows efficiency of the created fuzzy rule base for generation of desired fuzzy chaotic behavior to mask transmitted signals.

The corresponding lower and upper bounds of α-cuts of the obtained solution $(x_1(t), x_2(t), x_3(t))$ are shown in Figs. 1, 2 and 3 respectively.

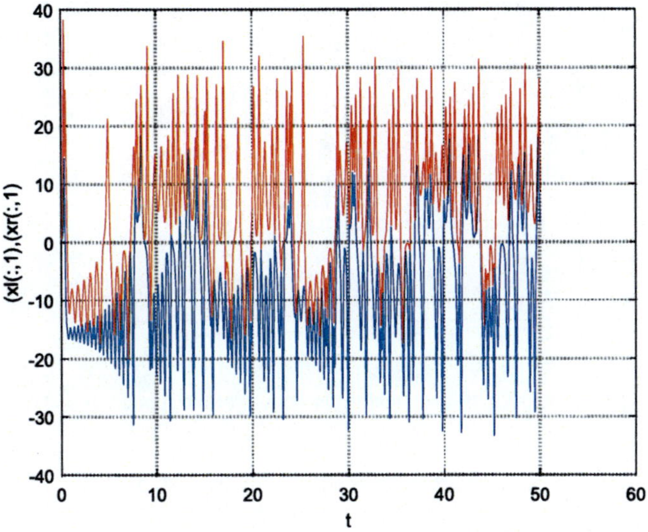

Fig. 1. Dynamics of lower and upper bounds of the support of $x_1(t)$

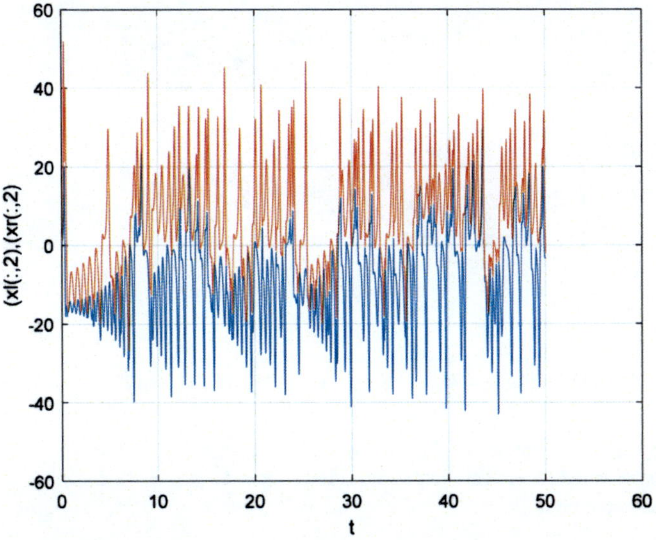

Fig. 2. Dynamics of lower and upper bounds of the support of $x_2(t)$

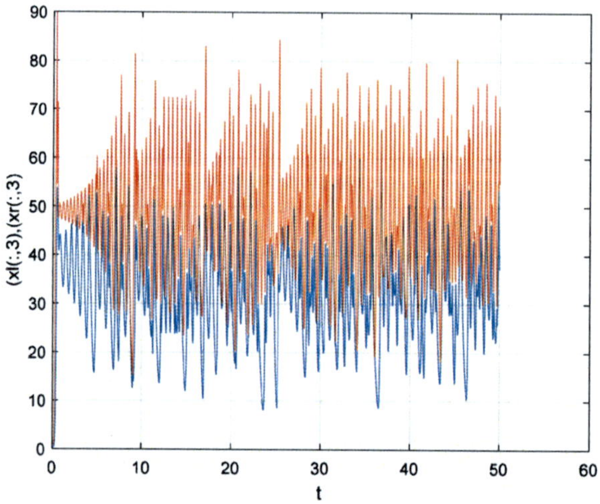

Fig. 3. Dynamics of lower and upper bounds of the support of $x_3(t)$

The corresponding phase trajectories are shown in Figs. 4, 5 and 6.

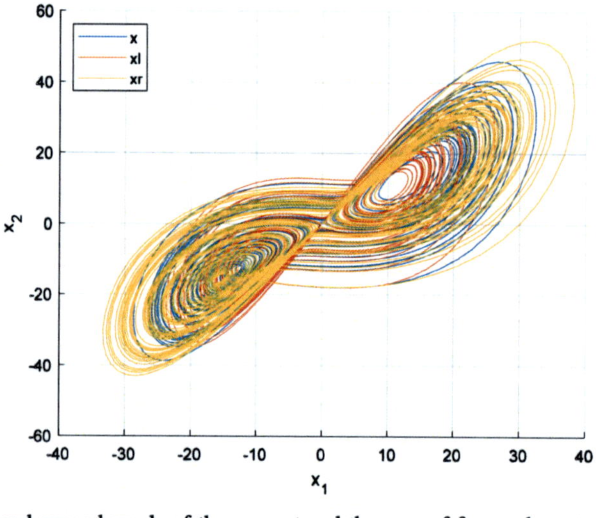

Fig. 4. Lower and upper bounds of the support and the core of fuzzy phase trajectories in the space of x_1 and x_2 variables (red curve describes lower bound, blue curve describes core, yellow curve describes upper bound)

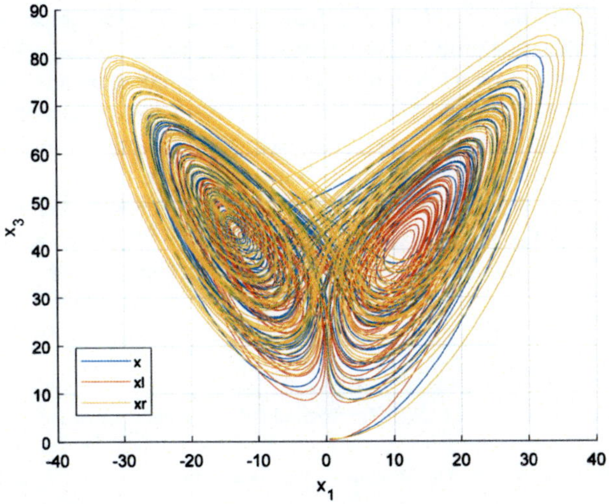

Fig. 5. Lower and upper bounds of the support and the core of fuzzy phase trajectories in the space of x_1 and x_3 variables (red curve describes lower bound, blue curve describes core, yellow curve describes upper bound)

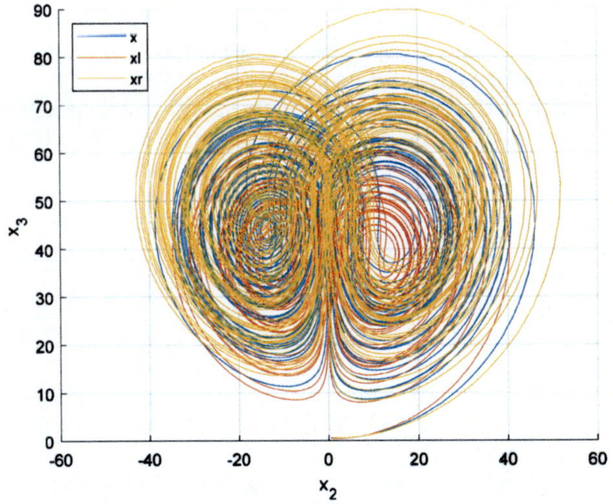

Fig. 6. Lower and upper bounds of the support and the core of fuzzy phase trajectories in the space of x_2 and x_3 variables (red curve describes lower bound, blue curve describes core, yellow curve describes upper bound)

5 Conclusion

A generation of fuzzy-chaotic behavior of secure communication system with transmitter and receiver under uncertainty of parameters is considered in this paper. Consideration of fuzzy-chaotic behaviors is only the beginning of secure communication issues. To respond to this challenging problem, we proposed an approach to linguistic description of Lorenz system expressed as system of fuzzy differential equations with uncertain parameters and initial conditions. For solving of the considered system, the generalized Hukuhara derivative is used. Even small changes in fuzzy parameters of the system lead to qualitative change in system dynamics. Huge simulation results allowed to create database that includes information on Lyapunov exponents and Lyapunov dimension. More complex fuzzy chaotic signal generated on the basis of the proposed approach can be applied to high-level communication security, file encryption simulation etc.

References

1. Babanli, K.M.: Construction of device for fuzzy chaos signal generation. In:14th International Conference on Theory and Application of Fuzzy Systems and Soft Computing ICAFS-2020, pp. 30–37, Springer, Budva, Montenegro (2021). https://doi.org/10.1007/978-3-030-64058-3_14
2. Li, S.Y., Ge, Z.M.: Generalized synchronization of chaotic systems with different orders by fuzzy logic constant controller. Expert Syst. Appl. **37**(3), 1357–1370 (2011). https://doi.org/10.1016/j.eswa.2010.08.018
3. Chen, H.-K., Lee, C.-I.: Anti-control of chaos in rigid body motion. Chaos Solit. Fractals. **21**(4), 957–965 (2004)
4. Wang, S.-P., Lao, S.-K., Chen, H.-K., Chen, J.-H., Chen, S.-Y.: Implementation of the fractional-order Chen-Lee system by electronic circuit. Int. J. Bifurcation chaos **23**(02), 1350030 (2013). https://doi.org/10.1142/S0218127413500302
5. Chen, G.H.: Controlling chaos and chaotification in the Chen-Lee system by multiple time delay. Chaos Solit. Fractals **36**(4), 843–852 (2008). https://doi.org/10.1016/j.chaos.2006.10.049
6. Chiu, C.-S., Lian, K.-Y.: Fuzzy model based chaotic cryptosystems. Integr. Fuzzy Logic Chaos Theory **187**, 507–525 (2006). https://doi.org/10.1007/3-540-32502-6_18
7. Silvio, A.B., Salgado, L.F., Danilo, M.P., Felipo, A.V.: Fuzzy differential equation in a RL type circuit. Int. J. Adv. Appl. Math. Mech. **6**(1), 43–48 (2018)
8. Tam, L.-M., Chen, H.-K., Sheu, L.-J., Lao, S.-K.: Alternative implementation of the chaotic Chen-Lee system. Chaos Solit. Fractals **41**(4), 1923–1929 (2009). https://doi.org/10.1016/j.chaos.2008.07.053
9. Castillo, O., Melin, P.: Theory of fuzzy chaos for the simulation and control of nonlinear dynamical systems. Integr. Fuzzy Logic Chaos Theory **187**, 391–414 (2006). https://doi.org/10.1007/3-540-32502-6_14
10. https://www.sciencedirect.com/science/article/abs/pii/S0020025512003180
11. Aliev, R.A., Aliev, R.R.: Soft Computing and Its Applications. World Scientific, New Jersey, London, Singapore, Hong Kong (2001). https://doi.org/10.1142/4766
12. Bede, B., Gal, S.G.: Generalizations of the differentiability of fuzzy-number-valued functions with applications to fuzzy differential equations. Fuzzy Sets Syst. **151**, 571–599 (2005)
13. Stefanini, L.: A generalization of hukuhara difference. In: Didier Dubois, M., Lubiano, A., Prade, H., Gil, M.Á., Grzegorzewski, P., Hryniewicz, O. (eds.) Soft Methods for Handling Variability and Imprecision, pp. 203–210. Springer, Heidelberg (2008). https://doi.org/10.1007/978-3-540-85027-4_25
14. Lakshmikantham, V., Mohapatra, R.N.: Theory of fuzzy differential equations and inclusions. New York, Taylor & Francis, London (2003).https://doi.org/10.1201/9780203011386

15. Lorenz, E.N.: Deterministic non-periodic flows. J. Atmos. Sci. **20**(2), 130–141 (1963). https://doi.org/10.1175/1520-0469(1963)020%3c0130:DNF%3e2.0.CO;2
16. Sprott, J.C.: Chaos and Time-Series Analysis. Oxford University Press, New York (2003)
17. Kaplan, J.L., Yorke, J.A.: Chaotic behavior of multidimensional difference equations. In: Peitgen, H.-O., Walther, H.-O. (eds.) Functional Differential Equations and Approximation of Fixed Points, pp. 204–227. Springer, Heidelberg (1979). https://doi.org/10.1007/BFb0064319

Exploring of the Problematic Industry 4.0 and Platform-Based Economic Development

Nadirbek Yusupbekov[1](✉), Shuhrat Gulyamov[1], Nargiza Usmanova[2], and Jamshid Khoshimov[2]

[1] Tashkent State Technical University, University Street, Tashkent, Uzbekistan2
[2] Tashkent University of Information Technologies, Amir Temur Street, Tashkent, Uzbekistan108

Abstract. The implementation of the digital economy is linked closely with the provisioning pillars of 'Industry 4.0' (e.g. functional man-machine compatibility while communicating directly via the Internet; transparency of information and the ability of systems to create a virtual copy of the physical world; technical support of machines to humans with combining large amounts of data and performing a number of specific tasks; the ability of systems to independently and autonomously make decisions), and being realized through the digital enterprises and production. The development of related technologies (i.e. Internet of Things, 5G, Cloud computing, artificial intelligence, etc.) and digital platforms ensured the emergence of open information systems and global industrial networks interacting each other and going beyond the boundaries of a single enterprise. Such systems and networks are having a transformative impact on all sectors of today's economy and businesses and transfer industrial automation to a new stage of industrialization. Considering and discussing the features of Industry 4.0 ecosystem, the authors emphasize the significance of cyber physical system and platform based manufacturing implementation (along with describing the idea of using fuzzy cognitive map) as conceptual and systematic bases for the use in the context of provisioning elements of the 'Industry 4.0' ecosystem and developing of integrative mechanisms to stimulate and improve overall system performance.

Keywords: Industry 4.0 · Digital economy · Digital manufacturing · Internet of Things · IoT platform

1 Introduction

As world practice shows, the widespread use of information and communication technologies (ICT) in various areas of the economy in next decades is bringing the physical and digital world closer and deeply integrates with the comprehensive transformation of traditional industries while stimulating the development of the digital economy.

The modern economy is developing under the influence of various trends, the most important of which can be noted as 'servicization' and digitalization, which are technologically and logically interconnected and should be studied on the basis of special

R. A. Aliev et al. (Eds.): ICSCCW 2021, LNNS 362, pp. 412–419, 2022.
https://doi.org/10.1007/978-3-030-92127-9_56

approaches and methods. Given the importance of such interconnection and interdependence, it is crucial to consider these trends from the standpoint of 'platformization' as a methodological approach that combines the concept of the platform economy as a social structure and technological content of the digital economy from the standpoint of supporting any type of services and communications.

The experience of global development shows that industrialization remains the main path to successful step forward, allowing countries to create and develop competitive skills and capabilities, thereby allowing them to succeed in a new technological paradigm. According to the results of the analysis (http://inecprom.spbstu.ru/blog/), digital manufacturing technologies, when applied to industrial production, offer enormous potential for economic growth and improving human well-being, as well as protecting the environment and achieving the goals of the Agenda in areas of sustainable development for the period up to 2030.

The implementation of the digital economy is linked closely with the provisioning pillars for fourth industrial revolution, 'Industry 4.0', and it being developed through/by the functioning of the digital manufacturing, enterprises and digital production [1]. The fourth stage of the industrial revolution is characterized mainly by the introduction of cyber-physical systems (CPS) into manufacturing processes. It is assumed that these systems will be integrated into one network, communicating with each other in real time, being autonomously managed and react to environment. This in turn, make them allow to adapt to new consumer demands. The concept of the Industry 4.0 in turn, can be described based on the followings: functional compatibility of man and machine - the ability to contact directly via the Internet; information transparency and the ability of systems to create a virtual copy of the physical world; technical assistance of machines to humans - combining large amounts of data and performing a number of specific tasks; the ability of systems to independently and autonomously make decisions. These aspects are logically gathered within the notion of Industry 4.0 platform, as a distinctive concept of the platform economy [2] that is essential today in all economic sectors, while having a platform one can run analytics and manage applications in order to get value from the Internet of Things.

Motivated by the complex and uncertain nature of digital manufacturing platform while implementing Industry 4.0, this paper explores the role of cyber-physical systems associated with digital manufacturing platforms, and proposes an approach for representing the functionality of platform on attracting the behavior of its components.

2 Overview of Problematics

The digital economy is established as a main driver of global growth whereas many countries around the world are determining measures to combine efforts to further development of the digital economy; at the same time, the digital economy will ensure the growth of gross domestic product by significant percentage [3]. Economic value is closely related with the production of goods and services. New business models emerged with the digital economy, driving value creation on platformization and the monetization, while rapidly expanding volume of digital data.

Overall technological development within 'all digital' context is changing the nature of the economies, both for country-scale and regional. Industrial competition is growing, markets are expanding as well. Obviously, the growth of national economies becomes evident and realistic. The digital economy changes the traditional models of businesses and markets, while determining the trends in growing of companies, industries and national economies in general. Industry 4.0 herewith links a number of key technologies such as Industrial Internet of Things, Big Data Analytics, Cybersecurity, Additive Manufacturing and Augmented Reality [4]. In turn, developed technological platform being the distinctive feature of Industry 4.0, can be defined by specific attributes, like using high-tech machines and equipment, ICT solutions, and cyber-physical systems that allow system integration.

From the research and development (R&D) perspective, the possibility of developing cyber-physical systems with wide range of smart components having abilities to sense the environment and adapt to changes, cause the issues of processing the ever increasing amount of production data, unpredictable due to the external and internal factors of the nature of functioning the components of industrial enterprises, while special models, methods and tools of artificial intelligence (AI) allow to take into account various features and innovations in the organization of the production process [5]. It is obvious to note this and related R&D are evolving with tremendous pace while attracting the interest of scientists and engineers all around the world. As an example, one can refer to [6–9], to name a few.

For the most comprehensive definition of CPS we refer to one given by the NIST in [10], wherein CPS are defined as smart systems including engineered interacting networks of physical and computational components. These highly interconnected and integrated systems provide new functionalities to improve quality of life and enable technological advances in different application areas. At the same time, CPS are closely related with Industrial Internet, Internet of Things (IoT), machine-to-machine, smart cities, and others to describe similar or related systems and concepts.

From the context of this research, the CPS are characterized by the close interaction between computational processes and physical processes and elements that constantly receive data from the environment and use them to arrange better the control processes. So far, the operation nature of CPS can be described via deep relationship between physical and computational elements while the AI and other technologies support the processing data from sensors in the real world, analyze this data and use it to further control physical elements. This interaction gives the CPS the ability to work effectively in changing conditions. Moreover, the cycle 'management - data acquisition - data processing – management' with the well-functioning system should yield positive results each time and create new value.

The paper [11] gives a good review slice for digital manufacturing platforms and systems within current research domain, and almost all of the platforms have the challenging issue of components to be effectively coordinated, among other significant technological and functional features (e.g. levels of integration in manufacturing, organizational, managerial, and technical implications, etc.). Moreover, the concerns of functional aspects of digital manufacturing platforms cover mainly manageability, measurability, monitorability, performance, physical context, uncertainty as the most attractive to the overall system capabilities [10]. Managing the effects of uncertainties can be seen as a fundamental challenge in CPS. Indeed, the uncertainty of data in sensing and actuation caused by manufacturing processes leads to incomplete knowledge on assessing the system behavior and achieving required performance.

The main contribution of this paper thus to establish the conceptual and systematic approach in gaining knowledge from the components of CFS which is represented as distributed environment with respective functional properties for nodes and entities.

3 Methodology

Any manufacturing system, in its nature, can be considered as a group of interacting or interrelated components (e.g. industrial equipment, software for various purposes, technological equipment, production facilities, etc.), that should be identified, be able to exchange information and interact while performing technological and production processes [11, 12]. Developing and implementing the industrial cyber-physical platform is aimed to collecting of data on the production facilities, the processes performed and the resources used for their implementation, as well as the processing of the information received and creating of knowledge bases.

Industrial cyber-physical systems, in simplified view, can be represented as layered structure:

1) physical layer containing all the physical objects involved in the production system;
2) virtual space consisting of interconnected digital 'twins' of the objects: digital twins record historical data about physical counterparts and provide it to the cloud.
3) cloud of services containing various software applications provided by equipment manufacturers and service providers.

From the modeling perspective (to attract the features of this complex and uncertain environment), this structure is considered as a distributed information state space, representing a collection of objects associated with or related to processes. Thus, methodologically, the decomposition in relation to the components and the representation of the state space for various components and information is performed in such a way that appropriate connections can be defined in the model. Such representation allows to define and detail a set of internal structures of a distributed environment, its components, their connections and possible interactions between components, including the properties, by specifying the corresponding structural elements. The set of objects that form the distributed space is determined depending on the task, and the behavior of objects, the establishment of relations between objects, the interfaces for each object and the implementation of the object are assigned by the rules. To specify the behavior (attract properties) of distributed objects, fuzzy systems relationships are used (for modeling causal relationships in a dynamic system), which are widely implemented in modeling of complex systems.

The knowledge about the system behavior entered into the fuzzy modeling system is reflected by the structure of nodes and relationships (positive, negative) in the structure (Fig. 1). In other words, existing knowledge on the behavior of the system will be reflected in the structure of nodes and relationships in the map, whereas fuzzy cognitive map is substantiated for solving the problem of optimal control of the behavior of a concept node (*i-th* entity) and the development of management decisions.

Each concept node represents one of the key factors of the system, the values of the weights indicate the value (strength) of influence of the entity C_i on C_j. The intensity of the influence of one concept-entity on another is an indicator with the help of which the transition from the classical relation to the fuzzy relation W is carried out, the elements w_{ij} of which characterize the direction and degree of intensity or the weight of the influence between the concepts C_i and C_j, i.e. $w_{ij} = w\ (C_i,\ C_j)$ when w_{ij} defines the characteristic function of the ratio W.

For *n* concepts, the fuzzy cognitive map is mathematically given by *1xn* state vector V with values of *n* concepts, and *nxn* by a matrix of weights (cognitive matrix) W. The value of V_i of each concept C_i is determined by the operation: $V^t = f(V^{t-1}W + V^{t-1})$, which computes the new value of the state vector as a result of interaction between concepts.

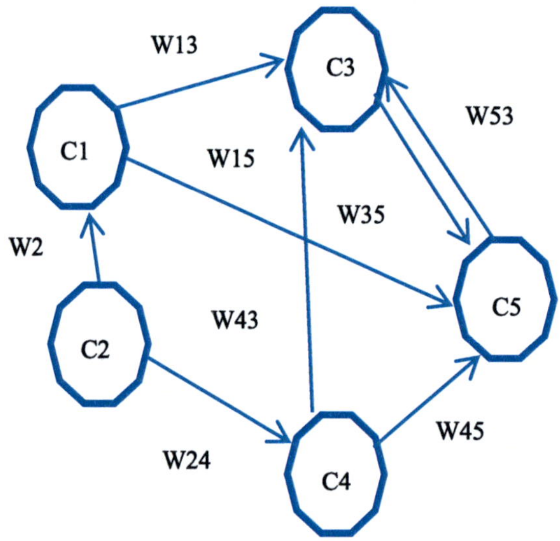

Fig. 1. Fuzzy cognitive map

The object (component) can be described based on the actual functionality it carries, including fragmented behavior, object allocation, resource sharing, etc., and based on the specific requirements of entities that may exist in production life cycle. By using cognitive maps in different scenarios, it is possible to describe the behavior and investigate various aspects of distributed functionality, features of the behavior of distributed objects by reflecting the properties of modeling elements in a multi-dimensional representation of structure and functionality.

Although the goal of this paper is not to get into the details of thorough modeling (this task is within the broader scope for further research), authors state the abovementioned provisioning of fuzzy relations and multi-agent systems being a separate issue as the decision-making process in fuzzy conditions: in particular, how close the behavior of a distributed object is to a particular property expected from interaction of components, the degree of such 'closeness', etc. in narrow view, and the possibilities of improving physical production processes by using information obtained by monitoring objects of the production system – in broad view of considered task.

4 IoT Platforms and Use Case Demo

For collecting and processing data from sensors in Smart Manufacturing solutions, it is necessary to use IoT platform. There are many IoT platforms functioning nowadays, such as IBM Watson, ThingWorx, Google Cloud Platform, Azure IoT of Microsoft, Amazon Web Services (AWS) IoT, ThingSpeak [13]. They are developed by top leading IT companies to ease IoT services meanwhile contributing its development. They have different functionality and mainly provide monitoring and/or controlling [14–18]. E.g. Amazon announced the AWS IoT platform [16] that supports features such as Registry for recognizing devices, Software Development Kit for devices, etc. IoT platform ThingSpeak provides analytic service that allows to aggregate, visualize, and analyze live data streams in the cloud [17]. Microsoft provides Microsoft Azure Intelligent System Service as an integrated platform and services to build IoT systems and applications by gathering, storing and processing data [18].

Based on ThingSpeak platform as appropriate for developing IoT platform (due to free service for monitoring; providing catchy and illustrative graphs for users; supporting Python) the Temperature and Pressure Remote Monitoring System is developed. For gateway, existing approaches could be Arduino, Raspberry Pi, BeagleBone, etc. Wi-Fi technology is used as a communication medium, which is one of the most widely-distributed technology especially for the smart environment. Moreover, it is possible to use different type of sensors for gathering temperature, humidity or pressure. For this project following components are used:

1. *Raspberry Pi* is a small-sized computer, can supply a complete Linux OS and can provide all the expected abilities that implies, at a low-power consumption level.
2. *BMP180* is a barometric pressure sensor with an I2C interface. It measures the absolute pressure of the air. This pressure varies with both the weather and altitude, allowing monitor changes in the weather, measure altitude, or any other tasks.
3. IoT platform – ThingSpeak.com. This IoT platform allows monitoring the condition in environment. To transmit data to the ThingSpeak Raspberry Pi 3 is used, to which BMP180 sensor is connected. ThingSpeak IoT platform is for remote monitoring. For this purpose, it is necessary to register on the site and create the channel. Then after appropriate variables are being set.

ThingSpeak is an application platform for the Internet of Things. ThingSpeak allows building an application with data collected by sensors. ThingSpeak processes the data, and then refers to applications to retrieve the data. Thus, different scenarios can be implemented and the performance be evaluated.

In Fig. 2 the final state of system is depicted as retrieved from ThingSpeak platform where data to be gathered for this project. This page/channel is accessible from any devices via following link: https://thingspeak.com/channels/957258.

As it is seen, there are different possibilities to explore the IoT platform and related environment (including analytical research in Matlab for expanded version), with getting parameters and characteristics of components.

Fig. 2. Page of project in ThingSpeak platform

5 Conclusion

An approach is considered to explore the features of information technology platform to support the production processes using technologies of industrial cyber-physical systems. The ThingsSpeak platform based use-case demonstrates the possibility to attract the specific task while implementing the solution. It is interesting to research such realization in close-to-practice conditions of real manufacturing or production systems, assessing the performance and determining the influence on overall system efficiency, including possibility to extract data for feasibility analysis. Taking into account the necessity to develop and deploy the Industry 4.0 solutions, additional challenge in technologies may arise, e.g. dealing with Big Data, Deep Learning, which in turn, brings more sophisticated R&D efforts in this domain. Authors see the future research direction in formalizing the mathematical basis of IoT platform including tools and methods applying to solve the tasks of development the national IoT platform (and thus to contribute to the scientific foundations and theoretical prerequisites for sustainable industrial development with creating more dynamic, sustainable, innovative and consumer-oriented Industry 4.0 ecosystem).

References

1. What is digital economy? Unicorns, transformation and the Internet of Things. www.deloitte.com/mt/en/pages/technology/articles/mt-what-is-digital-economy.html. Accessed 2020

2. Pillar, F.: Platforms for the industrial IoT: choosing the operating system of your digital factory of the future. https://theleadershipnetwork.com/article/platforms-for-the-industrial-iot-choosing-the-operating-system-of-your-digital-factory-of-the-future. Accessed 18 June 2018
3. Brent, R.M.: GDP and the digital economy: Keeping up with the Changes. In: Understanding the Digital Economy: Data, Tools, and Research. pp. 34-48, Cambridge, Mass. [u.a.]: MIT Press (2000), ISBN 0-262-02474-8. https://doi.org/10.7551/mitpress/6986.003.0004
4. Digital economy report 2019: Value creation and capture: implications for developing countries. United Nations: Conference on Trade and Development, UNCTAD, Geneva (2019). http://digitallibrary.un.org/record/3833647
5. Industrial Development Report 2020: Industrializing in the Digital Age. United Nations: Industrial Development Organization, Vienna (2019). https://www.unido.org/resources-publications-flagship-publications-industrial-development-report-series/idr2020
6. Bonomi, F., Milito, R., Natarajan, P., Zhu, J.: Fog computing: a platform for Internet of Things and analytics. In: Bessis, N., Dobre, C. (eds.) Big Data and Internet of Things: A Roadmap for Smart Environments. SCI, vol. 546, pp. 169–186. Springer, Cham (2014). https://doi.org/10.1007/978-3-319-05029-4_7
7. Arsénio, A., Serra, H., Francisco, R., Nabais, F., Andrade, J., Serrano, E.: Internet of Intelligent Things: bringing artificial intelligence into things and communication networks. In: Xhafa, F., Bessis, N. (eds.) Inter-cooperative Collective Intelligence: Techniques and Applications, pp. 1–37. Springer Berlin Heidelberg, Berlin, Heidelberg (2014). https://doi.org/10.1007/978-3-642-35016-0_1
8. Kuzlu, M., Fair, C., Guler, O.: Role of artificial intelligence in the Internet of Things (IoT) cybersecurity. Disc. Internet of Things **1**(1), 1–14 (2021). https://doi.org/10.1007/s43926-020-00001-4
9. Yusupbekov, N.R., Gulyamov, S.M., Kasimov, S.S., Usmanova, N.B.: Knowledge-based planning for industrial automation systems. In: Aliev R.A., et al. (eds.) 13th International Conference on Theory and Application of Fuzzy Systems and Soft Computing – ICAFS-2018, AISC vol. 896, 873–879 (2019). Springer, Cham. https://doi.org/10.1007/978-3-030-04164-9_115
10. Edward, R.G., Christopher, G., David, A.W., Martin, J.B.: Framework for cyber-physical systems. Overview/Natl. Inst. Stand. Technol. Spec. Publ. **1500–201**, 1 (2017). https://doi.org/10.6028/NIST.SP.1500-201
11. Gerrikagoitia, J.K., Unamuno, G., Urkia, E., Serna, A.: Digital manufacturing platforms in the Industry 4.0 from private and public perspectives. Appl.Sci. **9**(14), 2934 (2019). https://doi.org/10.3390/app9142934
12. Rüßmann, M.: Industry 4.0: The future of productivity and growth in manufacturing industries. Boston Consulting Group (2015). www.bcg.com/publications/2015/engineered_products_project_business_industry_4_future_productivity_growth_manufacturing_industries.as. Accessed Feb 5 2018
13. Top 20 IoT Platforms in 2018. Review by Santosh Singh. https://internetofthingswiki.com/top-20-iot-platforms/634/. Accessed Mar 8 2016
14. Nakhuva, B., Champaneria, T.: Study of various Internet of Things platforms. Int. J. Comput. Sci. Eng. Surv. (IJCSES). **6**(6), 61–74 (2015). https://doi.org/10.5121/ijcses.2015.6605
15. Google Cloud IoT platform. https://cloud.google.com/solutions/iot/
16. Amazon IoT platform. https://aws.amazon.com/iot/
17. ThingSpeak IoT platform. https://thingspeak.com/
18. Microsoft Azure IoT platform. https://azure.microsoft.com/en-us/overview/iot/

Toward an Analysis of Pairwise Comparison Matrices with Z-number-valued Elements

R. R. Aliyev[1], O. H. Huseynov[2(✉)], and Javanshir Guliyev[2]

[1] Department of Mathematics, Eastern Mediterranean University, via Mersin 10, Gazimağusa, Turkish Republic of Northern Cyprus, Turkey
rashad.aliyev@emu.edu.tr

[2] Research Laboratory of Intelligent Control and Decision Making Systems in Industry and Economics, Azerbaijan State Oil and Industry University, 20 Azadlig Avenue, AZ1010 Baku, Azerbaijan
oleq.huseynov@asoiu.edu.az, bozkurt.kuruc@neu.edu.tr

Abstract. Pairwise comparison matrices (PCMs) encode preferences in decision problems. Decision-relevant information may be characterized by partial reliability. In this paper, PCMs with Z-number-valued elements are considered. It is proposed to analyze such matrices in terms of consistency and informativeness. An example is used to illustrate that these two requirements may be conflicting.

Keywords: Z-number · Pairwise comparison matrix · Consistency · Coverage measure · Specificity measure

1 Introduction

Pairwise comparison matrices (PCMs) describe preferences over alternatives and criteria. Formally, pairwise comparison matrix (PCM) is a matrix (a_{ij}), where an entry a_{ij} denotes a degree to which an i-th alternative (criterion) is preferred to j-th one [1]. Natural conditions used for a_{ij} are $a_{ii} = 1$ and $a_{ji} = 1/a_{ij}$ (reciprocity), $\forall\, i, j = 1, \ldots, n$. The necessary condition for PCM is consistency. Traditionally, it is based on multiplicative transitivity (though different constructs are also used [2, 3]):

$$a_{ij}a_{jk} = a_{ik}, \forall i, j, k. \tag{1}$$

This implies that degree of preference a_{ik} is equal to the product of preferences degrees staying at all possible ways from i to k through j. Fulfillment of this condition is often problematic due to low computational abilities of a human brain. An inconsistency is related to violation of this multiplicative transitivity condition. In view of this, various indices were introduced to measure inconsistency. An inconsistency index ('consistency index' term is also used) is a function that maps a set of PCMs $\mathcal{A}$ to a set of real numbers $\mathcal{R}$, $I : \mathcal{A} \to \mathcal{R}$. The most famous consistency index was proposed by Saaty [4]. It is based on an affine transformation of maximal eigenvalue of a matrix. Among well known indexes are those proposed by Kou and Lin [5], Cavallo and D'Apuzzo [6, 7],

R. A. Aliev et al. (Eds.): ICSCCW 2021, LNNS 362, pp. 420–427, 2022.
https://doi.org/10.1007/978-3-030-92127-9_57

Koczkodaj and colleagues [8–13] and others. A systematic overview and analysis of inconsistency indexes is given in [1].

Initially, degrees of preference were described as crisp numbers from a predefined scale. Further, fuzzy sets were used to account for uncertainty of preference degrees [14–20]. Due to imperfect information and other restrictions, a decision maker (DM) finds difficulties to provide exact and accurate preference degrees. In this case, fuzzy degrees of preference are used to account for imprecision. Being imprecise, real-world preferences are also partially reliable. The reasons are restricted competence of a DM, complexity of alternatives, imperfect decision-relevant information, psychological biases etc. So, in line with describing imprecision, it becomes relevant to account of partial reliability of preferences. Up to day, no works have been proposed on consistency of partially reliable preferences. In order to handle imprecision and partial reliability, Zadeh introduced the Z-number concept [21]. A Z-number $Z = (A, B)$ is an ordered pair of fuzzy numbers used to describe a value of a random variable X. A is a fuzzy restriction on values of X. B is a fuzzy reliability of A, defined as a value of probability measure of A. A series of works is devoted to theoretical studies and practical applications of Z-numbers [22–45]. In [46] they propose a new approach for construction of consistency-driven partially reliable preferences described by pairwise comparisons matrix with Z-number-valued entries. The problem is formulated as a complex optimization problem involving membership functions and probability distributions. In view of this, a differential evolution (DE) optimization technique-based solution approach is proposed.

Along with consistency measure, such measures as specificity, coverage and entropy are important to analyze quality of uncertain granules. Specificity of Z-numbers is considered in [47]. Several works are devoted to entropy and uncertainty of Z-numbers (see [48, 49] and references herein).

In this work we consider a problem of analysis of PCMs with Z-number-valued elements in terms of consistency and some information quality indices. An example from decision making in country selection is used. It is shown that attaining consistency of PCMs may lead to loss of information contained in initial Z-number-valued elements.

2 Preliminaries

The concept of Z-number is used to describe a value of real-valued random variable X under combination of fuzzy and probabilistic uncertainties. Such kind of combination is usually referred to as a bimodal distribution [24, 26].

Definition 1 A continuous Z-number [3, 5]. A continuous Z-number is an ordered pair $Z = (A, B)$. A is a continuous fuzzy number with membership function $\mu_A(x)$ playing a role of a fuzzy restriction on a value that a random variable X may take:

Value of X is A.

In other words, A is used to describe imprecise information about a value of X. A degree of reliability of A is described as a value of probability measure $P(A) = \int_R \mu_A(x)p(x)dx$, where p is probability distribution of X. If p is precisely known, the $P(A)$ is a crisp number. However, in real-world problems an actual p may not be precisely known, and one has to consider a set of distributions. This requires to deal with fuzzy restriction on a value of $P(A)$. Thus, to describe uncertainty related to p, a continuous fuzzy number B with a membership function $\mu_B : [0, 1] \to [0, 1]$ is used as a fuzzy restriction:

$$\text{The value of } P(A) \text{ is } B \tag{2}$$

This implies that fuzzy uncertainty related to p induces fuzziness of a value of $P(A)$. Thus, due to fuzziness and probabilistic uncertainty of information, a value of random variable X can be described as a Z-number $Z = (A, B)$: A is a fuzzy estimation of a value and B is a fuzzy reliability of this estimation.

Definition 2 A Z-number-valued pairwise comparisons matrix. A Z-number-valued pairwise comparisons matrix (Z_{ij}) is a square matrix of Z-numbers:

$$(Z_{ij} = (A_{ij}, B_{ij})) = \begin{pmatrix} Z_{11} = (A_{11}, B_{11}) \ldots Z_{1n} = (A_{1n}, B_{1n}) \\ . \quad\quad \ldots \quad\quad . \\ Z_{n1} = (A_{n1}, B_{n1}) \ldots Z_{nn} = (A_{nn}, B_{nn}) \end{pmatrix}. \tag{3}$$

A Z-number $Z_{ij} = (A_{ij}, B_{ij}),\ \ i, j = 1, \ldots, n$ describes partially reliable information on degree of preference for *i-th* alternative (criterion) against *j*-th one.

Definition 3 A Distance between Z-numbers. As a Z-number $Z = (A, B)$ is characterized by fuzzy number A, fuzzy number B and underlying set of probability distributions G, we propose to define distance between Z-numbers $D(Z_1, Z_2)$ as follows.

Distance between A_1 and A_2 is computed as

$$D(A_1, A_2) = \sup_{\alpha \in (0,1]} D(A_1^\alpha, A_2^\alpha) \tag{4}$$

$$D(A_1^\alpha, A_2^\alpha) = \left| \frac{A_{11}^\alpha + A_{12}^\alpha}{2} - \frac{A_{21}^\alpha + A_{22}^\alpha}{2} \right| \tag{5}$$

A_1^α and A_2^α denote α-cuts of A_1 and A_2 respectively, $A_{11}^\alpha, A_{12}^\alpha$ denote lower and upper bounds of A_1^α ($A_{21}^\alpha, A_{22}^\alpha$ are those of A_2^α). Distance between B_1 and B_2 is computed analogously.

We also have to find distance between the sets G_1 and G_2 of probability distributions p_1 and p_2 underlying Z_1 and Z_2. The distance between p_1 and p_2 can be expressed as

$$D(G_1, G_2) = \inf_{p_1 \in G_1, p_2 \in G_2} \left\{ (1 - \int_R (p_1 p_2)^{\frac{1}{2}} dx)^{\frac{1}{2}} \right\} \tag{6}$$

Given $D(A_1, A_2)$, $D(B_1, B_2)$ and $D(G_1, G_2)$, the distance for Z-numbers is defined as

$$D(Z_1, Z_2) = \beta D(A_1, A_2) + (1 - \beta) D_{total}(B_1, B_2), \tag{7}$$

where $D_{total}(B_1, B_2)$ is a distance for reliability restriction computed as

$$D_{total}(B_1, B_2) = wD(B_1, B_2) + (1 - w)D(G_1, G_2). \tag{8}$$

$\beta, w \in [0, 1]$ are DM's assigned importance degrees.

Below we propose a definition of inconsistency index for Z-number-valued matrix adopted from the index introduced for real-valued matrix in [46].

Definition 4 An inconsistency index for Z-number-valued pairwise comparisons matrix. An inconsistency index K for Z-number-valued pairwise comparisons matrix (Z_{ij}) is defined as follows:

$$K((Z_{ij})) = \max_{i<j<k} \min\left\{D\left(Z(1), \left(\frac{Z_{ik}}{Z_{ij}Z_{jk}}\right)\right)D\left(Z(1), \left(\frac{Z_{ij}Z_{jk}}{Z_{ik}}\right)\right)\right\}, \tag{9}$$

where the components of Z-number $Z(1) = (A, B)$, are fuzzy singletons $A = 1$ and $B = 1$. The operations of multiplication and division of Z-numbers in (9) are defined as in [5].

3 Statement of the Problem

Analysis of pairwise comparisons matrix (Z_{ij}) may be based on consistency, coverage, specificity, entropy measures. Consistency is necessary requirement of any PCM. For consistency, we use index (9).

For (Z_{ij}), measure Q may be defined as an aggregation of measure values of PCM entries above main diagonal:

$$Q((Z_{ij})) = Agg_{i=1,\ldots,n;j>i}(Q(Z_{ij})) \tag{10}$$

As an aggregation operator, *min* or some averaging operators may be used.

Specificity, $Sp(Z_{ij})$ can be found as [47]:

$$Sp(Z_{ij}) = \lambda Sp(A_{ij}) + (1-\lambda)Sp(B_{ij}), \lambda \in (0, 1) \tag{11}$$

where λ is an importance degree assigned by a DM.

Coverage measure is used to analyze an extent to which information contained in uncertain granules is preserved. $Cov(Z_{ij})$ may be computed as convex combination of $Cov(A_{ij})$, $Cov(B_{ij})$. In essence is Sp is an inverse of Cov.

A problem of comparative analysis of PCMs may be considered as a multiattribute decision making problem on the basis of measures Q_s, $s = 1, \ldots, S$.

4 An Application. Analysis of PCMs with Z-number-valued Elements in a Market Selection Problem

Let us analyze two PCMs considered in [46] by using consistency and coverage measures. Those are related to preferences over multiple criteria in a problem of country selection for doing business [50]. We will deal with three criteria that describe a series of economical and institutional characteristics: Institutional Proximity, C_1, Economic Proximity, C_2, and Social and Cultural Proximity, C_3. Criterion C_1 represents governance performance and economic freedom. Criterion C_2 describes both domestic development (such as socioeconomic progress, household's standard of living) and global competitiveness issues. Criterion C_3 concerns cultural characteristics (an extent to which collectivism or individualism-based behavior is intrinsic to a country, issues related to communication with customers etc.). Due to complexity of the considered criteria, a DM's preferences may be characterized by fuzziness [51, 52] and partial reliability. So we can use Z-numbers [53–55] for description of preferences. In view of this, we consider partially reliable preference degrees of the Saaty scale to represent comparative importance of criteria [46]:

$$\begin{pmatrix} Z_{11} = ((0.93, 0.95, 0.97), (0.95, 0.98, 1)) & \ldots & Z_{13} = ((3.92, 4, 4.08), (0.7, 0.8, 0.9)) \\ Z_{21} = ((2.94, 3, 3.06), (0.7, 0.8, 0.9)) & \ldots & Z_{23} = ((8.82, 9, 9.02), (0.6, 0.7, 0.8)) \\ Z_{31} = ((0.245, 0.25, 0.255), (0.7, 0.8, 0.9)) & \ldots & Z_{33} = ((0.93, 0.95, 0.97), (0.95, 0.98, 1)) \end{pmatrix}$$

Z_{ij} denotes a Z-number-valued degree to which the i-th criterion is preferred to the j-th one. For example, $Z_{21} = ((2.94, 3, 3.06), (0.7, 0.8, 0.9))$ is a Z-number-valued degree to which C_2 is preferred to C_1. The value of inconsistency index K for the considered matrix (Z_{ij}) is $K((Z_{ij})) = 0.31$(computed in the basis of Definition 4). The value of Cov is found as $Cov(Z_{ij}) = 0.5Cov(A_{ij}) + 0.5Cov(B_{ij})$. The overall coverage obtained by using (10) is $Cov((Z_{ij})) = 0.14$.

Another Z-number-valued matrix (Z'_{ij}) is consistency driven one:

$$\begin{pmatrix} Z_{11} = ((1, 1, 1), (1, 1, 1)) & \ldots & Z_{13} = ((3.65, 3.651, 3.6512), (0.997, 0.9973, 0.9973)) \\ Z_{21} = ((2.4827, 2.483, 2.483), (0.78, 0.78, 0.99)) & \ldots & Z_{23} = ((9.06, 9.06, 9.064), (0.99, 0.99, 0.995)) \\ Z_{31} = ((0.28, 0.28, 0.278), (0.0078, 0.008, 0.50)) & \ldots & Z_{33} = ((1, 1, 1), (1, 1, 1)) \end{pmatrix}$$

The computed value of K is $K((Z'_{ij})) = 0.003$. The overall $Cov((Z'_{ij})) = 0.035$ and is quite lower than $Cov((Z_{ij})) = 0.14$. Thus, when obtaining consistent matrix Z'_{ij}, one may loss information contained in initial inconsistent (Z_{ij}).

5 Conclusion

Real-world preferences are characterized by imprecise and partially reliable information. A Z-number is a formal construct used to describe such kind of information. In this paper, we consider some issues related to consistency and informativeness of Z-number-valued pairwise comparisons matrix. A problem of foreign market selection is considered. The results show that it is needed to develop a technique to reduce loss of information when deriving consistent PCMs with Z-number-valued elements.

References

1. Brunelli, M.: A survey of inconsistency indices for pairwise comparisons. Int. J. Gen. Syst. **47**(8), 751–771 (2018). https://doi.org/10.1080/03081079.2018.1523156
2. Lin, L.Z., Yeh, H.R.: A perceptual measure of mobile advertising using fuzzy linguistic preference relation. Iran. J. Fuzzy Syst. **10**(5), 25–46 (2013). https://doi.org/10.22111/IJFS.2013.1205
3. Pekala, B.: Comparing uncertainty data in epistemic and ontic sense used to decision making problem. Iran. J. Fuzzy Syst. **16**(2), 45–55 (2019). https://doi.org/10.22111/IJFS.2019.4541
4. Saaty, T.L.: A scaling method for priorities in hierarchical structures. J. Math. Psychol. **15**(3), 234–281 (1977). https://doi.org/10.1016/0022-2496(77)90033-5
5. Kou, G., Changsheng, L.: A cosine maximization method for the priority vector derivation in AHP. Eur. J. Oper. Res. **235**(1), 225–232 (2014). https://doi.org/10.1016/j.ejor.2013.10.019

6. Cavallo, B., D'Apuzzo, L.: A general unified framework for pairwise comparison matrices in multicriterial methods. Int. J. Intell. Syst. **24**(4), 377–398 (2009). https://doi.org/10.1002/int.20329
7. Cavallo, B., D'Apuzzo, L.: Characterizations of consistent pairwise comparison matrices over Abelian linearly ordered groups. Int. J. Intell. Syst. **25**(10), 1035–1059 (2010). https://doi.org/10.1002/int.20438
8. Koczkodaj, W.W., Szwarc, R.: On Axiomatization of inconsistency indicators for pairwise comparisons. Fundam. Inform. **132**(4), 485–500 (2014). https://doi.org/10.3233/FI-2014-1055
9. Koczkodaj, W.W., Herman, M.W., Orlowski, M.: Managing null entries in pairwise comparisons. Knowl. Inf. Syst. **1**(1), 119–125 (1999). https://doi.org/10.1007/BF03325094
10. Koczkodaj, W.W., et al.: On normalization of inconsistency indicators in pairwise comparisons. Int. J. Approx. Reason. **86**, 73–79 (2017). https://doi.org/10.1016/j.ijar.2017.04.005
11. Koczkodaj, W.W., Szybowski, J., Wajch, E.: Inconsistency indicator maps on groups for pairwise comparisons. Int. J. Approx. Reason. **69**, 81–90 (2016). https://doi.org/10.1016/j.ijar.2015.11.007
12. Koczkodaj, W.W., Urban, R.: Axiomatization of inconsistency indicators for pairwise comparisons. Int. J. Approx. Reason. **94**, 18–29 (2018). https://doi.org/10.1016/j.ijar.2017.12.001
13. Koczkodaj, W.W.: A new definition of consistency of pairwise comparisons. Math. Comput. Model. **18**(7), 79–84 (1993). https://doi.org/10.1016/0895-7177(93)90059-8
14. Khorramrouz, F., Pourmahdi Kajabadi, N., Rahiminezhad Galankashi, M., Mokhatab Rafiei, F.: Application of fuzzy analytic hierarchy process (FAHP) in failure investigation of knowledge-based business plans. SN Appl. Sci. **1**(11), 1–13 (2019). https://doi.org/10.1007/s42452-019-1394-3
15. Krejči, J.: Pairwise Comparison Matrices and their Fuzzy Extension: Multi-criteria Decision Making with a New Fuzzy Approach, p. 291. Springer, Cham (2018). https://doi.org/10.1007/978-3-319-77715-3
16. Lee, W., Lau, H., Liu, Z.-Z., Tam, S.: A fuzzy analytic hierarchy process approach in modular product design. Expert Syst. **18**, 32–42 (2001). https://doi.org/10.1111/1468-0394.00153
17. Liu, Y., Eckert, C.M., Earl, C.: A review of fuzzy AHP methods for decision-making with subjective judgements. Expert Syst. Appl. **161**, 113738 (2020) https://doi.org/10.1016/j.eswa.2020.113738
18. Shapiro, A.F., Koissi, M.-C.: Fuzzy logic modifications of the Analytic Hierarchy Process. Insurance Math. Econ. **75**, 189–202 (2017). https://doi.org/10.1016/j.insmatheco.2017.05.003.
19. Van Laarhoven, P.J.M., Pedrycz, W.: A fuzzy extension of Saaty's priority theory. Fuzzy Sets Syst. **11**, 199–227 (1983). https://doi.org/10.1016/S0165-0114(83)80082
20. Wang, Y.-M., Chin, K.-S.: Fuzzy analytic hierarchy process: a logarithmic fuzzy preference programming methodology. Int. J. Approx. Reason. **52**(4), 541–553 (2011). https://doi.org/10.1016/j.ijar.2010.12.004
21. Zadeh, L.A.: A note on Z-numbers. Inform. Sci. **181**, 2923–2932 (2011). https://doi.org/10.1016/j.ins.2011.02.022
22. Aliev, R.A., Alizadeh, A.V., Huseynov, O.H.: The arithmetic of discrete Z-numbers. Inform. Sci. **290**, 134–155 (2015). https://doi.org/10.1016/j.ins.2014.08.024
23. Aliev, R.A., Alizadeh, A.V., Huseynov, O.H., Jabbarova, K.I.: Z-number-based linear programming. Int. J. Intell. Syst. **30**, 563–589 (2015). https://doi.org/10.1002/int.21709
24. Aliev, R.A., Huseynov, O.H., Aliyev, R.R., Alizadeh, A.V.: The Arithmetic of Z-Numbers: Theory and Applications, p. 316. World Scientific, Singapore (2015). https://doi.org/10.1142/9575

25. Aliev, R.A., Huseynov, O.H., Serdaroglu, R.: Ranking of Z-numbers and its application in decision making. Int. J. Inform. Tech. Decis. Making **15**(6), 1503–1519 (2016). https://doi.org/10.1142/S0219622016500310
26. Aliev, R.A., Huseynov, O.H., Zeinalova, L.M.: The arithmetic of continuous Z-numbers. Inform. Sci. **373**, 441–460 (2016). https://doi.org/10.1016/j.ins.2016.08.078
27. Aliev, R.A., Huseynov, O.H.: Decision Theory with Imperfect Information, p. 444. World Scientific, Singapore (2014)
28. Aliev, R.A., Perdycz, W., Huseynov, O.H.: Hukuhara difference of Z-numbers. Inform. Sci. **466**, 13–24 (2018). https://doi.org/10.1016/j.ins.2018.07.033
29. Aliev, R.A.: Uncertain Computation-Based Decision Theory. World Scientific, Singapore (2017). https://doi.org/10.1142/10679. Accessed Jan 2018
30. Aliev, R.A., Alizadeh, A.V., Huseynov, O.H., Jabbarova, K.I.: Z-number based linear programming. Int. J. Intell. Syst. **30**(5), 563–589 (2015). https://doi.org/10.1002/int.21709
31. Allahviranloo, T., Ezadi, S.: Z-advanced numbers processes. Inform. Sci. **480**, 130–143 (2019). https://doi.org/10.1016/j.ins.2018.12.012
32. Allahviranloo, T., Hooshangian, L.: A method to find fuzzy eigenvalues and fuzzy eigenvectors of fuzzy matrix. Neural Comput. Appl. **23**, 1159–1167 (2013). https://doi.org/10.1007/s00521-012-1062-7
33. Banerjee, R., Pal, S.K.: Z*-numbers: augmented Z-numbers for machine-subjectivity representation. Inform. Sci. **323**, 143–178 (2015). https://doi.org/10.1016/j.ins.2015.06.026
34. Das, S., Dhalmahapatra, K., Maiti, J.: Z-number integrated weighted VIKOR technique for hazard prioritization and its application in virtual prototype based EOT crane operations. Appl. Soft Comput. **94**, 106419 (2020). https://doi.org/10.1016/j.asoc.2020.106419
35. Ezadi, S., Allahviranloo, T., Mohammadi, S.: Two new methods for ranking of Z-numbers based on sigmoid function and sign method. Int. J. Intell. Syst. **33**, 1476–1487 (2018). https://doi.org/10.1002/int.21987
36. Ezadi, S., Allahviranloo, T.: New multi-layer method for Z-number ranking using hyperbolic tangent function and convex combination. Intell. Autom. Soft Comput. (2017). https://doi.org/10.1080/10798587.2017.1367146
37. Hendiani, S., Bagherpour, M., Mahmoudi, A., Liao, H.: Z-number based earned value management (ZEVM): a novel pragmatic contribution towards a possibilistic cost-duration assessment. Comput. Ind. Eng. **143**, 106430 (2020) https://doi.org/10.1016/j.cie.2020.106430
38. Jiang, W., Cao, Y., Deng, X.: A novel Z-network model based on Bayesian network and Z-number. IEEE Trans. Fuzzy Syst. **28**(8), 1585–1599 (2020). https://doi.org/10.1109/TFUZZ.2019.2918999
39. Kang, B., Chhipi-Shrestha, G., Deng, Y., Hewage, K., Sadiq, R.: Stable strategies analysis based on the utility of Z-number in the evolutionary games. Appl. Math. Comput. **324**, 202–217 (2018). https://doi.org/10.1016/j.amc.2017.12.006
40. Kang, B., Deng, Y., Hewage, K., Sadiq, R.: A method of measuring uncertainty for Z-number. IEEE Trans. Fuzzy Syst. **27**(4), 731–738 (2018). https://doi.org/10.1109/TFUZZ.2018.2868496
41. Li, Y., Garg, H., Deng, Y.: A new uncertainty measure of discrete Z-numbers. Int. J. Fuzzy Syst. **22**(3), 760–776 (2020). https://doi.org/10.1007/s40815-020-00819-8
42. Shen, K., Wang, J.: Z-VIKOR method based on a new comprehensive weighted distance measure of Z-number and its application. IEEE Trans. Fuzzy Syst. **26**(6), 3232–3245 (2018). https://doi.org/10.1109/TFUZZ.2018.2816581
43. Yager, R.R.: On a view of Zadeh's Z-numbers, advances in computational intelligence. In: Greco, S., Bouchon-Meunier, B., Coletti, G., Fedrizzi, M., Matarazzo, B., Yager, R.R. (eds.) Advances in Computational Intelligence. IPMU 2012. Communications in Computer and Information Science, vol. 299, pp. 90–101. Springer, Heidelberg (2012). https://doi.org/10.1007/978-3-642-31718-7_10

44. Yager, R.R.: On Z-valuations using Zadeh's Z-numbers. Int. J. Intell. Syst. **27**, 259–278 (2012). https://doi.org/10.1002/int.21521
45. Zadeh, L.A., Aliev, R.A.: Fuzzy Logic Theory and Applications. Part I and Part II, p. 610. World Scientific, Singapore (2019). https://doi.org/10.1142/10936
46. Aliev, R., Guirimov, B., Huseynov, O., Aliyev, R.: A consistency-driven approach to construction of Z-number-valued pairwise comparison matrices. Iran. J. Fuzzy Syst. **18**(4), 37–49 (2021). https://doi.org/10.22111/ijfs.2021.6175
47. Valiev, A.A., Abdullayev, T.S., Alizadeh, A.V., Adilova, N.E.: Comparison of measures of specificity of Z-numbers. Procedia Comput. Sci. **120**, 466–472 (2017). https://doi.org/10.1016/j.procs.2017.11.265
48. Qiao, D., Wang, X.-K., Wang, J.-Q., Chen, K.: Cross entropy for discrete Z-numbers and its application in multi-criteria decision-making. Int. J. Fuzzy Syst. **21**(6), 1786–1800 (2019). https://doi.org/10.1007/s40815-019-00674-2
49. Li, Y., Wu, T., Mao, J., Guo, H., Yao, A.: A method of uncertainty measurements for multidimensional Z-number and their applications. Math. Probl. Eng. **2020**, 1–16 (2020). https://doi.org/10.1155/2020/8407830
50. Hortacsu, A., Tektas, A.: Modeling the country selection decision in retail internationalization. W. Acad. Sci. Eng. Tech. **31**, 1068–1075 (2009). https://doi.org/10.5281/zenodo.1061134
51. Adilova, N.E.: Quality criteria of fuzzy if-then rules and their calculations. In: Aliev, R.A., Kacprzyk, J., Pedrycz, W., Jamshidi, Mo., Babanli, M., Sadikoglu, F.M. (eds.) ICAFS 2020. AISC, vol. 1306, pp. 55–62. Springer, Cham (2021). https://doi.org/10.1007/978-3-030-64058-3_7
52. Gardashova, L.A., Aliyev, B.F.: Dental disease detection using fuzzy logic. In: Aliev, R.A., Kacprzyk, J., Pedrycz, W., Jamshidi, M., Babanli, M., Sadikoglu, F.M. (eds.) ICAFS 2020. AISC, vol. 1306, pp. 126–132. Springer, Cham (2021). https://doi.org/10.1007/978-3-030-64058-3_16
53. Jabbarova, K., Alizadeh, A.V.: Z-decision making in human resources department. In: Aliev, R.A., Kacprzyk, J., Pedrycz, W., Jamshidi, M., Babanli, M., Sadikoglu, F.M. (eds.) ICAFS 2020. AISC, vol. 1306, pp. 498–507. Springer, Cham (2021). https://doi.org/10.1007/978-3-030-64058-3_62
54. Jabbarova, K., Huseynov, O.H.: Solution of a decision making problem under risk with z-information. In: Aliev, R.A., Kacprzyk, J., Pedrycz, W., Jamshidi, M., Babanli, M., Sadikoglu, F.M. (eds.) ICAFS 2020. AISC, vol. 1306, pp. 781–786. Springer, Cham (2021). https://doi.org/10.1007/978-3-030-64058-3_98
55. Jabbarova, K.I., Jabbarova, A.I.: Z-information based PROMETHEE method. In: Aliev, R.A., Yusupbekov, N.R., Kacprzyk, J., Pedrycz, W., Sadikoglu, F.M. (eds.) WCIS 2020. AISC, vol. 1323, pp. 287–293. Springer, Cham (2021). https://doi.org/10.1007/978-3-030-68004-6_37

Development of an Automatic Parking Algorithm Based on Fuzzy Logic

A. B. Sultanova[1,2(✉)]

[1] Azerbaijan State University of Oil and Industry, Baku, Azerbaijan
saxira@mail.ru
[2] Institute of Control Systems of ANAS, Azerbaijan, AZ1010 Baku, Azerbaijan

Abstract. Intelligent Transportation System (ITS) has generated a lot of research interest in the last few decades. Many problems have been identified in this area, among which autonomous parking has attracted more attention. Automatically parking vehicles in a confined space - an area where there are moving and stationary obstacles - is a challenge. Therefore, the development of parking systems based on new technologies, the placement of a more optimal number of cars in a limited space has become an urgent problem at a time when there is a large number of ground vehicles. The article has developed an automatic parking algorithm based on fuzzy logic. A new model of the parking lot was proposed, as well as the development of an automatic parking control system for passenger cars in the proposed parking lot. To solve this problem, it is proposed to divide the parking space by methods of the theory of fuzzy sets and fuzzy logic.

Keywords: Intelligent transport system · Automatic parking system · Fuzzy logic · Fuzzy controller · Ultrasonic sensors

1 Introduction

With the rapid growth of the urban population and an increase in the standard of living, the number of vehicles has increased dramatically. The rapid growth in the number of owners of city cars not only increases the burden on urban transport, but also deepens the problem of lack of parking spaces. Today in our city there is a sufficient density of cars, which negatively affects the way of life of people. To eliminate this problem, it is necessary to develop a parking system based on a new technology.

Many automotive industries are working on the development of autonomous vehicles, introducing artificial intelligence into machines and machine learning. Autonomous vehicle parking is also one of the parts of a fully autonomous vehicle, where the vehicle is parked independently, determining close conditions [1, 2].

Parking is the process of temporarily storing a car in a parking lot on demand. Nowadays, when a driver approaches a parking space, he receives information about the availability of free space in the existing parking space, and then places the car in a certain place. It is able to help drivers automatically drive a vehicle in difficult conditions where

R. A. Aliev et al. (Eds.): ICSCCW 2021, LNNS 362, pp. 428–436, 2022.
https://doi.org/10.1007/978-3-030-92127-9_58

a lot of attention and experience is required, thereby saving the driver from complex parking procedures and helping to park the car more efficiently.

The main problem with automatic parking is the formation of the trajectory of movement, tracking the trajectory of movement, the optimal distribution of the available area. In general, one of the main issues when parking is the placement of a large number of cars on the existing territory.

Currently, intelligent vehicles are the main direction of the development of the automotive industry and the object of research of the largest domestic and foreign automakers and research institutes. All over the world, there are many problems associated with parking cars in large cities. It also requires a lot of attention to the automatic parking technology.

The autonomous intelligent parking system mainly includes four stages;

- receiving an object;
- road planning;
- track the path;
- human-machine interface (HMI);

In general, simple autonomous parking systems are performed in the following sequence: empty parking space detection, trajectory planning, trajectory monitoring and control, speed control.

Currently, due to the rapid growth in the number of cars in the world, the problem of organizing permanent storage of cars is urgent. The current problems in the intelligent transport system have led to a lot of research over the past few decades. Intelligent traffic tracking and control system is the main technology of intelligent vehicles.

Nowadays, intelligent recognition of parking spaces is associated with machine vision and target detection and recognition technology. The growing interest in automated transport systems and the corresponding involvement of many researchers in this field in recent years have led to the development of various algorithms [3].

Automatic parking [4], widely discussed. Currently, there are two main methods of studying automatic parking systems: research methods based on ultrasonic sensors and methods based on visual sensors. Studies of automatic parking systems based on ultrasonic sensors were started relatively early. Currently, ultrasonic sensors still play an important role in the development of automatic parking systems. In [5], parking technology and the results of his research in the field of parking are published and it is proposed to use the dynamic optimization method to optimize maneuver planning. In [6], investigates parking management methods based on fuzzy control theory. On the basis of fuzzy rules control method, the function of automatic parking on the model of the car is proposed. In [7], an algorithm for automatic traffic control during parking was developed, based on the restrictions of the parking place and the kinematics of the vehicle. In [7, 8], studies were developed on the planning and tracking of a Park Road, and a method was proposed for planning a road based on an appropriate map and tracking the road using a controller based on the distance traveled. With the rapid development of Image technology, some parking specialists began to apply image processing technology for automatic parking systems.

Intelligent recognition of parking spaces is also associated with machine vision technology, target detection and recognition technology. Common research in this field includes fuzzy PID regulation, fuzzy inference, neural network, or methods integrated with various methods [9, 10].

Today, various automatic parking systems are created and applied. But there are still problems in this area. There are many factors that affect the parking process, such as: the length of the car, the steering angle, the size of the tires, etc.

In addition, there are a number of problems with the parking process. For example, parking between two cars, between poles, against a wall, etc. can be. It should be borne in mind that not every car has the same length and width. Therefore, there is no default variable for the parking system, and there is no standard parking space in the world. According to European standards, cars are divided into different classes according to their size (see Table 1).

Table 1. Classification of automobiles and Euro Car Segment scheme

N	Classification by grades	Length (m)	Width (m)	Examples
1	**Class A** (mini)	2.7–3.7	1.60–1.70	Fiat Panda, Smart EQ fortwo cabrio Fiat500, Toyota Aygo, Volkswagen Up!, Hyundai i10, Kia Picanto, Peugeot 108, Citroën C1, Suzuki Ignis. Renault Twingo
2	**Class B** (small)	3.7–4.1	1.50–1.70	Toyota Yaris, Volkswagen Polo, Dacia Sandero, Ford Fiesta, Citroën C3, Mini, Renault Zoe, Skoda Fabia, Hyundai i20, Seat Ibiza, Audi A1
3	**Class C** (medium)	4.2–4.6	1.70–1.85	Mercedes-Benz A, Mazda 3, Honda Civic, Toyota Corolla Sedan etc.
4	**Class D** (large)	4.4–4.7	1.70–1.95	BMW 3 Series, Volkswagen Passat, Tesla Model 3, Mercedes-Benz C-Class, Audi A4/S4/RS4, Škoda Superb, Volvo S60/V60, Peugeot 508, Audi A5/S5/RS5, Ford Mondeo, BMW 4 Series, Volkswagen Arteon, Toyota Camry
5	**Class E** (Executive)	4.7–4.9	1.70–1.85	BMW 5-Series, Audi A6, Mercedes-Benz E-Class, Jaguar XF, Lexus GS, Chrysler 300
6	**Class F** (luxury)	4.9–5.0	1.85–2.0	BMW 8 Series, BMW 7 Series, Mercedes-Benz S-Class, Porsche Panamera, Audi A8/S8
7	**Class J** (Sport)	>5.0	>1.8	Kia Sorento, Mazda CX-5,Nissan Juke, Chevrolet Trax, Suzuki Jimny

(*continued*)

Table 1. (*continued*)

N	Classification by grades	Length (m)	Width (m)	Examples
8	**Class** M (Multi)	>5.5	>1.95	Fiat 500L, Ford B-Max, Hyundai ix20, Opel/Vauxhall Meriva, Kia Venga
9	**Class** S (Sport)			Not considered

Note. Classification is carried out according to various standards. In some states, cars are classified by volume.

Sometimes it is very difficult to find a suitable place during the parking process. Thus, despite the fact that this is an empty space, this area does not correspond to the dimensions of the car, and as a result, the driver is forced to leave this parking lot, settling in another compartment

2 Problem Statement

Let's assume that you want to place as many cars as possible in a parking area the size of NxM. To solve the problem, an algorithm was proposed using fuzzy set theory, fuzzy logic.

1. First of all, with the help of ultrasonic sensors, the width and length of the parking space are determined and the vehicle is localized.
2. A fuzzy logic controller can receive input data from various sensors. Ultrasonic sensors can determine the distance between the car and obstacles. With fuzzy inference, the end result allows the car to distinguish between different environments and perform any behavior.
3. In order to ensure the normal opening of the doors on both sides of the car, the system usually determines the width of its parking by at least 60–70 cm along the width of the car. If the width of the car is equal to C, then (C + 0.70) the distance should be observed.
4. If the width of the parking space exceeds the minimum width of the required parking space, the task of planning the parking path and regulating the final state of the parking lot is performed.
5. If the distance measured by the ultrasonic measuring module is less than the distance required for normal parking, it turns out that there are not as many places as required, and the driver offers a place in the new parking lot. The article offers two-and four-zone parking.

The fuzzy controller is used to ensure the optimal distribution of parking space. A fuzzy logic controller (FLC) with autonomous parking in this paper. Two ultrasonic sensors are used to detect the parking space (see Fig. 1).

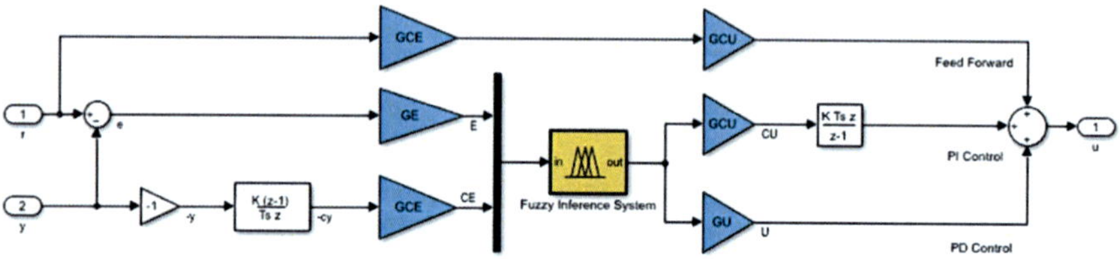

Fig. 1. Fuzzy controller

The fuzzy controller is a Mamdani-type fuzzy logic output system. A fuzzy model implements some real function $f : D_{X_1} \times \cdots\cdots \times D_{X_n} \to D_Z$ that is determined by the procedure for executing fuzzy rules. The Mamdani algorithm: Let us first note that the mechanism of fuzzy conclusions used in various expert and control systems is based on a knowledge base formed by specialists of the subject area in the form of a set of fuzzy predicate rules of the form:

R1: IF X is A1, THEN Z is B1,
R2: IF X is A2, THEN Z is B2,
....
RN: IF X is AN, THEN Z is BN

where x is an input variable (a name for known data values), z is an output variable (a name for the data value to be calculated); Ai and Vi are fuzzy sets defined on X and Z, respectively, using the membership and (z) functions [2].

To solve this problem, a corresponding base of rules of the fuzzy inference system is created, containing the rules of fuzzy production. To solve this problem, the following parameters are used:

– Input parameters

1. The size of the car entering the park or what class it belongs to-XN,
2. The size of the empty area in the parking lot- P1,P2,P3,P4

– Output parameters:

1. Decision on the placement of the car-PR

The fuzzy PID rule-base is a mapping between the inputs and the output; it contains normal heuristic control rules of controlling a plant. A sample of the rules has the following form:

Rule1*. IF the size of the car is large, and the remaining free space in the first zone is not large, and there is more space in the second zone than the size of the car, THEN it is necessary to park the car in the second zone.*

Rule 2*. IF the size of the car is small, and the remaining space in the first field is large, THEN the car should be parked in the first field.*

Since we know the minimum and maximum sizes of cars, we can determine the intervals of change of the linguistic variable for each input (Table 1).

The input linguistic variables "Distance to the obstacle" have limits of acceptable values from 0 to 5 m. (**Class** E, **Class** D). For other classes of cars, this size varies.

Each interval defined in this way is divided into regions: zero, close, middle, far, very far (see, Table 1).

A fuzzy logic controller can receive input data from various sensors. Ultrasonic sensors can determine the distance between the car and obstacles. The end result by means of fuzzy output allows the car to distinguish different environments and perform more expedient behavior. The fuzzy inference process shows that according to the fuzzy rule base and the database that was created, the fuzzy input data is processed to generate the corresponding control input data and control strategy.

3 Simulation and Performance Test of the System

In this, information related to the simulation of the system and the computer program and simulation results for these scenarios will also be given.The program consists of the integration of several programming languages, markup languages and hypertext design. These languages include: html və css and javascript. (the jQuery library is widely used). These languages were only used to create part of the view. The fuzziness in the system has been corrected to determine the size of the parking space in the parking space when parking cars. Thus, when comparing the size of the car and the size of the remaining parking space, specific terms such as "there is a place" or "there is no place" are not used. At this time, the use of expressions such as "little space "or" a lot of space" made it possible to use fuzziness.

The program allows the user to position the car and the parking point in the desired position. The main window of the simulation program is shown in Fig. 2.

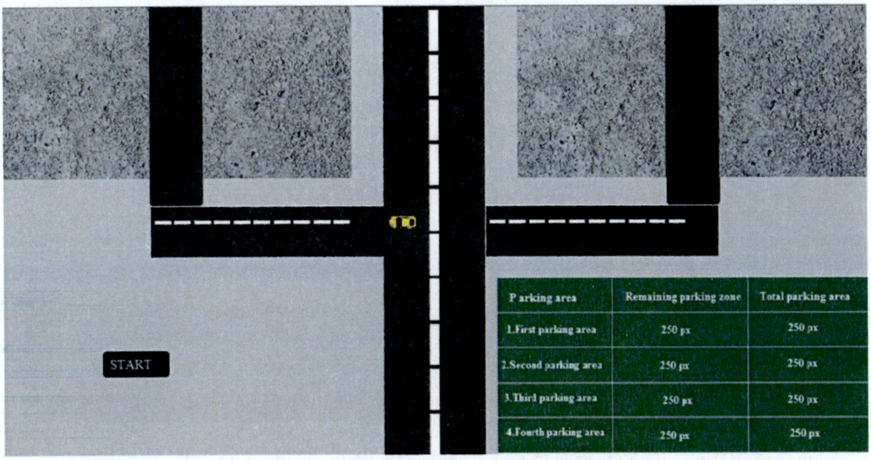

Fig. 2. Preview of the parking space

The table records the largest space remaining in each parking zone and the total free space remaining in this part. These locations change as each car is parked in this part. At the very beginning of the program, the largest space left empty and the total space left empty become the same in the packring areas where no cars are parked (see Figs. 3, 4, and 5).

Parking area	Remaining parking zone	Total Parking area
1.First parking area	250 px	250 px
2.Second parking area	250 px	250 px
3.Third parking area	250 px	250 px
4.Fourth parking area	250 px	250 px

Fig. 3. Table showing the size of parking

If there is not enough space in the parking zone in the first part, then it is allowed to move to the second zone.But when a car of the appropriate size approaches the entrance, it is placed in the place that remained in its first part, and the first parking area is filled. Access to the second parking area continues.

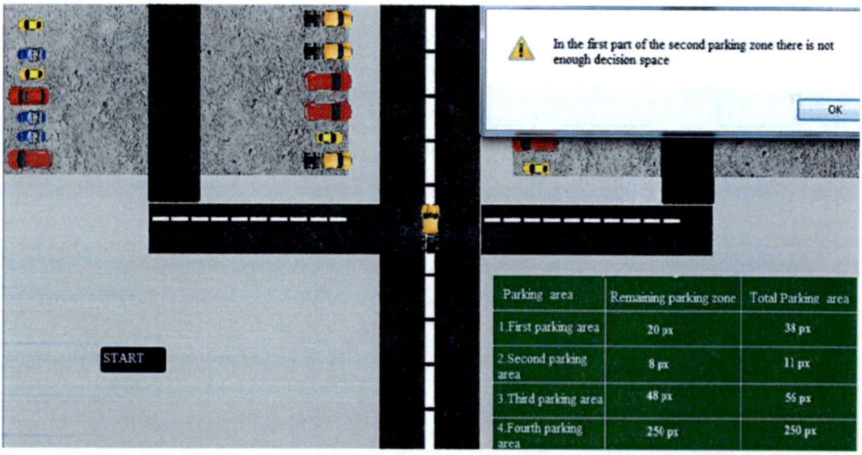

Fig. 4. In the first part of the second parking zone there is not enough decision space

Since the remaining territory at the second stop does not correspond to the size of the next car, it is allowed to go to the third stop. Thus, this process continues until the smallest area in the entire park area remains.

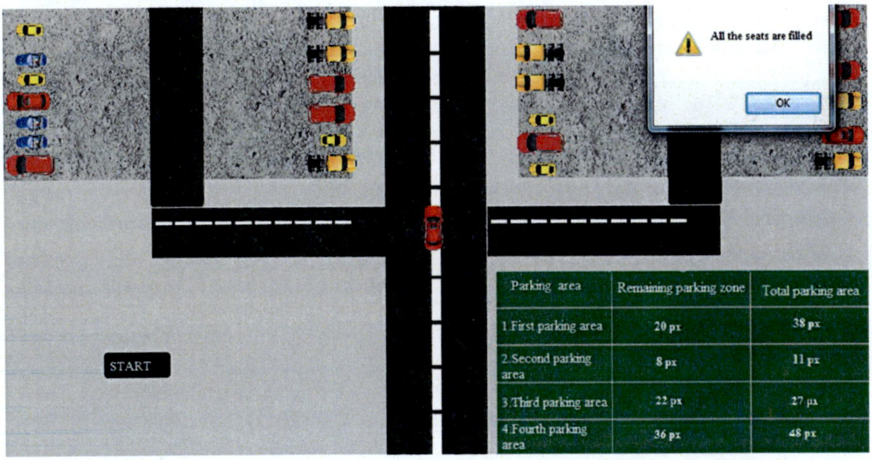

Fig. 5. Results of the program execution: all parking spaces are filled.

4 Conclusion

The paper develops an algorithm for automatic parking based on fuzzy logic. Here is a new model of a parking lot. The issue of developing an automatic parking control system for a passenger car in the proposed parking lot is being considered. To solve the problem, it is proposed to divide the parking space using the methods of fuzzy set theory and fuzzy logic.

When writing the practical part of his work, the following was done:

- the creation of an object was vaguely modeled taking into account its dynamic characteristics, such as size, shape, as well as speed, rotation angle;
- using the theory of fuzzy sets, the optimal use of parking space was ensured. (comparing the size of the machine and the size of the empty space).

The program is designed to work with the user in the interactive dialog mode.

References

1. Aliev, R.A., Fazlollahi, B., Aliev, R.R.: Soft Computing and its Applications in Business and Economics. Springer, Heidelberg (2004). https://doi.org/10.1007/978-3-540-44429-9
2. Wang, W., Song, Y., Zhang, J., Deng, H.: Automatic parking of vehicles: a review of literatures. Int. J. Automot. Technol. **15**(6), 967–978 (2014). https://doi.org/10.1007/s12239-014-0102-y

3. Bagloee, S.A., Tavana, M., Asadi, M., Oliver, T.: Autonomous vehicles: challenges, opportunities, and future implications for transportation policies. J. Mod. Transp. **24**(4), 284–303 (2016). https://doi.org/10.1007/s40534-016-0117-3
4. Lv, Z., Zhao, L., Liu, Z.: A path-planning algorithm for parallel automatic parking. In: Third International Conference on Instrumentation, Measurement, Computer, Communication and Control (IMCCC), pp. 474–478 (2013). https://doi.org/10.1109/IMCCC.2013.108
5. Li, B., Wang, K., Shao, Z.: Time-optimal maneuver planning in automatic parallel parking using a simultaneous dynamic optimization approach. IEEE Trans. Intell. Transp. **17**(11), 3263–3274 (2016). https://doi.org/10.1109/TITS.2016.2546386
6. Yang, W., Zheng, L., Li, Y., Ren, Y.: A trajectory planning and fuzzy control for autonomous intelligent parking system. Conference: WCX™ 17: SAE World Congress Experience 2017-01-0032 (2017). https://doi.org/10.4271/2017-01-0032
7. Degerman, P., Pohl, J., Sethson, M.: Hough transform for parking space estimation using long range ultrasonic sensors. In: SAE 2006 World Congress and Exhibition. Detroit, USA (2006). https://doi.org/10.4271/2006-01-0810
8. Gomez-Bravo, F., Cuesta, F., Ollero, A.: Parallel and diagonal parking in nonholonomic autonomous vehicles. Eng. Appl. Artif. Intel. **14**(1), 419–434 (2001). https://doi.org/10.1016/S0952-1976(01)00004-5
9. Yusnita, R., Fariza, N., Norazwinawati, B.: Intelligent parking space detection system based on image processing. Int. J. Innov. Manag. Technol. **3**(3) (2012). https://doi.org/10.1109/ICMIP.2017.4
10. Yan, G., Yang, W., Rawat, D.B., Olariu, S.: SmartParking: a secure and intelligent parking system. IEEE Intell. Transp. Syst. Mag. **3**(1), 18–30 (2011). https://doi.org/10.1109/MITS.2011.940473

Staff Selection with a Fuzzy Analytical Hierarchy Process in the Tourism Sector

Ahmad Valiyev[1], Rahib Imamguluyev[2(✉)], and Ilkin Gahramanov[1]

[1] Odlar Yurdu University, Baku AZ1072, Azerbaijan
ilkin.gahramanov@wbc.com
[2] Department of IT and Engineering, Odlar Yurdu University, Baku AZ1072, Azerbaijan

Abstract. As production in residential enterprises is based on labor, there is a need for more workers than in other sectors. In the tourism sector, there is a need for certain work in areas common to the customer, the need to properly perform services for the first time, as well as qualified personnel for production and consumption. Employees working in the tourism sector are more interested in customers than employees in other sectors. In enterprises in the tourism sector, employees satisfy customers and provide quality and efficient service; choosing the right human resources is very important. The choice of human resources also requires a very critical decision. In this study, fuzzy logic was applied in the selection of human resources in the tourism sector. The 3 candidates selected as examples in the study were evaluated within a total of 20 sub-criteria, which constitute the main criteria of External Criteria, Internal Criteria, Professional Competence, and Responsibility. As a result of the assessment, it was determined that the most important criterion is professionalism. The most important sub-criteria for candidates is experience and knowledge to work in the hotel business.

Keywords: Fuzzy logic · Fuzzy set · Tourism · Staff selection · Fuzzy analytical hierarchy process

1 Introduction

Human resources considered the most valuable asset of an enterprise, play an important role in the success of the enterprise. Due to the intense national and international competition between the enterprises and the uncertainties that arise with the market conditions, a precise and understandable information flow cannot be provided in some cases [1]. Especially in a structure where verbal variables are predominant, a blur may occur due to ambiguity and indecision. In such fuzzy environments, decision-making with fuzzy numbers is preferred [1].

Hotels, which are part of accommodation businesses, which are one of the most important building blocks of the tourism sector, are obliged to meet the mental needs of individuals who want to receive services, such as accommodation, food, and beverage, as well as their basic needs. Travel agencies are another part of accommodation businesses [2].

R. A. Aliev et al. (Eds.): ICSCCW 2021, LNNS 362, pp. 437–444, 2022.
https://doi.org/10.1007/978-3-030-92127-9_59

Travel agencies are the places where all the requests and needs of those who want to travel for service are answered in the promotion, information, and purchasing process of any touristic product. While travel agencies ensure that hotels are preferred by the service providers, hotels also provide product diversity for service providers who prefer travel agencies [3].

In accommodation establishments, more workers are needed compared to other sectors, since production is based on intensive labor. In the tourism sector, some features such as the presence of certain works in common areas with customers, the necessity of performing the services correctly at the first time, simultaneous production, and consumption require qualified personnel [3]. Due to the labor-intensive and seasonal nature of the tourism sector, there are some difficulties in the management of human resources.

Those in the field of management have realized that the employees are important and that there is no business without employees. One of the most important duties of the human resources department is the right selection of the employees applying for the job. Whether an employee will be selected in an enterprise, and if so, which criteria should be selected should be done with human resources planning. If a suitable employee for the job is not selected, problems may arise between the service recipient and the provider. Therefore, it is necessary to select the appropriate employee for the appropriate job [4].

Employees in the tourism sector are more interested in customers than employees in other sectors. In the enterprises in the tourism sector, in order for the employees to satisfy their customers and to provide a quality and efficient service; choosing the right human resources is of great importance. As a result of the developments in globalization, technology, and communication since the beginning of the 21st century, the level of competition of enterprises in the tourism sector has been increasing. In an increasingly competitive environment, businesses can only survive by making the selection of the employees they serve more effectively [5].

After the development of technology, the effects of globalization can affect even the smallest business. This effect increases the importance of human resources planning, especially in the service sector and businesses where the workforce is at the forefront. In this study, a personnel selection model has been created for the planning of human resources in the tourism sector due to the difficulty of choosing the right service provider. In the study, the application of the Fuzzy Analytical Hierarchy Process has been made in the selection of personnel in the tourism sector [6].

2 Fuzzy Set Theory

Fuzzy logic shows the relationship between individuals when making decisions with approximation rather than certainty [7, 8]. The most important part of fuzzy logic is that individuals prefer approximate decision terms in situations that may vary from person to person while making decisions. It is impossible for individuals to reach definite decisions in these uncertain and fuzzy environmental situations. As Lotfi A. Zadeh stated in his work titled “Fuzzy Sets”, uncertainties can be eliminated using fuzzy set theory [9, 10]. The real problem with the degree of membership in fuzzy set theory is represented by conventional mathematical expressions. Fuzzy set theory is expressed as the tendency of indecision to rationality due to inconsistency or uncertainty [11, 12].

The mathematical expression of fuzzy numbers is defined as convex fuzzy sets with membership degree between 0–1. The membership function of a triangular fuzzy number must have three basic properties [13–15].

3 Fuzzy Analytical Hierarchy Process

The Analytical Hierarchy Process (AHP), which was introduced for the first time by Thomas L. Saaty in 1971, is preferred in cases where qualitative decisions need to be made in multi-criteria problems [15]. AHP is based on decision makers' rating criteria using a pairwise comparison scale. The pairwise comparison scale used in the grading of the criteria is made with exact values between 1 and 9. However, in cases where the decisions made with the values used are uncertain or unstable, it is preferred to use the numbers by converting them to fuzzy numbers. In such cases, instead of the AHP method, the Fuzzy Analytical Hierarchy Process (FAHP) is used in uncertain and complex situations [16].

Modeling can be done with the preference of fuzzy numbers in the uncertainty and indecision of decision-makers in the face of situations. FAHP, based on the fuzzy preference values of the decision-makers, the general priorities values are obtained from the special priority combination. FAHP completes the decision-making phase more precisely than the AHP method. Pairwise comparison numbers in the weight matrix used in fuzzy field applications are chosen from fuzzy numbers [16]. Weight vectors and total scores for each alternative are calculated using the procedures in Table 1.

In the FAHP method, first, all verbal expressions (criteria, sub-criteria, and alternatives) are converted to fuzzy numbers. If there is no effect of the criteria on the sub-criteria among the criteria, the effect of the criteria on the alternatives is created with the help of triangular fuzzy numbers, and then the fuzzy priority matrix is obtained [16].

The fuzzy significance levels used in the FAHP method differ from the pairwise comparison matrix used in the AHP method (Wang and Wang, 2010: 8518). While the significance levels and triangular fuzzy numbers are being created, the data must meet this condition $\tilde{x}_{ij} = \left(1_{ij}, a_{ij}, b_{ij}\right) \rightarrow \tilde{x}\ \ 1_{ij} = (1/a_{ij}, 1/b_{ij}, 1/1_{ij})$. Table 2 shows the fuzzy significance scale.

Table 1. Fuzzy significance scale in FAHP

Explanation of binary comparison	Fuzzy significance	The inverse of fuzzy significance
Equally Important	(1, 1, 1)	(1, 1, 1)
Intermediate Value	(1, 2, 3)	(1/3, 1/2, 1/1)
A Little More Important	(2, 3, 4)	(1/4, 1/3, 1/2)

(continued)

Table 1. (*continued*)

Explanation of binary comparison	Fuzzy significance	The inverse of fuzzy significance
Intermediate Value	(3, 4, 5)	(1/5, 1/4, 1/3)
Strongly Important	(4, 5, 6)	(1/6, 1/5, 1/4)
Intermediate Value	(5, 6, 7)	(1/7, 1/6, 1/5)
Very Strongly Important	(6, 7, 8)	(1/8, 1/7, 1/6)
Intermediate Value	(7, 8, 9)	(1/9, 1/8, 1/7)
Extremely Important	(8, 9, 9)	(1/9, 1/9, 1/8)

4 Application of Staff Selection in the Tourism Industry with the Fuzzy Analytical Hierarchy Process

In the study, a suitable hotel business was determined by using the quota sampling method among the hotels operating in Azerbaijan. All managers who influenced the personnel selection in the selected hotel were accepted as decision-makers. The most important limitation of the research is that this research was carried out only in line with the opinions of 6 different department managers in the hotel business in Azerbaijan in terms of the best use of time and the availability of financial opportunities.

4.1 Implementation of Fuzzy Algorithm

In order to express the FAHP method and to divide the problem into sub-problems, a hierarchical structure is created and shown in Fig. 1. The criteria (sub-criteria) determined by the decision-makers to make comparisons; External Criteria (Military Service, Gender, Driving License, Marital Status, and Travel Status), Internal Criteria (Persuasion Ability, Communication Skill, Problem Solving Ability, Coping with Stress and Using Time Effectively), Professional Competence (Experience, Educational Status, General Program Knowledge, Program Knowledge, and Foreign Language) and Responsibility (Leader Trait, Customer Focus, Result Orientation, Taking Responsibility and Team Harmony).

For the main criteria in the hierarchical structure shown in Fig. 1, the group decision-making method of paired comparisons made by 6 experts in the hotel business was used. In practice, the calculations made in the pairwise comparison matrix created only for the comparison of the main criteria are shown in detail in Table 2.

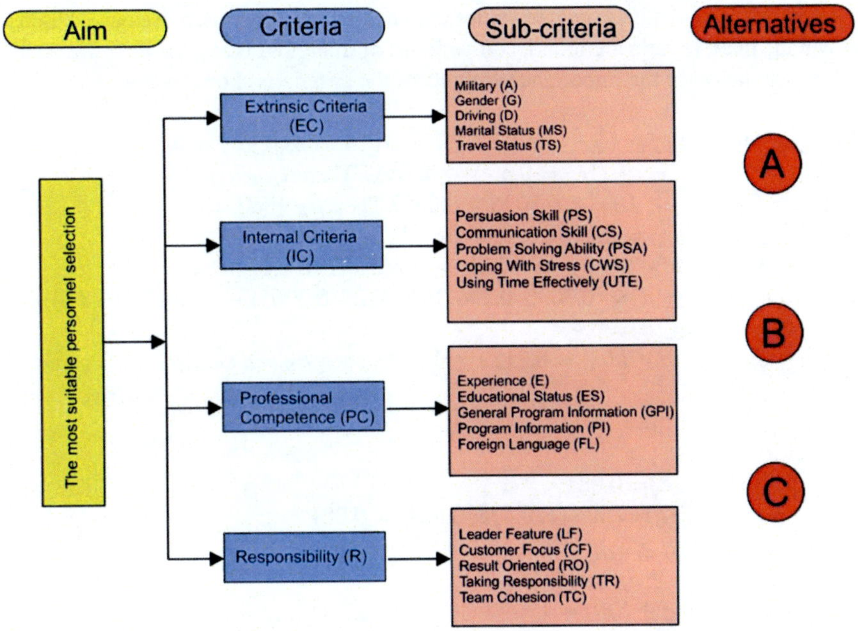

Fig. 1. Hierarchical structure of personnel selection problem

Table 2. Fuzzy evaluation matrix of main criteria

	EC			IC			PC			R		
EC	**1**	**1**	**1**	1/8	1/7	1/6	1/9	1/9	1/8	1/6	1/5	1/4
IC	6	7	8	**1**	**1**	**1**	1	1	1	1	1	1
PC	8	9	9	1	1	1	**1**	**1**	**1**	1	2	3
R	4	5	6	1	1	1	1/3	1/2	1	**1**	**1**	**1**
W_A=(0.00, 0.31, 0.64, 0.05)^T												

The calculation of fuzzy synthetic degree values for pairwise comparisons between the main criteria is shown below, respectively.

$S_{EC}= (1.41, 1.46, 1.53) \times (27.75, 31.94, 35.55)^{-1}$
$S_{EC} = (1.41, 1.46, 1.53) \times (0.02815, 0.03131, 0.03606) = (0.0396, 0.0456, 0.0554)$
$S_{IC} = (9, 10, 11) \times (27.75, 31.94, 35.55)^{-1}$
$S_{IC} = (9, 10, 11) \times (0.02815, 0.03129, 0.03606) = (0.2533, 0.3129, 0.3967)$
$S_{PC} = (11, 13, 14) \times (27.75, 31.94, 35.55)^{-1}$
$S_{PC} = (11, 13, 14) \times (0.02815, 0.03129, 0.03606) = (0.3096, 0.4067, 0.5049)$
$S_R = (6.34, 7.51, 9) \times (27.75, 31.94, 35.55)^{-1}$
$S_R = (6.34, 7.51, 9) \times (0.02815, 0.03129, 0.03606) = (0.1733, 0.2346, 0.3246)$

After the fuzzy synthetic degree values obtained as a result of pairwise comparisons between the main criteria are found, comparisons of triangular fuzzy numbers are made and the calculation of the importance weights of the criteria is shown below.

If V($S_{EC} \geq S_{IC}$) = $L_{IC} \geq U_{EC}$ = 0.2533 ≥ 0.0557 Then $\mu_{SEC}(d) = 0$
If V($S_{EC} \geq S_{PC}$) = $L_{PC} \geq U_{EC}$ = 0.3094 ≥ 0.0557 Then $\mu_{SEC}(d) = 0$
If V($S_{EC} \geq S_{R}$) = $L_{R} \geq U_{EC}$ = 0.3096 ≥ 0.0557 Then $\mu_{SEC}(d) = 0$
V($S_{EC} \geq S_{IC}, S_{PC}, S_{R}$) = min (0, 0, 0) = 0
If V($S_{IC} \geq S_{EC}$) = $M_{IC} \geq M_{EC}$ = 0.3131 ≥ 0.0456 Then $\mu_{SIC}(d) = 1$
V($S_{IC} \geq S_{PC}$) = $\mu_{SIC}(d)$= (0.3096–0.3967)/((0.3129–0.3965) – (0.4069–0.3096))
V($S_{IC} \geq S_{PC}$) = $\mu_{SIC}(d)$= 0.482
If V($S_{IC} \geq S_{R}$) = $M_{IC} \geq M_{EC}$ = 0.3129 ≥ 0.2346 Then $\mu_{SIC}(d) = 1$

.
.
.

V($S_{R} \geq S_{PC}$) = $\mu_{S_R}(d)$= 0.081
V($S_{R} \geq S_{EC}, S_{IC}, S_{PC}$) = min (1, 0.478, 0.081) = 0.081
d'(A_{EC}) = min V($S_{EC} \geq S_{K}$) = 0
d'(A_{IC}) = min V($S_{IC} \geq S_{K}$)) = 0.482
d'(A_{PC}) = min V($S_{PC} \geq S_{K}$)) = 1
d'(A_{R}) = min V($S_{RC} \geq S_{K}$)) = 0.081

When the above operations are performed sequentially, the next step is to obtain the weight vector (W′). The result obtained; $W' = (0, 0.482, 1, 0.081)^T$. The normalization of this vector is completed by sequential division of the obtained data by 1.561 (0 + 0.482 + 1 + 0.081). As a result of this process $W_A = (0, 0.32, 0.65, 0.06)^T$ weight vector is obtained.

For other pair comparison matrices, the priority vectors were calculated using the same algorithm above. Based on the pair matrices and calculated weight vectors obtained as a result of the assessments, the following result was obtained (Table 3).

Table 3. Overall conclusion

Criteria	Extrinsic	Internal	Professional competence	Responsibility	Importance weights
Weights	0,00	0,32	0,65	0,06	
Alternatives					
A	0,351	0,254	0,005	0,322	
B	0,352	0,215	0,399	0,262	
C	0,300	0,534	0,599	0,419	
Total	1,000	1,000	1,000	1,000	
A	0,00	0,079	0,004	0,017	0,1

(continued)

Table 3. (*continued*)

Criteria	Extrinsic	Internal	Professional competence	Responsibility	Importance weights
Weights	0,00	0,32	0,65	0,06	
Alternatives					
B	0,00	0,067	0,256	0,014	0,337
C	0,00	0,166	0,384	0,022	0, 572*
Total	1,000				

When personnel selection alternatives are examined according to the fuzzy logic model, alternative C is preferred primarily with a value of 0.572. Alternative B with 0.337 and Alternative A with 0.1 prevail.

5 Conclusion

The study aims to select the most suitable staff in an enterprise operating in Azerbaijan by applying the FAHP method, which is one of the most criteria-based decision-making methods. Criteria were evaluated by studying the opinions of 6 managers (specialists) who decided to select staff for a hotel job set for this purpose. After evaluating the opinions of the experts, a selected group of students who graduated from Odlar Yurdu University with a degree in "Organization of Tourism Work" were included in the model.

Considering the opinions of experts, the most important feature that applicants pay attention to is experience and knowledge to work in the hotel business. When planning staff, second-time applicants are required to have the most convincing skills, communication, and problem-solving skills. The distinctive feature of the applicants is that the candidates are leaders in taking responsibility.

If education in vocational schools is carried out in the practical and business areas, rather than in theory, it will be easier for graduates to move to work life. In future research, multi-criteria decision-making methods may be used to select different staff in different regions by adding additional criteria or using other criteria related to human resource planning.

References

1. Tyagi, S., Chambers, T., Yang, K.: Enhanced fuzzy-analytic hierarchy process. Soft Comput. **22**(13), 4431–4443 (2017). https://doi.org/10.1007/s00500-017-2639-y
2. Karczmarek, P., Pedrycz, W., Kiersztyn, A.: Fuzzy analytic hierarchy process in a graphical approach. Group Decis. Negot. **30**(2), 463–481 (2021). https://doi.org/10.1007/s10726-020-09719-6
3. Saaty, T.L.: Decision making — the analytic hierarchy and network processes (AHP/ANP). J. Syst. Sci. Syst. Eng. **13**, 1–35 (2004). https://doi.org/10.1007/s11518-006-0151-5

4. Javanbarg, M.B., Scawthorn, C., Kiyono, J., Shahbodaghkhan, B.: Fuzzy AHP-based multi-criteria decision making systems using particle swarm optimization. Exp. Syst. Appl. **39**(1), 960–966 (2012)
5. Huang, Z., Gardoni, M.: Concept evaluation based on fuzzy analytic hierarchy process. In: Conference: 2020 IEEE International Systems Conference (SysCon) (2020). https://doi.org/10.1109/SysCon47679.2020.9275886
6. Mahad, N.F., Mohamed Yusof, N., Ismail, N.F.: The application of fuzzy analytic hierarchy process (FAHP) approach to solve multi-criteria decision making (MCDM) problems. In: 12th Seminar on Science and Technology. https://doi.org/10.1088/1742-6596/1358/1/012081
7. Zadeh, L.A.: Fuzzy sets. Inf. Control **8**(3), 338–350 (1965)
8. Aliev, R., Tserkovny, A.: Fuzzy logic for incidence geometry. In: Kosheleva, O., Shary, S.P., Xiang, G., Zapatrin, R. (eds.) Beyond Traditional Probabilistic Data Processing Techniques: Interval, Fuzzy etc. Methods and Their Applications. SCI, vol. 835, pp. 49–85. Springer, Cham (2020). https://doi.org/10.1007/978-3-030-31041-7_4
9. Aliev, R.A., Pedrycz, W.: Fundamentals of a fuzzy logic based generalized theory of stability. IEEE Trans. Syst. Man Cybern. B Cybern. **39**(4), 971–988 (2009)
10. Zadeh, L.A., Aliev, R.: Fuzzy Logic Theory and Applications: Part I and Part II, p. 61 (2018). https://doi.org/10.1142/10936
11. Zadeh, L.A.: Toward a theory of fuzzy information granulation and its centrality in human reasoning and fuzzy logic. Fuzzy Sets Syst. **90**, 111–127 (1997)
12. Imamguluyev, R.: Determination of correct lighting based on fuzzy logic model to reduce electricity in the workplace, September 2020. In: Conference: International Conference on Eurasian EconomiesAt: Baku, Azerbaijan (2020). https://doi.org/10.36880/C12.02456
13. Imamguluyev, R.: Optimal solution of indoor lighting system based on fuzzy logic, May 2021. In: Conference: 2nd International Baku Conference on Scientific Research at: Baku, Azerbaijan (2021)
14. Valiyev, A., Imamguluyev, R., Ilkin, G.: Application of fuzzy logic model for daylight evaluation in computer aided interior design areas. In: Aliev, R.A., Kacprzyk, J., Pedrycz, W., Jamshidi, M.O., Babanli, M., Sadikoglu, F.M. (eds.) ICAFS 2020. AISC, vol. 1306, pp. 715–722. Springer, Cham (2021). https://doi.org/10.1007/978-3-030-64058-3_89
15. Imamguluyev, R.: Application of fuzzy logic model for correct lighting in computer aided interior design areas. In: Kahraman, C., Cevik Onar, S., Oztaysi, B., Sari, I.U., Cebi, S., Tolga, A.C. (eds.) INFUS 2020. AISC, vol. 1197, pp. 1644–1651. Springer, Cham (2021). https://doi.org/10.1007/978-3-030-51156-2_192
16. Li, J.: Analytic Hierarchy Process Automatically (2021). https://doi.org/10.1145/3469213.3470276
17. Wang, G.: Application of fuzzy analytic hierarchy process in the optimization of unproduced reserves. In: Lin, J. (ed.) IFEDC 2020. SSGG, pp. 2296–2305. Springer, Singapore (2021). https://doi.org/10.1007/978-981-16-0761-5_217

Type-2 Fuzzy Logic Approach for Port Selection

Kamala Aliyeva(✉)

Department of Instrument Engineering, Azerbaijan State Oil and Industry University, 20 Azadlig Avenue, AZ1010 Baku, Azerbaijan

Abstract. To assess the competitiveness of ports, an algorithm is proposed for choosing a solution based on situational algorithms for processing fuzzy information about factors that significantly affect their economic and economic activities. It is important to select the corresponding port for shipping. Different approaches have been accepted by carrying operators to estimate, analyze, and choose ports of calls. Most of these techniques are totally based on evaluating various port performance factors involved. The ports performance factors are related and are of a complex characteristic. Comparative analyzing of ports by various estimating technologies provides a basis for performing strategies for port and shipping operators. This paper goals to present interval type-2 fuzzy sets for choosing suitable port for transportation. The proposed hybrid approach under interval type-2 fuzzy sets is used environment to cover more of the vagueness of experts' opinions and expression in decision-making.

Keywords: Type-2 fuzzy logic · Port selection · Multicriteria decision making

1 Introduction

Currently, the ports are forced to work in an environment where their financial well-being depends primarily on the competitiveness of the port. Determining the indicators of the competitiveness of ports and ways to improve them are becoming the most important tasks, the solution of which is directly related to the efficiency of the ports. A comparative new concept of the competitiveness of ports in practice has appeared in the conditions of a transitional economy has not yet received a clear and unambiguous definition. Therefore, the task of determining the competitiveness of ports can be reduced to the task of comparing them with each other and choosing the best one according to common indicators that are of the greatest interest from cargo suppliers and carriers.

The peculiarities of solving such comparison problems are not only in the absence of a single base for assessing the economic and economic uncertainty of factors such as, geographic proximity to the places of production and consumption of products delivered by sea transport, possibility of timely delivery of cargo to any port of the world, technological and technical preparation of the port, effective use of the port's production facilities, quality and reliability of port services, financial and economic condition of the port. The emulating atmosphere of ports has made many explorers to regard the vitality of the most effective criteria in their performance estimation.

R. A. Aliev et al. (Eds.): ICSCCW 2021, LNNS 362, pp. 445–449, 2022.
https://doi.org/10.1007/978-3-030-92127-9_60

A lot of these literatures have based their analysis on the factors advised by the experts. Different articles have been founded on the selection of port rivalry criteria. Researches founded by Ha [1], Lirn et al. [2], and Chou [3] comprise factors leading to ports' performance evaluation such as delivered time, reliability, shipping costs, port expenditures, service characteristics, port size, and accessibility. Su et al. [6] have attempted to create a exhaustive performance measuring system by introducing a new management tool so called the balanced scorecard. Considering the character of commercial port activities, they have developed a lot of qualitative performance criteria, especially, criteria's such as finance, customer services, internal business process, training, and growth perspectives. This research classified the most defining factors through this concept.

Song and Yeo [5] have attached an AHP method to compare various attributes of the container terminal operations for ports situated at China and USA. They have selected the most determining criteria based on the expert's thinking in the region. Tai and Hwang [7], Chou [8] have estimated and compared the suitability of some models for port selection as a determining issue in the present transborder vessel commerce and transportation market. His articles have focused on the possibilities of decreasing transportation costs. In some article's authors determine parameters for port performance estimation have been provided. These articles have principally analyzed the influence of different characteristics on port estimation.

This paper designed as follows. Section 2 introduces basic definitions of the fuzzy Type-2 methodology that used in this problem. In Sect. 3, proposed method of the fuzzy Type-2 methodology for the port selection problem. Section 4 presents the basic results developed in this article.

2 Preliminaries

Definition 1. A Type-2 fuzzy set $\widetilde{A}$ on the universe of discourse X can be characterized by its fuzzy membership function $\mu_A(x, u)$ [9].

$$A = \{((x, u, \mu_{\widetilde{A}}(x, u))|\forall_x \in X, \forall_u \in J_x \subseteq [0, 1]\},\ 0 \leq \mu_{\widetilde{A}}(x, u) \leq 1 \quad (1)$$

Definition 2. An interval Type-2 fuzzy sets (IT2FS) $\widetilde{A}$ on the universe of discourse X is characterized by its upper A and lower membership function A.

$$\widetilde{A_\alpha} = \{(x, U)|\mu_{\widetilde{A}}(x, U) \geq a, x \in X, U \in J_x\} \quad (2)$$

Definition 3. The computational model of interval Type-2 fuzzy numbers extended to Type-2 fuzzy numbers. $\widetilde{A}_1$ and $\widetilde{A}_2$ are positive interval Type-2 fuzzy numbers. Sum of them is determined as follows:

$$\widetilde{A_1} + \widetilde{A_2} = \langle A_1^U + A_2^U; A_1^L + A_2^L\rangle =$$

$$= a_{11}^U + a_{21}^U, a_{12}^U + a_{22}^U, a_{13}^U + a_{23}^U, a_{14}^U + a_{24}^U, \frac{h(A_1^U) \cdot \|A_1^U\| + h(A_2^U) \cdot \|A_2^U\|}{\|A_1^U\| \cdot \|A_2^U\|}; \rangle \quad (3)$$

$$a_{11}^L + a_{21}^L, a_{12}^L + a_{22}^L, a_{13}^L + a_{23}^L, a_{14}^L + a_{24}^L, \frac{h(A_1^L) \cdot \|A_1^L\| + h(A_2^L) \cdot \|A_2^L\|}{\|A_1^L\| \cdot \|A_2^L\|}; \rangle \quad (4)$$

Where

$$\left\|A_j^U\right\| = \frac{a_{j1}^U + a_{j2}^U + a_{j3}^U + a_{j4}^U}{4} \; and \; \left\|A_j^L\right\| = \frac{a_{j1}^L + a_{j2}^L + a_{j3}^L + a_{j4}^L}{4} (j = 1, 2) \quad (5)$$

Definition 4. Let $\widetilde{A}_i = \langle A_i^U; A_i^L \rangle$ be two positive interval Type-2 fuzzy the expected value of A is defined as.

$$\widetilde{E}_\lambda(\tilde{A}) = \frac{E_\lambda(A^U) + E_\lambda(A^U)}{2} \quad (6)$$

$\lambda = 0.7$, where $\lambda \in [0; 1]$ is the index of optimism.

3 Statement of the Problem

The offered method of the fuzzy TOPSIS is used for the choice of the ports. For the selection of the best port, three alternatives are selected: A_1, A_2 and A_3. Four basic criteria have been used: C_1 – port performance, C_2 – port accessibility, C_3 – port tradition, C_4 –government assistance. The comparative weights of the factors were defined as $w_1 = 0.35, w_2 = 0.3, w_3 = 0.20, w_4 = 0.15$, respectively. Assume that a linguistic performance rating and codebook are represented in next form. (Table 1):

Very high = $< 0.8, 0.9, 1, 1, 1; 0.8, 0.9, 1, 1, 0.9 >$
High = $< 0.7, 0.9, 1, 1, 1; 0.7, 0.9, 1, 1, 0.8 >$
Average = $< 0.4, 0.5, 0.6, 0.7, 1; 0.4, 0.5, 0.6, 0.7, 0.6 >$

Table 1. Linguistic performance rating.

Attribute	C_1 performance	C_2 accessibility	C_3 tradition	C_4 gov. assistance
A_1	High	High	High	Average
A_2	Average	Very high	High	High
A_3	Avarage	Very high	Average	Very high

The problem of selecting best port is to determine on alternative A, $i = 1, 2, 3$ (port 1, port 2, port 3) with maximal value of the utility function U.

$$U(A^*) = \max U(A_i); \; A_1 \in A \quad (7)$$

Step 1: State decision matrix for port selection problem:

$$d_{11} = d_{12} = d_{13} = d_{23} = d_{24} = < 0.7, 0.9, 1, 1, 1; 0.7, 0.9, 1, 1, 0.8 >$$

$$d_{22} = d_{32} = d_{34} =< 0.8,\ 0.9, 1, 1, 1;\ 0.8,\ 0.9, 1, 1,\ 0.9 >$$

$$d_{14} = d_{21} = d_{31} = d_{34} =< 0.4,\ 0.5,\ 0.6,\ 0.7, 1;\ 0.4,\ 0.5,\ 0.6,\ 0.7,\ 0.6 >$$

Step 2. Normalize decision matrix D by (X-min)/(max-min):

$$d_{11} = d_{12} = d_{13} = d_{23} = d_{24} =< 0.4,\ 0.7,\ 0.7,\ 0.8, 1;\ 0.4,\ 0.7,\ 0.7,\ 0.8,\ 0.8 >$$

$$d_{22} = d_{32} = d_{34} =< 0.7,\ 0.9, 1, 1, 1;\ 0.7,\ 0.9, 1, 1,\ 0.9 >$$

$$d_{14} = d_{21} = d_{31} = d_{34} = < 0, 0.16, 0.33, 0.5, 1; 0, 0.16, 0.33, 0.5, 0.6 >$$

Step 3. Determine utility functions for all 3 ports:

$$U_{A_1} = d_{11}C_1 + d_{12}C_2 + d_{13}C_3 + d_{14}C_4$$

$$U_{A_2} = d_{21}C_1 + d_{22}C_2 + d_{23}C_3 + d_{24}C_4$$

$$U_{A_3} = d_{31}C_1 + d_{32}C_2 + d_{33}C_3 + d_{34}C_4$$

At last, we get next utility grades for different ports

$$U_{A_1} = < 0.47,\ 0.81,\ 0.86,\ 0.95, 1;\ 0.47,\ 0.81,\ 0.86,\ 0.95,\ 0.801 >$$

$$U_{A_2} = < 0.38,\ 0.60,\ 0.67,\ 0.81, 1;\ 0.38,\ 0.60,\ 0.67,\ 0.81,\ 0.851 >$$

$$U_{A3} = < 0.21,\ 0.44,\ 0.45,\ 0.53, 1;\ 0.21,\ 0.44,\ 0.45,\ 0.53,\ 0.803 >$$

Step 4. For determining best port, we make ranking of $U_{A_1}, U_{A_2}, U_{A_3}$. Assume that decision making is near optimistic and $\lambda = 0.7$. Expected values for different alternatives are

$$\widetilde{E}_\lambda(\widetilde{A}) = \frac{E_\lambda(A^L) + E_\lambda(A^U)}{2} \tag{8}$$

$$E_\lambda\left(A_1^U\right) = 1[(1 - 0.7)\,(0.47 + 0.81)\ + 0.7(0.86 + 0.95)]/2 = 0.82$$

$$E_\lambda\left(A_1^L\right) = 0.8[(1 - 0.7)\,(0.47 + 0.81)\ + 0.7(0.86 + 0.95)]/2 = 0.66$$

$$EV_{U_{A_1}} = 0.74$$

$$E_\lambda\left(A_1^U\right) = 1[(1 - 0.7)(0.38 + 0.60) + 0.7(0.67 + 0.81)]/2 = 0.66$$

$$E_\lambda\left(A_2^L\right) = 0.85[(1-0.7)(0.38+0.60)+0.7(0.67+0.81)]/2 = 0.56$$

$$EV_{U_{A_2}} = 0.61$$

$$E_\lambda\left(A_3^U\right) = 1[(1-0.7)\,(0.21+0.44)\,+0.7(0.45+0.53)]/2 = 0.44$$

$$E_\lambda\left(A_3^L\right) = 0.8[(1-0.7)\,(0.21+0.44)\,+0.7(0.45+0.53)]/2 = 0.35$$

$$EV_{U_{A_3}} = 0.40$$

$$EV_{U_{A_1}} = 0.74,\ EV_{U_{A_2}} = 0.61, EV_{U_{A_3}} = 0.40$$

After calculations and ranking we determine that best port is A_1.

4 Conclusion

Port selection problem can be characterized with different factors such as port performance, port accessibility, port tradition, government assistance. To make decision on port selection we have used fuzzy Type-2 methodology to take more profoundly vagueness appropriate the elaborate problem.

References

1. Ha, M.S.: A comparison of service quality at major container ports. Implications for Korean Ports J. Trans. Geography. (11), 131–137. Energy. (5), 1–14 (2003). https://doi.org/10.1016/S0966-6923(02)00069-8
2. Lirn, T.C., Thanopoulou, H.A., Beynon, M.J., Beresford, A.K.C.: An application of AHP on transhipment port selection, Global Perspect. Maritime Econ. Logist. (6), 70–91 (2004). https://doi.org/10.1057/palgrave.mel.9100093
3. Chou, C.C.: Application of FMCDM model to selecting the hub location in the marine transportation, Case Study Southeastern Asia Math. Comput. Modell. (51), 791–801 (2010). https://doi.org/10.1016/j.mcm.2009.10.015
4. Lirn, T.C., Thanopoulou, H.A., Beynon, M.J., Beresford, A.K.C.: An application of AHP on transhipment port selection, Global Perspect. Maritime Econ. Logis. (6), 70–91 (2004). https://doi.org/10.1057/palgrave.mel.9100093
5. Song, D., Yeo, K.: An evaluation of container ports in China and Korea with the analytic hierarchy process, J. Eastern Asia Soc. Trans. Stud. (5), 726–741 (2003)
6. Su, D., Liang, G., Liu, C.: A study on integrated port performance comparison based on the concept of Balanced Scorecard, J. Eastern Asia Soc. Trans. Stud. (5), 609–624 (2003)
7. Tai, H., Hwang, C.: Analysis of hub port choice for container trunk lines in East Asia. J. Eastern Asia Soc. Trans. Stud. **6**, 907–919 (2005)
8. Chou, C.: A comparative study of models for port choice. Proc. Eastern Asia Soc. Trans. Stud. **5**, 608–616 (2005)
9. Mendel, J.M.: Computing derivatives in interval type-2 fuzzy logic systems. IEEE T Fuzzy Syst. 12, 84–98 (2004). https://doi.org/10.1109/TFUZZ.2003.822681

Developing the PP Balance Model Scale and Fuzzy Logic Evaluation of the Latent Variable

Konul Memmedova(✉) and Guldal Kan Şebnem

Department of Psychological Counselling and Guidance, Near East University, North Cyprus, Turkey
{konul.memmedova,sebnem.gurdal}@neu.edu.tr

Abstract. The principal focal point of Positive Psychotherapy (PP) is to specify the "positive" and positively reinterpret the symptoms/disorders and help the person lead a balanced personal life. Human life is considered within four- balance dimensions; the body (senses), achievement (mind), relationships (traditions), and the future/spirituality. Therefore, any individual experiencing conflicts is inclined to cope with them through different dimensions. This study aimed to develop a PP Balance model scale to be administered to university academicians. The survey participants were picked through a random sampling method, in which 291 participants out of 1200 were interviewed considering confidence range and sampling error. The validity and reliability of 23 items were calculated.

In this study, the fuzzy logic approach is proposed to find a latent variable (Balance) in measurable variables (body, achievement, relationships, and spirituality. The application of the fuzzy logic approach allows increasing the precision of computing the personal life balance index.

Keywords: Positive psychotherapy · Fuzzy logic · Likert scale · Balance model · The Cronbach's Alpha test

1 Introduction

1.1 Theories of Positive Psychotherapy

There are four main theories to define "personality": personal characteristics theory, psychodynamic theory, humanistic theory, and unifying theory [1]. These theories attempted to explain an individual's behaviours in terms of instinct and drive. In later years, several psychotherapy movements such as individual psychology, existentialist, individual-oriented, gestalt, behaviourist, cognitive behaviourist, reality, short-term solution-oriented, narrative, and feminist [2]. One of these movements is positive psychotherapy, firstly developed in Germany [3]. When personality is at stake, positive psychology focuses more on well-being, subjective well-being, life satisfaction, hope, optimism, Flow, developing capacity, forgiveness, awareness, morality, responsibility, altruism, professional ethics, and happiness [4]. There have been developments in positive interferences through this movement which put forth individuals' positive feelings,

R. A. Aliev et al. (Eds.): ICSCCW 2021, LNNS 362, pp. 450–458, 2022.
https://doi.org/10.1007/978-3-030-92127-9_61

thoughts, and behaviours. These activities helped individuals feel better and more powerful [5]. In this respect, individuals should systematically and effectively exhibit positive feelings, thoughts, and behaviours, bringing about positive psychotherapy models. Studies in positive psychotherapy proved to be effective in preventive, constructive, and curative issues. For example, in studies [6] done, it has been observed that individuals receiving positive psychotherapy have more powerful therapeutic bonds, felt more efficient, and suffered fewer mental disorders. Positive psychotherapy is a therapeutic approach with its peculiar principles and interference based on source-tendency and clash-solution. Clash-solution therapeutic approach is a natural result of social interaction, defined as the opposition, disagreement, and resistance between two people [7]. Resistance affects individuals' way of living. The most effective way of solving conflicts is to know the reasons before anything else. At this point, the reasons for behaviours should be considered rather than interactions among individuals [8]. Job motivation can be associated with contextual elements in cultural and individual inclinations and be defined as the willingness in high-level efforts to achieve organizational targets [9]. Positive psychotherapy mainly focuses on specifying what is positive, reinterprets the symptoms/problems, and helps the involved have a balanced life. Besides a therapeutic method in a short-term consultation process, there is a long-term psychotherapy method [10].

2 Positive Psychotherapy Balance Model

Positive Psychotherapy examines human life in four dimensions; body (senses), achievement, relationships (traditions), and future/spirituality. These four dimensions influence the life of a positive individual and are explained by one's learning skill and knowledge, which are the indications of an individual's self and environmental perception as well as how the realities are interpreted. Every individual tries to cope with conflicts in different dimensions and adapts to new inclinations. While some react by developing physical symptoms, other head towards work/performance, some avoid social or establish social interaction, and some have fantasies [11].

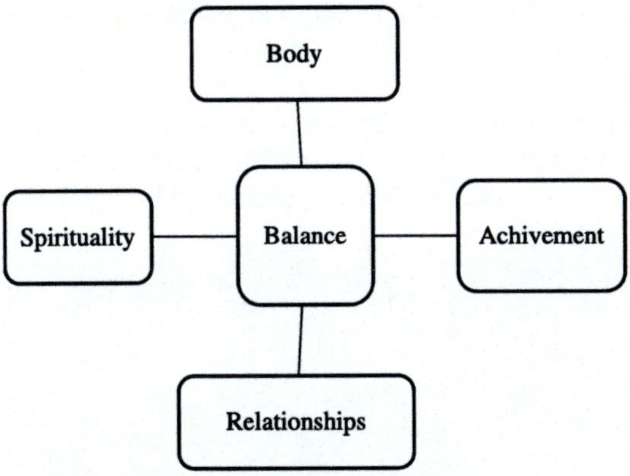

Fig. 1. Balance model in positive psychotherapy

Except for this study, research to investigate PP Balance Model (see Fig. 1) among academicians has not yet been done. In PP applications, besides their personality traits, balance principles are effectively practised making individuals healthy and productive [11].

3 Methodology

3.1 Data Collection and Sampling

The survey participants were 300 academicians from one of the universities of North Cyprus. The participants, specified through the random sampling method, were 300 academicians at a university in North Cyprus in the 2018–2019 academic years with a %95 reliability gap and % five sampling- errors. Totally 291 participants returned completed questionnaires. According to gender, %59.24 were female and %40.76 were male, %32.70 30 years of age and below, %41.23 between 31–39, and %26.07 over 40 years of age, %67.77 married, %32.23 single, %54.98 with children, and % 45.02 with no children. The descriptive statistic method [12] was conducted to design a Likert scale to measure PP Balance model. The data was processed using the Statistical Package for Social Sciences (SPSS-24.0) and structural equation Modelling package AMOS 24.0 [13]. The commentary and confirmatory factor were used for the validity analysis of the PP Balance Model Scale. The fitness of the data set to multi-variable normal distribution was overviewed, and Kaiser-Meyer-Olkin (KMO) coefficient and Bartlett's global test were administered. Firstly, the data set went through a commentary factor analysis and then specified the scale's factor structure, followed by a confirmative analysis for structural validity. The factor loads higher than 0, 70 were chosen. The reliability of the scale was tested using The Cronbach's Alpha test. The internal consistency test was administered to examine items correlations.

3.2 Design Likert Scale

In total 51 questions related to Achievement, Relationships, Body, and Imagination, on the PP Balance Model Scale were subjected to a confirmatory factor analysis test for the validity analysis. The 23 questions were selected as valid questions to measure the balance.

The questions were written in 5-Likert type and answered with scales as "Never" (N), "Rarely" (R), "Sometimes" (S), "Usually" (U), "Always" (A).

Table 1 presents the results of the analysis performed by SPSS software.

Table 1. Factor structure and factor loads of PP balance model scale

Items	Factors			
	F1	F2	F3	F4
Imagine myself somewhere different when stressed	0,90			
I'm very hopeful for my future	0,88			
Sometimes I reflect my imaginations to my real life	0,88			
Belief is partial success for me	0,84			
I always think of my future	0,84			
I like thinking of my future	0,80			
I can cope with problems		0,78		
I'm a perfectionist		0,78		
I believe I'm achieving my aims in my profession		0,75		
Success is my priority		0,75		
I care for my self-development		0,71		
Prestige in my profession is important for me		0,70		
I'm careful about my job responsibilities		0,69		
One's intelligence is important for me		0,66		
I believe we work in a positive environment			0,82	
I have good relationship with the staff			0,81	
One should express himself in a comfortable environment			0,80	
I'm happy with my work environment			0,80	
Relationships among juniors and seniors is important for me			0,79	
I'm careful about my physical appearance				0,90
I usually feel energetic				0,87
I'm pleased with my physical appearance				0,86
I spare time for physical activities				0,83

Notations: F1- Future/Spirituality; F2-Achievement; F3-Relationship; F4-Body. The Cronbach alfa test was administered, and item-total correlations were studied to confirm the reliability of the subject questions (see Table 2).

Table 2. The Cronbach Alpha test results

Factors	α
Body	0, 910
Achievement	0, 884
Relationship	0, 874
Future/Spirituality	0,937

Structural equation modelling was established using the AMOS software (see Fig. 2).

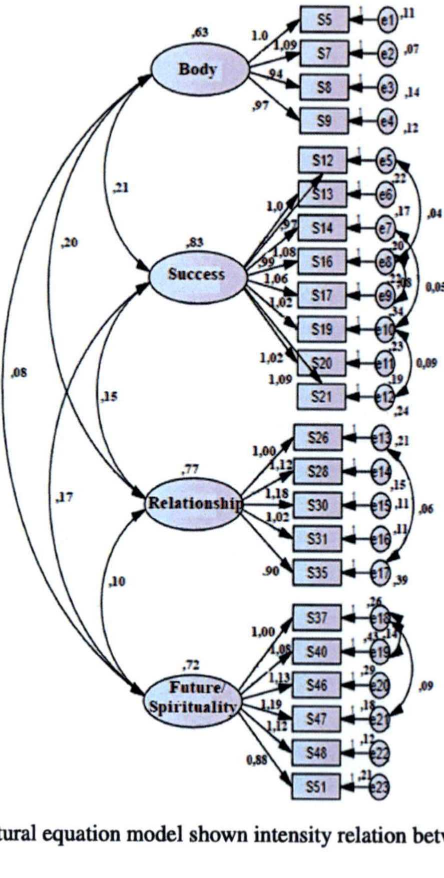

Fig. 2. Structural equation model shown intensity relation between of items

Table 3 presents the final version of the questionnaire. It is based on the statistical analysis and structural equation modelling results (N-Never; R-Rarely; S-Sometimes; U-Ususly; A-Always).

Table 3. The final version of the questionnaire

No.	Questions	N 1	R 2	S 3	U 4	A 5
1	I'm careful about my physical appearance					
2	I usually feel energetic					
3	I'm pleased with my physical appearance					
4	I spare time for physical activities					
5	I believe we work in a positive environment					
6	I have good relationship with the staff					
7	One should express himself in a comfortable environment					
8	I'm happy with my work environment					
9	Relationships among juniors and seniors is important for me					
10	I can cope with problems					
11	I'm a perfectionist					
12	I believe I'm achieving my aims in my profession					
13	Success is my priority					
14	I care for my self-development					
15	Prestige in my profession is important for me					
16	I'm careful about my job responsibilities					
17	One's intelligence is important for me					
18	I imagine myself somewhere different when stressed					
19	I'm very hopeful for my future					
20	Sometimes I reflect my imaginations to my real life					
21	Belief is partial success for me					
22	I always think of my future					
23	I like thinking of my future					

4 Fuzzy Logic Evaluation Results of Measurement

As shown in Table 3, the measurement scales have categorized (fixed) values: for Never-1; Rarely-2; Sometimes-3; Usually 4; Always-5. In the classical approach, respondents have chosen only one level from 5 levels Never, Rarely, Sometimes, Usually, Always. But in the real world, selecting only one categorized level does not reflect a genuine attitude of the respondent in expressing their opinion. As shown in [14], Likert scales with categorized intervals lack and do not handle the subjective error and uncertainty introduced by the respondent. For this reason, instead of the conventional Likert scale, the fuzzy logic approach is used to evaluate the latent variable. Fuzzy logic evaluation allows evaluating the opinion of survey participants using the partial reliability for two's neighbours' levels. The fuzzy logic evaluation approach involves in steps has been shown in Fig. 4.

Fuzzification

Fuzzification is a transformation of the Likert scales levels to the overlapping membership functions for each question. A standard triangular membership function is used in the present study. Figure 5 shows an example of fuzzification of the Input "Belief is a partial success for me".

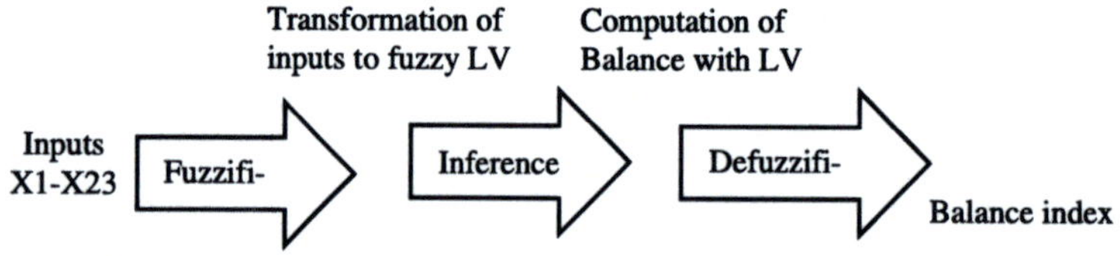

Fig. 4. Fuzzy evaluation steps

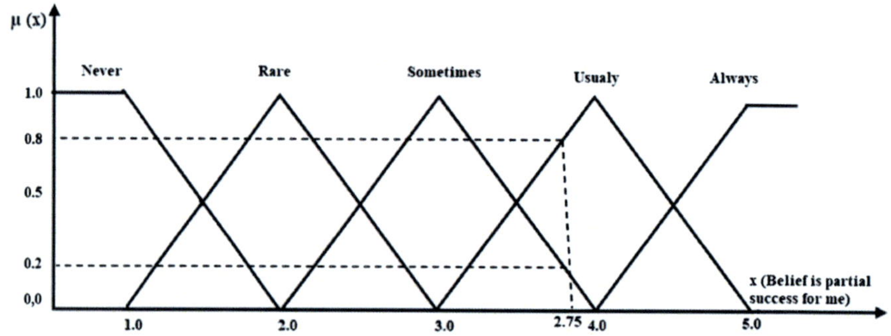

Fig. 5. Fuzzification of the item "Belief is a partial success for me"

In classical Likert scales two membership degrees are used: 1- belonging to level and 0-not belonging. The membership degree $\mu(x)$ of variables x varies between [0–1]. As shown in Fig. 5, the value for $x = 2.5$ falls on two fuzzy linguistic variables (LV): Sometimes and Usually with the degrees:

$$\mu_{\text{usually}}(x) = 0.8; \mu_{\text{sometimes}}(x) = 0.2$$

Thus conventional Likert scales have exceptional cases of fuzzy set that linguistic variables have membership degree between [0–1].

Fuzzy Inference Using If-Then Rules

After grouping the questions into four factors (Body, Achievement, Relationship and Spirituality), we reduce inputs up to 4. Thus the system has four inputs (B, A, R, and SP). Each input has five linguistic variables and one output (Balance) with five linguistic variables. The linguistic variables for inputs and output are expressed as Never; Rarely; Sometimes; Usually; Always. The FuzzyTech software [15] is used for modelling. The

If-Then rules are used to define a relationship between inputs and output. The fuzzy inference identifies the rules regarding the current inputs and computes the output linguistic variable. The 30 significant rules were chosen, with the expert's opinions. Below is given the fragment of the selected rules:

1. If (B is U and A is S and R is U and SP is U), Then Balance is U
2. If (B is S and A is U and R is U and SP is R), Then Balance is S
3. If (B is R and A is S and R is U and S is S), Then Balance is S
4. If (B is U and A is S and R is S and SP is U), Then Balance is U
5. If (B is A and A is U and R is A and SP is U), Then Balance is A

Aggregation is based on Mamdani MIN-MAX method [16].

The Medium -of- Maximum (MOM) method was used for defuzzification. For example, for rule 4, the fuzzy Balance is defined as

$$\mu_{\mathrm{Balance}} = (\mu_N = 0.00,\ \mu_R = 0.00,\ \mu_S = 0.35,\ \mu_U = 65,\ \mu_A = 0.00).$$

and the result of defuzzification gives Balance index = 0.8

5 Conclusion

In this study the four-dimension PP Balance model Scale for Academicians has been developed. The scale reliability measured by Gronbach alfa is 0. 99 for the body, 0.88 for achievement, 0.87 for relationship and 0.93 for spirituality. The proposed questionnaire consists of 23 items evaluated by 5-scales. Fifty-one were selected in terms of factor structure and load (factor loads > 0.7). The validity and reliability of the scale were tested using Statistical Software Packages for Social Sciences SPSS-24.0 and Structural Equation Modelling AMOS 26.

The categorized intervals such as (Never, Rarely, Sometimes, Usually and Always) in the questionnaires yield the vagueness and inaccuracy related to data acquisition. Unfortunately, the statistical approach used today for data processing does not consider this imprecision inherent in input data.

The application of fuzzy logic evaluation allows to handle the imprecision and vagueness of input data and compute Balance in terms of measurable variables of Body, Achievement, Relationship and Spirituality. The basic steps of fuzzy logic modelling, including fuzzification of psychological variables, inference and defuzzification, are considered using the FuzzyTech software.

References

1. Nelson, D.L., Quick, J.C.: Organizational Behaviour, 4th edn. Thomson, USA (1997)
2. Murdock, N.L.: Theories of Counselling and Psychotherapy. Pearson, London (2017)
3. Peseschkian, N.: Positive Psychotherapy. Sterling Publishers, New Delhi (2000)
4. Seligman, M.E., Csikszentmihalyi, M.: Positive psychology: an introduction. AM Psychol. **55**(1), 5–14 (2000). https://doi.org/10.1037//0003-066x.55.1.5

5. Layous, K., Chancellor, J., Lyubomirsky, S., Wang, L., Doraiswamy, P.M.: Delivering happiness: translating positive psychology intervention research for treating major and minor depressive disorders. J. Altern. Complement. Med. **17**(8), 675–83 (2011). https://doi.org/10.1089/acm.2011.0139
6. Diener, B.R., Kashdan, T.B., Minhas, G.: A dynamic approach to psychological strength development and intervention. J. Posit. Psychol. **6**, 106–118 (2011). https://doi.org/10.1080/17439760.2010.545429
7. Ricaud-Droisy, H., Zaouche-Gaudron, C.: Interpersonal conflict resolution strategies in children: a father-child co-construction. Eur. J. Psychol. Educ., 157–169 (2003). https://doi.org/10.1007/BF03173482
8. Zengin, F.: Negotiation (problem solution) and mediation education program to conflict solution styles and assertiveness levels of primary school 4th and 5th class students. MSc Thesis, Dokuz Eylül University, Izmir, Turkey (2008)
9. Latham, P.G., Pinder, C.C.: Work Motivation theory and research at the down of the twenty-first century. Annu. Rev. Psychol. **56**(1), 485–516 (2005). https://doi.org/10.1146/annurev.psych.55.090902.142105
10. Cope, T.: The inherently integrative approach of positive psychotherapy. J. Psychother. Integr. **20**(2), 203–250 (2010). https://doi.org/10.1037/a0019769
11. Peseschkian, N.: Positive Psychotherapy of Everyday Life (2016)
12. Brace, N., Richard K., Snelgar, R.: SPSS for Psychologists,7th edn. Macmillan Education, UK (2020)
13. Byrne, B.M.: Structural Equation Modelling with AMOS: Basic Concepts, Applications, and Programming. Lawrence Erlbaum Associates, Mahwah (2001)
14. Memmedova, K.: Quantitative analysis of effect of Pilates exercises on psychological variables and academic achievement using fuzzy logic. Qual. Quant. **52**(1), 195–204 (2017). https://doi.org/10.1007/s11135-017-0601-9
15. Von Altrock, C.: Fuzzy Logic and Neuro-fuzzy Applications in Business and Finance. Prentice-Hall International, Upper Saddle River (1997)
16. Mamdani, E.H., Assilian, S.: An experiment in linguistic synthesis with a fuzzy logic controller. Int. J. Man Mach. Stud. **7**(1), 1–3 (1975). https://doi.org/10.1016/S0020-7373(75)80002-2

Application of ANFIS to Predict Adhesive Wear Rate in Different Fiber-Reinforced HDPE Matrix Composite Materials

M. B. Babanli and K. H. Ismayilova(✉)

Department of Materials Science and Processing Technologies, Azerbaijan State Oil and Industry University, 20 Azadliq Avenue, Az 1010, Baku, Azerbaijan

Abstract. In this article, the results of adhesive wear tests in body fluid of different fiber-reinforced HDPE matrix polymer composite materials were investigated mathematically with the help of adaptive neuro-fuzzy infernece system (ANFIS). 25 different data were used in the calculations. 2/3 of these data were used for training and 1/3 for NN test, and it was determined that the data obtained as a result of the calculations largely overlapped with the experimental data.

Keywords: ANFIS · Fuzzy set · Composite materials · HDPE · Adhesive wear

1 Introduction

A composite material is basically a material system consisting of a mixture or combination of two or more components that do not mix with each other and have the composition of at least two different materials. Composite materials can be obtained by physically combining these materials under certain conditions and rate in order to get a clear feature from at least two separate substances that are not used directly for the desired purpose.

Composite materials have advantages such as high strength, lightness, design elasticity, dimensional stability, high dielectric resistance, corrosion resistance, ease of molding, surface applications, high heat resistance, high chemical resistance, as well as the characteristics of the manufacturing materials.

The use of artificial intelligence in the production of composite materials and determining their different properties is increasing rapidly.

As a result of the literature review, it has been seen that various databases have been created and widely used by scientists for the prices and different parameters of existing materials [1–3].

According to Rakkugila et al. [4] machine learning algorithms are used to synthesize and characterize new hybrid grade materials, and they think this method could replace experimental ones in the future.

Modeling in the field of materials science and the application of different mathematical methods, especially fuzzy logic, have been found to be effective in the development of new materials and improvement of the results [5–8]. Due to the complexity and uncertainty of the experimental data, the use of fuzzy logic to model different properties of

R. A. Aliev et al. (Eds.): ICSCCW 2021, LNNS 362, pp. 459–464, 2022.
https://doi.org/10.1007/978-3-030-92127-9_62

composite materials showed positive affect to the development of the properties of these materials [1, 5–7]. Particularly, it can be used to derive models from big data on material properties and production process parameters.

In this article, ANFIS is used to predict the adhesive wear rate in body fluid in different fiber-reinforced HDPE matrix composite materials.

The article is organized as follows. In Sect. 2, the preliminary material used in the paper is given. In Sect. 3, Neural Network is explored to predict the adhesive wear rate of composite materials. Section 4 concludes the paper.

2 Preliminaries

Definition 1. A fuzzy number. A fuzzy number is a fuzzy set A on R which possesses the following properties: a) A is a normal fuzzy set; b) A is a convex fuzzy set; c) α-cut of A, A^{α} is a closed interval for every $\alpha \in (0, 1]$; d) the support of A, supp (A) is bounded.

Definition 2. The sigmoidal membership function used here depends on the two parameters a and c and is given by

$$f(x; a, c) = \frac{1}{1 + e^{-a(x-c)}}$$

The membership function dsigmf depends on four parameters, a_1, c_1, a_2, and c_2, and it is the difference between two of these sigmoidal functions. $f_1(x; a_1, c_1) - f_2(x; a_2, c_2)$ The parameters are listed in the order: $[a_1, c_1, a_2, c_2]$. In MATLAB environment sigmoidal membership function is described as $y = \text{dsigmf}(x, [a_1, c_1, a_2, c_2])$.

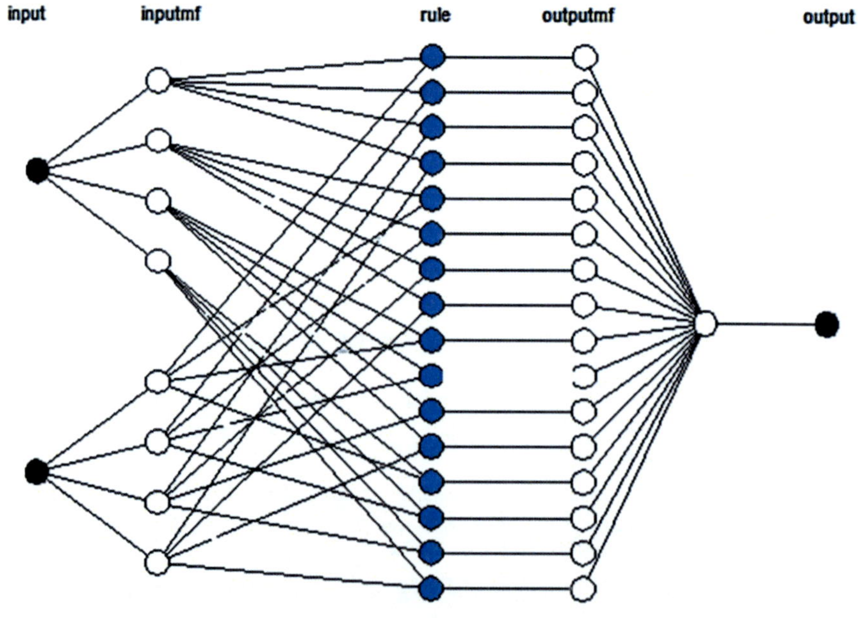

Fig. 1. Structure of neural network

Structure of the Neural network is given in Fig. 1. Described neural network consist of two inputs, one hidden layer and one output. The network is trained on the basis of a hybrid method.

3 An Application

As clustering method is used grid partition method, inference method Sugeno, membership functions dsigmf -sigmoid membership function (1), implication -min, aggregation -max, defuzzification -center of gravity method are selected. In calculations 25 data are used. From these data 2/3 were used for training, and 1/3 data were used for testing of NN. Fragment of the initial data is given in Table 1.

Table 1. Fragment of the initial data

Composite examples	Hardness (Shore D)	Load (N)	Wear volume (mm^3)
5,1	64,45	50	6,32
5,2	67,216	50	24
5,3	68,6	50	21,71
5,4	68,9	50	22,11
5,5	71,0625	50	13,21
5,6	72,1	50	18,89
5,7	62,74	50	19,69
5,8	62,94	50	21,24
5,9	63,2	50	19,02
5,10	63,485	50	29,07

Membership functions of the input variables are represented in Fig. 2.

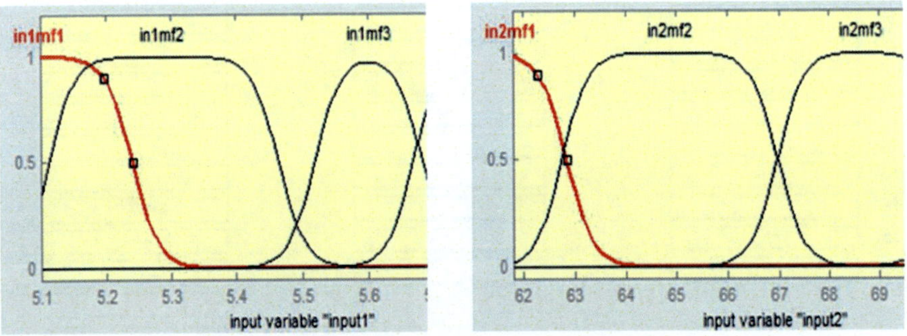

Fig. 2. Membership functions of the input variables

Using grid partition method 16 rules are formulated (Fig. 3) and result for determining output is given in Fig. 4.

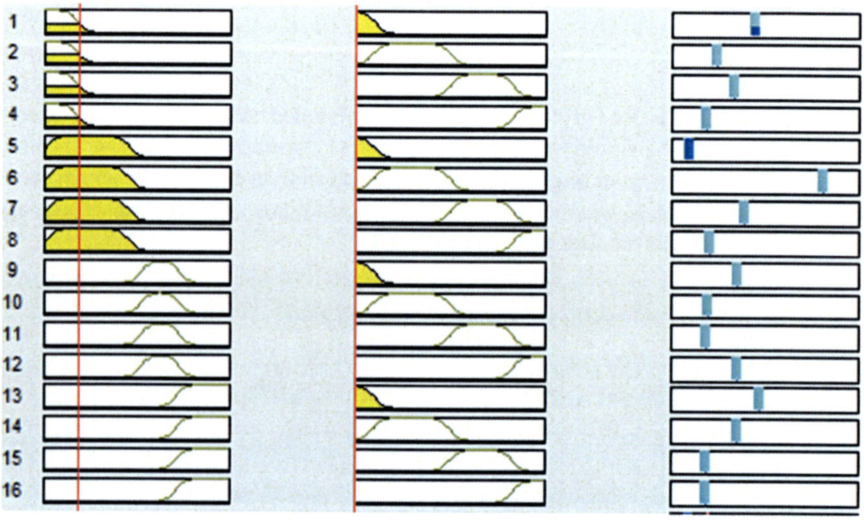

Fig. 3. Representation of the knowledgebase

According to values of input1 and input 2 is defined output as 35.2 (using neural network) and Fragment of tested data are given in Table 2.

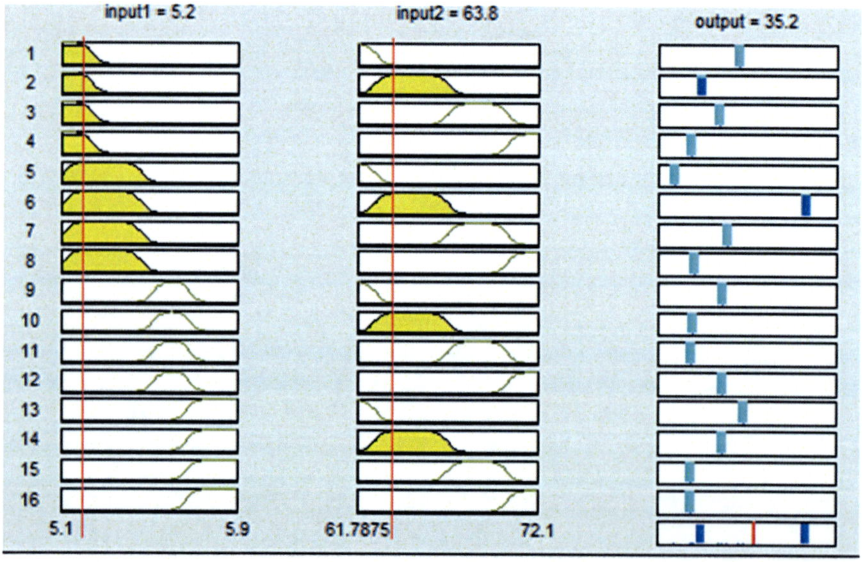

Fig. 4. The results of the networks grid partition method.

Table 2. Fragment of tested data.

X_1	X_2	Output	Defined	
5,19	64,375	31,67	33,5	3,3489
5,2	63,7625	34,76	35,2	0,1936
5,21	64,9	35,81	37,8	3,9601
5,11	64,942	23,17	24,5	1,7689
5,23	64,5625	44,18	41,4	7,7284
5,24	63,2375	42,99	41,8	1,4161
5,25	61,7875	1,35	1,92	0,3249
			RMSE	4,329076

The results are comprised with results of the practical data Root Mean Square Error (RMSE) defined as 4.32%. All calculations are made in MATLAB environment.

Obtained results give the basis to approve, that forecasts made by using adaptive neuro-fuzzy inference system are more reliable and exact.

4 Conclusion

The results of the ANFIS used to predict adhesive wear rates in composite materials with different fibrous HDPE matrices are the result of practical data from the Root Mean Square Error (RMSE), defined as 4.32%. The results confirm that predictions made with fuzzy neural networks are more reliable and more accurate.

References

1. Jabbarov, T.G., Gurbanov, N.A.: Synthesis of optimal technological parameters of "iron-cast-glass" grinding composite materials using fuzzy logic and big data concepts. In: Aliev, R.A., Kacprzyk, J., Pedrycz, W., Jamshidi, M., Babanli, M., Sadikoglu, F.M. (eds.) ICAFS 2020. AISC, vol. 1306, pp. 254–259. Springer, Cham (2021). https://doi.org/10.1007/978-3-030-64058-3_31
2. Babanli, M.B.: Synthesis of new materials by using fuzzy and big data concepts. In: 9th International Conference on Theory and Application of Soft Computing with Wods and Perception, ICSCCW 2017, Budapest, Hungary, 24–25 August 2017 (2017). https://doi.org/10.1016/j.procs.2017.11.216. Procedia Computer Science 120, pp. 104–111
3. Gurbanov, N.A., Abbasov, I.I., Ismayilova, K.H., Hasanova, N.A.: Production of polypropylene matrix polymer composites with hazelnut shell fillings, physical and mechanical properties. Nanosistemi Nanomater. Nanotehnologii **16**(3), 559–565 (2018)
4. Raccuglia, P., et al.: Machine-learning-assisted materials discovery using failed experiments. Nature **533**, 73–76 (2016). https://doi.org/10.1038/nature17439
5. Babanli, M.B., Huseynov, V.M.: Z-number-based material selection problem. In: 12th International Conference on Application of Fuzzy System sand Soft Computing, ICAFS 2016, Vienna, Austria, 29–30 August 2016 (2016). https://doi.org/10.1016/j.procs.2016.09.387. Procedia Computer Science 102, pp. 183–189

6. Chen, D., Li, M., Wu, S.: Modeling of microstructure and constitutive relation during super plastic deformation by fuzzy-neural network. J. Mater. Process. Technol. **142**, 197–202 (2003). https://doi.org/10.1016/S0924-0136(03)00598-3
7. Chen, S.-M.: A new method for tool steel materials selection under fuzzy environment. Fuzzy. Sets Syst. **92**, 265–274 (1997). https://doi.org/10.1016/S0165-0114(96)00189-3
8. Aliyarov, R.Y., Gardashova, L.A., Hasanli, N.I.: Predicting porosity through fuzzy logic based methods from South Caspian Basin Data. In: Aliev, R.A., Kacprzyk, J., Pedrycz, W., Jamshidi, M., Babanli, M., Sadikoglu, F.M. (eds.) ICAFS 2020. AISC, vol. 1306, pp. 268–274. Springer, Cham (2021). https://doi.org/10.1007/978-3-030-64058-3_33

Some Aspects of Fuzzy Decision Making in Digital Marketing Analysis

Gunay E. Imanova[1](✉) and Gunel Imanova[2]

[1] Azerbaijan State Oil and Industry University, 20 Azadlyg Avenue, Baku, Azerbaijan
[2] Department of Business Administration, Near East University, Mersin 10, Nicosia, North Cyprus, Turkey

Abstract. Advanced technology provides many opportunities to be utilized. Nowadays, window of opportunity is opened through digital marketing, which is a great tool to gain a competitive advantage over rivals. However, the ineffective use of digital marketing analytics and tools based on the insufficient or the wrong decisions will not bring any benefit to the companies. Especially during pandemic outbreak, digital marketing has gained a great popularity and become very useful for both companies and customers. In this state-of-art research we analyze the existing literature associated with digital marketing that is based on fuzzy logic. Several important research and their methodology were analyzed to provide a view of the existing studies that associate the digital marketing with the fuzzy logic. The main problem related to analyzed literature is the lack of certainty and reliability of the data provided. At the end, a proposed model under Z-environment is given for the decision-making process related to digital marketing analytics and tools.

Keywords: Fuzzy logic · Digital marketing · Z-environment · Decision making · Social media marketing · e-commerce

1 Introduction

Because of the several situations such as economic crisis and inflation, expiring industries and sectors, advanced technology, etc., marketing role and strategy have dramatically changed in the last decades. This diversification shifts marketing executives towards being more market-driven i.e., customer oriented. Providing satisfaction for todays' customers somehow related to acquiring timely and appropriate information about customer preferences, marketplace, and external environment. As a result of advanced technology and use of internet, digital marketing has become so popular. The companies such as Amazon, Alibaba, and Google have radically changed the way of selling something through using internet [1]. Use of digital marketing provides great advantages for the governments to be explored [2].

R. A. Aliev et al. (Eds.): ICSCCW 2021, LNNS 362, pp. 465–473, 2022.
https://doi.org/10.1007/978-3-030-92127-9_63

As the consumers are constantly interacting with the web and social media channels across the globe, buying preferences and decisions tend to change. Digital marketing technologies are implemented to keep pace with the changing demand of the customers [3]. In one of the researches [3] managers were asked to rate the benefits provided through digital marketing. As a result, information exchange, ease of use, knowledge acquisition, promotion of external and internal relationship, supporting decision making, are the main factors for using digital marketing, respectively.

It can be concluded that technological improvements may provide better customer satisfaction and competitive advantage through deep understanding of customer perceptions, customer engagement and building long-term customer relationship [4].

There are several effective digital marketing technologies to be utilized by the companies. The major problem is to find the best digital marketing tool to imply in order to gain the competitive edge over rivalry. Unfortunately, there are limited number of research about this evaluation and selection process. Some of the mentioned researches have used fuzzy-AHP (Analytical Hierarchy Process) approach for multi-criteria decision making (MCDM) which is proposed for the decision makers to ease their digital marketing tool selection process and to achieve their strategic objectives [5], AHP and CORPAS integration approach where, AHP is used for weight of criteria determination and CORPAS is used for alternative evaluation and ranking [6], fuzzy-TOPSIS and fuzzy-AHP integration were used to evaluate the alternatives and select the best digital marketing tool based on the given criteria, and ranking [7]. Here, the main problem is related to uncertainty of the considered variables and the ignored reliability.

The main purpose of this study is to analyze the existing literature related to use of fuzzy logic in digital marketing field and find out the existing gaps in this research field. The remaining of the paper is structured as the following. Section 2 involves preliminary information, and it is followed by the background information about the existing researches. Moreover, Sect. 4 involves suggestion for the gap in the literature. Finally, in Sect. 6 conclusion points are presented.

2 Preliminaries

Definition 1. Fuzzy numbers [8]**:** In the real-world environment, decision-making process is usually uncertain considering possibility of the values of interest. Moreover, the possible degrees of this vagueness or uncertainty are also unknown to DMs. Assigned degree for each possible value of x, in a fuzzy set is $\mu(x)\epsilon[0, 1]$.

$$x_0 \in R, \text{ where } \mu_M(x_0) = 1$$

$$\text{for any } 0 \leq \alpha \leq 1, A_\alpha = [x, \mu_{A_\alpha}(x) \geq \alpha] \text{ is a closed interval, where F(R)} \quad (1)$$

Definition 2. Random variables [8]**:** Random variables consist of two types, which are classified as continuous random variables and discrete random variables. Continuous variable which can be indicated by X, can take endless numerical outcomes of possible values x. Conversely, for a discrete random variable, X can only take countable distinct values.

Definition 3. Triangular fuzzy numbers [9]: A triangular fuzzy number is indicated by $\tilde{A}$ as a triplet, shown as (a_1, a_2, a_3), where the membership is defined using the formula given below (2).

$$\mu_{\tilde{A}}(x) = \begin{cases} 0 & \\ \dfrac{x - a_1}{a_2 - a_1} & x\epsilon(-\infty, a_1) \\ & x\epsilon[a_1, a_2] \\ \dfrac{c - x}{a_3 - a_2} & x\epsilon[a_2, a_3] \\ & x\epsilon(a_3, +\infty) \\ 0 & \end{cases} \quad (2)$$

Definition 4. Z-numbers [10, 11]: Acquired information or data must be reliable in order to make a sound decision. Z-number principle is based on this reality. Z-number is defined as a theory which comprises two parts of A and B, or Z = (A, B). Within this ordered pair of representation, A stands for a fuzzy constraint, or a restriction based on the real values that X a random variable, can take; whereas the second part B, stands for the fuzzy value for the probabilistic measure of reliability on the identified first part, A. Furthermore, both A and B are indicated in natural language-NL, which are the fuzzy numbers.

Definition 5. Z-valued matrix [12]: Z-valued matrix is a matrix for Z-number elements, that indicates the partial reliability of the data based on the random variable values of $X_{ij}, i, j = 1, \ldots, n$:

$$(Z_{ij} = (A_{ij}, B_{ij})) = \begin{pmatrix} Z_{11} = (A_{11}, B_{11})) \ldots Z_{1n} = (A_{1n}, B_{1n})) \\ . \quad \ldots \quad . \\ Z_{n1} = (A_{n1}, B_{n1})) \ldots Z_{nn} = (A_{nn}, B_{nn})) \end{pmatrix} \quad (3)$$

3 Background Information About Digital Marketing and Its Practices

There are various researches about the selection of the best digital marketing tool; such as acquiring the information about the perspectives of job seekers related to necessary digital marketing tool usage for the recruitment [13]. In this study some social media-based digital marketing tools are analyzed using a structured questionnaire based on non-probability sampling method. The main disadvantages or restrictions of this study are that the information is not generalizable to the target population and measuring the representativeness of the sample is not possible due to incalculable sampling error estimates.

In a research [14], some factors and actions in digital marketing environment important for attracting and retaining the users are analyzed using the Delphi method through in-dept interviews. Limitations of this research is related to number of subjective expert decisions, which decreases the reliability. In another research [15], digital marketing technologies are evaluated in terms of adoption of the most appropriate technology by startups, based on the existing literature. In this research, some variables such as ease of use, perceived benefits, cost and so forth are considered for the evaluation. In [16], researchers took the experts opinions based on the research of finding the best digital marketing method to be utilized. Here, interviews are used during a specific period of time, however, again the decision makers' opinions are crisp, not fuzzy. In another research [17] based on evaluation of digital marketing management, dynamic performance modelling was used. TOPSIS method was used for this decision-making process. Furthermore, study states that the use of this method proposes more practical abilities than one-time model construction based on the identical data in fuzzy method.

Fuzzy-TOPSIS and fuzzy-AHP integration were used by the researches to evaluate the alternatives and select the best digital marketing tool based on the given criteria, and ranking. According to results of this research, remarketing advertising is selected to be the best alternative digital marketing tool to be used which is followed by social media ad, ad-per-impression, ad-per-click, affiliate ad, banner marketing and email-ad [7].

3.1 Decision Making for Social Media Marketing (SMM) Practices

Social media marketing as a part of digital marketing, has gained a great importance with increase in access to internet and availability of smartphones. Social media marketing enables the share of information and knowledge, and marketing of goods and services, through effective communication [4]. Today, social media is used by marketing executives as a part of their marketing strategy. Social media can provide a wide array of information about the consumer behavior, their perceptions, their preferences toward brands and campaigns, past preferences, and the personality traits of the customers, and the brand loyalty. This bulk of information is the crucial one for understanding the target customer for better serving [18]. More information about the customers and marketing environment provides better understanding of the needs, wants of the customers and the rivalry. As a result, better marketing strategies with more informed decisions can be made [19]. Social media provides us the information about the perceptions, thoughts, and demographics of the customers through the shared contents. The mostly used 5 social media channels worldwide, in 2021 are listed in Table 1 [20].

Table 1. Most popular social networks worldwide, 2021 (Statista)

Social media	Active users (in millions)
Facebook	2797
YouTube	2291
WhatsApp	2000
Facebook Messenger	1300
Instagram	1287

By considering the above Table 1, marketers can gather the customer-related information from these social media channels and use it in marketing decision making process, constructing marketing strategies, and marketing mix (4Ps).

Another popular social media network is Twitter, with the 396 million active users [20]. There are lots of researches about analyzing the tweets and database of Facebook to gain information. Sentiment analysis, social bots and artificial intelligence and neural network-based techniques can be used to analyze the huge amount of data.

A research related to use of fuzzy logic in social media analysis suggest a digital marketing tool through combination of fuzzy logic and social bots with data mining for an effective targeting and communication ability, which can be very useful for customer relationship management.

Another research uses a new set of fuzzy rules consisted of multiple lexicons and datasets to analyze the sentiment of the social media posts shared by the target audience. In this research Natural Language Processing techniques and Word Sense Disambiguation is integrated through the proposed fuzzy system for the classification of social media posts as neutral, negative or positive sentiments [21].

Hybrid model research about optimal social media channel selection problem suggests the use ANP and fuzzy logic for determination of importance of social media channel criteria weights, in the fuzzy environment. Later, CORPAS-G method is utilized to analyse the alternatives and find out the ranking for them in order to select the best social media platform [22].

3.2 Fuzzy Logic and e-commerce

There are several research that apply the fuzzy logic to e-commerce platforms that involves B2B and B2C marketing. For instance, in research [23] fuzzy logic-based highly adaptive recommender system especially for B2C e-commerce platforms is proposed that offer: catalogue service for product categorization; product selection service and rule-based knowledge learning service.

In research [24] fuzzy logic is applied to trust evaluation in e-commerce sector. Fuzzy logic is used because of its ability to consider the uncertainties within e-commerce data, as the trust is often explained in linguistic terms.

In another research [25] fuzzy logic is applied to customer satisfaction index and a new model is presented. In [26], fuzzy hierarchical TOPSIS based on ES-QUAL is proposed for the evaluation of e-commerce websites.

4 Suggestion and Statement of the Problem

According to analyzed researches we suggest the method that is based on consistency driven approach for determining weights of the criteria of interest and identification of distance between ideal positive and negative solution under Z-number for ranking of alternatives related to digital marketing decisions such as selection of best digital marketing tool. Proposed method is given below. An example of pairwise comparison for each criterion is performed. Pairwise comparison matrix is deployed in Table 2.

Step 1: Consistency of DM's preferences is evaluated using consistency-driven approach to Z-valued preferences. Criteria weights are identified by a decision maker. The decision maker assigns a specific certainty degree on B part, in linguistic terms, based on the first part A, which is defined in triangular fuzzy numbers.

Step 2: Crisp maximum Eigenvalue and Eigenvectors are calculated.

Step 3: Ideal positive and negative solutions related to criteria weights acquired from the former step are formulated using TOPSIS model.

Step 4: Ranking of the alternatives is found.

Table 2. Pairwise comparison matrix

	C1	C_2	C_3	C_4
C_1	(1, 1, 1) (0.8, 0.9, 1.0)	(1/4, 1/3, 1/2) (0.6, 0.7, 0.8)	(1/4, 1/3, 1/2) (0.6, 0.7, 0.8)	(1/3, 1/2, 1/1) (0.6, 0.7, 0.8)
C2	(2, 3, 4) (0.6, 0.7, 0.8)	(1, 1, 1) (0.8, 0.9, 1.0)	(1, 2, 3) (0.6, 0.7, 0.8)	(1, 2, 3) (0.6, 0.7, 0.8)
C3	(1, 2, 3) (0.6, 0.7, 0.8)	(1/3 ,1/2, 1/1) (0.6, 0.7, 0.8)	(1, 1, 1) (0.8, 0.9, 1.0)	(2, 3, 4) (0.6, 0.7, 0.8)
C_4	(1, 2, 3) (0.6, 0.7, 0.8)	(1/3, 1/2, 1/1) (0.6, 0.7, 0.8)	(1/4, 1/3, 1/2) (0.6, 0.7, 0.8)	(1, 1, 1) (0.8, 0.9, 1.0)

5 Finding Eigenvalue and Eigenvectors

Solving the following equation gives the Maximum Eigenvalue, $\det(A-\lambda I) = 0$. Matrix A involves only medium values of fuzzy elements. DE objective function is minimized, $CF(\lambda) = D^2(\lambda) - \varepsilon\lambda$, where λ is the parameter to optimize, $D(\lambda) = \det(A - \lambda I)$ is a function dependent of λ, ε is a constant which may need to be mended for concrete problem (in our case it was set to 1, DE search space dimension is 1). A new DE objective function is set for finding Eigenvectors: CF(X) = VectorDistance (AX, λX). Furthermore, parameters to search are (N – 1) elements of vector X, (except the first one, which is set to 1) and DE search space dimension is N – 1.

Step 1. Finding fuzzy Eigenvalues. Here, Fuzzy A matrix is used. A new DE objective function is set as following: $CF(\lambda_L, \lambda_R)$ = FuzzyDistance (AX, $(\lambda_L, \lambda_M\ \lambda_R)$X), where "FuzzyDistance" is the total of fuzzy distances between corresponding elements of vectors. Vector X comprises fuzzy singletons that is produced from crisp Eigenvectors in step 1. Only left and right elements of fuzzy Eigenvalues are the subject for optimization (Medium component is taken from first step). DE search space dimension is 2.

Step 2. Finding fuzzy Eigenvectors. Fuzzy A matrix and fuzzy eigenvalues are used to construct DE objective function as follows: CF(X) = FuzzyDistance (AX, λX). Parameters to search are N elements of fuzzy vector X. Only left and right components are searched, and middle component is taken from the first step. Furthermore, DE search space dimension is 2 * N.

MAX LAMBDA = 5.14605236150391
Eigen Vector = 1 3.23272232962209 2.5401079704875 1.59222967350729
Correctness Check: A * X = 5.14594 16.63572 13.07148 8.19362
Lambda * X = 5.14605 16.63576 13.07153 8.19370
Fuzzy LAMBDA = [3.59896476396572, 5.14605236150391, 7.08349591173713]
Fuzzy Eigen Vectors = [0.98021 1.00000 1.19852] [2.83861 3.23272 3.62010] [2.46467 2.54011 2.89949][1.52681 1.59223 1.89307]
Correctness Check: A * X = [3.52774 5.14594 8.48971] [10.21605 16.63572 25.64297] [8.87025 13.07148 20.53854][5.49494 8.19362 13.40958]
Lambda * X = [3.52774 5.14605 8.48971] [10.21605 16.63576 25.64297] [8.87025 13.07153 20.53854] [5.49494 8.19370 13.40958].

Then, to find the criteria weights we normalize eigenvectors using the following formula:

$$Z = \frac{Xmax - X}{Xmax - Xmin} \tag{4}$$

6 Conclusion

Digital marketing has become an important part of marketing practices. In our research we analyze the existing literature associated with digital marketing that is based on fuzzy logic. The main problem related to analyzed literature is the lack of certainty and reliability of data. So, in this paper, a method for decision making process related to digital marketing usage is suggested for the future practices. Use of the proposed Z-number based method may improve the decision-making capability of marketing experts or DMs in a more robust and a certain way, by considering the vagueness of the decision-making environment.

References

1. Bala, M., Verma, D.: A critical review of digital marketing. Int. J. Manag. IT Eng. **8**(10), 321–339 (2018)
2. Munshi, A., Munshi, S.: Digital marketing: a new buzz word. ZENITH Int. J. Bus. Econ. Manag. Res. **2**(7), 190–196 (2012). https://doi.org/ijor.aspx?target=ijor:zijbemr&volume=2&issue=7&article=017
3. Tiago, M.T.P.M.B., Veríssimo, J.M.C.: Digital marketing and social media: why bother? Bus Horiz. **57**(6), 703–708 (2014). https://doi.org/10.1016/j.bushor.2014.07.002
4. Kotler, P., Armstrong, G., Opresnik, O.M.: Harlow: Principles of Marketing, England, Pearson (2018)
5. Leung, K.H., Mo, D.Y.: A fuzzy-AHP approach for strategic evaluation and selection of digital marketing tools. In: IEEE International Conference on Industrial Engineering and Engineering Management, (IEEM), Macao, pp. 1422–1426 (2019). https://doi.org/10.1109/IEEM44572.2019.8978797
6. Mukul, E., Büyüközkan, G., Güler, M.: Evaluation of digital marketing technologies with MCDM methods. In: 6th International Conference on New Ideas in Management, Economics and Accounting, France, Paris pp. 36–50 (2019)

7. Şengül, Ü., Eren, M.: Selection of digital marketing tools using fuzzy AHP-fuzzy TOPSIS. In: Fuzzy Optimization and Multi-Criteria Decision Making in Digital Marketing, vol. 5, pp. 97–126. IGI Global (2016). https://doi.org/10.4018/978-1-4666-8808-7
8. Aliev, R.A., Alizadeh, A., Aliyev, R.R., Huseynov, O.H.: Arithmetic of Z-Numbers, the: Theory and Applications. World Scientific (2015)
9. Aliev, R.A., Aliev, R.R.: Soft Computing and Its Application. World Scientific (2001)
10. Zadeh, L.A.: A note on Z-numbers. Inf. Sci. **181**(14), 2923–2932 (2011). https://doi.org/10.1016/j.ins.2011.02.022
11. Valiev, A.A., Abdullayev, T.S., Alizadeh, A.V., Adilova, N.E.: Comparison of measures of specificity of Z-numbers. Procedia Comput. Sci. **120**, 466–472 (2017). https://doi.org/10.1016/j.procs.2017.11.265
12. Aliyeva, K.: Eigensolution of 2 by 2 Z-matrix. In: Aliev, R., Kacprzyk, J., Pedrycz, W., Jamshidi, M., Babanli, M., Sadikoglu, F. (eds.) ICSCCW 2019. AISC, vol. 1095, pp. 758–762. Springer, Cham (2020). https://doi.org/10.1007/978-3-030-35249-3_98
13. Anute, N.B., Kabadi, S., Ingale, D.: A Study on perception of job seekers about digital marketing tools used for recruitment process. Int. J. 360 Manag. Rev. **07**(01) (2019). ISSN 2320-7132
14. López García, J.J., Lizcano, D., Ramos, C.M., Matos, N.: Digital marketing actions that achieve a better attraction and loyalty of users: an analytical study. Future Internet **11**(6), 130 (2019). https://doi.org/10.3390/fi11060130
15. Teixeira, S., Martins, J., Branco, F., Gonçalves, R., Au-Yong-Oliveira, M., Moreira, F.: A theoretical analysis of digital marketing adoption by startups. In: Mejia, J., Muñoz, M., Rocha, Á., Quiñonez, Y., Calvo-Manzano, J. (eds.) CIMPS 2017. AISC, vol. 688, pp. 94–105. Springer, Cham (2018). https://doi.org/10.1007/978-3-319-69341-5_9
16. Gaitniece, E.: Digital marketing performance evaluation methods. In: CBU International Conference Proceedings, vol. 6, pp. 135–140 (2018). https://doi.org/10.12955/cbup.v6.1145
17. Wątróbski, J., Jankowski, J., Ziemba, P.: Multistage performance modelling in digital marketing management. Econ. Sociol. **9**, 101–125 (2016). https://doi.org/10.14254/2071-789X.2016/9-2/7
18. Tuten, T., Solomon, M.: Social Media Marketing, 1st edn., Pearson, Harlow, Essex, UK (2014)
19. Howells, K., Ertugan, A.: Applying fuzzy logic for sentiment analysis of social media network data in marketing. Procedia Comput. Sci. **120**, 664–670 (2017). https://doi.org/10.1016/j.procs.2017.11.293
20. Statista: Most popular social networks worldwide as of April 2021, ranked by number of active users, April 2021. https://www.statista.com/statistics/272014/global-social-networks-ranked-by-number-of-users/
21. Vashishtha, S., Susan, S.: Fuzzy rule based unsupervised sentiment analysis from social media posts. Expert Syst. Appl. **138**, 112834 (2019). https://doi.org/10.1016/j.eswa.2019.112834
22. Tavana, M., Momeni, E., Rezaeiniya, N., Mirhedayatian, S.M., Rezaeiniya, H.: A novel hybrid social media platform selection model using fuzzy ANP and COPRAS-G. Expert Syst. Appl. **40**(14), 5694–5702 (2013). https://doi.org/10.1016/j.eswa.2013.05.015
23. Kong, F., Liu, H.: Applying fuzzy analytic hierarchy process to evaluate success factors of e-commerce. Int. J. Inf. Syst. Sci. **1**(3–4), 406–412 (2005)
24. Nefti, S., Meziane, F., Kasiran, K.: A fuzzy trust model for e-commerce. In: Seventh IEEE International Conference on E-Commerce Technology (CEC 2005), pp. 401–404. IEEE (2005). https://doi.org/10.1109/ICECT.2005.4

25. Liu, X., Zeng, X., Xu, Y., Koehl, L.: A fuzzy model of customer satisfaction index in e-commerce. Math Comput. Simul. **77**(5–6), 512–521 (2008). https://doi.org/10.1016/j.matcom.2007.11.017
26. Kang, D., Jang, W., Park, Y.: Evaluation of e-commerce websites using fuzzy hierarchical TOPSIS based on ES-QUAL. Appl. Soft Comput. **42**, 53–65 (2016). https://doi.org/10.1016/j.asoc.2016.01.017

Modeling of Number of Small and Medium Enterprises in Rostov Region Using a Modified Model Fuzzy Time Series

Taras Bogachev(✉), Tamara Alekseychik, Natalia Vovchenko, and Yana Kurinova

Rostov State University of Economics, Bolshaya Sadovaya Street, 69, 344002 Rostov-on-Don, Russia
bogachevTV@rsue.ru

Abstract. This paper proposes a method for predicting the number of small and medium enterprises per 10,000 people of the region's population using fuzzy logic. The variability of economic and industrial relations leads to incomplete data; therefore, the basis of the proposed method is fuzzy time series. To construct a modified time series of the indicator under research, the trend line equation is first constructed based on data for 2000–2017, which is then used to determine the forecasted value of the indicator for 2018 using fuzzy time series. When developing a forecasting model, the minimum and maximum values of the indicator are determined, because of which its universal set is determined and further divided into intervals. At these intervals, linguistic variables are introduced that characterize the qualitative state of the value of the indicator under study. Then its forecasted value is determined, which can also be described by the corresponding linguistic variable. Verification of the forecasted values obtained has shown the adequacy of the proposed modification of the fuzzy time series model. The proposed approach can be used to study various economic indicators.

Keywords: Small business · Fuzzy set theory · Forecasting · Fuzzy time series · Small and medium enterprises

1 Introduction

The sustainable state of the economic systems of countries is largely determined by favorable conditions for the development of small and medium enterprises (SMEs). The creation of SMEs contributes to increasing the economic activity of the population, reducing the unemployment rate, increasing the volume of goods produced, developing tourism, and increasing the number of consumer services. Therefore, support programs for SMEs are implemented not only by government agencies of various countries, but also by non-governmental organizations. The activity of small businesses is of particular importance for countries with developing economies [1, 2]. In small businesses, new environmental strategies are often more actively implemented and the problems that arise in this case are reflected in [3].

R. A. Aliev et al. (Eds.): ICSCCW 2021, LNNS 362, pp. 474–479, 2022.
https://doi.org/10.1007/978-3-030-92127-9_64

The development of SMEs is influenced by many different factors, among which it is necessary to identify more important and significant ones. In this regard, there is a need to develop methods for studying the state of SMEs using modern mathematical models and software tools. Various econometric models, particularly, time series models, are often used for such analysis [4].

Due to various reasons, the data used in the analysis of various economic processes are received in insufficient volume or in a distorted form. The variability of economic and industrial relations, the incompleteness of the legal framework regulating these relations, leads to the fact that the time series representing socio-economic processes may be incomplete. In connection with the above factors, it becomes necessary to develop forecasting methods using fuzzy logic, which would provide adequate forecasts in this situation [5]. To analyze the activities of small businesses, the preparation of relevant data is a rather serious problem, especially recently due to the restrictions associated with the coronavirus pandemic.

In this paper, we will analyze and determine the forecasted values of such an important indicator as the indicator of the number of SMEs per 10000 people population based on data for 2000–2018 in the Rostov region [6] using fuzzy time series.

2 Method

Fuzzy time series can be represented using fuzzy sets. A fuzzy set A, defined on a universal set U, has the form $= \langle u_r, f_A(u_r)\rangle$, $r = 1, .., n$. Here u_r is the element of the fuzzy set A, $f_A(u)$ is the membership function of the fuzzy set A, $f_A(u) : U \to [0, 1]$, $f_A(u_r)$ determines the grade of membership of the element u_r to the fuzzy set A.

Let $F(t)$ is a set of functions $f_i(t)$, $i = 1, 2, \ldots$ given on a universal set of the following form: $Y(t)$, $t = .., 0, 1, 2, \ldots$ Then $F(t)$ is called a fuzzy time series on the universal set $Y(t)$ [7, 8].

We put $F(t) = F(t-1) \circ R(t, t-1)$ where $R(t, t-1)$ is a fuzzy relation, and $\circ$ is the max-min composition operator. If $F(t)$ depends on $F(t-1), F(t-2), \ldots, F(t-k)$ then the fuzzy logical relationship is represented as $F(t-k), \ldots, F(t-2), F(t-1) \to F(t)$ and is called a k-th-order forecast model based on fuzzy time series.

Consider a forecasting model for $k = 1$. In this case, the fuzzy logical relationship has the form $F(t-1) \to F(t)$.

We represent the fuzzy values of the time periods i and $i + 1$ as fuzzy sets A_j and A_k (here the period is one year), defined on the universal set U. Thus, the fuzzy logical relationship can be represented in the following form $Aj \to A_k$, where A_j is the current state of the fuzzy logical relationship, and A_k is the next state.

Let T_i $(i = 1, 2, \ldots, n)$ are observed values for some factor of the time series. In order to be able to verify the model, we will find the equation of the trend line from the first $n - l$ values of the indicator under research, and from this equation we will calculate l approximate values for the corresponding time periods. After that, we replace the last l values of the time series (which will be used later to calculate the errors) with the found values.

The universal set U for indicator values is defined as follows:

$$U = [D_{min} - D_1,\ D_{max} + D_2] \tag{1}$$

where D_{min} is the minimum value of the indicator, and D_{max} is the maximum value of the indicator based on the known time series data, D_1, D_2 – are non-negative numbers that are selected in such a way that the set U can be divided into n equal intervals.

We define fuzzy sets A_r, $r = 1, \ldots, n$, in the form: $A_1 = 1/u_1 + 0.5/u_2 + 0/u_3 + \ldots + 0/u_{n-1} + 0/u_n$, $A_2 = 0.5/u_1 + 1/u_2 + 0.5/u_3 + \ldots + 0/u_{n-1} + 0/u_n$, $A_3 = 0/u_1 + 0.5/u_2 + 1/u_3 + 0.5/u_4 + \ldots + 0/u_{n-1} + 0/u_n$, …, $A_n = 0/u_1 + 0/u_2 + \ldots + 0/u_{n-2} + 0.5/u_{n-1} + 1/u_n$.

We perform fuzzification of the data by determining the intervals u_r, $r = 1,.., n$, which belong to the values of the researched indicator of number of SMEs per 10000 people. Based on the observed data of the indicator, fuzzy logical relationships are formed, from which s groups are then formed, each of which has the form $A_j \to A_{j_1}, \ldots, A_j \to A_{j_m}$. The forecast value of the indicator is found as the arithmetic mean of the middle points w_{j_1}, …, w_{j_m}, of the intervals $U_{j_1}, \ldots, U_{j_m}$, respectively, on which the membership functions of the corresponding fuzzy sets take the maximum value.

Therefore, the predicted value is found by the formula [8]:

$$F_{i+1} = \frac{w_{ji} + \cdots + w_{jm}}{m} \tag{2}$$

The relative forecast error is calculated using the following formula:

$$\frac{|F_i - T_i|}{T_i} * 100\% \tag{3}$$

Here F_i is the forecasted value for the i time period, T_i is the observed value.

3 Results

We regard the data on the number of SMEs in the Rostov region per 10000 people for 2000–2018 (Table 1). Using the data in Table 1 and econometric methods [4], we construct a trend line equation for the indicator number of SMEs per 10000 people:

$$y = 5.6536t + 52.038 \tag{4}$$

where t is the time period, $t = 1, 2, \ldots, 18$ (the values of the time periods correspond to the years from 2000 to 2017).

Based on (4), we will determine the forecasted value of the indicator for 2018. We will use this value when constructing a model of a fuzzy time series. Using observed data, we will calculate the relative error to verify the forecasted value. The problem was solved using a program written in Python [9] using the Pandas data analysis library [10].

We illustrate the model and results obtained for $n = 9$.

Using formula (4), we find that at $t = 19$ (corresponding to 2018), the value of y is 159. Then $D_{min} = 55$, $D_{max} = 159$. Based on (1), let $D_1 = 0.104$, $D_2 = 0$.

The universal set U for the values of the researched indicator was divided into 9 intervals: $U_1 = (54.896, 66.556]$, $U_2 = (66.556, 78.111]$, $U_3 = (78.111, 89.667]$, $U_4 = (89.667, 101.222]$, $U_5 = (101.222, 112.778]$, $U_6 = (112.778, 124.333]$, $U_7 = (124.333, 135.889]$, $U_8 = (135.889, 147.444]$, $U_9 = (147.444, 159]$. The values of the studied indicator at these intervals can be described by linguistic variables [8]: A_1 – “very little”, A_2 – “not very little”, A_3 – “not many”, A_4 – “not too many”, A_5 – “many”, A_6 – “many many”, A_7 – “very many”, A_8 – “too many”, A_9 – “too many many”.

We define fuzzy sets $A_1, \ldots, A_9$: $A_1 = 1/u_1 + 0.5/u_2 + 0/u_3 + \ldots + 0/u_8 + 0/u_9$, $A_2 = 0.5/u_1 + 1/u_2 + 0.5/u_3 + \ldots + 0/u_8 + 0/u_9$, $A_3 = 0/u_1 + 0.5/u_2 + 1/u_3 + 0.5/u_4 + \ldots + 0/u_8 + 0/u_9$, …, $A_9 = 0/u_1 + 0/u_2 + \ldots + 0/u_7 + 0.5/u_8 + 1/u_9$. The values of the indicator with the corresponding fuzzy sets are presented in Table 1.

Table 1. Values of the indicator of the number of SMEs per 10000 people with fuzzy sets

Year	The number of SMEs per 10000 people	Fuzzy set
2000	56	A_1
2001	57	A_1
2002	57	A_1
2003	55	A_1
2004	59	A_1
2005	68	A_2
2006	73	A_2
2007	75	A_2
2008	79	A_3
2009	110	A_5
2010	87	A_3
2011	103	A_5
2012	132	A_7
2013	129	A_7
2014	128	A_7
2015	113	A_6
2016	120	A_6
2017	128	A_7

The fuzzy logical relationship $A_i \rightarrow A_k$ means “If the value of the indicator of time period i is A_i, then that of time period $i + 1$ is A_k” [8]. Using the data in Table 1 we get the relationships $A_1 \rightarrow A_1, A_1 \rightarrow A_2, A_2 \rightarrow A_2, A_2 \rightarrow A_3, A_3 \rightarrow A_5, A_5 \rightarrow A_3$, $A_5 \rightarrow A_7, A_7 \rightarrow A_7, A_7 \rightarrow A_7, A_7 \rightarrow A_6, A_6 \rightarrow A_6, A_6 \rightarrow A_7$. The grouped fuzzy relationships are shown in Table 2.

We present an algorithm for constructing a forecast of an economic indicator using a modified model of fuzzy time series using the example of constructing a forecast of the value of the researched indicator for 2018.

Table 2. Fuzzy relationship groups for the indicator of the number of SMEs per 10000 people.

Fuzzy relationship groups	Fuzzy relationship
Group 1	$A_1 \to A_1; A_1 \to A_2$
Group 2	$A_2 \to A_2; A_2 \to A_3$
Group 3	$A_3 \to A_5$
Group 4	–
Group 5	$A_5 \to A_3; A_5 \to A_7$
Group 6	$A_6 \to A_6; A_6 \to A_7$
Group 7	$A_7 \to A_6; A_7 \to A_7$

1. The value of the indicator for 2017 belongs to the fuzzy set A_7 (see Table 1). We determine from Table 2 that for A_7 there are two fuzzy logical relations ($A_7 \to A_6$, $A_7 \to A_7$).
2. The maximum values of membership of fuzzy sets A_5, A_6, A_7, belong to the intervals U_6, U_7, respectively, where $U_6 = (112.778, 124.333]$, $U_7 = (124.333, 135.889]$.
3. The midpoints of the intervals U_6 and U_7 are 118.556 and 130.111, respectively. Using the formula (2), we determine the predicted value of the indicator in 2018: $0.5(118.556 + 130.111) = 124.334$. This value corresponds to the fuzzy set A6 described by the logistic variable "many many".
4. Assessment of the quality of the forecasted value of indicator.

4 Discussion

We evaluate the quality of the obtained forecast of the number of SMEs per 10000 people in the Rostov region for 2018 using both the equation of the trend line and a fuzzy time series model with calculating the relative errors according to formula (3) (see Table 3).

Obviously, the relative error of the value of the number of SMEs in the Rostov region, obtained using the fuzzy time series model, are much lower than the value obtained using the trend line equation.

Table 3. Forecasted values of the indicator number of SMEs in the Rostov region in 2018 and their relative errors

Characteristics of the forecasted value of the indicator	2018
Observed value	125
Forecasted values calculated using the trend line equation	159
Forecasted values calculated using a fuzzy time series model	124.334
Relative error of the value calculated using the trend line equation (%)	27.2
Relative error of the value calculated using the fuzzy time series model (%)	0.53

5 Conclusion

The paper investigates data on the number of SMEs in the Rostov region per 10000 people for 2000–2018. Using data for 2000–2017, a trend line equation for this indicator was constructed using econometric methods. This equation is then used to construct a modified time series. Next, a fuzzy time series model is built, which is used to determine the forecasted value of the indicator for 2018. Using observed data, relative errors are calculated to verify the forecasted values.

Modification of the well-known model of fuzzy time series made it possible to find adequate values of the researched indicator. Consequently, this method of constructing fuzzy time series can be used to analyze and forecast various economic indicators.

References

1. Qi, S., Nguyen, D.D.: Government connections and credit access around the world: evidence from discouraged borrowers. J. Int. Bus. Stud. **52**(2), 321–333 (2021). https://doi.org/10.1057/s41267-020-00341-x
2. Maksimov, V., Wang, S.L., Luo, Y.: Reducing poverty in the least developed countries: the role of small and medium enterprises. J. World Bus. **52**(2), 244–257 (2017). https://doi.org/10.1016/j.jwb.2016.12.007
3. Martín-Tapia, I., Aragón-Correa, J.A., Rueda-Manzanares, A.: Environmental strategy and exports in medium, small and micro-enterprises. J. World Bus. **45**(3), 266–275 (2010). https://doi.org/10.1016/j.jwb.2009.09.009
4. Anderson, T.W.: The Statistical Analysis of Time Series. Wiley, Hoboken (1994). https://doi.org/10.1002/9780470316894.oth1
5. Bogachev, T., Alekseychik, T., Pushkar, O.: Analysis of indicators of the state of regional freight traffic by method of fuzzy linear regression. In: Aliev, R., Kacprzyk, J., Pedrycz, W., Jamshidi, M., Babanli, M., Sadikoglu, F. (eds.) ICSCCW 2019. AISC, vol. 1095, pp. 632–638. Springer, Cham (2020). https://doi.org/10.1007/978-3-030-35249-3_81
6. The ROSSTAT: Region of Russia. Socio-economic Indicators. Statistical Compendium. https://gks.ru/bgd/regl/b19_14p/Main.htm. (in Russian)
7. Song, Q., Chissom, B.S.: Fuzzy time series and its models. Fuzzy Set Syst. **54**(3), 267–277 (1993). https://doi.org/10.1016/0165-0114(93)90372-O
8. Chen, S.-M.: Forecasting enrollments based on fuzzy time series. Fuzzy Set Syst. **81**, 311–319 (1996). https://doi.org/10.1080/019697202753306479
9. Python 3.9.5 Documentation. https://docs.python.org
10. Pandas Documentation. https://pandas.pydata.org/pandas-docs/stable

Decision Making on Firm's Software Selection with Type-2 Fuzzy Data

Nihad Mehdiyev(✉)

Research Laboratory of Intelligent Control and Decision Making Systems in Industry and Economics, Azerbaijan State Oil and Industry University, Baku, Azerbaijan

Abstract. Goal of this paper software selection problem for firms which is a multicriteria decision making problem whereas it maintains discrepant and various factors. In this article, decision making method using AHP and Interval type-2 fuzzy sets is used on firm's software selection problem with reliability, usability, vendor, flexibility factors and four alternatives. The eccentricity of this paper comes from applying interval type-2 fuzzy sets in the decision-making problem.

Keywords: Software selection · Type-2 fuzzy numbers · Multicriteria decision making · Fuzzy AHP

1 Introduction

In the development and use of information systems, the correct choice of software plays an important role, such as project management systems, programming systems, database management systems, etc. At the same time, various approaches are used to make a decision on the choice of software. at present, this direction is intensively developing, but an exhaustive solution has not yet been found. For example, for project management systems issues of integration with other systems are extremely important. For information technology projects, integration with code version control systems is important. Also, there is often a need to integrate management systems with accounting automation systems, and the scenarios for such integration can vary greatly.

Many factors influence the integration requirements: the scope of the company, the set of software tools used, the amount of data. Choosing software for firms is not a fast process. A lot of company owners and managers required a new system but are postpone by the number of competing products, the amount of time needed to estimate each one, or a fright of selecting the incorrect solution. The choosing process can be designed as a multi factor decision making problem which concentrates on problems with predetermined decision alternatives. Multicriteria decision making methods use various and discrepant factors such as behavioral patterns, technical and organizational abilities. In this article, a AHP method is applied to choose the suitable software due to its advantage, capability to separate a decision-making problem into its constitutive pieces and backing for group decision making via consent. In the typical expressing of multicriteria decision making problems, experts thinking is showed as exact numbers.

R. A. Aliev et al. (Eds.): ICSCCW 2021, LNNS 362, pp. 480–487, 2022.
https://doi.org/10.1007/978-3-030-92127-9_65

The fuzzy set theory [1] is improved to mathematically perform vagueness and afford designed instruments for dealing with the inexactness in decision-making problems. The fuzzy sets are able of processing vagueness and uncertainties, because of some drawback, notion of a type-2 fuzzy set was offered by Zadeh [2] as an expansion of standard fuzzy sets. Type-2 fuzzy sets, determined as fuzzy sets whose membership estimates are themselves type-1 decision making fuzzy sets, and these are useful for imparting model uncertainty and including linguistic uncertainties [3, 4]. The originality of the article comes from expanding the present works [5] on interval type-2 fuzzy AHP technique. Using interval type-2 fuzzy sets for various type decision making problems a more flexible and more intelligent results are determined interval type-2 fuzzy sets can indicate vagueness better than classical type-1 fuzzy sets [6, 7].

This article designed as follows. Section 2 is given preliminaries. Illustrative example presents applying AHP interval type-2 fuzzy method for the software selection problem in Sect. 3. Finally, Sect. 4 represents solution and the main results developed in the article.

2 Preliminaries

Definition 1. An interval Type-2 fuzzy sets (IT2FS) $\widetilde{A}$ on the universe of discourse X is characterized by its upper A and lower membership function *A*.

$$\widetilde{A_\alpha} = \{(x, U)|\mu_{\widetilde{A}}(x, U) \geq a, x \in X, U \in J_x\}$$

Figure 1 shows represents interval Type-2 fuzzy sets where upper and lower fuzzy numbers are drawn.

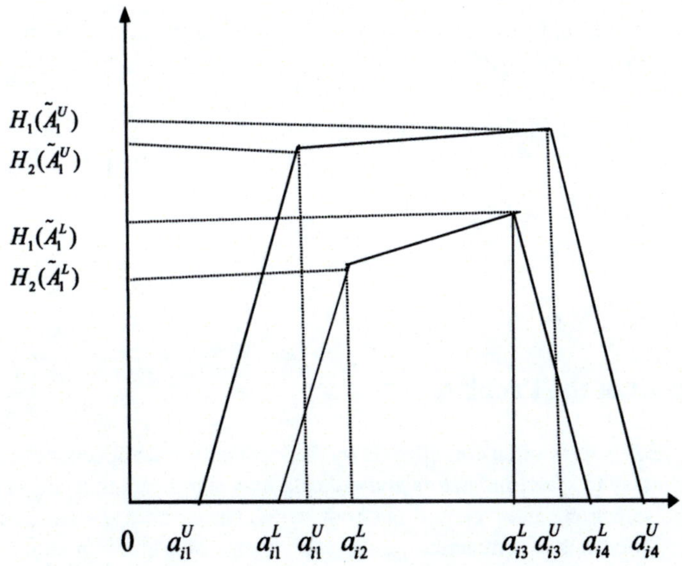

Fig. 1. The upper and the lower trapezoidal membership function *of* interval type-2 fuzzy sets.

Definition 2. Ranking value of the upper trapezoidal membership function of the interval type-2 fuzzy sets is determined by next function [7]:

$$Rank\left(\tilde{A}_i^U\right) = \frac{1}{n(n-1)}\left[\sum\nolimits_{k=1}^{n} p\left(\tilde{A}_i^U \geq \tilde{A}_k^U\right) + \frac{n}{2} - 1,\right] \tag{1}$$

where $Rank\left(\tilde{A}_i^U\right) = 1, \quad 1 \leq i \leq n.$

The ranking value $Rank\left(\tilde{A}_i^L\right)$ of the lower trapezoidal membership function $\tilde{A}_i^L$ of the interval type-2 fuzzy sets also determined as

$$Rank\left(\tilde{A}_i^L\right) = \frac{1}{n(n-1)}\left[\sum\nolimits_{k=1}^{n} p\left(\tilde{A}_i^L \geq \tilde{A}_k^L\right) + \frac{n}{2} - 1\right] \tag{2}$$

where $Rank\left(\tilde{A}_i^L\right) = 1, \quad 1 \leq i \leq n.$

Ranking value of the interval type-2 fuzzy sets $\tilde{\tilde{A}}_i$ determined as

$$Rank\left(\tilde{\tilde{A}}_i\right) = \frac{Rank\left(\tilde{A}_i^U\right) + Rank\left(\tilde{A}_i^L\right)}{2} \tag{3}$$

satisfying $\sum_{i=1}^{n} Rank\left(\tilde{\tilde{A}}_i\right) = 1.$

Definition 3. The ranking values of interval type-2 fuzzy sets are normalized by next functions,

$$r_{ij}^L = \frac{Rank\left(\tilde{\tilde{A}}_{ij}^L\right)}{\sqrt{\sum_{i,j=1} Rank\left(\tilde{\tilde{A}}_{ij}^{L2}\right)}}, \quad i = 1, 2, 3, \cdots, n, \ \ j = 1, 2, 3, \cdots, n \tag{4}$$

$$r_{ij}^L = \frac{Rank\left(\tilde{\tilde{A}}_{ij}^U\right)}{\sqrt{\sum_{i,j=1} Rank\left(\left(\tilde{\tilde{A}}_{ij}^U\right)^2\right)}}, \quad i = 1, 2, 3, \cdots, n, \ \ j = 1, 2, 3, \cdots, n \tag{5}$$

$r_{ij} = \frac{r_{ij}^L + r_{ij}^U}{2}$, where $\sum r_{ij} = 1.$

3 Statement of the Problem

The choice of software is often complicated by the fact that in each specific comparison task it is necessary to select indicators and indicate how significant they are in a given situation. The selection can be carried out both on the basis of the assessments of one expert, and on the basis of agreement on the assessments of a group of experts group discussion by company specialists can speed up decision-making, but this is a laborious process, and the presence of all specialists in a general discussion is not always possible. In addition, when organizing a group discussion, one may encounter the fact that the opinion of one expert or a group of experts differs greatly from the opinion of the majority. It is necessary to provide for the calculation and agreement general assessment based on data received from specialists. When evaluating a large number of software options, not all professionals may have competencies for each option. Lack of confidence in the assessment of the indicator, most experts tend to average estimates, which can prevent from choosing the solution that is optimal in terms of quality. Thus, the software comparison method must satisfy the following requirements: independence of the calculation method from the number and composition of indicators; correct processing of subjective assessments of experts; the ability to use the weights of indicators; the possibility of using an additional calculation parameter to determine the weight of the expert's assessment; high visibility and moderate computational complexity. Among the existing selection methods, the decision-making methods are of the greatest interest [8, 9]. Consider that multifactor decision making problem contains 4 factors -C_1, C_2, C_3, C_4 and 4 alternatives -A_1, A_2, A_3, A_4. C_1-Reliability, C_2-Usability, C_3-Vendor, C_4-Flexibility. For making decision is used the linguistical performance in terms of Type-2 fuzzy numbers that are represented in Table 1.

Table 1. Linguistic performance rating

Attribute	A_1	A_2	A_3	A_4
C_1-(Reliability)	Very high	High	Average	High
C_2-(Usability)	High	Very high	High	Average
C_3-(Vendor)	High	High	High	Average
C_4-(Flexibility)	Average	Average	Very high	Very high

4 Solution of the Problem

Step-1: The linguistic variables of AHP crisp number are converted to the interval type-2 fuzzy number for constructing matrix of factors and alternatives (Table 2 and Table 3).
Step-2: The average decision matrix of different criteria is determined by using next formula.

$$\tilde{\tilde{f}}_{ij} = \left(\frac{\tilde{\tilde{f}}^{1}_{ij} \oplus \tilde{\tilde{f}}^{2}_{ij} \oplus \ldots \oplus \tilde{\tilde{f}}^{p}_{ij}}{k} \right) \tag{6}$$

Table 2. AHP crisp number judgement matrix of criterion.

	C1	C2	C3	C4
C1	1	3	5	7
C2	0.33	1	5	7
C3	0.2	0.2	1	9
C4	0.15	0.15	0.11	1

Table 3. Interval type-2 fuzzy judgment matrix criterion.

	C1	C2	C3	C4
C1	((0, 1, 1, 2, 1), (0, 1, 1, 2, 0.9))	((2, 3, 3, 4, 1,), (2, 3, 3,4, 0.9))	((3, 4, 5, 6, 1,), (3, 4, 5, 6, 0.9))	((5,6,7, 8, 1,), (6,7,7,8, 0.9))
C2	((0.2,0.35,0.35,0.4, 1), (0,2,0.3,0.3, 0.4, 0.9))	((0, 1, 1, 2, 1), (0, 1, 1, 2, 0.9))	((4, 5, 5, 6, 1,), (4, 5, 5, 6, 0.9))	((5,6,7, 8, 1,), (5,6,7,8, 0.9))
C3	((0.1, 0.2,0.2,0.25, 1), (0,1,0.2,0.2, 0.3, 0.9))	((0.1,0.2,0.2,0.4, 1), (0,1,0.2,0.2,0.4, 0.9))	((0, 1, 1, 2, 1,), (0, 1, 1, 2, 0.9))	((8,9,9,11,1,), (8,9,9,11,0.9))
C4	((0.1,0.15,0.15,0.2, 1), (0,1,0.15,0.15,0.2,0.9))	((0.1, 0.15,0.15,0.2, 1), (0,1,0.15,0.15,0.2,0.9))	((0,0.12,0.12,0.15,1), (0,0.12,0.12,0.15, 0.9))	((0,1,1,2, 1,), (0,1,1,2, 0.9))

where

$$\tilde{\tilde{f}}^{L}_{C_1} = \begin{pmatrix} (\frac{0+2+3+5}{4}), (\frac{1+3+4+6}{4}) \\ (\frac{1+3+5+7}{4}), (\frac{2+4+6+8}{4}); 1, 1 \end{pmatrix} = [2.5, 3.5, 4, 5; 1]$$

$$\tilde{\tilde{f}}^{U}_{C_1} = \begin{pmatrix} (\frac{0+2+3+5}{4}), (\frac{1+3+4+6}{4}) \\ (\frac{1+3+5+7}{4}), (\frac{2+4+6+8}{4}); 1, 1 \end{pmatrix} = [2.75, 3.75, 4, 5; 0.9]$$

$$fc_1 = ((2.5, 3.5, 4, 5; 1, 1), (2.75, 3.75, 4, 5; 0.9, 0.9))$$

$$fc_2 = ((2.3, 3.08, 3.33, 4.1; 1, 1), (2.3, 3.07, 3.3, 4.1; 0.9, 0.9))$$

$$fc_3 = ((2.05, 2.6, 2, 6, 3.4; 1, 1), (2.05, 2.05, 2.6, 3.42; 0.9, 0.9))$$

$$fc_4 = ((0.05, 0.35, 0, 35, 0.63; 1, 1), (0.05, 0.35, 0, 35, 0.63; 0.9, 0.9))$$

Step-3: The weighted decision matrix for each criterion and alternatives can be obtained by using formula (8):

$$\tilde{\tilde{W}}_C = \left(\sum_{i=1, f \in F_i^L}^{n} \left(\tilde{\tilde{D}}_1 \tilde{\tilde{D}}_2 \cdots \tilde{\tilde{D}}_n\right) \otimes \begin{bmatrix} \tilde{\tilde{f}}^U_{C_1} \\ \tilde{\tilde{f}}^U_{C_2} \\ \tilde{\tilde{f}}^U_{C_3} \\ \tilde{\tilde{f}}^U_{C_4} \end{bmatrix} \right) \tag{7}$$

Let the importance of linguistic variables are:

$$D = \left(\tilde{\tilde{D}}_1 \tilde{\tilde{D}}_2 \cdots \tilde{\tilde{D}}_n\right) \tag{8}$$

$$D = \begin{matrix} 2.5,\ 3.5,\ 4,\ 5,\ 1,1; 2.75,\ 3.75,\ 4,\ 5,\ 0.9, 0.9 \\ 2.3,\ 3,08,\ 3.33,\ 4.1, 1, 1; 2.3,\ 3.07,\ 3.3,\ 4.1,\ 0.9, 0.9 \\ 2.05,\ 2.6,\ 2.6,\ 3.4, 1, 1; 2.05,\ 2.05,\ 2.6,\ 3.42,\ 0.9, 0.9 \\ 0.05,\ 0.35,\ 0.35,\ 0.63, 1, 1; 0.05, 0.35, 0.35, 0.63, 0.9, 0.9 \end{matrix}$$

$$W = \{0.35,\ 0.3,\ 0.25,\ 0.2\}$$

$$\tilde{\tilde{W}}_C = \begin{bmatrix} 2.5,\ 3.5,\ 4,\ 5,\ 1,1; 2.75,\ 3.75,\ 4,\ 5,\ 0.9, 0.9 \\ 2.3,\ 3,08,\ 3.33,\ 4.1, 1, 1; 2.3,\ 3.07,\ 3.3,\ 4.1,\ 0.9, 0.9 \\ 2.05,\ 2.6,\ 2.6,\ 3.4, 1, 1; 2.05,\ 2.05,\ 2.6,\ 3.42,\ 0.9, 0.9 \\ 0.05,\ 0.35,\ 0.35,\ 0.63, 1, 1; 0.05, 0.35, 0.35, 0.63, 0.9, 0.9 \end{bmatrix}$$
$$* [\,0.35,\ 0.3,\ 0.25,\ 0.1\,]$$
$$= \begin{bmatrix} 2.5,\ 3.5,\ 4,\ 5,\ 1,1; 2.75,\ 3.75,\ 4,\ 5,\ 0.9, 0.9 \\ 2.3,\ 3,08,\ 3.33,\ 4.1, 1, 1; 2.3,\ 3.07,\ 3.3,\ 4.1,\ 0.9, 0.9 \\ 2.05,\ 2.6,\ 2.6,\ 3.4, 1, 1; 2.05,\ 2.05,\ 2.6,\ 3.42,\ 0.9, 0.9 \\ 0.05,\ 0.35,\ 0.35,\ 0.63, 1, 1; 0.05, 0.35, 0.35, 0.63, 0.9, 0.9 \end{bmatrix}$$
$$* [\,0.35,\ 0.3,\ 0.25,\ 0.1\,]$$
$$= \begin{bmatrix} 0.87, 1.05, 1, 0.5;\ 1,\ 1;\ \ 0.96, 1.12, 1, 0.5; 0.9, 0.9 \\ 0.8\ ,\ 0.92, 0.\ 83, 0.41;\ 1, 1;\ \ 0.8, 0.92, 0.8, 0.4; 0.9, 0.9 \\ 0.71, 0.78, 0.65, 0.34\ ;\ \ 1, 1;\ \ 0.71,\ \ 0.61, 0.65, 0.34; 0.9, 0.9 \\ 0.01, 0.1, 0.08, 0.06\ ;\ \ 1, 1;\ \ 0.02, 0.1, 0.08, 0.06; 0.9\ ,\ \ 0.9 \end{bmatrix}.$$

Step-4: The ranking values of decision matrices of interval type-2 fuzzy sets are determined by using formulas (3), (4), and (5). The matrix of upper interval is determined as

$$\begin{matrix} 0.96 & 1.12 & 1 & 0.5 \\ 0.8 & 0.92 & 0.83 & 0.41 \\ 0.71 & 0.78 & 0.65 & 0.34 \\ 0.01 & 0.1 & 0.08 & 0.06 \end{matrix}$$

$$Rank(A^U_{C_1}) = (1/(4(4-1))(.96 + 1.12 + 1 + 0.5) + 4/2 - 1) = \mathbf{0.083} * (3.58 + \mathbf{1}) = 0.38$$

$$Rank(A^L_{C_1}) = (1/(4(4-1))(.87 + 1.05 + 1 + 0.5) + 4/2 - 1)$$
$$= \mathbf{0.083} * (3.42 + \mathbf{1}) = 0.36$$

Final rank will be normalized and employing the same models, we get next ranking values:

$$Rank\left(A^U_{C_2}\right) = 0.33,\quad Rank\left(A^L_{C_2}\right) = 0.32;$$
$$Rank\left(A^U_{C_3}\right) = 0.29,\quad Rank\left(A^L_{C_3}\right) = 0.27;$$
$$Rank\left(A^U_{C_4}\right) = 0.12,\quad Rank\left(A^L_{C_4}\right) = 0.1;$$

Step-5. The ranking values of IT2FS are normalized using the following TOPSIS equation to obtain weight relative by Eq. (6), Eq. (7) and Eq. (8).

$$r_{C_1}^{L} = \frac{Rank\left(\tilde{\tilde{A}}_{ij}^{L}\right)}{\sqrt{\sum_{i,j=1} Rank\left(\tilde{\tilde{A}}_{ij}^{L}2\right)}} = \frac{0.36}{1.06} = 0.34$$

$$r_{C_1}^{U} = \frac{Rank\left(\tilde{\tilde{A}}_{ij}^{U}\right)}{\sqrt{\sum_{i,j=1} Rank\left(\tilde{\tilde{A}}_{ij}^{U}2\right)}} = \frac{0.38}{1.1} = 0.345$$

$$r_{C_1} = (r_{C_1}^{U} + r_{C_1}^{L}) = (0.345 + 0.34)/2 = 0.342$$

$$r_{C_2}^{U} = 0.32, \quad r_{C_2}^{L} = 0.31; \quad r_{C_2} = 0.315$$

$$r_{C_3}^{U} = 0.26, \quad r_{C_3}^{L} = 0.25; \quad r_{C_3} = 0.255$$

$$r_{C_4}^{U} = 0.09, \quad r_{C_4}^{L} = 0.08; \quad r_{C_4} = 0.085$$

Step-6: For determining relative weight and ranking the alternatives is used model (9).

$$W_{ij} = \sum w_i A_{ij} \tag{9}$$

W_{ij} is normalized weight for factor j. A_{ij} is normalized weights aggregated matrix for factor j with considering alternatives i. The relative weight and rank are represented in Table 4.

Table 4. Priorities ranks and weights of software selection problem.

Basic factors					
	C1	C2	C3	C4	
Weights of Cn	0.345	0.315	0.255	0.085	Priorities weight
Alternatives					
A_1	0.31	0.33	0.26	0.045	0.24
A_2	0.27	0.28	0.10	0.036	0.17
A_3	0.24	0.21	0.17	0.03	0.16
A_4	0.017	0.10	0.16	0.05	0.08

From calculations and ranking we get that the best software is alternative A_1 (0.24) followed by alternative A_2 (0.17), alternative A_3 (0.16), and finally alternative A_4 (0.08).

5 Conclusion

The purpose of this article is software selection problem using interval type-2 fuzzy AHP technique with the feature of rank normalization, expert's weights. After solving this problem, we make decision that A_1 with the higher grade (24%) compared to the others alternatives.

References

1. Zadeh, L.A.: Fuzzy sets. Inf. Control **8**(3), 338–353 (1965). https://doi.org/10.1016/S0019-9958(65)90241-X
2. Zadeh, L.A.: The concept of a linguistic variable and its application to approximate reasoning – I. Inf. Sci. **8**(3), 199–249 (1975). https://doi.org/10.1016/0020-0255(75)90036-5
3. Karnik, N.N., Mendel, J.M.: Centroid of a type-2 fuzzy set. Inf. Sci. **132**, 195–220 (2001)
4. Kahraman, C., Sari, I.U., Turanoglu, E.: Fuzzy analytic hierarchy process with type-2 fuzzy sets. In: Proceedings of the 19th International FLINS Conference 26–29 August 2012, pp. 201–206 (2012). https://doi.org/10.1142/9789814417747_0033
5. Sari, I.U., Oztaysi, B., Kahraman, C.: Fuzzy AHP using type II fuzzy sets: an application to warehouse location selection. In: Doumpos, M., Grigoroudis, E. (eds.) Multicriteria Decision Aid and Artificial Intelligence: Links, Theory and Applications, pp. 258–308. Wiley, Hoboken (2013)
6. Chen, S.M., Lee, L.W.: Fuzzy multiple attributes group decision-making based on the interval type-2 TOPSIS method. Expert Syst. Appl. **37**, 2790–2798 (2010). https://doi.org/10.1016/j.eswa.2009.09.012
7. Chen, S.M., Lee, L.W.: Fuzzy multiple attributes group decision-making based on the ranking values and the arithmetic operations of interval type-2 fuzzy sets. Expert Syst. Appl. **37**, 824–833 (2010). https://doi.org/10.1016/j.eswa.2009.06.094
8. Aliev, R.A., Huseynov, O.H., Aliev, R.R., Alizadeh, A.V.: The Arithmetic of Z-Numbers: Theory and Applications. World Scientific (2015)
9. Aliev, R.A., Gardashova, L.A.: Z-set based approach to control system design. In: Aliev, R.A., Kacprzyk, J., Pedrycz, W., Jamshidi, M., Babanli, M., Sadikoglu, F.M. (eds.) ICAFS 2020. AISC, vol. 1306, pp. 10–21. Springer, Cham (2021). https://doi.org/10.1007/978-3-030-64058-3_2

Mathematical Model of Production and Exchange: Bifurcation Analysis and Management

V. Bratishchev Alexander[1], A. Batishcheva Galina[2](✉), I. Zhuravleva Maria[2], and Rogozhin Sergei[2]

[1] Don State Technical University, Gagarin Square, 1, 344000 Rostov-on-Don, Russia
[2] Economics, B. Sadovaya Street, 69, 344002 Rostov-on-Don, Russia

Abstract. The article provides a complete bifurcation analysis of the mathematical model of the dynamic system "Production and exchange" proposed by V. P. Milovanov. The behavior of trajectories at infinity is investigated using the Poincare transform. Using the theoretical analysis and numerical experiment in the Matlab+Simulink package, a scheme for dividing the phase space of the system into cells and the asymptotic behavior of the trajectories in each cell is obtained. The system was rough in the first quarter of the phase plane. By the method of analytical design of aggregated regulators, additive control of product production is designed, a class of permissible achievable states with such control and fixed system parameters is identified. If it is permissible to change one of the parameters, the control is constructed to achieve an arbitrary predetermined dynamic equilibrium, regardless of the initial state of the system. The numerical experiments conducted indicate the stability of such an equilibrium.

Keywords: Autonomous system · Equilibrium state · Phase portrait · Bifurcation analysis · Synergetic control theory

1 Introduction

In the monograph [1], the following mathematical model of the formation of the cost of manufacturing a product is proposed. Let $x(t)$ – the amount of money of the enterprise for the organization of production of a certain product; $x_i(t)$, $i = 1, \ldots, n$ – the number of available parts of type i at time t that are used in the manufacture of this product; a_i – constant inflow of type i parts per unit of time; a_1 – constant inflow of funds to the enterprise per unit of time; $c_i x(t) x_i(t)$ – expenditure of funds in order to integrate a type i part into the product being created; $b_0 x(t)$– expenses of the enterprise per unit of time not related to the manufacture of the product; $d_i x(t) x_i(t)$ – the required number of type i parts per unit of time for the manufacture of the product; $b_i x_i(t)$– outflow of type i parts

R. A. Aliev et al. (Eds.): ICSCCW 2021, LNNS 362, pp. 488–495, 2022.
https://doi.org/10.1007/978-3-030-92127-9_66

per unit of time from the sphere of production due to marriage, physical or moral wear and tear. The equation of the model has the form

$$\begin{cases} \frac{dx}{dt} = a_0 - b_0 x(t) - \sum_{k=1}^{n} {}_k x(t) x_k(t) \\ \frac{dx_i}{dt} = a_i - b_i x_i(t) - d_i x(t) x_i(t), \quad i = 1, \ldots, n \end{cases}.$$

All parameters a_i, b_i, c_i, d_i are considered positive.

The purpose of the simulation is to find the law of control of production processes and the choice of values of the system parameters, in which, regardless of the initial funds and the number of parts over time, the enterprise will always reach a stable production of a given number of products and a given number of parts for them.

2 Method

In the case of two parts, we have such a system:

$$\begin{cases} \frac{dx}{dt} = a_0 - b_0 x - c_1 x x_1 - c_2 x x_2 \\ \frac{dx_1}{dt} = a_1 - b_1 x_1 - d_1 x x_1 \\ \frac{dx_2}{dt} = a_2 - b_2 x_2 - d_2 x x_2 \end{cases}.$$

For specific numerical values of the parameters, the control law can be found. However, the study of this system is generally hindered by the complex form of the characteristic polynomial of the Jacobi matrix on the right side of the system. Therefore, in the article, we consider the general case for one detail, namely, for a system of the form:

$$\begin{cases} \frac{dx}{dt} = a_0 - b_0 x - c_1 x x_1 \\ \frac{dx_1}{dt} = a_1 - b_1 x_1 - d_1 x x_1 \end{cases}.$$

Note that the equations of the system are symmetric. Therefore, it can be considered as a mathematical model of the interdependent production of two products $x_1(t), x_2(t)$ and write the system in a symmetric form

$$\begin{cases} \frac{dx_1}{dt} = a_1 - b_1 x_1 - c_1 x_1 x_2 =: f_1 \\ \frac{dx_2}{dt} = a_2 - b_2 x_2 -{}_2\, x_1 x_2 =: f_2 \end{cases} \quad (1)$$

We use bifurcation analysis [2] and synergetic control theory [3] as the mathematical apparatus of the solution.

3 Results

At the first stage, it is necessary to find the equilibrium states of the system (1). Equating the right-hand sides of the equations to zero, we get two states:

$$S_{1,2} = \begin{pmatrix} \frac{a_1c_2 - a_2c_1 - b_1b_2 \pm \sqrt{(a_1c_2 - a_2c_1 - b_1b_2)^2 + 4a_1b_1b_2c_2}}{2b_1c_2}, \\ \frac{-a_1c_2 + a_2c_1 - b_1b_2 \pm \sqrt{(-a_1c_2 + a_2c_1 - b_1b_2)^2 + 4a_2b_1b_2c_1}}{2b_2c_1} \end{pmatrix}.$$

Since in each of the fractions the root is greater than the modulus of the first term, the state S_1 (with a + sign) for any positive parameter values lies in the first quarter of the phase plane, and the state S_2 (with a - sign) for any positive parameter values lies in the third quarter of the phase plane. To determine the topological nature of these states, we calculate the characteristic polynomial of the Jacobi matrices of the right parts of the system (1):

$$p(\lambda) := \begin{vmatrix} \lambda + b_1 + c_1 x_2 & c_1 x_1 \\ c_2 x_3 & \lambda + b_2 + c_2 x_1 \end{vmatrix} = \lambda^2 + (b_1 + c_1 x_2 + b_2 + c_2 x_1)\lambda$$
$$+(b_1 b_2 + b_1 c_2 x_1 + b_2 c_1 x_2) = 0.$$

The determinant of the Jacobi matrix

$$b_1 b_2 + b_1 c_2 x_1 + b_2 c_1 x_2 = \pm\frac{1}{2}\Big(\sqrt{(a_1 c_2 - a_2 c_1 - b_1 b_2)^2 + 4a_1 b_1 b_2 c_2}$$
$$+\sqrt{(-a_1 c_2 + a_2 c_1 - b_1 b_2)^2 + 4a_2 b_1 b_2 c_1}\Big) \neq 0$$

is not zero for any positive parameter values. It is easy to check that the trace of the matrix is not equal to zero. Therefore, S_2 is a saddle, and S_1 is a stable rough node for any positive parameter values.

This result is confirmed by a numerical experiment on the S-model [4] of the system (1) in the MATLAB-SIMULINK package for any numerical parameter values. Choose, for example, $a_1 = 1, b_1 = 2, c_1 = 3, a_2 = 3, b_2 = 2, c_2 = 1$. For these parameters, the equilibrium states are $S_1 = (0.3508, 0.4254)$, $S_2 = (-2.8508, -1.1754)$ (see Fig. 1, 2).

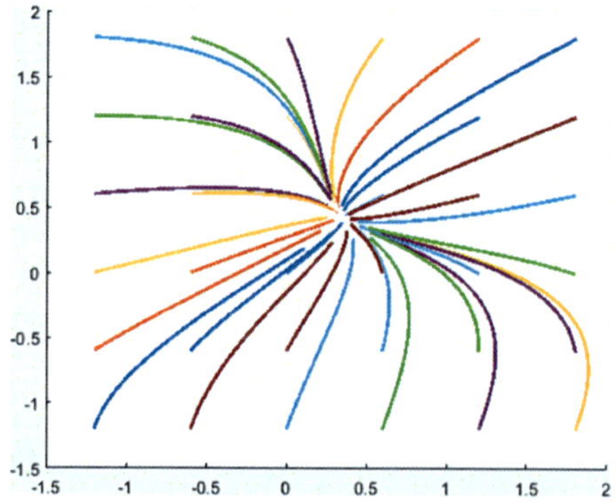

Fig. 1. Neighborhood of state S_1.

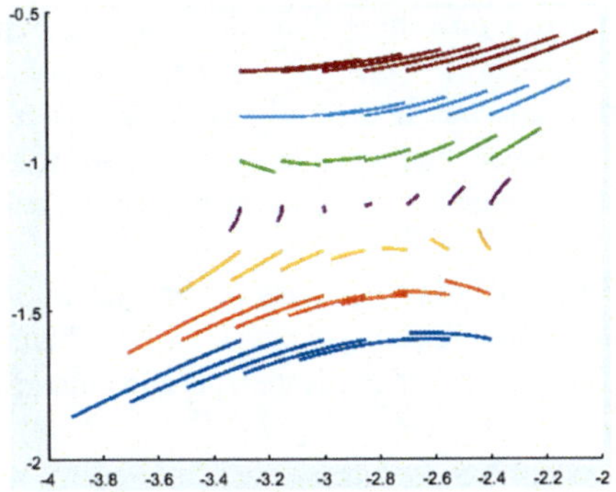

Fig. 2. Neighborhood of state S_2.

The phase portrait of the neighborhood S_1 is characteristic of the stable node, and the phase portrait of the neighborhood S_2 is characteristic of the saddle.

A complete bifurcation analysis of polynomial autonomous systems involves [2] the search and analysis of equilibrium states at infinity. The latter is carried out using the Poincare transform.

To determine the nature of possible equilibrium states ($\pm\infty$, 0), the following substitution of phase variables is used [2]: $x = \frac{1}{u}, y = \frac{v}{u}$. The system (1) is then transformed into the system:

$$\begin{cases} \frac{du}{dt} = b_1u^2 + c_1uv - a_1u^3 \\ \frac{dv}{dt} = -c_2v + a_2u^2 + c_1v^2 + (b_1 - b_2)uv - a_1u^2v \end{cases}.$$

Let's solve the system of equations by equating the right-hand sides to zero. The two equilibrium states lying on the coordinate axes are: $S_1^0 = (0, 0), S_2^0 = \left(0, \frac{c_2}{c_1}\right)$.

The state S_1^0 is a multiple of [2], since one of the eigenvalues of the corresponding Jacobi matrix is zero. According to the theorem on the nature of a multiple state of equilibrium, it is a saddle-node with a stable node sector.

The state S_2^0 has positive eigenvalues. Therefore, according to the first approximation stability theorem [5], or according to the Poincare criterion [2], it is an unstable node.

The other two equilibrium states have non-zero coordinates and are therefore images of the equilibrium states of the original system. The nature of the latter is previously established.

To determine the nature of the possible equilibrium states (0, $\pm\infty$), the Poincare transform $x = \frac{v}{u}, y = \frac{1}{u}$ is used. System (1) then switches to the system:

$$\begin{cases} \frac{du}{dt} = b_2u^2 + c_2uv - a_2u^3 \\ \frac{dv}{dt} = -c_1v + a_1u^2 + c_2v^2 + (b_2 - b_1)uv - a_2u^2v \end{cases}.$$

We are interested in the state $S_3^0 := (0, 0)$, since the nature of the other states has already been studied earlier. Just like S_1^0, it turned out to be a saddle node with a stable node sector.

We check the presence of cycles in system (1) using the Bendikson criterion [2]. Function $D(x_1, x_2) := (a_1 - b_1x_1 - c_1x_1x_2)'_{x_1} + (a_2 - b_2x_2 - c_2x_1x_2)'_{x_2} = -b_1 - c_1x_2 - b_2 - c_2x_1$ the sign is constant in the half-plane $c_2x_1 + c_1x_2 + (b_1 + b_2) > 0$, in particular in the first quarter, since the parameters are assumed to be positive. Therefore, according to the criterion, there are no cycles in the half-plane, and therefore in the first quarter.

According to the criterion of the roughness of the system, it is rough in the first quarter: its phase portraits are topologically equivalent. The obtained information about the nature of all the equilibrium states of the system (1) and numerical experiments allow us to obtain such a phase portrait of it (see Fig. 3).

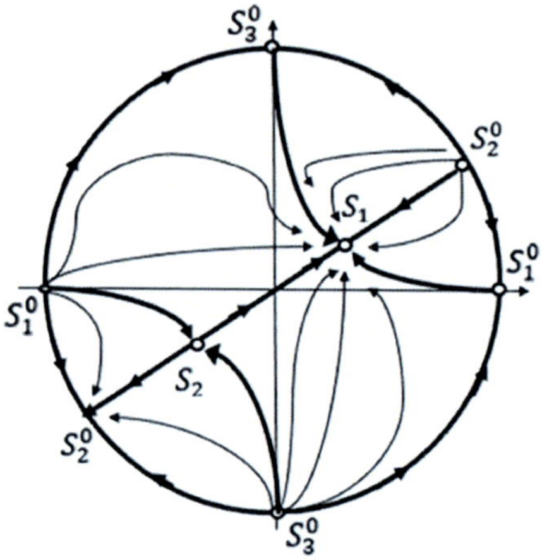

Fig. 3. The scheme of dividing the phase portrait of the system (1) into cells.

It can be seen that it consists of 5 elementary cells, in each of which the trajectories behave the same asymptotically.

4 Discussion

Consider the following economic problem: find an additive law for controlling the rate of production of products at any of the enterprises, for example, at the first one, in which the production of products at both enterprises is stabilized at a given quantity, regardless of their quantity at the initial time. From a mathematical point of view, it is necessary to find such a function $u(x_1, x_2)$, that the corresponding controlled system has one pre-set equilibrium state in the first quarter and the region of attraction of it would contain the first quarter.

We use the method of analytical construction of aggregated regulators (ACAR) [3], according to which we need to choose a suitable curve $\Psi(x_1, x_2) = 0$, which will be an invariant asymptotically stable set for the constructed controlled system and on which the predetermined state of equilibrium will be located. Since the second equation of the system (1) will not change, this state should also be on the hyperbola $a_2 - b_2x_2 - c_2x_1x_2 = 0$ (see Fig. 4).

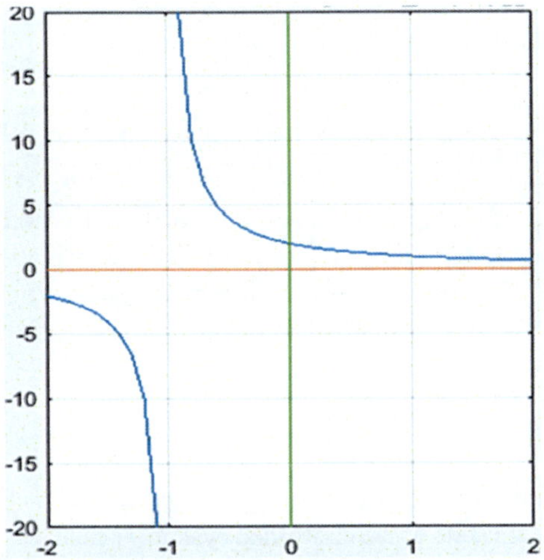

Fig. 4. Hyperbola graph.

That is, if the parameters are unchanged, only the points of this curve can be the states of the desired controller. Each such point lies at the intersection of a hyperbola and a suitable invariant set. Thus, we need to choose a family of curves that is transversal to the hyperbola. Figure 4 shows that this can be a bundle of lines $x_2 = \alpha x_1, \alpha > 0$, or a family of vertical lines $x_1 = \alpha, \alpha > 0$, and others. Among them, you need to choose a family that ensures the stability of the equilibrium states.

Let's say $\Psi(x_1, x_2) = x_1 - \alpha, \alpha > 0$. According to [3], the function Ψ, called the aggregate variable, must satisfy the differential equation $\Psi_t{}' = -\Psi$ on the solutions of the designed controller. This condition ensures the stability of the invariant curve $\Psi(x_1, x_2) = 0$, in the Kolesnikov sense: $\Psi(x_1(t), x_2(t)) \to 0$ on the trajectories. Expand the left part on the solutions of the controller:

$$\Psi'_{x_1}x'_{1t} + \Psi'_{x_2}x'_{2t} = x'_{1t} = -\Psi.$$

Combining this equation with the second equation of systems (1), we obtain the equation of the synergetic regulator:

$$\begin{cases} \frac{dx_1}{dt} = -x_1 + \alpha = f_1 + u \\ \frac{dx_2}{dt} = a_2 - b_2x_2 - {}_2x_1x_2 \end{cases}$$

Thus, the additive control is equal to:

$$u(x, y) := -x_1 + \alpha - a_1 + b_1 x_1 + c_1 x_1 x_2.$$

This system has a single equilibrium state:

$$\left(x_1^0, x_2^0\right) = \left(\alpha, \frac{a_2}{b_2 + 2\alpha}\right).$$

Let's check it for stability using the criterion [6]:

$$f'_{2x_2}(x_1^0, x_2^0) - \frac{\Psi'_{x_2}(x_1^0, x_2^0)}{\Psi'_{x_1}(x_1^0, x_2^0)} f'_{2x_1}(x_1^0, x_2^0) = -b_2 - c_2 x_1^0 = -b_2 - c_2\alpha.$$

Since the right-hand side is negative for any value of $\alpha > 0$, according to the criterion of the equilibrium state, the families of regulators will be asymptotically stable.

Since the function $\Psi(x_1, x_2) = x_1 - \alpha$ is linear, then according to remark 2 of the paper [7] invariant lines $x_1 = \alpha$ will be stable in general in the sense of Zubov [8].

5 Conclusion

It is proved that any point of the curve $a_2 - b_2 x_2 - c_2 x_1 x_2 = 0, x_1 > 0$ can be chosen as a predetermined equilibrium state of the controlled system. If such a point is selected, that is, $a_2 - b_2 x_2^0 - c_2 x_1^0 x_2^0 = 0$, then as the parameter α we take $\alpha := x_1^0$ and $\Psi(x_1, x_2) = x_1 - x_1^0$. If the parameters a_2, b_2, c_2 of the system (1) are fixed, then only the hyperbola points from the first quarter can be taken as the equilibrium states of the controlled system.

For example, the parameter (constant flow of parts to the second enterprise) a_2 can be changed. We fix an arbitrary point $\left(x_1^0, x_2^0\right)$ in the first quarter and select the parameter α so that it turns out to be the equilibrium state of the regulator. It must satisfy the equation $-x_1^0 + \alpha = 0$, where $\alpha := x_1^0$, and the equation $a_2 - b_2 x_2^0 - c_2 x_1^0 x_2^0 = 0$, in which the unknown is a_2. Obviously, $a_2 := b_2 x_2^0 + c_2 x_1^0 x_2^0 > 0$. The stability of the point $\left(x_1^0, x_2^0\right)$ is proved as before.

The fact that the region of attraction of the equilibrium state $\left(x_1^0, x_2^0\right)$ contains the first quarter of the phase plane was verified by numerical experiments with the S-model of the controlled system:

$$\begin{cases} \frac{dx_1}{dt} = -x_1 + x_1^0 \\ \frac{dx_2}{dt} = b_2 x_2^0 + 2x_1^0 x_2^0 - b_2 x_2 - 2x_1 x_2 \end{cases}$$

For example, let $b_1 = 2, c_1 = 3, a_2 = 3, b_2 = 2, c_2 = 1$, and the desired equilibrium state $\left(x_1^0, x_2^0\right) = (1, 2)$. Let us construct a numerical phase portrait of this system in the square $\{0 \le x_1 \le 100, 0 \le x_2 \le 100\}$ (see Fig. 5).

It can be seen that all trajectories for the selected network of initial data converge to the equilibrium state (1, 2).

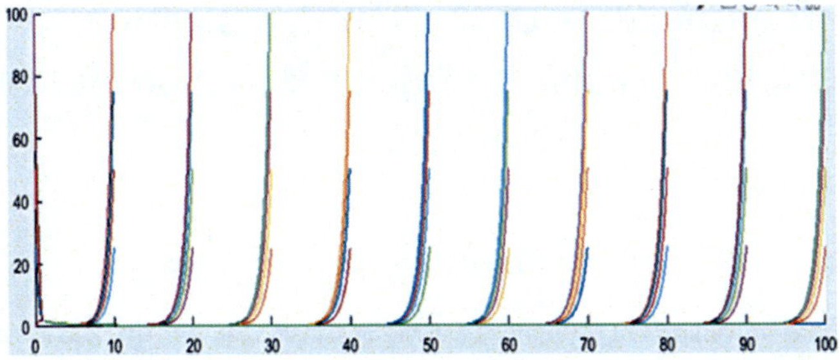

Fig. 5. The phase portrait of the regulator in the vicinity of the equilibrium state.

References

1. Milovanov, V.: Non-Equilibrium Socio-Economic Systems: Synergetics and Self-Organization, pp. 1–264. Editorial URSS, Moscow (2001). (in Russian)
2. Bautin, N., Leontovich, E.: Methods and Techniques of Qualitative Research of Dynamical Systems on the Plane, pp. 1–496. Nauka, Moscow (1976). (in Russian)
3. Kolesnikov, A.: Synergetic Methods of Managing Complex Systems. The Theory of System Analysis, pp. 1–240. Komkniga, Moscow (2006). (in Russian)
4. Dabney, J., Harman, T.: Mastering Simulink 4. Prentice Hall. Upper Saddle River, New Jersey 07458, pp. 1–400 (2001)
5. Malkin, I.: Theory of Motion Stability. Nauka, Moscow, pp. 1–532 (1966). (in Russian)
6. Bratishchev, A.: Mathematical theory of controlled dynamical systems. In: Introduction to Concepts and Methods, pp. 1–292. DSTU Publishing Center, Rostov-on-Don (2015). (in Russian)
7. Bratishchev, A.: On the characteristic polynomial of balance an autonomous system that has an attractive invariant manifold. Differ. Equ. Control Process. **2**, 15–23 (2017). (in Russian)
8. Zubov, V.: Stability of Movement. Lyapunov's Methods and Their Application, 2nd edn., pp. 1–232. Vysshaya shkola, Moscow (1984). (in Russian)

Identification of the Multiconnection Uncertainty Dynamic Objects Based on Neural Networks

S. M. Jafarov[1] and A. S. Aliyeva[1,2](✉)

[1] Azerbaijan State University Oil and Industry, Azadlig Avenue 20, AZ1010 Baku, Azerbaijan
[2] Institute of Control Systems of ANAS, B.Vahabzade Street 9, AZ1141 Baku, Azerbaijan

Abstract. In the proposed paper the suggested identification method affords to evaluate "black box" model of uncertain and non-stable dynamic objects with sufficient precision in any short time. One of the key advantages of the proposed method is letting to determine the adapted "black box" model while state variables of *m*-th order object are not measured and to create an adaptive control system while the derivatives of the structural variables are not measurable. "Black box" model neural networks (NN) for the two-dimensional 2-nd order object was trained according to error back propagation algorithm.

Keywords: Etalon transformers-filters · Neural network · Multiconnection uncertainty dynamic object · "Black box" model

1 Introduction

In novel manufacture processes the control objects work in intensity and under uncertainty. In the control systems of this kind of dynamic objects the requirements regarding to the quality are exacerbated. Therefore, to get a prior and posterior information on such objects, such as, obtaining (defining) dynamic characteristics, is necessary and actual.

So that, being known of the structure and values for the mathematical model parameters of the object are important in order to design control system with optimum quality index [1–6]. However, in some cases, the structure and parameters for the mathematical model of the object are unknown.

High-quality control of dynamic objects, which its uncertainty can be described as non-stationarity, could be achieved through adaptive systems [2–4]. However, in this kind of adaptive control systems state variables of the object should be fully observed or in other word, *n*-th order derivatives of the output variables for the object and itself must be measurable [2, 3].

But this type of mathematical models has sufficient nonlinearities and uncertainties [6–8]. For the experimental study of such models, it is necessary to observe (measure) the first, second and third order derivatives [6–8].

R. A. Aliev et al. (Eds.): ICSCCW 2021, LNNS 362, pp. 496–504, 2022.
https://doi.org/10.1007/978-3-030-92127-9_67

In the real objects it is not possible to measure the n-th order derivatives of input and output variables physically. Therefore, for the high-quality control of such uncertain dynamic objects, using the ethalon transformers, identification methods, tools of artificial intelligence-neural technology are necessary. For example, it is necessary to suggest more efficient methods and tools for designing intellectual and adaptive control systems without using of direct m-th order derivatives of real-state variables for the object [1, 5]. I.e. the identification method should be able to continuously evaluate the parameters of the mathematical model based on observations of input and output variables, or on the basis of the "black box" model, during the normal operation of an unknown object [9, 10].

In view of the foregoing, the present work proposes the identification method for the multidimensional- multiconnection dynamic objects with uncertainty according to neural technology and etalon transformers.

2 Statement of the Problem

Let's assume that, in generally, the movement of the dynamic object with uncertainty could be written by vector differential equations or by matrix transfer function:

$$\dot{y}(t) = \psi(\vec{a}(t), y(t), u(t)), \tag{1}$$

Here $y(t), u(t)$ - are n output and steering-input vectors, also are measurable (observable); $\dot{y}(t)$-$m - th$ order derivatives vector of the certain multidimensional-multiconnection dynamic object and it is not observed.

It is assumed, for multidimensional-multiconnection dynamic object the additivity is met. Then, (1) object could be written by operator equations as follows:

$$\left.\begin{aligned} Y_1(s) &= W_{11}(t,s)U_1(s) + W_{12}(t,s)U_2(s) + \cdots + W_{n1}(t,s)U_n(s) \\ Y_2(s) &= W_{21}(t,s)U_1(s) + W_{22}(t,s)U_2(s) + \cdots + W_{n2}(t,s)U_n(s) \\ &\vdots \\ Y_n(s) &= W_{n1}(t,s)U_{n1}(s) + W_{n2}(t,s)U_{n2}(s) + \cdots + W_{nn}(t,s)U_n(s) \end{aligned}\right\} \tag{2}$$

Here U_j, $\left(j = \overline{1,n}\right)$ j-th control input effect, $W_{ij}(t,s)$– are relevant transmission functions of the object through directly straight ($i = j$) and cross ($i \neq j$) channels, and transmission functions with $n \times n$ dimension are elements of matrix (3):

$$W_{ob}(t,s) = \begin{vmatrix} W_{11}(t,s) & W_{12}(t,s) & \ldots & W_{1n}(t,s) \\ W_{1n}(t,s) & W_{22}(t,s) & \ldots & W_{2n}(t,s) \\ \vdots & \vdots & & \vdots \\ W_{n1}(t,s) & W_{n2}(t,s) & \ldots & W_{nn}(t,s) \end{vmatrix} \tag{3}$$

Assume, the movement of the multidimensional- multiconnection dynamic object with uncertainty described by the transmission function of each $i, j = \overline{1, n}$ channel is as follows

$$W_{ij}(t, s) = {k_{ij}}\Big/{\sum_{l=0}^{m} a_{m-l,ij}(t)s^l} \tag{4}$$

Here k_{ij} , $a_{m-l,ij}(t)$ – are unknown non-stationary parameters of the object with uncertainty described by the (3) expressions of straightforward and cross effect channels.

According to the knowledges about identified dynamic object (our a prior information) we assume, values of $a_{l,ij}$, $l = \overline{1, m}$,k_{ij}, $i, j = \overline{1, n}$ parameters in (4) mathematical model could be limited, i.e.

$$a_{l,ijmin} \leq a_{l,ij}(t) \leq a_{l,ijmax}, \forall l \in L, L = \left[\overline{0, m}\right], \; k_{ij}, \; i, j = \overline{1, n}. \tag{5}$$

3 Solving of the Identification Problem Based on Neural Technology

Identification of the multidimension-multiconnection l dynamic object with uncertainty, i.e. in order to evaluate $a_{l,ij}$, $l = \overline{1, m}$, k_{ij}, $i, j = \overline{1, n}$ parameters, m-th order derivatives $y_j^l(t), j = \overline{1, n}, l = \overline{1, m}$ of the output variables must be measurable. However, in a case of considering, m-th order derivatives are not observed (measured). Therefore, using linear transformers-etalon filters [5, 8] of the measurable (observable) $y(t)$-output and $u(t)$-input vectors (signals) is proposed to evaluate (define) dynamic parameters $a_{l,ij}$, $l = \overline{1, m}$, k_{ij}, $i, j = \overline{1, n}$ of the object with uncertainty.

Let's assume, input variables vector ***u(t)*** are not periodic, i.e. they meet following conditions:

$$\boldsymbol{u(t)} \neq \mathbf{u}\big(t + T_p\big) - \infty < t < \infty, \; 0 < T_p < \infty, \tag{6}$$

(6), (2), (5) $y(t)$-output and $u(t)$-input variables vectors of the multiconnected dynamic object (1) with uncertainty and their derivatives $y_j^l(t)$, $u_j^l(t)$, $j = \overline{1, n}$, $l = \overline{1, m}$ are not periodic functions. Linear converters –(ethalon transformers)-$W_{FYj}(s)$ və $W_{FUj}(s), j = \overline{1, n}$ are added to the outputs of the each output and input for the object. m-th order derivatives $s^l Y_{Fj}(s)$, $s^l U_{Fi}(s)$, $l = \overline{0, m}$ of outputs for the ethalon filters $Y_{Fj}(s)$, $U_{Fj}(s)$ are observable (see Fig. 2).

Note that, for each of the outputs and inputs, $W_{FYj}(s)$ və $W_{FUj}(s), j = \overline{1, n}$ transmission function structures of the m-th order linear converters (ethlon filters) could be selected as follows:

$$W_{FYj}(s) = {Y_{Fj}(s)}\Big/{y_j(s)} = {1}\Big/{\sum_{l=0}^{m} d_{m-l,Fj}(t)s^l},$$

$$W_{FUi}(s) = {U_{Fi}(s)}\Big/{u_i(s)} = {1}\Big/{\sum_{l=0}^{m} d_{m-l,Fi}(t)s^l} \tag{7}$$

If we will equally ($W_{FYj}(s) = W_{FUi}(s) = W_F(s), j, i = \overline{1, n}$) and steadily select the structures and parameters of the transmission functions (7) for the linear converters (-electron filters) of the input and output variables, then using the operator model of the identification system (see Fig. 2 and 3) could be written as follows:

$$\left.\begin{array}{l} W_F(s)Y_1(s) = W_F(s)W_{11}(t,s)U_1(s) + \cdots + W_F(s)W_{n1}(t,s)U_n(s) \\ W_F(s)Y_2(s) = W_F(s)W_{21}(t,s)U_1(s) + \cdots + W_F(s)W_{n2}(t,s)U_n(s) \\ \vdots \qquad \vdots \qquad \vdots \qquad \vdots \\ W_F(s)Y_n(s) = W_F(s)W_{n1}(t,s)U_1(s) + \cdots + W_F(s)W_{nn}(t,s)U_n(s) \end{array}\right\}$$

If we consider $W_F(s)Y_j(s) = Y_{Fj}$, $W_F(s)U_i(s) = U_{Fi}$, then the relation between transformed variables Y_{Fj} and U_{Fi} could be described as follows:

$$Y_{Fj}(s) = W_{obij}(t, s)U_{Fi}, i, j = \overline{1, n} \tag{8}$$

Here $W_{obij}(t, s)$ - is transmission function "of j-th output relative to i–th input".

Thus, in a case of (6) conditions are met, state variables of the object are not observed, we have shown the identification possibility of the multidimensional dynamic object by measuring $s^l Y_{Fj}(s)$, $s^l U_{Fj}(s)$, $l = \overline{0, m}$ derivatives of the etalon transformers-filters, i.e. we have justified that.

If during the quasi-expiration period of the object needs parametric identification, then, $a_{l,ij}, l = \overline{1, m}, k_{ij}, i, j = \overline{1, n}$ could be evaluated by using the methods proposed in [5.8] or LSM (least squares method). However, in the synthesis and technical realization of adaptive control systems for multidimensional dynamic object the use of such models has some difficulties, and its practical efficiency may not be sufficient [1, 5, 8, 9].

But according to existing approach designed mathematical models have enough nonlinearity and uncertainties or nonstationarity [8–10]. Control systems that have the ability to change the object parameters should be intellectual or adaptive. Adaptive control system controllers should be adapted according to object information in either online or offline mode through the correction of the adjustment parameters. In such systems, to ensure optimization of control system and its stability, computer simulation of intelligent ACS by using –application of existing tools is important issue and its effective solution is actual [9].

In the article, according to the above mentioned, the determination issue of the "black box" model Neural Network technology of the multidimensional dynamic object with uncertainty and linear transformers under uncertainty of the structure and parameters of mathematical model is solved.

In the most identification method of dynamic objects $\psi(\cdot)$-structure of (1) mathematical model is accepted as the total sum of certain functions (for example, linear and nonlinear functions) [1–10]. If initial information and investigations about dynamic objects is not sufficient and the structure of identified mathematical model is not selected by the right way (sufficiently selected), then the solution for the synthesis of control will be complex and inefficient [9, 10].

Real dynamic model of the object (1a) could be described as formal reduction form, i.e.

$$y(t) = F(a(t), \dot{y}(t), u(t)), y \subset R^n, u \subset R^n \quad (9)$$

After each input and output variable has been converted through the $W_F(s)$ ethalon transformers (linear transformers), by measuring $s^l Y_{Fj}(s), s^l U_{Fj}(s), l = \overline{0, m}$ derivatives it is possible to define "black box" NN of the multiconnected dynamic object with uncertainty.

Let us assume that, the "black box" type mathematical model of object based on neural network could be defined by NN with r intermediate layers through the each i-th input j—th output channel:

$$Y_{Ni} = \phi_i\left(w_{ij,r}, \dot{y}(t), u(t)\right), w_{ij,r} \in W_{ij,r}, i, j = \overline{1, n}, r = \overline{1, g}, \quad (10)$$

Here $w_{ij} \in W_{ij}$, $i \in [1, n]$, $r \in \left[1, g\right]$, $i-$ th are weight coefficient of the neurons in the $g-$th intermediate layers through the $j-$ th output channel and can get values in the $W_{ij,r}$ region.

The defining issue of "black box" (neural network) model of dynamic object with uncertainty could be formulated as follows:

It is needed to define "black box" model with NN of the object which is observed (measured) all state variables of real dynamic object $Y_F(t) = [Y_{F1}(t), Y_{F2}(t), \ldots, Y_{Fn}(t)]$ and $U_F(t) = [U_{F1}(t), U_{F2}(t), \ldots, U_{Fn}(t)]^{\mathrm{T}}$ input effects, i.e. as (8) model, that the square of the difference (error) between outputs of NN and of the real object (state variables) should be minimum for $t_k = T_0 k$ ($T_0 = const, k = 1, 2, 3, \ldots, K$):

$$\begin{aligned} \mathbb{E} &= \frac{1}{2}\sum_{k=1}^{K} (E^k)^2 = \frac{1}{2}\sum_{k=1}^{K} [Y_F(t_k) - Y_N(t_k)]^2 \\ &= \frac{1}{2}\sum_{k=1}^{K}\sum_{i=1}^{n}\left[Y_{Fi}^k - \phi_i\left(w_{ij}, \dot{Y}_{Fi}^k, U_{Fj}^k\right)\right]^2 \Longrightarrow \min \end{aligned} \quad (11)$$

If one model, i.e. $i = 1$, is enough and if possible value-accuracy of error ε is given, then for (11) optimization could be satisfied that $\lim\limits_{k \to \infty} \frac{1}{2}\sum_{k=1}^{K} (E^k)^2 = \varepsilon$ is less.

The modeling system as "black box" of the multiconnected dynamic object with real uncertainty based on NN could be described as Fig. 1:

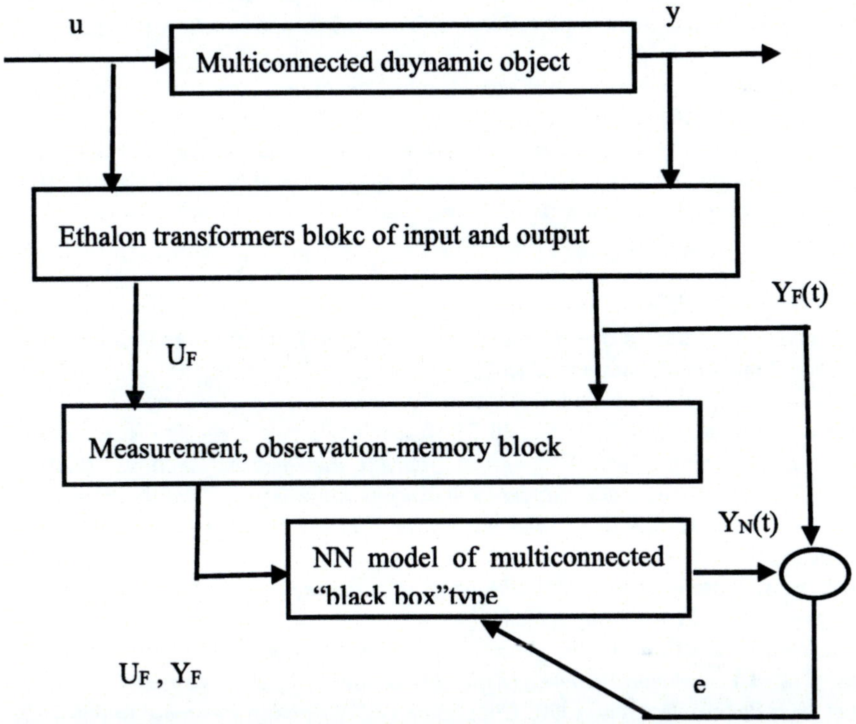

Fig. 1. The structure of modeling system as "black box" based on NN of the multiconnected dynamic object with uncertainty which transformed sate variables are observed

The solution of above formulated identification issue is $\boldsymbol{Y_N} = \phi\big(\boldsymbol{w}, \dot{\boldsymbol{Y}}_{\boldsymbol{F}}(t), \boldsymbol{U_F}(t)\big)$ function. These solutions are known as input-output pairs $(\boldsymbol{Y}_{\boldsymbol{F}}^1, \boldsymbol{Y}_N^1)$, $(\boldsymbol{Y}_{\boldsymbol{F}}^2, \boldsymbol{Y}_N^2)$, …, ($\boldsymbol{Y}_{\boldsymbol{F}}^{\boldsymbol{K}}, \boldsymbol{Y}_N^{\boldsymbol{K}}$) (formulated from the measurements). Identification process, i.e. NN training directly synthesize such $\phi(\cdot)$ function in $\boldsymbol{n}$ number, that could be close to $\boldsymbol{F}(\cdot)$ function realized by the object. Thus, training Neural Networks is becoming multidimensional optimization issue.

Since (11) function could be arbitrary, in general, training of $\boldsymbol{n}$ Neural Network is multiextremal nonconvex optimization issue. A number of iterative algorithms given in [8, 9] could be used to solve the above-mentioned (11) optimization problem. Currently the most widely used method of NN training (minimizing the mean square (quadratic) error - (11) function) is an iterative gradient algorithm.

This training algorithm, mentioned in most scientific literature [2, 9], was called error back propagation. Therefore, there is no need to give detailed explanation on error back propagation method. However, the identification of the multiconnected dynamic object with uncertainty as above formulated n-number "black box" for the mechatronic device realization based on Neural Networks Toolbox to provide "M" language program of MATLAB and training procedures with an explanation as follow is considered purposeful.

Through the single channel in order to determine "black box" model with explanation of NN training procedures can be summarized as follows.

In subsystems - object output, etalon transformers - filter variables input and output in input and output, Δt (Δt = 0.002 s) intervals in K (K = 4000, researcher selects) number, observation measurements, i.e., input matrices

$\boldsymbol{P} = [\boldsymbol{X}] = \Big[\boldsymbol{u}_1; \boldsymbol{u}_2; \boldsymbol{s}^2\boldsymbol{Y}_{F1}; \boldsymbol{s}\boldsymbol{Y}_{F1}; \boldsymbol{y}_2; \boldsymbol{s}^2\boldsymbol{Y}_{F2}; \boldsymbol{s}\boldsymbol{Y}_{F2}; \boldsymbol{s}^2\boldsymbol{U}_{F1}\boldsymbol{s}^2\boldsymbol{U}_{F1}; \boldsymbol{s}\boldsymbol{U}_{F1}; \boldsymbol{U}_{F1}; \boldsymbol{s}^2\boldsymbol{U}_{F2};$

$\boldsymbol{s}\boldsymbol{U}_{F2}; \boldsymbol{U}_{F2}; \Big]$ are formed:

Observational matrices of state variables $y_1^k; y_2^k$ $k = \overline{1, K}$ ($K = 4000$, *researcher selects*) of the real muliconnected object

$\boldsymbol{T} = [\boldsymbol{y}] = [\boldsymbol{y}_1; \boldsymbol{y}_2]$ are formed:

net = newff(min max(P), [300, 300, 2], {'purelin', 'logsig', 'tansig', 'traingd'}); *Creation and training of first NN (with appropriate transmission functions 'tansiq', 'logsiq', 'purlin') named **d** with 3 layer (300 neurons in two layers, 2 neurons in output layer). Number of neuron layers and etc. parameters are selected iteratively by the constructor).*

net.trainParam.goal = 0.001; *-the value of accuracy omission error of the NN training-* $\boldsymbol{e}^k = (\boldsymbol{e}_1^k$ *(heuristically selected by the constructor).*

net.trainParam.show = 50;- *from one of the fifty examples Param the first NN output* $y_1^k, k = \overline{1, K}$ *should be shown (defined heuristically by the constructor).*

net.trainParam.epoch = 1000; *Circulation of NN training -**'epocha'** (number of cycles selected heuristically by the constructor).*

net.train(net,P,T); *Launching NN training procedure.*

view(net);- *automatic installation of a "black box" model object in the Simulink and description of the flow chart (See* Fig. 2).

gensim(net); **ordisplay(net)**; – Decription of information on NN in MATLAB:- method of training (for example, training method "Gradient Decent"), criteria ("mse"), number of iterations ("epoch") and others [2, 9].

Dependence of the training error on the number of iterations (the "epoch"), the intended target - the "thousand" value of the error (a), and the description of the "black box" as subsystem (b) in the Simulink was given in Fig. 3.

After the training of NN the graphs (a, b) of its outputs (for example, $y_{N1}(t) = y_{N1}(kT_0) = y_{N1}^k, k = \overline{1, 4000}, T_0 = 0.01s$) and the transition processes of the real object (for example, $y_1(t) = y_1^k, k = \overline{1, 4000}, T_0 = 0.01$s) as well as, for $u_1(t) = 1[t] = 1[kT_0]$, $u_2(t) = 0$ values of input-control effects was shown in Fig. 3.

According to comparative analysis of the transient processes in Fig. 3, $\boldsymbol{y_N(t)}$ -NN output and $\boldsymbol{y(t)}$ derivative is exactly same as nonlinear dynamic object with uncertainty, i.e. $y(t) \cong y_N(t)$. Thus, in the issue of control influence or regulator synthesis, the possibility of using the trained NN ("black box" model of the real dynamic object) for the dynamic object with uncertainty in "offline" mode has been confirmed.

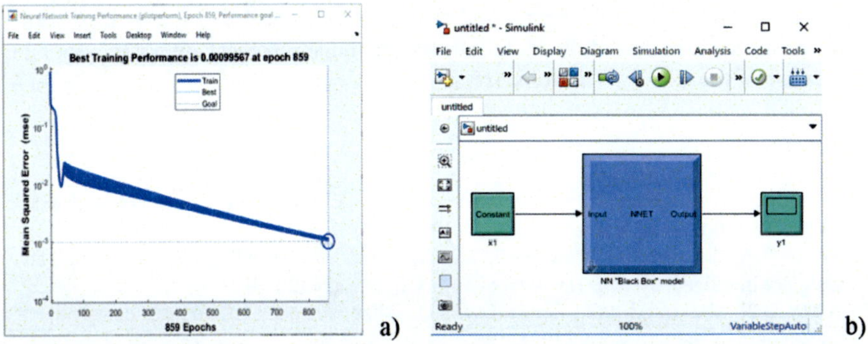

Fig. 2. Dependence graph for the cycles (epochs) of the training error and description of NN as "black box" in Simulink (b)

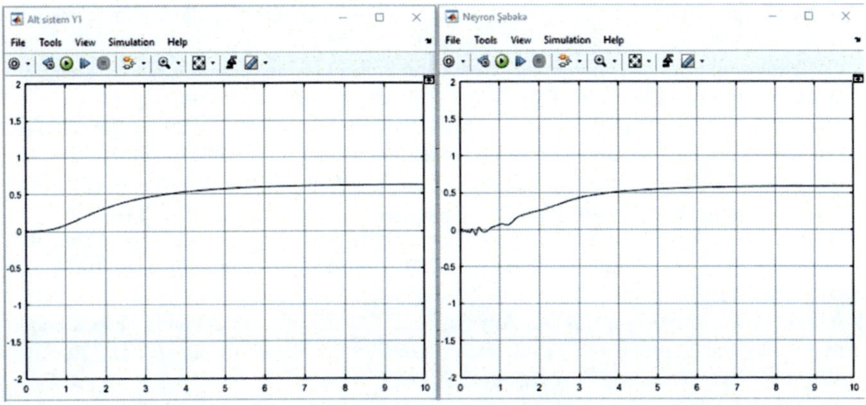

Fig. 3. The first transition processes Y_{N1} of the first NN with Y_{F1} output signal of the first ethalon filters in object output

4 Conclusion

A theoretical method for solving the determining problem of "black box" model for the multiconnected dynamic object with uncertainty based on neural technology and etalon transformers-filters, as well practical realization in MATLAB have been developed when information is not sufficient and state variables are not measurable. It has been proved that the modeling of real multidimensional uncertainty dynamic objects based on transformations of input and output variables is correct and efficient. "Black box" model NN for the two-dimensional 2-nd order object was trained according to error back propagation algorithm, and in order to overlap measured state variables vectors U_{Fi}, Y_{Fj}, $(i, j = 1, 2)$ of the 2-nd order ethalon transformers with state variables of NN output $y_N(t)$ the structure of NN (layers-3), number of the neurons in the layer (300, 300, 1), transmission functions, accuracy of training (0.001) and etc. parameters were identified.

One of the advantages of the proposed method is that it affords parametric identification when state variables of the multidimensional dynamic object are not measurable and creation of the adaptive control system.

References

1. Djafarov, S.M.: Design of adaptive control system for one class of dynamic objects with delay (in Russian). Proc. USSR Acad. Sci. **2**, 81–190 (1979)
2. Aliyev, R.A., Jafarov, S.M., Babayev, M.C., Huseynov, B.Q.: Principles of construction and design of intelligent systems (in Azerbaijan). Nargiz, Baku (2005)
3. Petrov, B.N., Rutkovski, V.Y., Zemlyakov, S.D.: Adaptive control of non-stationary objects (in Russian). Nauka, Moscow (1981)
4. Jafarov, P.S., Jafarov, S.M., Zeynalov, E.R.: Design of control system described by fuzzy differential equation. In: Fifth International Conference on Soft Computing, Computing with Words and Perceptions in System Analysis Decision and Control, pp. 298–302, Famagusta, North Cyprus, 2–4 September (2009)
5. Jafarov, S.M., Jafarov, P.S., Zeynalov, E.R., Zeynalova, L.M.: İdentification of fuzzy models of nonlinear dynamic objects with observability of input and output. In: Fifth International Conference on Soft Computing with Words and Perceptions in System Analysis, Decision and Control, Famagusta, North Cyprus, 2–4 September, pp. 302–305 (2009)
6. Jafarov, P.S., Zeynalov, E.R., Jafarov, S.M., Mustafayeva, A.M.: The analytical method of synthesis of a controller with a fuzzy TS model for control of a filexible joint robot arm. In: Sixth International Conference on Soft Computing with Words and Perceptions in System Analysis, Decision and control. Antalya, Turkey, 1–2 September, pp. 107–113 (2011)
7. Jafarov, S.M., Aliyeva, A.S.: Synthesis of intelligent control system with coordinate and parametric feedback. Procedia Comput. Sci. **120**, 554–560 (2017). https://doi.org/10.1016/j.procs.2017.11.278
8. Jafarov, S.M., Mustafayeva, A.M., Aliyeva, A.S.: Identification of dynamic objects under uncertainty and non-stationarity using Etalon transformers. Procedia Comput. Sci. **120**, 561–567 (2017). https://doi.org/10.1016/j.procs.2017.11.279
9. Aliyeva, A.S., Jafarov, S.M., Ismayilzade, S.E.: Modeling and intelligent control of dynamic objects with uncertainty based on NN. Sci. Pedagogical News Odlar Yurdu Univ. (Azerbaijan). **51**, 19–21 (2019)
10. Aliyev, T.: Noise Control of the Beginning and Development Dynamics of Accidents. Springer, Cham (2019). https://doi.org/10.1007/978-3-030-12512-7

Application of Digital Twin Theory for Improvement of Natural Gas Treatment Unit

Nodirbek Yusupbekov[1], Farukh Adilov[2], and Arsen Ivanyan[2(✉)]

[1] Tashkent State Technical University, Tashkent, Uzbekistan
[2] LLC "XIMAVTOMATIKA", Tashkent, Uzbekistan
{Farukh.Adilov,Arsen.Ivanyan}@himavtomatika.uz

Abstract. This paper describes fundamental principles of Digital Twins theory and provides exact investigation results in application of Digital Twins theory in upstream branch of oil and gas industry, namely based on example of natural gas treatment plant's performance increase. As one of key process units of natural gas treatment which allows to implement most powerful functions of digital twin gas sweetening unit with set of membranes is considered as object of investigation. On the base of membrane technology manipulated variables are defined as inputs to digital twin model. Some theoretical results as well as real references of model's engine calculations are reflected in the paper. Details of technical dashboards to visualize calculated results of running model based on manipulated variables are presented including monthly key performance indicators report dashboard, process flow diagram dashboard and high-level management dashboard. Paper also demonstrates data flow between digital twin model and real process unit and also inside digital twin model.

Keywords: Digital Twins theory · Key performance indicator (KPI) · Artificial Intelligence (AI) · Machine Learning (ML) · Hydrocarbons (HC) Loss · Model accuracy · Dashboard

1 Introduction

The importance of the issue of creating "digital twins" is confirmed by the increasing number of publications on this topic, but all of them are related with the creation of models of individual units of equipment, devices or compact package units [1–4]. Creating a full-fledged twin of a complex technological process, which involves large-scale industrial installations, involves the creation of a mathematical model and associated software and network "cloud" infrastructure of an incredible level of complexity [5, 6].

From another side there is number of traditional challenges for operation of plant like:

- Longer, harsher operating cycles between turn arounds, at peak performance
- Differing outcomes across shifts, units, plants and enterprise
- Lack of integrated insights across the operating spectrum

R. A. Aliev et al. (Eds.): ICSCCW 2021, LNNS 362, pp. 505–512, 2022.
https://doi.org/10.1007/978-3-030-92127-9_68

Such multi-factored set of permanent challenges which flexibly changes depending on different real process modes of operation definitely cannot be resolved by set of standardized off-line solutions even the root causes of several typical problems are well-known. It requires applying of much more tightly interacted solution which could "live" together with real plant and ensure the data exchange with plant in real-time mode. Such online solutions are being developed under theory of "digital twins" which explores and reflects such scientific directions as advanced monitoring, analytical and predictive capabilities, allowing 24-h online monitoring of plant data, and ongoing operational health checks and recommendations to close performance gaps [7].

Digital twin is a digital model of physical assets, processes and systems that can be used to optimize the operation, maintenance and reliability of physical assets, systems and manufacturing processes.

Digital twin incorporates technologies like Artificial Intelligence (AI), Machine Learning (ML), data science and software analytics with information to make live digital simulation models which are being updated and changed with any change in their physical counterpart.

Digital Twins theory becomes more and more important for industrial enterprises because:

- It unifies existing data silos into a virtual entity.
- It federates data across different applications to drive end-to-end integration.
- It ensures understanding how a plant responds to changes to drive increased reliability and optimization using process simulation technology.
- It enables those with strong domain expertise to diagnose and troubleshoot.

2 Description of Scientific-Technical Solution

To demonstrate all capabilities and new level of improvement ways using digital twin technology we investigated relatively simple technology from oil and gas upstream sector – gas sweetening process unit of natural gas treatment plant.

Carbon dioxide (CO_2) and hydrogen sulphide (H_2S) are common impurities in natural gas. These impurities need to be removed to achieve required concentration levels to meet each specific process requirements.

The most common methods for removal of CO_2 and/or H_2S from the gas is via amines, physical solvents, membranes or adsorbents. The choice of technology to be used depends on the levels of impurities to be removed. Membranes offer specific advantages over other basic processes including [8]:

- Low operator requirement
- Elimination of solvents from plant environment
- Simple operation
- Ideal for onshore and offshore applications
- No moving parts

Gas sweetening system is one of core technologies and optimization of this process has big influence on the total effectiveness of natural gas treatment plant [9].

To create digital twin of gas sweetening unit the informational-control system (ICS) of this unit should be connected to cloud. In the cloud digital twin is deployed and hosted in analytical platform and leverages a digital twin of the membrane unit (which is most complicated part of gas sweetening unit) to create a tuned model of the membrane.

The plant data from process unit are collected to cloud historian system and the solution performs gross error detection and parse the information into a project specific asset model. The data is then reconciled and analyzed using the tuned model. The model is periodically retuned based on the reconciled data to ensure continued accuracy over time.

The model is used to infer unmeasured parameters, forecast performance and determine optimal operating scenarios.

The purpose of the solution is to enable minimizing of hydrocarbon loss (or maximize hydrocarbon recovery) from the CO_2 removal process through improved execution and optimal control of the production process.

The solution displays the key variables that impact the performance and reliable operation of the unit. The solution provides targets for the key manipulated variables to recommend optimal operating conditions while maintaining the constraints of the process. It is forecasted the performance of each bank over year and quantifying the difference in performance against the fresh bank in same service. The system notifies the user when the difference has exceeded certain threshold value to help identify when and which banks should be replaced.

The following variables are manipulated using digital twin model to optimize the objective:

- Flow per bank
- Feed temperature
- Permeate pressure
- Number of banks online
- Feed pressure

Baseline for model accuracy is based on the reconciled data set. Reconciled data will be calculated by performing a material balance linear reconciliation across the entire membrane system.

The model accuracy is defined as:

$$1 - ABS(Model\ Prediction - Reconciled\ Value)/(Reconciled\ Value)\% \quad (1)$$

The accuracy of the solution will be judged on the ability of the model to predict the hydrocarbon loss in the permeate stream based on the definition of accuracy expressed above. The accuracy of the predictions and recommendations will be subject to the accuracy of the incoming data. Only reconciled data that passes the global gross error detection test (based on a 95% confidence) will be considered to calculate accuracy [10, 11].

Hydrocarbon loss is defined as:

$$(Permeate\ flow) * (100\% - \%CO2 - \%N2\ in\ permeate) \quad (2)$$

Current HC Loss can be expressed as:

$$\begin{aligned} &\textit{Current HC Loss} = (\textit{Reconciled Permeate flow}) \\ &*(100\% - \textit{Reconciled Permeate \%CO2} - \textit{Reconciled Permeate \%N}) \end{aligned} \quad (3)$$

Optimal HC Loss can be expressed as:

$$\begin{aligned} &\textit{Optimal HC Loss} = (\textit{Optimal Permeate flow}) \\ &*(100\% - \textit{Optimal Permeate \%CO2} - \textit{Optimal Permeate \%N}) \end{aligned} \quad (4)$$

Baseline HC Loss can be expressed as:

$$BaselineHCLoss = \frac{(a * ResidueFlow + b)}{100} * \left(\frac{\left(\frac{Feed\%CO2 - 22.5}{Feed\%CO2} \right)}{\left(\frac{28-22.5}{28} \right)} \right)^{c} * ResidueFlow, \quad (5)$$

where coefficients a,b,c are based on historical data from data historian system of plant and correlating Feed CO_2, Residue Flow and Hydrocarbon Loss.

Current Savings can be expressed as:

$$\begin{aligned} &\textit{Current Savings} = \textit{MAX}\,((\textit{Avg Baseline HC Loss}[1H] - \textit{Avg Current HC Loss}[1H]) \\ &* \textit{Heating Value of HC} * \textit{Sales Gas Price},\ 0) \end{aligned} \quad (6)$$

Optimal Savings can be expressed as:

$$\begin{aligned} &\textit{Optimal Savings} = \textit{MAX}\,(\textit{MAX}\,((\textit{Avg Baseline HC Loss}[1H] \\ &-\textit{Avg Optimal HC Loss}[1H]) * \textit{Heating Value of HC} \\ &* \textit{Sales Gas Price},\ (\textit{Avg Current HC Loss}\,[1H] \\ &- \textit{Avg Optimal HC Loss}[1H]) * \textit{Heating Value of HC} \\ &* \textit{Sales Gas Price}),\ 0) \end{aligned} \quad (7)$$

Forecasted Savings can be expressed as:

$$\begin{aligned} &\textit{Forecasted Savings} = \textit{Savings Forecast Avg}\,[\textit{Savings YTD}/\textit{Day of the Year}] \\ &*\textit{Num Days Remaining in Year} * 0.98 + \textit{Savings YTD} \end{aligned} \quad (8)$$

Current CO_2 Emissions in the model are assumed as:

$$\begin{aligned} &\textit{Current Emissions} = (\textit{Reconciled Permeate flow} \\ &* (C1[Rec] * 1 + C2[Rec] * 2 + C3[Rec] * 3 \\ &+ IC4[Rec] * NC4[Rec] * 4 + IC5[Rec] * 5 \\ &+ NC5[Rec] * 5 + C6[Rec] * 6 + N2[Rec] * 1) \\ &* 1/0.8364 * 0.98 * 44/100) * 1 \\ &+ (\textit{Reconciled Permeate flow} * C1[Rec] * 1/0.8364 \\ &* 0.02 * 16/100) * 25 \end{aligned} \quad (9)$$

Optimal CO_2 Emissions in the model are assumed as:

$$\begin{aligned} Optimal\,Emissions &= (Optimal\,Permeate\,flow \\ &* (C1[Opt] * 1 + C2[Opt] * 2 + C3[Opt] * 3 \\ &+ IC4[Opt] * NC4[Opt] * 4 + IC5[Opt] * 5 \\ &+ NC5[Opt] * 5 + C6[Opt] * 6 + N2[Opt] * 1) \\ &* 1/0.8364 * 0.98 * 44/100) * 1 \\ &+ (Optimal\,Permeate\,flow * C1[Opt] * 1/0.8364 \\ &* 0.02 * 16/100) * 25 \end{aligned} \quad (10)$$

Baseline CO_2 Emissions in the model are assumed as:

$$\begin{aligned} Baseline\,Emissions &= (ga*Feed\,\%\,CO2 + gb) \\ &* \left(gc * Residue\,Flow + gd * \sqrt{Residue}\,Flow\right) + ge, \end{aligned} \quad (11)$$

where ga, gb, gc, gd, ge coefficients are based on historical data from data historian system of plant and correlating Feed CO_2, Residue Flow and Greenhouse Gas Emissions.

Forecasted Emission Reductions in the model are assumed as:

$$\begin{aligned} &Forecasted\,Emission\,Reductions \\ &= Emission\,Reductions\,Forecast\,Avg\left[Emissions\,Reductions\,YTD/Day\,of\,the\,Year\right] \\ &*Num\,Days\,Remaining\,in\,Year + Emissions\,Reductions\,YTD \end{aligned} \quad (12)$$

To reflect functionality of digital twin and its' interaction with real plant specially programmed dashboards are used.

Technical dashboard displays to process engineers and operators key performance indicators of the gas sweetening process. It includes:

- Trend of current hydrocarbon loss and optimized hydrocarbon loss (MMSCFD)
- Trend of current production/nomination rate (MMSCFD).
- Current Operating Conditions vs Optimal/Recommended conditions
- Current performance/Conditions vs Limits
- Current and Model Predicted Conditions

Monthly KPI Report Dashboard provides a monthly summary of the overall unit performance.

Process Flow Diagram Dashboard provides a process flow diagram of the CO_2 removal membrane system. The key measured (reconciled) values as well as important non-measured, model predicted values are displayed in the PFD Dashboard. In addition, the PFD Dashboard also shows the recommended values of the manipulated variables from the optimizer run results.

High Level Management Dashboard provides a high-level view of the current system performance, provides visibility into the potential and actual benefit gains and provides recommendations on when to changeout a bank of elements. It includes:

- Current Hydrocarbon Loss
- Post Implementation Performance Gains
- Greenhouse Gas Emissions
- Post Implementation Greenhouse Gas Emission Reductions
- Operating Status
- Performance Rating
- Changeout Recommendation

Overall data integrated architecture based on real reference plant is illustrated in Fig. 1 below.

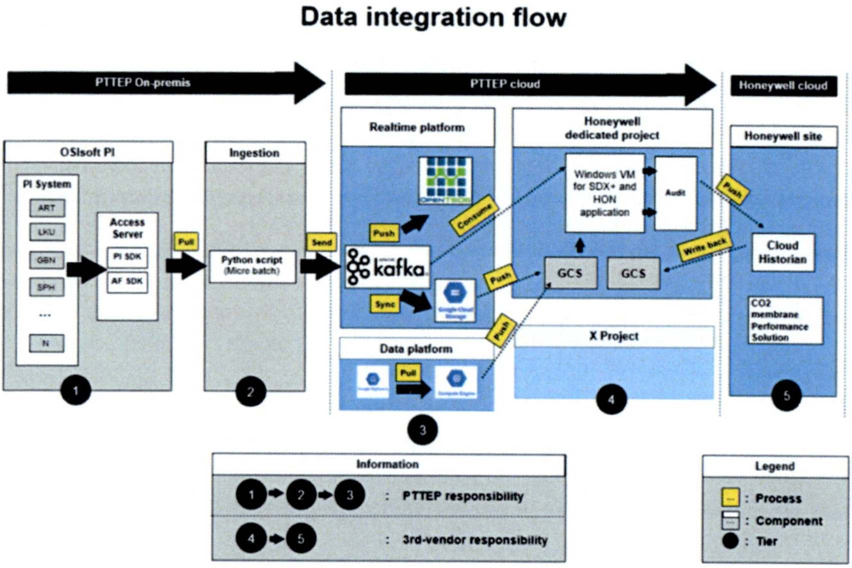

Fig. 1. Data integration flow between real natural gas treatment plant and Digital Twin in cloud.

3 Conclusion

Implementation of digital twin technology to gas sweetening process unit of natural gas treatment plant ensures achievement of following targets:

1. Minimizing hydrocarbon loss via permeate stream.
2. Minimizing annual cost of membrane bank operations.
3. Evaluation membrane performance and identification of optimal maintenance/replacement schedule.

This research is just one of examples how digital twin model works in conjunction with real process. So even complexity of the process unit chosen for this research is not so high it is obviously demonstrated that based on digital twin technology the efficiency of the process can be raised to a new level of optimization, where the standard methods of operated industrial facility optimization do not work anymore.

References

1. Varshney, S., Rohit, S.S., Sunidhi, Jamkhandi, A.G., Thanya, D., Parathodiyil, M.: Cyber-Physical Control and Virtual Instrumentation. In: Auer, M., Ram, B.K. (eds.) Cyber-physical Systems and Digital Twins. Lecture Notes in Networks and Systems, vol. 80, pp. 19–27. Springer, Cham (2019). https://doi.org/10.1007/978-3-030-23162-0-2
2. Redelinghuys, A.J.H., Kruger, K., Basson, A.: A six-layer architecture for digital twins with aggregation. In: Borangiu, T., Trentesaux, D., Leitão, P., Giret Boggino, A., Botti, V. (eds.) Service Oriented, Holonic and Multi-agent Manufacturing Systems for Industry of the Future. SOHOMA 2019. Studies in Computational Intelligence, vol. 853, pp. 171–182. Springer, Cham (2019). https://doi.org/10.1007/978-3-030-27477-1-13
3. Nagar, S.V., Chandrashekar, A.C., Suvarna, M.: Optimized additive manufacturing technology using digital twins and cyber physical systems. In: Auer, M., Ram, B.K. (eds.) Cyber-physical Systems and Digital Twins. Lecture Notes in Networks and Systems, vol. 80, pp. 65–73. Springer, Cham (2019). https://doi.org/10.1007/978-3-030-23162-0-7
4. Xie, F., Zhou, Z.: Discussion on the application of digital twins in the wear of parts. In: Liu, Q., Liu, X., Li, L., Zhou, H., Zhao, H.H. (eds.) Proceedings of the 9th International Conference on Computer Engineering and Networks. Advances in Intelligent Systems and Computing, vol. 1143, pp. 145–153. Springer, Singapore (2020). https://doi.org/10.1007/978-981-15-3753-0-15
5. Yusupbekov, N.R., Abdurasulov, F.R., Adilov, F.T., Ivanyan, A.I.: Application of cloud technologies for optimization of complex processes of industrial enterprises. In: Aliev, R., Kacprzyk, J., Pedrycz, W., Jamshidi, M., Sadikoglu, F. (eds.) 13th International Conference on Theory and Application of Fuzzy Systems and Soft Computing – ICAFS-2018. Advances in Intelligent Systems and Computing, vol. 896, pp. 852–858. Springer, Cham (2019). https://doi.org/10.1007/978-3-030-04164-9
6. Yusupbekov, N., Abdurasulov, F., Adilov, F., Ivanyan, A. Application of advanced process control technologies for optimization of polymers production processes. In: Aliev, R., Kacprzyk, J., Pedrycz, W., Jamshidi, M., Babanli, M., Sadikoglu, F. (eds.) 10th International Conference on Theory and Application of Soft Computing, Computing with Words and Perceptions – ICSCCW-2019. Advances in Intelligent Systems and Computing, vol. 1095, pp. 345–351. Springer, Cham (2020). https://doi.org/10.1007/978-3-030-35249-3
7. Honeywell Process Solutions: Real-Time Process Performance Monitoring: Powerful Analytic Tool Enables Smart Operations. Whitepaper, WP-18-02-ENG. Honeywell Process Solutions, Houston (2018). https://www.honeywellprocess.com/en-US/pages/default.aspx
8. He, X.: Membranes for natural gas sweetening. In: Drioli, E., Giorno, L. (eds.) Encyclopedia of Membranes, pp. 44–72. Springer, Heidelberg (2016) https://doi.org/10.1007/978-3-642-40872-4_1334-6
9. Matvienko, E.V., Adilov, F.T., Ivanyan, A.I.: Intelligent information technologies for integrated management systems of enterprises with a complex scheme of gas-extraction and processing. J. Multimedia Inf. Syst. **2**(3), 249–254 (2015). https://doi.org/10.9717/JMIS.2015.2.3.249

10. Crowe, C.M., Garcia Campos, Y.A., Hrymak, A.: Reconciliation of process flow rates by matrix projection. Part I Linear Case. J. Am. Inst. Chem. Eng. (AIChE) **29**, 881–888 (1983). https://doi.org/10.1002/aic.690290602
11. Crowe, C.M.: Reconciliation of process flow rates by matrix projection. Part II Nonlinear Case. J. Am. Inst. Chem. Eng. (AIChE) **32**, 616–623 (1986) https://doi.org/10.1002/aic.690320410

Determination of the Boundaries of Solid Solutions in the $MnTe$-Sb_2Te_3 and $SnTe$-Sb_2Te_3 Systems

E. N. Orujlu[1], A. E. Seidzade[1], A. N. Mammadov[1,2](✉), D. B. Tagiev[1], and M. B. Babanly[1]

[1] Nagiyev Institute of Catalysis and Inorganic Chemistry of ANAS, Baku, Azerbaijan
[2] Azerbaijan Technical University, Baku, Azerbaijan

Abstract. The boundaries of stable and metastable states of solid solutions of the $MnTe$-Sb_2Te_3 and $SnTe$-Sb_2Te_3 systems were determined by solving the thermodynamic equations of phase equilibria using the multipurpose genetic algorithm (MGA). As initial information, we used a small amount of experimental data from DTA and X-ray diffraction analysis, and thermodynamic functions of formation of MnTe, SnTe, Sb_2Te_3, $MnSb_2Te_4$, $MnSb_4Te_7$ $SnSb_2Te_4$, and $SnSb_4Sb_7$ compounds. The temperature-concentration dependences of the Gibbs energy of the formation of solid solutions from binary compounds, as well as from simple substances in quasi-binary sections, were determined using a subregular model of solutions of nonmolecular compounds as applied to telluride systems. The coordinates of the boundaries of metastable and stable solid solutions, in contrast to the entropy, are not very sensitive to the error in determining the enthalpy of formation of binary and ternary tellurides of manganese, tin, and antimony. It was revealed that the boundaries of β (Sb_2Te_3) solid solutions in $MnTe$-Sb_2Te_3 system extend to 10 mol% MnTe, and to 20 mol% SnTe in $SnTe$-Sb_2Te_3 system. γ ($MnSb_2Te_4$), δ ($MnSb_4Te_7$), γ ($SnSb_2Te_4$) and δ ($SnSb_4Te_7$) solid solutions are stable in the 50–55 mol% Sb_2Te_3 and 67–72 mol% Sb_2Te_3 concentration ranges, respectively. Metastable solid solutions areas based on the $SnSb_2Te_4$ and $SnSb_4Te_7$ compounds have been identified.

Keywords: Solid solutions · MnTe(SnTe)-Sb_2Te_3 systems · Thermodynamics · Multipurpose genetic algorithm

1 Introduction

Ternary tetradymite-like compounds $A^{IV}B^{V}{}_2Te_4$, $A^{IV}B^{V}{}_4Te_7$, $A^{IV}B^{V}{}_6Te_{10}$, etc., formed in the $A^{IV}Te$-$B^{V}{}_2Te_3$ (A^{IV}-Mn, Ge, Sn, Pb; BV-Sb, Bi) systems, are thermoelectric materials with low thermal conductivity [1]. These compounds belong to a new generation of inorganic substances possessing topological insulator properties and are very promising for use in spintronics and the creation of new devices for medicine and security systems [2]. Among them, manganese contained layered phases are a particularly interesting research platform for studying the interplay between magnetism and

R. A. Aliev et al. (Eds.): ICSCCW 2021, LNNS 362, pp. 513–521, 2022.
https://doi.org/10.1007/978-3-030-92127-9_69

topology [3, 4]. More recently, this situation becomes more interesting with the discovery of the first antiferromagnetic topological insulator – $MnBi_2Te_4$ [5].

In [6, 7] works, it is reported that ternary $SnSb_2Te_4$, $SnSb_4Te_7$, $MnSb_2Te_4$, and $MnSb_4Te_7$ compounds are formed in MnTe-Sb_2Te_3 and SnTe-Sb_2Te_3 systems. There are also solid solutions based on these ternary, as well as binary MnTe, SnTe, and Sb_2Te_3 compounds. However, up to date, the boundaries of the stable and metastable solid solutions in both MnTe-Sb_2Te_3 and SnTe-Sb_2Te_3 systems have not been determined.

The search for an optimal thermodynamic model of phase equilibria in quasi-binary systems is a difficult task in materials science due to the complex interaction of components in solid and liquid alloys. For thermodynamic optimization of the phase diagrams of binary and multicomponent systems, the Multipurpose Genetic Algorithm (MGA) was successfully applied in [8–13].

The aim of this work is thermodynamic calculation and modeling of the solid solution boundaries in the MnTe-Sb_2Te_3 and SnTe-Sb_2Te_3 systems.

2 Thermodynamic Equations

2.1 Equilibrium Between Condensed Phases

To determine the composition of the two phases in heterogeneous equilibrium, the equations obtained in [12] were used:

$$x_2^l = \frac{1 - \exp(F_1(T))}{\exp(F_2(T)) - \exp(F_1(T))} \tag{1}$$

$$x_2^s = x_2^l \cdot \exp(F_2)) \tag{2}$$

$$F_i(T) = \left(\Delta S_i^o - \Delta H_i^o/T\right)/R + \Delta\overline{G}_i^{exs,I/II}/RT \tag{3}$$

Here I and II represent phases that are in equilibrium; x_i^I and x_i^{II} are molar fractions of phase components; $x_1^I = 1 - x_2^I; x_1^{II} = 1 - x_2^{II}$; $\Delta\overline{G}_i^{exs,I/II} = \left(\Delta\overline{G}_i^{exs,I} - \Delta\overline{G}_i^{exs,II}\right)$ - difference between excess Gibbs free energies of the components of I and II phases. ΔS_i^o and ΔH_i^o are enthalpy and formation entropy of MnTe, SnTe, Sn_2Te_3, $MnSb_2Te_4$, $MnSb_4Te_7$, $SnSb_2Te_4$, and $SnSb_4Te_7$ compounds.

Equations (1–3) are used to determine the boundaries of the various crystalline modifications of MnTe, α (SnTe), and β (Sb_2Te_3) solid solutions (Fig. 1 and 2). The functions of internal stability were used for determination of the phase boundaries of the γ ($MnSb_2Te_4$), δ ($MnSb_4Te_7$) γ ($SnSb_2Te_4$), and δ ($SnSb_4Te_7$) solid solutions [14]:

$$\Psi = x(1 - x)\left(\partial^2 \Delta G^0/\partial x^2\right)_{P,T} > 0 \tag{4}$$

In (4), ΔG^0 is the integral molar Gibbs free energy of formation of solid solutions.

2.2 Integral Gibbs Free Energy of Solutions

To calculate the integral molar Gibbs free energy of formation of solid solutions of the γ ($SnSb_2Te_4$) and δ ($SnSb_4Te_7$) solid solutions, a modified version of the regular solutions model was used, which has been successfully tested in [15, 16]:

$$\Delta G_T^0 = x\left(\Delta H_1^{s,0} - T\Delta S_1^{s,0}\right) + (1-x)\left(\Delta H_2^{s,0} - T\Delta S_2^{s,0}\right) + ax^m(1-x)^n \\ + RT[xlnx + (1-x)\ln(1-x)] \quad (5)$$

Here: 1- $MnSb_2Te_4$ or $SnSb_2Te_4$; 2- $MnSb_4Te_7$ or $SnSb_4Te_7$; $x = x_2$.

$MnSb_2Te_4$-$MnSb_4Te_7$ and also $SnSb_2Te_4$-$SnSb_4Te_7$ system (Fig. 3) are internal quasi-binary systems of MnTe–Sb_2Te_3 and SnTe–Sb_2Te_3 systems, respectively. As a result, the thermodynamic conditions of phase equilibria can also be applied to the $MnSb_2Te_4$-$MnSb_4Te_7$ and $SnSb_2Te_4$-$SnSb_4Te_7$ systems. Equation (5) represents the determination of the Gibbs free energy of formation of $(MnSb_2Te_4)_{1-x}(MnSb_4Te_7)_x$ and $(SnSb_2Te_4)_{1-x}$ $(SnSb_4Te_7)_x$ solid solutions from pure Mn, Sn, Sb, and Te components. However, this equation can be written in (6) form if we use values of $MnSb_2Te_4$, $MnSb_4Te_7$, $SnSb_2Te_4$, and $SnSb_4Te_7$ compounds instead of elementary components.

$$\Delta G_T^0 = ax^m(1-x)^n + RT[xlnx + (1-x)\ln(1-x)] \quad (6)$$

In (6), the first term represents the enthalpy of mixing of solid solutions in an asymmetric version of the regular solutions model; for solid solutions, decomposing into two phases has the mixing parameter $a > 0$; the second term represents the configurational entropy of mixing solid solutions. In all equations: $R = 8.314$ J mol^{-1} K^{-1}.

The thermodynamic functions of the formation of MnTe, SnTe, Sb_2Te_3, and ternary compounds are listed in Table 1.

Table 1. Formation enthalpies and entropies of manganese, tin, and antimony tellurides.

Variable	Units	Value
$\Delta H^o_{298}(MnTe)$	kJ·mol−1	−108.7 ± 0.3
$\Delta H^o_{298}(MnSb_2Te_4)$	kJ·mol−1	−189.4 ± 1.6
$\Delta H^o_{298}(MnSb_4Te_7)$	kJ·mol^{-1}	−263.7 ± 3.1
$\Delta H^o_{298}(SnTe)$	kJ·mol−1	−60.7 ± 0.8
$\Delta H^o_{298}(Sb_2Te_3)$	kJ·mol^{-1}	−56.5 ± 0.8
$\Delta H^o_{298}(SnSb_2Te_4)$	kJ·mol−1	−140.4 ± 1.6
$\Delta H^o_{298}(SnSb_4Te_7)$	kJ·mol^{-1}	−215.7 ± 3.1
$\Delta S^o_{298}(MnTe)$	J·mol^{-1}K	11.3 ± 0.8
$\Delta S^o_{298}(MnSb_2Te_4)$	J·mol^{-1}K	25.8 ± 0.5
$\Delta S^o_{298}(MnSb_4Te_7)$	J·mol^{-1}K	21.7 ± 0.5

(continued)

Table 1. (*continued*)

Variable	Units	Value
$\Delta S^o_{298}(SnTe)$	J·mol^{-1}K	0.3 ± 0.1
$\Delta S^o_{298}(Sb_2Te_3)$	J·mol^{-1}K	7.1 ± 0.2
$\Delta S^o_{298}(SnSb_2Te_4)$	J·mol^{-1}K	16.8 ± 0.5
$\Delta S^o_{298}(SnSb_4Te_7)$	J·mol^{-1}K	13.8 ± 0.5

Values of the formation enthalpies and entropies of MnTe, SnTe, Sb_2Te_3, $SnSb_2Te_4$ compounds are taken from [17, 18]. Using the calculation method described in [19], the thermodynamic parameters of $MnSb_2Te_4$, $MnSb_4Te_7$, $SnSb_2Te_4$ and $SnSb_4Te_7$ compounds, were determined in this work.

2.3 Method for Solving Thermodynamic Equation

Optimization of the phase diagrams of MnTe–Sb_2Te_3 and SnTe–Sb_2Te_3 systems (Fig. 1 and 2) by using Eq. (1–3) was carried out according to the following scheme using the MGA multipurpose genetic algorithm [8]: initially, the search range for each parameter is determined for the MGA using the experimental data of DTA. The initial function is (3), which includes the values of entropy, enthalpy, excess Gibbs free energies of phases that are in thermodynamic equilibrium in the temperature range under consideration. Further, considering that sum of molar fractions is equal to one, using Eqs. 1 and 2, the compositions of the components in the considered region of phase equilibrium are determined. The boundary of each phase-type is generated by separate liquidus and solidus curves. However, these curves must fit together (Fig. 1 and 2). The genetic algorithm solves this problem based on the experimental error in determining the molar fraction (0.001 mol fraction) and temperature (±5). The deviations of the liquidus and solidus from linearity are related to the difference in the partial excess free energies of the MnTe, SnTe, and Sb_2Te_3 compounds in the phases at equilibrium. This parameter is determined based on the experimental DTA values. Thus, the entire phase diagrams for the MnTe-Sb_2Te_3 and SnTe-Sb_2Te_3 systems are generated with the accuracy of experimental determination of the composition and temperatures. The liquidus and solidus curves shown in Fig. 1 and 2 are determined on the basis of the absolute values of the enthalpy and entropy of formation of binary and ternary compounds in Table 1. Taking into account the error in determining these values, the liquidus and solidus represent a band of uncertainty, as in the case of the YbTe-SnTe system [11]. A detailed description of the MGA method as applied to modeling the boundaries of phase equilibrium is available in [12].

To determine the parameters of the enthalpy of mixing of solid solutions in Eq. (6), an asymmetric version of the regular solutions model was used, according to which m $\neq$ n $\neq$ 1 [16]. The following conditions were used to carry out the iterative process for the MGA:

$$x = 0 \div 1; \mathrm{a} > 0; \mathrm{m} > \mathrm{n} > 1;\ 300\,\mathrm{K} < \mathrm{T} < 800\,\mathrm{K}$$

As a result of calculations, dependence (6) was obtained in the form:

$$\Delta G_T^0(J/mol) = 40 * 10^7 * \mathrm{x}^8 * (1-\mathrm{x})^{10.5} + 8.314 * \mathrm{T} * (\mathrm{x} * \ln(\mathrm{x}) + (1-\mathrm{x}) * \ln(1-\mathrm{x})) \quad (7)$$

From (7) for the stability function of solid solutions based on $SnSb_2Te_4$ and $SnSb_2Te_4$ we obtain:

$$\Psi/4 = \left(100 * 10^8 * x^9 * (1-x)^9.5 - 168 * 10^8 * x^8 * (1-x)^1 0.5 + 56 * 10^8 * x^7 * (1-x)^1 1.5\right) + 1663 \quad (8)$$

Here x- mole fraction $SnSb_4Te_7$ in solid solution $(SnSb_2Te_4)_{1\text{-}x}(SnSb_4Te_7)_x$. Equations (7) and (8) are presented in the computer version.

The optimized phase diagrams of MnTe–Sb_2Te_3, SnTe–Sb_2Te_3 systems and of the Gibbs free energy of mixing, the stability function of solid solutions $(SnSb_2Te_3)_{1\text{-}x}(SnSb_4Te_7)_x$ are shown in Fig. 1, 2 and Fig. 3, respectively.

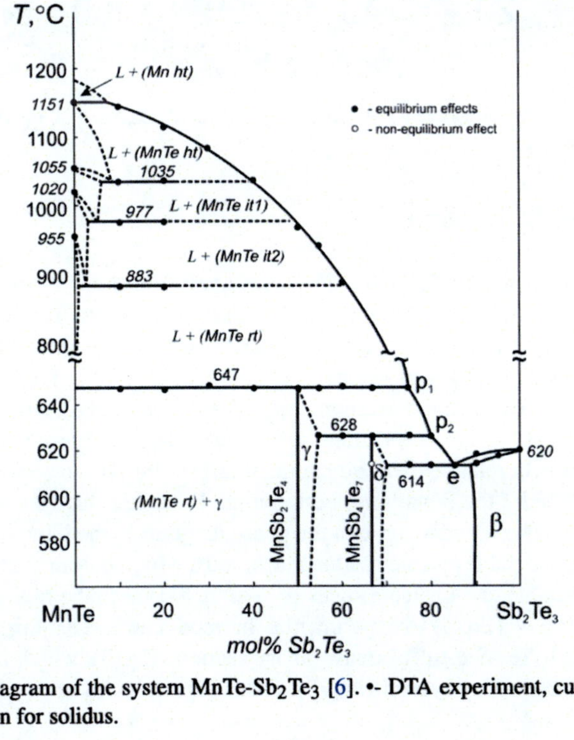

Fig. 1. Phase diagram of the system MnTe-Sb_2Te_3 [6]. •- DTA experiment, curves - thermodynamic calculation for solidus.

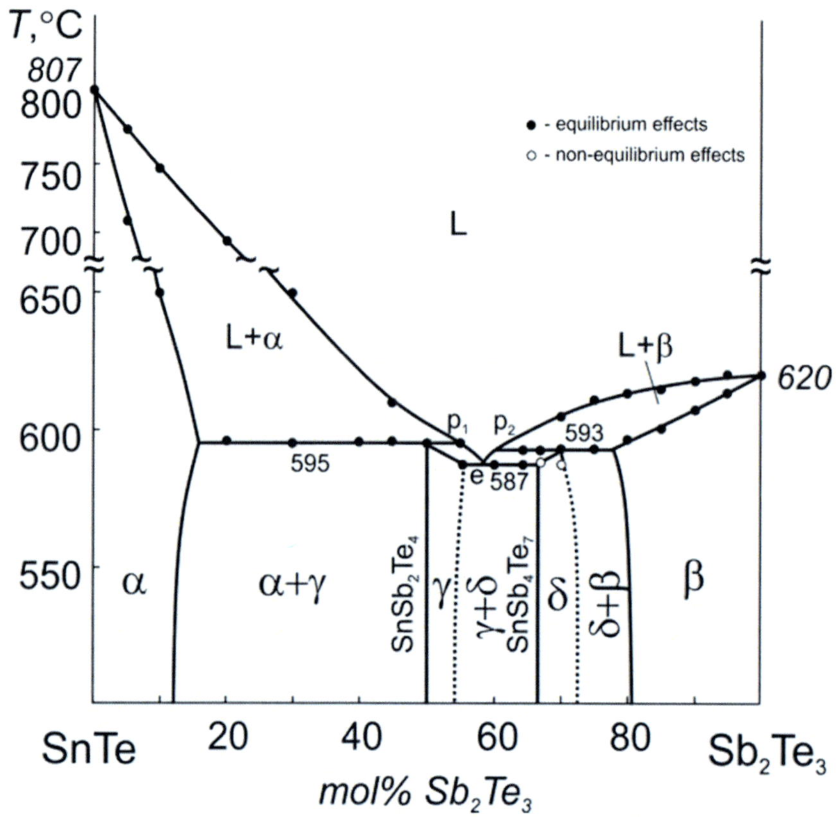

Fig. 2. Phase diagram of the system SnTe-Sb_2Te_3 [7]. •- DTA experiment, curves - thermodynamic calculation solidus.

3 Results and Discussion

The phase diagrams of the MnTe(SnTe)–Sb_2Te_3 systems (Fig. 1 and 2) are characterized by the formation of the peritectic compounds $MnSb_2Te_4$, $MnSb_4Te_7$ $SnSb_2Te_4$, and $SnSb_4Te_7$, solid solutions based on these compounds, as well as based on SnTe, Sb_2Te_3 compounds. Solid solutions γ ($MnSb_2Te_4$), δ ($MnSb_4Te_7$), γ ($SnSb_2Te_4$), δ ($SnSb_4Te_7$), α (SnTe) and β (Sb_2Te_3) in the corresponding concentration ranges (Fig. 1 and 2) are thermodynamically stable. However, along with stable phases, as a result of fluctuations, metastable phases with relative thermodynamic stability (Eq. 4) can be formed. As can be seen, in Fig. 1 and 2 the boundaries between the homogeneous and heterogeneous phases are binodal. The binodal curve in the phase diagram of the binary system is the boundary between the stable and metastable states, and the spinodal curve is the boundary between the metastable and unstable states of solutions. Due to the high sensitivity of the stability function ψ (Eq. 4) to experimental thermodynamic data, this function was applied to the $SnSb_2Te_4$-$SnSb_4Te_7$ quasi-binary section (Fig. 3), which is the internal section of the SnTe-Sb_2Te_3 system (Fig. 2). From Fig. 3, based on the values of the

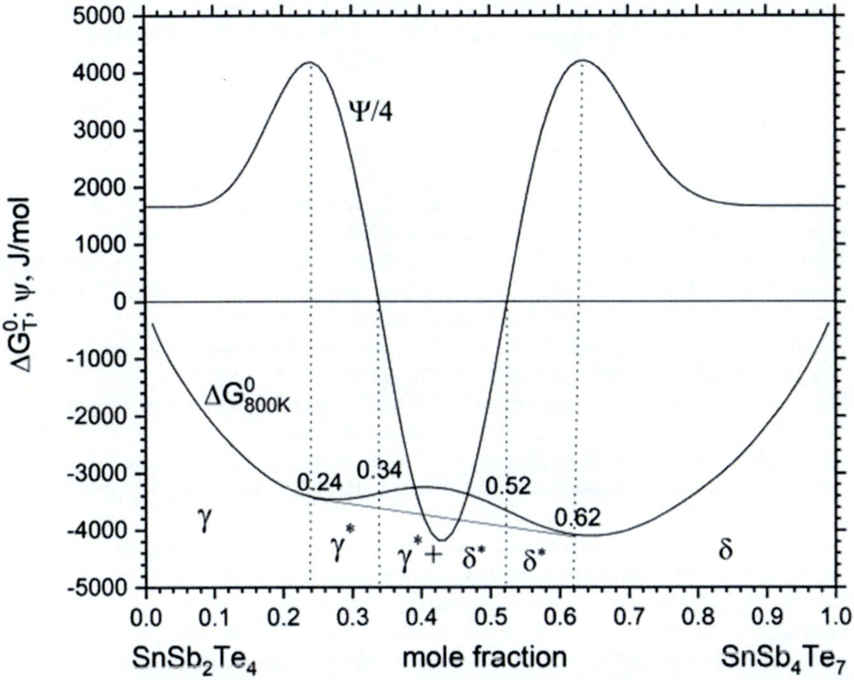

Fig. 3. Dependences of the Gibbs free energy of mixing (Eq. 7) and the stability function (Eq. 8) of solid solutions $(SnSb_2Te_3)_{1-x}(SnSb_4Te_7)_x$ on the composition at 800K.

stability function ($\psi > 0$), it follows that, in addition to stable solid solutions of phases γ and δ, metastable phases γ*($SnSb_2Te_4$) and δ*($SnSb_4Te_7$) can form in the system, which has relative stability.

In the case of decomposition of metastable phases, infinitesimal fluctuations lead to an increase in the regions of coexistence of final stable phases, in particular, nanoscale γ ($MnSb_2Te_4$), δ ($MnSb_4Te_7$) γ*($SnSb_2Te_4$) and δ*($SnSb_4Te_7$) solid solutions.

4 Conclusions

MGA made it possible to determine the parameters of the analytical dependences of the Gibbs free energy of formation of liquid and solid solutions of various modifications on the composition and temperature in the MnTe-Sb_2Te_3 and SnTe-Sb_2Te_3 systems. The boundaries of the equilibrium liquid-solid alloys, stable, metastable, and unstable state of solid solutions in the MnTe-Sb_2Te_3, SnTe-Sb_2Te_3, and in the internal section of this system $SnSb_2Te_4$-$SnSb_4Te_7$ have been determined. Uncertainty bounds stable and metastable solid solution are most sensitive to the values of entropy of formation MnTe, SnTe, Sb_2Te_3, $MnSb_2Te_4$, $MnSb_4Te_7$, $SnSb_2Te_4$, and $SnSb_4Te_7$.

The information obtained is intended to determine the optimal conditions for the synthesis of nanolayers of solid solutions based on two and three-component manganese, tin, and stibium tellurides for obtention of perspective topological insulator phases.

Acknowledgments. This work was supported by the Science Development Foundation under the President of the Republic of Azerbaijan – Grant № EIF-GAT-5-2020-3(37)-12/02/4-M-02.

References

1. Kanatzidis, M.G.: Discovery-synthesis, design, and prediction of chalcogenide phases. Inorg. Chem. **56**(6), 3158–3173 (2017). https://doi.org/10.1021/acs.inorgchem.7b00188
2. Babanly, M.B., Chulkov, E.V., Aliev, Z. S., Shevelkov, A.V., Amiraslanov, I. R.: Phase diagrams in materials science of topological insulators based on metal chalcogenides. Russ. J. Inorg. Chem. **62**(13), 1703–1729 (2017) https://doi.org/10.1134/S0036023617130034
3. Tokura, Y., Yasuda, K., Tsukazaki, A.: Magnetic topological insulators. Nat. Rev. Phys. **1**, 126–143 (2019). https://doi.org/10.1038/s42254-018-0011-5
4. Mong, R.S.K., Moore, J.E.: Magnetic and topological order united in a crystal. Nature, **576**, 390–392 (2019). https://doi.org/10.1038/d41586-019-03831-7
5. Otrokov, M.M., Klimovskikh, I.I., Bentmann, H., et al.: Prediction and observation of an antiferromagnetic topological insulator. Nature, **576**, 416–422 (2019). https://doi.org/10.1038/s41586-019-1840-9
6. Orujlu, E.N., Aliev, Z.S., Amiraslanov, I.R. Babanly M.B.: Phase equilibria of the MnTe-Sb_2Te_3 system and synthesis of novel ternary layered compound – $MnSb_4Te_7$. Phys. Chem. Solid St. **22**(1), 39–44 (2021). https://doi.org/10.15330/pcss.22.1.39-44
7. Seidzade, A., Orujlu, E., Doert, T., Amiraslanov, I., Aliyev, Z., Babanly, M.: An updated phase diagram of the SnTe-Sb_2Te_3 system and the crystal structure of the new compound $SnSb_4Te_7$. J. Phase Equilibria Diff. **42**, 373–378 (2021). https://doi.org/10.1007/s11669-021-00888-8
8. Preuss, M., Wessing, S., Rudolph, G., Sadowski. G.: Solving phase equilibrium problems by means of avoidance-based multiobjectivization. In: Springer Handbook of Computational Intelligence. Evol. Comput., Part E.58, pp. 1159–1171. Springer, Heidelberg (2015). https://doi.org/10.1007/978-3-662-43505-2_58
9. Stan, M., Reardon, B.J.: A Bayesian approach to evaluating the uncertainty of thermodynamic data and phase diagrams. Calphad **27**(3), 319–323 (2003). https://doi.org/10.1016/j.calphad.2003.11.002
10. Duong, T.C., Hackenberg, R.E., Landa, A., Honarmandi, A., Talapatra A.: Revisiting thermodynamics and kinetic diffusivities of uranium–niobium with Bayesian uncertainty analysis. Calphad **55**(2), 219–230 (2016). https://doi.org/10.1016/j.calphad.2016.09.006
11. Mammadov, A.N., Aliev, Z.S., Babanly, M.B.: Study of the uncertainty heterogeneous phase equilibria areas in the binary YbTe-SnTe alloy system. In: Aliev, R., Kacprzyk, J., Pedrycz, W., Jamshidi, M., Sadikoglu, F. (eds.) 13th International Conference on Theory and Application of Fuzzy Systems and Soft Computing - ICAFS-2018/Advances in Intelligent Systems and Computing, vol 896, pp. 815–822. Springer, Cham. (2019). https://doi.org/10.1007/978-3-030-04164-9_107
12. Mammadov, A.N., Alverdiev, I.D., Aliev, Z.S., Tagiev, D.B., Babanly, M.B.: Thermodynamic modeling of the phase diagram for Cu_2SnS_3-Cu_2SnSe_3 system. In: Aliev, R., Kacprzyk, J., Pedrycz, W., Jamshidi, M., Babanli, M., Sadikoglu, F. (eds.) 10th International Conference on Theory and Application of Soft Computing, Computing with Words and Perceptions - ICSCCW-2019. Advances in Intelligent Systems and Computing, 2020, vol. 1095, pp. 888–895. Springer, Switzerland (2020). https://doi.org/10.1007/978-3-030-35249-3
13. Imamaliyeva, S.Z., Alekberzade, G.I., Mamedov, A.N., Tagiev, D.B., Babanly, M.B.: Modeling the phase diagram of the Tl_9SmTe_6-Tl_4PbTe_3-Tl_9BiTe_6 system. In: Aliev, R., Kacprzyk,

J., Pedrycz, W., Jamshidi, M., Babanli, M., Sadikoglu, F. (eds.) 14th International Conference on Theory and Application of Fuzzy Systems and Soft Computing – ICAFS-2020. Advances in Intelligent Systems and Computing, 2021, vol 1306, pp. 480–489. Springer, Cham. https://doi.org/10.1007/978-3-030-64058-3_60
14. Stolen, S., Grande, T.: Chemical Thermodynamics of Materials: Macroscopic and Microscopic Aspects. Wiley, England (2004) https://doi.org/10.1002/0470092688
15. Mamedov, A.N., Tagiev, E.R., Aliev, Z.S., Babanly, M.B.: Phase boundaries of the $(YbTe)_x(PbTe)_{1-x}$ and $(YbTe)_x(SnTe)_{1-x}$ solid solutions series. Inorg. Mater. **52**(6), 543–545 (2016). https://doi.org/10.1134/S002016851606008X
16. Asadov, S.M., Mamedov, A.N., Kulieva, S.A.: Composition and temperature-dependent thermodynamic properties of the Cd, Ge||Se, Te system, containing $CdSe_{1-x}Te_x$ solid solutions. Inorg. Mater. **52**(9), 876–885 (2016). https://doi.org/10.1134/S0020168516090016
17. Guseinov, F.N., Seidzade, A.E., Yusibov, Y.A., Babanl, M.B.: Thermodynamic properties of the $SnSb_2Te_4$ compound. Inorg. Mater. **53**(4), 354–357 (2017). https://doi.org/10.1134/S0020168517040057
18. Iorish, V.S., Yungman, V.S.: Database of thermal constants of substances (2006). http://www.chem.msu.ru/cgi-bin/tkv.pl
19. Gurbanov, G.R., Adygezalova, M.B., Mamedov, A.N., Gulieva, S.A.: Syntesis, crystal structure, and thermodynamic functions of $GeSnSb_4Te_8$. Moscow Univ. Chem. Bull. **74**(3), 134–137 (2019). https://doi.org/10.3103/S0027131419030064

Check for updates

Numerical Determination of Gas and Oil Reserves

I. Y. Bayramov, A. N. Gurbanov, O. M. Mirzayev, and I. Z. Sardarova(✉)

Azerbaijan State Oil and Industry University, Baku, Azerbaijan

Abstract. The article discusses some of the problems of modeling and managing complex oil and gas production systems under uncertainty, provides a detailed technology for calculations at risk and uncertainty, as well as practical calculations at objects and processes in the oil and gas industry under uncertainty and optimization issues are discussed, in which case inaccurate data (interval, fuzzy) are used in static and dynamic models. Specific practical examples show the advantages of using fuzzy set theory in solving problems of management and control of gas fields and gas production system facilities in conditions of uncertainty.

Keywords: Management · Uncertainty · Fuzzy set · Affiliation function · Oil and gas industry · Numerical methods · Static models · Dynamic models

1 Introduction

In the development of the gas industry in modern times, the most pressing issues are the improvement of methods for calculating the amount of gas reserves in gas fields and the process of exploration of natural gas fields by rational methods. Thus, the improvement of these methods is the rapid discovery of natural gas fields and their development at low cost.

Calculation methods and mining studies of wells and reservoirs should allow to determine the parameters of aerated aquifers based on their areas and thicknesses. Giving accurate forecast of field development, regulating the processing process, finding the optimal design solution, etc. without having reliable knowledge about the parameters of the layers is not possible. The accuracy of the solutions to these issues depends to a large extent on the strategy for extracting gas reserves and the efficient use of capital investment for infrastructure and additional drilling.

The uncertainty of the situation for gas fields as a whole is "porosity", "thickness" and so on. mainly depends on the ambiguity of the concepts. Thus, the measurement of these parameters is carried out at different points in the field, but the heterogeneous environment of these quantities varies significantly throughout the field. The application of average and measured average values of parameters for calculations can lead to the acquisition of significantly mixed accurate values of parameters.

As a result of the research, it became clear that there are different methods of estimating reserves, the results of the calculations are inconsistent due to significant errors in the initial data, as the redundancy of information should allow to accurately estimate

R. A. Aliev et al. (Eds.): ICSCCW 2021, LNNS 362, pp. 522–528, 2022.
https://doi.org/10.1007/978-3-030-92127-9_70

reserves. The study also found that the cost of reserves, which depends on the accuracy of the installation of isobar cards, gas layer, porosity, layer's power and the amount of gas produced, reaches 25–35%.

2 Statement of the Problem

Since the errors of various devices and sensors indirectly affect the values of a number of parameters, it became necessary to calculate the coefficients of inaccurately given parameters and equations. Substituting deterministic, i.e. exact, quantities with inaccurately given quantities significantly complicates the calculation procedure, so it is necessary to use their iterative selection to obtain the accepted result. In this case, it is also difficult to estimate the error of the result. Often the parameters and coefficients of the equation also include various quantities of uncertainty (interval, fuzzy, stochastic, heuristic (based on expert estimates), etc.). Therefore, there is a need to describe all information in the language of a single formal fuzzy set, in which case the inaccurately given characteristic quantity can be expressed as a membership function.

Let's use the known equation to calculate gas reserves by the volume method and note the fuzzy given quantities. Let's describe it with the following expression:

$$V = \alpha \cdot m \cdot h \cdot S \cdot \frac{P_1 T_0}{P_0 T_1 Z} = \alpha \cdot m \cdot h \cdot S \cdot C$$

Here:
α - gas reserve factor;
h – layer's thickness, m;
S - gas area, m^2;
Z - gas compression ratio;
P_1 - formation pressure, Pa;
T_1 – layer's temperature, K;
V - gas reserves brought to normal conditions when P_0 and T_0, m^3.

Unlike existing practice, it is more efficient to use their membership functions $\mu(m)$, $\mu(\alpha)$, $\mu(h)$, $\mu(S)$ than the various averaging methods used to obtain accurate estimates for each parameter in advance.

We obtain the resultant function of gas reserves from Eq. (1) based on the definition of algebraic operations:

$$\mu_0(\alpha) = \max_U[\mu(\alpha) \wedge \mu(\alpha) \wedge \mu(h) \wedge \mu(S)]$$

Here: $U = \{(\alpha, m, h, S) | \alpha \cdot m \cdot h \cdot S \cdot C = V\}$.

It is extremely difficult to find the quantity $\mu_0(V)$ from the formula (2) by analytical methods. Therefore, numerical methods are used to solve this type of practical problems. The simplest of them is the opposite method. In this case, the resulting membership function is calculated by sequentially applying this algebraic operation (see Fig. 1).

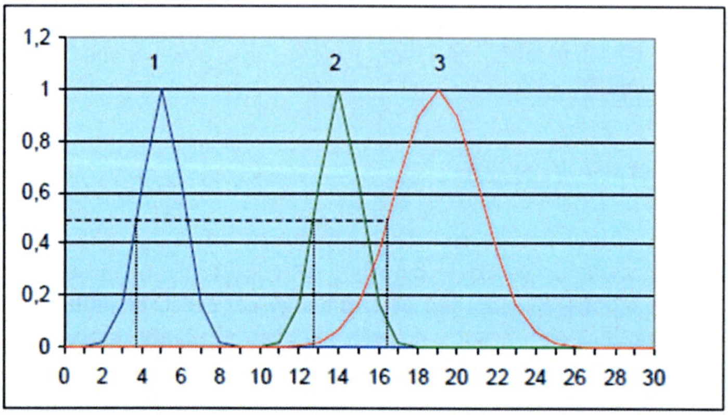

Fig. 1. The resulting membership functions. 1 - μ_1(x) first membership function, 2 - μ_2(y) second membership function, 3 - μ_3 (z) membership function of result.

The algorithm for finding the resultant membership function $\mu_3(z)$ by the inverse method as a result of algebraic operations on two arbitrary membership functions, such as $\mu_1(x)$ and $\mu_2(y)$, consists of the following steps:

1) *For the functions* $\mu_{1(x)} \cdot [0, \alpha], \mu_2(y) \cdot [0, \beta], \alpha \leq 1, \beta \leq 1$ *we divide the piece [0,* $\wedge\beta]$ *into parts by the points ni ϵ [0,*$\alpha \wedge \beta$*] (i = 1, 2, ..., n).*
2) For every m_i, we determine the n_i set of corresponding levels from equations $\mu_1(x) = n_i$, $\mu_2(y) = n_i$

$$\sigma_{n_i}(\widehat{x}) = [x_i^{'}, x_i^{''}];$$

$$\sigma_{n_i}(\widehat{y}) = [y_i^{'}, y_i^{''}].$$

$$\sigma_{n_i}(\widehat{z}) = \sigma_{n_i}(\widehat{x}) * \sigma_{n_i}(\widehat{y}) = [z_i^{'}, z_i^{''}]$$

3) We find the resolution functions from the ni leveled set. Note that symbol * is the corresponding interval operation.

As an example, we use the algorithm for calculating gas reserves by the volume method for any hypothetical field. The data for the calculation are obtained as a result of the processing of geological and mining information (Table 1), taking into account the measurement error of each variable.

Table 1. Preliminary data for the calculation of gas reserves

Parameters	Minimum rate of parameter	Similar to reality rate of parameter	Maximum rate of parameter
α	0,4	0,6	0,8
m	0,2	0,3	0,4
h	10	30	70
S	$1{,}9 \cdot 10^9$	$2 \cdot 10^9$	$2{,}1 \cdot 10^9$

Based on the data in the table above, we build a triangular membership function for each parameter:

$$\mu(\alpha) = \begin{cases} 3.33(\alpha - 0.3), \ 0.3 \le \alpha \le 0.6, \\ -5(\alpha - 0.8), \ 0.6 \le \alpha \le 0.8, \end{cases}$$

$$\mu(m) = \begin{cases} 10(m - 0.2), \ 0.2 \le m \le 0.3, \\ -10(m - 0.4), \ 0.3 \le m \le 0.4, \end{cases}$$

$$\mu(h) = \begin{cases} 0.05(h - 10), \ 10 \le h \le 30, \\ -0.025(\alpha - 70), \ 30 \le h \le 70, \end{cases}$$

$$\mu(S) = \begin{cases} 0.01(S - 1900), \ 1900 \le S \le 200, \\ -0.01(S - 2100), \ 2000 \le S \le 2100 \end{cases}$$

Membership functions can be set more accurately. In addition, rock survey methods, measurement errors of appropriate devices, survey data for all wells, etc. should be used. However, it should be noted that this numerical method used by us allows to work with any type of function.

Based on this algorithm, an n-level set is calculated for the $\mu_0(V)$ function (Table 2). In this case, the membership function takes the value 1, so the fraction [0,1] is divided into n - levels. This is presented in the first column of Table 2.

The results of the second stage of this algorithm for member functions are given in columns 2, 3, 4, and 5, respectively, of Table 2. The results are presented at sixth column, taking into account that the final result is multiplied by the constant C.

Table 2. N-level sets calculated for the function $\mu_0(V)$

N	$\sigma_n(\widehat{\alpha})$	$\sigma_n(\widehat{m})$	$\sigma_n(\widehat{h})$	$\sigma_n(\widehat{S})$	$\sigma_n(\widehat{V})$
0	[0.3, 0.8]	[0.2, 0.4]	[10, 70]	$[1.9, 2.1]10^9$	$[0.145, 5.97]10^{12}$
0,5	[0.45, 0.7]	[0.25, 0.35]	[20, 50]	$[1.95, 2.05]10^9$	$[0.557, 3.19]10^{12}$
1	0.6	0.3	30	$2.0 \cdot 10^9$	$1.37 \cdot 10^{12}$

As a result of the application of the inverse numerical method, the graph of the $\mu_0(V)$ function is shown in the figure below (Fig. 2).

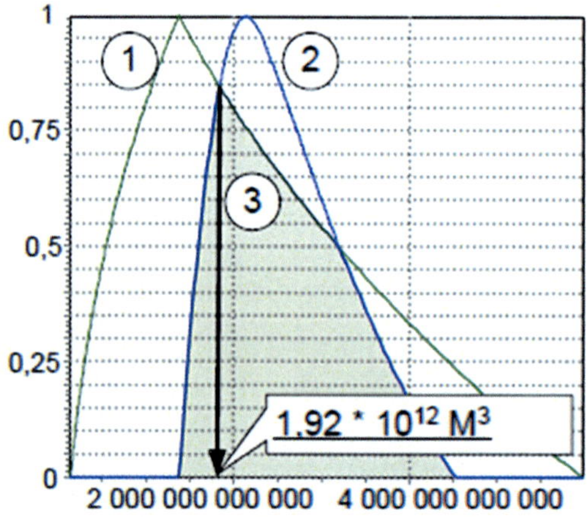

Fig. 2. Membership functions for gas reserves: 1 - prices of gas reserves by the volume method; 2 - prices of gas reserves by the balance method; 3 - prices agreed by both methods

The resulting reserves are estimated by analogy to formula (1). Note that only in this case the multiple factor is used, as it actively characterizes the type of collector used. The resultant coefficient of gas flow is defined as the fuzzy quantitative ratio of output and balance reserves (Fig. 3).

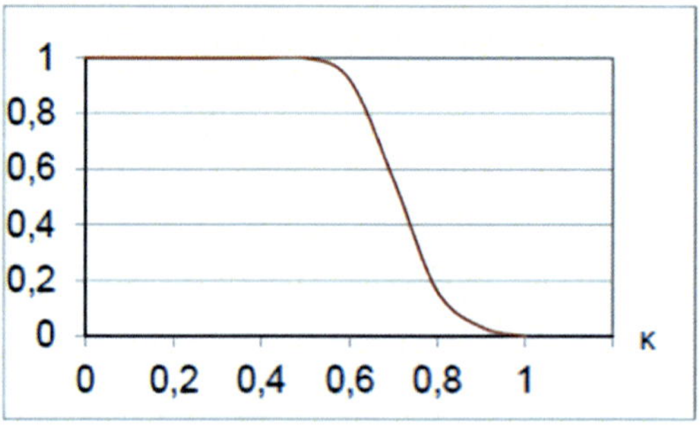

Fig. 3. Membership function of gas transfer coefficient

Figure 4 shows the estimation of gas reserves using the triangular function and the Monte Carlo method. Analysis of the results shows that the Monte-Carlo method cuts off some of the distribution and narrows the area of uncertainty in reserve prices.

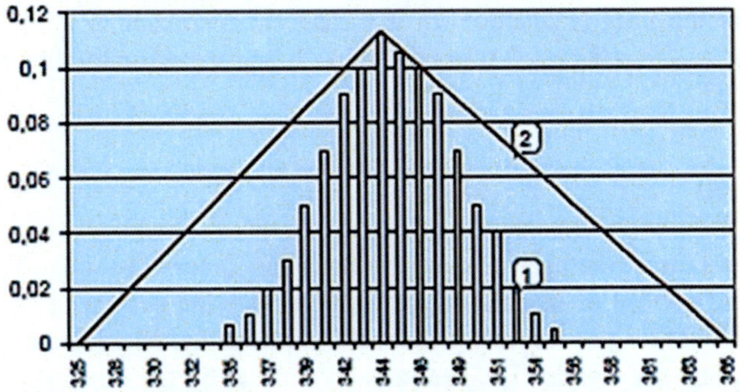

Fig. 4. Comparison of gas reserves estimation: 1 - frequency histogram by Monte-Carlo method; 2 - normalization of the triangular-type membership function

Now let's describe the process of calculating oil reserves based on fuzzy sets theory.

Let's use a known equation to calculate oil reserves by the volume method based on average prices, so that all or some quantities of this equation can be given fuzzily. This equation can be expressed as follows:

$$Q_e = F \cdot h_{e.n.l.h} \cdot k_{a.m.em.} \cdot k_{n.d.em.} \cdot \theta \cdot \rho$$

Where, Qe – primary oil reserves;
F – square of oil field;
he.n.h.l – effective oil line height;
ka.m.em – open porosity coefficient;
kn.d.em. – oil content coefficient;
θ – recalculation coefficient taking into account oil compression;
ρ – oil density under standard conditions.

We replace all calculated parameters with their corresponding membership functions $\mu(F)$, $\mu(h_{e.n.l.h})$,$\mu(k_{a.m.em..})$, $\mu(k_{n.d.em.})$.

The boundary points for the μ(F) membership function are determined by the internal and external contours of the oil or by the error of the value of the F-field. In lithological heterogeneity of deposits, the membership functions $\mu(h_{e.n.l.h})$,$\mu(k_{a.m.em..})$, $\mu(k_{n.d.em.})$ are determined by the wells that fall into the productive zone, the area F and, accordingly, the membership function μ(F) by the zone of the strata.

From the equation shown by performing algebraic operations on fuzzy quantities, we obtain the following for the resultant membership function for oil reserves:

$$\mu_0(Q_e) = \max_U[\mu(F) \wedge \mu(h_{e.n.l.h}) \cdot \mu(k_{a.m.em..}) \cdot \mu(k_{n.d.em.})]$$

$$U = \{(F, h_{e.n.l.h}, k_{a.m.em..}, k_{n.d.em.}) | F \cdot h_{e.n.l.h} \cdot k_{a.m.em..} \cdot k_{n.d.em.} \cdot \theta \cdot \rho = Q_e\}$$

For the solution of practical problems, $\mu_0(Q_e)$ membership function is defined by the inverse numerical method. To apply the algorithm, we build a triangular membership function for each parameter. Based on the inverse numerical method, an r-level set is calculated for the $\mu_0(Q_e)$ function. In this case, the membership functions $\mu(F)$, $\mu(h_{e.n.l.h}), \mu(k_{a.m.em..})$, $\mu(k_{n.d.em.})$ takes the value 1, and therefore the fraction [0,1] is divided by r-level.

3 Conclusion

The article shows the advantages of using of fuzzy sets theory in solving problems of management and control of objects of gas fields production system in conditions of uncertainty. Using of the gas resource price method was analyzed using the volume and balance method as an example. The analyzes carried out by the Monte-Carlo method revealed that during the distribution, the area of uncertainty of the fractional and reserve prices of a certain part is narrowed. It has also been found that used numerical method allows to work with any type of function.

References

1. Krasnov, O.S.: Theory and practice of probabilistic assessment geological risks and uncertainties in the preparation of oil and gas reserves. Oil and gas geology. Theory and practice: electr. Sci. J. VNIGRI, (2) (2007). http://www.ngtp.ru/rub/3/8_2009.pdf
2. Magomedov, I.A., et al.: Application of the theory of fuzzy sets to the control problems of non-stationary processes. In: Collection: Methods and systems of decision making. - Riga: RPI, pp. 60–65 (1984)
3. Mirzajanzade, A.K., Khasanov, M.M., Bakhtizin, R.N.: Modeling of Oil and Gas Production Processes. Nonlinearity, Disequilibrium, Uncertainty, p. 368. Institute of Computer Research, Moscow-Izhevsk (2005)
4. Semukhin, M.V.: Fuzzy estimates of oil reserves. In: Collection of Reports of the International Conference on Soft Computing. T.2. S.-Pb. Publishing House of ETU "LETI", p. 164–167 (2003)

Development of the Geoinformation Space Soft Model Considering Its Intelligent Properties

Vladimir V. Khramov(✉)

Southern University (IMBL), Rostov-on-Don, Russian Federation

Abstract. Among numerous methods of presenting and processing spatial information, special attention is given to those that are oriented towards consideration of the entire available set of NOT - factors characterizing real available data and approaches to their processing, having ambiguities, uncertainties and incomplete corresponding models and methods. In holistic established systems, their behaviour may be described by order parameters whose relationship to the system of known measures is generally non-linear and complex. However, these order parameters allow one to adequately describe the system, its stability, and the possibility of deterministic chaos.

In other words, hidden qualities of the environment objects, including those caused by human factor impact, can be present. In this paper, we suggest to extend the certain well-proven best practices in data analysis and transformation in the single training environment information space to solving the actual social and economic problems. The subject of this study is the information semantic system in terms of recording non-obvious properties of space images and prospects of their application in identifying extended objects of the Earth's surface. In addition, the paper provides approaches to revealing hidden and hideable artifacts, information bookmarks. A version of classification of this property manifestation types is given, as well as its description in soft model language.

Keywords: NOT-factors · Security · Intelligent systems · Semantic information · Intelligent security · Order parameters

1 Introduction

Human participation in the formation, perception, processing, and dissemination of information, in the form of certain intellectual products, raises a number of problems related to the elucidation of mechanisms of functioning of the individual, as an intellectual system and the interpretation of the results of such activities. Intellectual products often have discernible traits of developer's (or a man-machine development complex) individuality. For the last few years, such "intelligent bookmarks" have been rapidly transferring from the psychologists' area of interest to mathematical and technical areas.

It is supplemented by historically established system of measures for the properties of the physical environment and its separate components. In holistic established systems, their behaviour may be described by order parameters whose relationship to the

R. A. Aliev et al. (Eds.): ICSCCW 2021, LNNS 362, pp. 529–536, 2022.
https://doi.org/10.1007/978-3-030-92127-9_71

system of known measures is generally non-linear and complex. However, these order parameters allow one to adequately describe the system, its stability, and the possibility of deterministic chaos. In other words, hidden qualities of the environment objects, including those caused by human factor impact, can be [1, 2].

The methodology for studying a single geoinformation space involves a set of interrelated principles and approaches, including a digital description of a set of private representations of the studied space, created by a person in a virtual (computer) environment and its use in solving spatial problems and developing spatial solutions.

The analysis of known human models as a decision-maker (DM) revealed that they have certain common components that form structures related to the intelligent-hidden (intellatent). The concept of intellatency was introduced in [3] for solving problems of logical level and further extended [4] to the other, more complex intelligent problems.

Although such notions as "intelligence", "artificial intelligence", "intelligent systems" and others are widespread by now, they have no strict and unambiguous definitions. Without setting the task of their regular redefinition, let us use such an unambiguous notion as "metric intelligence" as basic.

2 Materials and Methods

In work [5], the system complexity is manifested through: a) complexity of its components; b) complexity of structure and architecture, c) complexity of behavior, systems, representing the subject of our study, have any combination of the conditions specified.

In terms of cognitive science, "studying and simulating principles of organization and operation of natural and artificial intelligent systems" [6], an intelligent model of the Single Geoinformation Space ((SGS) can get a special role. In the first turn, it happens because an object of manipulation (acceptance, processing, storage) in SGS [7] is semantic information, such as knowledge, a subject, a person, a decision maker. Therefore, the case is one of versions of the information semantic system, the functioning of which is aimed at achieving the certain goal, i.e., ability process information.

$$\Phi = \{\tau, \zeta, \gamma, \delta\}, \quad (1)$$

where τ, ζ, γ, δ is, respectively, text, sound, graphic and (thermo)dynamic information. Complex (or multimedia) forms, the number of which at two simultaneously used components is $C_4{}^2 = 6$, and at three components, number of forms is $C_4{}^3 = 4$, can be formed on the basis of homogeneous forms.

Let us assume the existence of some semantic information (knowledge) of the certain object (physical or abstract) in any form, available for formation and perception, and having the completed nature. Let us call this information primary. On practice, it may be performed in the form of a message about the location of a problem-oriented object on the certain territory. Assume that, as the result of moving in the information semantic system (ISS), analytical and synthetic and logical transformation, this information has changed in volume and/or presentation form. Let us call it resulting or secondary. Obviously, that its content cannot exceed primary, because loss is possible, and sometimes even inevitable.

Elements that form ISS are semantic objects (SOs), for example, man-machine systems. Numerous typical SOs in ISS can be described with an expression:

$$\ni = \{\lambda, \nu, \kappa\}, \tag{2}$$

where λis (are) a person (people), ν, a carrier (carriers) of semantic information of any form, κ are intelligent systems (computers). he component ν acts as a mediator, and components λ, κ are equal to the certain extent. Let us also note that, due to the fact that the SGS is a system with a purpose, the prerequisite for its existence is the principle possibility of reaching this goal. For this purpose, 1) SOs should be intercompatible; 2) ISS should have the "soft logic", i.e., process information and make decisions based on using the semantic interaction with real world objects, in the first turn, people [3].

The first requirement for semantic compatibility is achieved through conjunction in all components of any pair of semantic objects with non-zero structural relations. The second requirements may be fulfilled when using the fuzzy logic apparatus [8].

SGS efficient operation is achieved when the following principles are met:

1. *Principle of semantic topology. Semantic information cannot be reduced upon changing forms of its representation or their complexation* [4].

   In formal recording, the specified principle is presented as

$$(\forall\varphi_1^i(\varphi_1^i \in \Phi 1) \vee \ldots \vee (\forall\varphi_1^i(\varphi_n^j \in \Phi n)ad|SI|)), \tag{3}$$

   where $\varphi^i{}_k$, – semantic information representation form; $k = 1\ldots n$ - number of multimedia components; $i = 1\ldots m$ – number of types of SI homogeneous presentation forms; $j = 1\ldots C_i{}^k$ - number of types of SI complex presentation forms at i transfer channels and k complexation degrees; |SI| - semantic information, required and sufficient for unambiguous knowledge of an object; ad – adequacy ratio sign.

   In practical application for the SGS, this principle is expressed in possibilities for knowledge provision through various forms at constant meaning preservation. At the modern stage of SGS development, the principle can be fully implemented only with human participation with intuitive and heuristic methods. However, hardware and software have been already developed and successfully functioning [6], that provide automated procedures of transfer from text to sound form of semantic information presentation and back, and to graphical and graph-dynamic form, that deal with virtual images.
2. *Principle of inadequacy. "Full adequacy between an object and semantic information, that reflects it, is impossible"* [4].

   In formal recording, the specified principle is presented as

$$\forall\Omega(\Omega \Rightarrow \Phi) \to ad(\Omega \Rightarrow SI), \tag{4}$$

   where Ω - object; ⇒ - representation;

   $\Phi = \{\Phi_{ik}\}$, $i = 1..m$; $k = 1..n$– "totality of all possible forms of the object semantic information (SI) presentation; ad – inadequacy ratio. Therefore, the case is not absolute, but rather relative adequacy of primary semantic information to the cognitive object" [4].

Security as the Information System Property
In general case, security characterizes information "objects with conditions that are not hidden in a set of similar objects. From it, an object selects an actually used state, that may be changed with time. An observer has information on the set of possible conditions, 3 probabilistic characteristics, and may know previous actual conditions, but the current one is unknown to him/her" [3].

Any information objects of natural and artificial origin, without regard of destination, can be hidden. These are agricultural objects, mineral deposits, human diseases, movable robot coordinates, and so on.

2.1 Principles of Intelligent Security (Intellatency)

3. *Principle of direct intellatency. Secondary semantic information about the object coincides on the semantic source with primary information with accuracy to the subject intelligence index.*

   In formal recording, the specified principle is given as

$$\forall \sigma_{\mathrm{I}}(\sigma_{\mathrm{I}} \in SI)[SI_2 + \sigma_r \subset SI_1], \tag{5}$$

where σ_{I} - intelligence index, having dimensionality of semantic information and meeting the condition

$$\sigma = k_n(Sup(SI_2)) - Inf(SI_2)), \tag{6}$$

At this, *Sup* (SI_2), *Inf* (SI_2) are the upper and the lower edges of SI_2, respectively, and k_{n} is the nonlinear coefficient, in general case.

   On practice taking this principle into account provides for identification of k_{n} coefficient nonlinearity.

4. *Principle of secondary intellatency. Secondary semantic information about the object coincides on the semantic source of artificial origin with primary information with accuracy to the subjectivity index.*

In formal recording, the specified principle is similar to the previous one and has the form of

$$\forall \sigma_c(\sigma_c \in SI)[SI_2 + \sigma_c \subset SI_1], \tag{7}$$

where

$$\sigma_c = f(\sigma_{I1}, \Delta_1; \sigma_{I2}, \Delta_2; \sigma_{Ij}, \Delta_j; \ldots \sigma_{Im}, \Delta_m) \tag{8}$$

$\sigma_I{}^j$, Δ^j, $j = 1..\mathrm{m}$ are, respectively, intelligence indices of the semantic source components (for example, software) and their share (weight, significance coefficients) in this source system.

On practice of creating and setting of SGS with the high level of applied digital technologies, taking this principle into account provides for selection and development of such software that would provide variant recognition with the possibility of automatic or automated SGS adaptation to the test object, possibility, and permissibility to involve a person or artificial intelligence not only separately as a decision maker (DM), but hybrid intelligence, biont, too [8]. But, as the entire modern IS theory proved, “modern intelligent system should be capable of recording and simulation of the complex of NOT-factors, running through its interaction with the environment” [9]. This is also fully applied to space image analysis system.

A. NO Factors and Security

Real condition of SGS, as a man-machine system, provides for the known level of uncertainty, which is “accompanied by the hazard of non-achieving the goal of this system and requires risk, which increases uncertainty and gives rise to new hazards” [10].

Let us distinguish the main NOT-factors, related to security in the whole and to object-oriented security, in particular, and generate uncertainty, “connected with:

1) lack of knowledge *of the form of some deterministic functions*, their numerical characteristics and values of constants, describing intra- and intersystem processes;
2) lack of knowledge *of the certain factors(processes), affecting operation of intelligent systems,* including SGS;
3) *mathematic incommensurability of numerical estimates of values*, characterizing processes in the system (irrationality);
4) *quantum and mechanic and thermodynamic effects, originating from the complementarity principle*: they follow Heisenberg indeterminacy principle, which may considerably affect the functioning of conflicting systems;
5) *unknown deliberate action or behavior* (of a subject, for example, competitor). It is operative uncertainty, inherent to man-machine systems and leading to risk”; [9].

Totality of such security elements was revealed during the earth surface objects monitoring with SGS [7].

B. Means of Intellatency Estimation

It is usual to distinguish software and hardware means of estimation of any security, including intelligent security [3–5]. However, algorithms of revealing hidden and hideable (including active methods) intelligent bookmarks are primary in both cases. Direct solution of the detection problem is based on building binary trees of revealing hidden semantic objects. Prior to this case, even the security unit [11] was obtained that was called *diz* (number of binary measurements).

As a security measure R of numerous possible events

$$X = \{x_i\},\ I = 1..N_a,$$

where N_a is (a priori random value) the number of possible alternatives in search for the bookmark, path length expectation l_{av} is accepted from the binary tree root to the bookmark identification point. In other words, the case is the average number of steps during the search with regard of probabilities of their use:

$$R = m(l_i) = \sum\nolimits_{i=1}^{A} P(l_i),$$

where $m(\ l_i)$ is expectation;

$P\ (l_i)$ – probability of using the path l_i, in general case, at equal probability of their use, we may accept $P\ (\ li) = 1\ /\ Na$.

At such problem setting, the following security types may be differentiated [11, 12].

"Algorithmic security, determined by the required average number of binary measurements for revealing a bookmark at the certain set search algorithm. If interference is absent, it may be computed by the formula (8).

Potential security, determined by the required minimal number of diz for revealing a bookmark at all possible search algorithms $\sigma_i \in \Sigma\ S = min\ R(\sigma)$. At this, potential security becomes the objective characteristic of properties of the set (system) of semantic objects to resist the revealing of a bookmark, the measure of complexity of the search procedure organization.

Entropy security, potential security, computed by the formula defining Shannon entropy" [6]:

$$S = -\sum_{i=1}^{A} P_i\ log\ P_i \tag{9}$$

Imaginary security, (let us set it as Q) - a part of initial uncertainty (defined by the Eq. (9)), that cannot be removed with binary measurements due to interference and masking.

Complex security, (let us set it as $S = S + j(Q)$ – "complex number, a real part of which can be, for example, computed by formulas (8,9) and is subject to removal with binary measurements, and imaginary, to such measurements" [3] and not subject to removal, requiring special procedures for its identification.

The main part of "imaginary" security in all intellectual work products can be classified as Intel latency [3] and can be removed only with heuristic methods at the modern stage. In respect to knowledge, stored in the SGS base, as an intellectual work product, the probabilistic and entropy approaches, set forth above, are applicable with the significance level, defined with the so-called "imaginary" component.

3 Discussion

Let us assume that we obtained a set from n (scalar) parameters following the results of, in general case, inadequate tests. Then, integral estimate of hidden semantic information can be defined as a linear separating function for the fuzzy classification algorithm in the following form: $F(\lambda, \vec{x})$, where $\vec{x} = \{x_1, x_2, \ldots, x_n\}$ – vector of parameter values, revealing inner content of integral estimate of the information object security components; $\lambda = \{\lambda_1, \ldots, \lambda_m\}$ - weight vector.

It is possible to introduce characteristic functions for each described test of the information object x_i, with regard of initial fuzziness:

$$0 \leq \omega_{it}(x_t) \leq 1, \ \forall x_t \in X, \tag{10}$$

where $t \in \Xi$ is a sampling element.

The methodology of using the properties of a single information space to improve the efficiency of managing socio-economic systems has shown its consistency. In a number of publications, author suggests variants of using the proposed classification of the latent properties of information objects for pattern recognition in computer vision systems of robots [4, 13], for studying the Earth from space and assessing the state ecology [7, 14, 15], as well as to assess the state and management of education in a particular region [14]. Currently, research is being carried out to establish the measurability of relationships between individual NOT-factors of the model of a single information space with an assessment of their latency.

The task here is, by "varying the vector λ, values, define such value of the separating function that would correspond to the best, to the certain extent, object separation" [7] by security classes.

4 Conclusion

The obtained security characteristics allow to carry out the constructive qualitative analysis of information objects of the Earth's surface images by space intelligence. The research has focused on the SGS information semantic system in terms of considering the opaque properties of space images and prospects for their use in the identification of extended objects of Earth's surface.

The approaches proposed may be also used in the line of other applications, technical diagnostics, problems of intelligent analysis of states of objects with various origin.

References

1. Iskhakov, R.I.: On possibility to use security theory for determining the source (channel) of confidential information leak (in Russian). Bull. Udmurt Univ. 3, 70–75 (2010). https://cyberleninka.ru/article/n/o-vozmozhnosti-ispolzovaniya-teorii-skrytnosti-dlya-opredeleniya-istochnika-kanala-utechki-konfidentsialnoy-informatsii
2. Markova, D.G.: Human Factor in information security (in Russian). In: Salvedi, G.(eds). M. Mir, 1., 4p. (2019). https://cyberleninka.ru/article/n/chelovecheskiy-faktor-v-informatsionnoy-bezopasnosti/viewer
3. Khramov, V.V.: The principle of Intel latency and its use in recognition problems (in Russian). In: Collection of Papers: Thematic Research and Technology Collection. Pushchino-on-Oka, Materials of the International: NTC BC RAS: Spectral methods in research, pp. 62–66 (1994). https://www.elibrary.ru/item.asp?id=32838003
4. Khramov, V.V.: Features of majority processing of fuzzy information (in Russian). In Collection: Spectral Methods of Information Processing in Scientific Research. Reports of the I All-Russian Conference (Spectrum-2000). Russian Foundation for Basic Research, Institute of Mathematical Problems of Biology, Russian Academy of Sciences, pp. 136–138 (2000). https://www.elibrary.ru/item.asp?id=32656899

5. Kholodnaya, M.A.: Intelligence Psychology. Paradoxes and studies: Textbook for Bachelor's and Master's programme (in Russian), p. 334. Urait Publishing House (2019). https://aldebaran.ru/author/aleksandrovna_holodnaya_marina/kniga_psihologiya_intellekta_paradoksyi_issled/
6. Akperov, G.I.O., Alekperov, I.D.O., Khramov, V.V.: Intelligent information systems in the age of digital economy(in Russian), p. 113. Publishing House of South University (Institute of Management, Business and Law) (2020). https://www.elibrary.ru/item.asp?id=43829920
7. Akperov, I.G., Khramov, V.V.: Soft models of evaluating the condition of information ecology of the region single geoinformation space (in Russian). In: Collection of Papers: Innovative Technologies in Science and Education ("ITSE 2020" Conference). Collection of Research Papers of the VIII International Research-to-Practice Conference, with the Use of Remote Technologies. Rostov-on-Don, pp. 29–33 (2020). https://doi.org/10.23947/itno.2020.29-33
8. Kramarov, S.O., Khramov, V.V., Grebenyuk, E., Belyaev A.: Multiparametric selection of distance learning tools using fuzzy approaches (in Russian). In: Collection of Papers: E3S Web of Conferences. VIII International Scientific and Practical Conference "Innovative Technologies in Science and Education" (ITSE 2020), pp. 18024 (2020). https://doi.org/10.1051/e3sconf/202021018024
9. Nariniani, A.S.: NO factors: ambiguity and under definiteness: difference and interconnection (in Russian). Izv. RAS, Theo. Manage. Syst. **5**, 44–56 (2000). https://davaiknam.ru/text/ne-faktori-neodnoznachnoste-doformalenoe-issledovanie-a-s-nari
10. Zade, L.: Notion of linguistic variable and its application to making approximate decisions (in Russian). Mir (1976). https://scask.ru/a_book_zade.php
11. Khramov, V.V.: Concept of providing efficiency of organization and technical systems based on bionics and intelligent approach (in Russian). Bull. Rostov State Transp. Univ. **2**, 138–141 (2001). https://www.elibrary.ru/item.asp?id=11713661
12. Kanevsky, Z.M., Litvinenko, V.P., Makarov, G.V.: Security theory. Part 1, Fundamentals of Security Theory (in Russian), p. 92. Voronezh State Technical University, Voronezh (2003). https://www.studmed.ru/kanevskiy-z-m-litvinenko-v-p-makarov-g-v-osnovy-teorii-skrytnosti_a2717828d3e.html
13. Khramov, V.V.: A method for describing objects in the scene of a robot vision system by means of fuzzy Delaunay triangulation using an adaptive mesh. Int. Sci. J. Mod. Inf. Technol. IT Educ. [S.l.], **16**(4), ISSN 2411-1473 (2021). http://sitito.cs.msu.ru/index.php/SITITO/article/view/702
14. Khramov, V.V.: The concept of connectivity model measurements of geographic and information space of the region. In: Aliev, R.A., Yusupbekov, N.R., Kacprzyk, J., Pedrycz, W., Sadikoglu, F.M. (eds.) WCIS 2020. AISC, vol. 1323, pp. 553–558. Springer, Cham (2021). https://doi.org/10.1007/978-3-030-68004-6_72
15. Akperov, I.G., Khramov, V.V.: The concept of a unified geoinformational space of the region: ecological aspect. In: Collection of papers: E3S Web of Conferences. VIII International Scientific and Practical Conference "Innovative technologies in science and education" (ITSE 2020), p. 09006 (2020). https://doi.org/10.1051/e3sconf/202021009006

Use of Defuzzification Technique in a Pulse Echo Method for Cable Fault Locating

Salahaddin I. Yusifov, Farid H. Aghayev, and Rza S. Safarov(✉)

Azerbaijan State Oil and Industry University, Azadliq Avenue 1010, Baku, Azerbaijan

Abstract. A pulse echo method is widely used for cable fault locating. This approach is one of time domain reflectometry methods as well as arc reflection, and impulse current. Some advantages and disadvantages of a pulse echo method are listed. The main disadvantage of methods using an oscilloscope is difficulty of defining the returning moment in the result of distorting of the front of the pulse. Then a possibility of use of adaptive algorithms for cable fault locating was analyzed. The disadvantage of use of adaptive algorithms is the complexity of calculations. So, a new computational algorithm using centroid defuzzification method is offered for cable fault locating. Firstly, we proved that time diagrams (reflectograms) of the impulse reflected from the fault location can be considered as a fuzzy set. This new approach will be able to use in all problems connected with signals processing, not only in this problem. Then it is offered to define the fault location on basis of a defuzzification operation. We have chosen the centroid defuzzification method. In comparison, with methods using an oscilloscope, in the offering method, the distorting of the front of the pulse returning from the fault location does not disturb precision so much. In comparison, with adaptive algorithms, calculations carried out with the offered method are simpler.

Keywords: Pulse echo method · Time domain reflectometry · Adaptive algorithm · Defuzzification · Centroid method

1 Introduction

A pulse echo method is widely used for cable fault locating. This approach is one of time domain reflectometry methods as well as arc reflection, and impulse current. A pulse echo method technique uses a combination of a receiver and a transmitter. These devices are parts of a time domain reflectometer [1, 2].

According to this method a sounding signal from a transmitter is a short duration pulse signal that has a high frequency and a low voltage. This signal enters onto a cable. Traveling along the surface of the cable this signal's energy reflects from disruptions and comes back. The energy can reflect partially or in full. The reflected energy depends on a value of impedance changing in disruptions.

R. A. Aliev et al. (Eds.): ICSCCW 2021, LNNS 362, pp. 537–544, 2022.
https://doi.org/10.1007/978-3-030-92127-9_72

That time interval between a sounding pulse and a reflected pulse is measured by means of a time domain reflectometer. The distance until a fault location is calculated on basis of this time interval:

$$l = \frac{v \cdot t}{2} \tag{1}$$

where l is the distance from the time domain reflectometer until a fault location; v is a speed of traveling (propagation velocity) of the sounding pulse along the cable; t is the measured time interval between the sounding and reflected pulses. Dividing by 2 is because of the pulse goes until a disruption and as well as reflects from that during the time t.

A pulse echo method has some advantages. For example, the ability to determine damage that is unstable and distinguishing the places of each of several simultaneous damages. A fault location can be found fast and easily. The method uses pulses of small voltage that is safe for operating personnel and prevents stress to the cable.

Nevertheless, this method has some disadvantages. A part of systems based on this method includes an oscilloscope. With the help of an oscilloscope, the time between the sounding and reflected pulses is determined. As it is known, a real pulse has a finite duration and has an unlimited spectrum. Due to the limited bandwidth of the cable, the reflected pulse loses its high-frequency components. Since the steepness of the pulse fronts is due to the presence in their spectrum of components with frequencies that are many times higher than the fundamental frequency, the front of the reflected signal is distorted. In addition, as it travels through the cable, the pulse is attenuated and interfered with noise. As a result, it is difficult to determine accurately the beginning of the reflected pulse on the oscilloscope.

Adaptive [3, 4] and correlation [5] algorithms realized on computers are devoid of this disadvantage because, they do not just work with the impulse front. The values of the coefficients they calculate are most affected by the middle part of the pulse. The pulse's middle part has more energy (amplitude) and is less distorted. That middle part of the pulse consists of low-frequency components not high-frequency ones.

In [3, 4], it was analyzed an adaptive algorithm for cable fault locating. The cable has a multitude of conditional sections. Their lengths are equal to the distance to which the pulse can pass during the time τ, where the τ is the length of the sounding pulse. The algorithm is mathematically shown as follows:

$$C_I(J) = C_I(J-1) + \frac{A}{A+J}\left\{Y(J) - \sum_{K=1}^{N} C_K(J-1)X_K(J)\right\}X_I(J) \tag{2}$$

where $C_I(J)$ is the value of the coefficient C_I in the Jth moment, "A" is any constant coefficient, Y (J) is the value of the reflected pulse Y in the Jth moment, $X_I(J)$ is the value of the delayed pulse X_I of the sounding pulse X in the Jth moment.

The sounding pulse is repeatedly delayed for a time τ (the length of the sounding pulse). Index "I" shows the serial number of the sounding pulse delay. In general $I = \overline{1, N}$, where N is the number of conventional cable sections and delayed pulses. Delayed pulses are mutually orthogonal. The end of one delayed pulse coincides with the beginning of the next one.

The essence of the method based on the adaptive algorithm is that at 100% time coincidence of the pulse reflected from the damage with the Ith delayed pulse X_I (case 1), the value of the Ith coefficient C_I should noticeably exceed values of other coefficients C_M. Here $M = \overline{1, N}$ and $I \neq M$. In this case, localization of damage takes place in the Ith section.

In case of partial time coincidence of the pulse reflected from the damage with two subsequent delayed pulses X_I and X_{I+1} (case 2), similarly, the values of the coefficients C_I and C_{I+1} must noticeably exceed values of other coefficients C_K. Here $K = \overline{1, N}$, where $K \neq I$ and $K \neq I + 1$. Moreover, that coefficient of C_I and C_{I+1} has the greatest value, which corresponds to the delay of the sounding pulse, the most coinciding in time with the pulse Y reflected from the damage. In this case, the location of the damage is determined within sections I and I + 1.

The disadvantage of this method is the complexity of the calculations according to formula (2). In addition, it is necessary to imagine that a cable consists of many conventional parts. After finding the C_I coefficient corresponding to each Ith part, it is necessary to compare all C_I coefficients to determine the largest one of them (case 1) or the largest two of them (case 2). In both cases (case 1 and case 2), the coming moment of the pulse returning from the location of damage is determined by the index of the largest coefficient C_I: $t = I \cdot \tau$.

In this paper, a new computational algorithm using centroid defuzzification method is offered for cable fault locating.

2 Method

In this paper, a new calculation algorithm for cable fault locating on basis of pulse echo method is offered. The algorithm is based on using centroid defuzzification method. Here we mainly use defuzzification as fuzzy technology. To do this, we must firstly prove that the time diagrams (reflectograms) of the impulse reflected from the fault location can be considered as a fuzzy set.

First of all, let us remember what a fuzzy set is.

Definition of fuzzy sets [6, 7]. Suppose that x elements belong to the universal set X, i.e. $x \in X$. If we mark entering of x elements to a set A with $\mu_A(x)$, then a collection of pairs $(\mu_A(x), x)$ is a fuzzy set A in a universal set X. The quantity $\mu_A(x)$ is called the membership degree of the fuzzy set A.

Reflectograms show the time dependence of the amplitude of the pulse returning from the fault location (Fig. 1). In other words, it shows exactly when the impulse returning from the fault location exists. We are just interested in that moment. If we know this moment, we can determine the fault location according to formula (1). Let us imagine a fuzzy set called "the returning moment of the impulse from the fault location". In this case, the all time axis, i.e. $t \in (0, \infty)$, plays the role of a universal set. Therefore, different moments of time are elements of the fuzzy set that we define. The amplitude of the impulse returning from the fault location plays the role of a membership degree. Indeed, in the absence of the pulse returning from the fault location, the amplitudes on the reflectogram are either zero or very close to zero. In the presence of the pulse returning from the fault location, the amplitude differs sharply from the background of total zero (or close to zero). In other words, in those moments, the membership degree gets higher in the fuzzy set, which we call "the returning moment of the impulse from the fault location". Then we can say that time moments, during that the impulse returning from the fault location exists really, enter the fuzzy set we defined with larger membership degrees.

Fig. 1. A reflectogram of the pulse returning from the fault location.

It should also be noted that the shape of the impulse returning from the fault location is similar to the Gaussian membership function. This is due to the fact that according to theory of signals, fronts of the impulse returning from the fault location are more distorted. This is because the pulse fronts are formed by the high-frequency components of the signal. Because the bandwidth of the cable and the receiver is limited, some of these high-frequency components are cut off. In other words, the pulse is smoothed. We talked about this above. The resemblance of the impulse returning from the fault location to a Gaussian membership function encourages us to consider time diagrams as a fuzzy set. It should also be noted that the Gaussian membership function belongs to convex membership functions. Fuzzy theory prefers the use of convex membership functions.

There are two possible objections to considering the time diagrams of the pulse returning from the fault location as a fuzzy set. Firstly, as accepted in fuzzy theory, membership degree varies in the range [0, 1]. In the fuzzy set we set, the amplitude of the impulse, which plays the role of membership degree, can be greater than 1. The second objection may be that the membership degree is not negative. The amplitude of the pulse returning from the fault location can be negative, as shown in the Fig. 2. In short, the amplitude of the pulse returning from the fault location can exceed the range [0, 1] in both directions.

Fig. 2. A negative sign of the pulse returning from the fault location.

In response to the first objection, we can say that amplitudes of the pulse returning from the fault location in the time diagram can be divided by the amplitude of the sounding pulse. Normally, the amplitude of the pulse returning from the cable is not larger than the amplitude of the sounding pulse. This is due to the energy of the pulse being damped during propagation in the cable. If the amplitude of the sounding pulse is equal to the amplitude of the returning pulse, then we obtain a normal fuzzy set after the dividing. If the amplitude of the sounding pulse is greater than the amplitude of the pulse returning from the fault location, then we obtain a subnormal fuzzy set. By dividing amplitudes of the returning pulse in the time diagram by the maximum amplitude of the returning pulse, then again, we get a normal fuzzy set. Therefore, we respond to the first objection by a normalizing operation.

In response to the second objection, we can say that the polarity of the impulse returning from the fault location may be both positive and negative. A positive polarity indicates a break in the cable line, and a negative polarity indicates a short circuit. Thus, the polarity of the returning impulse provides useful information about the type of damage. Once we have this information about the type of damage (breakage or short circuit), we can get rid of the negative sign by operating with the modulus of the amplitudes in the course of calculations. Also, we can keep the negative polarity as it is. We will show below that it does not disturb the result of calculations (finding the moment of return).

So we proved that the time diagrams of the impulse returning from the fault location can be considered as a fuzzy set.

It can also be proved that time diagrams of the pulse returning from the fault location can be considered as a fuzzy number. That is a fuzzy number indicating the time. The fuzzy number itself is a fuzzy set, but under certain conditions.

First of all, let us remember what a fuzzy number is.

Definition. Fuzzy numbers [6, 7]. If a fuzzy set is normal, convex, it has the restricted support and each of its alpha-cuts is a closed interval then such a fuzzy set is a fuzzy number.

Let us examine whether the conditions set out in this definition are met. Above it was mentioned the possibility of normalization in response to the first objection. As can be seen from the Fig. 1, the time diagram of the pulse returning from the fault location is in the form of a convex function. In the fuzzy set, which we call "the returning moment of the impulse from the fault location", for any three different time moments t1, t2, t3 (where $t1 < t2 < t3$) the condition $\mu(t2) > \min[\mu(t1), \mu(t3)]$ is met. The fact that the shape of the impulse returning from the fault location is close to the Gaussian membership function means that support is restricted and each of its alpha-cuts is a closed interval.

As you can note, all the conditions are met to consider time diagrams as a fuzzy number. So we have proved that the time diagrams of the pulse returning from the fault location can also be considered a fuzzy number.

Now we can apply the defuzzification operation to the time diagrams (fuzzy sets) of the pulse returning from the fault location.

Suppose that the fuzzy set we obtain is any aggregated fuzzy set. Let us apply the defuzzification operation to that fuzzy set. We choose the centroid defuzzification method as the type of defuzzification. This method is one of the most popular ones. If the universal set X is finite then the formula of the defuzzification centroid method is given below:

$$\frac{\sum_{j=1}^{k} x_j \cdot \mu_A(x_j)}{\sum_{j=1}^{k} \mu_A(x_j)} \tag{3}$$

Here x_j are elements of a fuzzy set, and $\mu_A(x_j)$ are membership degrees of the inclusion of those elements in any fuzzy set "A". We write time moments t_j instead of x_j and Y_j instead of $\mu_A(x_j)$. Then we obtain for the centre of the impulse returning from the fault location:

$$t_{centre} = \frac{\sum_{j=1}^{k} Y_j \cdot t_j}{\sum_{j=1}^{k} Y_j} \tag{4}$$

The use of defuzzification technique in a pulse echo method for cable fault locating implies the use of formula (4).

We stated above that saving a negative polarity as it is in case of a short circuit does not disturb the result of calculations (finding the moment of return). Indeed, if the values Y_j are less than zero, both the numerator and the denominator in expression (4) are negative. As a result, the moment of return of the impulse from the fault location remains positive.

Here is a description of the designed algorithm.

1. A sounding signal from a transmitter enters onto a cable. That signal reflects from the fault location and comes back.
2. All the trace of the reflected signal (reflectogram) is sampled and measured as values Y (J) in moments J. The sampling interval Δt is chosen to provide some samples within the pulse duration t_{pulse}. So in (4), $t_j = J \cdot \Delta t$.
3. A normalizing operation is applied to values Y (J) and as it is proven above, we obtain a fuzzy set.
4. Formula (4) is applied to the fuzzy set and we find the centre of the impulse returning from the fault location.
5. The returning moment (beginning) of the impulse from the fault location is found as following:

$$t_{reflection} = t_{centre} - \frac{t_{pulse}}{2} \tag{5}$$

Let us carry out simulation for theoretical proving of above mentioning. Suppose that the maximal length of the cable is 750 m. There is a break from the beginning of the cable in the 600 m distance. We assume the speed of spreading of an electromagnet wave in the cable as $3{\cdot}10^8$ m/sec. Then the returning moment of the impulse from the fault location according to formula (1) is

$$t_{reflection} = 4 \cdot 10^{-6} s.$$

A resolution equal to 0.9 m is accepted. So the pulse duration is

$$t_{pulse} = 6 \cdot 10^{-9}\ s.$$

We obtain the following values from (4) and (5)

$$t_{centre} = 4.0030e - 006$$

$$t_{reflection} = 4.0000e - 006$$

as a result of simulation. It corresponds to primary data and formula (1). All of these show that calculation is true.

3 Results

Time diagrams (reflectograms) of the impulse reflected from the fault location can be considered as a fuzzy set.

1. Time diagrams (reflectograms) of the impulse reflected from the fault location can also be considered as a fuzzy number.
2. Defining the fault location on basis of the centroid defuzzification method is offered.
3. Research carried out with Matlab program package has showed that the coming moment of the pulse returning from the fault location was found with high precision.

4 Discussion

Research carried out in this paper and its results are important. A new calculation method for cable fault locating based on pulse echo method with application of a defuzzification method is analysed. The centroid defuzzification method is used for this purpose. The returning pulse was considered as a fuzzy set (number) for the first time for this purpose.

Calculations carried out with the offered method are simpler than calculations on basis of adaptive algorithms.

Carried out calculations showed high precision of the results.

5 Conclusion

A method for cable fault locating based on pulse echo method with application of the centroid defuzzification method is offered for the first time. This approach has a meaning also for synthesizing of the signals theory with the fuzzy theory. Idea of possibility of considering the pulse returning from the fault location as a fuzzy set is put forward. This new approach will be able to use in all problems connected with signal processing, not only in this problem.

A disadvantage (difficulty of defining the returning moment in the result of distorting of the front of the impulse) of methods using an oscilloscope for fault locating has been shown.

In the offering method, the distorting of the front of the pulse returning from the fault location does not disturb precision so much, because the centre of the returning pulse is found. Since the centre of the returning pulse is more precisely found, the coming back moment of the pulse according to formula (6) is defined with high precision, too.

It is necessary to note that in methods using an oscilloscope, a small duration of the sounding pulse is required to provide a high resolution. But there is no need to take so little time duration of the sounding pulse in the offering method, because the centre of the returning pulse is found with high precision. This is explained with two causes. In the first place, the centroid defuzzification method defines the centre of the returning pulse with high precision. Secondly, in accordance with signals theory, the centre of the returning pulse is less distorted.

A disadvantage (difficulty of calculations) of methods using adaptive algorithms have also been shown. A comparison of formulae (2) and (4) shows that the offering method demands simpler calculations.

References

1. Introduction to Cable Fault Locating. www.kep.ua
2. Giaquinto, N., Scarpetta, M., Spadavecchia, M.: Algorithms for locating and characterizing cable faults via stepped-frequency waveform reflectometry. IEEE Trans. Instrum. Meas. **69**, 7271–7280 (2020). https://doi.org/10.1109/TIM.2020.2974110
3. Abdullayev, I.M., Yusifov, S.I., Safarov, R.S.: On computer modeling an adaptive algorithm recognising the fault location and character in a communication cable. In: Scientific Papers 1996, no. 3, pp. 200–204, Baku (1996)
4. Safarov, R.S.: On computer modeling adaptive algorithms recognising the fault location and character in a communication cable. In: Scientific Papers 1998, Baku, no. 1, pp. 183–188 (1998)
5. Safarov, R.S.: A correlation approach for definition of the fault location in communication cables for control systems in oil industry. In: Proceedings of the Azerbaijan Academy of Sciences. Series of Physicotechnical and Mathematical Sciences 1998, Baku, vol. XVIII, no. 1, pp. 34–36 (1998)
6. Ross, T.J.: Fuzzy Logic with Engineering Applications. Wiley (2010). https://doi.org/10.1002/9781119994374
7. Belohlavek, R., Klir, G.J. (eds.): Concepts and Fuzzy Logic. The MIT Press (2011). https://doi.org/10.1007/s11225-014-9586-8

Application of the Fuzzy Cobb-Douglas Model to Predict the Final Product of the Region

V. J. Akhundov(✉)

Research Laboratory of Intelligent Control and Decision Making Systems in Industry and Economics, Azerbaijan State Oil and Industry University, 20 Azadlig Avenue, AZ1010 Baku, Azerbaijan

Abstract. The economic development of the region is significantly influenced by a set of factors. The aim of the study is, taking these factors into account, the construction of a fuzzy model that allows calculating the fuzzy value of the gross regional product ($\tilde{Y}$). The object of analysis is the regions of the Azerbaijan Republic. It should be noted that the work studied statistics for 10 regions. In the article, the problem of forecasting the regional final product is solved using the Cobb-Douglas fuzzy model.

Keywords: Econometric model · Gross regional product · Fuzzy number · Fuzzy Cobb-Douglas model · Confidence intervals

1 Introduction

Clark's rigid accelerator model [1] is one of the first attempts to explain investment behavior at the aggregate level. He suggested that the only reason firms invest is attaining the optimal level of capital changes, which in turn is proportional to the output. An attempt to provide such a justification using a neoclassical approach is contained in the work of Grossman (1972) [2] which draws attention to the fact that the marginal return on investment can positively depend on the volume of investment. In the models discussed above, the aggregate demand was assumed to be known. However, uncertainty in aggregate demand can affect aggregate investment. Therefore, other authors propose a model that includes both definite and uncertain demand.

In paper [3] it is shown an application of fuzzy modeling in solving a problem of scenario forecasting of regional development, based on the analysis of expert evaluation data and the data of regional statistics.

This paper is devoted to modeling gross regional products of Azrebaijan regions by using fuzzy Cobb-Douglas function.

The second section of the article presents an econometric model explaining the gross regional product of the regions of the Republic of Azerbaijan from the point of view of investments in fixed capital. According to the constructed model, the variances of random deviations were determined, and dependence of Gross Regional Product (GRP) on investments is considered.

In the third section of the article, fuzzy values for the final product are calculated for 10 regions of the country using the fuzzy Cobb-Douglas model.

R. A. Aliev et al. (Eds.): ICSCCW 2021, LNNS 362, pp. 545–551, 2022.
https://doi.org/10.1007/978-3-030-92127-9_73

2 Regional Econometric Specifications

Investment activity stimulates business development and leads to an increase in the received profit. One of the important areas of investment is investment in fixed assets, which represent the cost of construction and reconstruction of fixed assets, leading to an increase in their initial cost. Thus, the goal in the first stage of the study is to use an econometric model that allows explaining the gross regional product by investments in fixed assets. The object of analysis is the Aran economic region of the Azerbaijan Republic (Table 1).

Table 1. Indices of Aran economic region of the Republic of Azerbaijan

	2010	2011	...	2018	2019	2020
Fixed industrial-production funds (K) mln.man	1699.321	2067.173	...	4712.208	4772.049	5220.694
Population in comes (L) mln.man	3393.2	4104.95	...	6793.32	7216.63	6987.4
Gross Regional Product (Y) mln.man	2973.691	3874.433	...	6926.287	7521.157	7075.981
Fixed capital investments (I) mln.man	294226.2	510780.3	...	708680.7	846430.3	755489

It is worth noting that the work studied statistics for 10 regions.

As an initial specification of the econometric model, the paper proposes using a regression equation [4]:

$$Y_i = \boldsymbol{a}_0 \cdot X_{1,i}^{a} \cdot \varepsilon_i \quad (1)$$

where $\acute{\iota}$ is the index of observation of the region, varies from 1 to 10; $Y_{\acute{\iota}}$ – gross regional product for the subjects of the Azerbaijan Republic; $X_{1,\acute{\iota}}$ – investments in fixed capital for $\acute{\iota}$ region per year; parameter $\boldsymbol{a}$ is the coefficients of elasticity of GRP for investments in fixed assets.

The values of the size of the gross regional product, depending on investments in fixed assets, demonstrate that with the growth of investments costs, the GRP also grows.

However, it should be noted that with similar costs for technological innovation, the regions have different GRPs, which may indicate the influence of other factors, such as the location of the region, its resource availability and others. Using the actual data on "investment in fixed assets" and "production volume" for the Aran economic region, we check the assumption of homoscedasticity using the Goldfeld-Quandt test (Table 2). This testing led to satisfactory results and the adoption of the null hypothesis of the homoscedasticity of random disturbances in the model, since the p-value for each regressor turned out to be greater than the specified significance level.

The final part of this frequent research is to construct a prediction of the variable ***Yi***. Consequently, model (1) can be recognized as adequate to the actual dependences of GRP on investments in fixed assets. In order to evaluate the statistical data for K, L, Y,

Table 2. The results of the Goldfeld-Quandt test for the Aran economic region

Explaining variable	p-value
$X_{1 \cdot inew}$	0.23

I the Goldfeld-Quandt test is used. At this stage of the study, confidence intervals (Ỹ−; Ỹ+) are determined.

3 Forecasting the Final Product by Using the Fuzzy Cobb-Douglas Model

In the context of an imperfect information about ongoing regional investment projects, the use of fuzzy sets becomes relevant. With this factor in mind, when forecasting the final product for the regions of the Republic of Azerbaijan, a fuzzy model of Cobb-Douglas was used [5, 6]:

$$\tilde{Y} = \tilde{A} \cdot \tilde{K}^{\tilde{\alpha}} \cdot \tilde{L}^{\tilde{\beta}} \tag{2}$$

where the values of $\tilde{Y}'$, K̃, L̃, and parameters A˜, α˜, β˜ are described by fuzzy numbers. Using fuzzy data, we will calculate fuzzy values for the end product of the region based on fuzzy estimates of the fixed assets and the average number of employees.

Table 3 shows fuzzy estimates (in form of triangular fuzzy numbers, TFN) of the average annual cost of fixed assets and the average number of employees in the regions, taking into account the degree of probable difference in 2010–2020.

Using this table, construction of Cobb-Douglas production function under fuzzy information is based on a solution to the following optimization problem:

$$\sum_{years} d\left(\tilde{Y}, \tilde{Y}'\right) \rightarrow \min., \ \text{s.t.} \underline{\tilde{A}} \leq \tilde{A} \leq \overline{\tilde{A}}, \ \underline{\tilde{\alpha}} \leq \tilde{\alpha} \leq \overline{\tilde{\alpha}}. \tag{3}$$

$\tilde{Y}$ denotes fuzzy reduced values of end product (Table 3), $\tilde{Y}'$ is a fuzzy value of function (2), $\tilde{A}$, $\tilde{\alpha}$ and $\tilde{\beta}$ are the fuzzy parameters for the fuzzy Cobb-Douglas production function, $\overline{\tilde{A}}$, $\underline{\tilde{A}}$ and $\underline{\tilde{\alpha}}$, $\overline{\tilde{\alpha}}$, $\underline{\tilde{\beta}}$, $\overline{\tilde{\beta}}$ are the boundary conditions of the ranges of these parameters. Distance between fuzzy numbers d is adopted from comparison of fuzzy numbers ($\leq$ is considered as formulated in [7]). The problem is to find such values of $\tilde{A}$, $\tilde{\alpha}$ and $\tilde{\beta}$ that minimize distance d between fuzzy data and fuzzy values of Cobb-Douglas production function(distance d is used as in [8]). $\tilde{A}$, $\tilde{\alpha}$ and $\tilde{\beta}$ parameters are taken for the studied Aran region production as follows: $\underline{\tilde{A}}$ = (0.98; 1; 1.02), $\overline{\tilde{A}}$ = (294.12; 300; 306); $\underline{\tilde{\alpha}}$ = (0.098; 0.1; 0.102), $\overline{\tilde{\alpha}}$ = (0.882; 0.9; 0.918); $\underline{\tilde{\beta}}$ = (0.098; 0.1; 0.102), $\overline{\tilde{\beta}}$ = (0,882; 0,9; 0,918).

By solving the problem, the fuzzy parameters $\tilde{A}$ and $\tilde{\alpha}$ and $\tilde{\beta}$ for the investigated Aran region production were found as follows: $\tilde{A}$ = (0.538; 0.549; 0.56); $\tilde{\alpha}$ = (0.882; 0.9; 0.918); $\tilde{\beta}$= (0.171; 0.174; 0.1779).

Table 3. Fuzzy estimates of Y, K, L for the Aran region (2010–2020).

Years	The probable difference rate of Y	Average annual cost of fixed assets, (thousand dollars) $\tilde{K}$	Average number of employees (thousands people) $\tilde{L}$	Cost of extracted oil and gas (thousands of dollars) (the reduced value of the product) $\tilde{Y}'$
2010	23%	(1648.3;1699,3;1750,3)	(3291;3393,2; 3495)	(2417; 2973.7; 3657,6)
2011		(2005.2; 2067.2; 2129.2)	(3981.8;4104.9;4228)	(3150;3874.4; 4765.5)
2012		(2250.5; 2320.2; 2389.8)	(4514.1;4653.6;4793)	(3531.7; 4344; 5343.2)
2013		(2640.3; 2722; 2803.6)	(4767.6;4915;5062.5)	(3933.7;4838; 5951.3)
2014		(3109.5; 3273.2; 3436.8)	(4910; 5168.3; 5426.7)	(3941.5; 4848; 5963.2)
2015		(2950.8; 3106.1; 3261.4)	(5062.7;5329.1;5595.6)	(3976; 4890.5; 6015.3)
2016		(3663.2; 3856; 4048.77)	(5578.8; 5872.4;6166.1)	(4140.8; 5093; 6264.6)
2017		(5197.5; 5471; 5744.6)	(6026.1; 6343.2;6660.4)	(5251.3; 6459; 7944.8)
2018		(4476.6; 4712.2; 4947.8)	(6453.6; 6793.3;7132.9)	(5631;6926.2; 8519.3)
2019		(4533.5; 4772.1; 5010.7)	(6855.8; 7216.6;7577.4)	(6114.7; 7521; 9251)
2020		(4855.3; 5220.7; 5010.7)	(6498.2; 6987.4;7476.5)	(5752.8; 7076; 8703.4)

The obtained value of (2) of the solution was 218,65 thousand manats, which is of a high accuracy. Thus, the obtained results can be used for modeling of production in the regions with acceptable accuracy. The corresponding values of Cobb-Douglas production function $\tilde{Y}'$ are shown in Table 4.

Table 4. Fuzzy Cobb-Douglas function values

Years	TFN-based values of production function, $\widetilde{Y'}$
2010	(2413.806; 3021.495; 3781.879)
2011	(2952.476; 3711.003; 4664.042)
2012	(3363.683; 4239.12; 5341.984)
2013	(3627.449; 4578.583; 5778.658)
2014	(3828.292; 4946.906; 6386.376)
2015	(3898.143; 5039.027; 6507.704)
2016	(4406.518; 5710.457; 7393.302)
2017	(5007.27; 6505.926; 8445.192)
2018	(5185.32; 6742.071; 9278.503)
2019	(5481.088; 7134.708;927545.25)
2020	(5289.831; 7039.869; 9343.943)

The fuzzy values of end product and those of Cobb-Douglas production function $\widetilde{Y'}$ are shown graphically below (Fig. 1, the lower bound, the core and the upper bound of the TFNs are shown).

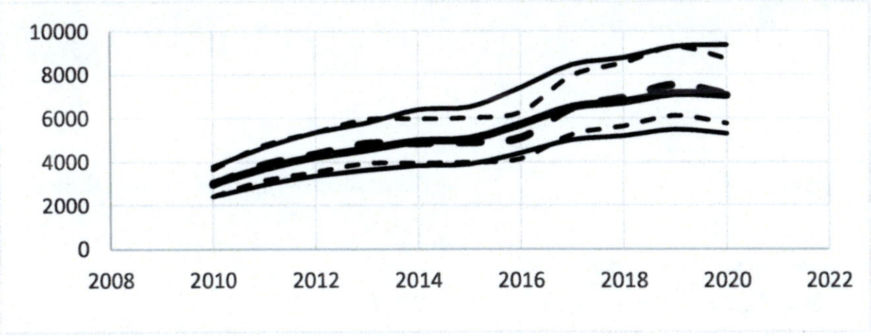

Fig. 1. Fuzzy Cobb-Douglas function graph (solid curve) and imprecise data (dashed curve).

Likewise, values for $\tilde{A}$, $\tilde{\alpha}$, $\tilde{\beta}$ were determined for the 10 economic regions that make up the economy of Azerbaijan, and the results are presented in Table 5 below.

As can be seen from the table, no positive effect of production scale ($\tilde{\alpha} + \tilde{\beta} > 1$) was observed in any of the studied regions (10 regions), except for Aran economic region, Nakhchivan Autonomous Republic, Absheron economic region.

Table 6 compares the accuracy of modeling for all regions, calculated with the defined values $\tilde{A}$, $\tilde{\alpha}$, $\tilde{\beta}$ in the research process.

Table 5. Determined values of $\tilde{A}$, $\tilde{\alpha}$, $\tilde{\beta}$ by regions

	Regions	Ã	$\tilde{\alpha}$	$\tilde{\beta}$
1	Baku city	(202.9; 207; 211.4)	(0.457; 0.466; 0.48)	(0.098; 0.1; 0.102)
2	Absheron economic region	(0.119; 0.12; 0.124)	(0.437; 0.446; 0.45)	(0.882; 0.9; 0.918)
3	Ganja-Gazakh economic region	(43.5; 44.41; 45.3)	(0.35; 0.357; 0.364)	(0.262;0.267;0.273)
4	Sheki-Zagatala economic region	(4.35; 4.44; 4.53)	(0.71; 0.723; 0.738)	(0.098; 0.1;0.102)
5	Lankaran economic region	(14.7; 15; 15.3)	(0.098; 0.1; 0.102)	(0.543; 0.554; 0.564)
6	Guba-Khachmaz economic region	(2.05; 2.1; 2.137)	(0.87; 0.88; 0.91)	(0.098; 0.1; 0.102)
7	Aran economic region	(0.538; 0.549;0.56)	(0.882; 0.9; 0.918)	(0.171; 0.174; 0.178)
8	Karabakh economic region	(2.19; 2.24; 2.28)	(0.098; 0.1; 0.102)	(0.81; 0.83; 0.843)
9	Mountainous-Shirvan economic region	(1.67; 1.7; 1.73)	(0.098; 0.1; 0.102)	(0.83; 0.846; 0.863)
10	Nakhchivan Autonomous Republic	(0.49; 0.5; 0.51)	(0.882; 0.9; 0.918)	(0.289; 0.295; 0.301)

Table 6. Mean square deviation of $\widetilde{Y'}$ from the data

	Regions	The standard deviation of $\widetilde{Y'}$ from the data
1	Baku city	9.56%
2	Absheron economic region	22.1%
3	Ganja-Gazakh economic region	5.57%
4	Sheki-Zagatala economic region	16.80%
5	Lankaran economic region	13.07%
6	Guba-Khachmaz economic region	10.78%
7	Aran economic region	5.53%
8	Karabakh economic region	4.47%
9	Mountainous-Shirvan economic region	22.68%
10	Nakhchivan Autonomous Republic	7.27%

As can be seen from the table, the greatest deviation was in the Mountainous-Shirvan economic region. A large number of impact factors and their influence on production led to a more squared deviation (22.68%) of fuzzy values for the final product in the region.

4 Conclusion

In order to get a more adequate forecast of end product it is necessary to take into account imprecise information. In this paper fuzzy Cobb-Douglas model to account for fuzzy values of end product, capital, labor and production parameters is used. The obtained results show adequacy of the used approach.

References

1. Clark, J.: Business acceleration and the law of demand: a technical factor in economic cycles. J. Polit. Econ. **25**, 217–235 (1917). https://doi.org/10.1086/252958
2. Grossman, H.: A choice-theoretic model of an income-investment accelerator. Am. Econ. Rev. **62**, 630–641 (1972). https://doi.org/10.2307/1806104
3. Luchyk, V., et al.: Forecasting of cross-border regions development: scenario approach. Editura Universitatii "Stefan cel Mare", Suceava (2017). https://doi.org/10.31617/Lknute.2019-10-10.44
4. Baltagi, B.H., Fingleton, B., Pirotte, A.: Estimating and forecasting with a dynamic spatial panel data model. Oxford Bull. Econ. Stat. **76**, 112–138 (2014). https://doi.org/10.1111/obes.12011
5. Aliev, R.A., Fazlollahi, B., Aliev, R.R.: Soft Computing and Its Applications in Business and Economics. Springer Berlin Heidelberg, Berlin, Heidelberg (2004)
6. Akhundov, V.J.: Modeling of oil and gas production for an enterprise by using fuzzy Cobb-Douglas function. In: 14th International Conference on Theory and Applications of Fuzzy Systems and Soft Computing – ICAFS-2020, pp. 731–737, Budva, Montenegro (2020). https://doi.org/10.1007/978-3-030-64058-3_91
7. Zhang, X., Chen, W.M.: New similarity of triangular fuzzy number and its application. Adv. Inf. Technol. 1–7 (2014). https://doi.org/10.1155/2014/215047
8. Zhang, G.Q.: Group Decision Making Method with Fuzzy Preference of Decision Makers. Faculty of Information Technology University of Technology, Sydney, P.O. Box 123, Broadway, NSW 2007, Australia (2007)

Applying Multi-layers Feature Fusion in SSD for Detection of Small-Scale Objects

Elbrus Imanov[1] (✉) and Zubair Shah[2]

[1] Department of Computer Engineering, Near East University, via Mersin 10, North Cyprus, Turkey
elbrus.imanov@neu.edu.tr

[2] Research Assistant Area of Computer Engineering, Sakhı Chesma, Peshawar, Pakistan
zubair.peshawar@gmail.com

Abstract. At present, object detection using convolution neural networks occupies a dominant position. However, the structure of the competent neural network faces hereditary difficulties: high-level networks have large receptive areas and meaning information has high capacity to illustrate, but the solution is diminished, the calculations are weak. In grassroots networks there are relatively small acceptable areas, and it has strong geometric detail information representation capability. SSD (single shot multi-box detection) object prediction of multi-level feature uses mAP, through predicting things with large fields of reception and small fields of prediction of conditions, it also uses high-layer facilities to do. Based on the analysis and introduction of classic SSD algorithms, in this paper, we aim to find small-scale complain at a fast run, through presenting an approach which adapts the SSD detector in relation to trading with precision and speed. An MSSD (Modified single shot multi-box detection) procedure of multistage components combination is offered to supply old knowledge in SSD. The VGGl6 and deep residual networks are used by the MSSD. With Residual network, this article uses the FPN-based network architecture to integrate high and low layers and improves the traditionally sampled structure. The high-layer semantic information is integrated into the low-layer network feature information, and the multi-scale feature mAPs for predicting relapse location case and classification work input are enriched to develop the disclosure efficiency. Experimental results show that these two fusion modules obtain better mAP on PASCAL VOC2007 and Logo datasets than base-line SSD, especially on some small objects categories.

Keywords: Small object detection · Single shot multi-box detector · Feature pyramid networks · Real-time · Mean average precision · Convolution neural networks

1 Introduction

Advanced technology brings computer hardware and software from artificial intelligence, unmanned driving, intelligent transportation, travel to identification and reuse of many other applications. Computers should be used to detect and identify Real-time

R. A. Aliev et al. (Eds.): ICSCCW 2021, LNNS 362, pp. 552–559, 2022.
https://doi.org/10.1007/978-3-030-92127-9_74

tracking, rapid tracking which makes it important as a fundamental, and long-standing subject. In computer vision, the detection of object has always been in practice for decades, and it dominates an important and active area of research.

Recently, the techniques of deep learning have emerged as efficient and powerful methods for automatic learning features representation from [1, 2]. Specifically, these techniques had already provided significant improvement for object detection as never before, which is a practice that has mobilized large number of attentions recently. This problem is extremely relevant in many of today's challenging research applications like detecting traffic signs, pedestrians and cars on roads and a so on. Artificial Neural Network has become a popular direction in the field of artificial intelligence research. First, the method abstracts the neural network of living objects from the direction of information analysis and processing, and then builds some simple models. Finally, these models are combined according to different connected methods to form different network structures. McCulloch and Pitts established the MP model in the early 1940s [3]. Subsequent years of investigation and development, the theoretical structure of artificial neural networks has achieved excellent results in a wide range of research areas. At the same time, a large number of neural network structural models are proposed, such as perceptions, feed-forward neural networks [4], Boltzmann machines convolution neural networks, and so on [5]. Whilst relevant feature has to be manually extracted for traditional machine learning, which in deep learning certain features are learned by using raw images as input. A CNNs working on a layer of sends and receives also several hidden layers, convolution, pooling and entirely linked films are examples like invisible layers. Although CNNs have been proved efficient and effective on object detection, the reliable and accurate detection for small objects, due to their limited information and resolution in images deemed a quite challenging task [6],which the small object activations become forwardly smaller with each pooling layer, as passing an image throughout a standard CNN architecture like ResNet or VGG16 [7] where current methods often cannot detect small size objects effectively as well as they have done for the large objects [8]. The challenges involved in the detection of small objects are multiple folds, but the biggest challenge stems from the comparatively small size of an object compared to its background in an image, e.g., the small object of interest occupies only 1–5% percent of pixels in an image. In addition, the input size for all these networks is indicated by in the place of storage over GPUs due to the enormous network running memory requirements. For instance, an SSD detection model based on VGG16 needs more than 10 gigabytes to handle only a single image with a 2048 R 2048 input size [9, 10]. Simplifying the network, e.g., using a shallow one, with a tradeoff of performance degradation is the only way to overcome the above-mentioned problem. This research work is based on existing SSD methods, receptive fields and low-level information about the characteristics used to predict small objects. When a person is at the specific level smaller than SSD detection it may have adverse effects on objects. In order to solve this problem with top level iconic detail, little features facts must be combined. In our study, the solved object detection problems accuracy and increased speed classification are realized. It has important practicability in artificial intelligence, unmanned driving, intelligent transportation, face recognition and other fields. The rest of the study is structured as follows. Part 2 involves system and methodology used for our research, and it is followed by part 3, which involves the experimental results and analysis. Moreover, in part 4 conclusion points are presented.

2 System Design and Methodology

Single Shot Multi-Box Detector Network is one of the architectures used most commonly for detecting objects. The network is fully convolution and can therefore be used for images with any resolution. Two architectures are proposed in the original paper: 300 × 300, input resolution architecture and 512 × 512-pixel input resolution architecture. Our baseline model is SSD300 because its default model and original document is described [11]. Furthermore, other papers also use this network as a baseline [12]. The SSD network is called Single-shot since both localization and classification of objects were carried out through the network within a single feed-forward. This is contrast to, for example, the Faster-RCNN network [13], from which it differs since it does not have a separate regional proposal network. The network consists of three parts, a base network, SSD layers and prediction layers attached to multiple feature maps in the network. These first layers are called the base network and this base network consists of stacked convolutions to decrease size. The purpose of this base network is to provide response maps that enable detection of different sizes. The base network can be seen in Fig. 1 and is represented the convolutions conv1 till conv5. Here, we use a truncated VGG16 base network and initialize those layers with image-net weights. However, as mentioned by the authors one could replace the base network by any standard or non-standard architecture, e.g., inception or resent [14, 15]. Network architecture is shown in Fig. 1.

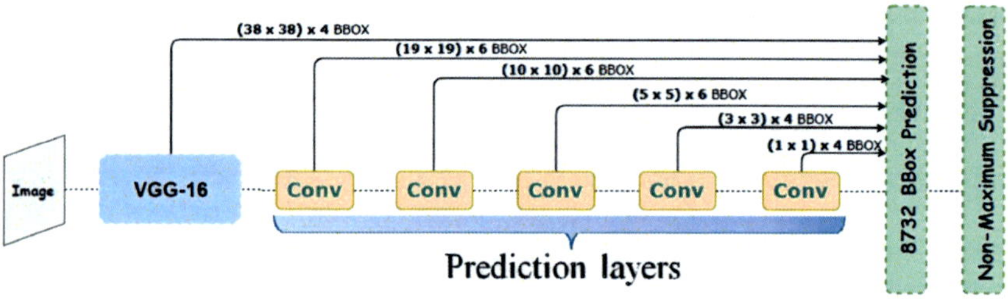

Fig. 1. SSD network architecture.

Modified single shot multi-box detector model from the SSD model is introduced and the knowledge of the deep residual network, the researcher found that with the deepening of the depth convolution network level. In image classification functions, outstanding networks have proven to be better than VGGS because it provides escape links between blocks. In fact, ResNets typically can go up to 101 layers whereas VGG networks can only go up to 16. Because deep nets are often better for sorting images, ResNets are generally more accurate than VGG. Here we will apply our idea on both of the VGG and ResNet networks showing the result that we got in our experiment later. The shallower layers (conv-3) suffering from lacking of semantic information, so in order to compensate that lack we inject contextual information from other layer and come out with new design model named element-sum and concat-sum model. Network structure based a deep VGG16 network base network to obtain feature maps in order to feed them forward into next detection layers. In MSSD proposed model, instead of classify the normal ConvNet feature map, we exploit the pyramidal feature hierarchy in convolution layers before feeding to the detection layers. We take advantage of the appropriate Conv-layers to provide helpful contextual information as extra-large receptive field often would absolutely introduce useless, large noise in the background. For large objects in deeper layers, we don't use the feature fusion module in order to keep up speed as long as SSD uses their shallower layers, as in the case of conv4 3, to predict small objects.

3 Experimental Results and Analysis

The purpose of this research is to train the network and to accurately classify and locate the objects. While the better object detection algorithm should have higher detection accuracy and detection speed, the detection accuracy is measured by mean Average Precision and the disclosure run is consistent by Frames Per-Second (FPS). In order to verify the effect of the feature pyramid network layer module and prediction model on detection performance, trained a VGG16 model with an input image 300 * 300 and a ResNetl01-SSD model with an input image of 321 * 321 for both Elt-sum and concat-sum models which lead to a slight improve in accuracy at the expense of speed as it will be seen later. The source of the experimental data, i.e., the source of the object detection image, is the VOC2007/2012 dataset and logo dataset. Test results under VOC2007/2012 and logo dataset, models have trained on the union of PASCAL VOC2007/2012 and Logo datasets which includes 20 categories and 8 classes respectively. The proposed MSSD model is implemented on the basis of the original SSD network built on the foundation of VGG16 and ResNet architectures, all of which are available on the website. The baseline SSD has trained with a 300 × 300 input size and batch size of 16 [16]. Both of the two MSSD feature fusion models, concat-sum and elt-sum, are fine-tuned on the well-trained SSD baseline for 10K iterations. In our case changing the learning rate is a must and we gave 10^{-3} for the first 60K iterations and then reduce it to 10^{-4} and 10^{-5} at 70K iterations. Experiment of MSSD based on deep VGG16 network, the appropriate layers are explored which will be the best to fuse with results of experiment. We choose two appropriate layers, conv4-3 and conv5-3, to fuse, because fc6 has a greater receptive field than conv5-3 for tiny objects that could add much more noise in the background. The two MSSD, concat-sum and element-sum, methods are both enhanced with regard to overall object detection compared to their SSD baseline. The MSSD with concat-sum module gets 77.6 mAP, while the element-sum module gets 77.4 mAP, which are 0.3 and 0.1 higher than the current SSD baseline. We test MSSD models on the logo dataset which contain 8 classes. The MSSD is implemented count on original SSD built on the VGG16 architectures and caffe 5 frameworks. The MSSD with concat-sum module gets 77.0 mAP, while the element-sum module gets 76.7 mAP, which are 0.6 and 0.3 higher than the current SSD baseline. Dataset includes 20 classes and each class might have small objects, we choose 181 images. Figure 2 shows the evaluation results of 3 models-SSD300, MSSD with concate-sum and concate-sum methods.

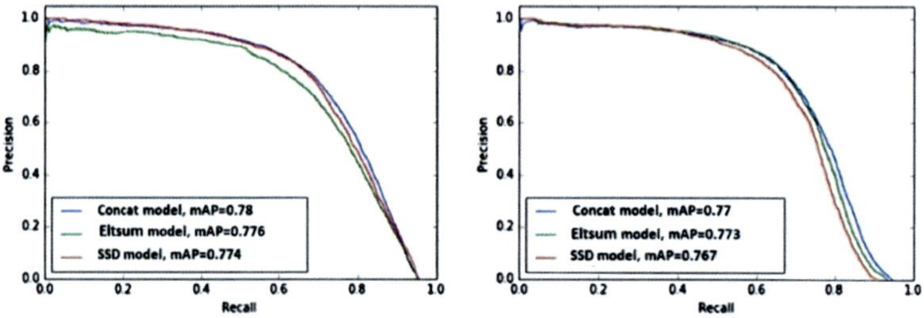

Fig. 2. Evaluation results of models.

As it is shown in Fig. 3 we began our training from a pre-trained checkpoint instead of starting from scratch. The absolute accident character actuate but exhibit reducing behavior overall. Figure 4 demonstrates overall mAP development for 80,000 steps at 0.5 IoU. The mAP values are classified for the recognition deadest at 0.5 IoU. As it seems from Fig. 4, around 6800 measures, mAP is witnessing a tremendous increase to 67.8%. The mAP then increases slightly more and in 19,000 steps reaches closer to 77.6%. The mAP value remains relatively constant with minor fluctuations after 19,000 moves.

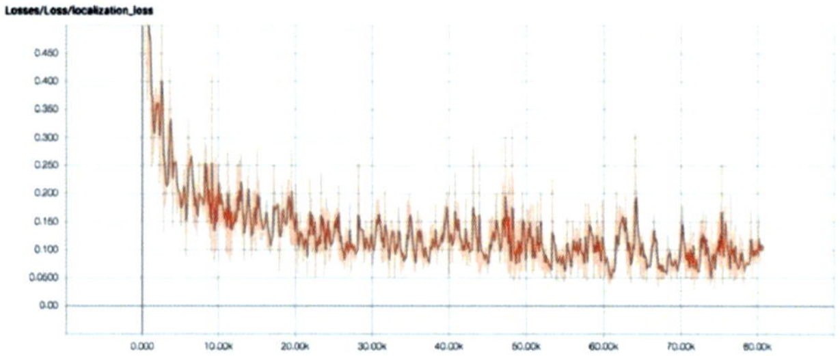

Fig. 3. Decline of total loss when concate-sum.

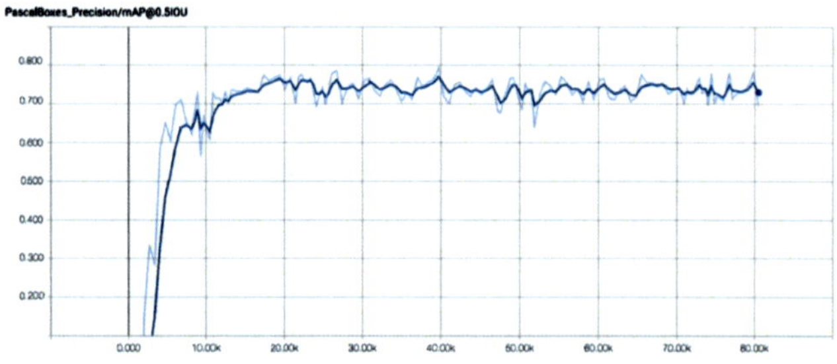

Fig. 4. Development of overall mAP when concate-sum.

Because the mAP character height out after 19,000 moves, the model is exported at various training stages and checked on the test dataset. The MSSD figure is adjust for 100,000 steps as shown in Figs. 5 and 6. Figure 6 demonstrates the downward trend in total loss during the entire training phase and shows overall mAP development at 0.5 IoU per 100,000 steps. For the validation dataset the mAP character is classify at 0.5 IoU. As is evident from Fig. 6 in around 9000 stages, mAP is experiencing a tremendous increase up to 53%. The mAP then increases slightly more and in 33,000 steps reaches nearer to 73%.

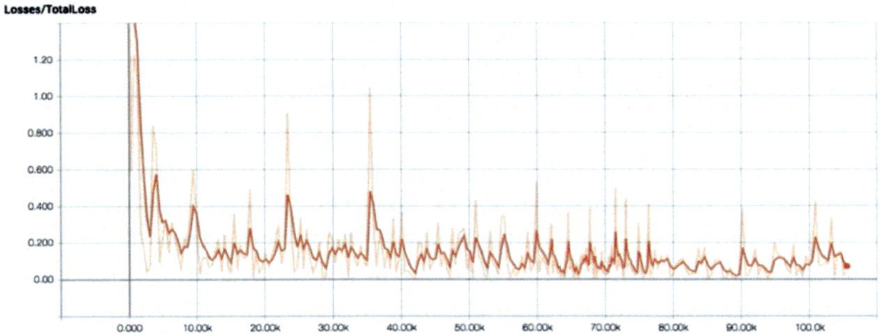

Fig. 5. Decline of total loss when Eltsum-ResNet-101 on PASCAL VOC2007 dataset.

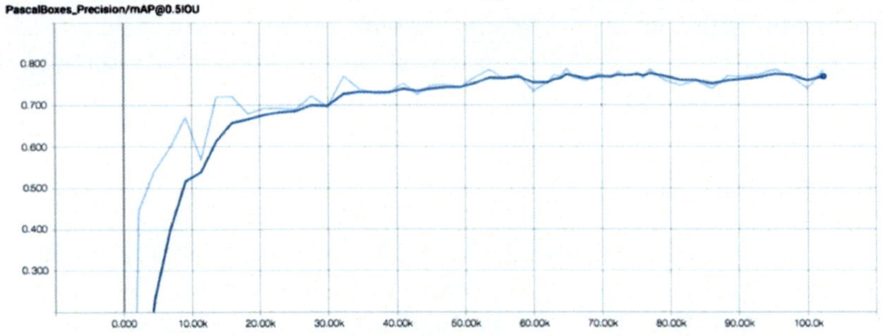

Fig. 6. Development of overall mAP when Eltsum-ResNet-101.

Value increased little after 19,000 measures and settled down with minor fluctuations of 77.4%. On PASCAL VOC 2007 and Logo test datasets, running time for those both fusion methods has evaluated, as it can be seen in Table 1. The two fusion methods, concat-sum and element-sum modules, have detection speed of 40 FPS and 43 FPS, respectively, in PASCAL VOC 2007 test dataset, while it is 13.12 FPS in Logo test dataset. Unfortunately, because of the additional feature fusion layers, both became slower than original SSD model. Nevertheless, MSSD fusion methods is also still achieving a real-time detection. Different models are shown in Table 1.

Table 1. The running time illustration of different models.

Dataset	Method	Network	mAP	FPS
PASCAL VOC 200	SSD300	VGG16	77.3	50
	DSSD321	Residual-101 [11]	78.6	13.6
	MSSD concat model	VGG16	77.6	40
	MSSD Eltsum model	VGG16	77.4	43
	MSSD Eltsum model	Residual-101 [11]	77.4	13.12
LOGO test	SSD300	VGG16	76.4	50
	MSSD concat model	VGG16	77	40
	MSSD Eltsum model	VGG16	76.7	43
	MSSD Eltsum model	Residual-101 [11]	76.9	13.12

Module has used 2 convolution layers in each layer, with 384 kernels and the concat-sum model used 3 convolution layers in each layer with 512 kernels. Which is why the model with the element-sum approach is faster than the model with the concat-sum approach by 3 FPS. Performance of MSSD modules based on VGG16 network, we show a comparison in performance of the pre-trained SSD300 model and the fine-tuned MSSD models on both datasets, and the experimental results demonstrate that helpful contextual information proves the existence of small objects. Performance of SDD Modules is based on residual network. The reason why the better detection result cannot be obtained is that the SSD model itself is based on a deep convolution neural network, and the high-level feature information with a large receptive field is used to predict a large object, and the low-level feature information with a small receptive field is used to predict a small object.

4 Conclusion

This article mainly studies the process of item disclosure based on SSD model. An improved SSD object detection algorithm MSSD is proposed. The high-layer linguistic knowledge is embedded in the low-level network's feature information, and the multi-scale feature map of the forecast relapse position pack and the classification task input is enriched to improve the detection accuracy. The VGG16 network used for SSD training was used with a deep residual network to optimize the feature maps of candidate box regression and classification task input, while showing the experimental results and corresponding advantages and disadvantages. This research uses the FPN-based network structure to fuse the upper and lower layers and improves the traditional up-sampling structure. The high linguistic knowledge is embedded in the low network's feature information, and the multi-scale feature map of the prediction regression position pack and the coordination work absorption is enriched to improve the detection accuracy. Correlated to the case of the craft small object detector, experiments show that the MSSD model is more improved than the classical SSD model in detection accuracy but not detection speed. In conclusion, we observe that although the SSD framework is scale-invariant, it can still benefit from the feature fusion architecture to find complain of different capacity.

The SSD method together with the feature fusion architecture can be adapted to work for the object detection task.

References

1. Hinton, G.E., Salakhutdinov, R.R.: Reducing the dimensionality of data with neural networks. Science **313**(5786), 504–507 (2006)
2. Lecun, Y., Bengio, Y., Hinton, G.: Deep learning. Int. J. Nature **521**(7553), 436–444 (2015)
3. Eggert, C., Brehm, S., Wincel, A.: Small object detection in faster R-CNN. In: IEEE International Conference on Multimedia and Expo (ICME), pp. 421–426 (2017)
4. Krishna, H., Jawahar, C.: Improving small object detection. In: 4th IAPR Asian Conference on Pattern Recognition (ACPR), pp. 340–345 (2017)
5. Redmon, J., Divvala, S., Girshick, R.: Unified, real-time object detection. In: Proceedings of the IEEE Conference on Computer Vision and Pattern Recognition, pp. 779–788 (2016)
6. Li, J., Liang, X., Wel, Y.: Perceptual generative adversarial networks for small object detection. In: Proceedings of the IEEE Conference on Computer Vision and Pattern Recognition, pp. 1222–1230 (2017)
7. Bell, S., Lawrence, Z.C., Bala, K.: Detecting objects in context with skip pooling and recurrent neural networks. In: Proceedings of the IEEE Conference on Computer Vision and Pattern Recognition, pp. 2874–2883 (2016)
8. Liu, W., Aguelov, D., Erhan, D.: Single shot multibox detector. In: European Conference on Computer Vision, pp. 21–37 (2016)
9. Liu, M., Dong, J., Dong, X.: Segmentation of lung nodule in CT images based on mask R-CNN. In: 9th International Conference on Awareness Science and Technology, pp. 1–6 (2018)
10. Simonyan, K., Zisserman, A.: Very deep convolution networks for large-scale image recognition. arXiv preprint, pp. 1409–1556 (2014)
11. Liu, L., Ouyang, W., Wang, X.: Deep learning for generic object detection: a survey. Int. J. Comput. Vision **1809**, 1–58 (2018)
12. He, K., Zhang, X., Ren, S.: Deep residual learning for image recognition. In: Proceedings of the IEEE Conference on Computer Vision and Pattern Recognition, pp. 770–778 (2016)
13. Ren, S., He, K., Girshick, R.: Towards real-time object detection with region proposal networks. In: Advances in Neural Information Processing Systems. pp. 91–99 (2015)
14. Ioffe, S., Szegedy, C.: Accelerating deep network training by reducing internal covariate shift. arXiv preprint, p. 1502.03167 (2015)
15. He, K., Zhang, X., Ren, S.: Spatial pyramid pooling in deep convolution networks for visual recognition. IEEE Trans. Pattern Anal. Mach. Intell. **37**(9), 1904–1916 (2015)
16. Jia, Y., Shelhamer, E., Donahue, J.: Convolutional architecture for fast feature embedding. In: Proceedings of the 22nd ACM International Conference on Multimedia, pp. 675–678 (2014)

Fuzzy Information Granulation Methodology for Identification of Class-Conscious Speech

Farida Huseynova(✉)

Azerbaijan State Oil and Industry University, Baku AZ 1010, Azerbaijan

Abstract. In recent years, great advances have been made in NLP for the accuracy and coverage of given word, appearing in a certain context. However, there are some problems, with processing of the meaning representation in automatic word sense disambiguation systems, and it becomes incomprehensible for different social classes to understand expressive language. We try to explore and determine the correct words and expressions usage in "Bag of words analysis" and give some recommendations for future successes in interpretation systems. Our aim is to apply the fuzzy information granulation method that can improve these systems without any discussion.

Keywords: Social class · Speech · Education · Fuzzy set · Granulation

1 Introduction

Despite the fact that social class concept is appearing with greater intensity in all environments, for most people it sounds very tough today. However, the whole universe exists and acts as a hierarchic system and there is a society, environment of human beings, to which they belong. Most people are increasingly embarrassed about the class-consciousness, and do their best to deny or disguise it. Nevertheless, they do not avoid calling anyone or anything 'working class' – resorting to polite euphemisms such as 'low-income groups', 'less privileged', 'ordinary people', 'less educated', 'the man in the street', 'tabloid readers', 'blue collar', 'state school', 'council estate' and 'popular'.

It is obvious that, there are different ways of doing things for different classes, we mean that they have rules about; food, mealtimes, dress, greetings, trade, hospitality, joking, status-differentiation, etc. The rules governing unofficial code of conduct comes across class problems and affects people's speech too. In many cases, people are usually unaware of the language they use. In different social groups, people have different characteristic mannerisms for using and interpretation of meaning in a language. Accordingly, people of different classes are differently sensitive in the mental sphere for using the words or understanding it, as they decode the meanings with internal highly sensitive social positioning sensors as – consciousness, cognition, and intelligence by which they communicate.

R. A. Aliev et al. (Eds.): ICSCCW 2021, LNNS 362, pp. 560–566, 2022.
https://doi.org/10.1007/978-3-030-92127-9_75

People with high income, hence, privileged to higher social classes and comprehends the world and himself in the world differently. This inequality has also led to major changes in a range of institutions, education being one of the more important ones. Education becomes a form of elite education, which is associated with family income, hence, privileged to higher social classes. In addition, upper class and middle class parents manage to obtain for their children the best education that works for them. Educated' speech is more upper and upper-middle speech; it lacks the haw-haw tones, vowel swallowing/drawling and pronunciation differences. Although upper-class speech as a whole is not necessarily intelligible than lower-class speech, it must be not only in mispronunciation of certain words, as it is often a lower-class signal, indicating a less-educated speaker. For example: saying 'nucular' instead of 'nuclear' is a common mistake.

Upper class people almost unconsciously try to achieve some degree of symmetry or balance in their conversations by using proper words, hence the person feels obliged, if only out of reflex politeness, to reciprocate with a comparably personal disclosure. As it is, even the introductions and greetings can be different in different layers: The most common greetings can be "hello" or "hi" nowadays, as "How do you do?" for some people can be uncomfortable, clumsy and inelegant, because it is now regarded by many as somewhat archaic, and is no longer the universal standard greeting. For some 'How do you do?' is a real question about health or well-being, and the 'Nice day, isn't it?' exchange must, however, be understood in the same light, and not taken literally, as it' is not a real question about the weather.

When we think about how computers may tackle the problem of social class' language, as there are so many words and expressions, which have identical meanings, but can be replaced by one-another depending on goals and consequences. It is too hard to understand the speech according to different social classes algorithmically, but the Semantic Web needs to be able to deal with such kind imprecise data and knowledge in a useful manner.

In the paper, we tried to determine some examples which are used as queries to the applications to understand how ambiguity, uncertainty, vagueness affects poorly on the performance of the NLP systems and analyzed many researches on the purpose. According Prof. Rafik Aliev all the decisions with imperfect information are made under risky, ambiguity or uncertain situations [1]. We think that the application of fuzzy logic in Computing with words (CW) is needed to use, as it provides a much more expressive language for knowledge representation and fuzzy Systems closely resembles human decision making with the ability to solve problems and generate solutions which are precise". Besides fuzzy models are transparent and easily adaptable, and theyuse the sub-symbolic information available in the universe of discourse providing the context to the given model and solution [2].

The paper is organized as follows: Next section describes the analyses of different researchers' discussion on depth of class' speech. Section 3 describes the methods of analyses of resolving the problem by fuzzy granulation method in CWW (Computing with words).

2 Social Class Research

According to Weber's [3] notion of status, as "the effective claim to social esteem in terms of positive or negative privileges is typically founded on (a) style of life, (b) formal education, which may be empirical training or rational instruction, and the corresponding forms of behavior, and (c) hereditary or occupational prestige".

William Labov [4] is perhaps the best known exponent of bringing class into sociolinguistic analysis, most notably in his classic book. From the establishment of an individual's class position, he followed the reasoning that one's social class position led to particular speech patterns, and this allowed him to argue that one can identify people as working class or upper middle class via how they speak.

He is mainly concerned of codes of conduct that comes across behavior influencing the distinctions of the language, especially, the way of communication, the words they use and how they pronounce them [5].

Meanwhile, in the United Kingdom, Basil Bernstein [6–8] framed the interrelationship between social class and language in terms of social structures in society and a theory of socialization, according to which children come to master different ways of communication in their day-to-day interactions in family settings, interactions among peers, and in formal education.

Shirley Brice Heath [9] in his classic Ways with Words compared and contrasted the language socialization practices of the residents of two communities in the southeastern part of the United States.

Ben Rampton in his book [10], Language in Late Modernity developed an informed and in-depth discussion of class. He outlined two levels at which class is a phenomenon: 1. material conditions-different people in different times, places and networks 2. "meta-level" representations: ideologies, images, and discourses about social groups. According to Jewitt, C. "approaches that understand communication and representation to be more than about language, and in which language is seen as one form of communication" [11]. Researchers like Snell [12, 13] and Emma Moore [14] have combined their interest in linguistic variation, multimodal repertoires, and stylization and stance with an interest in social class, focusing on adolescent speech patterns and forms in northeast England, respectively.

Moore found out the important point that about how working class speech and culture may be seen, not just as an alternative to middle-class speech, but as embodying fundamentally different ways of being and communicating. Moore noted, "Upper class men gossip just as much as working class, both classes devoted the same amount of conversation time (about 65 per cent) to social topics such as personal relationships; in another, the difference was found to be quite small, with gossip accounting for 55 per cent of upper class conversation time and 67 per cent of working class time. As business, sport and leisure have been shown to occupy about 10 per cent of conversation time, discussion of football could well account for the difference. In fact, research has revealed only one significant difference, in terms of content, between the class gossips: upper class spend much more time talking about themselves. Of the total time devoted to conversation about social relationships, working class spend two-thirds talking about their own relationships".

Kate Fox underlines that 'Cultural capital' is a more helpful concept, but the only form of cultural capital that functions as a near-infallible class indicator in real life (as opposed to surveys) is 'linguistic capital' [15].

3 Research Method

Signs of using the language, in particular ways a word, term, notation, sign, symbol, phrase, sentence, or any other form used for communication, which could be ambiguous for different social classes. The aim is to identify how can the language be interpreted in more than one way by different people and apply it in NLP systems? Zadeh's computational theory of perceptions (CTP), based on the methodology of computing with words (CW), may come to play an important role in the conception, design and utilization of information systems. According to Zadeh "Exploitation of the tolerance for imprecision is an issue of central importance in CW and CTP" [16, 17]. Besides he stated that all human concepts are context dependent, they can be divided into a granular structures. Following Zadeh's other recommendation "the degree of associated meaning and fuzziness of perceptions is the main sense of the human brain".

A granule is the object (points) drawn together by indistinguishability, similarity, proximity or functionality. Formal granule is the cluster of words (points) defined by generalized constraint. X is R. Here X is a constrained variable, is copula, R constraining relation. The concept of CW is rooted in which the concepts of a linguistic variable and granulation can be introduced [18, 19]. Simple examples of linguistic characterization of perceptions drawn from everyday experiences are given below:

he was deadly serious.
He looked very solemn.
the sincerity of his beliefs is unquestionable.
the deadly earnestness of the conversation.

So, the upper class are probably more acutely sensitive than any working class to the distinction between 'serious' and 'solemn', between 'sincerity' and 'earnestness', which may be a linguistic characterization of a perception. At the most basic level, in all above-mentioned sentences the main semantic meaning underlies "honesty". Therefore, if we translate this word as "serious" during the detection process, but the document uses "solemn", "sincerity" or "earnestness" version, during comparison of the texts we can get an incorrect result, because of value difference and perception difference. If we check all synonym versions of the word according to fine line distinction in "Bag of words", n – count of terms, which have synonym terms, m – average count of synonyms of terms- the various levels, and types of semantic information involved are not clearly understood. Acute sensitivity to the between seriousness and solemnity, may occur because of disproportionate class-consciousness.

The importance of speech in the context may point to another characteristic: our love of words. We are particularly 'tactile' or physically expressive in expressing the thoughts by reliance on linguistic signals. Words are our preferred medium, so it is perhaps significant that they should be our primary means of signaling and recognizing social status. William Hanson, an 'etiquette' expert suggested "More class-based conceptual work needs to be done" and he provided a guideline to 'U' and 'Non-U' words, guessed, the U stands for 'upper class'[12]. He considers educating people and first stop using the word "posh" is necessary, and instead of it to say "smart" as it is the norm by upper class standards. The other class experts suggest, that the upper classes don't need to borrow words that may be in fashion for a few months, so the more traditional repartee is preferable" and they think it would be more literal to use 'old' instead of 'vintage'. More traditional words for upper classes are given in the table (Table 1).

Table 1. Traditional and common words

Traditional words	Common words
Champagne	Prosecco-Bubbly/Fizz
Napkin	Serviette, doily, overlay
Basement	Lower Ground, foundation, base, groundwork, footing
Film	Movie, flick, pic, flicker, show
Jacuzzi	Hot Tub, vortex, swirl, whirlwind, whirl, eddy, curl

We consider, in order to choose the correct word CW and the close relationship between fuzzy information granulation must be used is as it is discussed in [20]. Fuzzy information granulation (fuzzy IG) may be viewed as a human way of employing data compression for reasoning and, more particularly, making rational decisions in an environment of imprecision, uncertainty and partial truth. In a very broad sense, granulation involves a partitioning of whole into parts. Modes of information granulation (IG) in which granules are crisp play important roles in a wide variety of methods, approaches and techniques. In a very broad sense, granulation involves a partitioning of whole into parts. Fuzziness of granules, their attributes and their values is characteristic of ways in which human concepts are formed, organized and manipulated. fuzzy information granulation (fuzzy IG) may be viewed as a human way of employing data compression for reasoning and, more particularly, making rational decisions in an environment of imprecision, uncertainty and partial truth. Information granulation (IG) can be: a. possibilistic (X is A partial knowledge); b. probabilistic (X is P partial certainty); c. veristic (X is V partial truth); d. generalized (X is R hybrid).

A basic generic problem in CW is: We are given a collection of propositions expressed in a natural language which constitute the initial data set (IDS). From the initial data set we wish to infer an answer to a query expressed in a natural language. The answer, also expressed in a natural language, is referred to as the terminal data set (TDS). In using feature extraction process fuzzy c (fuzzy clustering) can be used. In this example, what we have done, in effect, amounts to a derivation of a linguistic characterization of the joint probability distribution of X and Y starting with linguistic characterizations of the probability distribution of X and the probability distribution of Y.

4 Conclusion

Natural language processing systems which can handle fuzzy logic based statements must be encouraged. CW – which is based on fuzzy logic – can be used for much more expressive language for knowledge representation, reasoning and computation. This gives more scalability, flexibility and efficiency to the natural language processing system. In CW, a granule is a fuzzy set of points drawn together by similarity, where a word is a description of a constraint on a variable. By fuzzy logic incorrect reasoning can be prevented with measurements of mental perceptions.

References

1. Aliev, R.A., Fazlollahi, B., Aliev, R.R.: Soft Computing and Its Applications in Business and Economics. Springer, Heidelberg (2004). https://doi.org/10.1007/978-3-540-44429-9
2. Zadeh, L.A.: Fuzzy sets. Inf. Control **8**, 338–353 (1965). https://doi.org/10.1016/S0019-9958(65)90241-X
3. Weber, M.: Economy and Society: An Outline of Interpretative Sociology, vol. 1. Bedminster Press, New York (1968)
4. Labov, W.: The Social Stratification of (r) in New York City Department Stores. In: Coupland, N., Jaworski, A. (eds.) Sociolinguistics. MLS, pp. 168–178. Macmillan Education UK, London (1997). https://doi.org/10.1007/978-1-349-25582-5_14
5. Labov, W.: The Social Stratification of English in New York City. Center for Applied Linguistics, Washington, DC (1966)
6. Wootton, A.J., Bernstein, B.: Class, codes and control, volume I: theoretical studies towards a sociology of language. Br. J. Sociol. **24**(1), 120–121 (1973). https://doi.org/10.2307/588808
7. Wootton, A.J.: Class, codes and control (Vol. 2, Applied Studies Towards a Sociology of Language). Sociology **8**(2), 334–336 (1974). https://doi.org/10.1177/003803857400800217
8. Eggleston, J., Bernstein, B.: Class, codes and control. Volume 3. Towards a theory of educational transmissions. Br. J. Educ. Stud. **25**(1), 91 (1977). https://doi.org/10.2307/3120051
9. Heath, S.B.: Ways with Words Language, Life, and Work in Communities and Classrooms. Cambridge University Press, Cambridge (1983)
10. Rampton, B.: Language in Late Modernity: Interaction in an Urban School. Cambridge University Press, Cambridge (2006)
11. Jewitt, C.: An introduction to multimodality. In: Jewitt, C. (ed.) Handbook of Multimodal Analysis, pp. 14–27. Routledge, London (2009)
12. Snell, J.: From sociolinguistic variation to socially strategic stylization. J. Sociolinguist. **14**(5), 630–656 (2010). https://doi.org/10.1111/j.1467-9841.2010.00457.x
13. Snell, J.: Dialect, interaction and class positioning at school: from deficit to difference to repertoire. Lang. Educ. **27**, 110–128 (2013). https://doi.org/10.1080/09500782.2012.760584
14. Moore, E.: The social life of style. Lang. Lit. **21**, 66–83 (2012)
15. Fox, K.: Watching the English: The Hidden Rules of English Behaviour. Hodder & Stoughton, London (2005)
16. Zadeh, L.A.: From computing with numbers to computing with words: from manipulation of measurements to manipulation of perceptions. IEEE Trans. Circuits – I **46**(1), 105–119 (1999). https://doi.org/10.1109/81.73
17. Zadeh, L.A.: Outline of a new approach to the analysis of complex system and decision processes. IEEE Trans. Syst. Man Cybern. **SMC-3**, 28–44 (1973). https://doi.org/10.1109/TSMC.1973.5408575

18. Zadeh, L.A.: Outline of a computational approach to meaning and knowledge representation based on the concept of a generalized assignment statement. In: Winter, H. (ed.) Artificial Intelligence and Man-Machine Systems, pp. 198–211. Springer-Verlag, Berlin/Heidelberg (1986). https://doi.org/10.1007/BFb0006964
19. William Hanson – The UK's Leading Etiquette Coach & Expert. https://www.williamhanson.co.uk
20. Zadeh, L.A.: Toward a theory of fuzzy information granulation and its centrality in human reasoning and fuzzy logic. Fuzzy Sets Syst. **90**, 111–127 (1997). https://doi.org/10.1016/s0165-0114(97)00077-8

Possibility Optimal Risk Management in Magistral Gas and Oil Transportation Systems

S. A. Agammedova and I. A. Habibov(✉)

Azerbaijan State Oil and Industry University, Azadliq Street 20, Baku, Azerbaijan

Abstract. In this paper, the problem of risk management in trunk gas and oil transportation systems (MGNTS) with limited resources for the elimination of accidents is formulated in the form of a fuzzy decision-making model with a nonlinear income function and linear fuzzy constraints. The problem is also formulated in the form of a semi-Markov decision-making model for a controlled Markov process in continuous time with a nonlinear objective function of income and linear inequalities-constraints with coefficients expressing the costs of each type of funds for a certain type of accidents and given in the form of random fuzzy variables. With the help of a feasibility approach to the problem of fuzzy optimization and level maximization under line-by-line constraints according to the expected possibility, the formulated problem is reduced to an equivalent deterministic nonlinear programming problem with Boolean variables. An algorithm for solving the problem is proposed. On the basis of optimization methods, the solution of the problem is reduced to a deterministic nonlinear Boolean programming problem. Based on the methods of possible optimization, the solution of the problem is reduced to a deterministic nonlinear Boolean programming problem.

Keywords: Semi-Markov decision-making process · Renewal theory · Pseudo-Boolean programming methods

1 Introduction

The paper considers the problem of risk management in gas and oil transportation systems (MGNTS) with limited amounts of funds allocated for measures to prevent or eliminate the consequences of accidents, and the known distribution functions of the uptime of the type and distribution functions of the recovery time, using each solution from a given a set of measures for accidents of a certain type. The problem is also formulated in the form of a semi-Markov decision-making model for a controlled Markov process in continuous time with a nonlinear objective function of income and opportunity constraints on costs.

Using the possibility approach to the problem of fuzzy optimization and maximization of the level under line-by-line constraints on the expected possibility [1], the formulated problem is reduced to an equivalent deterministic nonlinear programming problem with Boolean variables. In the case of clear coefficients of linear constraints, this problem was solved in [2].

R. A. Aliev et al. (Eds.): ICSCCW 2021, LNNS 362, pp. 567–574, 2022.
https://doi.org/10.1007/978-3-030-92127-9_76

2 Formulation of the Problem

Let each state $i \in S$, $S = \{0, 1, 2, ..., N\}$, of the considered MGOTS, be assigned a finite set K_i of solutions (various complexes of actions in case of accidents and support of the normal functioning mode in the absence of accidents), the elements of which we denote $k = 0, 1, 2, ..., k_i$. If the system is in the state $i \in S$ and the decision $k \in K_i$ is made, then its further behavior is determined by the probabilistic law:

$$Q_i^{(k)}(t) = p_{ij}^{(k)} F_{ij}^{(k)}(t), \ \ j \in S, \tag{1}$$

where $p_{ij}^{(k)}$ - is the probability of the system transition from state i to state j when making a decision k; $F_{ij}^{(k)}(t)$ - is the distribution function of the time spent by the system in the state when making a decision k and provided that the next transition occurs to state j.

The risk management problem for MGOTS should include variables x_{kj} ($j = 0, 1, ..., N$; $k = 0, 1, ..., m$), where m is the total number of complexes of actions provided for the elimination of accidents from the set $\tilde{S} = S\backslash\{0\}$ and forming a set $K = U_{j \in S} K_j$, where K_j is a set of complexes of measures provided for the elimination of accidents of type j from the set $\tilde{S}$. At $j = 0$, a known decision $x_{00} = 1$ is made, and at $j \neq 0$, a previously unknown solution with components or ($k = 1,..., m$).

The goal function f_0 of the considered problem, which determines the total income of the system, depends on all components of the vector $x = (x_{00}, x_{11}, ..., x_{m1}, ..., x_{1N}, ..., x_{mN})$, while the cost constraints depend on all components of the vector $\tilde{x} = (x_{11}, ..., x_{m1}, ..., x_{1N}, ..., x_{mN})$ of dimension $n = mN$. We will assume that cost constraints are represented by linear functions of the fuzzy variables $a_{kj}(\gamma)$ and $b_k(\gamma)$ in the form [1]

$$f_k(x, \gamma) = \sum_{j=1}^{N} a_{kj}(\gamma) x_{kj} - b_k(\gamma), \ \ (k = 1, ..., m) \tag{2}$$

We will assume that the following conditions are met:

1. State $i = 0$ corresponds to the normal functioning of MGOTS, and $i \neq 0$ - to emergency (or pre-fail) situations.
2. The functions $F_{0j}^{(k)}(t)$ and $F_{0j}^{(k)}(t)$, $j \in \tilde{S} = S\backslash\{0\}$, $k \in K_j$, together with their first derivatives, are continuous for $t > 0$, with the possible exception of a finite number of points, and increase at infinity no faster than an exponential function, which ensures the existence of the Laplace-Stiltjes integral for these functions [3–5].
3. For a unit of time spent in state i, if decision k is made, the average income is paid $r_i^{(k)}$ (for $i \neq 0$, the number $r_i^{(k)}$ is negative and equal to the system's costs per unit of time spent in state i, provided that the state is exited using solution k).
4. The quantities $\left| r_i^{(k)} \right|$ are bounded for all, $i \in S$, $k \in K_i$ and the probabilities $p_{ij}^{(k)}$ satisfy the relations $\sum_{j \in S} p_{ij}^{(k)} = 1, \ \ i \in S, \ k \in K_i; \ \ p_{ij}^{(k)} \geq 0, \ i, j \in S, \ k \in K_i$.

Thus, in each state there are k_i solutions from the finite set K_i. The choice of some solution k from K_i in the state $i \in S$ means setting the values $Q_{ij}^{(k)}(t)$, $P_{ij}^{(k)}$, $F_{ij}^{(k)}(t)$, $r_{ij}^{(k)}$, $j \in S$. For $i = 0$, $K_0 = \{0\}$; $p_{0j}^{(0)} \neq 0 (j \in S)$ is the probability of transition to state j from state $i = 0$. $\overline{p}_{00}^{(0)}$ - is the probability of the system staying in the state $i =$ 0, in this case $F_{00}^{(0)}(t)$- the distribution function of the time accumulated by the moment t of staying in the state $i = 0$. The probabilities $p_{0j}^{(0)}$ $(j \in \tilde{S})$ are calculated in practice as the proportion of accidents of type j in the total set of accidents of various types of MGOTS based on the data bank of the process history. In this case, the function $F_{00}^{(0)}(t)$ is the function of the distribution of the time of failure-free operation between adjacent accidents of type j. For $i = 1,\ldots,N$ for any $k \in K_i p_{i0}^{(k)} = 1, p_{ij}^{(k)} = 0$ $(j \neq 0)$; $F_{j0}^{(k)}(t)$- the distribution function of the recovery time of the MGOTS operability using the solution k in case of an accident of type j. Due to the continuity in time of the process under study, we will use an exponential overestimation with the norm α, i.e. if at some point in time a unit income is paid, then after time t this income will already $e^{-\alpha t}$ cost units. Then, if r_i is income per unit of time, then the total income for time t has the form $\int_0^t r_i e^{\alpha\tau} d\tau = \frac{r_i}{\alpha}(1 - e^{-\alpha t})$. Let us denote by i_n the state of the system after the n-th transition, u_n is the decision made, and τ_n is the time spent in this state ($n = 0, 1, 2,\ldots,$ i_0 is the initial state). An admissible strategy π for a controlled MGOTS is defined as the sequence $\{\pi_0, \pi_1, \pi_2,\ldots\}$, where $\pi_n(\bullet/z_n)$ is a probabilistic measure focused on the function of constraints *U(S)* on the decisions (controls) made, determined by the system of of equations

$$\Pi\{\{f_k(x, \gamma) = 0\} \geq \alpha_k\}, k = \overline{1, m} \tag{3}$$

(Π - is a possible measure, a_{kj} - are level sets of possible quantities is equivalent to a system of linear a_{kj}, $k = \overline{1, m}, j = \overline{1.N}$) and depending on the history of the controlled system at the time of the n-th transition $z_n = (i_0, u_0, \tau_0, \ldots, i_{n-1}, u_{n-1}, \tau_{n-1}, i_n)$. The measure $\pi_n(\bullet/z_n)$ defines a randomized rule for choosing a solution u_n based on the information z_n. We denote by $w_i(t, \alpha, \pi)$ the total income over time t of the system making a decision according to the strategy π with the overestimation rate α, provided that the system starts functioning at time $t = 0$ from state i. Here α (0 < α < 1) is the coefficient of discounting income (in normal operation) or costs (in emergency situations); coefficient α is determined by inflation (at inflation on 10% α = 0.1). By $v_i(t, \alpha, \pi) = w_i(t, \alpha, \pi)/t$ we denote the total average income of the system over time t under the same conditions. It is necessary to find an α-optimal nonrandomized Markov strategy π* that maximizes the average income $v_i(\alpha, \pi)$ for an arbitrary initial distribution of the process, $\vec{\alpha} = (\alpha_0, \alpha_1, \ldots, \alpha_n)$ where $\sum_{i \in S} \alpha_i = 1$, $\alpha > 0$, $i \in S$. Without loss of generality, we will take a $\vec{\alpha} = (1, 0, \ldots, 0)$ vector as the initial distribution, i.e. the initial state of the system is normal operation.

3 Method for Solving the Problem

Using a matrix $Q^{(k)}(t) = [Q_{ij}^{(k)}(t)]$ called the matrix of transition distributions, the function $H_i^{(k)}(t) = \sum_{j \in S} Q_{ij}^{(k)}(t), \ i \in S, \ k \in K_i$, is determined which is the distribution function of the time spent by the process in state i when making a decision $k \in K_i$. Let, $h_i^{(k)}(\alpha) = h_i^{(k)}(s)/s = \alpha$ where $h_i^{(k)}(s)$ is the Laplace-Style-tyes transformation of the function $H_i^{(k)}(t)$. Using the standard for the renewal theory [5] reasoning as $t \to \infty$ in the Laplace-Stieltjes image field, we obtain the following equation

$$\mathrm{v_i}(\alpha) = \sum_{k \in K_i} d_i^{(k)} \left(\rho_i^{(k)}(\alpha) + \sum_{j \in S} q_{ij}^{(k)} \mathrm{v_j}(\alpha) \right) \tag{4}$$

where $\mathrm{v_i}(\alpha)$ is a short entry $v_i(\alpha, \pi)$; $q_{ij}^{(k)}(\alpha) = L_{s=\alpha}^{*}\{Q_{ij}^{(k)}(t)\}$, $L^*(s)$ - Laplace-Stilties transformation; $d_i^{(k)}(k \in K_i)$ - strategy system in state i. Let be

$$\rho_i(\alpha) = \sum_{k \in K_i} d_i^{(k)} \rho_i^{(k)}(\alpha), \ \rho_i^{(k)}(\alpha) = \frac{r_i^{(k)}}{\alpha}\left(1 - h_i^{(k)}(\alpha)\right) \tag{5}$$

and $\rho(\alpha) = (\rho_0(\alpha), \ldots, \rho_N(\alpha))^T$, $\mathrm{v}(\alpha) = (\mathrm{v_0}(\alpha), \ldots, \mathrm{v_N}(\alpha))^T$. Then

$$\mathrm{v}(\alpha) = \rho(\alpha) + q(\alpha)\mathrm{v}(\alpha), \ q(\alpha) = [q_{ij}(\alpha)], \ q_{ij}(\alpha) = \sum_{k \in K_i} d_i^{(k)} q_{ij}^{(k)}(\alpha) \tag{6}$$

from (4) we find

$$\mathrm{v}(\alpha) = [I - q(\alpha)]^{-1}\rho(\alpha). \tag{7}$$

where I is the $(N + 1) \times (N + 1)$ identity matrix. Equality (7) is valid, since the matrix $[I - q(\alpha)]^{-1}$ is nondegenerate. Multiplying both sides of equality (5) by the vector $\vec{a}$ of the initial distribution, we obtain

$$\vec{a} \cdot \mathrm{v}(\alpha) = \sum\nolimits_{i \in S} \sum\nolimits_{j \in S} \sum\nolimits_{k \in K_i} q_i \mu_{ij}(\alpha) \rho_j^{(k)}(\alpha) d_j^{(k)}, \ [\mu_{ij}(\alpha)] = [I - q(\alpha)]^{-1} \tag{8}$$

The quantities $\mu_{ij}(\alpha)$ depend on $d^{(k)}, k \in K_i, i \in S$, since the elements of the matrix $[I - q(\alpha)]$ are expressed in terms of $d^{(k)}, k \in K_i, i \in S$. Let $\{d_j^{(k)}\}$ $(k \in K_j)$ a nonrandomized stationary Markov strategy be a system in state j: $d_j^{(k)} \in \{0, 1\}, \ \sum_{j \in S} d_j^{(k)} = 1$.

We put $x_{00} = 1$ and $x_{kj} = d_j^k, \ k \in K_j, \ j \in \tilde{S}$. For linear functions (2) by methods of interval mathematics [6] it can be proved [7] that the system

$$\begin{cases} \sum_{j=1}^{N} a_{kj}^{-} \cdot x_{Rj} \le b_k^{+}, \\ \sum_{j=1}^{N} a_{kj}^{+} \cdot x_{Rj} \ge b_k^{+}, \\ x_{Rj} \in X, k = 1, ..., m, \end{cases} \tag{9}$$

where a_{kj}^{-}, a_{kj}^{+}, b_k^{-}, b_k^{+} are the boundaries of closed intervals $[a_{kj}^{-}, a_{kj}^{+}]$, $[b_k^{-}, b_k^{+}]$, representing α_k- level sets of possible quantities $a_{kj}(\gamma)$ and $b_k(\gamma)$, respectively.

Maximization of the total average income (8) with functions of cost constraints of the form (3) is formulated as the following optimization problem for Boolean variables $x = (x_{kj})$, $k \in K_j, j \in \tilde{S}$ or $j \in J_k$, $k \in K$ which is all the same:

$$f(\alpha, x) = \sum_{i \in S} \sum_{j \in S} \sum_{k \in K_j} a_i \mu_{ij}(\alpha, x) \rho_j^{(k)} x_{kj} \to \max \tag{10}$$

$$\sum_{k \in K_j} x_{kj} = 1, \quad j \in \tilde{S} \tag{11}$$

$$\sum_{j \in J_k} a_{kj}^{-} \cdot x_{kj} \le b_k^{+}, \quad (k = 1, ..., m) \tag{12}$$

$$\sum_{j \in J_k} a_{kj}^{+} \cdot x_{kj} \ge b_k^{-}, \quad (k = 1, ..., m) \tag{13}$$

$$x_{kj} \in \{0, 1\}, \quad k \in K_j, \quad j \in \tilde{S}, \tag{14}$$

where a_{kj}^{-}, a_{kj}^{+}, b_k^{-}, b_k^{+} boundaries of intervals $[a_{kj}^{-}, a_{kj}^{+}]$, $[b_k^{-}, b_k^{+}]$, representing α_k-level sets of possible quantities $a_{kj}(\gamma)$ and $b_k(\gamma)$, respectively.

To construct solutions to such a system under additional conditions (14), one can use the approach proposed in [8], based on finding basic solutions. However, in practice, in inequalities (12) and (13), the right-hand sides b_k^{-} and b_k^{+} are determined by the known coefficients a_{kj}^{-} and a_{kj}^{+}, respectively. We will assume that, a_{kj}^{-}, a_{kj}^{+} and b_k^{-}, b_k^{+}, are triangular possible quantities [1].

Based on the above, the following main conclusions can be drawn. The optimal solution to problem (10)–(14) is obtained by choosing such a solution to system (11)–(14), at which the maximum value of the objective function (10) is achieved. This value is obtained by a simple enumeration of the values of function (10) on the set of particular solutions of system (11)–(14) which are determined which using the pseudo-Boolean methods of bivalent programming.

4 Numerical Example

Consider the operation of a gas pipeline under the risk of two possible accidents: accident 1 (j = 1) - blockage; emergency 2 (j = 2) - valve skip.

Alternative measures in case of accident 1: 1) Stopping the gas pipeline and restoring normal conditions. 2) Gas pipeline purging. 3) Application of a viscoelastic cleaning system.

Alternative measures in case of accident 2: 1) Stopping the gas pipeline and restoring the normal state. 2) Replacement of the valve.

We assume that in both cases the first activity is carried out at the same time T = 1 h. Let us denote the measures to eliminate accident 1 through K = 1, K = 2 and K = 3, and in the case of accident 2 through K = 1 and K = 4. Then the number of system elements will be N = 2, and the total number of different measures for liquidation in the aggregate of accidents will be m = 4.

$$F_j(t) = 1 - e^{-\lambda_j t}, G_j^{(k)}(t) = 1 - e^{-\mu_j^{(k)} t}$$

where $\lambda_j = 1\big/T_{j1}$; $\mu_j^{(k)} = 1\big/T_{j2}^{(k)}$; $\mu_j^{(k)} = 1\big/T_{j2}^{(k)}$; T_{j1} and $T_{j2}^{(k)}$ is the average lifetime of the system between two successive accidents of type j and the average recovery time of the system after a crash of type j when solving k. Let be a non $d_i^{(k)}$ - randomized stationary strategy of the system in a state when $i (i \in S)$ solving k (i.e., the probability of making a decision k in state i): $d_i^{(k)} \in \{0, 1\}$, $\sum\limits_{k \in K_i} d_i^{(k)} = 1$, $i \in S$. Then the distribution law of the lifetime and the distribution law of the recovery time of the system as a whole will be written in the form $F_j(t) = 1 - e^{-\lambda t}$, $G(t) = 1 - e^{-\mu t}$, where $\lambda = \sum\limits_{j=1}^{N} \lambda_j$, $\mu = \sum\limits_{j=1}^{N} \sum\limits_{k \in K_j} d_j^{(k)} \mu_j^{(k)}$. Let us $T_{11} = 8h$, $T_{21} = 8h$, $T_{12}^{(1)} = 1h$, $T_{12}^{(2)} = 2h$, $T_{12}^{(3)} = 1h$, $T_{22}^{(1)} = 1h$, $T_{22}^{(4)} = 0,5h$.

Then $F_0(t) = F(t) = 1 - e^{-0,25t}$, $F_j(t) = 1 - e^{-0,25t}$, $j = 1, 2$; $G_1^{(1)}(t) = 1 - e^{-t}$, $G_1^{(2)}(t) = 1 - e^{-0,5t}$, $G_1^{(3)} = 1 - e^{-t}$; $G_2^{(1)}(t) = 1 - e^{-t}$, $G_2^{(4)}(t) = 1 - e^{-2t}$.

Let us assume that the throughput of the gas pipeline is 7 106 m^3/day and the cost of 1000 m^3 of gas is 230 USD. Then the costs from the downtime of the gas pipeline during the time T = 1 h will be 7·103·30/24 = 66850 USD. Let us denote by the average cost of action k in the event of an accident j. Then, neglecting the costs of restoration work in the event $k = 1$, we get the $c_{11} = c_{12} = 66850$ c.u.

Let, further, the average costs for other activities are equal, $c_{21} = 300$, $c_{31} = 400$, $c_{42} = 600$ cu. We will assume that in the state $i = 0$ the only decision ($k = 0$) has been made - to continue normal functioning - and in this state the following probability distribution is given:

$$P_{00}^{(0)} = 0,7,\ P_{01}^{(0)} = 0,1,\ P_{02}^{(0)} = 0,2.$$

Using the Laplace-Stieltjes transformation, we find that the matrix $[I - q(\alpha, x)]^{-1} = [\mu_{ij}(\alpha, x)]$ $(i, j \in S)$ has elements:

$$\mu_{00}(\alpha, x) = 0,\ \mu_{01}(\alpha, x) = \frac{1}{D(\alpha, x)} \cdot \frac{0,0125}{\alpha + 0,125},\ \mu_{02}(\alpha, x) = \frac{1}{D(\alpha, x)} \cdot \frac{0,025}{\alpha + 0,125},$$

$$\mu_{1,0}(\alpha, x) = \frac{1}{D(\alpha, x)}\left(\frac{x_{11}+x_{31}}{\alpha+1} + \frac{x_{21}}{\alpha+0,5}\right),$$

$$\mu_{11}(\alpha, x) = \frac{1}{D(\alpha, x)}\left(1 - \frac{0,175}{\alpha+0,25} - \frac{0,025}{\alpha+0,125}\left(\frac{x_{12}}{\alpha+1} + \frac{2x_{42}}{\alpha+2}\right)\right),$$

$$\mu_{12}(\alpha, x) = \frac{1}{D(\alpha, x)} \cdot \frac{0,025}{\alpha+0,125}\left(\frac{x_{11}+x_{31}}{\alpha+1} + \frac{x_{21}}{\alpha+0,5}\right),$$

$$\mu_{20}(\alpha, x) = \frac{1}{D(\alpha, x)} \cdot \left(\frac{x_{12}}{\alpha+1} + \frac{2x_{42}}{\alpha+2}\right),$$

$$\mu_{21}(\alpha, x) = \frac{1}{D(\alpha, x)} \cdot \frac{0,0125}{\alpha+0,125}\left(\frac{x_{12}}{\alpha+1} + \frac{2x_{12}}{\alpha+2}\right),$$

$$\mu_{22}(\alpha, x) = \frac{1}{D(\alpha, x)}\left(1 - \frac{0,175}{\alpha+0,25} - \frac{0,0125}{\alpha+0,125}\left(\frac{x_{11}+x_{31}}{\alpha+1} + \frac{x_{21}}{\alpha+0,5}\right)\right)$$

where $D(\alpha, x) = -\frac{0,025}{\alpha+0,125}\left(\frac{x_{12}}{\alpha+1} + \frac{2x_{12}}{\alpha+2}\right) + 1 - \frac{0,175}{\alpha+0,25} - \frac{0,0125}{\alpha+0,125}\left(\frac{x_{11}}{\alpha+1} + \frac{x_{21}}{\alpha+0,5}\right)$- determinant of a matrix $[I - q(\alpha, x)]$. The quantity $\rho_i^{(k)}$ in (8) is written in the form:

$$\rho_0^{(0)}(\alpha) = \frac{66850}{\alpha}\left(1 - \frac{0,175}{\alpha+0,25} - \frac{0,0375}{\alpha+0,125}\right),\ \rho_1^{(1)}(\alpha) = -\frac{66850}{\alpha}\left(1 - \frac{1}{\alpha+1}\right),$$

$$\rho_1^{(2)}(\alpha) = -\frac{150}{\alpha}\left(1 - \frac{0,5}{\alpha+0,5}\right),\ \rho_1^{(3)}(\alpha) = -\frac{400}{\alpha}\left(1 - \frac{1}{\alpha+1}\right),$$

$$\rho_2^{(1)}(\alpha) = -\frac{66850}{\alpha}\left(1 - \frac{1}{\alpha+1}\right),$$

$$\rho_2^{(4)}(\alpha) = -\frac{1200}{\alpha}\left(1 - \frac{2}{\alpha+2}\right).$$

Assuming the levels α_k in (3) to be equal $\alpha_1 = \alpha_2 = \alpha_3 = \alpha_4 = 0,8$ and the fuzzy coefficient of the quantity c_{kj} and b_k (to simplify the calculations) equal to 1, after applying the above algorithm for the numerical solution of the system of inequalities (11)–(14), we obtain four solutions with the following values of Boolean variables $\{x_{kj}\}$:

1) $x_{11} = 1,\ x_{42} = 1$; 2) $x_{21} = 1,\ x_{12} = 1$; 3) $x_{21} = 1,\ x_{42} = 1$; 4) $x_{31} = 1,\ x_{12} = 1$;

5) $x_{31} = 1,\ x_{42} = 1$ (not specified variables in each solution are equal to zero).

The objective function $f(\alpha, x)$ for $\alpha = 0,1$ and $a = (1, 0, ..., 0)$ the initial distribution takes $f_5 = -175,2$ the largest value when solving 5) $x_{11} = 0, x_{21} = 0, x_{31} = 1, x_{12} = 0, x_{42} = 1$.

5 Conclusion

The problem of managing technical and technological risks in the main gas and oil pipelines with possible restrictions on the funds allocated for the elimination of possible accidents is considered in the form of a semi-Markov decision-making model with the criterion of the maximum average discounted income. Provided that both the coefficients and the right-hand sides of the constraints are disconnected, the problem is reduced to a deterministic optimization problem with Boolean variables, a nonlinear objective function and constraints in the form of a system of linear inequalities, which is solved using pseudo-Boolean methods of bivalent programming, based on the methods of feasibility optimization.

References

1. Yazenin, A.V.: Optimization and Decision Making Methods for Fuzzy Data (in Russian). Tver (1995). http://tekhnosfera.com/metody-optimizatsii-i-prinyatiya-resheniy-pri-nechetkih-dannyh
2. Dyshin, O.A., Azizov, I.A.: Semi-Markov risk management models in Magistral gas and oil pipeline systems (in Russian). Electron. Model. **32**(2), 15–30 (2010)
3. Beichelt, F., Franken, P.: Reliability and Maintenance (in Russian). Mathematical Approach - M.: Radio and Communication, 392 p. (1988)
4. Mine, H., Osaki, S.: Markov Decision-Making Processes (in Russian), 176 p. Nauka, Moscow (1977). https://bookree.org/reader?file=445877
5. Cox, D.R., Smith, V.L.: Renewal Theory, Science Paperbacks, SP58, London, 142 p. (1967). https://openlibrary.org/books/OL5663880M/Renewal_theory
6. Alefeld, G., Herzberger, Yu.: Introduction to Interval Computations (in Russian), Mir, 356 p. (1987). https://search.rsl.ru/ru/record/01001380077
7. Yazenin, A.V.: On the problem of maximizing the possibility of achieving a fuzzy goal (in Russian), News of the Academy of Sciences. Theory Contr. Syst. **4**, 120–123 (1999)
8. Gorodetsky, V.I., Lebedev, A.I., Ponomarev, V.V.: Pseudo-Boolean methods of bivalent programming in problems of structural synthesis of ACS (in Russian). Autom. Contr. Syst. Publ. House Leningrad State Univ. **4**, 50–55 (1977)

Modern Approaches to the Formation of Management Methodology in the Conditions of Digital Transformation

Akperov Imran Gurru Ogly(✉)

Southern University (IMBL), Rostov Region, Rostov-on-Don, Russia
rector@iubip.ru

Abstract. Modern conditions for the formation of the digital economy imply the use of new approaches to organizing information flows in the management of organizations of all types. The article discusses approaches that are typical for management processes in the Russian Federation, but have common properties with the well-known trends in the world economy. Based on Russian and foreign experience, the country has formed the "Program for the Development of the Digital Economy in the Russian Federation until 2035", the main provisions of which presuppose a combination of several key components of the digital transformation in management of the modern socio-economic development of Russia. These include digital big data technologies, soft models and computations based on fuzzy sets and fuzzy logics, complex chaotic Systems of Systems, as well as intellectual data mining, including machine learning and neural network technologies. The proposed study examines the issues of the practical application of these approaches within a particular region. The main examples are given within the framework of the specifics of the southern region of the Russian Federation. The main feature of the work is that the existing NON-factors are taken into account to the maximum extent.

Keywords: Soft models · NON-factors · Artificial neural networks · Data mining · Big data · Chaos · System of systems

1 Introduction

Modern global trends of VUCA-world and digital transformation determine deep tectonic changes in socio-economic and cultural-political practices of a person and necessarily raise new questions for the organization and management of human activity. The unstable, multi-vector, hyper-fast modern world is characterized by an increase in information flows, the constant emergence of new technologies, and the formation of ecosystems. In such conditions, many researchers speak about the need to form a new management paradigm [1].

The emergence of big data as a new type of strategic resource in an organization requires the search for new methods of effective management. These methods are connected with the collecting of the necessary information, processing, transformation into

R. A. Aliev et al. (Eds.): ICSCCW 2021, LNNS 362, pp. 575–579, 2022.
https://doi.org/10.1007/978-3-030-92127-9_77

knowledge, appropriate packaging and further transfer. The methodology of these items is practically absent in the modern process of making managerial decisions [2]. In this regard, the data concept – Data Driven Management (DDM) that has appeared in recent years, "aimed at changing the business model and organizing management decision-making processes" [3], solves a number of problems. But the formation of a corporate culture of working with data remains an open and complex issue.

Thus, it can be stated that in the world practice there is a trend of changes in the theory and methods of management associated with the total introduction of digital technologies that determine transformational processes. The search for a methodological basis for a new management paradigm becomes relevant and necessary.

According to the developers of the "Program for the development of the digital economy in the Russian Federation until 2035", "the conceptual digital model of managing socio-economic systems will be based on such principles as: obtaining real-time data; managing economic processes based on automated analysis of big data; the digital ecosystem of the state, business and citizens" [4].

In turn, a specialist in systems thinking and modeling of business systems, J. Gharajedaghi notes that in modern conditions, an organizational model is being formed [5] as a social model of a multi-intelligent system, and a person is required to have a coherent, holistic thinking designed to help see through chaos and complexity as well as to cope with a set of interrelated variables in a changing external environment [6].

2 Chaos Theory as the Basis of Management Methodology

Analyzing the strategies of digital business transformation and the directions of changes in management approaches in the digital economy, McKinsey specialists distinguish "three levels of changes: new ecosystems, business architecture and foundation" [7]. At the same time, the external level of organizational changes is determined by the activity of ecosystems and the transformation of markets. The internal level of the organization is based on changes in business models and values. The third level depends on the convergence of digitalization processes and the corresponding processes of corporate culture within the organization. To optimize management in the conditions of digital transformation, it seems heuristic and relevant to form a methodology based on such approaches as chaos theory, fuzzy set methods and neural network models.

Chaos theory in the desired management methodology represents the external environment of an organization as an asymmetric, nonlinear dynamically changing space in which any element acquires an emergent value [8, 9]. The managerial effect of applying chaos theory is the ability to use nonlinear, chaotic processes to achieve the goals of the organization, determining stable parametric data areas and states of "self-organized criticality of the controlled object" [10].

In the mathematical formulation, the problem is reduced to the study of a system of differential equations with nonlinear feedbacks in the form of:

$$\frac{\partial}{\partial t}X_i(t) = F_i(X_1, X_2, ..., X_n; t), \quad i = 1, 2, .., n, \tag{1}$$

where X_i are the phase variables that determine the state of the object under consideration at a given time t.

The number of phase variables "characterizing the socio-economic state of the region may include the average income level of the population, the index of social tension, the number and characteristics of sources of potential tension, quantitative and qualitative characteristics" [11] of forces and logistical, medical, anti-epidemic and other means aimed at preventing emergency situations (emergencies) and minimizing their consequences. "The system of Eq. (1) will describe the behavior of an object in case of emergency, if its right-hand sides of equations with nonlinear feedbacks are known. The presence of positive feedback in the system is always potentially dangerous and can cause loss of controllability. An example of such systems are systems with a variable delay" [12]. Their mathematical models have the form of:

$$\frac{dx}{d} = F(x(t), \quad x(t-\tau)), \tag{2}$$

where τ – the delay.

The simplest example of a model with a delay is the Hutchinson equation:

$$\frac{dx}{dt} = \alpha \ (x(t) \cdot (1 - x(t-\tau)), \tag{3}$$

When the delay time is long, this equation describes rare, periodically occurring, giant bursts that can be interpreted as catastrophes" [13]. Thus, the change in the delay time is one of the sources of chaos and, accordingly, unpredictability, fuzziness of complex systems.

As it was already mentioned above, the modern VUCA world is characterized by such a fundamental principle as uncertainty, "which is based on objective and subjective factors". "The objective factor is due to the very existence of objects and phenomena of the external world with ambiguous characteristics. The subjective factor is related to the individual characteristics of the perception of these phenomena by different people" [11].

The mathematical theory of fuzzy sets proposed by L. Zadeh allows describing fuzzy concepts and knowledge, operating with this knowledge [12] and making fuzzy conclusions. "The presence of mathematical means of reflecting the fuzziness of the initial information allows us to build a model that is adequate to reality and to process large amounts of data in a relevant way. In the problem of choosing alternatives to data processing, an important place is occupied by the analysis of situations in which qualitative characteristics, rather than quantitative, are decisive. For these purposes, L. Zadeh created a theory of a linguistic variable, the value of which is characterized not by a numerical value, but by a word (or a set of words) of a natural language. Thus, using the theory of fuzzy sets in management, qualitative information can be used in formalized analysis procedures" [13].

The use of neural networks in management in recent decades has found application in such functional areas as project management, inventory management, and risk management. However, today, a person who is considered in the organization as a valuable asset that needs to be preserved and developed comes to the fore in the digital economy. The digital footprint left by a person in the context of its functioning is determined by a large amount of data, the processing of which is the task of self-learning neural networks. Based on the data coming from different sources, the neural network can identify problem areas in the management system.

The proposed methodological platform, instrumentally consisting of elements of chaos management, neural networks and fuzzy sets, is based on the "Theory O", in which the object of management is the organizational culture, and the subject of management is the joint interaction of people in decision–making and proactivity as a reaction to predicted changes [14, 15].

3 Conclusion

Thus, it can be noted that the problems of management in the conditions of transformation mostly lie in formation of the digital culture of the organization. The structural elements of this culture can be considered: flexibility and adaptability in the search and selection of information; making decisions based on data using Big Data technologies; digital consciousness focused on the priority of using digital technologies. They are the result of effective management of digital transformation based on the methodological integration of chaos theory, fuzzy sets and neural networks.

References

1. Akperov, I.G., Khramov, V.V.: Management of social and economic systems of the region – formation of digital economy. Intellectual Resources – to the Regional Development. **2**, pp. 36–47 (2020). https://www.elibrary.ru/item.asp?id=43033189 (in Russian)
2. Markova, V.D., Kuznetsova, S.A.: Management development in digital economy: analytical research review. Vestnik NSU. Serias: Social and Economic Sciences. **3** (2020). https://cyberleninka.ru/article/n/razvitie-menedzhmenta-v-tsifrovoy-ekonomike-analiticheskiy-obzor-issledovaniy (in Russian)
3. Antonov, V.G., Samosudov, M.V.: Problems and perspectives of digital management development. E-Management. **2**, 38–48 (2018) https://www.elibrary.ru/item.asp?id=37291115 (in Russian)
4. Program of Digital Economy Development in the Russian Federation Until 2035 (2017). https://d-russia.ru/wp-content/uploads/2017/07/programma-tsifrov-econ.pdf (in Russian)
5. Akperov, I.G., Khramov, V.V.: The concept of a unified geoinformational space of the region: ecological aspect. In: E3S Web of Conferences. VIII International Scientific and Practical Conference "Innovative Technologies in Science and Education" (ITSE 2020), p. 09006 (2020). https://doi.org/10.1051/e3sconf/202021009006
6. Gharajedaghi, J.: System thinking: how to manage chaos and difficult processes. Grevtzov Books, Minsk (2010). https://lib-bkm.ru/13915
7. Akperov, I.G., Khramov, V.V.: Soft models of functional connectivity of measurements of geoinformational space of the region. Intellectual Resources – to the Regional Development. **2**, pp. 54–59 (2020) https://www.elibrary.ru/item.asp?id=43033191 (in Russian)
8. Kulagin, V., Sukharevskaya, A., Meffert, J.: Digital@Scale: Desk Book of Business Digitalization. Intellektualnaya literatura (2020). https://iknigi.net/avtor-vladimir-kulagin/179818-digitalscale-vladimir-kulagin/read/page-1.html (in Russian)
9. Savchenko, A.V.: "Chaos theory" by Stiven Mann and corporative management. Management. **1**(15) (2017). https://cyberleninka.ru/article/n/teoriya-haosa-stivena-manna-i-korporativnoe-upravlenie (in Russian)

10. Akperov, I.G., Khramov, V.V.: Soft evaluation models of the state of informational ecology in unified geoinformational space of the region. In: Collection: Innovative Technology in Science and Education (Conference "ITNO 2020"). Collection of Scientific Articles of VIII International Scientific and Practical Conference, With the Use of Distant Technologies. Rostov-on-Don, pp. 29–33. (2020) (in Russian) https://doi.org/10.23947/itno.2020.29-33
11. Imran, A., Vladimir, K., Viktor, L., Olga, M.: Fuzzy methods and algorithms in data mining and formation of digital plan-schemes in earth remote sensing. Procedia Comput. Sci. **120**, 120–125 (2017). https://doi.org/10.1016/j.procs.2017.11.218
12. Zadeh, L.A.: The concept of a linguistic variable and its application to approximate reasoning. In: Learning Systems and Intelligent Robots, pp. 1–10. https://doi.org/10.1007/978-1-4684-2106-4_1
13. Romanenko, G.N.: Mathematical methods in tasks of management and decision-making. Izvestiya VUZ. South-Caucuses Region. **6**(164), 154–156 (2011) https://cyberleninka.ru/article/n/matematicheskie-metody-v-zadachah-upravleniya-i-prinyatiya-resheniy (in Russian)
14. Beer, M., Noria N.: Resolving the Tension between Theories E and O of Change//Breking the Code of Change. Harvard Business School Press, Boston, pp. 1–33 (2000) https://dis.ru/library/558/27084/ (in Russian)
15. Akperov, I.G., Khramov, V.V.: Development of instruments of fuzzy identification of extended objects based on the results of satellite monitoring. In: Advances in Intelligent Systems and Computing, pp. 325–332 (2018). https://doi.org/10.1007/978-3-030-04164-9_44

Synthesis of the Polymer Bitumen Compositions for Pavement Using Neural Network

D. S. Mamed Hasan-zade, A. I. Babayev, and G. S. Hasanov(✉)

Azerbaijan State Oil and Industry University, Azadlig Ave. 20, Baku, Azerbaijan

Abstract. The paper presents the synthesis of the polymer bitumen compositions by adding wastes of high-pressure polyethylene production – clods. Modified bitumen composition, which contains clods, rubber crumbs, fuel oil with improved properties for pavement, compared to bitumen composition complying with the requirements of bitumen BND 60/90 according to GOST 22245-90, has been produced. It was found that addition of a bitumen composition as a modifying additive of 2% polymer waste from the production of high-pressure polyethylene – clod, along with rubber crumb, enables obtaining a bitumen composition of improved quality. Using Fuzzy neural network the factors have an impact on the composition quality were determined due to traditional forecasting methods cannot take into account uncertainty factor completely. It was shown that the clod in the composition has the strong effect on the mass extensibility, and the bitumen in the mass has an impact on their softening temperature and needle penetration.

Keywords: Binding paving bitumen · Modification · Polymer wastes · Clod (polyethylene wastes) · Softening · Extensibility · Multiple regression

1 Introduction

Chemical composites have been ranked highest in innovations per employee, leaving behind even metallurgical production, pharmaceutics and automotive industry [1]. It especially concerns to the production of such products of oil and gas chemistry like modern polymers. Nowadays, the most important polymer materials are polyether and polyamide fibers, polyethylene, polypropylene, and the production of polycarbonates and polyurethanes will noticeably increase. The increase in the consumption of these materials rises amount of their wastes that leads to an additional ecological pressure in the environment. Therefore, the secondary treatment of polymer wastes, which are practically valuable raw materials for the production of new materials and compositions, is an serious problem and allows reducing environmental tensions. From this point of view, the use of various polymers or their wastes in obtain of paving bitumen compositions along with plasticizing and stabilizing additives, is one of the promising and widely used by improving their quality [2–6]. Addition of modifying polymer additives into bitumen composition may purposefully modify the structural and mechanical properties of polymer-bitumen compositions, thereby giving them polymer properties such as elasticity, heat resistance, transparency, freeze resistance, etc.

R. A. Aliev et al. (Eds.): ICSCCW 2021, LNNS 362, pp. 580–590, 2022.
https://doi.org/10.1007/978-3-030-92127-9_78

Due to the polymer additive, bitumen acquires a number of valuable physical and mechanical properties and high resistance to polymer aging, which makes the production of polymer-bitumen compositions most cost-effective. In the production of polymer-bitumen compositions, achievement of desired results requires a good combination of bitumen and polymer additive, which in turn depends on the chemical nature of the polymers and bitumen [7, 8]. The molecular weight of the polymer and the technology of preparing polymer-bitumen mixture play a decisive role in this process [9, 10].

Aim of the present work is to obtain polymer-bitumen compositions with improved properties for road pavement by adding wastes of high-pressure polyethylene into their composition and to study the dependence of their structural and mechanical properties on production conditions and content of the compositions. Thus, the generation of the multiple and standard regression equations also enabled establishing the dependencies of the main properties of the composition on the percentage of mass components.

2 Statement of the Problem

The aim of this work is to obtain polymer bitumen compositions for pavement. Fuzzy neural network is selected for analyzing this problem, because it is characterized by the following strengths: curve fitting, generalization ability, adaptation, approximation ability and learning.

Proceeding from this, it is provided to solve the following problems:

- Selection of effective clustering and inference method for fuzzy neural networks;
- Comparison of the received results with results of traditional methods.

For the decision of these problems the architecture of fuzzy recurrent neural networks with two inputs, with one hidden layer and one output has been selected. The network is trained based on a hybrid method. Such neural network is characterized by smaller computing complexity and good ability to training.

In calculations 63 data are used. From these data 2/3 were used for training, and 1/3 data were used for testing of FRNN. Fragment of the initial data is given on Table 1.

Table 1. Fragment of the initial data

	x1	x2	x3	y1
1	1,425	1,52	73	11,73421
2	1,44	1,534	73,118	11,70977
3	1,455	1,548	73,236	11,68534

(continued)

Table 1. (*continued*)

	x1	x2	x3	y1
	...	...	...	...
24	1,665	1,744	74,888	50,53533
25	1,68	1,758	75,006	50,58349
26	1,695	1,772	75,124	50,63165
27	1,71	1,786	75,242	50,67981
28	1,725	1,8	75,36	50,72797
	...	...	...	...
58	2,235	2,276	79,372	142,1238
59	2,25	2,29	79,49	142,2923
60	2,265	2,304	79,608	142,4607
61	2,28	2,318	79,726	142,6292
62	2,295	2,332	79,844	142,7977

3 Experimental Part

Wastes of high pressure polyethylene (HPP) – clods with a molecular weight of 700–11000 – were used as polymer additives while conducting researches. Amount of polyethylene in the product was found to be 99%, the concentration of double bonds was 2–10%, thermographic value was 150 °C. High-melting bitumen characterized by softening temperature by ring and ball ("R and B") method – 110 °C (GOST 11506-73) and needle penetration depth at 25 °C 16 × 0,1 mm (GOST 11501-73) and extensibility at 25 °C – 3 cm (GOST 11505-75) has been used as a raw material. Based on literature data [11, 12], a rubber crumb with diameter of 0.06 mm (GOST-1293), which is a waste of rubber products, has been used as a modifier.

Use of rubber crumb as a modifier of road bitumen is due to the fact that addition of rubber crumb into the bitumen composition enables increasing adhesion of types with a coating, and this allows preventing bitumen bleeding from coating at high air temperatures and reduces dust formation. To obtain a homogeneous polymer-bitumen composition, bitumen is first mixed with fuel oil M-40 (GOST 10585-99) at 12 °C (Table 2). Further, HPP and resin crumb are added into the mixture at a specified temperature and intense stirring. While stirring, the temperature of bitumen mixture was raised up to 170 °C and remained for another 30–40 min.

Table 2. Physical and mechanical parameters of used fuel oil M-40 according to GOST 10585-99

№	Name of parameters	Parameters
1	Kinematic viscosity, at 80 °C, mm^2/c	43
2	Density at 20 °C, kg/m^3	922,1
3	Chilling point, °C	−12
4	Flash point, °C	184
5	Ignition residue, %	0,0228
6	Mass share of sulfur, %	0,23
7	Mass share of water, %	N/A

4 Results and Discussion

The studies show the possibility of obtaining polymer-bitumen composition with improved quality by adding HPP into the bitumen composition. In particular, with an amount of specified polymer waste of 1.8% per unit weight of bitumen, melting point of the composition is reduced to 57 °C by "R and B" method which is 53 °C versus 110 °C for the raw material. At the same time, the penetration depth of a needle is increased from 16 × 0,1 mm to 100 × 0,1 mm, as well as the extensibility from 3 to 8 cm at 25 °C. Thus, based upon softening temperature, the resulting polymer-bitumen composition complies with the requirements of bitumen BND 60/90 (Table 3). For comparison, the effect of adding one polymer waste – clod in the amount of 2,0 and 4,0% mass into bitumen has been studied.

Table 3. Physical and mechanical properties of polymer-bitumen compositions

№	Name	Softening temperature by R and B method, °C, GOST 11506-73	Penetration depth of needle at 25 °C × 0,1 mm, GOST 11501-73	Extensibility cm, at 25 °C n.n GOST 11505-75
	Bitumen BND 60/90 GOST 22245-90	n.n. 47	61–69	55
	High melting bitumen (raw material)	110	16	3

(*continued*)

Table 3. (*continued*)

№	Name	Softening temperature by R and B method, °C, GOST 11506-73	Penetration depth of needle at 25 °C × 0,1 mm, GOST 11501-73	Extensibility cm, at 25 °C n.n GOST 11505-75
Polymer-bitumen composition				
I	Bitumen – 60,6% Clod – 1,8% Fuel oil – 36,4% Resin crumb – 1,2%	53	100	8
II	Bitumen – 49,02% Clod – 1,96 Fuel oil – 49,2%	46	71	13
III	Bitumen – 48,07% Clod – 3,85% Fuel oil – 48,08%	80	51	4
IV	Bitumen – 77,5% Rubber crumb – 1,6% Clod – 1,5% Fuel oil – 19,4%	54	134	9
V	Bitumen – 73,5% Rubber crumb – 2,2% Clod – 2,2% Fuel oil – 22,1%	47	139	13,5
VI	Bitumen – 77,6% Rubber crumb – 3,1% Clod – 7,7% Fuel oil – 11,6%	43	143	3,0

As the results show in Table 3, a decrease in the softening temperature of polymer-bitumen composition is observed at all concentrations of added polymer waste. The softening temperature of the composition is reduced from 110 to 46 °C while adding 1.95% of clod, and with addition of 4% it is reduced to 80 °C. Penetration depth of a needle at 25 °C increases from 16 × 0,1 mm to 71 × 0,1 mm and 51 × 0,1 mm, and the extensibility indicator rises up to 8 and 13 cm, respectively.

It should be noted that one of the ways of producing compositions with given operational properties is the modification of bitumen with various polymer additives. Polyethylene is able to form a single homogeneous spatial structure with viscous road bitums, which causes an increase of softening temperature of their resistance to aging of the organic binder, as well as an increase of its chemical resistance under the effect of acids and alkalis of liquid fuel. In addition, polyethylene imparts elasticity to the coating, widens the temperature range, reduces the glass transition temperature of bitumen and increases softening temperature. Additionally, this prevents the formation of cracks, as well as the softening of coating under the sun [13].

Addition of a polymer residue – clod to the bitumen composition sharply changes its qualitative characteristics, that requires determining their optimum amount, since the composition also includes fuel oil and rubber crumbs. As Table 2 shows that the sample III of polymer and bitumen composition characterized by extensibility at 25 °C and softening temperature by "R and B" method concedes to the sample II. The sample II exhibits the best parameters at 25 °C on extensibility (13.5 cm) and at softening temperature (47 × 0.1 mm) (Table 3), while using bitumen BND60/70 46–54 × 0.1 mm that makes a difference 47 × 0.1 mm.

Comparison of the data given in Table 3 shows that the present composition has improved performance, namely, the softening temperature was achieved by "R and B" method and a needle penetration depth is found to be at 25 °C, that comply with bitumen BND 60/90 requirements, as well as the extensibility is increased at 25 °C, allowing the composition to be used under different climatic conditions, while the prototype was analyzed for only one parameter.

By increasing the percentage of polymer waste from 1.5 to 2,2% in polymer-bitumen compositions, the amount of fuel oil from 19,4 to 22,1% and rubber crumb from 1.6 to 2.2%, an increase of extensibility from 9 cm to 13,5 cm and accordingly an improvement in the elasticity of the composition were achieved (Table 3). At the same time, further increase in the number of added components led to a deterioration in parameters. In particular, with an increase in the amount of polymer additive in the composition, the production of a homogeneous bitumen composition is complicated. Table 3 shows that in this case compositions (sample 3) with extensibility at 25 °C – 3 cm and softening temperature 43 °C by "R and B" method have been produced. From Table 4 it also follows that sample 2 (Table 3) of the polymer-bitumen composition was the best among three samples, where the extensibility at 25 °C reached a maximum value of 13,5 cm, which, together with the softening temperature indicator, corresponds to the construction bitumen BND 60/90.

To study the dependencies of the main properties of the composition on the percentage of mass components, the equations of multiple and standard regression were generated. For this purpose, table of experimental data was first prepared. The percentage share of the total mass of the clod (X_1), rubber (X_2) and bitumen (X_3) crumbs were selected as independent variables. Mass extensibility (Y_1), softening temperature (Y_2) and needle penetration depth (Y_3) (Table 4) were taken as dependent variables.

Table 4. Planning matrix of experiments

Number of measurements	X_1	X_2	X_3	Y1	Y2	Y3
1	1,5	1,6	77,5	9	54	134
2	2,2	2,2	73,5	13,5	47	139
3	8	4	69	10	52,5	75
4	10	3	72,5	13,5	47	139
5	10	4	70	13	43	143
6	12	1,6	76	9	54	105
7	12	2	78,5	7	61,5	134
8	16	8	75,5	5	61	135

Further, according to the known procedure [14, 15], calculations were made on the generation of regression equations and the determination of equation parameters to clarify the effect of components on certain properties of the composition. As a result, the multiple regression equation for mass extensibility was obtained:

$$Y_1 = 62.1785 - 0.1255X_1 - 0.702X_2 - 0.6581X_3$$

and standardization form of regression equation:

$$y_1 = -0.198x_1 - 0.475x_2 - 0.716x_3$$

Mean square error (MSE) for Y1 is 1.8%.

According to the maximum coefficient $\beta_1 = -0.198$, we conclude that the X_1 factor, which complies with the clod percentage in the mass, has the greatest effect on the extensibility of the mass. Polymer structures include straight, radial, cross-linked and uneven-special fuzzy branching distribution that have a noticeable effect on both the physical-mechanical-chemical properties of the polymer and the quality of polymer-bitumen composite materials, including their chemical composition, structure, types of glutinating and technological features of the production process. This is due to the high amount of polymers in the polymer-bitumen composition.

Similar equations for the dependence of softening temperature of the mass on the percentage of constituent components are as follows:

$$Y_2 = -63.3052 + 0.2803X_1 + 1.0015X_2 + 1.4851X_3$$

$$y_2 = 0.21x_1 + 0.322x_2 + 0.767x_3$$

MSE for Y2 is 0.4%.

Maximum coefficient $\beta_3 = 0.767$ shows that the X3 factor, which corresponds to the bitumen percentage in the mass, has the greatest effect on softening temperatures. The asphalt composition with a high molecular weight includes relatively insoluble and non-volatile compounds and polar molecules with lower molecular weight known as asphaltenes, as well as low molecular weight compounds called maltenes. Asphaltenes, which are the main component of asphalt, give hardness, and maltenes provide plasticity and facilitate adhesion. Maltenes consist mainly of oils (aromatic and saturated) and resins, which explains the change in the softening temperature of the mass from the percentage of constituent components of bitumen.

These equations for the dependence of the needle penetration depth on the percentage of constituent components is as follows:

$$Y_3 = -99.566 - 1.2178X_1 + 3.3364X_2 + 3.0376X_3,$$

$$y_3 = -0.257x_1 + 0.302x_2 + 0.442x_3$$

MSE for Y3 is 3.5%.

The maximum coefficient $\beta_3 = 0.442$ shows that the factor X_3 ha the largest effect on needle penetration, that corresponds to bitumen percentage in the mass. When used in the production process of an oil composite, the polymer swells that enables to be distributed more evenly in the mass, forming a relatively homogeneous medium. This partially affects the X_3 parameter. As the analysis results of the equations show the addition of polymers into asphalt binders in polymer-modified asphalt leads to the modification of some key physical properties, such as elasticity, tensile strength, sensitivity to high and low temperatures, viscosity, and adhesion and cohesion.

Using neural network [16–18] is defined Y1, Y2 and Y3 (Fig. 1). Fragments of the obtained results are given in Figs. 2, 3 and 4.

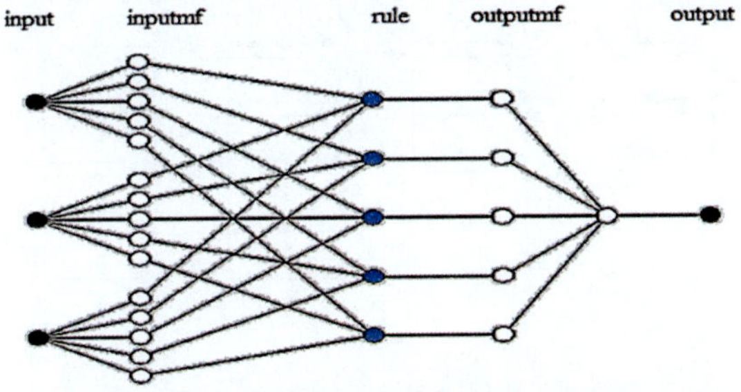

Fig. 1. Structure of the neural network

These parameters help to increase the durability of pavement achieved by reducing fatigue and thermal cracking, as well as reducing susceptibility to high temperatures (for example, wear rutting, pushing, rockiness) and increased retention of aggregates in applications, for example, chip breakers. These indicators constitute a set of mutually confirming and mutually exclusive factors for complex systems, capable of aggregating fuzzy representations, the mechanism of which is closely related to both the quantitative ratio of components and the technology of their production.

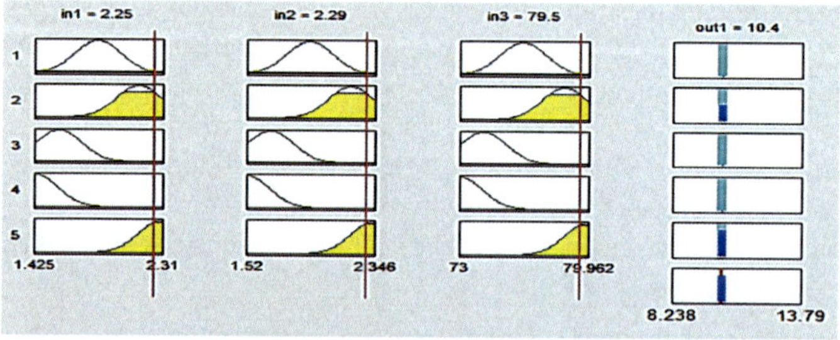

Fig. 2. Mass extensibility (Y_1)

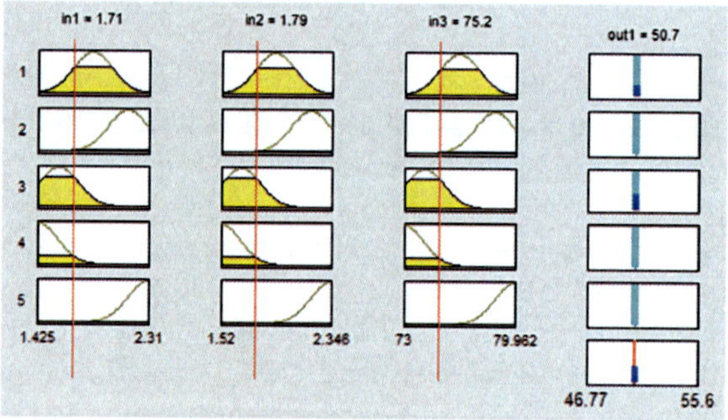

Fig. 3. Softening temperature (Y_2)

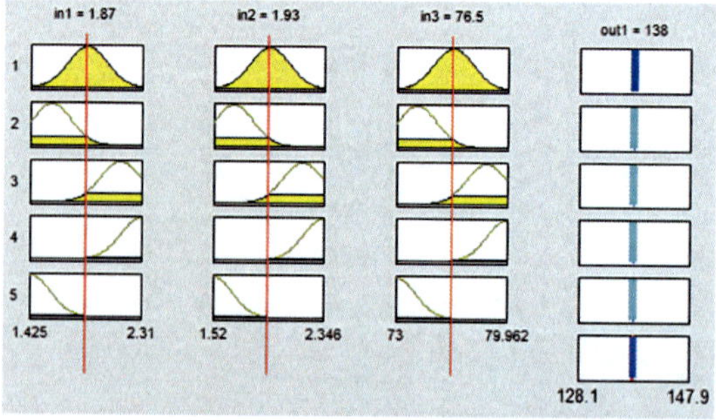

Fig. 4. Penetration depth (Y_3)

5 Conclusion

The results are comprised with results of the traditional method regression. All calculations are made in MATLAB soft.

For regression method Mean Square Error (MSE) has made: For Y1 1.8%, for Y2 0.4%, for Y3 3.5%. For fuzzy neural network: on Y1 0.45%, on Y2 0.1%, on Y3 0.8%

The received results give the basis to approve, that for today forecasts made by using of fuzzy recurrent neural networks are more reliable and exact.

Thus, a series of studies found that the addition of polymer waste of the production of high-pressure polyethylene to bitumen as a modifying additive – clod, along with rubber crumbs, allows obtaining a bitumen composition with improved quality and complies with the requirements of GOST 22245-90 for construction bitumen PPB 60/90. In addition, the possibility of recycling polymer waste of polyethylene production, along with crumbs of treated tires, helps to reduce the cost of roadways and solve the environmental problem. By comparing the coefficients of regression equations, we may conclude that clod percentage in the composition has the greatest effect on mass extensibility, and their softening and needle penetration temperatures are influenced by the proportion of bitumen in the mass.

It has been found that polymer-bitumen composition as a set of interacting factors in a complex system, is capable of aggregating fuzzy representations, the mechanism of which is closely related both to the quantitative ratios of components and the technology of their production, considering the dynamics and nature of synergy of these factors.

References

1. Babkin V.V., Uspenskiy D.D.: New strategy: chemistry 2030 high limits of a raw material. Clustering. Chemization of Industry. RF/Babkin V.V., Uspenskiy D.D. Publ. "Lica", 222 c. (2015) (in Russian)
2. Study and Production of Petroleum Bitums. Collection of Scientific Works. Center on Scientific-Research Information Institute and Technical-Economical Studies Petroleum Chemistry. Under Editorship of Fyodorov V.V., pp. 77–124 (1981) (in Russian)
3. Samadova, F.I., Allahverdiyeva, A.A. Technology of producing bitums. Elm 158 (2007) (in Russian)
4. Fyodorov, V.V., Siroejko, A.M., Begak, O.Y., Proskuyakov, V.A., Borovikov, G.I.: J. Appl. Chem., **75**(6) 1027 (2002). (in Russian) https://doi.org/10.1134/S0044460X
5. Pozdnyaeva, L.V.: Petroleum polymer resin as a modifier of petroleum pavement bitums. Col. Sci. Pap. **8**, 42–55 (1996) (in Russian)
6. Leonenko, V.V., Safarov, G.A.: Chemistry and technology of oil fuels **5**, 43 (2001) (in Russian)
7. Nekhoroshev, V.N., Rossel, L.P.: Oil and gas of West Siberia. In: Collection of Papers of All Russian Scientific-Technological Conference, pp. 268–269 (2005) (in Russian)
8. Huseynov, F.I., Bilalov, M.M.: Modified polymer compositions. Study Guide, p. 216 (2001) (in Russian)
9. Rosenthal, D.L., Tobolina, M.S., Fedosova, V.A.: Modification of bitums with polymer additives. Oil treatment. Subject Survey **6**, 15–42 (1988) (in Russian)
10. Allahverdiyeva, A.A.: Polymer-bitumen composition. Azerbaijan Oil Ind. **5**, 46 (1993) (in Russian)
11. Khristoforova, A.A., Filippov, S.E., Gogolev, I.N.: E-journal. Eng. J. Don **4**, 347–350 (2011) (in Russian)
12. Khristoforova, A.A., Sokolova, M.D., Potlov, S.N.: Materials. Technol. Instrum. **13**(1), 73–76 (2008). (in Russian)
13. Renner, A.G., Aralbayeva, G.G., Zinovyova, O.A.: Correlation-regression analysis. Methodological Guidance on Laboratory Practicals, 26 p. (2002) (in Russian)

14. Protsenko, M.Y.: The effect of low-molecular polyethylene on the properties of petroleum pavement bitumen. Irkutsk 140–148 (2018) (in Russian)
15. Dreyper, N., Smith, G.: Applied Regression Analysis, 3rd Issue, 912 p. (2016)
16. Aliev, R.A., Fazlollahi, B., Aliev, R.R.,Guirimov, B.G.: Linguistic time series forecasting using fuzzy recurrent neural network. J. Knowledge. "Education" Society of Azerbaijan Republic. Series Business 3–13 (2005)
17. Gardashova, L.A.: Benchmark fuzzy time – series forecasting using fuzzy recurrent neural network. J. Knowledge. "Education" Society of Azerbaijan Republic. Series Business 38–44 (2005). https://doi.org/10.1007/s00500-007-0186-7
18. Aliev, R.A., Fazlollahi, B., Aliev, R.R.: Soft Computing and Its Applications in Business and Economics. Springer Verlag, 450 p. (2004). https://doi.org/10.1007/978-3-540-44429-9

Assessment of the University Competitiveness in the Paradigm of the Humanistic System Behavior

Misir Mardanov[1], Ramin Rzayev[2(✉)], and Zeynal Jamalov[2]

[1] Institute of Mathematics and Mechanics of ANAS, Vahabzadeh str. 9, 1141 Baku, Azerbaijan
[2] Institute of Control Systems of ANAS, Vahabzadeh str. 9, 1141 Baku, Azerbaijan

Abstract. The decision-making process is the most recognized type of human intellectual activity, which is presented as a continuous chain of multi-criteria choices developed and implemented under the influence of numerous factors of a diverse nature. According to the Bellman-Zadeh principle, decision-making in the fuzzy environment is a multi-criteria choice of the best alternative that simultaneously satisfies fuzzy goals (quality criteria) and fuzzy constraints. In the article, taking into account the presence of uncertainties of the human factor, each university is considered as a humanistic educational system, and the assessment of the level of its competitiveness and the calculation of the corresponding rating are presented as a weakly structured problem that has not only quantitative but also qualitative characteristics. As a weakly structured problem, the comprehensive assessment of the competitiveness of universities requires a multidisciplinary approach. Therefore, to solve it, methods and models of mathematical statistics, elements of fuzzy logic, theories of fuzzy sets and neural networks, as well as hybrid (neuro-fuzzy) systems were used.

Keywords: University · Humanistic system · Multi-criteria choice · Expert system · Fuzzy set · Neural network · Decision-making

1 Introduction

Zadeh's non-standard ideas relative to significance and role of linguistic variables in approximate decision making reflected his new approaches to describing weakly structured humanistic systems (i.e., implying human participation), providing new opportunities for their analysis and problem solving in the paradigm of sensible human reasoning. Owing to the tools of fuzzy logic, it is possible to formalize all kinds of expert verbal models that implicatively reflect the cause-effect relations between input and output linguistic characteristics. As a result, along with quantitative (metrizable) data, it became possible to involve qualitative (non-metrizable, weakly structured) categories to the computational process [1–5].

Now in the world there are a lot of specialists from different fields who actively apply the fuzzy logic to solve various applied problems. In particular, over the past five years, we have developed a methodological basis for the system of information and

R. A. Aliev et al. (Eds.): ICSCCW 2021, LNNS 362, pp. 591–598, 2022.
https://doi.org/10.1007/978-3-030-92127-9_79

analytical support for assessing the competitiveness of universities and calculating their ratings. The starting point here was that the problem itself is weakly structured, and each university is presented as the humanistic system. L. Zadeh has always opposed a tradition that reinforces the growing tendency to analyze humanistic systems as if they were mechanistic systems described in terms of difference, differential, or integral equations. In particular, L Zadeh believed that the accurate quantitative analysis has not a significant practical importance for solving really social, economic and other problems caused by the active participation of a person.

2 Problem Definition

As a rule, the methods of scoring- analysis suffer from the fact that they do not take into account the contextual cause-effect relation. In contrast, analytical methods provide the comparison of alternatives by an index reflecting the total influence of the finite number of factors $\{x_k\}_{k=\overline{1,n}}$ in the form of the corresponding function $F = F(x_1, x_2, \ldots, x_n)$. However, due to the absence of the sufficient number of necessary statistical data that form the notion of indicators and evaluation criteria, and the means of formalizing the preferences of decision-makers, the construction of an adequate econometric model of type F becomes impossible. Therefore, from the point of view of a multi-criteria assessment of a humanistic system behavior, it is reasonable to represent a working model for calculating the university rating in the form of a "black box" with input and output characteristics in the form of linguistic variables анализ.

3 Comprehensive Assessment of the University Competitiveness

The important step in the development and application of fuzzy logic was reflected in [6], devoted to decision-making in the fuzzy environment. This work was published in 1970 by L. Zadeh in collaboration with R. Bellman, where authors, in particular, illustrate a fundamentally new approach using examples of multistep decision-making processes, in which key goals and/or basic constraints are sets defined in a weakly structured form, i.e., they are of a fuzzy nature. Subsequently, this approach began to be called the Bellman-Zadeh principle in decision-making under uncertainty, or, more specifically, in solving the problem of multicriteria optimization in the presence of fuzzy information. This problem is formulated as follows:

$$\min_{\vec{x}}\{f_1(\vec{x}), f_2(\vec{x}), \ldots, f_k(\vec{x})\},\ \vec{x} = (x_1, x_2, \ldots, x_n)^T \in S,$$

where $f_k: R^n \to R\ (k \geq 2)$ is the objective functions that describe the criteria of satisfaction and, as a rule, mutually contradict each other; S is a nonempty domain.

According to the Bellman-Zadeh principle, decision-making in a fuzzy environment is a multi-criteria choice of the best alternative that simultaneously satisfies fuzzy goals (quality criteria) and fuzzy constraints (Fig. 1). In this case, no distinction is made between purpose and limits. If there is such a difference, then it is conditional.

As weakly structured problem the comprehensive assessment of the university competitiveness is required a multidisciplinary approach. Therefore, to solve it, we used

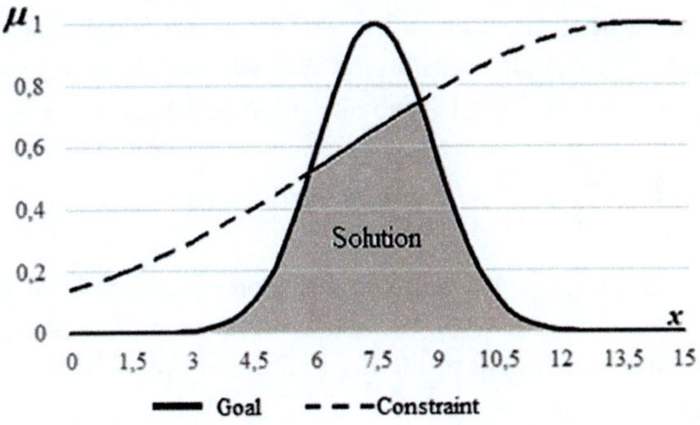

Fig. 1. Illustration of decision-making according to the Bellman-Zadeh principle

methods and models of mathematical statistics, tools and elements of fuzzy logic, fuzzy set theories and neural networks, as well as hybrid (neuro-fuzzy) systems. Based on the results of our research, nine articles and one monograph were published.

In 2015, to estimate the marketing space of educational services and the university's competitive position in the labor market, the appropriate models were proposed, developed based on the application of fuzzy inference [7]. In the context of this study, a fuzzy approach to the formation of the flexible and universal system for assessing the competitiveness of universities has proposed. In the same year, the leading Russian journal "Control Science" (Moscow) published the article [8], where it was proposed the method for assessing the competitiveness of universities with their subsequent ranking, based on the application of fuzzy inference system. Here, the analysis of the competitiveness of universities includes multi-criteria assessments of the marketing environment and the competitive position of universities in the educational services market. Thus, the marketing environment of universities was assessed according to 13 indicators: x_1 – state support, x_2 – competition of applicants, x_3 – connection with secondary schools, x_4 – popularization by media structures, x_5 – the condition of the national economy, x_6 – socio-demographic situation, x_7 – socio-cultural position of society, x_8 – political and legal support of society, x_9 – target orientation of the university, x_{10} – applied education technology, x_{11} – organizational structure of the university, x_{12} – the level of the teaching staff, x_{13} – the number of competing universities. The competitive positions of university specialties have estimated by two indicators: y_1 – the market share of the university in given specialty, y_2 – the rate of growth of the market share of the university in given specialty. In both cases, to describe the qualitative evaluation criteria (as terms of linguistic variables x_i ($i = 1 \div 13$) and y_j ($j = 1, 2$)) by appropriate fuzzy sets $\tilde{A}$, the following Gaussian membership function was used:

$$\mu_{\tilde{A}}(u) = \exp[-(u - u_0)^2/\sigma^2] \quad (1)$$

where u is the indicator of the university for the current position; u_0 is the average value of the indicators of all universities for the current position; σ is the mean-square deviation.

Value judgements relative to the marketing environment and the competitive position of universities were constructed in the form of terms of output linguistic variables described by the corresponding fuzzy subsets of the discrete universe $J = \{0, 0.1, 0.2, \ldots, 1\}$. In particular, $\forall j \in J$ such terms are: F = FAVORABLE, $\mu_F(j) = j$; MF = MORE THAN FAVORABLE, $\mu_{MF}(j) = j^{(1/2)}$; VF = VERY FAVORABLE, $\mu_{VF}(j) = j^2$; P = PERFECT,

$$\mu_P(j) = \begin{cases} 1, & \text{if } j = 1, \\ 0, & \text{if } j < 1; \end{cases} \quad UF = \text{UNFAVORABLE}, \ \mu_{UF}(j) = 1 - j.$$

Approbation of this approach was carried out on the example of five hypothetical universities u_k ($k = 1 \div 5$). As a result, the sought general functional solution, for example, regarding the assessment of the marketing environment of universities, looks like the following matrix

$$R = \left[\begin{array}{c|ccccccccccc} & 0 & 0.1 & 0.2 & 0.3 & 0.4 & 0.5 & 0.6 & 0.7 & 0.8 & 0.9 & 1 \\ \hline u_1 & 0.8701 & 0.9701 & 0.9870 & 0.9870 & 0.9679 & 0.8679 & 0.7679 & 0.6679 & 0.5679 & 0.4679 & 0.3679 \\ u_2 & 0.3874 & 0.4874 & 0.5874 & 0.6874 & 0.7874 & 0.7904 & 0.7904 & 0.7904 & 0.7904 & 0.7904 & 1.0000 \\ u_3 & 0.9588 & 0.9817 & 0.9817 & 0.9441 & 0.8441 & 0.7441 & 0.6441 & 0.5441 & 0.4441 & 0.3441 & 0.2441 \\ u_4 & 0.6827 & 0.7827 & 0.8827 & 0.9827 & 0.9870 & 0.9870 & 0.8894 & 0.7994 & 0.6994 & 0.5994 & 0.4994 \\ u_5 & 0.8946 & 0.9777 & 0.9777 & 0.9777 & 0.9777 & 0.9777 & 0.9777 & 0.9777 & 0.9777 & 0.8788 & 0.7788 \end{array}\right]$$

where each row is the essence of the values of the membership function $\mu_{Ei}(j)$, $i = 1 \div 5$; $j = 0, 0.1, 0.2, \ldots, 1$, which restore the corresponding fuzzy sets, reflecting the fuzzy conclusions relative to the assessment of the marketing space of the corresponding universities.
To defuzzify the fuzzy evaluative conclusions E_i, the following formula had used:

$$F(E_i) = \frac{1}{\alpha_{\max}} \int_0^{\alpha_{\max}} M(E_{i\alpha}) d\alpha, \ i = 1 \div 5, \tag{2}$$

where $E_{i\alpha}$ is the α-level set ($\alpha \in [0, 1]$) defined as $E_\alpha = \{j | \mu_{Ei}(i) \geq \alpha, j \in I\}$; $\alpha_{\max}$ is maximum value on $E_{i\alpha}$; $M(E_{i\alpha}) = (1/n) \sum_{k=1}^{n} x_k$ is the cardinal number of the set $E_{i\alpha}$.

In 2016, to aggregate the estimates of key indicators of universities according to the QS version an alternative approach was proposed based on the application of the fuzzy inference [9]. The world universities ranked in the first ten places from the list according to the QS WUR version were selected as the evaluated alternatives. Based on the results of the calculations, the ranking of the top universities was obtained, which is somewhat different from the QS WUR version, based on the weighted aggregated assessment of key indicators.

A year later, to assess and subsequent rank of universities relative to the quality of the provided educational services two approaches were applied: *statistical*, implying the weighted estimates of key indicators, and *verbal*, implying the application of the fuzzy inference system [10, 11]. Based on the results of using these methods to estimating the arbitrarily chosen universities as hypothetical alternatives, similar aggregate results (ratings) and the corresponding two ranking methods were obtained. It was noted that in the case of covering the larger number of universities, for example, as in the case of the QS WUR methodology, the quality of fuzzification of evaluation criteria significantly improves and, as a result, the accuracy of the desired estimates increases.

In 2017, in the German publisher "Palmarium Academic Publishing" published the book "Fuzzy methodology for multi-criteria ranking of universities" [12], which outlines fuzzy methods for multi-criteria choice of alternatives under uncertainty, as well as new approaches to solving problems associated with comprehensive assessment of educational services, ranking of universities and optimization of their organizational structures.

In the next 2018-th and 2019-th years, the fuzzy methods for multi-criteria assessment of alternatives were considered under insufficient information [13, 14]. To provide analytical support of the decision-making process, new approaches to the compilation of heuristic knowledge were formulated by adequately conversion the external representation of information relative to weighted integral estimates of alternatives obtained by expert judgements into the adequate internal representations by neural network based fuzzy inference system.

Finally, in 2019, it was proposed a methodology for constructively assessing the effectiveness of universities' activities using the variety of criteria adjusted for regional specifics [15]. Given set consists of 18 criteria and is divided into three sections – according to the key indicators: "Potential", "Activity and quality of education", and "International recognition". Some of them are calculated as the proportion of objects or results of the university's activities of a certain quality relative to their total number or volume. Other indicators are purely qualitative in nature, such as attractiveness among applicants and/or reputation among employers. The number of indicators is calculated according to various schemes, such as the level of university participation in international programs. The rest are absolute measurement values. The proposed methodology takes into account the fact that various indicators of universities' performance have different weights in determining their ranking, and their priorities should change depending on the specifics of the university, its size, age, and the number of other characteristics. As a result of such analysis, the ranking of universities is carried out, but also the mechanism of its progressive development is planned. Based on these considerations, within the framework of this methodology, the approach to the multi-criteria assessment of educational services and the ranking of universities were proposed, based on the joint use of expert judgements and neural network modeling.

A preliminary expert analysis of the relative influence of factors reflecting the quality of educational services included: 1) selection of influences; 2) identification of generalized weights for the selected factors, based on their relative influence; 3) formation of key indicators of educational services and, accordingly, weighted total indices as university ratings.

To establish the rank estimates of the factors x_{ki} $(i = 1 \div n)$, experts are involved, who are invited to rank the variables x_{ki} according to the following principle: the most important is to be indexed with the number "1", the next, less important, with the number "2" and then in descending order of expert preference. The obtained rank scores are summed and evaluated for consistency. To determine the degree of consistency of expert opinions relative to the priority x_{ki}, the following Kendall's concordance coefficient is used, which demonstrates the multiple rank correlation of expert opinions:

$$W = 12 \cdot S / \left[m^2 \left(n^3 - n \right) \right], \tag{3}$$

where m is the number of involved experts; n is the number of influences; S is the deviation of expert opinions from the mean value of the ranking of the influences x_i ($i = 1 \div n$), which is calculated as:

$$S = \sum_{i=1}^{n} \sum_{j=1}^{m} \left[r_{ij} - \frac{m(n+1)}{2} \right]^2, \tag{4}$$

where $r_{ij} \in \{1, 2, \ldots, n\}$ is the rank of the i-th influence, established by the j-th expert.

At the preliminary stage of the independent questionnaire, each of the involved experts is also invited to set the values of the normalized estimates of the generalized weights of x_{ki} in the form α_{ij}. Further, the group estimates of the influences and the degree of competence of the experts are calculated. To calculate the average value α_i over the i-th group of normalized estimates of generalized weights of x_{ki}, the following equality is applied:

$$\alpha_i(t+1) = \sum_{j=1}^{m} w_j(t) \alpha_{ij}, \tag{5}$$

where $w_j(t)$ is a coefficient characterizing the degree of competence of the j-th expert ($j = 1 \div m$) at time t. As can be seen from (5), the process of finding the group estimates of the normalized values of the generalized weights of the influences is iterative and completes under condition

$$\max \{ |\alpha_i(t+1) - \alpha_i(t)| \} \leq e, \tag{6}$$

where ε is the permissible accuracy of calculations.

Competence indicators of experts $w_j(t)$ ($j = 1 \div m$) at time t are calculated based on the following equalities:

$$\begin{cases} w_j(t) = \dfrac{1}{\eta(t)} \displaystyle\sum_{i=1}^{n} \alpha_i(t) \cdot \alpha_{ij} \ (j = \overline{1, m-1}), \\ w_m(t) = 1 - \displaystyle\sum_{j=1}^{m-1} w_j(t), \ \sum_{j=1}^{m} w_j(t) = 1. \end{cases} \tag{7}$$

where $w_j(t)$ is the indicator of the j-th expert's competence in the t-th approximation, and $\eta(t)$ is the normalizing factor calculated by the formula:

$$\eta(t) = \sum_{i=1}^{n} \sum_{j=1}^{m} \alpha_i(t) \alpha_{ij}. \tag{8}$$

The knowledge obtained on the basis of expert conclusions, as external representations of the weighted total assessments of universities, are compiled into effective internal representations of them in the logical basis of a three-layer feedforward neural network inducing output signals as follows [15]:

$$z_j = \sum_{p=1}^{m} c_p \varphi[\sum_{i=1}^{r} w_{pi} x_{ij} - \theta_p] \ (j = 1 \div 50), \tag{9}$$

where r is the number of components of the input vector; m is the number of nonlinear neurons in the "hidden" layer, selected by the user during the simulation by trial and error; w_{pi} and c_p are the weights of the input and output connections, respectively; θ_p is the threshold of the p-th nonlinear neuron from the "hidden" layer; $\varphi(\cdot)$ is the activation function of the nonlinear neuron from the "hidden" layer, for example, sigmoidal type $\varphi(t) = 1/(1 + e^{-t})$.

4 Conclusion

The main message of L. Zadeh, directed to the entire scientific community, is that in humanistic systems, reasoning and decision-making are based not only on measurements (quantitative assessments), but also on linguistic assessments. Owing to fuzzy logic technology, it is possible to involve both quantitative and qualitative characteristics in the computational process.

The comprehensive indicator of the quality of educational services at the university reflects a very wide range of heterogeneous parameters characterizing the degree of compliance of educational programs, material and technical support of the educational process, scientific and methodological base, staff, etc. According to J. von Neumann, the desire to obtain an adequate model for assessment of the weakly structured system, which is undoubtedly the educational process at a university, loses its meaning, since model complexity becomes commensurate with the complexity of the process itself. Obviously, the use of such model cannot allow a relatively simple and clear interpretation of the mechanism of the educational process, use of any formal procedures for studying its characteristics and synthesizing a control system for it. Moreover, for the educational process, as a complex, open and dynamically developing system, the principle of incompatibility of L. Zadeh [3] is applicable, according to which it is impossible to obtain accurate and at the same time practical judgments about its current and future states. Therefore, many rating agencies carry out a statistical calculation of key indicators of the quality of educational services by weighted summation of their components with pre-selected weights. To identify these weights, expert systems are used, where the heuristic knowledge of specialized specialists is the main resource.

References

1. Zadeh, L.A.: Fuzzy sets. Inf. Control **8**, 338–353 (1965). https://doi.org/10.1016/S0019-9958(65)90241-X
2. Zadeh, L.A.: Probability measures of fuzzy events. J. Math. Anal. Appl. **10**, 421–427 (1968). https://doi.org/10.1016/0022-247X(68)90078-4
3. Zadeh, L.A.: Outline of a new approach to the analysis of complex systems and decision processes. IEEE Trans. Syst. Man Cybern. **SMC-3**(1), 28–44 (1973). https://doi.org/10.1109/TSMC.1973.5408575
4. Zadeh, L.A.: Fuzzy sets as a basis for a theory of possibility. Fuzzy Sets Syst. **1**, 3–28 (1978). https://doi.org/10.1016/S0165-0114(99)80004-9
5. Zadeh, L.A.: A Theory of approximate reasoning. In: Hayes, J., Michie, D., Mikulich, L.I. (eds.) Machine Intelligence, vol. 9, pp. 149–194. Halstead Press, New York (1979). https://doi.org/10.1016/B978-1-4832-1450-4.50047-X

6. Bellman, R.E., Zadeh, L.A.: Decision-making in a fuzzy environment. Manage. Sci. **17**(4), 141–164 (1970). https://doi.org/10.1287/mnsc.17.4.B141
7. Mardanov, M.J., Rzayev, R.R., Jamalov, Z.R., Khudadova, A.K.: Assessment of the university competitiveness based of a fuzzy analysis of its qualitative characteristics. News Azerbaijan Eng. Acad. **7**(4), 113–130 (2015) (in Russian)
8. Mardanov, M.J., Rzayev, R.R., Jamalov, Z.R., Hasanov, V.I.: An approach to the assessment of higher education institutions competitiveness. Control Sci. **6**, 23–34 (2015). (in Russian)
9. Mardanov, M.J., Rzayev, R.R.: Alternative aggregation of rating estimates within the QS-technology of ranking higher educational institutions. Math. Mach. Syst. **4**, 122–133 (2016). (in Russian)
10. Mardanov, M.J., Rzayev, R.R.: Two approaches to integrated assessment and ranking of universities. J. Autom. Inf. Sci. **4**, 36–55 (2017)
11. Mardanov, M.J., Rzayev, R.R., Jamalov, Z.R., Khudadova, A.K.: Integrated assessment and ranking of universities by fuzzy inference. Procedia Comput. Sci. **120**, 213–220 (2017). https://doi.org/10.1016/j.procs.2017.11.231
12. Mardanov, M.J., Rzayev, R.R., Jamalov, Z.R.: Fuzzy Methodology for Multi-criteria Ranking of Universities. Palmerium Academic Publishing, Saarbruchen (2017).(in Russian)
13. Mardanov, M.J., Rzayev, R.R.: Methods for multi-criteria assessment of alternatives in the logical basis of neural networks. Autom. Meas. Mech. Engs. **3**, 75–85 (2018). (in Russian)
14. Mardanov, M., Rzayev, R.: One approach to multi-criteria evaluation of alternatives in the logical basis of neural networks. In: Aliev, R.A., Kacprzyk, J., Pedrycz, W., Jamshidi, M., Sadikoglu, F.M. (eds.) ICAFS 2018. AISC, vol. 896, pp. 279–287. Springer, Cham (2019). https://doi.org/10.1007/978-3-030-04164-9_38
15. Mardanov, M.J., Rzayev, R.R., Ibragimov, N.S., Jamalov, Z.R.: Formation of university rankings based on the compilation of expert assessments in the logical basis of the neural network. Math. Mach. Syst. **1**, 107–123 (2019). (in Russian)

Fuzzy Model of Estimating the Specific Weight of Shadow Economy Being a Manifestation of Economic Crime on the Basis of the Utility Function

Mirali S. Kazimov[1,2], Nazim K. Aliyev[1,3], and Shahzada G. Madatova[1,4,5](✉)

[1] Training Centre of Ministry of Taxes of the Republic of Azerbaijan, Agha Nematullah str. 44, AZ1033 Baku, Azerbaijan

[2] Baku State University, Acad. Zahid Xalilov str. 23, AZ1148 Baku, Azerbaijan

[3] Institute of Law and Human Rights, ANAS, H.Javid Ave., 115, AZ1073 Baku, Azerbaijan

[4] The Academy of Public Administration under the President of the Republic of Azerbaijan, Lermontov str. 74, AZ1001 Baku, Azerbaijan

[5] Azerbaijan State Oil and Industry University, Azadlig ave. 34, AZ1000 Baku, Azerbaijan

Abstract. The article is devoted to the control measures of the tax authority against the cases of tax reduction that occur during the fulfillment of the tax liability of the economic entity and the mechanism for detecting tax evasion. On this basis, in the assessment of quantitative and qualitative indicators, using the methodology of fuzzy logic IF–THEN, it is possible to increase the declared income only by increasing the fines without paying attention to the usefulness of the collection, or by increasing the usefulness of the collection without changing the fines. The article assumed that declared profitability, specific weight of the collected tax and the specific weight of income remaining at the disposal of a taxpayer were taken as fuzzy variables. However, the analysis shows that profitability differs both by fields and by taxpayers operating in each field. Therefore, the declared profitability factor and the penalty for its concealment should be treated differently.

Keywords: Shadow economy · Economic crime · Utility function · Fuzzy model · Fuzzy approach

1 Introduction

The shadow economy is one of the main determinants of crime in the field of economic activity, especially its more organized manifestation forms. Therefore, ensuring the effective struggle against the "shadow economy" is considered one of the main directions of state policy of the Republic of Azerbaijan. Reforms carried out in this area in recent years, complex institutional measures have significantly reduced the characteristics of the "shadow economy", in particular, its widespread informal employment, conducting bookkeeping by double-entry. However, due to the large potential of the "shadow economy", the effective struggle against it remains its significance in terms of increasing state budget revenues [1].

R. A. Aliev et al. (Eds.): ICSCCW 2021, LNNS 362, pp. 599–606, 2022.
https://doi.org/10.1007/978-3-030-92127-9_80

The phenomenon of shadow economy is one of the problems forthcoming in modern economics, which requires scientific estimation and practical solutions. Analysis of the emergence process of this problem and review of existing scientific research shows that one of the main reasons affecting its emergence is related to the gaps in tax policy, inadequacy of tax legislation to economic policy, inadequate stimulation of entrepreneurship, etc.

The fuzzy model of the interaction between the two components of economic system as an economic entity in the example of the state and taxpayer in the form of a tax authority, the process of shadow economy emergence in this economic relationship and reducing its specific weight was proposed [2]. The article assumed that declared profitability, specific weight of the collected tax and the specific weight of income remaining at the disposal of a taxpayer were taken as fuzzy variables. However, the analysis shows that profitability differs both by fields and by taxpayers operating in each field. Therefore, the declared profitability factor and the penalty for its concealment should be treated differently.

2 The Mechanism of Joint Action of Tax Authority and Economic Entity

One of the main tasks of the tax authority, which represents the state and communicates directly with economic entities within the powers granted to it by legislation, by controlling their timely and full fulfillment of tax obligations, is to prevent tax evasion. Here, the state control function consists of cameral, mobile and operational inspections. As it is seen, the state (tax authority) and economic entity operate in a two-tier hierarchical relationship in the process of fulfilling the tax liability [3]. At the same time, state and economic entities involved in this relationship are characterized by certain, and sometimes significant, differences in corporate goals.

Each of them tries to get some benefit as a result of their activities. If in this case the profit of the state is expressed by the amount of tax mobilized to the state budget, the profit of the economic entity is expressed by the amount of income remaining at its disposal after fulfillment of the tax obligation [4].

On the other hand, if we consider that the state is socially responsible to the economic entity for the sustainability of its activities, then the second component of the state profit will appear, which is reflected in the income remaining at the disposal of the economic entity.

At the same time, the economic entity has a two-component utility function. In addition to trying to keep the income at its disposal for its social security, the economic entity should make appropriate allocations to the state budget in accordance with its corporate responsibility and legislation [5].

Thus, both of them have utility function, which depends on the disposable income and taxes paid to the state budget. In [6], A.B. Kostin calls the income remaining at the disposal of the economic entity as its consumption, and assumes that the utility function for both entities is Cobb-Douglas shaped. Note that in this study, we will also use this hypothesis.

At the same time, the state should also take control measures on the accuracy of the taxation base of the economic entity, the existence risk of shadow economy, and so on. As it is seen, it is very important to take into account numerous economic and social indicators in the process of studying the relationship between the state and economic entities. Here, the emergence of shadow economy and the study of the relationship of its measurement with other economic indicators is no exception.

3 The Utility Function of the State

The utility function of the state is written as follows:

$$Y_1 = A * V^{\alpha} * C^{\beta} \rightarrow \max_{r_0} \tag{1}$$

Where,
Y_1 – is the overall utility function of the state;
V – is tax revenue;
A – positive constant
C – is income remaining at the disposal of the economic entity;
r_0 – is the level of profitability declared by the economic entity;
α, β – accordingly, is the measure of the contribution of the tax mobilized to the state budget and the income remaining at the disposal of the economic entity and acting as a guarantee for its consumption to the total utility.

It is obvious that the income to the state budget is expressed in the tax calculated from the income determined in accordance with the declared profitability of the economic entity with the application of the tax rate fixed by the tax legislation:

$$V = G * u * r_0 \tag{2}$$

Where,
G – is the value of the assets declared by the economic entity;
$G * r_0$ – is the income declared by the economic entity;
u – is the tax rate.

The income (consumption) remaining at the disposal of the economic entity C will be calculated as follows:

$$C = (1 - u) * G * r_0 + G * (r - r_0) \tag{3}$$

Where,
r – is real profitability;
$r - r_0$ – is profitability kept in the shadow;
$G * (r - r_0)$ – is the income remaining in the shadow.

As it is seen, the income remaining at the disposal of an economic entity consists of two parts:

$(1 - u) * G * r_0$ – is income remaining after payment of tax calculated on declared income;

$G * (r - r_0)$ – is income hidden in full.

Thus, the logarithmic line of the utility function of the state can be

$$\ln Y_1 = \ln A + \alpha \ln V + \beta \ln C \quad (4)$$

or

$$D = \ln A + \alpha \ln(G * u * r_0) + \beta \ln[(1 - u) * G * r_0 + G * (r - r_0)] \quad (5)$$

Where, $D = \ln Y_1$.

In this case, if we indicate the forecast task on tax revenues by P, then we can write the mathematical model of the state in the example of the tax authority as follows:

$$D = \ln A + \beta \ln[G(1 - u)r_0 + G(r - r_0)] + \alpha \ln Gur_0 \rightarrow \max \quad (6)$$

$$Gur_0 + Gu(r - r_0)(1 - k) \geq P; \quad (7)$$

$$0 \leq r_0 \leq r; \alpha, \beta, u \in (0, 1); \; \alpha + \beta = 1; \; \alpha, \beta, k \geq 0 \quad (8)$$

Where, k – is the income remaining in the shadow (penalty for tax evasion).

The constraint indicates that the conditionally economically collected tax cannot be less than the tax forecast task. (8) indicates the necessary constraint conditions.

4 Utility Function of the Economic Entity

The utility function of the economic entity is also Cobb-Douglas function shaped:

$$Y_2 = B * V^{\alpha} * C^{\beta} \rightarrow \max_{r_0}$$

Where,

Y_2 – is utility function of the economic entity;

V – is the amount of tax paid by the economic entity;

C – is income remaining at the disposal of the economic entity.

α, β – accordingly, is the measure of the contribution of the tax mobilized to the state budget and the income remaining at the disposal of the economic entity and acting as a guarantee for its consumption to the total utility.

The tax paid by the economic entity to the state budget will be expressed as the tax calculated from the income determined in accordance with its declared profitability with the application of the tax rate fixed by the tax legislation:

$$V = G * u * r_0 \quad (9)$$

The income (consumption) remaining at the disposal of the economic entity C will be calculated as follows:

$$C = (1 - u) * G * r_0 + G * (r - r_0) * (1 - k) \quad (10)$$

Where,

k – is the penalty imposed by the state for the economic entity tending to remain in the shadow (the cost remaining in the shadow).

Thus, the logarithmic line of the utility function of the economic entity will be

$$\ln \mathrm{Y}_2 = \ln \mathrm{B} + \alpha \ln \mathrm{V} + \beta \ln \mathrm{C} \tag{11}$$

or

$$\mathrm{S} = \mathrm{B} + \alpha \ln(\mathrm{G} * \mathrm{u} * \mathrm{r}_0) + \beta \ln[(1 - \mathrm{u}) * \mathrm{G} * \mathrm{r}_0 + \mathrm{G} * (\mathrm{r} - \mathrm{r}_0) * (1 - \mathrm{k})] \tag{12}$$

Where, $\mathrm{S} = \ln \mathrm{Y}_2$.

In this case, if we indicate the tax potential of the taxpayer by VP, then we can write the mathematical model of the economic entity as follows:

$$\mathrm{S} = \beta \ln[\mathrm{G}(1 - \mathrm{u})\mathrm{r}_0 + \mathrm{G}(\mathrm{r} - \mathrm{r}_0)(1 - \mathrm{k})] + \alpha \ln \mathrm{Gur}_0 \rightarrow \max \tag{13}$$

$$\mathrm{Gur}_0 + \mathrm{G\,u}(\mathrm{r} - \mathrm{r}_0)(1 - \mathrm{k}) \leq \mathrm{VP}; \tag{14}$$

$$0 \leq \mathrm{r}_0 \leq \mathrm{r};\ \alpha, \beta, \mathrm{u} \in (0, 1);\ \alpha + \beta = 1;\ \alpha, \beta, \mathrm{k} \geq 0 \tag{15}$$

Here, (14) the constraint indicates that the conditionally economically the tax collected from the economic entity cannot be more than its tax potential. (15) indicates the necessary constraint conditions.

It should be noted that since the economic entity tends to remain in the shadow, its contribution to the total utility α, β varies depending on cost k remaining in the shadow.

5 Construction of IF–THEN Fuzzy Model

At the initial stage, as mentioned above, the economic entity declares a certain part (r_0) of its real profitability (r) and appropriates the remaining part ($\mathrm{r} - \mathrm{r}_0$). r is known to the tax authority at the level of the average value of the area to which the economic entity belongs. Therefore, in order to eliminate this uncertainty in the report of the economic entity, the tax authority will determine penalty of k_1 if $\mathrm{r}_0 \in (\mathrm{r}_{01}, \mathrm{r}_{02})$ and if $\mathrm{r}_0 \in (\mathrm{r}_{02}, \mathrm{r}_{03})$ it determines penalty of k_2, and if $\mathrm{r}_0 \in (\mathrm{r}_{03}, \mathrm{r})$, it determines penalty of k_3 and informs the economic entity about it. In this case, it is obvious that $\mathrm{k}_1 > \mathrm{k}_2 > \mathrm{k}_3$. The economic entity declares a new greater value of r_0 in order not to lose its income, taking into account the applied penalty, and so on. Let's consider profitability r_0 of economic activity acting as a governing factor, cost k remaining in the shadow, contribution of taxes and income to the total utility and evaluation of α, β with IF-THEN fuzzy approach [7, 8]. For this, based on the tax potential of the economic entity, taking into account the conditions (8) and (15) and accepting $\beta = \mathrm{k}$ and $\alpha = 1 - \beta$ for simplicity, we shall present fuzzy approach for the selection of appropriate k:

- fuzzify k and r_0 using the existing fuzzy methods;

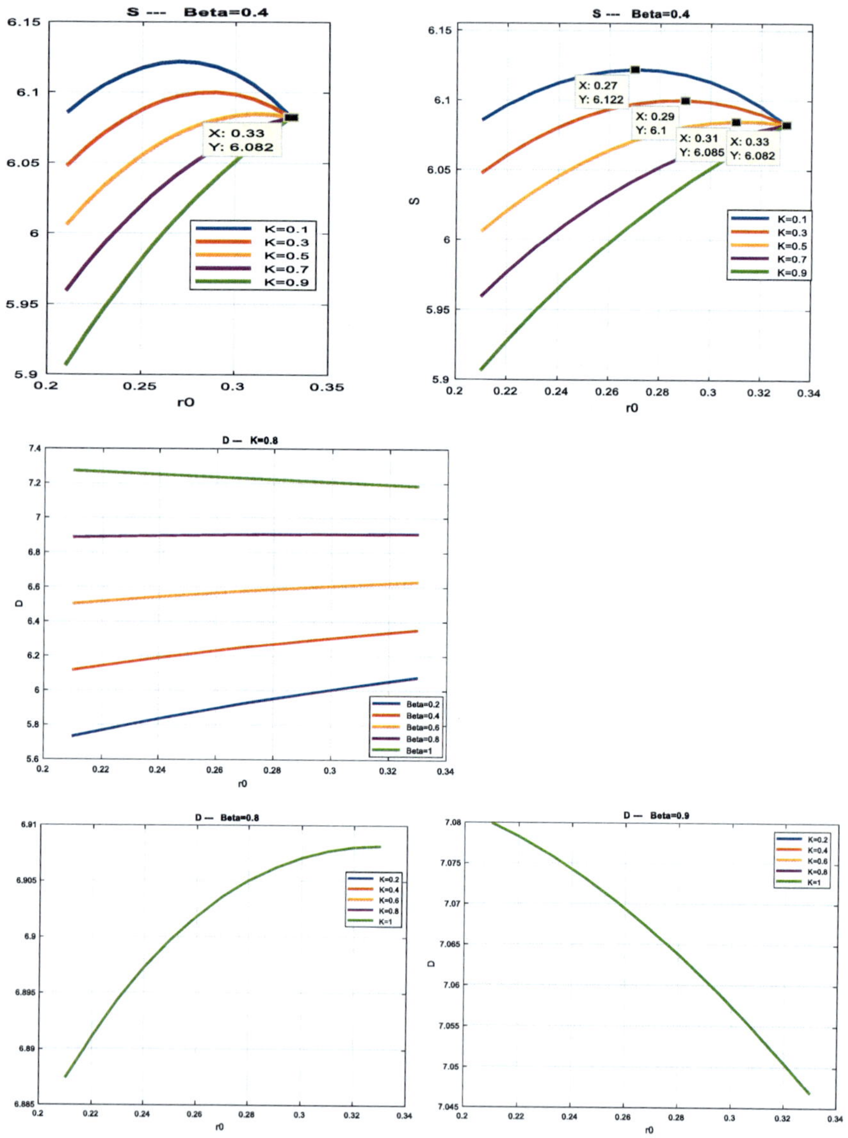

Fig. 1. In case of K = {0.1; …; 1} the change of D at different valuables of β

- determine the values of k and α with 3 linguistic quantities: "weak", "medium", "strong"
- also express the values of r_0 with three linguistic quantities in each interval $(r_{01}, r_{02,})$, (r_{02}, r_{03}) and (r_{03}, r):

"low", "medium", "upper"

According to these linguistic variables, we can write IF-THEN conditions as follows:

IF k = “weak” and α = “strong” Then r_0 = “upper”
IF k = “medium” and α = “medium” Then r_0 = “medium”
IF k = “weak” and α = “strong” Then r_0 = “upper”
IF k = “strong” and α = “weak” Then r_0 = “low”

The fragment of the calculations of the proposed model and fuzzy approach on the sample will be as follows:

Initial data: For example, G = 4000\$, P = 150\$, VP = 500\$, u = 0.2, r = 0.33, $r_{min} = 0.21$, $k = 0.2$

The increase in the value of the utility function of the state depending on the increase in profitability declared by the economic entity is shown in the graph below (see Fig. 1; Fig. 2a and b):

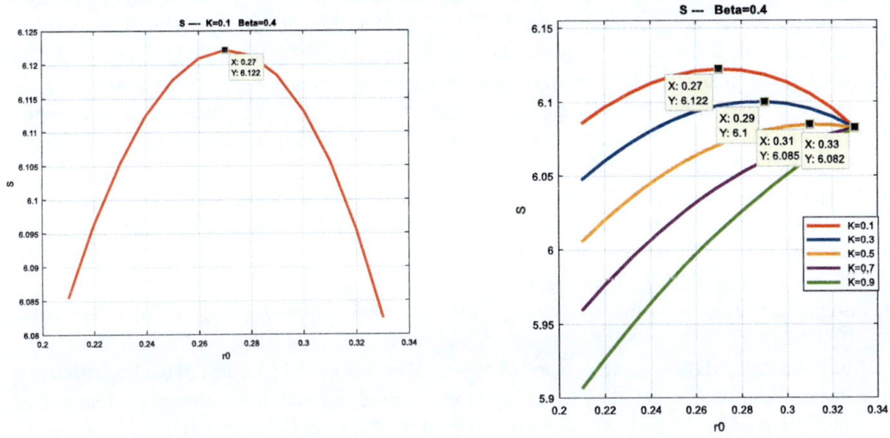

Fig. 2. (a) The change of S, where KK = 0.1 and β = 0.4 so that S∈[6.08;6.122]. (b) The change of S in case of K = {0.1; 0.3; 0.5; 0.7 and 0.9} and β = 0.4. Note: In Fig. 2a and b ▬The line represents the change in S at K = 0.1 and β = 0.4, the difference in convexity in the visual appearance of the graphs is due to the difference in scale on the ordinate axis.

6 Conclusion

- Shadow economy usually exists in areas where the tax burden is high, the state puts great pressure on business, and business transactions are possible to be realized;
- The obtained results show that the proposed instrumentation reflects the process of shadow economy emergence quite realistically;
- It is possible to increase the declared profitability (income) only by increasing the penalties, regardless of the utility of the collection;
- Declared profitability (income) can also be increased by increasing the profitability of the collection without changing the penalties;
- When analyzing these graphs, it is easy to observe that as the value of k penalty increases, the tendency of an economic entity to come out of the shadow increases;

- As the value of the declared profitability r_0 approaches the real profitability r, tax revenues to the state budget increase (Fig. 1.);
- The economic entity is obliged to declare to the state a higher value of r_0 in order not to reduce the value of the utility function (Fig. 2);
- At a certain value of shadow penalty k, the value of r_0 significantly approaches the real profitability r;
- As the value of k penalty increases, the shadow income decreases significantly.

References

1. Schneider, F.: Estimating the size of the danish shadow economy using the currency demand approach: an attempt. Scand. J. Econ. **88**(4), 643 (1986). https://doi.org/10.2307/3440435
2. Musayev, A.F., Kazimov, M.S., Rustamov, S.S., Aliyev, N.K., Madatova, S.G.: Fuzzy regulation model of the interaction between the state and economic subject within the shadow economy and tax field. In: Aliev, R.A., Kacprzyk, J., Pedrycz, W., Jamshidi, M., Babanli, M., Sadikoglu, F.M. (eds.) ICAFS 2020. AISC, vol. 1306, pp. 223–229. Springer, Cham (2021). https://doi.org/10.1007/978-3-030-64058-3_27
3. Musayev, A.F., Madatova, S.G., Rustamov, S.S.: Evaluation of the impact of the tax legislation reforms on the tax potential by fuzzy inference method. In: 12th International Conference on Application of Fuzzy Systems and Soft Computing/Procedia Computer Science, vol. 102, pp. 507–514 (2016). https://doi.org/10.1016/j.procs.2016.09.435
4. Zadeh, L.: Fuzzy sets. Inf. Control **8**, 338–353 (1965). https://doi.org/10.1016/S0019-9958(65)90241-X
5. Musayev, A.F., Madatova, S.G., Rustamov, S.S.: Evaluation of the impact of the tax legislation reforms on the tax potential by fuzzy inference method. In: 12th International Conference on Application of Fuzzy Systems and Soft Computing/Procedia Computer Science (in russian), vol. 102, pp. 507–514 (2016). https://doi.org/10.1016/j.procs.2016.09.435
6. Kostin, A.V.: A model of generation of the shadow economy in the process of interaction between the state and the population. Vestn. Novosib. State Univ. Ser. Socio-Econ. Sci. **14**(1), 15–26 (2014)
7. Musayev, A., Madatova, S., Rustamov, S.: Mamdani-type fuzzy inference system for evaluation of tax potential. In: Zadeh, L.A., Yager, R.R., Shahbazova, S.N., Reformat, M.Z., Kreinovich, V. (eds.) Recent Developments and the New Direction in Soft-Computing Foundations and Applications. SFSC, vol. 361, pp. 511–523. Springer, Cham (2018). https://doi.org/10.1007/978-3-319-75408-6_39
8. Rustamov, S., Musayev, A., Madatova, S.: Evaluation of the impact of state's administrative efforts on tax potential using sugeno-type fuzzy inference method. In: Aliev, R.A., Kacprzyk, J., Pedrycz, W., Jamshidi, M., Sadikoglu, F.M. (eds.) ICAFS 2018. AISC, vol. 896, pp. 352–360. Springer, Cham (2019). https://doi.org/10.1007/978-3-030-04164-9_47

Optimal Regimes of Cooling Processes of Plastic Details

Djahid A. Kerimov(✉)

Azerbaijan State Oil and Industry University, Azadlig 34, Baku, Azerbaijan

Abstract. The article studies the maximum difference in thermic voltages and volumetric deformation. In order to determine the functional dependence of the temperature drop on temperature stresses and volumetric deformations, the rate of change in the temperature drop is taken into account. In this paper, we consider finding the optimal mode for the sizes of plastic details often used in mechanical engineering. At the same time, we discuss some issues of fuzzy modeling of the considered dependence.

Keywords: Plastic details · Cooling regime · Thermic voltages · Volumetric deformation · Cylinder radius · Temperature drop

1 Introduction

For plastic details the cooling regime is a decisive factor [1] and the selection of their optimal values, ensuring the minimization of temperature voltages and volumetric deformation, preventing cracking of details, becomes a very important task [2–5]. In the work "Analytical determination of the quality indicator of plastic parts" in this book [5], we discussed issues of fuzzy modeling of dependence of plastic parts quality on its influential factors. In this article, we consider fuzzy modeling of the functional dependence of the temperature drop on temperature stresses and volumetric deformations. The maximum difference in thermic voltages and volumetric deformation is determined by the following formulas:

$$\overline{\sigma}_{rr} - \overline{\sigma}_{\theta\theta} = \frac{1+\mu}{1-\mu}\alpha a \Delta T_{in} \cdot \mathrm{A}\left[\frac{1}{r}\frac{\mu_1}{R_1} \cdot U_0'\left(\mu_1 \frac{r}{R_1}\right) - \frac{\mu_1^2}{R_1^2} U_0''\left(\mu_1 \frac{r}{R_1}\right)\right] \cdot \cos\mu_1 \frac{z}{n} \cdot e^{-\mathrm{A}_0 \cdot \frac{at}{R_1^2}}, \quad (1)$$

$$\theta = 2\frac{1+\mu}{1-\mu}\alpha a \Delta T_{in} \cdot \mathrm{A}\left[\left(\frac{\mu_1}{R_1}\right)^2 \cdot U_0''\left(\mu_1 \frac{r}{R_1}\right) + \frac{1}{r} U_0'\left(\mu_1 \frac{r}{R_1}\right) - \left(\frac{\mu_1}{n}\right)^2 \cdot U_0\left(\mu_1 \frac{r}{R_1}\right)\right] \cdot \cos\mu_1 \frac{z}{n} \cdot e^{-\mathrm{A}_0 \cdot \frac{at}{R_1^2}}, \quad (2)$$

where R_1, R_2 are radii of the inner and outer surfaces of the hollow cylinder, α_1, α_2 and α_3 are heat transfer coefficients. The other notations are explained in [5]. As information

R. A. Aliev et al. (Eds.): ICSCCW 2021, LNNS 362, pp. 607–614, 2022.
https://doi.org/10.1007/978-3-030-92127-9_81

on the values of the mentioned variables is often imprecise, we consider that values of $R_1, R_2, \alpha_1, \alpha_2$ and α_3, h are described by fuzzy numbers. Indeed, the imprecision of $\alpha_1, \alpha_2, \alpha_3$ and λ is conditioned by the fact that thermal conductivity and heat transfer depends significantly on composition of material (plastic details). In turn, information on composition is, as a rule, imprecise. So, fuzzy version of formulas (1)–(2) may be considered in terms of the extension principle [13].

For a complete study of the temperature drop on the cylindrical surfaces of the plastic parts, it is necessary to pay attention to the following circumstances, i.e. if the radius of the cylinder is sufficiently small, it is necessary to study the rate of the temperature drop. This is due to the fact that at small dimensions of the radius of the cylinder, the angle between the tangents, i.e. the angle between the tangential components of rates the temperature drop will be large enough. Therefore, in order to accurately determine the functional dependence of the temperature drop on temperature voltages and volumetric deformations, it is necessary to take into account the rate of change in the temperature drop.

The temperature drop fully characterizes thermic voltages and volumetric deformation. Further, if the radius of the cylinder is large enough, then the difference in temperature drop is infinitely small. Therefore, to study thermic voltages and volumetric deformation at sufficiently large values of the cylinder radius, it is necessary to introduce weight functions.

2 Investigation of the Temperature Drop on the Cylindrical Surfaces of Plastic Details

Based on these considerations, for the study of thermic voltages and volumetric deformation, the size range of details (cylinder radius) is divided into an interval. For this interval, the task of optimal choice of the criterion B_{i_1} (cooling gime), minimizing thermic voltages, is considered.

In this regard, we will consider finding the optimal regime for the sizes of plastic details often used in mechanical engineering.

For this interval, the problem of finding the optimal cooling regime $\left(B_{i_1}\right)$ is considered, at which the function (rate of thermic voltages) has the following form [3, 4]:

$$J_1 = \int_{R_1}^{R_2} \left\{ \frac{\partial}{\partial r} [\overline{\sigma}_{rr} - \overline{\sigma}_{\theta\theta}] \right\}^2 dr, \tag{3}$$

and the stationary value (rate of change of volumetric deformation)

$$J_2 = \int_{R_1}^{R_2} \left\{ \frac{\partial}{\partial r} [\overline{\sigma}_{rr} + \overline{\sigma}_{\theta\theta} + \overline{\sigma}_{zz}] \right\}^2 dr \tag{4}$$

takes a minimum value. In order to consider imprecise information, the fuzzy form of (3) and (4) can be dealt with as fuzzy integrals by using the methods proposed in [6–13].

The functions J_1 and J_2 practically mean the full square of the rate of the difference between thermic voltages and volumetric deformation, on a segment $[R_1, R_2]$, respectively.

Taking this into account in formulas (3) and (4), we transform the expression of the function J_1 and J_2 into a more convenient form:

$$\overline{\sigma}_{rr} - \overline{\sigma}_{\theta\theta} = \beta_0 \Delta T_{in} \cdot \mathrm{A}\left[\frac{1}{r}\frac{\mu_1}{R_1} \cdot U_0'\left(\mu_1 \frac{r}{R_1}\right) - \left(\frac{\mu_1}{R_1}\right)^2 U_0''\left(\mu_1 \frac{r}{R_1}\right)\right] \cdot \cos \mu_1 \frac{z}{n} \cdot e^{-\mathrm{A}_0 \cdot \frac{at}{R_1^2}}$$

$$\frac{\partial}{\partial r}[\overline{\sigma}_{rr} - \overline{\sigma}_{\theta\theta}] = \beta_0 \Delta T_{in} \cdot \mathrm{A}\left[-\frac{1}{r^2}\frac{\mu_1}{R_1} \cdot U_0'\left(\mu_1 \frac{r}{R_1}\right) + \frac{1}{r}\left(\frac{\mu_1}{R_1}\right)^2 U_0''\left(\mu_1 \frac{r}{R_1}\right) - \left(\frac{\mu_1}{R_1}\right)^3 U_0'''\left(\mu_1 \frac{r}{R_1}\right)\right] \cdot \cos \mu_1 \frac{z}{n} \cdot e^{-\mathrm{A}_0 \cdot \frac{at}{R_1^2}} = \frac{\beta_0 \mu_1}{R_1} \Delta T_{in} \cdot \mathrm{A}\left[\frac{1}{r}\frac{\mu_1}{R_1} U_0''\left(\mu_1 \frac{r}{R_1}\right) - \frac{1}{r^2} U_0'\left(\mu_1 \frac{r}{R_1}\right) - \left(\frac{\mu_1}{R_1}\right)^2 U_0'''\left(\mu_1 \frac{r}{R_1}\right)\right] \cdot \cos \mu_1 \frac{z}{n} \cdot e^{-\mathrm{A}_0 \cdot \frac{at}{R_1^2}} \tag{5}$$

$$\theta = \overline{\sigma}_{rr} + \overline{\sigma}_{\theta\theta} + \overline{\sigma}_{zz} = 2\beta_0 \Delta T_{in} \cdot \mathrm{A}\left[\left(\frac{\mu_1}{R_1}\right)^2 U_0''\left(\mu_1 \frac{r}{R_1}\right) + \frac{\mu_1}{R_1}\frac{1}{r} U_0'\left(\mu \frac{r}{R_1}\right) - \left(\frac{\mu_1}{n}\right)^2 U_0\left(\mu_1 \frac{r}{R_1}\right)\right] \cdot \cos \mu_1 \frac{z}{n} \cdot e^{-\mathrm{A}_0 \cdot \frac{at}{R_1^2}} \tag{6}$$

$$\frac{\partial}{\partial r}\theta = 2\beta_0 \Delta T_{in} \cdot \mathrm{A}\left[\left(\frac{\mu_1}{R_1}\right)^3 U_0'''\left(\mu_1 \frac{r}{R_1}\right) - \frac{1}{r^2} U_0'\left(\mu_1 \frac{r}{R_1}\right) \cdot \frac{\mu_1}{R_1} + \frac{1}{r}\left(\frac{\mu_1}{R_1}\right)^2 U_0''\left(\mu_1 \frac{r}{R_1}\right) - \left(\frac{\mu_1}{n}\right)^2 \frac{\mu_1}{R_1} U_0'\left(\mu_1 \frac{r}{R_1}\right)\right] \cdot \cos \mu_1 \frac{z}{n} \cdot e^{-\mathrm{A}_0 \cdot \frac{at}{R_1^2}} = 2\beta_0 \Delta T_{in} \cdot \mathrm{A}\frac{\mu_1}{R_1}\left[\left(\frac{\mu_1}{R_1}\right)^? U_0''\left(\mu_1 \frac{r}{R_1}\right) - \frac{1}{r^2} U_0'\left(\mu_1 \frac{r}{R_1}\right) + \frac{1}{r}\frac{\mu_1}{R_1} U_0''\left(\mu_1 \frac{r}{R_1}\right) - \left(\frac{\mu_1}{n}\right)^2 U_0'\left(\mu_1 \frac{r}{R_1}\right)\right] \times \tag{7}$$

$$\times \cos \mu_1 \frac{z}{n} \cdot e^{-\mathrm{A}_0 \cdot \frac{at}{R_1^2}} = 2\beta_0 \Delta T_{in} \cdot \mathrm{A}\frac{\mu_1}{R_1}\left\{\left[\left(\frac{\mu_1}{R_1}\right)^2 + \frac{1}{r}\frac{\mu_1}{R_1} U_0''\left(\mu_1 \frac{r}{R_1}\right) - \left[\frac{1}{r^2} + \left(\frac{\mu_1}{n}\right)^2\right] U_0'\left(\mu_1 \frac{r}{R_1}\right)\right]\right\} \cdot \cos \mu_1 \frac{z}{n} \cdot e^{-\mathrm{A}_0 \cdot \frac{at}{R_1^2}}$$

Then

$$J_1 = \int_{R_1}^{R_2} \left\{\frac{\beta_0 \mu_1}{R} \Delta T_{in} \cdot \mathrm{A}\left[\frac{1}{r}\frac{\mu_1}{R_1} U_0''\left(\mu_1 \frac{r}{R_1}\right) - \frac{1}{r^2} U_0'\left(\mu_1 \frac{r}{R_1}\right) - \left(\frac{\mu_1}{R_1}\right)^2 U_0'''\left(\mu_1 \frac{r}{R_1}\right)\right] \times \cos \mu_1 \frac{z}{n} \cdot e^{-\mathrm{A}_0 \cdot \frac{at}{R_1^2}}\right\}^2 dr = \left(\frac{\beta_0 \mu_1}{R} \Delta T_{in}\right)^2 \mathrm{A}^2 \int_{R_1}^{R_2} \left\{\left[\frac{\mu_1}{R_1} U_0''\left(\mu_1 \frac{r}{R_1}\right) - \frac{1}{r^2} U_0'\left(\mu_1 \frac{r}{R_1}\right) - \left(\frac{\mu_1}{R_1}\right)^2 U_0'''\left(\mu_1 \frac{r}{R_1}\right)\right] \cdot \cos \mu_1 \frac{z}{n} \cdot e^{-\mathrm{A}_0 \cdot \frac{at}{R_1^2}}\right\}^2 dr, \tag{8}$$

$$J_2 = \int_{R_1}^{R_2} \left\{ 2\beta_0 \Delta T_{in} \cdot \mathrm{A}\frac{\mu_1}{R} \left[\left(\left(\frac{\mu_1}{R_1} \right)^2 + \frac{1}{r}\frac{\mu_1}{R_1} \right)^2 U_0''\left(\mu_1 \frac{r}{R_1} \right) \right. \right.$$
$$\left. \left. - \left(\frac{1}{r^2} + \left(\frac{\mu_1}{R_1} \right)^2 \right) \cdot U_0\left(\mu_1 \frac{r}{R_1} \right) \right] \cdot \cos \mu_1 \frac{z}{n} \cdot e^{-\mathrm{A}_0 \cdot \frac{at}{R_1^2}} \right\}^2 dr$$
$$= 4\beta_0^2 (\Delta T_{in})^2 \mathrm{A}^2 \left(\frac{\mu_1}{R_1} \right)^2 \int_{R_1}^{R_2} \left\{ \left[\left(\left(\frac{\mu_1}{R_1} \right)^2 + \frac{1}{r}\frac{\mu_1}{R_1} \right) U_0''\left(\mu_1 \frac{r}{R_1} \right) \right. \right.$$
$$\left. \left. - \left(\frac{1}{r^2} + \left(\frac{\mu_1}{R_1} \right)^2 \right) U_0'\left(\mu_1 \frac{r}{R_1} \right) \right] \cdot \cos \mu_1 \frac{z}{n} \cdot e^{-\mathrm{A}_0 \cdot \frac{at}{R_1^2}} \right\}^2 dr \tag{9}$$

Taking into account thermic voltages J_1 and J_2 volumetric deformation, we will obtain [1, 5, 6]

$$J_2 = \frac{B_{i_1}^2}{\left(\mu_1^2 + B_{i_1}^2 \right)^2 \left[\frac{R_2}{R_1} U_0\left(\frac{R_2}{R_1}\mu_1 \right) - \frac{2}{\pi B_{i_1}} \right]^2} \cdot \omega \tag{10}$$

$$\omega = G4\beta_0^2 (\Delta T_{in})^2 \left(\frac{\mu_1}{R_1} \right)^2 \cdot \frac{B_{i_1}^2 \left(B_i^2 + \mu_1^2 \right)}{\mu_1^4 \left(B_{i_1}^2 + B_i + \mu_1^2 \right)} \int_{R_1}^{R_2} \left\{ \left[\left(\left(\frac{\mu_1}{R_1} \right)^2 + \frac{1}{r}\frac{\mu_1}{R_1} \right) U_0''\left(\mu_1 \frac{r}{R_1} \right) \right. \right.$$
$$\left. \left. - \left(\frac{1}{r^2} + \left(\frac{\mu_1}{R_1} \right)^2 U_0'\left(\mu_1 \frac{r}{R_1} \right) \right) \right] \cdot \cos \mu_1 \frac{z}{n} \cdot e^{-\mathrm{A}_0 \cdot \frac{at}{R_1^2}} \right\}^2 dr$$

$$J_1 = \left(\frac{\beta_0 \mu_1}{R_1} \Delta T_{in} \right)^2 \frac{4B_{i_1}^2}{\left(\mu_1^2 + B_{i_1}^2 \right)^2 \left[\frac{R_2}{R_1} U_0\left(\frac{R_2}{R_1}\mu_1 \right) - \frac{2}{\pi B_{i_1}} \right]^2}$$
$$\times \frac{4B_{i_1}^2 \left(B_i + \mu_1^2 \right)}{\mu_1^4 \left(B_{i_1}^2 + B_i + \mu_1^2 \right)} \cdot \int_{R_1}^{R_2} \left\{ \left[\frac{1}{r}\frac{\mu_1}{R_1} U_0''\left(\mu_1 \frac{r}{R_1} \right) - \frac{1}{r^2} U_0'\left(\mu_1 \frac{r}{R_1} \right) \right. \right.$$
$$\left. \left. - \left(\frac{\mu_1}{R_1} \right)^2 U_0'''\left(\mu_1 \frac{r}{R_1} \right) \right] \cdot \cos \mu_1 \frac{z}{n} \cdot e^{-\mathrm{A}_0 \cdot \frac{at}{R_1^2}} \right\}^2 dr \tag{11}$$

Since ω does not depend on B_{i_1} [6, 7], then.

$$\frac{dJ_2}{dB_{i_1}} = \left\{ \frac{4B_{i_1}^2}{\left(\mu_1^2 + B_{i_1}^2 \right)^2 \cdot \frac{R_2}{R_1} U_0\left(\frac{R_2}{R_1}\mu_1 \right) - \frac{2}{\pi B_{i_1}}} \right\} = 0,$$

where a fuzzy form of derivative may be considered in a generalized settings formulated in [10, 13]. The expression of this condition requires the fulfillment of the following equality

$$2B_{i_1}\left(\mu_1^2 + B_{i_1}^2\right)\cdot\left[\frac{R_2}{R_1}U_0\left(\frac{R_2}{R_1}\mu_1\right) - \frac{2}{\pi B_{i_1}}\right]^2 - B_{i_1}^2\left\{2B_{i_1}\left(\mu_1^2 + B_{i_1}^2\right)\right.$$

$$\left.\times\left[\frac{R_2}{R_1}U_0\left(\frac{R_2}{R_1}\mu_1\right) - \frac{2}{\pi B_{i_1}}\right]^2 + 2\left(\mu_1^2 + B_{i_1}^2\right)^2\cdot\left[\frac{R_2}{R_1}U_0\left(\frac{R_2}{R_1}\mu_1\right) - \frac{2}{\pi B_{i_1}}\right]\cdot\frac{2}{\pi B_{i_1}^2}\right\} = 0$$

Hence

$$2B_{i_1}^2 - \pi\frac{R_2}{R_1}U_0\left(\frac{R_2}{R_1}\mu_1\right)B_{i_1} + 4\mu_1^2 = 0 \tag{12}$$

Consequently, when solving (12), the function J_2 takes on an extreme value. Equation (12) has a solution

$$B_{i_1} = \frac{\pi\frac{R_2}{R_1}U_0\left(\frac{R_2}{R_1}\mu_1\right) \pm \sqrt{\frac{\pi^2}{R_1^2}R_2^2U_0^2\left(\frac{R_2}{R_1}\mu_1\right)^2 - 32\mu_1^2}}{4} \tag{13}$$

since $R_2 > R_1$, then $D = \frac{\pi^2}{R_1^2}R_2^2U_0^2\left(\frac{R_2}{R_1}\mu_1\right)^2 - 32\mu_1^2 \succ 0$, therefore, Eq. (11) has a actual solution. Then it can be easily shown that

$$J_2''\left(B_{i_1}^2\right) \prec 0$$

where

$$B_{i_1}^2 = \frac{\pi\frac{R_2}{R_1}U_0\left(\frac{R_2}{R_1}\mu_1\right) + \sqrt{\frac{\pi^2}{R_1^2}R_2^2U_0^2\left(\frac{R_2}{R_1}\mu_1\right)^2 - 32\mu_1^2}}{4}.$$

Consequently, at $B_{i_1} = B_{i_1}^2 \leftarrow$, the function J_2 takes on its smallest value.

Thus, the optimal regime $B_{i_1}^*$ for the first range of cylinder radius sizes is determined as follows:

$$B_{i_1}^* = \frac{\pi\frac{R_2}{R_1}U_0\left(\frac{R_2}{R_1}\mu_1\right) + \sqrt{\frac{\pi^2}{R_1^2}R_2^2U_0^2\left(\frac{R_2}{R_1}\mu_1\right)^2 - 32\mu_1^2}}{4}. \tag{14}$$

In this case, the minimum value of the function J_2 is determined by the formula:

$$J_2 = \frac{\left(B_{i_1}^*\right)^2}{\left[\mu_1^2 + \left(B_{i_1}^*\right)^2\right]\cdot\left[\frac{R_2}{R_1}U_0\left(\frac{R_2}{R_1}\mu_1\right) - \frac{2}{\pi B_{i_1}^*}\right]}.\omega \tag{15}$$

Now we transform ω into a convenient form for calculation:

$$\int_{R_1}^{R_2}\left\{\left[\left[\left(\frac{\mu_1}{R_1}\right)^2+\frac{1}{r}\frac{\mu_1}{R_1}\right]\cdot U_0''\left(\mu_1\frac{r}{R_1}\right)-\frac{1}{r^2}+\left(\frac{\mu_1}{n}\right)^2\cdot U_0'\left(\mu_1\frac{r}{R_1}\right)\right]\times\right.$$
$$\left.\times\cos\mu_1\frac{z}{n}\cdot e^{-\mathrm{A}_0\cdot\frac{at}{R_1^2}}\right\}^2dr=\int_{R_1}^{R_2}\left\{\left[\left[\left(\frac{\mu_1}{R_1}\right)^2+\frac{1}{r}\frac{\mu_1}{R_1}\right]\cdot U_0''\left(\mu_1\frac{r}{R_1}\right)-\left[\frac{1}{r^2}+\left(\frac{\mu_1}{n}\right)^2\right]\times\right.\right.$$
$$\left.\times U_0'\left(\mu_1\frac{r}{R_1}\right)\right\}^2dr\cdot\cos^2\mu_1\frac{z}{n}\cdot e^{-2\mathrm{A}_0\cdot\frac{at}{R_1^2}}=\int_{R_1}^{R_2}\left\{\left[\left(\frac{\mu_1}{R_1}\right)^4+2\left(\frac{\mu_1}{R_1}\right)^3\cdot\frac{1}{r}+\frac{1}{r^2}\left(\frac{\mu_1}{R_1}\right)^2\right]\times\right.$$
$$\times U_0''\left(\mu_1\frac{r}{R_1}\right)^2-2\left[\frac{1}{r^2}+\left(\frac{\mu_1}{R_1}\right)^2+\frac{\mu_1^4}{R_1^2h^2}+\frac{1}{r^3}\frac{\mu_1}{R_1}++\frac{1}{r}\frac{\mu_1^3}{R_1^2h^2}\right]\cdot U_0''\left(\mu_1\frac{r}{R_1}\right)\times$$
$$\left.\times U_0'\left(\mu_1\frac{r}{R_1}\right)+\left[\frac{1}{r^4}+2\frac{1}{r^2}\left(\frac{\mu_1}{n}\right)^2+\left(\frac{\mu_1}{n}\right)^4\right]\cdot U_0'\left(\mu_1\frac{r}{R_1}\right)^2\right\}dr\cdot\cos^2\mu_1\frac{z}{n}\cdot e^{-2\mathrm{A}_0\cdot\frac{at}{R_1^2}}.$$

In what follows, we introduce the notations. After these designations, the function J_2 is represented as:

$$J_2=\frac{\left(B_{i_1}^*\right)^2}{\left[\mu_1^2+\left(B_{i_1}^*\right)^2\right]\cdot\left[\frac{R_2}{R_1}U_0\left(\frac{R_2}{R_1}\mu_1\right)-\frac{2}{\pi B_{i_1}^*}\right]}\cdot 64\beta_0^2(\Delta T_{in})^2\cdot\left(\frac{\mu_1}{R_1}\right)^2$$
$$\times\frac{B_{i_1}^2\left(B_i^2+\mu_1^2\right)}{\mu_1^4\left(B_{i_1}^2+B_i+\mu_1^2\right)}\cdot\left[\left(\frac{\mu_1}{R_1}\right)^2\cdot a_1+2\left(\frac{\mu_1}{R_1}\right)^2\cdot a_2+\left(\frac{\mu_1}{R_1}\right)^2\cdot a_3\right.$$
$$-2\left(\frac{\mu_1}{R_1}\right)^2\cdot a_4-2\frac{\mu_1^4}{R_1^2h^2}\cdot a_5-2\frac{\mu_1}{R_1h^2}\cdot a_6-2\frac{\mu_1^3}{R_1h^2}\cdot a_7+a_8$$
$$\left.+2\left(\frac{\mu_1}{n}\right)^2\cdot a_9+\left(\frac{\mu_1}{R_1}\right)^4\cdot a_{10}\right]\cdot\cos^2\mu_1\frac{z}{n}\cdot e^{-2\mathrm{A}_0\cdot\frac{at}{R_1^2}} \tag{16}$$

3 Discussion and Conclusions

Thus, the minimum value of the volumetric deformation of a particular cylinder size depending on time is determined by formula (16).

Similarly, the limit value of the square of the rate of the difference in thermic voltages in the interval $[R_1, R_2]$ for any value of the height of the cylinder and for any value of time is determined.

Hence, from $J_{2\,\min}$ and $[J_1]$ the expressions we can get

$$[J_1]=\frac{\omega'}{\omega}J_{2\,\min} \tag{17}$$

Formula (17) characterizes the dependence of $J_{2\,\min}$ and $[J_1]$. In the future, it is possible to determine the total amount of deformation and, as a result, the total deformation of details must be used when calculating the strength of parts in production conditions.

Let us note, that a shift from classical description of the considered process to fuzzy representation requires delicate consideration of a lot of aspects. Namely, types of fuzzy derivatives [12, 14], approaches to numerical solutions of FDE and fuzzy equations [9, 21] etc. An important issue is reducing computational complexity when dealing with fuzzy models.

References

1. Kerimov, D.A.: Scientific bases and practical methods of optimization of parameters of quality of plastic details of the oil-field equipment. Dissertation of Doctor of technical sciences, 298 p. Baku (1985). (in Russian)
2. Kerimov, D.A., Mustafaev, A.D.: The Application of plastics in petroleum engineering. Нефть и газ, no.11, 12 (1964). (in Russian)
3. Plutalov, V.N.: Standardization of parameters and tolerances of metric thread for plastic parts. Standards and Quality, M., no. 5 (1966). (in Russian)
4. Badalova, T.A., Kerimov, D.A., Mustafaev A.D.: Investigation of the shrinkage process during the formation of threaded connections from plastics by the method of mathematical modeling. Trans. Azerbaijan Inst. Oil Chem. (6) (1973). (in Russian)
5. Gasanova, N.A.: Definition of mechanical indicators of plastic details of the oil-field equipment. Int. J. Innov. Res. Comput. Sci. Technol. (IJIRCST) **8**(4), 313–314 (2020). ISSN: 2347-5552. https://doi.org/10.2139/ssrn.3673168
6. Abbasbandy, S., Allahviranloo, T., Armand, A., Gouyandeh, Z.: A fuzzy solution of heat equation under generalized Hukuhara differentiability by fuzzy Fourier transform. Fuzzy Sets Syst. **309**, 81–97 (2017). https://doi.org/10.1016/j.fss.2016.04.010
7. Ahmady, N.: A fuzzy newton-cotes method for integration of fuzzy functions. J. Fuzzy Set Valued Anal. **2**, 171–178 (2015). https://doi.org/10.5899/2015/jfsva-00247
8. Bede, B., Gal, S.G.: Generalizations of the differentiability of fuzzy-number-valued functions with applications to fuzzy differential equations. Fuzzy Sets Syst. **151**, 581–599 (2005). https://doi.org/10.1016/j.fss.2004.08.001
9. Bede, B., Gal, S.G.: Quadrature rules for integrals of fuzzy-number-valued functions. Fuzzy Sets Syst. **145**, 359–380 (2004). https://doi.org/10.1155/2014/413570
10. Bede, B., Stefanini, L.: Generalized differentiability of fuzzy-valued functions. Fuzzy Sets Syst., 1–23. https://doi.org/10.1016/j.fss.2012.10.003
11. Costa, T.M.: Jensen's inequality type integral for fuzzy-interval-valued functions. Fuzzy Sets Syst., 327. https://doi.org/10.1016/j.fss.2017.02.001
12. Gomes, L.T., Barros, L.C.: Fuzzy calculus via extension of the derivative and integral operators and fuzzy differential equations. In: Annual Meeting of the North American Fuzzy Information Processing Society (NAFIPS), 6–8 August, pp. 1–5 (2012). https://doi.org/10.1109/NAFIPS.2012.6290965
13. Gomes, L.T., Barros, L.C.: Fuzzy differential equations with arithmetic and derivative via Zadeh's extension. Mathware Soft Comput. Mag. **20**, 70–75 (2013)
14. Mazandarani, M., Najariyan, M.: Differentiability of type-2 fuzzy number-valued functions. Commun. Nonlinear Sci. Numer. Simul. **19**, 710–725 (2014). https://doi.org/10.1016/j.cnsns.2013.07.002

15. Stefanini, L.: A generalization of Hukuhara difference and division for interval and fuzzy arithmetic. Fuzzy Sets Syst. **161**, 1564–1584 (2010). https://doi.org/10.1016/j.fss.2009.06.009
16. Stefanini, L.: A generalization of Hukuhara difference. In: Dubois, D., Lubiano, M.A., Prade, H., Gil, M.Á., Grzegorzewski, P., Hryniewicz, O. (eds.) Soft Methods for Handling Variability and Imprecision. ASC, vol. 48, pp. 203–210. Springer, Heidelberg. (2008). https://doi.org/10.1007/978-3-540-85027-4_25
17. Stefanini, L., Bede, B.: Generalized Hukuhara differentiability of interval-valued functions and interval differential equations. Nonlinear Anal. **71**, 1311–1328 (2009). https://doi.org/10.1016/j.na.2008.12.005
18. Wu, X.: Optimal fuzzy quadrature formula for classes of fuzzy-number-valued functions of Lipschitz type. Fuzzy Sets Syst., 1–9. https://doi.org/10.1016/j.fss.2012.09.001

The Problem of Measuring Human Capacity Using Ahangyol (Way to Harmony) Methods and the Need for Fuzzy Evaluation

Ahmad Gashamoglu(✉)

Institute of Philosophy and Sociology, Azerbaijan National Academy of Sciences, Baku AZ1073, Azerbaijan

Abstract. A range of methods are exploited to study human capacity (capacity of social groups, social systems). This article proposes a completely new approach. Already widely known to the scientific community, the Science of Ahangyol proposed by the author helps to define such an approach. According to Ahangyol, there are ways of collecting information when determining the capacity of a person, which is regarded as a "cosmo-bio-psycho-socio" being. At the same time, scientific analysis is carried out to determine each factor more accurately. According to specific indicators, it is possible to determine ways of selecting the more important influential factors from the collected information. These selected indicators can play an important role in developing recommendations for the future success of new hires or existing staff. It can be useful in studying, forecasting and managing social processes.

Keywords: Human capacity · Personnel selection · Fuzzy evaluation · Information reliability · Science of Ahangyol · Social management

1 Introduction

It is commonly known that the capacity of every being, including human being, plays a very important role in its activity. Depending on the environment, this capacity develops and manifests itself in different forms. Therefore, it is highly important from the scientific point of view to define the capacity of social systems, social groups and individuals in a more detailed and accurate manner. Various studies are currently conducted on this issue in the fields of philosophy, sociology, psychology, economics, management, etc. [1]. Issues of studying human capacity are also in the UN spotlight and are examined in its annual human development reports. The theoretical research of the well-known Indian scientist, Nobel Laureate Amartia Sen [2] in this field are widely known to the scientific community.

Scientific theories and methods based on fuzzy logic can help deepen this research. For example, we can mention the important works by L. Zadeh and R. Aliev, which have extensively attracted the attention of the scientific community [3]. Given that only a small part of human capacity is realized, these research results can significantly help identify

R. A. Aliev et al. (Eds.): ICSCCW 2021, LNNS 362, pp. 615–620, 2022.
https://doi.org/10.1007/978-3-030-92127-9_82

ways of removing restrictions and expanding opportunities. Although the application of the results obtained is considered more relevant for economic systems, it is also highly important for all other systems.

As many well-known scientists have pointed out, the more we learn about man and his/her capacity, the better we understand the world. This information provides an important impetus to the development of science in general [4]. The Science of Ahangyol [5] proposed by the author can contribute a great deal to research in this area. The Science of Ahangyol argues that it is important to look at man not only as a "bio-social" or "bio-psycho-social" being, but also as a "cosmo-bio-psycho-socio" one. It is logically clear that man is also a "cosmic" being. It should be noted that the concept of a "cosmic" being generally implies a being that is connected to the universe in various ways. It is known that the functions as a hierarchical system. There is hardly any doubt that the various processes taking place in space (such as magnetic vortices, the approach of planets, solar eclipses, etc.) affect the processes taking place in the human body. Even doctors often issue warnings about that.

Thus, a human being can be viewed as a being that exists and operates under the influence of four important factors:

$$H(c, b, p, s) \tag{1}$$

c – cosmic factor; b – biological factor; p – psychological factor; s - social factor. Clearly, the lower limit of the influence of this factor is greater than zero. Because if there is a human being, he/she definitely has a cosmic connection. Therefore, c is always >0. To determine the upper limit of the influence of this factor is beyond the intellectual capacity of man. According to the definition of "infinity" in the Science of Ahangyol, we should mark the upper limit as ∞. Therefore, $(0 < c < \infty)$, where c is different for everyone. In other words, different people have different levels of cosmic connection. Also, one and the same person has different levels of connection at different times. Everyone feels at dawn that their cosmic connections are very active. To put it in the vernacular, people feel that their "brain works well", "they have a peace of mind", "they are more inspired", "more enthusiastic", etc. Under these circumstances, the hierarchical systems in which man exists provide them with more energy and their connection with the spheres of the universe that affect them becomes stronger.

Let's clarify the concept of c – "cosmic connection" in more detail. As we know from philosophy, all beings in the world are interconnected. There are certain similarities in the existence of all beings. These similarities can be found even between humans and society, animals, birds, insects and plants. If this is the case, then shouldn't there by a similarity between man and the universe? All organs and glands of the human body are closely connected with a certain area of the brain. The sensory wires of this connection can be observed in the hands, feet and other parts of the body. Therefore, if we see the universe as an organism, we can think that there is an area in it that performs the function of the brain. Again, if we recall the principle of the existence and functioning of the human body, then we can say that every being on earth is closely connected with the part of the universe that performs the function of the brain. It is the influence of different parts that leads to the formation of certain characteristics in a person. Various beings and living things emerge under the influence of a world that performs this brain function.

The "spirit" factor, studied at the categorical level in the Science of Ahangyol, plays an important part in the formation of human being's cosmic relations. When a person has been in the mother's womb for about four months, the spiritual essence that enters the body from the universe plays an important part in the formation of that person's cosmic relations, their later life and activity.

The fact that every person has different abilities is also due to the nature of these cosmic connections. Let's formulate the question first. Why do people have different abilities? Imagine an area several hectares large. In this area, in close proximity to each other, there can be both clear drinking water and water rich in oil, iodine, sulfur, different muds and minerals. Let me note that there is such an area in Zykh settlement near the city of Baku, for example. Why are there fluids of different composition in such a close area? Science responds immediately. There are different horizons in the lower layers of the soil in that area, and each of the wells drilled there connects to different horizons. The same applies to man. If we imagine that each person represents a horizon, we can compare the fluid flowing from the well to that person's ability. The nature of the relationship with various layers of the universe plays the key role in the formation and emergence of human abilities. Of course, the "bio", "psycho" and "socio" factors of a person also have an important influence on the formation of this ability.

Thus, the "cosmic" factor has a serious impact on the determination of human capacity, and it is absolutely necessary to take this factor into account when studying the capacity of any person. Obviously, information about this factor cannot be obtained only through the four-factor traditional measurement (nominal, regulated, interval, quantitative). To do this, there is a need for a new scale proposed in the Science of Ahangyol, which is based on the principles of harmony and fuzzy evaluation. The c-factor can only be studied and taken into account with the aid of fuzzy evaluation. With the help of rational knowledge about the c-factor, it is also possible to identify non-rational factors.

There is no need to dwell on the "bio", "psycho" and "socio" (b, p, s) factors due to the availability of relatively detailed information about them in scientific literature.

H (c, ƅ, p, s) – we can proceed in the following sequence to study human potential using Ahangyol methods. It is clear from the previous explanations that along with being different for different people, each of the c, b, p, s factors is also different at different times. Therefore, when we study the capacity of a person, these factors may represent a combination of different levels in the following Capacity levels matrix.

$$\begin{array}{l} c1,\ c2 \ldots\ldots\ldots\ c\alpha \\ b1,\ b2 \ldots\ldots\ldots\ b\beta \\ p1,\ p2 \ldots\ldots\ldots\ p\mu \\ s1,\ s2 \ldots\ldots\ldots\ s\pounds \end{array} \tag{2}$$

Each line of this matrix represents different levels of c, b, p, s factors. α, β, μ, £ is the number of levels the researcher can engage. We mentioned above that c varies in the $(0 < c < \infty)$ interval. In other words, each person's cosmic connections can be at different levels at different times. However, the researcher's ability as a human being is limited. Let's say that we can take it up to α level. The same applies to the levels of other factors. Each line in Matrix 2 can be calculated and taken into account only with the help of fuzzy evaluation.

Different combinations of elements of this matrix may be the focus in each specific situation. If we are interested in a scientist's or a specialist's work on a research project during the week, then we will need to use this matrix with high precision. Let's suppose that a group of people has been asked to perform an important task in three days and three hours. Then the capacity of each of these individuals separately and of the group as a whole can be analyzed using Matrix 2. Depending on the nature of each task, different lines in this matrix may be more relevant.

The capacity in each person can vary as follows:

Cab – capacity by abilities. Everyone has a certain level of capacity by their abilities. They can have the abilities of a mathematician, singer, builder, etc.: ab1, ab2………. ab_{χ}. Here ab1 represents different areas of ability. c- "cosmic" factors play an important role in the formation of capacity by abilities. The other three factors also play a role in the formation and development of this capacity. There is also the capacity of the person according to their occupation: C occ- capacity by occupation. A person's education, work experience, working environment and other factors have a significant impact on the formation of this capacity. S-"socio" factors play a greater role here. However, as mentioned earlier, other factors also play a role in the formation and development of this capacity. If a person's occupation is aligned with their abilities and objectives, the capacity of this person gradually increases. The more a person is in harmony with their abilities, the more their capacity is revealed. Such capacity is likely to increase. The lower the degree of harmony, the lower the probability that the capacity may be revealed. The harmony of Cab and Cocc create the total human capacity Ct – the total capacity. Cab and Cocc can seriously affect each other.

There is another issue that needs to be addressed. An activity that may not seem to fit a person's abilities at first glance can sometimes help to develop that ability. Examples of this include great Azerbaijani thinkers such as A. Bakikhanov, M. Akhundov and a number of writers and scientists who, despite being military officers, produced serious philosophical and artistic works. If they had worked in the field of their specialization from the beginning, they would probably have become ordinary specialists. However, the knowledge they had gained in their field of work helped them show their talents even more. This depends on how much a person enjoys doing what they do. The more a person enjoys the field in which they are engaged, the more their capacity is revealed. Therefore, although the field in which a person is engaged does not seem to correspond to their capacity at first glance, the degree to which they are in harmony with it plays an important role. Therefore, the issue of analyzing and measuring the degree of harmony can be the object of scientific research.

The degree of harmony of any being with another being may be different. Let's take two strings on a musical instrument, for example. They can be tuned "to harmony with each other". Harmony can be "slightly disturbed", "noticeably disturbed", "very disturbed", and there can be "no harmony at all". Or, on a social level, let's take two people – a couple or business partners. Their characters may be: 1) completely harmonious. In this case, they inspire each other, and their joint life and activities increase the capacity of both; 2) there is a slight discrepancy. In this case, there may be some tensions between these people. 3) There is some incompatibility. In this case, there may be frequent conflicts, etc. The lower the degree of harmony, the less likely people are to continue working together. With the help of the "degree of harmony" proposed by the Science of Ahangyol, it is possible to determine the level of harmony in the coexistence and activity of different beings.

Now let's look at the issue of personnel selection. Let's imagine that a company needs to identify a person capable of working in a certain position and succeeding. In this case, it is necessary to pay attention to the following issues: 1) Company goals; 2) The extent to which the company is currently performing its functions. To what extent the company's performance is in order; 3) Responsibilities of the person to be selected; 4) The degree of importance of the lines and elements of Matrix 2 for this task; 5) What is the main area of the person's abilities – Cab; 6) The capacity of this person in this occupation – Cocc. 7) The degree of importance of the person's capacity in this area; prospects for developing or hampering Cocc.

Considering Matrix 2, we can say that when studying the capacity of an individual (social group, social system), we should first consider the existence of the following entropies:

1. H1 – entropy for "cosmo" relations
2. H2 – entropy for "bio" characteristics
3. H3 – entropy for "psycho" characteristics
4. H4 – entropy for "socio" characteristics
5. H5 – entropy for general capacity.

Data should be collected for each of these five directions, for specific indicators – g1, g2 – on the basis of available equipment and methodologies. This information may be measured both using the known scales and other methods (scale of harmony proposed in the Science of Ahangyol, etc.).

Then, on the basis of the methods proposed by R. Aliyev (1), the degree of reliability of data collected should be determined. The next step requires studying the nature of interaction between collected data and data ratios. Then, using the Gc graph built on the basis of these ratios, we can determine which indicators are "more relevant", "more central" in the determination of a person's capacity. Then, we should check how aligned these "relevant indicators" are with the task set for the person (social group, social system). If this capacity is adequate for the fulfillment of the task, this person may be appointed to this position. Otherwise, not. If a person is appointed to the position, then we should determine the direction in which "relevant indicators" should developed for a more successful future. Each of these may only succeed if there is strong mathematical assurance.

2 Final Provisions

It would be useful to use the Science of Ahangyol proposed by the author to determine the capacity of a person (social group, social system). To do this, we should consider humans as a "cosmo-bio-psycho-socio" being and collect data on these directions. This information may range in the interval of 0, ∞. Each of "cosmo", "bio", "psycho" and "socio" factors may have different values for different humans at different times. Therefore, there appears a need to use only the methods of Fuzzy Evaluation in this issue. It should also be noted that studying of each of these factors in greater detail requires a new scientific approach. The Science of Ahangyol may be of great help in this [5]. After checking the reliability of the data collected through special equipment and scientific methods, the study of mutual relationships between them enables the determination of "more relevant indicators" in ascertaining the capacity of a specific person (social group, social system). This, in its turn, allows the opportunity to determine how useful this employee may be for a specific position. It may also contribute to the determination of how apt a person serving in this position is. It is also helpful in developing recommendations for a more successful activity of this person in the future.

References

1. Ivanov, O.I.: Human Potential (Formation, development and use). St. Petersburg (2013)
2. Sen, A.: Rationality and Freedom, 752 p. Belknap Press (2004). https://doi.org/10.1521/siso.2007.71.1.59
3. Zadeh, L.A., Aliev, R.A.: Fuzzy Logic Theory And Applications: Part I and Part II, 435 p. World Scientifific (2019). ISBN: 13: 978-981-3238-17-6; 10: 981-3238-17-8
4. Spirkin, A.G.: Philosophy. Moscow. Gradiarka (2007). 978-5-9916-3260-7; 978-5-9692-1493-4
5. Gashamoglu, A.: Fuzzy mathematics & science of ahangyol. In: 12th International Conference on Application of Fuzzy Systems and Soft Computing, ICAFS 2016, 29–30 August 2016. Vienna, Austria (2016). 978-1-5108-3162-9

The Use of Mathematical Methods in the Analysis and Forecasting of Stock Indices

I. V. Fedorova(✉), L. K. Popova, T. N. Sharypova, and E. N. Lozina

Rostov State University of Economics, Rostov-on-Don, Russia

Abstract. In the modern world, there are many questions concerning the prediction of indicators related to the financial market. A model that can accurately predict the direction of market movement is currently very valuable and relevant. This paper presents a linear regression model for predicting the Dow Jones financial index in the stock market. The key issue in the work of the stock market is the prediction of periods of increase (growth) or decrease of the index function for a certain period of time. Models that can predict the direction of market movement with high accuracy can be built using the statistical theory of machine learning. To implement the predictive model, the initial data from open platforms were taken. Future values are predicted based on the obtained statistical data. In this paper, a predictive model is implemented to predict the closing price of the Dow Jones index on the stock market, using a linear regression model.

Keywords: Modeling · Fuzzy sets · Soft computing · Cloud technology · Fuzzification · Defuzzification

1 Introduction

Dow Jones index is one of the most famous and oldest indexes among the American market indexes [1]. Its task is to track the development of the industrial component of the American stock markets. The index covers the 30 largest US companies. Among them are: Microsoft Corp., Apple Inc., Coca-Cola Co.

2 Materials and Methods

The data is taken from the open platforms Quandl and AlphaVantage, which provide access to more than 9 million free datasets [3].

Figure 1 shows the data that the Dow Jones data set contains:

R. A. Aliev et al. (Eds.): ICSCCW 2021, LNNS 362, pp. 621–629, 2022.
https://doi.org/10.1007/978-3-030-92127-9_83

Date	Open	High	Low	Close	Adj Close	Volume
2001-01-01	10790.919922	11028.000000	10468.040039	10887.360352	10887.360352	5825960000
2001-02-01	10884.820313	11035.139648	10294.009766	10495.280273	10495.280273	4164820000
2001-03-01	10493.250000	10859.500000	9106.540039	9878.780273	9878.780273	6086470000
2001-04-01	9877.160156	10906.410156	9375.719727	10734.969727	10734.969727	5241100000
2001-05-01	10734.049805	11350.049805	10673.219727	10911.940430	10911.940430	4518100000
...	...	...	...	...	...	...
2020-01-01	28638.970703	29373.619141	28169.529297	28256.029297	28256.029297	6170770000
2020-02-01	28319.650391	29568.570313	24681.009766	25409.359375	25409.359375	6991990000
2020-03-01	25590.509766	27102.339844	18213.650391	21917.160156	21917.160156	15521140000
2020-04-01	21227.380859	24764.769531	20735.019531	24345.720703	24345.720703	9795600000
2020-05-01	24120.779297	25758.789063	22789.619141	25383.109375	25383.109375	7940630000

Fig. 1. Data on the Dow Jones index

During each trading day (Monday - Friday), the share price changes and is recorded in real time. Five values show the price change in one day and are the key trading indicators:

1. Date: Date;
2. Open: Opening price;
3. Close: Closing price;
4. High: the highest price;
5. Low: lowest price;
6. Volume: The total number of shares sold before the market closes.

When creating a model, we will rely on the method described in [6].

Let's list the signs used in predicting the closing price of the Dow Jones index:

- opening price;
- opening price for the previous day;
- closing price for the previous day;
- the highest price for the previous day;
- lowest price for the previous day;
- sales volume for the previous day;
  - average closing price for the previous five days;
  - average closing price for the previous month;
  - average closing price for the previous year;
  - the ratio between the average price for the previous week and the previous month;
  - the ratio between the average price for the previous week and for the previous year;
  - the ratio between the average price for the previous month and for the previous year;

- average volume for the previous five days;
- average volume for the previous month;
- average volume for the previous year;
- the ratio between the average volume for the previous week and the previous month;
- the ratio between the average volume for the previous week and for the previous year;
- the ratio between the average volume for the previous month and for the previous year;
- standard deviation of the closing price for the previous five days;
- standard deviation of the closing price for the previous month;
- standard deviation of the closing price for the previous year;
- the ratio between the standard deviation of prices for the previous week and for the previous month;
- the ratio between the standard deviation of prices for the previous week and for the previous year;
- the ratio between the standard deviation of prices for the previous month and the previous year;
- standard deviation of volumes for the previous five days;
- standard deviation of volumes for the previous month;
- standard deviation of volumes for the previous year;
- the ratio between the standard deviation of volumes for the previous week and for the previous month;
- the ratio between the standard deviation of volumes for the previous week and for the previous year;
- the ratio between the standard deviation of volumes for the previous month and for the previous year;

- total return of the previous day;
- total return of the previous week;
- total return of the previous month;
- total return of the previous year;

- moving average of the total return for the previous week;
- moving average of the total return for the previous month;
- the moving average of the total return for the previous year.

To predict the closing prices of the Dow Jones index on the transformed data, we use the standard SGDRegressor function from the sklearn library. To select the parameters, use the GridSearchCV function.

The result is shown in Fig. 2.
MSE: 50415.891
Accuracy on the training set: 0.96
Accuracy on the test set: 0.92.

To compare the performance of SGDRegressor, we use RandomForestRegressor and SVR. To select the function parameters, use GridSearchCV.

SGDRegressor {'alpha': 3e-05, 'eta 0': 0.03}:
MSE: 29876.811
Accuracy on the training set: 1.00
Accuracy on the test set: 0.89
RandomForestRegressor {'max_depth': 50, 'min_samples_split': 10}:
MSE: 46358.319
Correctness on the training set: 1.00
Accuracy on the test set: 0.85
SVR {'C': 10000, 'epsilon': 0.0001}:
MSE: 26895.198
Accuracy on the training set: 1.00
Accuracy on the test set: 0.91

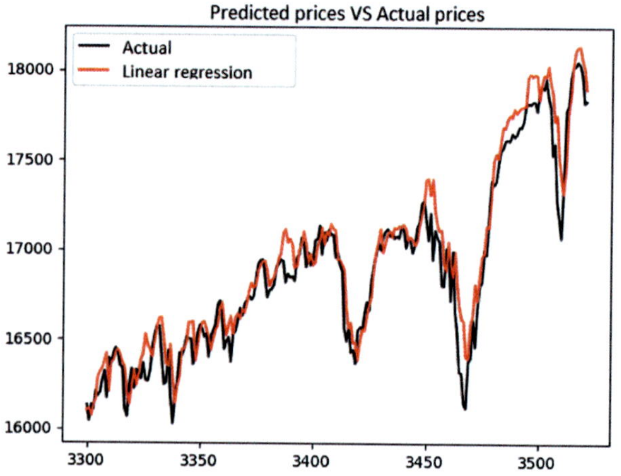

Fig. 2. Price comparison

Thus, the accuracy of the SGDRegressor method is lower than the SVR, but higher than the accuracy of the RandomForestRegressor method.

To increase the accuracy of the prediction of our model, we will check whether our data is stationary, since non-stationary data can lead to a deterioration in the regression.

A stationary time series is a time series whose statistical properties, such as mean, variance, and autocorrelation, are constant over time.

To check for stationarity, we use the extended Dickey-Fuller Test (ADF). ADF is a type of statistical test that determines whether a single root is present in time series data. Consider two hypotheses: the null hypothesis, H0: if it failed to reject, then there is a chance that the time series is non-stationary; the alternative hypothesis, H1: if H0 is rejected, then the time series is stationary.

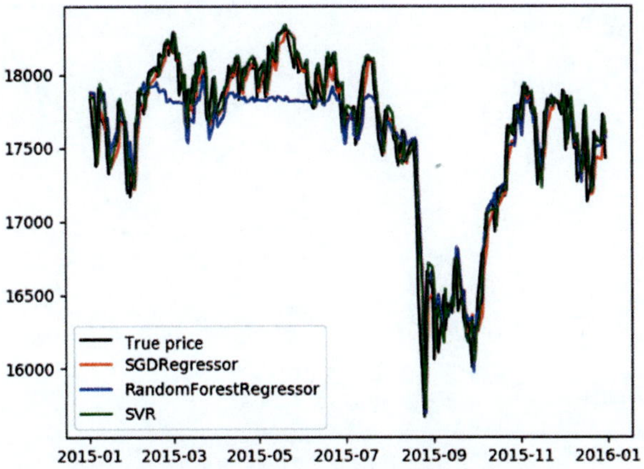

Fig. 3. Comparison of 3 methods with the true value

To accept or reject the null hypothesis, it is customary to use the test. When this is used, the following critical p-value is used:

$p > 0.05$: H0 cannot be rejected, so the data has a unit root and is non-stationary;

$p \leq 0.05$: we reject H0, therefore, the data does not have a unit root and is stationary.

Calculate the moving average and standard deviation in the variables df_mount and df_sdt, respectively, with a window period of one year.

```
df_settle = df[' Close'].resample('MS').ffill().dropna()
df_rolling = df_settle.rolling(12)
df_mean = df_rolling.mean()
df_std = df_rolling.std().
```

We will plot the moving average relative to the original time series (Fig. 4) and the moving standard deviation separately (Fig. 3). We will plot the moving average relative to the original time series (Fig. 4) and the moving standard deviation separately (Fig. 4).

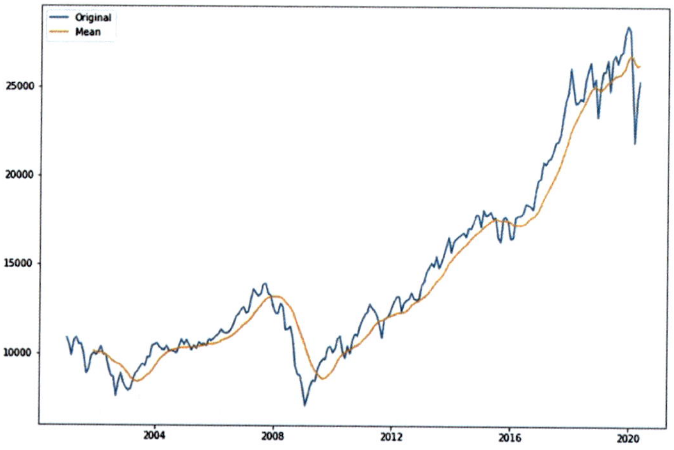

Fig. 4. Moving average chart

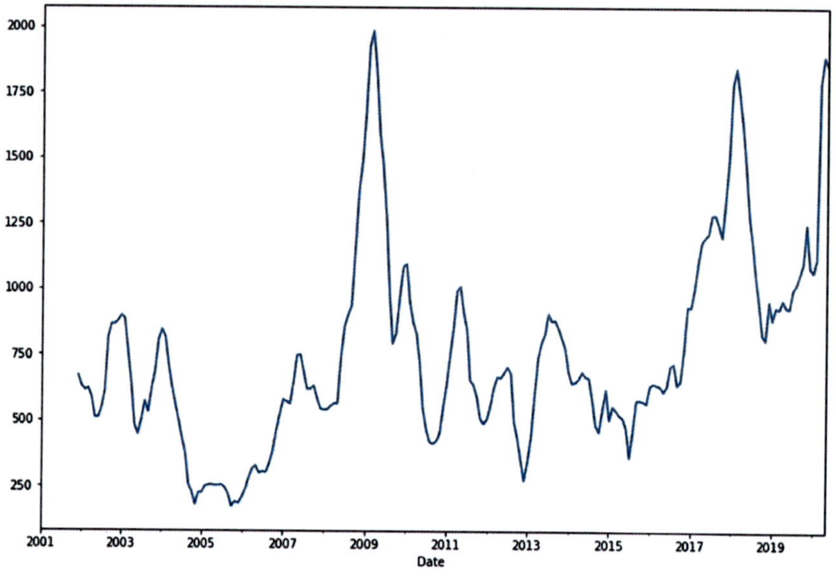

Fig. 5. Moving standard deviation chart

Figure 5 shows a graph of the movement of the standard deviation.

Using the statsmodels module, we perform an ADF test for our dataset using the adfuller() method.

```
from statsmodels.tsa.stattools import adfuller
result = adfuller(df_settle)
print('ADF statistic: ', result[0])
print('p-value:', result[1])
critical_values = result[4]
for key, value in critical_values.items():
print('Critical value (%s): %.3f' % (key, value))
```

Получаем следующие значения:

ADF statistic: 0.21558996270503175
p-value: 0.9731225564902721
Critical value (1%): -3.459
Critical value (5%): -2.874
Critical value (10%): -2.574

The statistical p-value of the ADF test is greater than 0.05. Thus, it is impossible to reject H0 about the existence of a unit root, hence the data is non-stationary.

The non-stationarity of the time series is affected by the trend or seasonality. In order to make a series stationary, you need to eliminate the trend and seasonality. To do this, use the following methods: detrending, differentiation, and decomposition. The obtained stationary data are suitable for statistical forecasting.

Let's detrend it. Detrending-the process of removing a trend line from non-stationary data (Fig. 6).

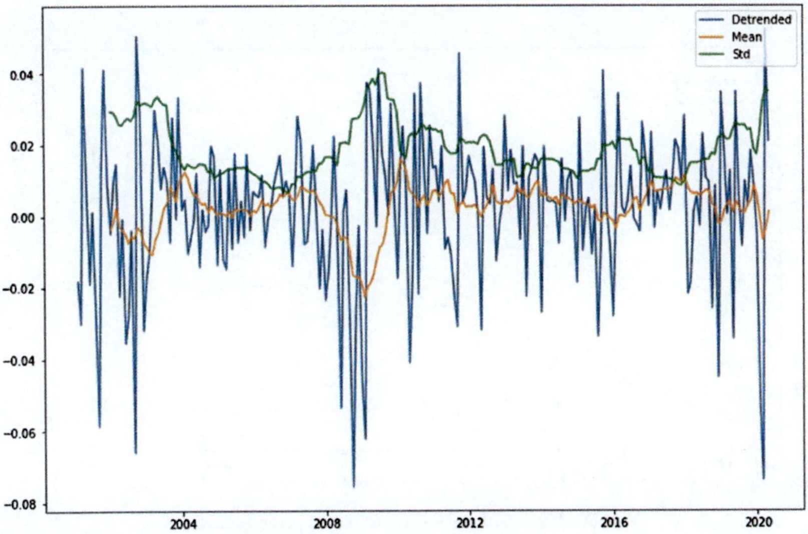

Fig. 6. Detrending result

Note that the mean and standard deviation do not have a long-term trend.

After running the ADF test, we get.

ADF statistic: −14.198292224311526
p-value: 0.00000
Critical value (1%): −3.459
Critical value (5%): −2.874
Critical value (10%): −2.573
p-value for this data is less than 0.05.

Thus, it is possible to reject H0, hence the data is stationary.

We predict the future value based on the obtained statistical data using the autoregressive integrated moving average (ARIMA) method. ARIMA is a prediction model for stationary time series based on linear regression.

The search for parameters for our model will be performed using the "grid search", also known as the hyperparameter optimization method [4]. Let's adjust the ARIMA model using the SARIMAX () function of the statsmodels module. At each iteration, the function returns an object of the Meresults class that contains the aic attribute to return the AIC value. The model with the lowest AIC value gives us the most suitable model that defines our parameters p, d, and q.

3 Results

The model of the seasonal component ARIMA (0, 1, 1, 12) will give us the lowest value of AIC - 3289.336. We will plot prices starting from 2008 (Fig. 7):

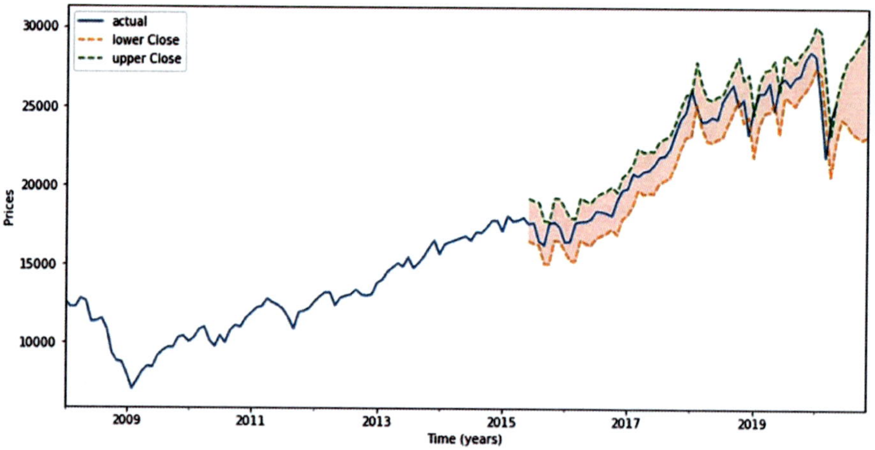

Fig. 7. The result of the Dow Jones price forecast

ARIMA (1, 1, 1) RMSE Error: 366.8729197
ARIMA (1, 1, 1) MSE Error: 134595.7392
MSE: 125080.536

The solid line shows the observed values, while the dashed lines show five-year moving forecasts. As the forecast for the next five months moves into the future, the confidence interval expands to reflect the loss of confidence in the forecast.

Thus, a predictive model is implemented to predict the closing price of the Dow Jones index on the stock market.

References

1. Shalev-Schwartz, Sh., Ben-David, Sh.: Machine Learning Ideas (2019)
2. Muller, G.: Introduction to Machine Learning with Python, pp. 59–71 (2017)
3. Documentation on the Panda library [Electronic resource]. https://pandas.pydata.org/docs/user_guide/index.html
4. Documentation on Quandl [Electronic resource]. https://docs.quandl.com
5. Weiming-Mastering Python for Finance Implementing Advanced modern Financial Statistics Applications Using Python-Packt Publishing (2019)
6. Documentation on Alpha Vantage [Electronic resource]. https://www.alphavantage.co/documentation

EFL Teachers' Competencies According to Student Opinions

Saide Sadıkoğlu[1](✉), Şahin Akdağ[1], Murat Tezer[1], and Rashad Khalilov[2]

[1] Near East University, 99138 Nicosia, Northern Cyprus
{Saide.sadikoglu,Sahin.akdag,Murat.tezer}@neu.edu.tr
[2] Azerbaijan University of Languages, Baku, Az 1010, Azerbaijan

Abstract. An important task facing education is to present fundamentally new requirements for the training of teaching staff. Flexible, individually oriented teaching and training should be a modern instructor who can provide the necessary competencies for innovative development and develop the creative abilities of students. The study aims to determine the proficiency of English teachers competencies who will teach English based on students' perceptions. The survey method was used in the research. According to the result obtained from the data collected with the questionnaire, the proficiency of the instructors working at the university, who gave English lessons with the distance education method during the pandemic period, was generally determined as good. In addition, students see the teaching staff's knowledge and application skills of English Teaching techniques as good. Students also consider the proficiency of the instructors to monitor and evaluate the learning progress in the teaching of English as a foreign language as very good. In addition, the students said that the proficiency of the instructors in the process of teaching English as a foreign language in terms of program and content knowledge competencies is very good.

Keywords: English teaching · Foreign language · Foreign language teaching · Teacher competence

1 Introduction

Man, who has been interacting with his environment for centuries, is a social being. He tried to meet the need to convey his feelings, thoughts, and dreams through speaking and writing and wanted them to be understood through listening and reading. Thus, the phenomenon called communication has emerged, and the necessity of communicating with the people around has emerged for the individual to survive [1]. All skill areas that make up the language are interrelated and can only develop in unity with each other. The primary responsibility of English teaching is to develop students' four basic language skills in a planned manner with a scientific understanding and to gain the sensitivity and habit of using our language correctly and consciously [2]. Fulfilling this responsibility is possible by maintaining the English language teaching, which starts from the first grade of primary school and lasts until the end of the education stages, effectively and

R. A. Aliev et al. (Eds.): ICSCCW 2021, LNNS 362, pp. 630–635, 2022.
https://doi.org/10.1007/978-3-030-92127-9_84

efficiently. There is a strong interaction between mother tongue education and foreign language teaching. It is not possible for individuals whose mother tongue skills are not developed to be successful in foreign language teaching [3]. It is necessary to understand what the mother tongue is before understanding what the foreign language is [4]. It is possible to say that teachers have similar problems regarding the acquisition of four basic language skills in both mother tongue education and foreign language teaching.

It is a reality accepted by everyone that students who take foreign language courses starting from primary education and taking foreign language courses during higher education experience serious problems despite the efforts to make innovations with the effect of different education policies implemented in Turkey over time [5]. The difficulties experienced in foreign language education in Turkey are due to more fundamental factors such as language policies and approaches to language teaching, as well as crowded classrooms, inadequate physical conditions, and dilemmas regarding qualified teacher training. In reality, the situation is aggravated by a large number of students and the mistakes made in language policies and contemporary language teaching approaches. Undoubtedly, the mass of teachers, who are directly affected by the decisions taken but are not directly or extensively involved in the decision-making processes on foreign language education, but are responsible for all in-class arrangements, represent a group that has difficulty in finding a place in this flow [6].

An education that ignores social and individual needs is doomed to fail [7]. Meeting these needs can be through improvement in education; Improvement in education can only be achieved by training better trainers. Because success in education cannot increase without equipped trainers [8]. Effective use of these competencies training by teachers and students in the field of foreign language education may result in a more successful educational process. Despite the competence of a student who acquires 21st-century skills, the expectations above the perceptions of a good teacher and a good education system are also changed [9]. For this purpose, 21st-century skills training should be given to teachers. Teacher competencies should be defined and results that can be associated with 21st-century skills should be produced. Due to the changing educational trends and the diversity of the society-individual, the Ministry of National Education determined fourteen teacher qualifications in 2002 and formed a teacher training strategy accordingly [10]. Self-efficacy perceptions of teachers Kanadlı [11] who have various ideas about whether they can do their jobs successfully, teachers' determination in the face of difficulties, it is one of the factors that affect success in education, as it can affect the efforts made within the scope of the determined goals and general education process [12]. Aktaş [13] summarized the importance of researching the subject in the context of Turkey by linking the difficulties of foreign language education in Turkey to teachers' perception of language competencies, students' interests and motivations, teaching methods, learning environment, and used materials. For these reasons, examining the personal and professional self-efficacy concepts of teachers in the 21st century, looking at the differences in self-efficacy perceptions between different groups within the profession will enable us to understand the targeted social and individual progress, and to understand the current position in the process and the success of the process in achieving these goals. In this way, necessary arrangements for the healthy functioning of the process and planning for the future will be possible [14]. The purpose of this research is to determine the proficiency of English teacher competencies who will teach students based on students' perceptions.

2 Research Method

In this research, which is in the type of descriptive research, the survey method was used. According to Büyüköztürk [15], survey studies aiming to describe a past or present situation as it exists and conducted on larger samples than other studies are known as survey studies. The survey method was used in this study because it reveals what the said event or situation is, what is wanted to be done, and solutions, and since the proficiency of the instructor's competencies who will teach English will be determined based on students' perceptions.

2.1 Participants

For the research, the participants of this research consist of 79 Dental Faculty students studying in the field of health of a private university serving in Northern Cyprus. After obtaining the necessary ethical permission and approval from the university due to the Pandemic, the students were informed about the questionnaire form via e-mails, and they were asked to fill it out with the Google Form tool. The average time to fill out the questionnaire is 20 min on average.

2.2 Data Collection Tool

As a data collection tool in the research, the questionnaire form "The competencies of instructors who will teach Turkish to foreigners according to students' perceptions" was used by adapting it and developed by Yıldız and Tepeli [16]. Only four parts of the developed questionnaire were used in this study. The questionnaire form was developed to determine the proficiency of the instructors who will teach/teach Turkish to foreigners according to student perceptions. In this study, the words "Turkish" were changed to "English" in the entire questionnaire form. Thus, in the questionnaire, there are questionnaire items regarding the proficiency of the instructors who will teach/teach English according to student perceptions. The questionnaire form consists of 36 items. Due to the change made in the questionnaire, the questionnaire form was evaluated in terms of content validity by 3 field experts. Necessary corrections were made as a result of the evaluation of the items by experts. In this study, the Cronbach Alpha reliability coefficient for 36 items of the questionnaire was found to be 0.95. Büyüköztürk, Böke, and Köklü [17] state that if the reliability coefficient is between 0.70–1.00, the reliability is good. The result obtained showed that the consistency of the questionnaire for the application was high. Thus, the questionnaire was adapted to the English language by the researchers.

The questionnaire items were created in a 5-point Likert type and the degree of agreement of the people to the items; It is classified as "Do not agree at all", "Do not agree", "Partially agree", "Agree" and "Completely agree".

2.3 Data Analysis

After the application of the questionnaire, the obtained data were subjected to statistical processing in a computer environment using SPSS 24 program; In this context, the frequency and percentage values of the data were calculated. In addition, to make a general judgment, the mean and standard deviation values are also calculated. The findings obtained at the end of these processes were interpreted together to ensure clarity.

2.4 Results

In this part of the research, the findings as a result of the analysis of the data obtained from the survey are included.

Considering the data on the "Effects of Teacher Competencies on Students in Teaching English as a Foreign Language" dimension, more than half of the students (more than 60%) said to "He/she knows English as a foreign language teaching methods/or foreign language teaching methods" (32.9% Completely agree and 45.6% agree; $\overline{X} = 4.03, Sd = .933$), "He/she is proficient in the field of grammar, He/she is advanced in Communication skills" ($\overline{X} = 4.27, Sd = .635$), "He/she is advanced in Writing skills" ($\overline{X} = 3.96, Sd = .912$), ""He/she has advanced Verbal skills" ($\overline{X} = 4.11, Sd = .906$), "He/she is advanced in Reading-comprehension skills" ($\overline{X} = 4.19, Sd = 0.863$), "He/she is advanced in Listening-comprehension skills" ($\overline{X} = 4.13, Sd = .790$), "He/she has a rich vocabulary" ($\overline{X} = 4.11, Sd = .816$), "He/she should have knowledge and values on scientific, legal and ethical requirements related to the teaching profession required by the age and laws" ($\overline{X} = 4.29, Sd = .865$). However, students are more likely about the teaching staff are less agree on "He/she has a good command of literature knowledge and theories, children's literature, world literature and folk literature in order to benefit in the field of teaching English to foreigners" data compared to other data ($\overline{X} = 3.43, Sd = 1.034$).

Considering the data belonging to the "English Teaching Techniques, Knowledge and Practical Skills" dimension, the students stated that they agreed with the English teaching techniques knowledge and application skills for the instructors. In other words, students see the teaching staff's knowledge and application skills of English teaching techniques as good ($3.40 < \overline{X} < 4.20$).

Considering the data for the "Competencies for Monitoring and Assessing Learning Development in the Process of Teaching English as a Foreign Language" dimension, the students answered "Agree" for the dimension of ability to monitor and evaluate learning development in the processes of teaching English as a foreign language for instructors. In other words, the students see the proficiency of the instructors in monitoring and evaluating the learning development in English as a foreign language teaching process as very good ($\overline{X} > 4.20$).

Considering the data belonging to the "Competencies on Program and Content Knowledge" dimension, the students answered "Agree" and "Completely agree" for the dimension of competencies regarding program and content information in the processes of teaching English as a foreign language for instructors. In other words, the students see the proficiency of the instructors in teaching English as a foreign language in terms of curriculum and content knowledge as very good ($3.40 < \overline{X} < 5$).

2.5 Discussion

Considering the findings of the effect of teacher proficiency in teaching English as a foreign language on students, more than half of the students stated that the proficiency of the instructors is good or very good. Thus, we can say that the competencies desired by the students are found in the instructors. However, the students also stated that they expect more from the instructors to have a command in the fields of literature knowledge and theories, children's literature, world literature, and folk literature to benefit from in the field of teaching English to foreigners. Similarly, in the study conducted by Yıldız and Tepeli [16], similar findings were obtained in terms of the effect of teacher competencies on students in teaching Turkish as a foreign language.

Students consider teaching staff to be good in English Teaching Techniques, Knowledge and Practical Skills. Other studies show that students in the university preparatory program have moderate levels of English self-efficacy beliefs and that their self-efficacy beliefs are not at the desired level. Students do not consider themselves very competent in learning the English language. They do not have a high level of determination and determination to learn English [18]. In this regard, the competence of the instructors is important. In other words, if the instructors have good knowledge and application skills of English Teaching techniques, it will also increase the self-efficacy beliefs of the students.

Students answered "Agree" for Competencies for Monitoring and Assessing Learning Development in the Process of Teaching English as a Foreign Language dimension. In other words, the students see the proficiency of the instructors in monitoring and evaluating the learning development in English as a foreign language teaching process as very good.

Students consider Competencies on Program and Content Knowledge of instructors to be very good in teaching English as a foreign language. Similarly, Yıldız and Tepeli [16] revealed in their research that foreign students have a high expectation from teachers to have the ability to prepare Turkish teaching programs for foreigners. Considering the results that the expectations from the teachers are generally high in the fields of proficiency in curriculum and content knowledge, it is stated that the instructors working in the field of teaching Turkish to foreigners are expected to have a good command of the program dimension. Bayır and Yeşil [19] stated in their research that students experience some teacher-related difficulties in the foreign language learning process.

3 Conclusion and Recommendations

As a result of the research, more than half of the students stated the proficiency of the instructors in terms of teaching English as good or very good. In addition, students see the teaching staff's knowledge and application skills of English Teaching techniques as good. The students also consider the proficiency of the instructors to monitor and evaluate the learning progress in the teaching of English as a foreign language as very good. In addition, the students said that the proficiency of the instructors regarding the program and content knowledge in the processes of teaching English as a foreign language is very good. For the instructors to teach the English language, both their knowledge of the field and their knowledge of new teaching techniques will provide significant advantages for the learners and they will learn the English language better by creating self-confidence in the students.

References

1. Aydın, İ.S.: Development of a speaking self-efficacy scale for teacher candidates. Akdeniz J. Educ. Res. **14**, 33–46 (2013)
2. Karademir, C.A.: Gorgoz, S: English teachers' problems encountered in teaching four basic language skills. Int. Educ. Stud. **12**(4), 118–127 (2019). https://doi.org/10.5539/ies.v12n4p118
3. Dolunay, S.K.: The aim and the importance of grammar teaching. J. Turkology Stud. (TÜBAR) **15**(27), 275–284 (2010)
4. Tosun, C.: Foreign language education problem. J. Lang. Linguist. Stud. **2**(1), 28–42 (2006)
5. Demirel, Ö.: Foreign Language Teaching. Pegem Publishing, Istanbul (2003)
6. Haznedar, B.: Foreign language teaching in Turkey: reforms, trends, and our teachers. In: International Conference on New Trends in Education and Their Implications, pp. 11–13 (2010)
7. Demirel, Ö., Budak, Y.: In-service training needs for teachers. Educ. Manag. Theory Pract. **33(1)**, 62–81 (2003). https://dergipark.org.tr/en/pub/kuey/issue/10364/126866
8. Seferoğlu, S.S.: Teacher competencies and professional development. Sci. Educ. Light Reason **58**, 40–45 (2004). http://yunus.hacettepe.edu.tr/~%20sadi/yayin/Seferoglu_Ogretmen_Yeterlikleri_BAAE_2004-58.pdf
9. Paine, L.W., Fang, Y.: Reform as hybrid model of teaching and teacher development in China. Int. J. Educ. Res. **45**, 279–289 (2006). https://doi.org/10.1016/j.ijer.2007.02.006
10. Görkaş, B.: Görkaş, B.: Examination of the special field qualifications of English teacher candidates in terms of qualification of the Council of Higher Education, the Ministry of National Education, The English Language Teaching undergraduate programs. Unpublished Master's thesis. Akdeniz University, Institute of Educational Sciences, Antalya (2017)
11. Kanadlı, S.: Prospective teachers' professional self-efficacy beliefs in terms of their perceived autonomy support and attitudes towards the teaching profession: a mixed methods study. Educ. Sci. Theory Pract. **7(5)**, 1847–1871 (2017)
12. Tschannon-Moran, M., Woolfolk-Hoy, A., Hoy, W.K.: Teacher efficacy: its meaning and measure. Rev. Educ. Res. **68**(2), 202–248 (1998). https://doi.org/10.3102/00346543068002202
13. Aktaş, T.: Communication competence in foreign language teaching. J. Lang. Linguist. Stud. **1**(1), 89–100 (2005)
14. Ariş, B.: Differences in self-efficacy perceptions of foreign language teachers in Turkey. Istanbul University, Social Sciences Institute, Master thesis (2019)
15. Büyüköztürk, Ş., Kılıç Çakmak, E., Akgün, Ö.E., Karadeniz, Ş., Demirel, F.: Scientific Research Methods, 6th edn. Pegem A Academy, Ankara (2010)
16. Yıldız, Ü., Tepeli, Y.: A study on teacher competencies in teaching Turkish as a foreign language. Int. J. Lang. Acad. **2**(4), 564–578 (2014). (in Turkish). https://dergipark.org.tr/en/pub/hunefd/issue/7791/101927
17. Büyüköztürk, Ş., Bökeoğlu, Ö.Ç., Köklü, N.: Statistics for the Social Sciences, 3rd edn. Pegem A Academy, Ankara (2008)
18. İlbeği, A.S., Çeliköz, M.: An investigation of English self-efficacy beliefs of university students attending English preparatory program. IBAD J. Soc. Sci. **8**, 14–34 (2020)
19. Bayır, M., Yeşil, R.: The levels of university students to experience foreign language learning difficulties. MANAS J. Soc. Res. **9(3)**, 1479–1493 (2020). https://dergipark.org.tr/en/pub/mjss/issue/56124/595584

On a View of Decision Making Under Z-information

Akif V. Alizadeh(✉)

Department of Electronics and Automation, Azerbaijan State Oil and Industry University, 20 Azadlig Avenue, Baku AZ1010, Azerbaijan
a.alizadeh@asoiu.edu.az

Abstract. In this paper we investigate multi-attribute decision making problem, where the attribute values and weight are described by Z-numbers. The proposed approach is based on the concepts of positive ideal and negative ideal solutions and distance between Z-valued vectors. Final decision alternative is selected on basis of a degree to the positive ideal solution. A numerical example on multi-attribute decision making for Web Services selection is given to illustrate the solution processes of the suggested approach.

Keywords: Z-number · Type-2 fuzzy number · Decision making · Ideal solution · Z-distance

1 Introduction

Reliability of information is a very important aspect in decision making. The Z-number concept was proposed by Zadeh [1] as a new way to treat uncertainty and reliability of information. Z-number $Z = (A, B)$, is composed of two parts, the first one is a restriction on a value of random variable, and the second part is the reliability of information. Compared with the classical fuzzy number, Z-number has more ability to describe knowledge of human. Aliev et al. [2] initiated study of the basic arithmetic operations of Z-numbers. Further, its application is evident in different fields of research such as multicriteria decision making problems [3, 5–8], linear programming problem [7] and other.

The concept of Z-numbers captures the fuzziness of information better than type-1 and interval type-2 fuzzy sets. In this paper, we explain the relation between Z-numbers, type-2 and type-1 fuzzy sets. Real-world problems usually are characterized by bimodal information and complexity. But this is not taken into account in existing works on the multi-attribute decision making (MADM) problems. The use of one or another MADM theory depends mainly on decision making situations. Unfortunately, there are scarce works on MADM with Z-information in existence. In this paper, a new approach to multi-attribute decision analysis using the Z-number concept is proposed.

The rest of the paper is structured as fallows. In Sect. 2, some prerequisite material on Z-numbers is presented. In Sect. 3, the statement of the problem and the suggested approach to MADM with Z-information is described. In Sect. 4 an application of the

R. A. Aliev et al. (Eds.): ICSCCW 2021, LNNS 362, pp. 636–644, 2022.
https://doi.org/10.1007/978-3-030-92127-9_85

suggested approach to a real-world investment problem is illustrated. Finally, conclusions are given Sect. 5.

2 Preliminaries

Definition 1. A discrete Z-number [1, 2]. A discrete Z-number is an ordered pair $Z = (A, B)$ where A is a discrete fuzzy number playing a role of a fuzzy constraint on values of a random variable X: X *is* A. Is a discrete fuzzy number with a membership function $\mu_B : \{b_1, \ldots, b_n\} \to [0, 1]$, $\{b_1, \ldots, b_n\} \subset [0, 1]$, playing a role of a fuzzy constraint on the probability measure of A: $P(A) = \sum_{i=1}^{n} \mu_A(x_i)p(x_i)$ *is* B.

Definition 2. Square of a discrete Z-number. Let us now consider computation of $Z_Y = Z_X^2 = Zsqr(Z_X)$. Let $Z_X^+ = (A_X, R_X)$ where R_X is represented as

$$p_X = p_X(x_1)\backslash x_1 + p_X(x_2)\backslash x_2 + \ldots + p_X(x_n)\backslash x_n.$$

Then the square of discrete Z^+-number Z_Y^+ is determined as follows:

$$Z_Y^+ = (A_Y, R_Y),$$

where $A_Y = A_X^2$ and R_Y is a discrete probability distribution defined as follows:

$$p_Y = p_Y(y_1)\backslash y_1 + p_Y(y_2)\backslash y_2 + \ldots + p_Y(y_m)\backslash y_m \tag{1}$$

such that

$$y_r = x_k^2 \text{ and } p_Y(y_r) = \sum_{y_r = x^2} p_X(x),\ r = 1, \ldots, m. \tag{2}$$

Next, we compute $\mu_{p_X}(p_{X,l}) = \mu_{B_X}\left(\sum_{k=1}^{n} \mu_{A_X}(x_k)p_{X,l}(x_k)\right)$. Now we realize that the fuzzy set of probability distributions p_X with membership function $\mu_{p_X}(p_{X,l})$ naturally induces the fuzzy set of probability distributions $p_{Y,l}$ with the membership function defined as $\mu_{p_Y}(p_{Y,l}) = \mu_{p_X}(p_{X,l})$, subject to (1)–(2).

Next, we should compute probability measure of A_Y given p_Y. Finally, given a fuzzy restriction on p_Y described by the membership function μ_{p_Y}, we construct a fuzzy set B_Y with the membership function μ_{B_Y} defined as follows:

$$\mu_{B_Y}(b_{Y,l}) = \sup(\mu_{p_Y}(p_{Y,l}))$$

subject to

$$b_{Y,l} = \sum_k p_{Y,l}(x_k)\mu_{A_Y}(x_k).$$

As a result, $Z^2 = Zsqr(Z)$ is obtained as $Z^2 = (A_Y, B_Y)$.

Definition 3. Square Root of a discrete Z-number.

Let us consider computation of $Z_Y = \sqrt{Z_X} = Zsqrt(Z_X)$. Let Z_X^+ and Z_X be the same as Definition 3. Then the discrete Z^+-number Z_Y^+ is determined as follows:

$$Z_Y^+ = (A_Y,\ R_Y),$$

Where $A_Y = \sqrt{A_X}, \sqrt{A_X}$ and R_Y is represented by a discrete probability distribution

$$p_{R_Y} = p_{R_Y}(y_1)\backslash y_1 + p_{R_Y}(y_2)\backslash y_2 + \ldots . p_{R_Y}(y_n)\backslash y_n, \tag{3}$$

such that

$$y_k = \sqrt{x_k} \text{ and } p_{R_Y}(y_k) = p_{R_X}(x_k), \tag{4}$$

Then we construct $\mu_{p_X}(p_{X,l}) = \mu_{B_X}\left(\sum_{k=1}^{n} \mu_{A_X}(x_k)p_{X,l}(x_k)\right)$ and recall that

$$\mu_{p_Y}(p_{Y,l}) = \mu_{p_X}(p_{X,l}),$$

subject to (3)–(4).

Next, we compute probability measure of A_Y and, given the membership function μ_{p_Y},we construct a fuzzy set B_Y analogously to that we did in Definition 2. As a result, $\sqrt{Z} = Zsqrt(Z)$ is obtained as $\sqrt{Z} = (A_Y, B_Y)$.

Definition 4. Absolute value of Z-number is defined as

$$|Z| = Zabs(Z) = Zsqrt(Zsqr(Z)) = \sqrt{Z^2}.$$

Definition 5. A distance between Z-numbers. The Z-distance between Z-numbers $Z_1 = (A_1, B_1)$ and $Z_2 = (A_2, B_2)$ is defined as

$$Zd(Z_1, Z_2) = Zabs(Z_1 - Z_2). \tag{5}$$

Procedure Conversion Z-number to crisp number:

Step 1. Conversion Z-number to type-2 fuzzy number. Suppose that the Z-number $Z = (A, B)$ is given. The probability distributions connecting A and B are represented by the matrix $p = (p_{ij})$.

Here, the index i corresponds to the index i of the elements x_i in the possibility and probability distributions $\mu_A : \{x_1, \ldots, x_m\} \to [0, 1]$, $\{x_1, \ldots, x_m\} \subset X$, $p_X = p_X(x_1)\backslash x_1 + p_X(x_2)\backslash x_2 + \ldots + p_X(x_m)\backslash x_m$, and the index j corresponds to the index j of the elements b_j in the membership function $\mu_B : \{b_1, \ldots, b_n\} \to [0, 1]$, $\{b_1, \ldots, b_n\} \subset [0, 1]$.

For a defuzzified value b_* of B and a probability distribution of $p_{:,j}$ corresponding to b_j, a membership function $\mu_j(x_i)$ is found so that $b_j = \sum_{i=1}^{m} \mu_j(x_i)p_j(x_i)$.

The $\mu_j(x_i)$ level membership functions constructed in this way can be considered as a type-2 fuzzy description of a given Z-number.

Step 2. Conversion of type-2 fuzzy number to type-1 fuzzy number. By defuzzifying the type-2 $\mu_j(x_i)$ membership function with respect to the levels corresponding to each x_i, we obtain type-1 fuzzy numbers.

Step 3. Conversion of type-1 fuzzy number to crisp number (defuzzification). Using one of the defuzzification methods, type-1 fuzzy numbers are converted to crisp numbers.

3 Statement of the Problem and a Solution Method

Assume that $\mathcal{A} = \{\mathcal{A}_1, \mathcal{A}_2, \ldots, \mathcal{A}_n\}$ is a set of alternatives and $C = \{C_1, C_2, \ldots, C_m\}$ is a set of attributes. Every attribute C_j, $j = \overline{1,m}$ is characterized by weight W_j assigned by expert or decision maker. As we deal with Z-information valued decision environment, characteristics of alternative $\mathcal{A}_i$, $i = \overline{1,n}$ on attributes C_j $(j = \overline{1,m})$ is described by the form

$$\mathcal{A}_i = \{(Z(A_{i1}, B_{i1}), Z(A_{i2}, B_{i2}), \ldots, Z(A_{ij}, B_{ij}), Z(A_{im}, B_{im})\} \tag{6}$$

where $Z(A_{ij}, B_{ij})$ is evaluation of an alternative $\mathcal{A}_i$ with respect to attribute C_j. Values of attributes and weights of attributes are usually derived from decision maker or experts and are with partial reliability. In this case, the weights W_j, $i = \overline{1,m}$ are represented as

$$W_j = \left\{Z(A_j^w, B_j^w)\right\}, j = \overline{1,m}, \tag{7}$$

where A_j^w is value of weight of j-th is attribute, B_j^w is reliability of this value.

Hence, we can represent decision matrix D_{nm} as (Table 1):

Table 1. Decision matrix D_{nm}

	C_1	C_2	...	C_m
$\mathcal{A}_1$	$[Z(A_{11}, B_{11})]$	$[Z(A_{12}, B_{12})]$	...	$[Z(A_{1m}, B_{1m})]$
$\mathcal{A}_2$	$[Z(A_{21}, B_{21})]$	$[Z(A_{22}, B_{22})]$	...	$[Z(A_{2m}, B_{2m})]$
⋮	⋮	⋮	⋮	⋮
$\mathcal{A}_n$	$[Z(A_{n1}, B_{n1})]$	$[Z(A_{n2}, B_{n2})]$	...	$[Z(A_{nm}, B_{nm})]$

The common approach in MADM is the use of the utility theories. This approach leads to transformation of a vector-valued alternative to a scalar-valued quantity. This transformation leads to loss of information. It is related to restrictive assumptions on preferences underlying utility models. In human decision it is not needed to use artificial transformation.

In this case we will use the concept of positive and negative ideal point in multi-attribute decision making. We present a positive ideal Z-point for attributes as

$$\mathcal{A}_p^{id} = \left(Z\left(A_{p1}^{id}, B_{p1}^{id}\right), Z\left(A_{p2}^{id}, B_{p2}^{id}\right), \ldots, Z\left(A_{pm}^{id}, B_{pm}^{id}\right)\right). \tag{8}$$

A negative ideal point will be described as

$$\mathcal{A}_N^{id} = \left(Z\left(A_{N_1}^{id}, B_{N_1}^{id}\right), Z\left(A_{N_2}^{id}, B_{N_2}^{id}\right), \ldots, Z\left(A_{N_m}^{id}, B_{N_m}^{id}\right)\right) \quad (9)$$

Solving of the stated decision-making problem, i.e. the process of choice of the best alternative among $\mathcal{A} = \{\mathcal{A}_1, \mathcal{A}_2, \ldots, \mathcal{A}_n\}$ consist of the following steps:

1. Weighted distance d_{ip} i-th alternative and positive ideal solution (8) is defined by (5).
2. Weighted distance d_{iN} between i-th alternative and negative ideal solution (9) is defined by (5).
3. Degree of membership r_i, $i = \overline{1, n}$, of each alternative belonging to the positive ideal solution is calculated by using (7):

$$r_i = \frac{d_{iN}}{d_{iN} + d_{ip}} \quad (10)$$

4. Convert r_i, $i = \overline{1, n}$ to crisp number by using procedure described in Sect. 2.
5. Finally, decision alternative is selected as $\max(r_i)$, $i = 1, n$

4 An Example

We consider multi-attribute decision making for Web services selection problem [4]. Today a wide variety of services are offered that can satisfy quality of services for agents. The options, i.e., web services, are $\mathcal{A}_1$, $\mathcal{A}_2$, ... $\mathcal{A}_8$. An agent has to make a decision taking into account 5 attributes C_1 (cost), C_2 (time), C_3 (reliability), C_4 (availability), C_5 (repetition). In this case all 8 alternatives are evaluated under 5 attributes by Z-numbers. Components of these Z-numbers are presented by triangular fuzzy numbers. The initial and normalized decision matrix are shown in Tables 2 and 3.

Table 2. Decision matrix

	C_1	C_2	C_3
$\mathcal{A}_1$	(0.45 0.5 0.55)(0.5 0.6 0.7)	(0.441 0.49 0.539)(0.5 0.6 0.7)	(0.621 0.69 0.759)(0.5 0.6 0.7)
$\mathcal{A}_2$	(0.126 0.14 0.154)(0.5 0.6 0.7)	(0.531 0.59 0.649)(0.5 0.6 0.7)	(0.423 0.47 0.517)(0.5 0.6 0.7)
$\mathcal{A}_3$	(0.225 0.25 0.275)(0.5 0.6 0.7)	(0.711 0.79 0.869)(0.7 0.8 0.9)	(0.27 0.3 0.33)(0.5 0.6 0.7)
$\mathcal{A}_4$	(0.612 0.68 0.748)(0.5 0.6 0.7)	(0.603 0.67 0.737)(0.5 0.6 0.7)	(0.378 0.42 0.462)(0.5 0.6 0.7)
$\mathcal{A}_5$	(0.333 0.37 0.407)(0.5 0.6 0.7)	(0.225 0.25 0.275)(0.5 0.6 0.7)	(0.522 0.58 0.638)(0.5 0.6 0.7)
$\mathcal{A}_6$	(0.432 0.48 0.528)(0.5 0.6 0.7)	(0.549 0.61 0.671)(0.5 0.6 0.7)	(0.621 0.69 0.759)(0.5 0.6 0.7)
$\mathcal{A}_7$	(0.738 0.82 0.902)(0.7 0.8 0.9)	(0.324 0.36 0.396)(0.5 0.6 0.7)	(0.522 0.58 0.638)(0.5 0.6 0.7)
$\mathcal{A}_8$	(0.531 0.59 0.649)(0.5 0.6 0.7)	(0.378 0.42 0.462)(0.5 0.6 0.7)	(0.648 0.72 0.792)(0.7 0.8 0.9)

Table 3. Decision matrix

	C_4	C_5
$\mathcal{A}_1$	(0.702 0.78 0.858)(0.5 0.6 0.7)	(0.126 0.14 0.154)(0.5 0.6 0.7)
$\mathcal{A}_2$	(0.585 0.65 0.715)(0.5 0.6 0.7)	(0.828 0.92 1.012)(0.7 0.8 0.9)
$\mathcal{A}_3$	(0.747 0.83 0.913)(0.7 0.8 0.9)	(0.576 0.64 0.704)(0.5 0.6 0.7)
$\mathcal{A}4$	(0.405 0.45 0.495)(0.5 0.6 0.7)	(0.342 0.38 0.418)(0.5 0.6 0.7)
$\mathcal{A}_5$	(0.351 0.39 0.429)(0.5 0.6 0.7)	(0.243 0.27 0.297)(0.5 0.6 0.7)
$\mathcal{A}_6$	(0.621 0.69 0.759)(0.5 0.6 0.7)	(0.702 0.78 0.858)(0.5 0.6 0.7)
$\mathcal{A}_7$	(0.216 0.24 0.264)(0.7 0.8 0.9)	(0.324 0.36 0.396)(0.5 0.6 0.7)
$\mathcal{A}_8$	(0.522 0.58 0.638)(0.5 0.6 0.7)	(0.252 0.28 0.308)(0.5 0.6 0.7)

For the simplicity, the weights of the attributes are: $W_1 = 0.3$, $W_2 = 0.2$, $W_4 = 0.18$ and $W_5 = 0.2$. The positive ideal alternative is as

$\mathcal{A}_p^{id} =$ ((0.738 0.82 0.902)(0.7 0.8 0.9), (0.711 0.79 0.869)(0.7 0.8 0.9), (0.648 0.72 0.792)(0.7 0.8 0.9), (0.747 0.83 0.913)(0.7 0.8 0.9), (0.828 0.92 1.012)(0.7 0.8 0.9)).

The negative ideal alternative is as

$\mathcal{A}_N^{id} =$ ((0.126 0.14 0.154)(0.5 0.6 0.7), (0.225 0.25 0.275)(0.5 0.6 0.7), (0.27 0.3 0.33)(0.5 0.6 0.7), (0.216 0.24 0.264)(0.5 0.6 0.7), (0.126 0.14 0.154)(0.5 0.6 0.7)).

According to (1)–(2), weighted distances between Z-vectors of alternatives and Z-vector of positive ideal solution are obtained as

Zd1P = (trimf(0.188, 0.32, 0.452), trimf(0.3919, 0.5066, 0.6392));
Zd2P = (trimf(0.584, 0.680.776), trimf(0.394, 0.5099, 0.6192));
Zd3P = (trimf(0.463, 0.57, 0.677), trimf(0.3919, 0.508, 0.6406));
Zd4P = (trimf(0.0036, 0.14, 0.29), trimf(0.3917, 0.5063,0. 0.6387));
Zd5P = (trimf(0.331, 0.45, 0.569), trimf(0.3913, 0.5071, 0.6287));
Zd6P = (trimf(0.21, 0.34, 0.47), trimf(0.392, 0.5066, 0.6393));
Zd7P = (trimf(0.0, 0.0, 0.164), trimf(0.5797, 0.6839, 0.8235));
Zd8P = (trimf(0.089, 0.23, 0.371), trimf(0.3913, 0.5064, 0.6389));

Analogously, we have obtained weighted distances between Z-vectors of alternatives and negative ideal Z- vector:

Zd1N = (trimf(0.296, 0.36, 0.424), trimf(0.3049, 0.4085, 0.5257));
Zd2N = (trimf(0.0, 0.0, 0.028), trimf(0.4512, 0.5114, 0.5898));
Zd3N = (trimf(0.071, 0.11, 0.149), trimf(0.3134, 0.4173, 0.5117));
Zd4N = (trimf(0.458, 0.54, 0.622), trimf(0.3046, 0.4091,0. 0.5257));
Zd5N = (trimf(0.179, 0.23, 0.281), trimf(0.3053, 0.4131, 0.5263));
Zd6N = (trimf(0.278, 0.34, 0.402), trimf(0.305, 0.4097, 0.5258));
Zd7N = (trimf(0.584, 0.68, 0.776), trimf(0.394, 0.5098, 0.6192));
Zd8N = (trimf(0.377, 0.45, 0.523), trimf(0.3047, 0.4093, 0.526));

The membership degrees $r_i,\ i = \overline{1, 8}$ calculated according to (7):

Zr1 = (trimf(0.3379, 0.5294, 0.876), trimf(0.08115, 0.1657, 0.2827));
Zr2 = (trimf(0.0, 0.0, 04795), trimf(0.3411, 0.4235, 0.4857));
Zr3 = (trimf(0.08596, 0.1618, 0.279), trimf(0.09074, 0.1787, 0.281));
Zr4 = (trimf(0.5022, 0.7941, 1.347), trimf(0.08127, 0.1656,0. 0.2829));
Zr5 = (trimf(0.2106, 0.3382, 0.551), trimf(0.0841, 0.1718, 0.2837));
Zr6 = (trimf(0.3188, 0.5, 0.8238), trimf(0.08135, 0.1657, 0.2825));
Zr7 = (trimf(0.6213, 1.0, 1.329), trimf(0.2164, 0.3148, 0.4066));
Zr8 = (trimf(0.4217, 0.6618, 1.122), trimf(0.08092, 0.1652, 0.2823));

Applying conversion of Z-number to type-2 fuzzy number, we obtain F2r1 (Fig. 1).

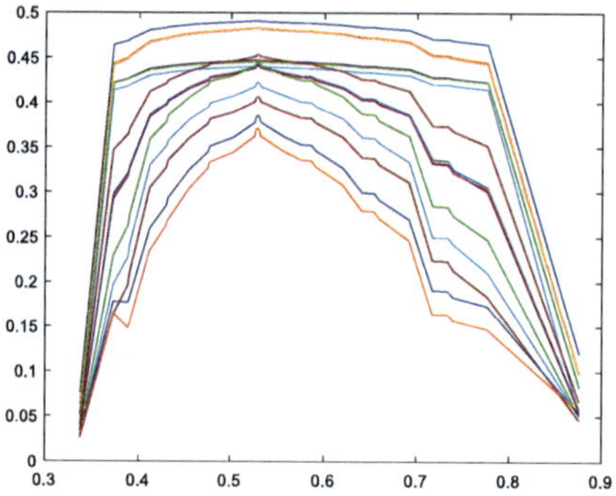

Fig. 1. Fuzzy type-2 number F2r1.

Similarly, we obtain F2r8 (Fig. 2):

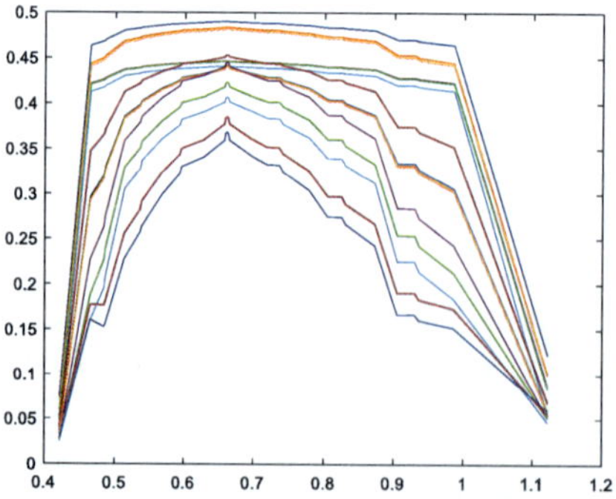

Fig. 2. Fuzzy type-2 number F2r8.

By converting type-2 fuzzy number to type-1 fuzzy number, we obtain F1r1 (Fig. 3):

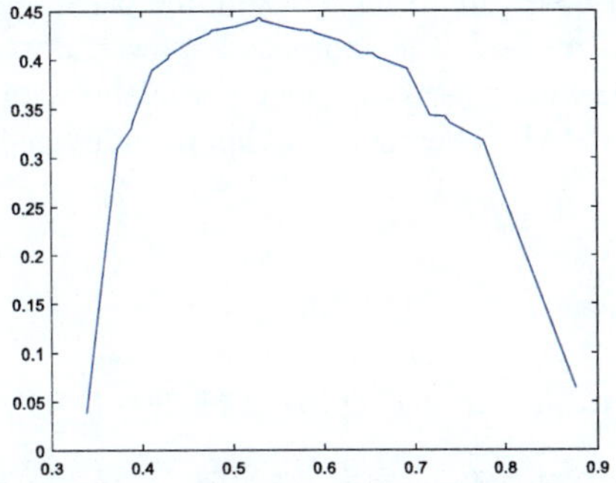

Fig. 3. Fuzzy type-1 number F1r1.

Similarly, we obtain F1r8 (Fig. 4):

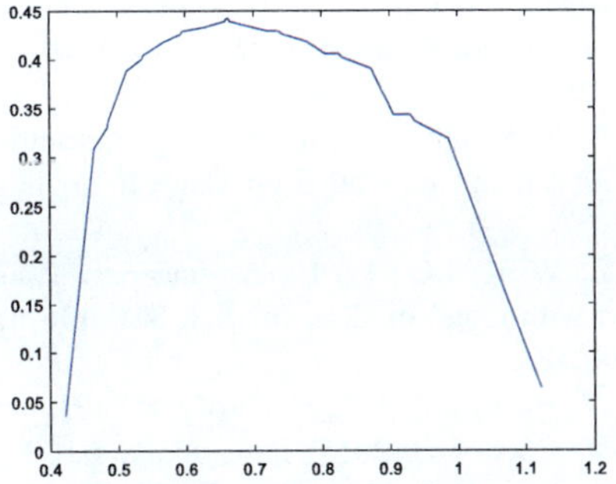

Fig. 4. Fuzzy type-1 number F1r8.

By defuzzifying type-1 fuzzy numbers, we obtain Cri, i = 1…8:

Cr1 = 0.552809486050035
Cr2 = 0.0173484461718796
Cr3 = 0.168706838634238
Cr4 = 0.835595045394880
Cr5 = 0.351927001340508
Cr6 = 0.521689234935137
Cr7 = 0.976116239204448
Cr8 = 0.693625370931505

The final decision is determined as $\max(r_1, r_2, r_3, r_4, r_5, r_6, r_7, r_8) = 0.976116239204448$. Thus, the best alternative is A_7.

5 Conclusion

This paper introduces a novel approach by using the capability of the concept of Z-number and the TOPSIS method. The approach is powerful to describe knowledge of human being and may be widely used for solving problems with uncertain information. A numerical example on MADM demonstrates applicability and efficiency of proposed method.

References

1. Zadeh, L.A.: A note on a Z-number. Inf Sci **181**, 2923–2932 (2011). https://doi.org/10.1016/j.ins.2011.02.022
2. Aliev, R.A., Huseynov, O.H., Aliyev, R.R., Alizadeh, A.A.: The arithmetic of Z-numbers: theory and application. World Scientific Publishing Co., Pte. Ltd., Singapore (2015).ISBN: 9789814675284 9814675288
3. Yaakob, A.M., Gegov, A.: Interactive TOPSIS based group decision making methodology using Z-numbers. Int J Comput Intell Sci. **9**(2), 311–324 (2016). https://doi.org/10.1080/18756891.2016.1150003
4. Aliev, R.A., Huseynov, O.H., Serdaroglu, R.: Ranking of Z-numbers and its application in decision making. Int J Inf Technol Decis Mak. **15**(6), 1503–1519 (2016). https://doi.org/10.1142/S0219622016500310
5. Kang, B., Hu, Y., Deng, Y., Zhou, D.: A new methodology of multicriteria decision-making in supplier selection based on Z-numbers. Math Probl Eng. (2016). https://doi.org/10.1155/2016/8475987
6. Peng, H.G., Zhang, H.Y., Wang, J.Q., Li, L.: An uncertain Z-number multicriteria group decision-making method with cloud models. Inf Sci. **501**, 136–154 (2019). https://doi.org/10.1016/j.ins.2019.05.090
7. Aliev, R.A., Alizadeh, A.V., Huseynov, O.H., Jabbarova, K.I.: Z-number-based linear programming. Int J Intell Syst. **30**(5), 563–589 (2015). https://doi.org/10.1002/int.21709
8. Aboutorab, H., Saberi, M., Asadabadi, M.R., Hussain, O., Chang, E.: ZBWM: the Z-number extension of best worst method and its application for supplier development. Expert Syst Appl. **107**, 115–125 (2018). https://doi.org/10.31181/dmame2003019d

Assessment of the Economic Efficiency of Investment Projects Using a Fuzzy Inference System

Naila Kalantarly[1], Inara Rzayeva[2(✉)], and Vasif Hasanov[3]

[1] Azerbaijan State Academy of Physical Education and Sport, F. Xoyski Avenue 98, AZ1072 Baku, Azerbaijan
nailakalantarli@sport.edu.az

[2] Azerbaijan State University of Economics, Istiglaliyat Street 6, AZ1101 Baku, Azerbaijan

[3] Ganja State University, Haydar Aliyev Street 429, AZ2001 Ganja, Azerbaijan

Abstract. Investment projects can be evaluated relative to many criteria: from the point of view of their social significance, the scale of their influence on the environment, the degree of involvement of labor resources, etc. However, the central place in these assessments belongs to the economic efficiency of the investment project. Analysis and estimation of the economic efficiency of investment projects is one of the difficult problems in the economy, production and decision-making. Its complexity is specified by presence of uncertainty and set of knowingly conflicting criteria that are both quantitative and qualitative nature. The paper proposes a fuzzy approach to assessing the economic efficiency of investment projects. This approach is based on a verbal model, which is implemented through the use of a fuzzy inference system on the example of six real alternative investment projects, financial and economic indicators of which are selected from open sources. The fuzzy inference system is formed on six logically consistent implications, which are implemented by Lukachevich's rule.

Keywords: Economic efficiency · Investment project · Verbal model · Fuzzy set · Fuzzy inference

1 Introduction

Creation of suitable conditions for sustainable economic growth and development of the territorial socio-economic system of the region as a whole is one of the main problems within the government assistance program. It is clear that in order to achieve high levels of regional development, it is necessary to carry out rather large targeted investments. But for this, at first it is necessary to clarify and justify the possibility and economic feasibility of implementing the selected investment projects. The complexity of assessing the economic efficiency of investment projects is due to the presence of uncertainty, since when deciding on investments, it is always necessary to foresee the future, and by the presence of many obviously contradictory criteria, which have both quantitative and qualitative nature.

R. A. Aliev et al. (Eds.): ICSCCW 2021, LNNS 362, pp. 645–653, 2022.
https://doi.org/10.1007/978-3-030-92127-9_86

As a rule, quantitative criteria are expressed in monetary terms and/or in conventional units on the scale of the unified interval. Qualitative criteria do not lend themselves to monetary or any other quantitative expression and cannot be aggregated in traditional ways with monetary estimates, but with all this, they must be taken into account in the framework of investment calculations to analyze the economic effectiveness of the proposed regional investment projects. Such comprehensive approach, capable of more accurately taking into account and assessing all aspects of the economic efficiency of investment project, can be a fuzzy inference system, which is able to identify the most significant and priority projects for the development of the region.

2 Problem Definition

In general, indicators of the economic efficiency of investment projects are qualitative (weakly structured) categories, which, as a rule, are specified in the form of statements such as "HIGH", "SUFFICIENT", etc., that is, in the form of terms of the corresponding linguistic variables. Based on this paradigm, it is necessary to form a fuzzy inference system capable of providing an objective selection and ranking of investment projects according to the aggregate indicators, which reflect the economic efficiency of the regional investment projects and provide decision-making support in terms of financing and, accordingly, choosing the best project.

3 Fuzzy Inference System for Assessing the Economic Efficiency of Investment Projects

In the context of a limited amount of public investment, the choice of the most rational investment project is the main prerogative of the regional authorities, which are guided by a system of criteria in accordance with the Decision of the Supervisory Board of the National Fund of Entrepreneurship Support (hereinafter Fund) in the Azerbaijan Republic [1]. In particular, to estimate the economic efficiency of investment projects realized in Azerbaijan, the Fund recommends the authorized credit institutions to use the criteria: x_1 – investment efficiency; x_2 – reliability of calculations of efficiency indicators; x_3 – return on investment; x_4 – investment profitability; x_5 – risk assessment. To estimate the economic efficiency of investment projects there are proposed to use following appropriate verbal model [2], which assumes that the criteria are ranked according to the importance degree:

e_1: "If the investment efficiency is high, reliability of calculations of efficiency indicators is sufficient, and return on investment is high, then the economic efficiency of project is satisfactory.";
e_2: "If the investment efficiency is high, reliability of calculations of efficiency indicators is sufficient, return on investment is high, and investment profitability is high, then the economic efficiency of project is more than satisfactory";
e_3: "If the investment efficiency is high, reliability of calculations of efficiency indicators is sufficient, return on investment is high, investment profitability is high, and the risk assessment is low, then the economic efficiency of project is perfect";

e_4: "If the investment efficiency is high, reliability of calculations of efficiency indicators is sufficient, return on investment is high, and the risk assessment is low, then the economic efficiency of project is very satisfactory";
e_5: "If the investment efficiency is high, reliability of calculations of efficiency indicators is sufficient, and return on investment is high, and the risk assessment is high, then the economic efficiency of project is still satisfactory";
e_6: "If the investment efficiency is low and return on investment is low, then the economic efficiency of project is unsatisfactory".

Proposed verbal model provides a multi-criterion estimation of alternative projects, where the criteria x_k ($k = 1 \div 5$) are input linguistic variables, whose values are the following corresponding terms: X_1 = HIGH (investment efficiency), X_2 = SUFFICIENT (reliability of calculations of efficiency indicators), X_3 = HIGH (return on investment), X_4 = HIGH (investment profitability), X_5 = LOW (risk assessment), and output linguistic variable y, which assigns terms: UNSATISFACTORY, SATISFACTORY, MORE THAN SATISFACTORY, VERY SATISFACTORY and PERFECT.

The adaptation of the verbal model to the fuzzy environment is carried out by the corresponding fuzzy inference system:

e_1: "If x_1 = HIGH and x_2 = SUFFICIENT and x_3 = HIGH, then y = SATISFACTORY";
e_2: "If x_1 = HIGH and x_2 = SUFFICIENT and x_3 = HIGH and x_4 = HIGH, then y = MORE THAN SATISFACTORY";
e_3: "If x_1 = HIGH and x_2 = SUFFICIENT and x_3 = HIGH and x_4 = HIGH and x_5 = LOW, then y = PERFECT";
e_4: "If x_1 = HIGH and x_2 = SUFFICIENT and x_3 = HIGH and x_5 = LOW, then y = VERY SATISFACTORY";
e_5: "If x_1 = HIGH and x_2 = SUFFICIENT and x_3 = HIGH and x_5 = HIGH, then y = SATISFACTORY";
e_6: "If x_1 = LOW and x_3 = LOW, then y = UNSATISFACTORY".

According to [3], the terms of linguistic variable y can be defined by appropriate fuzzy subsets on the discrete universe $J = \{0, 0.1, 0.2, \ldots, 1\}$. So, $\forall j \in J$ these fuzzy sets are restored by following corresponding membership functions: S = SATISFACTORY: $\mu_S(x) = x$; MS = MORE THAN SATISFACTORY: $\mu_{MS}(j) = j^{(1/2)}$; P = PERFECT: $\mu_P(j) = 1$ if $j = 1$ and $\mu_P(j) = 0$ if $j < 1$; VS = VERY SATISFACTORY: $\mu_{VS}(j) = j^2$; US = UNSATISFACTORY: $\mu_{US}(j) = 1-j$. Then, the implicative rules $e_1 \div e_6$ are represented in the symbolic form as follows:

e_1: $(x_1 = X_1)\ \&\ (x_2 = X_2)\ \&\ (x_3 = X_3) \Rightarrow (y = S)$;
e_2: $(x_1 = X_1)\ \&\ (x_2 = X_2)\ \&\ (x_3 = X_3)\ \&\ (x_4 = X_4) \Rightarrow (y = MS)$;
e_3: $(x_1 = X_1)\ \&\ (x_2 = X_2)\ \&\ (x_3 = X_3)\ \&\ (x_4 = X_4)\ \&\ (x_5 = X_5) \Rightarrow (y = P)$;
e_4: $(x_1 = X_1)\ \&\ (x_2 = X_2)\ \&\ (x_3 = X_3)\ \&\ (x_5 = X_5) \Rightarrow (y = VS)$;
e_5: $(x_1 = X_1)\ \&\ (x_2 = X_2)\ \&\ (x_3 = X_3)\ \&\ (x_5 = \neg X_5) \Rightarrow (y = S)$;
e_6: $(x_1 = \neg X_1)\ \&\ (x_3 = \neg X_3) \Rightarrow (y = US)$.

To approve the Fuzzy Inference System, six alternative investment projects u_i ($i = 1 \div 6$) are chosen: 2 smalls (up to \$100,000.00); 2 middle (from \$100,000.00 to

\$1,000,000.00), and 2 larges (over \$1,000,000.00). Concise description of these projects is given in [1]. Thus, for fuzzification the terms from left parts of rules $e_1 \div e_6$ the expert assessments of projects obtained at the initial stage of the study are applied. The averaged values of expert assessments of project economic efficiency relative to criteria are summarized in Table 1. In this case, $U = \{u_1, u_2, u_3, u_4, u_5, u_6\}$ is the discrete universe on the base of which there are described the terms of input characteristics by appropriate fuzzy sets with the corresponding Gaussian membership functions

$$\mu(t) = \exp\{-(t-4)^2/\sigma_k^2\}, \tag{1}$$

where ${\sigma_k}^2$ is the density of scattering which select subject to the priority of the corresponding criteria [2]. In our case, estimation of the project for each criterion is calculated in accordance with the value of corresponding Gaussian function from [0, 1]. For example, estimate u_1 by first criterion A_1 is $\mu_{A1}(3.6) = 0.9314$.

Table 1. Average expert assessments of project economic efficiency relative to criteria

Project	Influences				
	x_1	x_2	x_3	x_4	x_5
	Qualitative evaluation criteria				
	A_1 = HIGH	A_2 = SUFFICIENT	A_3 = HIGH	A_4 = HIGH	A_5 = LOW
u_1	3.6	3.6	3.8	3.8	2.4
u_2	3.4	3.8	3.6	3.6	3.0
u_3	2.8	3.2	2.8	2.6	2.2
u_4	2.6	3.2	2.4	2.6	1.6
u_5	3.4	3.4	3.4	3.2	2.6
u_6	3.2	3.8	3.0	2.8	2.2

Empirically, the values of parameters (root mean square deviations) σ_k in formula (1) were obtained as following: $\sigma_1 = 1.5$, $\sigma_2 = 1.75$, $\sigma_3 = 2$, $\sigma_4 = 2.25$ and $\sigma_5 = 2.5$ Then, adopted criteria can be described by following fuzzy sets [2]:

- HIGH (investment efficiency): $X_1 = \{0.9314/u_1, 0.8521/u_2, 0.5273/u_3, 0.4185/u_4, 0.8521/u_5, 0.7524/u_6\}$;
- SUFFICIENT (reliability of calculations of efficiency indicators): $X_2 = \{0.9491/u_1, 0.9870/u_2, 0.8114/u_3, 0.8114/u_4, 0.8891/u_5, 0.9870/u_6\}$;
- HIGH (return on investment): $X_3 = \{0.9900/u_1, 0.9608/u_2, 0.6977/u_3, 0.5273/u_4, 0.9139/u_5, 0.7788/u_6\}$;
- HIGH (investment profitability): $X_4 = \{0.9921/u_1, 0.9689/u_2, 0.6790/u_3, 0.6790/u_4, 0.8812/u_5, 0.7524/u_6\}$;
- LOW (risk assessment): $X_5 = \{0.6639/u_1, 0.8521/u_2, 0.5955/u_3, 0.3979/u_4, 0.7308/u_5, 0.5955/u_6\}$.

Further, for each left-hand side of rules $e_1 \div e_6$ the unified membership function $\mu_{Mi}(u)$ is established by intersection of corresponding fuzzy set. In particular, according to [4] we have:

e_1: $\mu_{M1}(u) = \min\{\mu_{X1}(u), \mu_{X1}(u), \mu_{X3}(u)\}$, $M_1 = \{0.9314/u_1, 0.8521/u_2, 0.5273/u_3, 0.4185/u_4, 0.8521/u_5, 0.7524/u_6\}$;
e_2: $\mu_{M2}(u) = \min\{\mu_{X1}(u), \mu_{X1}(u), \mu_{X3}(u), \mu_{X4}(u)\}$, $M_2 = \{0.9314/u_1, 0.8521/u_2, 0.5273/u_3, 0.4185/u_4, 0.8521/u_5, 0.7524/u_6\}$;
e_3: $\mu_{M3}(u) = \min\{\mu_{X1}(u), \mu_{X1}(u), \mu_{X3}(u), \mu_{X4}(u), \mu_{X5}(u)\}$, $M_3 = \{0.6639/u_1, 0.8521/u_2, 0.5273/u_3, 0.3979/u_4, 0.7308/u_5, 0.5955/u_6\}$;
e_4: $\mu_{M4}(u) = \min\{\mu_{X1}(u), \mu_{X1}(u), \mu_{X3}(u), \mu_{X5}(u)\}$, $M_4 = \{0.6639/u_1, 0.8521/u_2, 0.5273/u_3, 0.3979/u_4, 0.7308/u_5, 0.5955/u_6\}$;
e_5: $\mu_{M5}(u) = \min\{\mu_{X1}(u), \mu_{X1}(u), \mu_{X3}(u), 1 - \mu_{X5}(u)\}$, $M_5 = \{0.3361/u_1, 0.1479/u_2, 0.4045/u_3, 0.4185/u_4, 0.2692/u_5, 0.4045/u_6\}$;
e_6: $\mu_{M6}(u) = \min\{1 - \mu_{X1}(u), 1 - \mu_{X3}(u)\}$, $M_6 = \{0.0100/u_1, 0.0392/u_2, 0.3023/u_3, 0.4727/u_4, 0.0861/u_5, 0.2212/u_6\}$.

As a result, the rules $e_1 \div e_6$ can be rewritten in a more compact form:

e_1: $(x = M_1) \Rightarrow (y = S)$;
e_2: $(x = M_2) \Rightarrow (y = MS)$;
e_3: $(x = M_3) \Rightarrow (y = P)$;
e_4: $(x = M_4) \Rightarrow (y = VS)$;
e_5: $(x = M_5) \Rightarrow (y = S)$;
e_6: $(x = M_6) \Rightarrow (y = US)$.

To convert these rules the Lukasiewicz's implication [5, 6]

$$\mu_H(u, j) = \min_{u \in W}\{\, 1, 1 - \mu_M(u) + \mu_Y(j)\}$$

is applied. In this case, for each pair (u, j) it is obtained the fuzzy relation in the form of following matrixes:

$R_1 =$

	0	0.1	0.2	0.3	0.4	0.5	0.6	0.7	0.8	0.9	1
0.9314	0.0686	0.1686	0.2686	0.3686	0.4686	0.5686	0.6686	0.7686	0.8686	0.9686	1.0000
0.8521	0.1479	0.2479	0.3479	0.4479	0.5479	0.6479	0.7479	0.8479	0.9479	1.0000	1.0000
0.5273	0.4727	0.5727	0.6727	0.7727	0.8727	0.9727	1.0000	1.0000	1.0000	1.0000	1.0000
0.4185	0.5815	0.6815	0.7815	0.8815	0.9815	1.0000	1.0000	1.0000	1.0000	1.0000	1.0000
0.8521	0.1479	0.2479	0.3479	0.4479	0.5479	0.6479	0.7479	0.8479	0.9479	1.0000	1.0000
0.7524	0.2476	0.3476	0.4476	0.5476	0.6476	0.7476	0.8476	0.9476	1.0000	1.0000	1.0000

,

$$R_2 = \left[\begin{array}{c|ccccccccccc}
 & 0 & 0.3162 & 0.4472 & 0.5477 & 0.6325 & 0.7071 & 0.7746 & 0.8367 & 0.8944 & 0.9487 & 1 \\ \hline
0.9314 & 0.0686 & 0.3849 & 0.5159 & 0.6164 & 0.7011 & 0.7757 & 0.8432 & 0.9053 & 0.9631 & 1.0000 & 1.0000 \\
0.8521 & 0.1479 & 0.4641 & 0.5951 & 0.6956 & 0.7803 & 0.8550 & 0.9225 & 1.0000 & 1.0000 & 1.0000 & 1.0000 \\
0.5273 & 0.4727 & 0.7889 & 0.9199 & 1.0000 & 1.0000 & 1.0000 & 1.0000 & 1.0000 & 1.0000 & 1.0000 & 1.0000 \\
0.4185 & 0.5815 & 0.8977 & 1.0000 & 1.0000 & 1.0000 & 1.0000 & 1.0000 & 1.0000 & 1.0000 & 1.0000 & 1.0000 \\
0.8521 & 0.1479 & 0.4641 & 0.5951 & 0.6956 & 0.7803 & 0.8550 & 0.9225 & 0.9845 & 1.0000 & 1.0000 & 1.0000 \\
0.7524 & 0.2476 & 0.5638 & 0.6948 & 0.7953 & 0.8800 & 0.9547 & 1.0000 & 1.0000 & 1.0000 & 1.0000 & 1.0000
\end{array}\right],$$

$$R_3 = \left[\begin{array}{c|ccccccccccc}
 & 0 & 0 & 0 & 0 & 0 & 0 & 0 & 0 & 0 & 0 & 1 \\ \hline
0.6639 & 0.3361 & 0.3361 & 0.3361 & 0.3361 & 0.3361 & 0.3361 & 0.3361 & 0.3361 & 0.3361 & 0.3361 & 1.0000 \\
0.8521 & 0.1479 & 0.1479 & 0.1479 & 0.1479 & 0.1479 & 0.1479 & 0.1479 & 0.1479 & 0.1479 & 0.1479 & 1.0000 \\
0.5273 & 0.4727 & 0.4727 & 0.4727 & 0.4727 & 0.4727 & 0.4727 & 0.4727 & 0.4727 & 0.4727 & 0.4727 & 1.0000 \\
0.3979 & 0.6021 & 0.6021 & 0.6021 & 0.6021 & 0.6021 & 0.6021 & 0.6021 & 0.6021 & 0.6021 & 0.6021 & 1.0000 \\
0.7308 & 0.2692 & 0.2692 & 0.2692 & 0.2692 & 0.2692 & 0.2692 & 0.2692 & 0.2692 & 0.2692 & 0.2692 & 1.0000 \\
0.5955 & 0.4045 & 0.4045 & 0.4045 & 0.4045 & 0.4045 & 0.4045 & 0.4045 & 0.4045 & 0.4045 & 0.4045 & 1.0000
\end{array}\right],$$

$$R_4 = \left[\begin{array}{c|ccccccccccc}
 & 0 & 0.01 & 0.04 & 0.09 & 0.16 & 0.25 & 0.36 & 0.49 & 0.64 & 0.81 & 1 \\ \hline
0.6639 & 0.3361 & 0.3461 & 0.3761 & 0.4261 & 0.4961 & 0.5861 & 0.6961 & 0.8261 & 0.9761 & 1.0000 & 1.0000 \\
0.8521 & 0.1479 & 0.1579 & 0.1879 & 0.2379 & 0.3079 & 0.3979 & 0.5079 & 0.6379 & 0.7879 & 0.9579 & 1.0000 \\
0.5273 & 0.4727 & 0.4827 & 0.5127 & 0.5627 & 0.6327 & 0.7227 & 0.8327 & 0.9627 & 1.0000 & 1.0000 & 1.0000 \\
0.3979 & 0.6021 & 0.6121 & 0.6421 & 0.6921 & 0.7621 & 0.8521 & 0.9621 & 1.0000 & 1.0000 & 1.0000 & 1.0000 \\
0.7308 & 0.2692 & 0.2792 & 0.3092 & 0.3592 & 0.4292 & 0.5192 & 0.6292 & 0.7592 & 0.9092 & 1.0000 & 1.0000 \\
0.5955 & 0.4045 & 0.4145 & 0.4445 & 0.4945 & 0.5645 & 0.6545 & 0.7645 & 0.8945 & 1.0000 & 1.0000 & 1.0000
\end{array}\right],$$

$$R_5 = \left[\begin{array}{c|ccccccccccc}
 & 0 & 0.1 & 0.2 & 0.3 & 0.4 & 0.5 & 0.6 & 0.7 & 0.8 & 0.9 & 1 \\ \hline
0.3361 & 0.6639 & 0.7639 & 0.8639 & 0.9639 & 1.0000 & 1.0000 & 1.0000 & 1.0000 & 1.0000 & 1.0000 & 1.0000 \\
0.1479 & 0.8521 & 0.9521 & 1.0000 & 1.0000 & 1.0000 & 1.0000 & 1.0000 & 1.0000 & 1.0000 & 1.0000 & 1.0000 \\
0.4045 & 0.5955 & 0.6955 & 0.7955 & 0.8955 & 0.9955 & 1.0000 & 1.0000 & 1.0000 & 1.0000 & 1.0000 & 1.0000 \\
0.4185 & 0.5815 & 0.6815 & 0.7815 & 0.8815 & 0.9815 & 1.0000 & 1.0000 & 1.0000 & 1.0000 & 1.0000 & 1.0000 \\
0.2692 & 0.7308 & 0.8308 & 0.9308 & 1.0000 & 1.0000 & 1.0000 & 1.0000 & 1.0000 & 1.0000 & 1.0000 & 1.0000 \\
0.4045 & 0.5955 & 0.6955 & 0.7955 & 0.8955 & 0.9955 & 1.0000 & 1.0000 & 1.0000 & 1.0000 & 1.0000 & 1.0000
\end{array}\right],$$

$$R_6 = \left[\begin{array}{c|ccccccccccc}
 & 1 & 0.9 & 0.8 & 0.7 & 0.6 & 0.5 & 0.4 & 0.3 & 0.2 & 0.1 & 0 \\ \hline
0.0100 & 1.0000 & 1.0000 & 1.0000 & 1.0000 & 1.0000 & 1.0000 & 1.0000 & 1.0000 & 1.0000 & 1.0000 & 0.9900 \\
0.0392 & 1.0000 & 1.0000 & 1.0000 & 1.0000 & 1.0000 & 1.0000 & 1.0000 & 1.0000 & 1.0000 & 1.0000 & 0.9608 \\
0.3023 & 1.0000 & 1.0000 & 1.0000 & 1.0000 & 1.0000 & 1.0000 & 1.0000 & 0.9977 & 0.8977 & 0.7977 & 0.6977 \\
0.4727 & 1.0000 & 1.0000 & 1.0000 & 1.0000 & 1.0000 & 1.0000 & 0.9273 & 0.8273 & 0.7273 & 0.6273 & 0.5273 \\
0.0861 & 1.0000 & 1.0000 & 1.0000 & 1.0000 & 1.0000 & 1.0000 & 1.0000 & 1.0000 & 1.0000 & 1.0000 & 0.9139 \\
0.2212 & 1.0000 & 1.0000 & 1.0000 & 1.0000 & 1.0000 & 1.0000 & 1.0000 & 1.0000 & 0.9788 & 0.8788 & 0.7788
\end{array}\right].$$

Finally, a general functional solution is obtained by the intersection of the fuzzy relations $R_1, R_2, \ldots, R_6$ and it looks like as following matrix

$R_1 =$

	0	0.1	0.2	0.3	0.4	0.5	0.6	0.7	0.8	0.9	1
u_1	0.0686	0.1686	0.2686	0.3361	0.3361	0.3361	0.3361	0.3361	0.3361	0.3361	0.9900
u_2	0.1479	0.1479	0.1479	0.1479	0.1479	0.1479	0.1479	0.1479	0.1479	0.1479	0.9608
u_3	0.4727	0.4727	0.4727	0.4727	0.4727	0.4727	0.4727	0.4727	0.4727	0.4727	0.6977
u_4	0.5815	0.6021	0.6021	0.6021	0.6021	0.6021	0.6021	0.6021	0.6021	0.6021	0.5273
u_5	0.1479	0.2479	0.2692	0.2692	0.2692	0.2692	0.2692	0.2692	0.2692	0.2692	0.9139
u_6	0.2476	0.3476	0.4045	0.4045	0.4045	0.4045	0.4045	0.4045	0.4045	0.4045	0.7788

To estimate the economic efficiency of projects, the rule of compositional inference in the fuzzy environment is applied: $E_i = G_i \circ R$, where E_i is the degree of satisfaction of the i-th project ($i = 1 \div 6$); G_i is the imaging of the i-th project by fuzzy subset of U.

Then, according to $\mu_{E_i}(j) = \max\{\min(\mu_{G_i}(u), \mu_R(u))\}$, where $\mu_{G_i}(u) = \begin{cases} 0, & u \neq u_i; \\ 1, & u = u_i, \end{cases}$ whence it follows that $\mu_{E_i}(j) = \mu_R(u_i, j)$, i.e., E_i is the i-th row of the matrix R [5, 6].

To obtain the numerical estimates of fuzzy conclusions relative to the economic efficiency of projects the defuzzification procedure is applied. In particular, relative to economic efficiency of project u_6 the conclusion is described by following fuzzy set:

$$E_6 = \{0.2476/0, 0.3476/0.1, 0.4045/0.2, 0.4045/0.3, 0.4045/0.4, 0.4045/0.5, 0.4045/0.6, 0.4045/0.7, 0.4045/0.8, 0.4045/0.9, 0.7788/1\}.$$

First, it is necessary to calculate the α-level sets $E_{j\alpha}$ and the corresponding cardinal numbers $M(E_{j\alpha})$ by the formula: $M(E_{j\alpha}) = \sum_{i=1}^{n} x_i/n$:

- for $0 < \alpha < 0.2476$: $\Delta\alpha = 0.2476$, $E_{6\alpha} = \{0, 0.1, 0.2, \ldots, 0.9, 1\}$, $M(E_{6\alpha}) = 0.50$;
- for $0.2476 < \alpha < 0.3476$: $\Delta\alpha = 0.1$, $E_{6\alpha} = \{0.1, 0.2, \ldots, 0.9, 1\}$, $M(E_{6\alpha}) = 0.55$;
- for $0.3476 < \alpha < 0.4045$: $\Delta\alpha = 0.0570$, $E_{6\alpha} = \{0.2, 0.3, \ldots, 0.9, 1\}$, $M(E_{6\alpha}) = 0.60$;
- for $0.4045 < \alpha < 0.7788$: $\Delta\alpha = 0.3743$, $E_{6\alpha} = \{1\}$, $M(E_{1\alpha}) = 1$.

Thus, numerical estimate of the economic efficiency of project u_6 is obtained as

$$F(E_6) = \frac{1}{\alpha_{\max}} \int_0^{\alpha_{\max}} M(E_{6\alpha})d\alpha = \frac{1}{0.7788} \int_0^{0.7788} M(E_{6\alpha})d\alpha$$
$$= \frac{1}{0.7788}[0.5 \cdot 0.1238 + 0.55 \cdot 0.0550 + 0.6 \cdot 0.0342 + 1 \cdot 0.3743] = 0.7540.$$

Numerical estimates of the economic efficiency for the rest projects are established by similar actions: for project u_1 – 0.8556, u_2 – 0.9231, u_3 – 0.6612, u_4 – 0.4955, u_5 – 0.8605. From the point of view of economic efficiency, the most satisfactory investment project is the one with the highest numerical estimate. In our case, this is the project u_2. The rest of the projects are ranked in descending order of numerical estimates: u_5 (0.8605), u_1 (0.8556), u_6 (0.7540), u_3 (0.6612), u_4 (0.4955).

The obtained results are summarized in Table 2, which also presents the summary assessments of the experts.

Table 2. Comparison of the obtained results

Project	Fuzzy inference system	Order	Average expert rating	Order
u_1	0.8556	3	17.20	2
u_2	0.9231	1	17.40	1
u_3	0.6612	5	13.60	5
u_4	0.4955	6	12.40	6
u_5	0.8605	2	16.00	3
u_6	0.7540	4	15.00	4

4 Conclusion

As can be seen from the result, the approach proposed in the article can be successfully applied for operative assessment of the economic efficiency of regional investment projects that apply for budgetary funds allocated by the National Investment Fund for Support of Entrepreneurship in Azerbaijan in accordance with criteria x_k $(k = 1 \div 5)$. Obviously, the main advantageous feature of the fuzzy approach is its ability to adapt to the existing regional environment, to the requirements and preferences of specialists responsible for investment decision-making. More specifically, the advantage of the fuzzy approach is that the criteria for evaluating investment projects approved by the Fund (especially those that are qualitative in nature) are formulated on the basis of existing alternatives, which, in turn, significantly increases the objectivity of the choice process.

In comparison with the methods of scoring-analysis that do not reflect internal cause-effect relations, the fuzzy inference induces a more objective and balanced assessment of investment projects based on the bounded set of alternatives, even in the absence of accumulated heuristic knowledge in the subject area. In addition, the implication rules take into account the priority of the evaluation criteria, which provides the more balanced choice of the investment projects. Integral estimates of the considered six alternative projects are not absolute, since the fuzzy inference system was not subjected to structural and parametric optimization. Actually, this is not the main goal of this study. The resulting overall assessments make it possible to correctly rank investment projects that differ both in the amount of funding and in target orientation.

References

1. Entrepreneurship Development Fund of the Republic of Azerbaijan. http://www.anfes.gov.az/saffman/show.img.php?id=81
2. Rzayev, R.R., et al.: Estimation the effectiveness of regional investment projects by fuzzy conclusion method. In: MS'12 International Conference on Decision Making Systems in Business Administration, vol. 8. World Scientific Publishing Co. Pte. Ltd, Rio de Janeiro, pp. 37–53 (2013)

3. Andreichenkov, A.V., Andreichenkova, O.N.: Analysis, synthesis, planning decisions in the economy. Finance and Statistics, Moscow (2000).(in Russian)
4. Zadeh, L.A.: The concept of a linguistic variable and its application to approximate reasoning–I. Inf Sci **8**(3), 199–249 (1975). https://doi.org/10.1007/978-1-4684-2106-4_1
5. Rzayev, R.R.: Neuro-fuzzy modeling of economic behavior. Lambert Academic Publishing, Moscow (2012).in Russian
6. Rzayev, R.R.: Analytical support for decision-making in organizational systems. Palmerium Academic Publishing, Saarbruchen (2016).in Russian

Stable Algorithms for Solving the Problem of Determining the Weighting Coefficients of Neural Networks with Radial-Basis Activation Functions

Husan Igamberdiev, Azizbek Yusupbekov, Uktam Mamirov(✉), and Inomjon Abdukaxxarov

Tashkent State Technical University, 2 University, Tashkent, Uzbekistan

Abstract. The article proposes computational procedures for the formation and construction of stable algorithms for solving the problem of determining the weight coefficients of neural networks with radial-basis activation functions. The approximation properties of multilayer neural networks are analyzed. It was revealed that neural networks with radial-basis activation functions are universal approximators. It is noted that the problem of RBF networks is the choice of the number of radially based functions required for the approximation. When solving this problem for the purpose of quality approximation, the network output error is minimized. Regularized iterative algorithms for determining the weight coefficients of neural networks with radial-basis activation functions based on projection methods are proposed. To solve the extended regularized normal system of equations, the block version of the Kachmazh's algorithm and the pseudo-inversion of rectangular matrices proposed by Greville is used. To give stability to the procedure for determining the weight coefficients of neural networks, an algorithm for finding a pseudoinverse matrix was also used using the representation of a symmetric matrix in a diagonal form.

Keywords: Neural networks · Radial basis functions · Weights

1 Introduction

Currently, artificial neural networks are widely used to solve a variety of tasks related to data mining under conditions of a priori and current uncertainty. Artificial neural networks are necessary computational models for solving a wide variety of problems, such as pattern recognition, curve and function approximation, data compression, associative memory, modeling and control of nonlinear unknown objects [1–3]. Neural networks are distinguished by their computational efficiency and ease of hardware implementation. They have the property of generalization - the ability to correctly classify new images [4].

R. A. Aliev et al. (Eds.): ICSCCW 2021, LNNS 362, pp. 654–661, 2022.
https://doi.org/10.1007/978-3-030-92127-9_87

To date, a large number of different types of neural networks have been developed with their own distinctive features. Artificial neural networks differ in their architecture: the structure of connections between neurons, the number of layers, neuron activation function, learning algorithm [5].

In accordance with the Stone-Weierstrass [5] theorem, multilayer neural networks (MNN) approximate any continuous function with any accuracy depending on the number of layers and the number of basic processing elements in the layers. In this case, the approximation properties of the MNN have a number of advantages in practical applications over the approximation properties of polynomials and orthogonal functions. For example, when approximating continuous functions by polynomials, there is a scaling problem, and its solution requires a priori information about the properties of the function being approximated. The same problem exists when using families of orthogonal functions. Another problem is the implementation of the approximation properties of polynomials and orthogonal functions in real time. Customizable multilayer neural networks are largely free of these shortcomings and, in the light of the known results of studies of their approximation properties, can be considered a universal tool for arbitrarily accurate approximation of continuous functions of many variables in the form of a superposition of fairly simple and practically any nonlinear functions of one variable [5–7].

However, nonlinear multilayer neural networks also have associated disadvantages. Due to the nonlinear dependence of the network outputs on the tunable parameters, the approximation problem becomes multi-extremal, and along with the possible existence of a global minimum of the learning functional, there are also many local minima from which the network does not derive the standard learning algorithm. The interrelation of setting the weight coefficients of all basic elements of the network significantly increases the training time, which is critical for real-time systems, which include control systems for dynamic objects [5].

The neural networks with Radial Basis Networks (RBFN) considered below or further briefly RBF-networks are free of the mentioned disadvantages of trained MNN and in this sense are considered as an alternative to multilayer networks, although, like MNN, they serve as an analogue of biological neural networks. It is shown in [8, 9] that RBF - networks are universal approximators and, with easy constraints, can be applied to approximate any continuous function.

The specificity of RBF networks is that in the operating mode only the weight coefficients of the linear output layer are adjusted in them. The approximation error is calculated directly at the output of the network in the same way as in the MNN, but setting only a single layer that is linear in the setting parameters removes the problem of finding the global minimum of the training functional (usually quadratic) and contributes to the rapid convergence of the network training process [10].

2 Problem Definition

The problem of RBF networks is the choice of the number of radial basis functions required for approximation, and this circumstance becomes a critical factor in the use of RBF networks in identification and control problems, where the necessary information for determining the dimension of RBF networks is usually absent. It was noted in [11] that the number of required radial basis functions grows exponentially with the number of input variables, that is, in our case, with increasing n. Hence, it follows that RBF-networks are applicable in those problems where the dimension of the input set is bounded and known. In the general case, there is no such limitation in multilayer neural networks. Another problem with RBF networks is the need to preconfigure the hidden layer. Many works are devoted to this problem [5, 12].

Let the output layer contain p neurons, so that the output vector has the form:$Y_i = [y_{i_1}, y_{i_2}, ..., y_{i_p}]^T$.

To determine the weight of neurons in the output layer $w_{ij}, i = \overline{1, n}, \ j = \overline{1, p}$ of the network, the entire set of templates is presented, so that for all n input vectors we can write $FW = Y$, where

$$F = \begin{bmatrix} f_1(X_1)\, f_2(X_1) \ldots f_n(X_1) \\ f_1(X_2)\, f_2(X_2) \ldots f_n(X_2) \\ \vdots \quad \vdots \quad \vdots \quad \vdots \\ f_1(X_n)\, f_2(X_n) \ldots f_n(X_n) \end{bmatrix}, W = \begin{vmatrix} w_{11}\ w_{12} \ldots w_{1p} \\ w_{21}\ w_{22} \ldots w_{2p} \\ \vdots \quad \vdots \quad \vdots \quad \vdots \\ w_{n1}\ w_{n2} \ldots w_{np} \end{vmatrix}, Y = \begin{vmatrix} y_{11}\ y_{12} \ldots y_{1p} \\ y_{21}\ y_{22} \ldots y_{2p} \\ \vdots \quad \vdots \quad \vdots \quad \vdots \\ y_{n1}\ y_{n2} \ldots y_{np} \end{vmatrix}.$$

Then the weight matrix W can be calculated by the formula: $W = F^{-1}Y$, if the matrix F is invertible.

However, in practical problems, condition $m = n$ is usually unacceptable. In real cases, it is required to find an approximate solution to the approximation problem at $m << n$ [10].

Consider an RBF network with one output and m hidden neurons:

$$y = h(X) = \sum_{i=1}^{m} w_i f_i(X). \tag{1}$$

Let it be necessary to approximate the dependence given by the set of input-output data (training sample): $\{(X_1, y_1), (X_2, y_2), ..., (X_n, y_n)\}$.

For a qualitative approximation, it is required to minimize the network output error given by the formula

$$E = \sum_{i=1}^{n} (h(X_i) - y_i)^2 \tag{2}$$

We denote

$$F_j = [f_j(X_1) f_j(X_2) \ldots f_j(X_n)]^T, H = [h(X_1) h(X_2) \ldots h(X_n)]^T, Y = [y_1\, y_2 \ldots y_n]^T.$$

Then $F_j^T H = F_j^T Y, \ \ j = \overline{1, m}$, where $F = [F_1\ F_2 \ldots F_m]$.

Insofar as

$$H = \begin{bmatrix} h(X_1) \\ h(X_2) \\ \vdots \\ h(X_n) \end{bmatrix} = \begin{bmatrix} \sum_{j=1}^{m} w_j f_j(X_1) \\ \sum_{j=1}^{m} w_j f_j(X_2) \\ \vdots \\ \sum_{j=1}^{m} w_j f_j(X_n) \end{bmatrix} = \begin{bmatrix} f_1(X_1)\, f_2(X_1) \ldots f_m(X_1) \\ f_1(X_2)\, f_2(X_2) \ldots f_m(X_2) \\ \vdots \quad \vdots \quad \ddots \quad \vdots \\ f_1(X_n)\, f_2(X_n) \ldots f_m(X_n) \end{bmatrix} \begin{bmatrix} w_1 \\ w_2 \\ \vdots \\ w_m \end{bmatrix} = FW,$$

can be written $F^T FW = F^T Y$, and finally

$$W = (F^T F)^{-1} F^T H = F^+ H, \tag{3}$$

where F^+ – is a pseudoinverse matrix.

3 Solution

At the same time, here, in the implementation of the stabilization algorithm of the form (3), the pseudoinverse matrix $\tilde{F}^+$ is used. It is known [13–15] that the problem of calculating the pseudoinverse matrix is generally unstable with respect to errors in specifying the matrix. In practice, consideration of such disturbances may be prompted by the limited accuracy with which the observed phenomenon is described by quantitative information. The effect of round-off errors made during the numerical procedure can also be analyzed as if it caused perturbation of the input data.

We take the approximation conditions for Eq. (2) in the form

$$\left\|F - \overline{F}\right\| \le \eta, \quad \left\|H - \overline{H}\right\| \le \delta, \tag{4}$$

where $\overline{F}$ and $\overline{H}$ – the exact values of the matrix operator and the right-hand side of Eq. (3), respectively.

It is known [13, 15, 16] that the normal pseudo solution $W_* = F^+ H$ is the normal solution of the normal system of equations

$$F^T FW = F^T H \tag{5}$$

or $F^T r = 0$, where $r = H - FW$.

When solving Eq. (3), the A.N.Tikhonov regularization method is usually used [17]:

$$(F^T F + \alpha I)W = F^T H. \tag{6}$$

As shown in [14, 18], the regularized normal system of Eqs. (6) can be written in the form:

$$\left(\begin{array}{c|c} \omega I_m & F \\ \hline F^T & -\omega I_n \end{array}\right)\begin{pmatrix} L \\ H \end{pmatrix} = \begin{pmatrix} H \\ 0 \end{pmatrix} \Leftrightarrow \tilde{F}_\omega g = \tilde{H}, \tag{7}$$

where $y = \omega^{-1} r$, $\omega = \sqrt{\alpha}$.

The matrix $\tilde{F}_\omega$ system (7) for all $\alpha > 0$ is nondegenerate [14] and its only solution is the vector $g_* = \left(L_*^T,\ W_*^T\right)^T$ where

$$W_* = \left(F^T F + \alpha I_n\right)^{-1} F^T H, L_* = \omega^{-1} r_*, r_* = H - FW_*.$$

We write the extended system (7) in the form of a system of two equations:

$$\left(\omega I_m \vdots F\right) g = H, \tag{8}$$

$$\left(F^T \vdots - \omega I_n\right) g = 0. \tag{9}$$

To solve system (8), (9), we use the block version of the Kachmazh algorithm [14, 19]:

$$g_{i,1} = g_{i,0} - \left(\omega I_m \vdots F\right)^{+} \left[\left(\omega I_m \vdots F\right) g_{i,0} - H\right], \tag{10}$$

$$g_{i,t} = g_{i,t-1} - \beta_{i,t-1} \begin{pmatrix} f_{t-1} \\ -\omega e_{t-1} \end{pmatrix}, \tag{11}$$

$$\beta_{i,t-1} = \frac{(f_{t-1}^T - \omega e_{t-1}^T) g, t-1}{\left\| (f_{t-1}^T - \omega e_{t-1}^T)^T \right\|^2}, \quad t = 2, 3, \ldots, n+1, \quad g_{i+1,0} = g_{i,n+1}, i = 1, 2, \ldots, \tag{12}$$

where $(e_1, e_2, ..., e_n) = I_n$; $F = (f_1, f_2, ..., f_n)$.

If we denote $H = \left(h^{(1)}, ..., h^{(n)}\right)^T$, then taking into account that the vector $g = \left(L^T, H^T\right)^T$, the recurrent Eq. (11) can be written in the form of two recurrent equations:

$$\begin{aligned} l_{i,t} &= l_{i,t-1} - \beta_{i,t-1} f_{t-1}, \\ \beta_{i,t-1} &= \frac{f_{t-1}^T l_{i,\,t-1} - \omega w_{i,t-1}^{(t-1)}}{\|f_{t-1}\|^2 + \omega^2}, \end{aligned} \tag{13}$$

$$w_{i,\,t}^{(t-1)} = w_{i,\,t-1}^{(t-1)} + \omega \beta_{i,\,t-1}, \quad t = 2, 3, \ldots, n+1. \tag{14}$$

As $l = \omega^{-1} r$ where $r = H - FW$, recurrent Eqs. (13), (14) can be transformed to the form:

$$r_{i,t} = r_{i,t-1} - \rho_{i,t-1} a_{t-1}, \quad \rho_{i,t-1} = \frac{f_{t-1}^T r_{i,t-1} - \alpha w_{i,t-1}^{(t-1)}}{\|f_{t-1}\|^2 + \alpha^2}, \tag{15}$$

$$w_{i,t}^{(t-1)} = w_{i,t-1}^{(t-1)} + \rho_{i,t-1}, \quad t = 2, 3, \ldots, n+1. \tag{16}$$

Formulas (14)–(16) are some modifications of the algorithm for pseudo-inversion of rectangular matrices proposed by Greville [13, 20]. Obviously, to compute the matrices

$\left(\omega I_m \vdots F\right)_j^+$ any other known method can be used [13, 20]. In particular, to increase the stability of the computation process $\left(\omega I_m \vdots F\right)_k^+$ in relation to calculation errors, instead of the conditions for branching the algorithm specified in formula (12), the following conditions can be used:

$$\left\| \left(\omega I_m \vdots F\right)_j^k - g_j \left(\omega I_m \vdots F\right)_{k,j-1} \right\| > \eta \text{ and } \left\| \left(\omega I_m \vdots F\right)_j^k - g_j \left(\omega I_m \vdots F\right)_{k,j-1} \right\| \leq \eta,$$

where $\|\cdot\|$ is any vector norm, ah is a selection of regularization parameter selected taking into account the accuracy of specifying the matrix $\left(\omega I_m \vdots F\right)_k$ and the word length of the computer used.

When calculating the quantity z according to the equation $z = \left(\omega I_m \vdots F\right)^+ H$ (8) pseudoinverse matrix $\left(\omega I_m \vdots F\right)^+$ represent as follows:

$$\left(\omega I_m \vdots F\right)^+ = \left(\omega I_m \vdots F\right)^T \left[\left(\omega I_m \vdots F\right)\left(\omega I_m \vdots F\right)^T\right]^+ .$$

For brevity, given the definition $\left(\omega I_m : F\right)$, denotes:

$$Q = \left[\left(\omega I_m \vdots F\right)\left(\omega I_m \vdots F\right)^T\right],$$

where Q – symmetric, non-negative definite matrix of dimension $(m+n) \times (m+n)$.

According to [13], we represent the symmetric matrix in a diagonal form:

$$Q = TUT^T,$$

where

$$T = (t_1 \mid t_2 \mid \ldots \mid t_{m+n})$$

is a block orthogonal matrix and U is a diagonal matrix.

Then, according to [20], we can write:

$$Q^+ = \sum_{i=1}^{m+n} \lambda_i^{-1} t_i t_i^T,$$

where $\lambda_1 \geq \lambda_2 \geq \ldots \geq \lambda_{m+n} > 0$ Are the eigenvalues of the matrix Q.

In this way:

$$\left(\omega I_m \vdots F\right)^+ = \left(\omega I_m \vdots F\right)^T Q^+,$$

$$z = \left(\omega I_m \vdots F\right)^T Q^+.$$

4 Conclusion

Presented algorithms make it possible to obtain a stable solution to the problem of determining the weights of an RBF network and use them to solve issues of identification and control of linear indefinite dynamic objects. Based on the well-known theoretical provisions of the theory of regular estimation, it would be argued that the use of randomized and quasi-optimal variants of the regularized modification of the design algorithm can increase the convergence rate and obtain the desired solution with the required accuracy in the shortest possible time. Moreover, the efficiency of these algorithms, as is known, significantly increases with a decrease in the spectral condition number of a regularized extended system of the form (7), which causes an increased accuracy of the algorithm used.

Based on developed algorithmic support, created the software, which is used to solve automation and control problems for a specific industrial facility. The results of practical implementation of the developed system have shown their effectiveness [16].

References

1. Aliev, R.R.: Soft Computing and MIQ Rise, vol. 1. Scientific Notes of the Azerbaijan State Oil Academy (ASOA), in Russian (1997)
2. Aliev, R.A., Aliev, R.R.: The Theory Of Intelligent Systems And Its Application. Chashyoglu, Baku, in Russian (2001)
3. Yusupbekov, N.R., Aliev, R.A., Aliev, R.R., Yusupbekov, A.N.: Intelligent control and decision-making systems. State Scientific Publishing House "O'zbekiston Milliy Ensklopediyasi," Tanhkent (2014).in Russain
4. Neumerkel, D., Donne, J., Murray-Smith, R.: Neural network applications for process control. In: Proceedings, EUFIT '94, Second European Congress on Intelligent Techniques and Soft Computing: Aachen, vol 3, pp. 138–14. Elite Foundation, Aachen, Germany (1994)
5. Terekhov, V.A., Efimov, D.V., Tyukin, I.: Neural network control systems. High School, Moscow (2002).in Russian
6. Omatu, S., Khalid, M., Yusof, R.: Neuro-Control and Its Applications. Springer, London (1995)
7. Yusupbekov, N.R., Igamberdiev, H.Z., Mamirov, U.F.: Adaptive control system with a multilayer neural network under parametric uncertainty condition. In: Russian Advances in Fuzzy Systems and Soft Computing: selected contributions to the 8-th International Conference on Fuzzy Systems, Soft Computing and Intelligent Technologies (FSSCIT-2020), vol 2782, pp 228–234. CEUR Workshop Proceedings, Aachen, Germany (2020). http://ceur-ws.org/Vol-2782/paper_32.pdf

8. Park, J., Sandberg, I.W.: Approximation and radial-basis-function networks. Neural Comput. **5**, 305–316. Massachusetts Institute of Technology, USA (1993). https://doi.org/10.1162/neco.1993.5.2.305
9. Park, J., Sandberg, I.W.: Universal approximation using radial-basis-function networks. Neural Comput. **3**, 246–257. Massachusetts Institute of Technology, USA (1991). https://doi.org/10.1162/neco.1991.3.2.246
10. Faris, H., Aljarah, I., Mirjalili, S.: Evolving radial basis function networks using moth–flame optimizer. In: Samui, P., Sekhar, S., Balas, V.E. (eds.) Handbook of Neural Computation, pp 537–550. Academic Press (2017). https://doi.org/10.1016/B978-0-12-811318-9.00028-4
11. Handbook of Intelligent Control: Neural, Fuzzy and adaptive approaches. In: White, D.A., Sofge, D.A. (eds.) Van Nostrand Reinhold, New York (1992)
12. Golovko, V.A.: Neurointelligence: theory and application. Book. 1: Organization and Training of Neural Networks with Forward and Backward Connections. Publishing House BPI, Brest (1999)
13. Charles, L.L., Richard, J.H.: Solving Least Squares Problems. Prentice-Hall, New Jersey, Englewood Cliffs, USA (1986)
14. Il'in, V.P.: On the Kaczmarz iterative method and its generalizations. J. Appl. Ind. Math. **2**, 357–366 (2008). https://doi.org/10.1134/S199047890803006X
15. Igamberdiev, H.Z., Mamirov, U.F.: Regular Algorithms for the Parametric Estimation of the Uncertain Object Control. In: Aliev, R.A., Yusupbekov, N.R., Kacprzyk, J., Pedrycz, W., Sadikoglu, F.M. (eds.) WCIS 2020. AISC, vol. 1323, pp. 322–328. Springer, Cham (2021). https://doi.org/10.1007/978-3-030-68004-6_42
16. Mamirov, U.F.: Regular Synthesis of Systems for Adaptive Control of Uncertain Dynamic Objects. "Bilim va intellektual salohiyat", Tashkent, (in Russian) (2021)
17. Tikhonov, A.N., Arsenin, V.Ya.: Methods for Solving ill–Posed Problems. Nauka, Moscow, (in Russian) (1986)
18. Zhdanov, A.I.: Introduction to methods for solving ill-posed problems. Aerospace University, Samara (2006).in Russian
19. Zhdanov, A.I., Sidorov, Y.V.: Parallel implementation of a randomized regularized Kaczmarz's algorithm. Comput. Opt. **39**(4), 536–541 (2015). https://doi.org/10.18287/0134-2452-2015-39-4-536-541
20. Gantmacher, F.R., Brenner, J.L.: Applications of the Theory of Matrices. Courier Corporation (2005)

Conversion of Volatile Time Series into a Fuzzy Time Series by the Example of the Dow Jones Index Dynamics

Parvin Alizada(✉)

Baku State University, Z.Xalilov str. 23, AZ1148 Baku, Azerbaijan

Abstract. Fuzzy modeling of time (or dynamic) series is the independent direction of scientific and applied research. This direction includes a battery of problems, the methodology of which is based on the theory of fuzzy sets, fuzzy logic and fuzzy models (or fuzzy inference systems). The initial procedure for fuzzy modeling of time series is fuzzification of historical data obtained by observing on the basis of "soft measurements" of the behavior of a dynamic system for a certain period of time. The paper proposes a new rule of fuzzification of historical data, which is tested on the indicators of the Dow Jones Industrial Average, established by the results of daily trading on the US stock exchange by the usual arithmetic averaging of component indicators. The fuzzification procedure proposed in the given paper is implemented through a fuzzy inference system, which ensures that the membership functions of the corresponding fuzzy subsets of the discrete universe are found, covering the entire set of indicators of the Dow Jones index for more than a year.

Keywords: Dow Jones index · Fuzzy set · Fuzzy time series · Fuzzy inference

1 Introduction

Most of the existing algorithms for forecasting time series, sas a rule, work with real numbers. However, in practice, observation data on the behavior of a particular system should be considered as weakly structured, i.e., those about which their belonging to a certain class is known. The simplest way to formalize these data are fuzzy sets [1], which in a fairly acceptable way identify the values (terms) of linguistic variables through the construction of appropriate membership functions. Therefore, it is advisable to interpret weakly structured time series as fuzzy with all the ensuing consequences relative to their modeling and forecasting. So, over the past two decades, numerous publications are devoted to the study of fuzzy time series. The approaches to time series reconstruction described in [2–6] differ in the rules of fuzzification of historical data and/or defuzzification of fuzzy model outputs. It is clear that the reliability of the final predicts directly depends on the adequate description of weakly structured time series data by fuzzy sets and, accordingly, the interpretation of the defuzzified results.

R. A. Aliev et al. (Eds.): ICSCCW 2021, LNNS 362, pp. 662–670, 2022.
https://doi.org/10.1007/978-3-030-92127-9_88

2 Problem Definition

The object of research is a time series (TS): $x(t)\}$ $(t = 1 \div T)$, in which $x(t)$ is a weakly structured historical data (HD) or, in our case, a fuzzy set (FS) A_j $(j = 1 \div J)$, which is characterized by the tuple [1]: $\{x(t)/\mu_{Aj}[x(t)]\}$, $\mu_{Aj}[x(t)] \rightarrow [0, 1]$. It is necessary to develop a method for fuzzification of HD, which would allow to restore the fuzzy TS (FTS) more adequately and, thereby, increase the reliability of its forecasting on the basis of existing fuzzy models.

3 Fuzzification of HD Using Fuzzy Inference

It is known that in order to introduce fuzziness at the beginning it is necessary to determine the universe. In the case of TS, it is based on coverage of a range of HD. According to [6], the segment $d = [D_{\min} - D_1; D_{\max} + D_2]$, where $D_{\min}$ and $D_{\max}$ are the minimum and maximum values of the TS indicator, respectively; D_1 and D_2 are positive numbers selected by the user, for example, from the calculation of dividing the segment d into equal intervals u_j according to the number of preselected evaluation criteria. In particular, the terms C_1, C_2, ..., C_n of the linguistic variable (LV) "the value of the TS indicator" are understood as appropriate fuzzy subsets of the discrete universe $U = \{u_1, u_2, ..., u_n\}$, reflecting the qualitative criteria for assessing the value of HD as follows:

$$C_1 = \mu_{C1}(u_1)\,/u_1 + \mu_{C1}(u_2)/u_2 + \ldots + \mu_{C1}(u_n)/u_n,$$
$$C_2 = \mu_{C2}(u_1)\,/u_1 + \mu_{C2}(u_2)/u_2 + \ldots + \mu_{C2}(u_n)/u_n,$$
$$\ldots$$
$$C_n = \mu_{Cn}(u_1)/u_1 + \mu_{Cn}(u_2)/u_2 + \ldots + \mu_{Cn}(u_n)/u_n,$$

where $\mu_{Ci}(u_j) \in [0, 1]$ $(i, j = 1 \div n)$ are the values of the membership function (MF) of the interval u_j to FS C_i, i.e., if HD belongs to the interval u_j, then its fuzzy interpretation is the FS C_i. Therefore, it is necessary to correctly identify the corresponding MF $\mu_{Ci}(u)$.

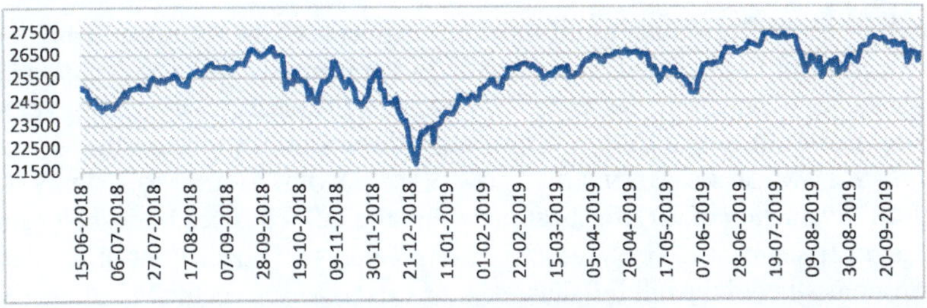

Fig. 1. Volatile DJI time series.

Figure 1 shows the TS reflecting the volatile dynamics of the Dow Jones Industrial Index (DJI) for the period from June 15, 2018 to October 10, 2019 [7], where the minimum and maximum values among 333 DJIs are the numbers $D_{\min} = 21792.2$ and $D_{\max} = 27359.2$, respectively. Choosing $D_1 = 21.2$ and $D_2 = 11.8$, the corresponding universe can be obtained as $U = [21771, 27371]$, which let us divide into eight equal intervals of 700 units in length: $u_1 = [21771, 22471]$, $u_2 = [22471, 23171]$, $u_3 = [23171, 23871]$, $u_4 = [23871, 24571]$, $u_5 = [24571, 25271]$, $u_6 = [25271, 25971]$, $u_7 = [25971, 26671]$, $u_8 = [26671, 27371]$. Then, according to traditional approach [1], the following terms are chosen as possible values of LV "DJI value":

- TOO LOW: $C_1 = 1/u_1 + 0.5/u_2 + 0/u_3 + 0/u_4 + 0/u_5 + 0/u_6 + 0/u_7 + 0/u_8$,
- VERY LOW: $C_2 = 0.5/u_1 + 1/u_2 + 0.5/u_3 + 0/u_4 + 0/u_5 + 0/u_6 + 0/u_7 + 0/u_8$,
- MORE THAN LOW: $C_3 = 0/u_1 + 0.5/u_2 + 1/u_3 + 0.5/u_4 + 0/u_5 + 0/u_6 + 0/u_7 + 0/u_8$,
- LOW: $C_4 = 0/u_1 + 0/u_2 + 0.5/u_3 + 1/u_4 + 0.5/u_5 + 0/u_6 + 0/u_7 + 0/u_8$,
- HIGH: $C_5 = 0/u_1 + 0/u_2 + 0/u_3 + 0.5/u_4 + 1/u_5 + 0.5/u_6 + 0/u_7 + 0/u_8$,
- MORE THAN HIGH: $C_6 = 0/u_1 + 0/u_2 + 0/u_3 + 0/u_4 + 0.5/u_5 + 1/u_6 + 0.5/u_7 + 0/u_8$,
- VERY HIGH: $C_7 = 0/u_1 + 0/u_2 + 0/u_3 + 0/u_4 + 0/u_5 + 0.5/u_6 + 1/u_7 + 0.5/u_8$,
- TOO HIGH: $C_8 = 0/u_1 + 0/u_2 + 0/u_3 + 0/u_4 + 0/u_5 + 0/u_6 + 0.5/u_7 + 1/u_8$.

In this case, the fuzzy interpretation of DJI is carried out taking into account the belonging of the interval of their localization u_j ($j = 1 \div 8$) to one or another FS C_j with the value of a rather trivial MF. Therefore, below another approach to fuzzification of DJIs is considered on the basis of verbal modeling. In other words, based on the previous reasoning, the following consistent and rather trivial statements were chosen:

e_1: "If the DJI is closer to the middle of the segment u_1, then its value is too low";
e_2: "If the DJI is closer to the middle of the u_2, then its value is very low";
e_3: "If the DJI is closer to the middle of the u_3, then its value is more than low";
e_4: "If the DJI is closer to the middle of the u_4, then its value is low";
e_5: "If the DJI is closer to the middle of the u_5, then its value is high";
e_6: "If the DJI is closer to the middle of the u_6, then its value is more than high";
e_7: "If the DJI is closer to the middle of the u_7, then its value is very high";
e_8: "If the DJI is closer to the middle of the segment u_8, then its value is too high".

Analysis of this reasoning, as information fragments of the common verbal model, allows us to identify one input characteristic in the form of LV x = "*Location of the* DJI", the values of which are the terms: CLOSER TO THE MIDDLE OF THE SEGMENT u_k ($k = 1 \div 8$), and one output LV y = "DJI *value*", the values of which are the terms: TOO LOW, VERY LOW, MORE THAN LOW, LOW, HIGH, MORE THAN HIGH, VERY HIGH, TOO HIGH. Evaluation of $x(t)$ localization within the DJI TS based on its belonging to one or another segment u_j ($j = 1 \div 8$) is interpreted as a fuzzy subset of the discrete universe, consisting of the full DJI dataset: $U = \{x(t)\}_{t=1 \div 333}$. As the MF of the FS, the Gaussian functions are chosen as $\mu(x) = \exp\left[-(x_t - u_{j0})^2/\sigma^2\right]$, where $x_t = x(t)$ is the DJI based on the results of the completion of trading on the stock exchange for the t-th day; u_{j0} is the middle of the interval u_j; $\sigma^2 = 2500000$ is the density parameter chosen as the same for all cases (see Fig. 2). Then, taking into account the midpoints of the segments u_j ($j = 1 \div 8$): $u_{10} = 22121$, $u_{20} = 22821$, $u_{30} = 23521$, $u_{40} = 24221$, $u_{50} = 24921$, $u_{60} = 25621$, $u_{70} = 26321$, $u_{80} = 27021$, localization factors of DJI $x(t)$ is interpreted as follows:

- "CLOSE TO 22121" as FS: $X_1 = 0.952181/x_1 + 0.958630/x_2 + 0.964640/x_3 + \ldots + 0.000793/x_{332} + 0.000472/x_{333}$;
- "CLOSE TO 22821" as FS: $X_2 = 0.643393/x_1 + 0.656883/x_2 + 0.670320/x_3 + \ldots + 0.006941/x_{332} + 0.004497/x_{333}$;
- "CLOSE TO 23521" as FS: $X_3 = 0.293758/x_1 + 0.304145/x_2 + 0.314743/x_3 + \ldots + 0.041079/x_{332} + 0.028958/x_{333}$;
- "CLOSE TO 24221" as FS: $X_4 = 0.090627/x_1 + 0.095155/x_2 + 0.099859/x_3 + \ldots + 0.164269/x_{332} + 0.125994/x_{333}$;
- "CLOSE TO 24921" as FS: $X_5 = 0.018892/x_1 + 0.020116/x_2 + 0.021408/x_3 + \ldots + 0.443858/x_{332} + 0.370415/x_{333}$;
- "CLOSE TO 25621" as FS: $X_6 = 0.002661/x_1 + 0.002873/x_2 + 0.003101/x_3 + \ldots + 0.810382/x_{332} + 0.735842/x_{333}$;
- "CLOSE TO 26321" as FS: $X_7 = 0.000253/x_1 + 0.000277/x_2 + 0.000304/x_3 + \ldots + 0.999750/x_{332} + 0.987728/x_{333}$;
- "CLOSE TO 27021" as FS: $X_8 = 0.000016/x_1 + 0.000018/x_2 + 0.000020/x_3 + \ldots + 0.833393/x_{332} + 0.895873/x_{333}$.

Fig. 2. The Gaussian MF of FS, reflecting the degree of DJI localization

The terms of the output LV "DJI value" can be described in the form of fuzzy subsets of the universe $I = \{0, 0.1, 0.2, \ldots, 1\}$, so that, according to [7], $\forall i \in I$ we have: $TL =$ TOO LOW, $\mu_{TL}(i) = \begin{cases} 0, & \text{if } i = 1, \\ 1, & \text{if } i < 1; \end{cases}$ $VL =$ VERY LOW: $\mu_{VL}(i) = (1\text{-}i)^2$; $ML =$ MORE THAN LOW: $\mu_{ML}(i) = (1 - i)^{(1/2)}$; $L =$ LOW: $\mu_L(i) = 1 - i$; $H =$ HIGH: $\mu_H(i) = i$; $MH =$ MORE THAN LOW: $\mu_{MH}(i) = i^{(1/2)}$; $VH =$ VERY HIGH: $\mu_{VH}(i) = i^2$; $TH =$ TOO HIGH, $\mu_{TL}(i) = \begin{cases} 1, & \text{if } i = 1, \\ 0, & \text{if } i < 1. \end{cases}$

Thus, taking into account the introduced formalisms and as a result of applying the Lukasiewicz's implication for each pair $(u, i) \in X \times Y$: $\mu_W(u, i) = \min\{1, 1 - \mu_X(u) + \mu_Y(i)\}$, the rules $e_1 \div e_8$ are transformed to the following corresponding fuzzy relations:

$$R_1 = \left[\begin{array}{c|ccccccccccc} & 1 & 0 & 0 & 0 & 0 & 0 & 0 & 0 & 0 & 0 & 0 \\ \hline 0.0294 & 1.0000 & 0.9706 & 0.9706 & 0.9706 & 0.9706 & 0.9706 & 0.9706 & 0.9706 & 0.9706 & 0.9706 & 0.9706 \\ 0.0374 & 1.0000 & 0.9626 & 0.9626 & 0.9626 & 0.9626 & 0.9626 & 0.9626 & 0.9626 & 0.9626 & 0.9626 & 0.9626 \\ 0.0699 & 1.0000 & 0.9301 & 0.9301 & 0.9301 & 0.9301 & 0.9301 & 0.9301 & 0.9301 & 0.9301 & 0.9301 & 0.9301 \\ \vdots & \vdots & \vdots & \vdots & \vdots & \vdots & \vdots & \vdots & \vdots & \vdots & \vdots & \vdots \\ 0.0008 & 1.0000 & 0.9992 & 0.9992 & 0.9992 & 0.9992 & 0.9992 & 0.9992 & 0.9992 & 0.9992 & 0.9992 & 0.9992 \\ 0.0005 & 1.0000 & 0.9995 & 0.9995 & 0.9995 & 0.9995 & 0.9995 & 0.9995 & 0.9995 & 0.9995 & 0.9995 & 0.9995 \end{array}\right],$$

$$R_2 = \left[\begin{array}{c|ccccccccccc} & 1 & 0.81 & 0.64 & 0.49 & 0.36 & 0.25 & 0.16 & 0.09 & 0.04 & 0.01 & 0 \\ \hline 0.1274 & 1.0000 & 1.0000 & 1.0000 & 1.0000 & 1.0000 & 1.0000 & 1.0000 & 0.9626 & 0.9126 & 0.8826 & 0.8726 \\ 0.1530 & 1.0000 & 1.0000 & 1.0000 & 1.0000 & 1.0000 & 1.0000 & 1.0000 & 0.9370 & 0.8870 & 0.8570 & 0.8470 \\ 0.2435 & 1.0000 & 1.0000 & 1.0000 & 1.0000 & 1.0000 & 1.0000 & 0.9165 & 0.8465 & 0.7965 & 0.7665 & 0.7565 \\ \vdots & \vdots & \vdots & \vdots & \vdots & \vdots & \vdots & \vdots & \vdots & \vdots & \vdots & \vdots \\ 0.0069 & 1.0000 & 1.0000 & 1.0000 & 1.0000 & 1.0000 & 1.0000 & 1.0000 & 1.0000 & 1.0000 & 1.0000 & 0.9931 \\ 0.0045 & 1.0000 & 1.0000 & 1.0000 & 1.0000 & 1.0000 & 1.0000 & 1.0000 & 1.0000 & 1.0000 & 1.0000 & 0.9955 \end{array}\right],$$

... ...

$$R_8 = \left[\begin{array}{c|ccccccccccc} & 0 & 0 & 0 & 0 & 0 & 0 & 0 & 0 & 0 & 0 & 1 \\ \hline 0.2252 & 0.7748 & 0.7748 & 0.7748 & 0.7748 & 0.7748 & 0.7748 & 0.7748 & 0.7748 & 0.7748 & 0.7748 & 1.0000 \\ 0.1913 & 0.8087 & 0.8087 & 0.8087 & 0.8087 & 0.8087 & 0.8087 & 0.8087 & 0.8087 & 0.8087 & 0.8087 & 1.0000 \\ 0.1160 & 0.8840 & 0.8840 & 0.8840 & 0.8840 & 0.8840 & 0.8840 & 0.8840 & 0.8840 & 0.8840 & 0.8840 & 1.0000 \\ \vdots & \vdots & \vdots & \vdots & \vdots & \vdots & \vdots & \vdots & \vdots & \vdots & \vdots & \vdots \\ 0.8334 & 0.1666 & 0.1666 & 0.1666 & 0.1666 & 0.1666 & 0.1666 & 0.1666 & 0.1666 & 0.1666 & 0.1666 & 1.0000 \\ 0.8959 & 0.1041 & 0.1041 & 0.1041 & 0.1041 & 0.1041 & 0.1041 & 0.1041 & 0.1041 & 0.1041 & 0.1041 & 1.0000 \end{array}\right].$$

As a result of the intersection $R_1 \cap R_2 \cap \ldots \cap R_8$, a common functional solution is obtained in the form of the following matrix:

$$R = \left[\begin{array}{c|ccccccccccc} & 0 & 0.1 & 0.2 & 0.3 & 0.4 & 0.5 & 0.6 & 0.7 & 0.8 & 0.9 & 1 \\ \hline x_1 = 25090.5 & 0.0114 & 0.1114 & 0.2114 & 0.3114 & 0.4114 & 0.5114 & 0.6114 & 0.5610 & 0.4610 & 0.3610 & 0.2610 \\ x_2 = 24987.5 & 0.0018 & 0.1018 & 0.2018 & 0.3018 & 0.4018 & 0.5018 & 0.6018 & 0.5094 & 0.4094 & 0.3094 & 0.2094 \\ x_3 = 24700.2 & 0.0193 & 0.1193 & 0.2193 & 0.3193 & 0.4193 & 0.5193 & 0.4878 & 0.3878 & 0.2878 & 0.1878 & 0.0878 \\ \vdots & \vdots & \vdots & \vdots & \vdots & \vdots & \vdots & \vdots & \vdots & \vdots & \vdots & \vdots \\ x_{332} = 26346.0 & 0.0002 & 0.0102 & 0.0402 & 0.0902 & 0.1602 & 0.1666 & 0.1666 & 0.1666 & 0.1666 & 0.1666 & 0.8357 \\ x_{333} = 26496.7 & 0.0123 & 0.0223 & 0.0523 & 0.1023 & 0.1041 & 0.1041 & 0.1041 & 0.1041 & 0.1041 & 0.1041 & 0.8740 \end{array}\right],$$

which reflects the cause-effect relations between the localization factors of DJI $x(t)$ and, in fact, the estimate of the DJI value. In this case, the fuzzy interpretation of the t-th DJI ($t = 1 \div 333$) is found by the rule of compositional inference: $A_t = G_t \circ R$, where G_t is a mapping of the t-th DJI in the form of the fuzzy subset of the universe U.

Choosing the composition rule as $\mu_{At}(i) = \max\{\min\{\mu_{Gt}(x), \mu_R(i)\}\}$ and supposing $\mu_{G_t}(x) = \begin{cases} 0, & \text{if } x \neq x_t, \\ 1, & \text{if } x = x_t, \end{cases}$ we have: $\mu_{At}(i) = \mu_R(x_t, i)$. This means that A_t is the fuzzy interpretation of the t-th DJI on the I, and the values of corresponding MF are located on the t-th row of the matrix R. In particular, the fuzzy analog of the DJI $x_1 = 25090.5$ is the FS (the 1st row of the R matrix): $A_1 = \{0.0114/0, 0.1114/0.1, 0.2114/0.2, 0.3114/0.3, 0.4114/0.4, 0.5114/0.5, 0.6114/0.6, 0.5610/0.7, 0.4610/0.8, 0.3610/0.9, 0.2610/1\}$.

Thus, based on the number of DJI TS data, fuzzy subsets are obtained, which, in fact, describe these DJIs (see Table 1). Obviously, this is too large a mix of fuzzy sets to be able to apply the known fuzzy predictive TS models on its basis. Therefore, it is necessary to significantly narrow the list of fuzzy sets.

Summarized in Table 1 the point estimates of fuzzy sets (FSPE) are defuzzified values of fuzzy analogs A_t ($t = 1 \div 333$) of DJIs. On the scale of the segment [0, 1] these FSPEs conditionally restore the configuration of the DJI TS, such as shown in Fig. 3. FSPEs are obtained according to the following reasoning. For the fuzzy subset A of the universe I, the α-level sets ($\alpha \in [0, 1]$) are defined in the form of $A_\alpha = \{i | \mu_A(i) \geq \alpha, i \in I\}$. Further, for each A_α the corresponding cardinal number $M(A_\alpha)$ is calculated by the formula: $M(A_\alpha) = \sum_{k=1}^{n} u_k/n$, where $u_k \in A_\alpha$. As a result, the point estimate of A is defined by the formula $F(A) = (1/\alpha_{\max}) \int_0^{\alpha_{\max}} M(A_\alpha) d\alpha$, where $\alpha_{\max}$ is the maximum value of the MF of the fuzzy set A. In particular, for fuzzy set A_1 we have:

- for $0 < \alpha < 0.0114$, $\Delta\alpha = 0.0114$, $A_{1\alpha} = \{0, 0.1, \ldots, 0.8, 0.9, 1\}$, $M(A_{1\alpha}) = 0.50$;
- for $0.0114 < \alpha < 0.1114$, $\Delta\alpha = 0.1$, $A_{1\alpha} = \{0.1, 0.2, \ldots, 0.9, 1\}$, $M(A_{1\alpha}) = 0.55$;
- for $0.1114 < \alpha < 0.2114$, $\Delta\alpha = 0.1$, $A_{1\alpha} = \{0.2, 0.3, \ldots, 0.9, 1\}$, $M(A_{1\alpha}) = 0.60$;
- for $0.2114 < \alpha < 0.2610$, $\Delta\alpha = 0.0495$, $A_{1\alpha} = \{0.3, 0.4, \ldots, 0.9, 1\}$, $M(A_{1\alpha}) = 0.65$;
- for $0{,}2610 < \alpha < 0.3114$, $\Delta\alpha = 0.0505$, $A_{1\alpha} = \{0.3, 0.4, \ldots, 0.8, 0.9\}$, $M(A_{1\alpha}) = 0.60$;
- for $0.3114 < \alpha < 0.3610$, $\Delta\alpha = 0.0495$, $A_{1\alpha} = \{0.4, 0.5, \ldots, 0.9\}$, $M(A_{1\alpha}) = 0.65$;
- for $0.3610 < \alpha < 0.4114$, $\Delta\alpha = 0{,}0505$, $A_{1\alpha} = \{0.4, 0.5, \ldots, 0.8\}$, $M(A_{1\alpha}) = 0.60$;
- for $0.4114 < \alpha < 0.4610$, $\Delta\alpha = 0.0495$, $A_\alpha = \{0.5, 0.6, 0.7, 0.8\}$, $M(A_{1\alpha}) = 0.65$;
- for $0.4610 < \alpha < 0.5114$, $\Delta\alpha = 0.0505$, $A_{1\alpha} = \{0.5, 0.6, 0.7\}$, $M(A_{1\alpha}) = 0.60$;
- for $0.5114 < \alpha < 0{,}5610$, $\Delta\alpha = 0{,}0495$, $A_{1\alpha} = \{0.6, 0.7\}$, $M(A_{1\alpha}) = 0.65$;
- for $0.5610 < \alpha < 0.6114$, $\Delta\alpha = 0.0505$, $A_{1\alpha} = \{0.6\}$, $M(A_{1\alpha}) = 0.60$.

$F(A_1) = [0.0114 \cdot 0.5 + 0.1 \cdot 0.55 + \ldots + 0.0495 \cdot 0.65 + 0.0505 \cdot 0.60]/0.6114 = 0.6062$.

Now, on the basis of FSs, compiled by the number of IDs, it is necessary to form the evaluation criteria for DJI. Suppose, that the number of such criteria is chosen as 10. Then all FS A_t ($t = 1 \div 333$) can be divided into 10 groups by the following rules ($k = 1 \div 10$):

"If the point estimate of A_t is from the $(0.1(k–1), 0.1k]$, then A_t is from the k-th group",

In this case, the k-th evaluation criteria can be considered as the intersection of all FSs belonging to the k-th group, i.e., FSs from the k-th group, which are united by the logical operation "AND". For example, the 1st group includes FS, whose point estimate is located in the interval (0, 0.1] (see Table 1). These are the sets: A_{131}, A_{132}, A_{133}, A_{134} and A_{139}. Information about these sets, including the values of their MFs and point estimates, as well as the 1st evaluation criterion $B_1 = A_{131} \cap A_{132} \cap A_{133} \cap A_{134} \cap A_{139}$ are summarized in Table 2. Other evaluation criteria are established by similar actions.

Table 1. Detailed fuzzy analogue of DJI TS

Date	FS	The MF values on the universe I					FSPE
		0	0.1	…	0.9	1	
15.06.2018	A_1	0.0114	0.1114	…	0.3610	0.2610	0.6062
18.06.2018	A_2	0.0018	0.1018	…	0.3094	0.2094	0.5939
19.06.2018	A_3	0.0193	0.1193	…	0.1878	0.0878	0.5330
… … …							
08.10.2019	A_{331}	0.0098	0.0198	…	0.2546	0.7791	0.8959
09.10.2019	A_{332}	0.0002	0.0102	…	0.1666	0.8357	0.9321
10.10.2019	A_{333}	0.0123	0.0223	…	0.1041	0.8740	0.9534

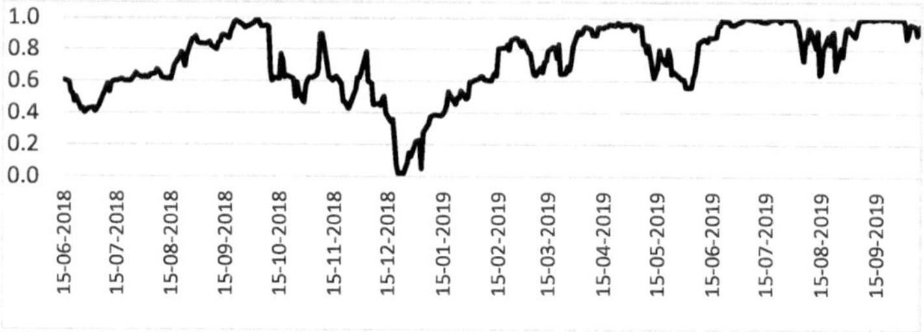

Fig. 3. DJI TS in notation of the FSPE

Table 2. The 1st criterion for assessing the DJI

FS	The values of MF for FS from the 1st group at the universe I						FSPE
	0	0.1	0.2	…	0.9	1	
A_{131}	0.8173	0.1960	0.1960	…	0.0106	0.0006	0.0785
A_{132}	0.9138	0.0412	0.0412	…	0.0412	0.0412	0.0226
A_{133}	0.9801	0.0423	0.0423	…	0.0423	0.0423	0.0216
A_{134}	0.8115	0.2051	0.2051	…	0.0113	0.0013	0.0821
A_{139}	0.8644	0.1200	0.1200	…	0.0172	0.0072	0.0514
B_1	0.8115	0.0412	0.0412	…	0.0106	0.0006	0.0210

Information about all generalized evaluation criteria B_k $(k = 1 \div 10)$ is summarized in Table 3. These criteria can be conventionally designated as: B_1 = TOO LOW, B_2 = VERY LOW, B_3 = MORE THAN LOW, B_4 = LOW, B_5 = BELOW THE AVERAGE, B_6 = AVERAGE, B_7 = ABOVE AVERAGE, B_8 = HIGH, B_9 = VERY HIGH, B_{10} = TOO HIGH. Thus, if the point estimate of FS A_t $(t = 1 \div 333)$ belongs to the k-th interval, then the corresponding DJI can be estimated by the quality criterion B_k. On the base of this approach the common fuzzy analogue of TS is obtained. This analogue is summarized in Table 4, where the criterion C_i ($i = 1 \div 7$) used FTS is also reflected.

Table 3. Generalized criteria for assessing the DJI

FS	The values of MF for FS B_k ($k = 1 \div 10$) at the universe I						FSPO
	0	0.1	0.2	…	0.9	1	
B_1	0.8115	0.0412	0.0412	…	0.0106	0.0006	0.0210
B_2	0.7193	0.2985	0.2985	…	0.0330	0.0230	0.1308
B_3	0.5875	0.4393	0.4393	…	0.1061	0.0031	0.2376
B_4	0.2361	0.3361	0.4361	…	0.1058	0.0000	0.3830
B_5	0.0466	0.1466	0.2466	…	0.1000	0.0000	0.4767
B_6	0.0000	0.1000	0.2000	…	0.1195	0.0195	0.5098
B_7	0.0000	0.1061	0.2014	…	0.3375	0.2375	0.6237
B_8	0.0009	0.0925	0.1225	…	0.4183	0.5679	0.7678
B_9	0.0079	0.0179	0.0479	…	0.2463	0.6597	0.8810
B_{10}	0.0001	0.0010	0.0010	…	0.0010	0.7883	0.9994

4 Conclusion

The proposed approach to fuzzification of ID was tested on the DJI, which are established based on the results of daily trading on the US stock exchange by the usual arithmetic averaging of the constituent indicators. This allows us to consider the daily readings of the DJI as weakly structured, and interpret the dynamics of its change in the form of the FTS. Therefore, the fuzzification is applied, which is realized using fuzzy inference system that provides the determination of the MF values for the corresponding fuzzy subsets of the discrete universe that covers the entire set of DJIs for more than a year. In article there are selected only ten criteria for assessing DJI. However, for the implementation of the known forecasting models in this particular case, the number of such criteria is not enough. Within the framework of the proposed approach, it is possible to easily and quickly increase the number of these evaluation criteria for a more detailed description of TS. Applying this approach, a more common fuzzy analogue of TS has obtained, which is summarized in Table 4. The Table 4 reflects the reconstruction of FTS using the criterion C_i ($i = 1 \div 7$).

Table 4. Fuzzy Time Series of DJI

Date	DJI	Interval	Fuzzy analogue of DJI		Fuzzy model of DJI TS:	
			Sign	FSPE	using C_i	using B_k
15.06.2018	25090.5	u_5	A_1	0.6062	C_5	B_7
18.06.2018	24987.5	u_5	A_2	0.5939	C_5	B_6
19.06.2018	24700.2	u_5	A_3	0.5330	C_5	B_6
20.06.2018	24657.8	u_5	A_4	0.5219	C_5	B_6
21.06.2018	24461.7	u_4	A_5	0.4723	C_4	B_5
………						
08.10.2019	26164.0	u_7	A_{331}	0.8959	C_7	B_9
09.10.2019	26346.0	u_7	A_{332}	0.9321	C_7	B_{10}
10.10.2019	26496.7	u_7	A_{333}	0.9534	C_7	B_{10}

References

1. Zadeh, L.A.: Fuzzy sets. Inf. Control **8**, 338–353 (1965). https://doi.org/10.1016/S0019-9958(65)90241-X
2. Song, Q., Chissom, B.S.: Fuzzy time series and its models. Fuzzy Sets Syst. **54**, 269–277 (1993). https://doi.org/10.1016/0165-0114(93)90372-O
3. Kumar, N., et al.: Fuzzy time series forecasting of wheat production. J. Comput. Sci. Eng. **2**(3), 635–640 (2010). https://doi.org/10.1063/5.0056946
4. Chen, S.M.: Forecasting enrollments based on fuzzy time series. Fuzzy Sets Syst **81**, 311–319 (1996). https://doi.org/10.1016/0165-0114(95)00220-0
5. Cheng, C.H., Chang, J.R., Yen, C.A.: Entropy-based and trapezoid fuzzification fuzzy time series approaches for forecasting IT project cost. Technol Forecast Soc Chang **73**, 524–542 (2006). https://doi.org/10.1016/j.techfore.2005.07.004
6. Poulsen JR (2009) Fuzzy time series forecasting–developing a new forecasting model based on high order fuzzy time series. AAUE: CIS 4
7. Rzayev, R.R., et al.: Time series modeling based on fuzzy data analysis. Fuzzy Syst. Soft Comput. **9**(1), 39–86 (2014). (in Russian)
8. Rzayev, R.R.: Neuro-fuzzy Modeling of Economic Behavior. Lambert Academic Publishing, Moscow (2012). (in Russian)
9. Rzayev, R.R.: Analytical Support for Decision-Making in Organizational Systems. Palmerium Academic Publishing, Saarbruchen (2016). (in Russian)
10. Dow Jones Industrial Index. https://ru.tradingview.com/symbols/DJ-DJI/

Multi-criteria Evaluation of Investment Projects Using the Fuzzy Method of Weighted Maximin Convolution

Elchin Aliyev(✉), Ramin Rzayev, and Adila Ali

Institute of Control Systems of ANAS, B. Vahabzadeh Street 9, AZ1141 Baku, Azerbaijan
elchin.aliyev@sinam.net

Abstract. The adoption of socio-economic, organizational and managerial decisions in most cases implies the implementation of a procedure for multi-criteria selection of alternatives, in which possible options for alternative solutions are assessed according to two or more criteria. Obviously, such decisions include a multi-criteria choice of investment strategies for certain projects. Actually, the justification of the economic feasibility, the scope of the planned work and the timing of their implementation, as well as a detailed description of the practical development of the planned capital investments constitute the subject and content of each investment project. The existing guidelines for assessing the feasibility and effectiveness of investment projects involve the use of scalar evaluation criteria based on financial and economic indicators. However, in real situations, these indicators are still weakly structured. Based on this premise, the article discusses the problem of evaluating investment projects, where the indicators of project performance are considered in the form of qualitative evaluation criteria with different degrees of importance. The fuzzy method of weighted maximin convolution of criteria was chosen as an assessment tool, which was tested on the example of ten hypothetical investment projects.

Keywords: Investment project · Efficiency indicator · Expert assessment · Fuzzy set · Maximin convolution

1 Introduction

Multi-criteria evaluation of investment projects is the primary procedure in the process of forming an "optimal" portfolio of investment projects. As a result, a comparative analysis of alternative investment solutions is carried out, which makes it possible to justify the financing of the most effective projects. In the mainstream of this topic, numerous developments are known. For example, in [1] a comprehensive approach to the comparative analysis of investment projects is proposed, which involves the sequential implementation of the following steps: 1) formation of a system of indicators characterizing an investment decision; 2) reasonable choice of the most effective projects using appropriate methods of analysis and evaluation; 3) establishment of an integral criterion for the final formation of the portfolio of investment projects.

R. A. Aliev et al. (Eds.): ICSCCW 2021, LNNS 362, pp. 671–679, 2022.
https://doi.org/10.1007/978-3-030-92127-9_89

In [2, 3], to evaluate investment projects, we used five financial and economic indicators, which are widely used in international calculation practice. These are: x_1 – NPV (Net Present Value); x_2 – PI (Profitability Index); x_3 – IRR (Internal Rate of Return); x_4 – PP (Payback Period); x_5 – ROI (Return on Investment). Used in [2] the fuzzy method of maximin convolution does not take into account the degree of importance of these indicators, that is, the specific weights of these indicators in deriving the overall assessment of the economic efficiency of the investment project. In the considered form, this method, reflecting the "pessimistic" approach, ignores the "positive" sides of the alternatives: it only identifies the best alternative, as having minimal disadvantages in all indicators. Therefore, based on this premise, it becomes obvious the importance and the relevance of a weighted multi-criteria evaluation of investment projects, which, in fact, predetermined the topic of this article.

2 Problem Definition

The choice of a method for evaluating investment projects and forming an investment portfolio is the prerogative of the investor. At the same time, the values of indicators relative to certain investment projects are provided to the investor in the form of averaged data, which do not always adequately reflect the economic efficiency of projects. In reality, the values of the financial and economic indicators of investment projects should be considered in the general case as weakly structured data, that is, those about which their belonging to a certain type is known. This paradigm formed the basis of this paper, which allows the use of fuzzy methods for multi-criteria evaluation of projects, provided that the indicators x_i $(i = 1 \div 5)$ are qualitative evaluation criteria. The task is to develop a mechanism for a weighted multi-criteria evaluation of investment projects and, accordingly, their ranking, using the fuzzy maximin convolution method. To demonstrate this approach, ten investment projects were selected, the averaged values of performance indicators of which are summarized in Table 1 (see [2]).

Table 1. Financial and economic indicators of investment projects

Project	x_1 NPV ($)	x_2 PI (%)	x_3 IRR (%)	x_4 PP (%)	x_5 ROI (%)
a_1	1,100,000.00	1.14	28	1.80	28
a_2	900,000.00	1.25	25	1.90	32
a_3	700,000.00	1.35	40	1.60	36
a_4	850,000.00	1.05	13	1.20	24
a_5	550,000.00	1.45	20	1.40	25
a_6	1,250,000.00	1.15	35	1.65	22
a_7	300,000.00	1.20	24	1.50	26
a_8	1,300,000.00	1.13	18	1.70	30
a_9	1,500,000.00	1.24	15	2.00	23
a_{10}	630,000.00	1.18	30	1.30	27

Representing the indicators x_i ($I = 1 \div 5$) within the system of comparative assessment of the effectiveness of investment decisions in the form of linguistic variables, the values of which are the weakly structured terms, it is necessary to adapt the fuzzy maximin convolution method to solving the problem of weighted multi-criteria evaluation, ranking and selection of the best investment project.

3 Estimation of Alternatives by the Fuzzy Method of Weighted Maximin Convolution

So, the indicators x_i ($i = 1 \div 5$) of the effectiveness of investment decisions will be considered as qualitative categories or linguistic variables of the form: x_1 = "*the net present value*", x_2 = "*the value of the profitability index*", x_3 = "*acceptability of the internal rate of return*", x_4 = "*payback period*", x_5 = "*the level of return on investment*", and the corresponding terms of the form: F_1 = LARGE, F_2 = ACCEPTABLE, F_3 = APPROPRIATE, F_4 = DESIRED, and F_5 = HIGH as qualitative evaluation criteria. At the same time, the numerical estimates of the considered alternatives are represented by the degrees of compliance with these evaluative concepts. Starting from these assumptions, let us discuss the essence of the applied method of weighted maximin convolution.

Let $A = \{a_1, a_2, \ldots, a_n\}$ is the set of alternatives, which will be considered as a discrete universe, and $F = \{F_1, F_1, \ldots, F_m\}$ is the set of qualitative evaluation criteria, which are fuzzy subsets of the discrete universe A. In general form these criteria can be represent as following fuzzy sets

$$F_i = \mu_{F_i}(a_1)/a_1 + \mu_{F_i}(a_2)/a_2 + \ldots + \mu_{F_i}(a_n)/a_n \ (i = 1 \div m). \tag{1}$$

Then, according to [3], the choice of the best alternative is based on the intersection of fuzzy sets

$$F = F_1 \cap F_2 \cap \ldots \cap F_m, \tag{2}$$

where, according to [4], the intersection operation is realized by following

$$\mu_F(a_j) = \min\{\mu_{F_1}(a_j), \mu_{F_2}(a_j), \ldots, \mu_{F_m}(a_j)\} \ (j = 1 \div n). \tag{3}$$

The best alternative is considered a_j ($j = 1 \div n$), for which the corresponding membership function has the greatest value, i.e.

$$\mu_F(a_j) = \max\{\mu_F(a_1), \mu_F(a_2), \ldots, \mu_F(a_n)\}. \tag{4}$$

In the decision-making process, the evaluation criteria *Fi*, as a rule, differ in degrees of their significance. In this case, the expression (2) should be replaced by the weighted intersection [3]:

$$F = F_1^{\alpha_1} \cap F_2^{\alpha_2} \cap \ldots \cap F_m^{\alpha_m}, \tag{5}$$

where αi is the weight of the *i*-th evaluation criterion. In this case, for the totality of all weights, the following condition must be fulfilled:

$$\sum\nolimits_{i=1}^{m} \alpha_i = 1. \tag{6}$$

4 Fuzzification of Quality Evaluation Criteria

The set of solutions to the problem is characterized by the set of quality criteria x_i ($i = 1 \div 5$) (see Table 1), which are described by fuzzy sets of the form (1). As the membership functions restoring these fuzzy sets, the following Gaussian functions are chosen:

$$\mu_i(u) = \exp\{-(u - u_i)^2/\sigma_i^2\} \quad u \in [0, u_i], \tag{7}$$

where u_i is the maximum value of the indicator x_i among the projects under consideration; $\sigma_i^2 = \sum_{i=1}^{51} (u - u_i)^2/51$ is the empirically selected mean-square deviation. Then, to estimate the investment projects the evaluation concepts (quality evaluation criteria) can be described by following fuzzy sets:

- LARGE (the net present value): $F_1 = \{0.8096/a_1, 0.6217/a_2, 0.4296/a, 0.5725/a_4, 0.3038/a_5, 0.9208/a_6, 0.1494/a_7, 0.9486/a_8, 1/a_9, 0.3682/a_{10}\}$;
- ACCEPTABLE (the value of the profitability index): $F_2 = \{0.8730/a_1, 0.9451/a_2, 0.9860/a_3, 0.7977/a_4, 1/a_5, 0.8806/a_6, 0.9155/a_7, 0.8653/a_8, 0.9396/a_9, 0.9021/a_{10}\}$;
- APPROPRIATE (the internal rate of return): $F_3 = \{0.7654/a_1, 0.6586/a_2, 1/a_3, 0.2584/a_4, 0.4759/a_5, 0.9546/a_6, 0.6217/a_7, 0.4072/a_8, 0.3134/a_9, 0.8306/a_{10}\}$;
- DESIRED (payback period): $F_4 = \{0.9707/a_1, 0.9926/a_2, 0.8880/a_3, 0.6217/a_4, 0.7654/a_5, 0.9130/a_6, 0.8306/a_7, 0.9354/a_8, 1/a_9, 0.6950/a_{10}\}$;
- HIGH (the level of return on investment): $F_5 = \{0.8636/a_1, 0.9640/a_2, 1/a_3, 0.7189/a_4, 0.7578/a_5, 0.6381/a_6, 0.7952/a_7, 0.9208/a_8, 0.6789/a_9, 0.8306/a_{10}\}$.

5 Identification of Indicator Weights Based on Peer Reviews

As noted, the efficiency of the investment project is the multifactorial category, which is characterized by financial and economic indicators x_i ($i = 1 \div 5$), which in varying degree affect the overall estimation of the project. The identification of the weights of these indicators implies: 1) ranking the indicators x_i for their priority; 2) the group assessment of the normalized values of the generalized weights of indicators x_i based on their relative influence on the level of project efficiency. A group of experts is involved in the implementation of these procedures. Each expert is invited to arrange the indicators x_i according to the principle: the most important indicator is designated by the number "1", the next less important one – by the number "2" and then in descending order of the expert's preference. The obtained rank estimates of the indicators x_i are entered into a single database. Further, to establish the degree of consistency of expert opinions, the Kendall's concordance coefficient is applied. This coefficient demonstrates the multiple rank correlation of expert opinions.

According to [5], the Kendall's concordance coefficient is calculated by the formula:

$$W = 12 \cdot S/[m^2(n^3 - n)], \tag{8}$$

where m is the number of invited experts; n is the number of indicators as influences on the total assessment of investment project; S is the deviation of expert opinions from the average value of the criteria ranking x_i, which is calculated by the formula:

$$S = \sum\nolimits_{i=1}^{n} [\sum\nolimits_{j=1}^{m} r_{ij} - m(n+1)/2]^2, \tag{9}$$

where $r_{ij} \in \{1, 2, 3, 4, 5\}$ is the rank of the i-th indicator established by the j-th expert.

Now, suppose that rank estimates r_{ij} of the priority of indicators x_i were obtained by the independent survey of 15 experts. These estimates are summarized in Table 2. In this case, according to formula (8) and $S = 1466$, calculated on the basis of (9) and data from Table 2, the value of the Kendall concordance coefficient will be as

$$W = 12 \cdot 1466/[15^2(5^3 - 5)] = 0.6516,$$

that, significantly exceeding the threshold value 0.6, indicates a fairly acceptable consistency of expert conclusions regarding the degrees of importance of the criteria x_i.

Table 2. Expert estimates of investment project indicators

Expert	Rank estimates (r_{ij}) and normalized values of the indicator weights (α_{ij})									
	x_1		x_2		x_3		x_4		x_5	
	r_{1j}	α_{1j}	r_{2j}	α_{2j}	r_{3j}	α_{3j}	r_{4j}	α_{4j}	r_{5j}	α_{5j}
e_1	5	0.05	2	0.30	4	0.10	3	0.15	1	0.40
e_2	5	0.10	1	0.30	3	0.20	4	0.15	2	0.25
e_3	4	0.10	2	0.25	5	0.05	3	0.15	1	0.45
e_4	5	0.05	2	0.25	3	0.20	4	0.15	1	0.35
e_5	5	0.05	4	0.10	3	0.15	2	0.30	1	0.40
e_6	4	0.10	2	0.25	5	0.05	3	0.20	1	0.40
e_7	3	0.15	5	0.05	2	0.20	4	0.10	1	0.50
e_8	5	0.10	2	0.25	4	0.15	3	0.20	1	0.30
e_9	4	0.10	3	0.15	2	0.30	5	0.05	1	0.40
e_{10}	5	0.10	3	0.20	4	0.15	2	0.25	1	0.30
e_{11}	4	0.10	2	0.25	3	0.20	5	0.05	1	0.40
e_{12}	5	0.05	2	0.20	4	0.10	3	0.15	1	0.50
e_{13}	4	0.15	2	0.25	3	0.20	5	0.10	1	0.30
e_{14}	5	0.05	3	0.15	1	0.40	4	0.10	2	0.30
e_{15}	5	0.05	2	0.25	4	0.10	3	0.20	1	0.40

Preliminary calculations for identifying the weights of the indicators x_i are carried out in the form of averaging over groups of normalized estimates using an iterative formula, which looks like

$$\alpha_i(t+1) = \sum\nolimits_{j=1}^{m} w_j(t)\alpha_{ij}, \tag{10}$$

where $w_j(t)$ is the degree of competence of the j-th expert ($j = 1 \div$ m) at time t, which is calculated by the following equalities:

$$\begin{cases} w_j(t) = \dfrac{1}{\eta(t)} \sum\nolimits_{i=1}^{n} \alpha_i(t)\alpha_{ij}, j = 1 \div (m-1); \\ w_m(t) = 1 - \sum\nolimits_{j=1}^{m-1} w_j(t);\ \sum\nolimits_{j=1}^{m} w_j(t) = 1, \end{cases} \tag{11}$$

where $\eta(t)$ is the normalizing factor, determined as

$$\eta(t) = \sum\nolimits_{i=1}^{n} \sum\nolimits_{j=1}^{m} \alpha_i(t)\alpha_{ij}. \tag{12}$$

The process of averaging over groups is completed under

$$\max_i\{|\alpha_i(t+1) - \alpha_i(t)|\} \le \varepsilon, \tag{13}$$

where ε is the calculation accuracy, which is accepted by user.

Assuming $\varepsilon = 0.001$ and the same degrees of expert competence at the initial stage $t = 0$ (as $w_j(0) = 1/15$), the average values for the groups of normalized estimates of the weights of the indicators x_i in the 1st approximation are obtained from (9) as

$$\alpha_i(1) = \sum_{j=1}^{15} w_j(0)\alpha_{ij} = \sum_{j=1}^{15} \alpha_{ij}/15 \ \ (i = 1 \div 5).$$

As a result, we get the corresponding numbers: $\alpha_1(1) = 0.0867$; $\alpha_2(1) = 0.2133$; $\alpha_3(1) = 0.1700$; $\alpha_4(1) = 0.1533$; $\alpha_5(1) = 0.3767$, which completely satisfy condition (6). Nevertheless, it is not difficult to see that requirement (13) for this approximation is not fulfilled. Therefore, to move to the next step it is necessary to calculate the normalizing factor $\eta(1)$ in the form:

$$\eta(1) = \sum\nolimits_{i=1}^{5} \sum\nolimits_{j=1}^{15} \alpha_i(1)\alpha_{ij} = 1.30 \cdot 0.0867 + 3.20 \cdot 0.2133 + 2.55 \cdot 0.1700 + 2.30 \cdot 0.1533 + 5.65 \cdot 0.3767 = 3.7097.$$

Based on (11) or, more specifically, based on equalities:

$$\begin{cases} w_j(1) = \dfrac{1}{\eta(1)} \sum_{i=1}^{14} \alpha_i(1)\alpha_{ij}, j = 1 \div 14; \\ w_{15}(1) = 1 - \sum_{j=1}^{14} w_j(1); \sum_{j=1}^{15} w_j(1) = 1, \end{cases}$$

the degrees of expert competences are updated, and they are assigned the corresponding numbers: $w_1(1) = 0.0698$; $w_2(1) = 0.0603$; $w_3(1) = 0.0709$; $w_4(1) = 0.0664$; $w_5(1) = 0.0668$; $w_6(1) = 0.0679$; $w_7(1) = 0.0704$; $w_8(1) = 0.0623$; $w_9(1) = 0.0674$; $w_{10}(1) = 0.0615$; $w_{11}(1) = 0.0686$; $w_{12}(1) = 0.0742$; $w_{13}(1) = 0.0616$; $w_{14}(1) = 0.0627$; $w_{15}(1) = 0.0690$.

Further, assuming $\alpha_i(2) = \sum_{j=1}^{15} w_j(1)\alpha_{ij}$, in the 2nd approximation, the average values for the groups of normalized estimates of the indicator weights are obtained in the form of the following numbers: $\alpha_1(2) = 0.0862$; $\alpha_2(2) = 0.2126$; $\alpha_3(2) = 0.1676$; $\alpha_4(2) = 0.1528$; $\alpha_5(2) = 0.3808$, which, as it is not difficult to see, satisfy condition (6).

Checking the obtained values of the weights for the fulfillment of condition (13), we make sure that it is not satisfied again:

$\max\{|\alpha_i(2) - \alpha_i(1)|\} = \max\{|0.0862 - 0.0867|; |0.2126 - 0.2133|; |0.1676 - 0.1700|; |0.1528 - 0.1533|; |0.3808 - 0.3767|\} = 0.0042 > \varepsilon$.

Therefore, it is necessary to calculate the following normalizing factor as

$$\eta(2) = \sum_{i=1}^{5} \sum_{j=1}^{15} \alpha_i(2)\alpha_{ij} = 1.30 \cdot 0.0862 + 3.20 \cdot 0.2126 + 2.55 \cdot 0.1676 + 2.30 \cdot 0.1528 + 5.65 \cdot 0.3808 = 3.7228.$$

In this case, the degrees of expert competences are the following corresponding numbers: $w_1(2) = 0.0699$; $w_2(2) = 0.0602$; $w_3(2) = 0.0710$; $w_4(2) = 0.0664$; $w_5(2) = 0.0669$; $w_6(2) = 0.0680$; $w_7(2) = 0.0706$; $w_8(2) = 0.0622$; $w_9(2) = 0.0674$; $w_{10}(2) = 0.0614$; $w_{11}(2) = 0.0686$; $w_{12}(2) = 0.0744$; $w_{13}(2) = 0.0615$; $w_{14}(2) = 0.0625$; $w_{15}(2) = 0.0691$.

According to (9) and assuming $\alpha_i(3) = \sum_{j=1}^{15} w_j(2)\alpha_{ij}$, for the 3rd approximation the average values for the groups of normalized estimates of the indicator weights are obtained as following numbers: $\alpha_1(3) = 0.0862$; $\alpha_2(3) = 0.2126$; $\alpha_3(3) = 0.1675$; $\alpha_4(3) = 0.1528$; $\alpha_5(3) = 0.3809$, which satisfy the basic condition (6). Moreover, as seen from:

$\max\{|\alpha_i(3) - \alpha_i(2)|\} = \max\{|0.0862 - 0.0862|; |0.2126 - 0.2126|; |0.1675 - 0.1676|; |0.1528 - 0.1528|; |0.3809 - 0.3808|\} = 0.0001$,

condition (13) is already satisfied. This means that $\alpha_1(3)$, $\alpha_2(3)$, $\alpha_3(3)$, $\alpha_4(3)$, $\alpha_5(3)$ are the final generalized weights of the corresponding indicators x_i.

6 Choosing the Best Investment Project and Ranking the Rest

According to (1) and (2), and taking into account various importance degrees for efficiency indicators of investment projects, the set of optimal alternatives A can be obtained by following intersecting of fuzzy sets $F_i(a)$: $F = F_1^{\alpha 1} \cap F_2^{\alpha 2} \cap \ldots \cap F_5^{\alpha 5}$, where, according to [4], the j-th project ($j = 1 \div 10$) is considered the best, because the condition $\mu_A(a_j) = \max\{\mu_A(a_1), \mu_A(a_2), \ldots, \mu_A(a_{10})\}$ is satisfied. In this case, the set of optimal alternatives will be:

$A = \{\min\{0.8096^{0.0862}, 0.8730^{0.2126}, 0.7654^{0.1675}, 0.9707^{0.1528}, 0.8636^{0.3809}\}; \min\{0.6217^{0.0862}, 0.9451^{0.2126}, 0.6586^{0.1675}, 0.9926^{0.1528}, 0.9640^{0.3809}\}; \min\{0.4296^{0.0862}, 0.9860^{0.2126}, 1^{0.1675}, 0.8880^{0.1528}, 1^{0.3809}\}; \min\{0.5725^{0.0862}, 0.7977^{0.2126}, 0.2584^{0.1675}, 0.6217^{0.1528}, 0.7189^{0.3809}\}; \min\{0.3038^{0.0862}, 1^{0.2126}, 0.4759^{0.1675}, 0.7654^{0.1528}, 0.7578^{0.3809}\}; \min\{0.9208^{0.0862}, 0.8806^{0.2126}, 0.9546^{0.1675}, 0.9130^{0.1528}, 0.6381^{0.3809}\}; \min\{0.1494^{0.0862}, 0.9155^{0.2126}, 0.6217^{0.1675}, 0.8306^{0.1528}, 0.7952^{0.3809}\}; \min\{0.9486^{0.0862}, 0.8653^{0.2126}, 0.4072^{0.1675}, 0.9354^{0.1528}, 0.9208^{0.3809}\}; \min\{1^{0.0862}, 0.9396^{0.2126}, 0.3134^{0.1675}, 1^{0.1528}, 0.6789^{0.3809}\}; \min\{0.3682^{0.0862}, 0.9021^{0.2126}, 0.8306^{0.1675}, 0.6950^{0.1528}, 0.8306^{0.3809}\}\} = \{0.9457, 0.9324, 0.9298, 0.7971, 0.8830, 0.8427, 0.8489, 0.8603, 0.8233, 0.9175\}$.

As a result, the desired solution relative to the estimates of investment project efficiency a_k ($k = 1 \div 10$) is obtained by equality (4). In particular, we have: $\max\{\mu_A(a_j)\} = \max\{0.9457, 0.9324, 0.9298, 0.7971, 0.8830, 0.8427, 0.8489, 0.8603, 0.8233, 0.9175\}$. As one can see, the project a_1 is the most effective with the final score of 0.9457. The rest of the projects are ranked in descending order: a_2 (0.9324), a_3 (0.9298), a_{10} (0.9175), a_5 (0.883), a_8 (0.8603), a_7 (0.8489), a_6 (0.8427), a_9 (0.8233) and a_4 (0.7971).

7 Conclusion

To make balanced decisions on the formation of the optimal investment portfolio, it is advisable to use different approaches, including using weighted estimates obtained on the basis of identifying the specific weights of project performance indicators. In [2], we considered the solution of the problem by both traditional methods of scoring analysis (Pareto and Bord) and fuzzy methods of analysis, including the fuzzy method of maximin convolution of criteria with the same degrees of importance. The results obtained by all assessment methods are summarized in Table 3. The estimates, obtained using fuzzy methods, differ from scoring relative to the best investment project. Significant differences in the ranking of projects dictate the need to harmonize expert assessments obtained at the stage of identifying the indicator weights with the process of forming the implicative rules of the fuzzy inference system. It is obvious that such harmonization can only take place within the framework of the unified expert group.

Table 3. Ranking of investment projects by different methods of analysis

Project	Pareto method	Bord method		Maximin convolution				Fuzzy inference	
				Same importance		Different importance			
	Order	Estimate	Order	Estimate	Order	Estimate	Order	Estimate	Order
a_1	3	32	3	0.7654	1	0.9457	1	0.8682	1
a_2	1	38	1	0.6217	3	0.9324	2	0.8038	3
a_3	2	38	2	0.4296	4	0.9298	3	0.7148	6
a_4	10	11	10	0.2584	9	0.7971	10	0.5763	9
a_5	7	23	8	0.3038	8	0.8830	5	0.6519	7
a_6	6	28	6	0.6381	2	0.8427	8	0.8243	2
a_7	8	21	9	0.1494	10	0.8489	7	0.5355	10
a_8	4	29	5	0.4072	5	0.8603	6	0.7687	5
a_9	5	32	4	0.3134	7	0.8233	9	0.7786	4
a_{10}	9	24	7	0.3682	6	0.9175	4	0.6498	8

References

1. Bystrov, O.F., et al.: Management of Investment Activities in the Regions of the Russian Federation. Publishing House INFRA-M, Moscow (2008)
2. Ali, A.E.: Investment projects portfolio analyses using fuzzy evaluation methods. Adv. Intell. Syst. Comput. **1306**, 685–693 (2021). https://doi.org/10.1007/978-3-030-64058-3_85
3. Andreichenkov, A.V., Andreichenkova, O.N.: Analysis, Synthesis, Planning Decisions in the Economy. Finance and Statistics, Moscow (2000).(in Russian)
4. Zadeh, L.A.: The concept of a linguistic variable and its application to approximate reasoning – I. Inf. Sci. **8**(3), 199–249 (1975). https://doi.org/10.1016/0020-0255(75)90046-8
5. Lin, A.S., Wu, W.: Statistical Tools for Measuring Agreement. Springer, New York (2012)

Investigating of Some Macroeconomic Indicators of the Republic of Azerbaijan in the Post-oil Period by Using Interval Analysis

Revana I. Davudova[1,2](✉)

[1] ANAS Institute of Economics, H. Javid Avenue, 115, AZ1143 Baku, Republic of Azerbaijan
[2] ANAS Institute of Control Systems, B.Vahabzadeh Street, 9, AZ1141 Baku, Republic of Azerbaijan

Abstract. The manuscript is devoted to the analysis of the economy of the Republic of Azerbaijan in the post-oil era on the basis of bilateral and multilateral relations of macroeconomic indicators in balanced and unbalanced open systems. For this purpose, the interrelationships of macroeconomic indicators such as consumption norm, investment norm, GDP structural efficiency and economic growth rate [1, 2] were assessed based on the monthly values of the post-oil period (2015–2020) that was obtained by Delphi survey method. This evaluation was implemented by using interval analysis. In order to realise calculations, software as M-files was created by using MathWoks Matlab 2017b. It was determined that the Azerbaijani economy can be studied as a balanced open economic system in March 2016, March, April and September 2020, and as an unbalanced open economic system in the other period of the research cycle. Based on the conducted analyses, the ideal interval values of the consumption norm, investment norm, and GDP structural efficiency for regulating balanced and unbalanced open perfectly economic systems of the Republic of Azerbaijan were determined.

Keywords: Consumption norm · Investment norm · GDP structural efficiency · Interval analysis · Post-oil period

1 Introduction

Azerbaijani economy can be divided into three stages in the independence period:

- the "transition period" (1991–2004) covers from the declaration of independence to the receipt of income from oil contracts;
- the "big oil revenues period" (2005–2014) covers from the initiating of access of oil revenues to budget, to the devaluation;
- the "post-oil revenues period" covers the post-devaluation period.

R. A. Aliev et al. (Eds.): ICSCCW 2021, LNNS 362, pp. 680–688, 2022.
https://doi.org/10.1007/978-3-030-92127-9_90

While President Ilham Aliyev assessing the fall of oil prices during his speech at a panel discussion on "The New Energy Equation" at the World Economic Forum in Davos, Switzerland on January 21, 2016 Azerbaijan economy was preparing following 20 years for post-oil period. He expressed attitude as "this was unforeseen to us and caused a difficult situation for our economy". This was a signal that Azerbaijan was experiencing a post-oil period.

In an interview with TASS (May 7, 2020), President Ilham Aliyev mentioned that "the sharp decline in oil prices in recent months marks the beginning of the post-oil period in the world economy," noting that his country is already moving towards independently from oil production and prices. He mentioned that industrial production in the non-oil and gas sector increased by 23% in the first quarter of 2020 and emphasized that we have to realize some crucial issues such as to reduce the level of shadow economy for stablishing robust economy, to develop infrastructure, to stimulate fields such as agriculture and tourism, to utilize the potential of information and communication technologies, to create suitable business environment for local and foreign investors, to support small and medium enterprises.

However, from the early 2020, the manifestations of the COVID-19 pandemic began to be observed in all countries, including the largest economies. The Republic of Azerbaijan, as a part of the global world, was not left out of this trouble. Near to the end of 2020, the scale of changes in the dynamics of some macroeconomic indicators of our country became even more noticeable. From this point of view, the analysis of macro indicators of our country is significant.

2 Methodology

This study is focused to the analysis of the dependence of macroeconomic indicators and norms expressing the effectiveness of norms and economic system in a balanced and unbalanced open economy for the Republic of Azerbaijan [2–4]. These built dependencies for macroeconomic indicators were analyzed by using interval analysis [5–8] of macro indicators based on expert estimates determined by the Delphi survey method.

During post-oil period considering the monthly prices in 2015–2020, the economy of the Republic of Azerbaijan was almost balanced ($Ex \approx Im$) in the 2016M3, 2020M3, 2020M4 and 2020M9 months, and in the remaining months it was unbalanced ($Ex \neq Im$). Therefore, some macroeconomic indicators can be assessed for balanced open economy system (BOES) using the mathematical apparatus given in [1–3]. Unbalanced open economy system (UOES) macroindicators can be assessed using the multilatperiodl relationships given in [4].

The estimations were implemented using the following expressions [3, 4] (Table 1), where $b = ex - im$, and $b = 0$ for BOES.

Table 1. Mathematical expressions of the macroeconomic indicators' interrelationships

	c	i	ψ	F
c	—	$c = 1 - 2i - \varepsilon b$	$c = 1 - \frac{1}{\psi} - \varepsilon b$	$c = 1 - 2\sqrt{F}$
i	$i = \frac{1-c-\varepsilon b}{2}$	—	$i = \frac{1}{2\psi}$	$i = \sqrt{F}$
ψ	$\psi = \frac{1-c-\varepsilon b}{2}$	$\psi = \frac{1}{2i}$	—	$\psi = \frac{1}{2\sqrt{F}}$
F	$F = \left[\frac{1-c-\varepsilon b}{2}\right]^2$	$F = i^2$	$F = \frac{1}{4\psi^2}$	—

3 Computational Experiments

The results of the calculations can be analyzed in two directions:

3.1 Perfect Balanced Economy System (PBOES)

The results of calculations using bilatperiodl relations of macroeconomic indicators for the months 2016M3, 2020M3, 2020M4 and 2020M9, when the economy of the Republic of Azerbaijan was a balanced open economy (BOE), are given in Table 2 and Fig. 1. In perfect economy system, being consumption norm ≈0.5, is considered acceptable. According to expert estimates, this norm was in the range of ≈[0.3,0.37] in March 2016, ≈[0.4,0.5] in March and April 2020, and [0.45,0.54] in September of the same year. However, being at the desired level of estimates which calculated on the basis of expert assessment of investment norm and the GDP structural efficiency is the fact of being at the desired level of macro indicators namely investment norm and GDP structural efficiency. However, the undesirable value of the calculated value can be explained by the discrepancy of economic growth rate. It has to be noted that calculated value was estimated by using $c_{perf} = c(F) = 1 - 2\sqrt{F}$ which expresses the dependency of the consumption norm services to the ratio of economic growth rate.

Table 2. Evaluated values of consumption expenditure norm according to expert and various macronorms, for PBOE

	c_{exp}		$c_{perf} = c(i)$		$c_{perf} = c(\psi)$		$c_{perf} = c(F)$	
	Low	Upp	Low	Upp	Low	Upp	Low	Upp
2016M03	0,31	0,37	0,49	0,60	0,51	0,62	−1,34	−1,10
2020M03	0,37	0,44	0,61	0,70	0,42	0,55	−1,07	−0,85
2020M04	0,42	0,50	0,45	0,57	0,16	0,35	−0,79	−0,61
2020M09	0,45	0,54	0,34	0,49	0,24	0,41	−1,15	−0,93

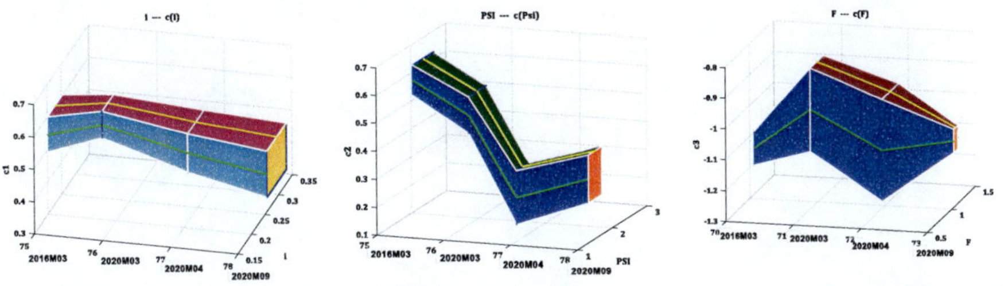

a) $c_{perf} = c(i)$ b) $c_{perf} = c(\psi)$ c) $c_{perf} = c(F)$

Fig. 1. Calculated values of consumption expenditure norm for PBOE depending on various indicators

Computed ratio of the investment norm depending on experts and macro indicators are given in Table 3 and Fig. 2. In perfectly open balanced economy system, 0.25 can be taken as the perfect value of the investment norm. According to the expression $i = \frac{1-c}{2}$, it can be said that macro indicators i and c are inversely related to PBOE, and the slight lower prices of the consumption expenditure norm have led to a slightly higher value of the investment norm. Calculated values of the investment norm based on the GDP structural efficiency demonstrates that the GDP structural efficiency is at a perfect level in 2016M03 and 2020M03 when the economy of the Republic of Azerbaijan is balanced, and it is at the level of ≈[1.8;2.5].

In the PBOES, the assessment based on the dependence $i_{perf} = i(F) = \sqrt{F}$ of the investment norm on the economic growth rate, suggests that a high economic growth rate causes the investment norm unfeasible prices that are too high.

Table 3. Evaluated values of investment norm according to expert and various macro norms, for PBOE

	i_{exp}		$i_{perf} = i(c)$		$i_{perf} = i(\psi)$		$i_{perf} = i(F)$	
	Low	Upp	Low	Upp	Low	Upp	Low	Upp
2016M03	0,20	0,25	0,32	0,35	0,19	0,24	1,05	1,17
2020M03	0,15	0,19	0,28	0,31	0,22	0,29	0,93	1,03
2020M04	0,21	0,27	0,25	0,29	0,32	0,42	0,80	0,90
2020M09	0,26	0,33	0,23	0,27	0,29	0,38	0,97	1,08

Estimations for the PBOES have shown that relatively low consumption expenditure norm and the highest economic growth rates can have undesirable consequences for the GDP structural efficiency. The values of the GDP structural efficiency calculated on the basis of the investment norm, are in the desirable and perfect limit (Table 4 and Fig. 3).

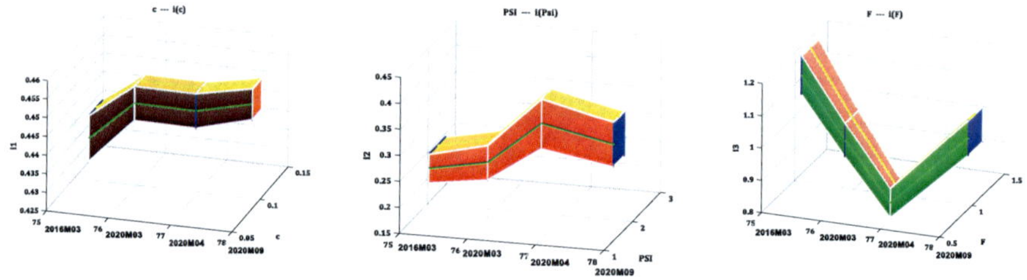

a) $i_{perf} = i(c)$ b) $i_{perf} = i(\psi)$ c) $i_{perf} = i(F)$

Fig. 2. Calculated prices of investment norm for PBOE depending on various indicators

Table 4. Evaluated values of the GDP structural efficiency according to expert and various macro norms, for PBOE

	ψ_{exp}		$\psi_{perf} = \psi(c)$		$\psi_{perf} = \psi(i)$		$\psi_{perf} = \psi(F)$	
	Low	Upp	Low	Upp	Low	Upp	Low	Upp
2016M03	2,06	2,66	1,45	1,59	1,96	2,51	0,43	0,48
2020M03	1,73	2,24	1,59	1,80	2,59	3,31	0,48	0,54
2020M04	1,19	1,54	1,72	2,00	1,82	2,33	0,56	0,62
2020M09	1,32	1,70	1,83	2,18	1,52	1,94	0,46	0,52

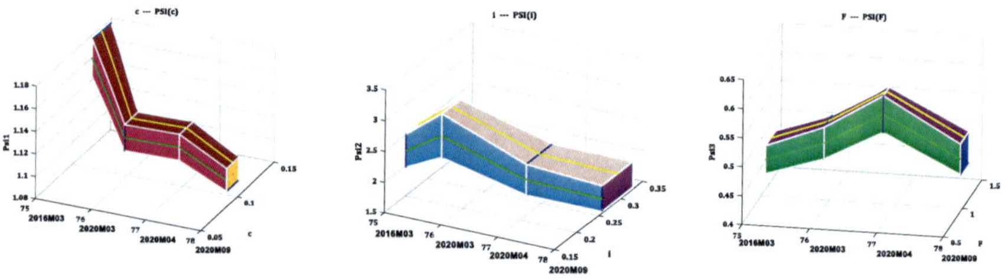

a) $\psi_{perf} = \psi(c)$ a) $\psi_{perf} = \psi(c)$ a) $\psi_{perf} = \psi(c)$

Fig. 3. Calculated values of the GDP structural efficiency for PBOE depending on various indicators

3.2 Perfect Unbalanced Open Economy System (PUOES)

It can be mentioned that the Republic of Azerbaijan had an unbalanced open economy (UOE) in 2015, 2016, 2017, 2018, 2019 and 2020. The perfect unbalanced open economy (PUOE) can be analyzed using multilatperiodl relationships of macro indicators [4]. The results of the calculations made for this purpose, are given at the Tables 5, 6 and 7, and they are expressed visually in Figs. 4, 5 and 6.

Evaluations for PUOE give opportunity to state that the investment norm and the GDP structural efficiency for this period are almost desirable, however the economic growth rate is very high. Thus, according to expert estimates, it can be considered perfect in 2020, according to the investment norm of consumption expenditure norm and the GDP structural efficiency, respectively.

Table 5. Evaluated values of consumption expenditure norm according to expert and various macro norms, for PUES

	c_{exp}		$C_{perf} = c(i)$		$C_{perf} = c(\psi)$		$C_{perf} = c(F)$	
	Low	Upp	Low	Upp	Low	Upp	Low	Upp
2015	0,29	0,52	0,20	0,46	0,16	0,43	−1,17	−0,95
2016	0,27	0,54	0,21	0,47	0,13	0,41	−1,14	−0,92
2017	0,34	0,56	0,29	0,53	0,21	0,48	−1,14	−0,92
2018	0,32	0,58	0,38	0,63	0,25	0,54	−1,12	−0,91
2017	0,35	0,57	0,35	0,59	0,22	0,49	−1,17	−0,95
2018	0,41	0,58	0,28	0,52	0,34	0,58	−0,96	−0,80

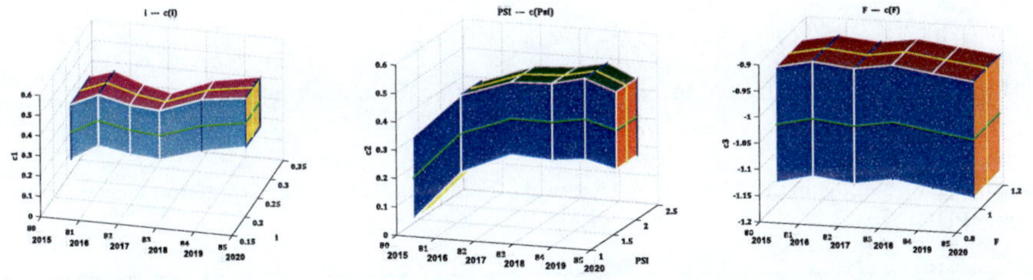

a) $c_{perf} = c(i)$ b) $c_{perf} = c(\psi)$ c) $c_{perf} = c(F)$

Fig. 4. Calculated values of consumption expenditure norm for PUOE depending on various indicators

The analysis of the evaluated values of the investment norm on the basis of dependence on other macro indicators in PUOE demonstrates that the values are relatively perfect that was calculated by expression $i_{perf} = i(c) = \frac{1-c-\varepsilon b}{2}$ and the assessed values can be considered perfect on the basis of the GDP structural efficiency.

Table 6. Evaluated values of consumption expenditure norm according to expert and various macro norms, for PUOES

	i_{exp}		$i_{perf} = i(c)$		$i_{perf} = i(\psi)$		$i_{perf} = i(F)$	
	Low	Upp	Low	Upp	Low	Upp	Low	Upp
2015	0,24	0,30	0,14	0,32	0,25	0,33	0,97	1,09
2016	0,22	0,28	0,11	0,31	0,24	0,31	0,96	1,07
2017	0,17	0,22	0,08	0,27	0,20	0,26	0,96	1,07
2018	0,16	0,21	0,11	0,32	0,21	0,27	0,95	1,06
2017	0,19	0,24	0,13	0,31	0,24	0,31	0,97	1,08
2018	0,24	0,31	0,16	0,29	0,21	0,28	0,90	0,98

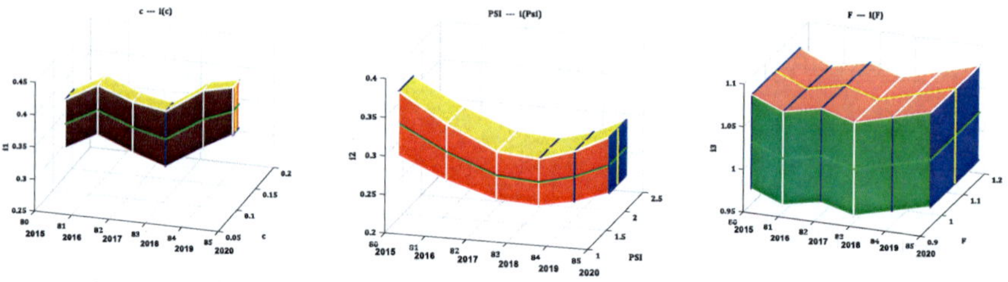

a) $i_{perf} = i(c)$ b) $i_{perf} = i(\psi)$ c) $i_{perf} = i(F)$

Fig. 5. Calculated values of investment norm for PUOE depending on various indicators

Table 7. Evaluated values of the GDP structural efficiency according to expert and various macro norms, PUOES

	ψ_{exp}		$\psi_{perf} = \psi(c)$		$\psi_{perf} = \psi(i)$		$\psi_{perf} = \psi(F)$	
	Low	Upp	Low	Upp	Low	Upp	Low	Upp
2015	1,54	1,99	1,55	3,47	1,65	2,11	0,46	0,51
2016	1,59	2,05	1,59	4,47	1,81	2,31	0,47	0,52
2017	1,95	2,52	1,88	6,02	2,29	2,92	0,47	0,52
2018	1,85	2,39	1,57	4,69	2,41	3,07	0,47	0,52
2017	1,63	2,11	1,62	3,76	2,05	2,62	0,46	0,51
2018	1,79	2,39	1,72	3,11	1,60	2,11	0,51	0,56

The expert assessment of the GDP structural efficiency of an UOES was higher in 2017 $\psi_{exp}\Big|_{2017}^{UOE} = [1.95; 2.52]$. However, taking into account the influence on other macro indicators, its perfect value is $\psi_{perf}\Big|_{2020}^{UOE} = [1.79; 2.39]$.

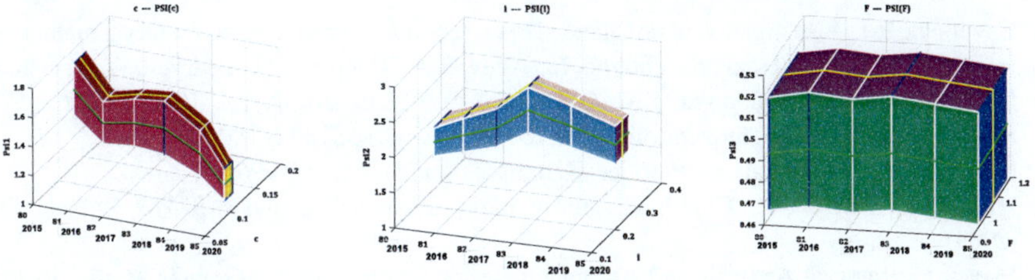

a) $\psi_{perf} = \psi(c)$ b) $\psi_{perf} = \psi(i)$ c) $\psi_{perf} = \psi(F)$

Fig. 6. Calculated values of the GDP structural efficiency for PUOE depending on various indicators

4 Conclusion

The economy of the Republic of Azerbaijan in 2015–2020, covering the post-oil period, was investigated on the basis of expert assessments of indicators by using interval analysis on the basis of bilatperiodl and multilatperiodl intperiodctions of macro indicators. The analysis revealed that economy in the case of balance, that means, *Ex* = *Im*, the perfect interval values of macro indicators can be adjusted by choosing as follows:

$$c_{perf}\left|\begin{matrix} BOE \\ 2020M09 \end{matrix}\right. = [0.45; 0.54],\ i_{perf}\left|\begin{matrix} BOE \\ 2020M04 \end{matrix}\right. = [0.21; 0.27],$$

$$\psi_{perf}\left|\begin{matrix} BOE \\ 2020M03 \end{matrix}\right. = [1.73; 2.24]$$

For PUOES the avperiodge annual perfect values of macroeconomic indicators can be regulated by selecting as follows:

$$c_{perf}\Big|_{2020}^{UOE} = [0.41; 0.58], \qquad i_{perf}\Big|_{2018}^{UOE} = [0.16; 0.21] \ \ və$$

$$\psi_{perf}\Big|_{2020}^{UOE} = [1.79; 2.39].$$

References

1. Vladimirov, S.A.: On the macroeconomic essence of the strategic development objectives of effective, balanced macroeconomy systems. Scientific works of the North-West Institute of Management, branch of RANEPA, **6**, 4(21), 105–116 (2015). (in Russian)
2. Musayev, A.F., Davudova, R.I., Musayeva, A.A.: Application of interval analysis in evaluation of macroeconomic impacts of taxes. In: Aliev, R.A., Kacprzyk, J., Pedrycz, W., Jamshidi, M., Sadikoglu, F.M. (eds.) ICAFS 2018. AISC, vol. 896, pp. 627–634. Springer, Cham (2019). https://doi.org/10.1007/978-3-030-04164-9_82
3. Musayev, A.F., Davudova, R.I., Musayeva, A.A.: Evaluatıon of interrelatıons of macroeconomıc ındıcators by usıng ınterval analysıs. Mag. ANAS News Econ. Ser. **4**, 5–18 (2018). (in Azerbaijani)

4. Davudova, R.I.: Investigation of unbalanced open economy system by using interval mathematics for the case of Azerbaijan Republic. In: Aliev, R.A., Kacprzyk, J., Pedrycz, W., Jamshidi, M., Babanli, M.B., Sadikoglu, F.M. (eds.) ICSCCW 2019. AISC, vol. 1095, pp. 565–572. Springer, Cham (2020). https://doi.org/10.1007/978-3-030-35249-3_72
5. Moore, R.: Interval Analysis. Prentice-Hall, upper Saddle River (1996)
6. Moore, R., Kearfott, R., Cloud, M.: Introduction to Interval Analysis. SIAM, Philadelphia (2009)
7. Mayer, G.: Interval Analysis and Automatic Result Verification. De Gruyter, Berlin, Boston (2017)
8. Alefeld, G., Herzberger, J.: Real interval arithmetic. Introduction to Interval Computation, pp. 1–9. https://doi.org/10.1016/b978-0-08-091636-1.50007-1

Problems of Fuzzy – Multiple Control of the "Smart House" System (Review)

G. Akperov Imran[1(✉)] and Sakharova Lyudmila[2]

[1] Private Educational Institution of Higher Education "Southern University (IMBL)", Rostov-on-Don, Russia
rector@iubip.ru
[2] Rostov State University of Economics, Rostov-on-Don, Russia

Abstract. A study of the available mathematical models of the "Smart Home" system, as a system that controls electrical loads, lighting, security, climate control, and media devices in households, has been carried out. The article presents an overview of the most relevant devices that carry out the technical configuration of the "Smart Home" system for controlling electrical appliances and the environment, including air temperature, humidity levels, light, noise, as well as air quality, security and alarms, cleaning and cleaning systems. It is noted that the control of all these devices and subsystems, as a rule, is based on fuzzy logic, as well as gamification as the use of fascinating elements that are usually present in games, in real or productive actions. An overview of the most relevant modern works devoted to fuzzy-multiple control of heating, energy saving, gates, lighting, watering, alarms, video surveillance is given; special attention is paid to the model of connected thermostats based on gamification, as well as to complex models of the "Smart Home" system. The second part of the review is devoted to the theoretical foundations of modeling the Internet of Things based on fuzzy-set models.

Keywords: Smart home · Fuzzy-multiple control · Internet of Things

1 Introduction

Smart home means a system that controls electrical loads, lighting, security, climate control, media devices. The control of all these devices and subsystems, as a rule, is based on fuzzy logic [1], as well as gamification ([2], Yu-KaiChou) as the use of fascinating elements that are usually present in games, in real or productive actions. The main advantage of the "Smart Home" system is that it simplifies the daily life of a person and increases the comfort of living. One notable example is that lights throughout the house can be turned off thanks to the voice assistant, without touching the switch. Almost limitless options for possible scenarios will allow you to put the home on the alarm, thereby de-energize all the sockets, leaving only a couple of necessary ones, turn on the music, turn off the gas supply, close the curtains, regulate the temperature and leave it with a calm sense of security.

R. A. Aliev et al. (Eds.): ICSCCW 2021, LNNS 362, pp. 689–696, 2022.
https://doi.org/10.1007/978-3-030-92127-9_91

2 Smart Home System and Its Components

Today, there are a number of representatives who provide a complete set of smart home.

Refrigerator Samsung Family Hub 2.0 – helps you manage the rest of your Samsung smart home appliances. The Samsung Connect app, integrated into Family Hub, expands these capabilities: the platform is able to connect and control IoT devices, regardless of whether they run on Android or iOS. Without interrupting cooking or chatting in the kitchen, users can use the Family Hub display to turn on the AddWashTM or instruct the POWERbotTM to start cleaning [3]. Climate technology is an area in which Wi-Fi remote control is gaining ground. For example, Haier's Lightera Premium, Lightera Crystal and Lightera DC-Inverter inverter air conditioners are equipped with this convenient and modern feature as standard. In addition, the Lightera, Elegant, Family and Tibio non-inverter split systems included in the 2017 model range can be retrofitted with a Wi-Fi module as an option [3]. Remote control of the REDMOND SkyWaterHeat 802S smart water heater is performed using the Ready for Sky mobile application. Working on a set schedule, the device will heat the water for you to wake up in the morning and return in the evening after work, or it will keep the water warm throughout the day if you have a day off and you spend time at home [3].

Smart kettle Xiaomi Mi Smart Kettle is an innovative electric kettle with the possibility of smart control via a smartphone via Bluetooth 4.0 connection. The device has a laconic, minimalistic design, a double body and a convenient handle with a button for opening the lid [4]. Xiaomi Mi Water Purifier White – Using reverse osmosis technology, not only pathogens and antibiotics are removed, but the water is also purified from impurities of heavy metals, scale. Thanks to a four-level filtration system and multi-layer purification, the likelihood of secondary pollution is eliminated [4]. Robot vacuum cleaner Xiaomi Xiaowa Robot Vacuum Cleaner Lite is a home assistant who will do the cleaning for you, and in addition, before you return home from work [4]. Smart curtains Xiaomi Aqara Smart Curtain – With just a light press on the dedicated wireless switch (not included), you can open and close the smart curtains. Using your smartphone, according to your rhythm of life, you can independently set the time when the curtains should open or close [4].

Ezviz Husky Air 1080p is a Wi-Fi surveillance camera that you can connect to from your smartphone. It transmits not only the image, but also the sound, and in two directions, so you can observe and talk [5]. Rubetek RS-3210 is an opening sensor that will notify you when a window or door is opened [5]. Chipolo Classic is a keychain tracker that will help you easily find a thing if it is lying around somewhere [5]. Xiaomi Mi LED Smart Bulb - this smart bulb is equipped with its own Wi-Fi module, so it can directly connect to a smartphone without a gateway. You just need to install the Mi Home App [5].

WiFi relay SonoFF World On - WiFi relay for controlling any electrical appliance or device from a smartphone via the Internet [6]. Smart GSM socket SC1-GSM. Allows you to turn on and off the equipment remotely via SMS commands, voice calls or buttons on the case [6]. The Broadlink A1 e-Air smart home control system (sensors) is a device for monitoring the quality of the living environment: air temperature, humidity, light, noise, and air quality [6]. Universal remote control for smart home BroadLink RM Pro, which allows you to control almost all household appliances via Wi-Fi or the Internet, using an Android or IOS device to control: heaters, air conditioners, TVs, audio systems and other devices with an IR remote control [6]. Smart wireless Wifi socket BroadlinkWiFi Smart Socket, which allows you to see the online status, as well as turn on/off any electrical device via the Internet, set timers for turning on/off [6]. Door stop alarm with a siren (signal stop for the door) Door Stop Alarm, which counteracts intruders trying to enter the house by means of an alarm [6]. Autonomous motion sensor with remotes and siren FL1770 detecting movement within a radius of 8 m around the house [6]. WiFi Ceiling SonoFF: lighting control via smartphone from anywhere in the world [6]. Autonomous water leakage sensor (water leakage alarm) [6]. Sapsan gas leakage control system designed for installation in rooms where gas stoves, gas boilers or any other gas equipment are used [6].

3 Management of "Smart Home" and Its Subsystems Based on Fuzzy Logic and Gamification

The control of all these devices and subsystems, as a rule, is based on fuzzy logic. This is due to the fact that models based on fuzzy logic allow us to more fully describe all aspects of uncertainty inherent in processes and things in the world around us. Let us give a brief overview of fuzzy models for the implementation of the Internet of Things using the example of the implementation of the "Smart Home" system. Let us consider the control models described in the literature both for individual units of the "Smart House" and for the entire system as a whole.

3.1 Control of Individual Smart Home Systems

Heating Control. The fuzzy-multiple model proposed in [7], which allows autonomously, without human intervention, to control the level of heat in the house. The system controls a device that changes the temperature and keeps it comfortable for life. To implement the model, the Visual Studio 2017 environment was used using AForge.NET Framework and the A Forge.Fuzzy library. AForge.NET Framework is a C# framework designed for developers in the fields of artificial intelligence, neural networks, genetic algorithms, and machine learning. The framework consists of various libraries designed to solve different problems. In particular, the AForge.Fuzzy library is designed to work with fuzzy computations.

Energy Saving. In [8], the systems of energy efficient control of building heating, based on fuzzy sets and ensuring the maintenance of the required comfortable air temperature in the premises, based on the assessment and processing of data from external and internal conditions, were considered.

Goal. In [9], a system of automatic control of a barrier using intelligent technologies of fuzzy inference and computer vision is proposed. The article is devoted to improving the quality of the vehicle access control and management system based on the technology of vehicle license plate recognition. The factors influencing the reliability of recognition and the quality of the system as a whole are determined. A structural diagram of an intelligent control system and algorithms for the joint functioning of computer vision and fuzzy logic systems are proposed.

Lighting. In [10], actual problems of intelligent control of lighting are considered, which have become possible to solve using modern light-emitting devices, computer technology and fuzzy logic.

Watering. In [11], on the basis of a comparative analysis in the field of automated watering of tea plantations, the task was set to create an intelligent control system for watering tea bushes. A fuzzy algorithm with its implementation on Fuzzy Logic (based on MATLAB) is proposed for modeling the irrigation control process.

Video Surveillance. The system of fuzzy-multiple video surveillance control is considered in [12].

Signaling. In [13], the authors considered the problem of assessing the security of wireless, including automobile, signaling from unauthorized access. The analysis of open sources showed that at present such methods do not exist. The authors of the article propose to use a simple and visual apparatus of fuzzy logic to carry out a quantitative and qualitative assessment of the security of the signaling radio channel from unauthorized access.

Thermostats. In [14], connected thermostats (CTS) are often considered to save less energy than predicted because consumers may not know how to use them and may not be saving energy. In addition, some models perform according to consumer expectations and are thus not used as intended. As a result, CTSs save less energy and are underutilized in households. To save energy, [14] proposes a system of gamification and serious games, adapted by a fuzzy logic system to motivate connected thermostat consumers. This intelligent gamification framework can be used to customize gamification and serious gaming strategy for each consumer so that fuzzy logic systems can be tailored to meet the needs of each consumer.

3.2 Complex Models of the "Smart House" System

The article [15] examines the main aspects of the reliability of the "smart home" system: the presence in the building of a unified control system for all engineering systems; availability of computing power sufficient for decision-making; elimination of all maintenance personnel of the building and transfer of control functions, decision-making and execution of actions to the "intelligent center" and its subsystems; the ability to instantly turn off computer control of a smart home and transfer control functions to a person; own automation, ensuring their safe operation when the central control is turned off; the system must be configured for expansion.

In [16], an algorithm is considered for optimizing information exchange processes in security and monitoring systems of the automated control system, based on the use of a closed exponential model of a queuing network and characterized by the ability to operate with fuzzy sets. Article [17] is devoted to the application of the theory of fuzzy sets in the problem of forming a portfolio of the "Smart House" project. The article [18] argues that smart home scenarios are pre-programmed behavior of the smart home system for a particular event. The work [19] discusses methods of building and controlling an intelligent home based on such functions as: optimization of lighting, heating, protection and control of the house remotely, using mobile devices and special key fobs. Article [20] is devoted to the creation of an integrated application solution for smart home control based on Andurino. In [21], general characteristics of specialized interfaces used in the design of "smart home" systems are given, as well as the possibilities of studying controllers, sensors and actuators working with these interfaces are considered.

In [22], an overview analysis of the current state of engineering systems, controls, platforms and data transfer protocols of home automation systems is given; the direction of development of home automation systems is determined by the introduction of speech control technologies and the use of the Z-Wave wireless data transfer protocol. In [23], various research papers on a wide range of energy management methods for smart homes are considered, aimed at reducing energy consumption and minimizing energy losses. The idea of a smart home is being developed, followed by a review of existing energy management methods, which, as a rule, work on the basis of fuzzy logic, heuristic methods, neural networks and evolutionary algorithms.

In [24], general principles for the implementation of a software and hardware complex for intelligent control of engineering equipment in the "smart home" system are proposed. The principles are in the use of standard equipment connected by open data transfer protocols, in the allocation of three levels of control (physical, logical and intellectual) and the use of a distributed knowledge base that allows remote control.The article [25] analyzes the loads generated by the consumers of the "smart home", considers the equipment for the implementation of the power supply control system for the "smart home", provides an analysis of possible circuit solutions for the power supply of the "smart home".

4 Theoretical Foundations of Modeling the Internet of Things Based on Fuzzy-Set Models

In [26], the process of designing complex technical systems (CTS) is considered, which is an iterative process and which is characterized by significant expenditures of resources (financial, labor, time), execution of a large amount of documentation, as well as many emerging risk situations. In [27], a methodology and an example of using the capabilities of the frame model with the use of the apparatus of fuzzy logic in procedures-daemons of frames is given. This technique can serve as a basis for the development of information support for an automated monitoring system. In [28], mathematical models and methods of multicriteria selection of the best solutions with fuzzy initial information for the "Smart House" system are considered. A method of representing an indistinctly described situation by a representative vector is given, which makes it possible to reduce the complexity of finding a standard for a given class of fuzzy situations, an example of finding a reference situation is considered.

In [29], algorithms for controlling electromechanical systems using the theory of fuzzy logic are considered, the main provisions of their synthesis are given, methods for analyzing their stability based on fuzzy Lyapunov functions are considered. The main methods of forming a logical solution used in the design of various types of regulators with fuzzy logic, proposed by Zade, Mamdani, Takagi, Sugeno and Mendel, are described; a typical block diagram of such regulators is given in general form.

In [30], the results of interaction of the classical and fuzzy regulators of an automated control system are presented. It is based on a classic PID controller and a fuzzy controller, the rule base of which can be generated in an automatic mode. A standard Mamdani fuzzy inference algorithm is considered, on the basis of which a hybrid algorithm for forming a base of rules of a fuzzy controller is proposed, which allows obtaining the desired control depending on the degree of complexity of the problems being solved. The key difference of the proposed hybrid algorithm is the change in the main stages of fuzzy inference, implemented in the standard Mamdani algorithm, in particular, the first stage is the fuzzification stage, the second is the stage of forming the base of the rules of the fuzzy regulator. In [31], an algorithm for tuning systems of the Mamdani type is proposed, which uses the principle of a proportional-integral controller with a limited integral component to adjust the rules of fuzzy inference. Since the classical methods of the theory of automatic control work well for a completely deterministic control object and a deterministic environment, the materials of this work [32] propose fuzzy control methods for systems with incomplete a priori information and high complexity of the control object. The block diagram of an automatic control system based on fuzzy logic is considered.

The goal and objectives of work [33] are to develop a fuzzy model for managing production facilities operating under conditions of a priori uncertainty, followed by an assessment of its accuracy. Article [34] is devoted to the consideration of the application of fuzzy logic controller in control systems and the analysis of existing research methods in the field of intelligent control technologies for solving the problems of adapting the applied models and algorithms to various objects and systems, in particular to systems for maintaining the microclimate parameters of the building's life support environment, and also determining the basic parameters for increasing the economic efficiency of using a fuzzy logic controller in a control system.

The work [35] describes the use of fuzzy control in various control systems for the parameters of technological processes, in particular, in systems where the objects of regulation are rooms in which it is necessary to regulate the microclimate parameters.

In the article [36], a system of automated control of modes of maintaining the desired climate in residential premises is considered. Comfortable conditions are mainly determined by the indoor air temperature. A fuzzy controller is proposed for efficient temperature control (heating, ventilation and air conditioning). Fuzzy logic air conditioning control has an advantage over traditional PID control on three main criteria, namely robustness, response speed and energy savings. In [37], a model of the ventilation system is considered, taking into account the unevenness of the heat load and the gas composition of the air. Methods for the development of intelligent control systems for electric drives of ventilation and air conditioning systems are proposed. As a result of work [38], transient processes were obtained, for which quality indicators were determined. Comparison and analysis of the results obtained made it possible to draw a conclusion about the advantages of one regulator over another.

5 Conclusion

At the present time, fuzzy logic has a great influence on our everyday life, simplifying it and making ordinary everyday things much more pleasant and convenient, so that today can be called the future, because it contains such systems and adaptations that simplify our life. But still, not everything is as simple as it seemed at first glance, all these devices and systems depend on their manufacturer, and if you have one application on a smartphone from each manufacturer, then over time the phone's memory will be crammed with various applications for a "smart home", therefore the best solution is to adjust one smart home interface with which the user can customize (change) and satisfy his needs and demands in one equipped system, which is equipped with a set of smart devices.

References

1. https://ru.wikipedia.org/wiki/Fuzzy_set – Fuzzy logic and fuzzy sets and their father Lutfi Zadeh
2. Chou, Y.-K.: Actionable gamification: beyond points, badges and leaderboards. Octalysis Media: Fremont. CA. Rev. Int. Organizaciones, (18), 137 (2016). https://doi.org/10.17345/rio18.137-144
3. http://irvispress.ru/catalog/v-dome-i-sadu/umnyj-dom/smart-home-smart-tech/
4. https://gooosha.ru/umnaya-kuxnya-top-15-bytovyx-priborov-kompanii-xiaomi-2018-goda/
5. https://www.mvideo.ru/obzor-gadzhetov-dlya-umnogo-doma
6. https://smartdomus.ru/umnyj-dom/umnaya-dacha.html
7. Kristalinsky, V.R., Golubentsov, A.A.: Fuzzy model of control control in a smart house. Fuzzy model of control of a control system in a smart home, pp. 47–52 (2015)
8. Ⱪuanyshbaeva, T.: Development and research of a system of intelligent control of energy supply of a building, P. 88 (2020)
9. Amosov, O.S., Baena, S.G., Ivanov, Y.S., Soe H.: Automatic barrier control system using fuzzy inference technologies and computer vision. Sci. Sci. Internet J. **9**(1) (2017)
10. Avanyan T.G., Zubov, D.V., Paramonov, A.: Actual problems of intelligent lighting control. Izvestia MGTU "Mami" **4**(22), 3, 57–58 (2014)
11. Mammadov, J.F., Narimanova, R.O.: Development of intellectual irrigation system for growing tea plantation. Vestnik Tambovskogo Gosudarstvennogo Tehnicheskogo Universiteta **26**(2), 234–242 (2020)
12. Bukharin, S.V., Melnikov, A.V., Chernikov, D.N.: Preliminary clustering of video surveillance equipment based on nonlinear and fuzzy multiple quality indicators. Bull. VI Ministry Internal Aff. Russ. **3**, 23–34 (2018)
13. Gavrishev, A.A., Burmistrov, V.A, Osipov, D.L.: Assessment of the security of wireless signaling from unauthorized access based on the concepts of fuzzy logic. Appl. Inf. **10. 4**(58), 62–69 (2015)
14. Ponce, P., Meier, A., Méndez, J.I.: Lawrence Berkeley national laboratory. J. Cleaner Prod.40 (2020)
15. Rychkova, V.A.: Reliability of the "smart house" system as the main criterion for the efficiency of the functioning of an object. Bull. Sci. Educ. **4–2**(58), 31–34 (2019)
16. Eremenko, V.T.: Informatics, Computer Science and Management, 146–151 (2017)
17. Application of the theory of fuzzy sets to the problem of forming a portfolio of projects
18. Berchuk, D.Y., Zhuravlev, D.V.: Application of fuzzy logic to control scenarios of the Smart House system. 254–255 (2016)

19. Averin A.I.: Intellectual house control. Smart House
20. Kuznetsov, I.M.: Iot and Smart Home control systems, 5
21. Shishov, O.V., Bobrov, M.A.: Specialized interfaces in the "Smart House" systems, 8
22. Alimuradov, A.K., Churakov, P.P., Kvitk, Y.S.: Voice control device for home automation system Z-Wave. 48–57 (2014)
23. https://Swsu.Ru/Sbornik-Statey/Metody-Regulirovaniya.Php
24. Baigozin, D.V., Pervukhin D.N., Zakharova G.B.: Development of the principles of intelligent management of engineering equipment in the "Smart House" system. 168–172 (2008)
25. Karnitsky, V.Y., Ershov, S.V., Ryumov, A.Y.: Features of power supply of the "Smart House" system. 134–140 (2017)
26. Kovalev, C.M., Kolodenkova A.E.: Building a knowledge base of an intelligent system for monitoring and preventing risk situations for the design stage of complex technical systems. 398–409 (2017)
27. Ermilov, A.E., Misevich P.V.: The use of a frame model and fuzzy logic as the basis for constructing tools for automated monitoring systems. 71–76 (2015)
28. Glushan, V.M., Karelin, V.P., Kuzmenko, O.L.: Fuzzy models and methods of multicriteria choice in intelligent decision support systems. 106–113 (2018)
29. Stzheletskia, R., Demidova, G.L., Lukichev, D.V., Polyakov, N.A., Abdullin, A.A., Lovlin, S.Y.. Control algorithms for electromechanical objects using controllers based on fuzzy logic. 14 (2019)
30. Ignatiev, V.V., Spiridonov, O.B.: Hybrid algorithm for forming the base of rules of a fuzzy controller. 177–185 (2018)
31. Golosovsky, M.S., Bogomolov A.V., Terebov D.S., Yevtushenko E.V.: Algorithm for tuning a fuzzy inference system of the Mamdani type. 19–29 (2018)
32. Finaev: Prospects for solving control problems under uncertainty. 79–86
33. Finaev, E.D., Sinyavskaya, I.V.: Fur. Fuzzy model of temperature control in the baking chamber. 149–159
34. Shilkina, S.V., Fokina, E.N.: The controller of fuzzy logic in the management of technological processes. Vestn. SibADI **15**(1), 106–114 (2018)
35. Yuran, S.I., Vershinin, M.N.: Improvement of the microclimate control system based on fuzzy logic Bulletin of Ngiei, **9**(100), 33–45 (2019)
36. Bulgakov, A.G., JuburiIssam, A.I., Ali, M.: Fuzzy control system of microclimate of premises of buildings. 31–37 (2009)
37. Blagodarov, A.A. Kostin, Y.M., Safonov, A.S.: Tarasov Intelligent control methods in ventilation and air conditioning systems. 4 (2013)
38. Nadezhdin, I.S., Goryunov, A.G.: Control systems for a non-stationary object based on the MRS-controller and a pid-controller with fuzzy logic. 50–75

A Fuzzy Logic Approach for Evaluating the Effectiveness of the Use of Information Technology in an Industrial Enterprise

Nusrat Muradov(✉)

UNEC, Baku, Azerbaijan

Abstract. One of the main tasks and directions of the "Strategic Road Map" of the Republic of Azerbaijan is the application of modern information technology (IT) to increase economic efficiency in industrial enterprises of the country. The choice of this direction is the active use of all sectors of the economy, including industry. Creates conditions for intensive development. The use of information technology in industrial enterprises increases their competitiveness, expands the opportunities to integrate their relations into the international economy, as well as improves the management system of social and economic processes at the macro and microeconomic levels. However, there are a number of internal reasons, including the weak legal framework, the unwillingness of many industries to join effective management technologies, as well as ICT, the effective market for goods and services, including a single information infrastructure and e-commerce. Lack of information support: low level of training in the field of creation and use of ICT: obstacles due to shortcomings in the regulation of economic activity when exporting industrial products to domestic and foreign markets: High monopoly within the network service leads to negative consequences. To overcome this problem, it is advisable to approach the assessment of the effectiveness of the use of information technology in an industrial enterprise with fuzzy logic.

Keywords: İnformation · Technology · Economy · IT · Fuzzy logic

1 Introduction

An integral approach to assessing the effectiveness of the use of IT in an industrial enterprise has been implemented in the following areas:

- calculation of weights with the help of Fishber school;
- calculation of differentiated indicators for individual groups;
- finding a generalized integral indicator for each individual group;
- determining the efficiency achieved in the use of information technology.

Recall that all calculations are for the mining industry, processing industry, energy sector, water and sewage treatment plants.

R. A. Aliev et al. (Eds.): ICSCCW 2021, LNNS 362, pp. 697–703, 2022.
https://doi.org/10.1007/978-3-030-92127-9_92

At the same time, information-analytical system (IAS), integration indicators of the effectiveness of a set of measures to ensure information security, etc.

We will first consider the essence of the above approaches [1].

In general, an information and analytical system (IMS) can be presented as a model for estimating the total value of property: [1]

$$TCO = \sum inv_k \tag{1}$$

TCO, where the IAS is common property costs; Group k inversion costs, which include: capital (purchase of equipment, integrated and applied software); management and ownership (internal and external activities); with hardware and software technical support); development of an application program by a firm or an external partner; about the application of the system [2].

Evaluating the effectiveness of IAS implementation is a system of quantitative and qualitative indicators. Quantitative indicators are defined as follows: income growth; reduction of production cycle; cost reduction; increase the efficiency of resource use; increase the level of customer service, reduce production defects, etc.

In addition to the measured indicators, quality effects are also observed in the implementation of the system: increasing the company's investment attractiveness; increase of organizational culture; formation of an IP, increasing the value of IT, etc.

Expert assessments can be used to assess quality indicators [2].

The analysis of the effectiveness of the IAS implementation project should be carried out by calculating the discounted DP, which allows to make a rational decision on the application of IP. A number of adjustments need to be made to the well-known methodology for calculating development indicators: as an investment, the total capital costs of the IMS and the timely delivery of the TCO to the project are taken into account. To assess the effectiveness of investment projects from these positions, the following discount indicators are the criteria for the integrated evaluation of IT projects: investment return index, internal rate of return, discounted return on investment [3].

An important step in this work is to use a single indicator of the effectiveness of a number of measures to ensure information security. To do this, we must assume that S is a significant number of security services for firm; N - number of security measures that can be applied to ensure IS; such as a1, a2,… – the importance of a separate security service.

This method allows the calculation of the general conditions of pre-reading/negligence of factors relative to each other in the weight of Fishburn and ensures the fulfillment of the following conditions:

For ∀j∈ {1,2, ... S}: a_j∈ [0,1] və Σ_ (j = 1) = 1

We show the process data in more detail. In particular, we describe the technical efficiency E_{ij}^{t}, which reflects a value that provides a j-th security service (I = 1,…, N; j = 1) at the required level of the i-th set of measures during this period…. S).

As a result of computational experience, the E_{ij}^{t} account can be obtained. According to this model, sets of protective measures weaken or eliminate IP vulnerabilities, which leads to a reduction in the likelihood of an attack, which in turn has a positive effect on the state of IS services [3].

After finding E_{ij}^{t} by solving the problem of choosing the optimal option, it is expedient to use an additional part of TE calculations for the i-th package of measures, which can be written in the following form:

$$E_{ij}^{t} = \sum\nolimits_{j=1}^{s} a_j E_{ij}^{t} \quad (2)$$

Such a choice of the FC criterion would be positive if the requirements for a minimum level of each of the important IB services are met. The set of restrictions that reflects these requirements, in fact, determines the number of acceptable options, among which it is necessary to find the optimal one [4].

By determining the TE indicators for all possible sets of PU security, it is necessary to determine the most optimal complex selection criteria from the EE point of view.

Adaptation of industrial production to market changes, as well as the formation of a synergistic effect in the form of progressive growth of the most important parameters is possible only through the introduction of innovations in the field of IT. However, innovative activities can only be carried out with sufficient financial resources and favorable conditions [4].

In modern conditions, investment opportunities are limited for individual firms and the state as a whole. Therefore, effective innovative activity in the industrial sector of the economy is under the force of large, multi-faceted vertically integrated combinations of holding structures. At the same time, such associations confirm their ability to create the necessary conditions for a synergistic impact in the implementation of investment activities.

Therefore, the actual creation of a quantitative assessment of the effectiveness of investment indicators of vertically integrated industrial structures in innovation investments [5].

In this regard, experience shows that the level of innovation efficiency of innovation activity (EINA) can generally be calculated on the basis of a model that determines the profitability of a firm in innovative products:

$$EInA = \frac{P_{\partial on}}{Z_{UH}} \quad (3)$$

Here, P is the additional income generated by the introduction of innovative products.

As can be seen, the features of organizational and management innovations allow us to evaluate EE very accurately. The types of importance of the firm are so diverse that indirectly allowing us to conduct a comprehensive assessment of the results of innovative changes in the SED database, which provides a single information space within the firm as an element of innovation, indirectly providing storage and retention of certain documents in AOC effect (see Table 1).

Table 1. Types of efficiency from the application of innovations based on EDS and the possibility of their evaluation.

Types of efficiency	Impact, appropriate indicator effectiveness		Quantitative evaluation
	Open/it is not clear	Directly/indirectly	
Personnel productivity	It is clear	Indirectly	Possible
Management costs ¬ Nosta	It is clear	Indirectly	Possible
Profit from management activities	It is clear	Indirectly	Possible
ESA control system	It is clear	Indirectly	Possible
Total profitability	It is clear	Indirectly	Possible
Business activity	It is clear	Indirectly	Possible
Management decision-making speed	Covered	Indirectly	Difficult
The effectiveness of ICT	Covered	Indirectly	Difficult

There is no universal formula for evaluating the effectiveness of new IT and organizational and management innovations based on them. Valuation procedures have both direct and indirect effects on income and expenses, respectively.

The economic sense is defined in many senses, even in the economic sense. Similar to the concept of "influence", there is a broad and narrow approach to the definition of its essence:

- achieve certain results at the lowest possible cost;
- purchase of the maximum volume of products from a certain number of resources;
- distribution of available resources between branches, as it is impossible to increase the production of any goods without interrupting the production of other products.

These approaches do not apply to forecasting, so it is necessary to use a more objective meaning of this concept. Efficiency refers to the relative impact of a process or robbery, defined as the ratio of costs to the cost of delivery of the receipt [5].

It should be noted that the implementation of CRM through IT is not short-lived. In particular, this is due to the lack of an appropriate system of measures as a result of the application of the OIE.

Thus, in order to assess the effectiveness of the use of information technology in an industrial enterprise, the opinions of experts were studied and modeled in the fuzzy section of the Matlab program.

2 Application of Fuzzy Logic

Fuzzy logic, fuzzy or fuzzy logic is a logic structure that was formed as a result of an article published by Lütfü Aliasker Zade in 1965 [6–9].

Fuzzy logic is based on fuzzy sets and subsets. In the classical approach, an object is either a member of the set or it is not [9, 10]. When expressed mathematically, it takes the value "1" if the object is a member of the set in terms of its member relationship with the set, and "0" if it is not a member of the set. Fuzzy logic is an extension of classical set theory [11–13]. In a fuzzy set, each object has a degree of membership. The membership degree of the object can be any value in the range (0, 1) and is represented by the membership function M(x) [13–15].

In this study, an assessment was made between the traditional method and the man-made method based on the Sugeno-type Fuzzy Inference System, and as a result, the advantage of the proposed model was chosen. In order to solve the problem, the MATLAB program developed a derivative system based on fuzzy logic with 3 inputs – 1 output type and 18 conditional constraint rules (Table 2).

Table 2. Expert opinions

Place	Experts		
	Expert1	Expert2	Expert3
Mining industry enterprises	Medium	Good	Important
Production facilities	Good	Medium	Useless
Energy sector enterprises	Useless	Medium	Effective
Water and waste treatment facilities	Very Good	Excellent	Very Good

The input parameters are:

1) "Expert1",
2) "Expert2"
3) "Expert3",

The output parameter is unique: Value.

Figure 1 shows the FIS-Editor editing window of the QSTS development phase.

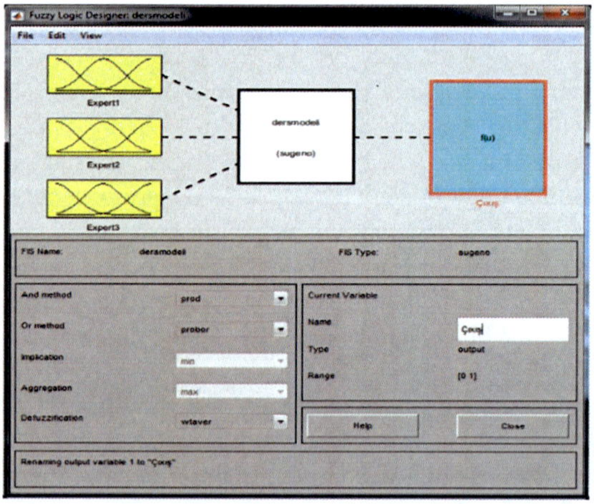

Fig. 1. FIS-Editor editing window

The following triangular fuzzy numbers are used for criteria evaluations:

Useless = Δ (0 0 15); Middle = Δ (0 15 30); Important = Δ (15 30 45); Effective = Δ (30 45 60); Good = Δ (45 60 75); Very Good = Δ (60 75 90); Excellent = Δ (75 100 100).

In order to achieve the output, the system needs to be trained according to the input data. Membership function parameter training should be included to obtain training information. The training process is stopped when the maximum epoch number is reached or the training error level is reached.

Thus, the following results were obtained with fuzzy logic (see: Table 3):

– for extractive industry enterprises – 0.810;
– for production enterprises – 0.376;
– for enterprises in the energy sector – 0.651;
– for water and sewage treatment plants – 0.549.

Table 3. Phase III. Evaluation of the effectiveness of information technology research in industrial enterprises

Industrial enterprises	Find the generalized integral index for each group (Y)		
	Integral indicator (Yi)	Number of group indicators (d)	Number of experts (weight ratio) (sn)
Mining industry enterprises	0,810	4,049	0.2
Production facilities	0376	2,114	0.178
Energy sector enterprises	0.651	4.20	0.155
Water and waste treatment facilities	0.549	2.892	0.190

As can be seen, the closest integration processes in the field of information technology are in the enterprise industry (0.810), which is explained by the use of the most modern innovative technologies along with the technological processes of the extractive industry.

References

1. Kipervar, E.A.: Conference: International Conference on Finance, Entrepreneurship and Technologies in Digital Economy. (2021). https://doi.org/10.15405/epsbs.2021.03.23
2. Golik, V.I., Dmitrak, Y.V., Gabaraev, O.Z., Kozhiev, K.K.: Innovative technologies at the uranium industry enterprises. News Tula State Univ. Sci. Earth **3**(1), 131–140 (2020). https://doi.org/10.46689/2218-5194-2020-3-1-131-140
3. Karamyshev, A.N., Isavnin, A.G., Zaytseva, Z.I.: Technology management of large industrial enterprises. Helix **9**(5), 5375–5378 (2019). https://doi.org/10.29042/2019-5375-5378
4. Kotliarova, V.H., Bratishko, Y.S.: Metrological management at an industrial enterprise. Bus. Inform **3**(518), 212–219 (2021). https://doi.org/10.32983/2222-4459-2021-3-212-219
5. Rakhisheva, A.B., Shaimagambetova, A.C., Ostapenko, E.I., Abylkasova, Z.B.: Management Of industrial enterprises. National Acad. Sci. Republic Kazakhstan **3**(325), 161–165 (2019). https://doi.org/10.32014/2019.2224-5294.107
6. Aliev, R.A., Fazlollahi, B., Aliev, R.R.: Soft Computing and its Applications in Business and Economics. Springer-Verlag, Berlin Heidelberg (2004)
7. Imamguluyev, R.: Determination of correct lighting based on fuzzy logic model to reduce electricity in the workplace. In: International Conference on Eurasian Economies (2020). https://doi.org/10.36880/c12.02456
8. Imamguluyev, R.: Optimal solution of indoor lighting system based on fuzzy logic. In: 2nd International Baku Conference on Scientific Research. Baku, Azerbaijan (2021)
9. Valiyev, A., Imamguluyev, R., Ilkin, G.: Application of fuzzy logic model for daylight evaluation in computer aided interior design areas. In: Aliev, R.A., Kacprzyk, J., Pedrycz, W., Jamshidi, M.O., Babanli, M., Sadikoglu, F.M. (eds.) ICAFS 2020. AISC, vol. 1306, pp. 715–722. Springer, Cham (2021). https://doi.org/10.1007/978-3-030-64058-3_89
10. Aliev, R., Tserkovny, A.: Fuzzy logic for incidence geometry. In: Kosheleva, O., Shary, S.P., Xiang, G., Zapatrin, R. (eds.) Beyond Traditional Probabilistic Data Processing Techniques: Interval, Fuzzy etc. Methods and Their Applications. SCI, vol. 835, pp. 49–85. Springer, Cham (2020). https://doi.org/10.1007/978-3-030-31041-7_4
11. Imamguluyev, R., Suleymanli, T., Umarova, N.: Evaluation of the effectiveness of integration processes in production enterprises based on the fuzzy logic model. In: Aliev, R.A., Kacprzyk, J., Pedrycz, W., Jamshidi, M.O., Babanli, M., Sadikoglu, F.M. (eds.) ICAFS 2020. AISC, vol. 1306, pp. 133–139. Springer, Cham (2021). https://doi.org/10.1007/978-3-030-64058-3_17
12. Imamguluyev, R.: Application of fuzzy logic model for correct lighting in computer aided interior design areas. In: Kahraman, C., Cevik Onar, S., Oztaysi, B., Sari, I.U., Cebi, S., Tolga, A.C. (eds.) INFUS 2020. AISC, vol. 1197, pp. 1644–1651. Springer, Cham (2021). https://doi.org/10.1007/978-3-030-51156-2_192
13. Zadeh, L.A., Aliev, R.A.: Fuzzy logic theory and applications (2018). https://doi.org/10.1142/10936
14. Zadeh, L.A.: Toward a theory of fuzzy information granulation and its centrality in human reasoning and fuzzy logic. Fuzzy Sets Syst. **90**(2), 111–127 (1997). https://doi.org/10.1016/s0165-0114(97)00077-8
15. Kilic, K., Abdullayev, T., Alakbarov, R., Kilic, N.: Processing of fuzzy queries and software implementation to a relational database of wholesale and retail commercial enterprises. Procedia Comput. Sci. **102**, 490–494 (2016). https://doi.org/10.1016/j.procs.2016.09.432

Development of Methods for Processing Acoustic Emission Signals of Sensors for the Compressor-Pump Station's Control

Valeh I. Bakhshali(✉), Nail T. Mardanov, Aygun A. Bekirova, and Ismayil A. Ismayil

Azerbaijan Technical University, H. Javid Avenue, 25, AZ1073 Baku, Azerbaijan
{v.bakhshali,ismayil_i}@aztu.edu.az

Abstract. The durable and reliable work of compressor machines and their technical security have great practical importance in reducing the environmental impact of hydrocarbon transport. Investigation and control of the vibration processes of these machines by sensors are current problems for oil and gas transportation. Piston and centrifugal machines are heavy-duty systems that are still failure-prone due to high stresses and wear proses. Specialized methods will be developed for the influence of impacts, vibrations, temperature, film lubrication and unified system model. The research efforts will lead to more reliable and efficient design alternatives for reciprocating machines which are used for the transport of oil and gas and will thus contribute directly to the competitiveness of a key industry in the Caspian Sea - Black Sea Region. The durable and reliable work of compressor-pump stations used in the oil and gas industry has great importance for practice. The article investigated the main causes of failures through the collection and analysis of data from different parts of the machines are used for transportation the oil and gas resources using sensors. The need for planned technical diagnostics of aggregates was sufficiently justified. The main causes of malfunctions of these machines and their causes were identified. It is noted that the statistical unbalance is characterized by a high amplitude of vibration signals. Finally, we explain how we can determine the existence of a statistical unbalance using graphs constructed using these vibration signals.

Keywords: Dynamics · Compressor · Vibration · Sensors · Diagnosis · Reliability · Stability

1 Introduction

Different heavy loaded machines and equipment (compressors and pumps) are used in the pipelines used for transportation of oil and gas resources. The durable and reliable work of machines and their technical security have great importance for reducing the environmental impact of hydrocarbon transport and exploitation of oil and gas resources. Piston and centrifugal machines are heavy-duty systems that are still failure-prone due

R. A. Aliev et al. (Eds.): ICSCCW 2021, LNNS 362, pp. 704–710, 2022.
https://doi.org/10.1007/978-3-030-92127-9_93

to high stresses and wear proses. Specialized methods will be developed for the influence of impacts, vibrations, temperature, and film lubrication and unified in a system model. The research efforts will lead to more reliable and efficient design alternatives for reciprocating machines which are used for the transport of oil and gas and will thus contribute directly to the competitively of a key industry in the Caspian Sea - Black Sea Region, for example in Baku-Tbilisi-Ceyhan (BTC) Main Oil Pipeline, there are 8 pump-compressor stations (2 stations in Azerbaijan, 2 stations in Georgia and 4 stations in Turkey territory). Besides BTC today the region is functioning the Trans Adriatic Pipeline (TAP) which transported Caspian natural gas to Europe (Fig. 1). Connecting with the Trans Anatolian Pipeline (TANAP) at the Greek-Turkish border, TAP will cross Northern Greece, Albania, and the Adriatic Sea before coming ashore in Southern Italy to connect to the Italian natural gas network.

Fig. 1. The pipelines for transport of oil and gas in the Caspian Sea - Black Sea Region.

The energy security of the Black Sea - Caspian Sea Region depends on the technical conditions -reliability and durability of the machines and equipment used in oil and gas pipelines in Region. The overall goal of project is development of integrated model of the piston machine system that is than used to evaluate and improve current design and increase the environment conditions. Providing the technical security of oil and gas pipelines is one of the very important problems of exploitation and transport in oil and gas industry.

2 The Mechanical Characteristics Analysis of the Compressor-Pump Stations

It is well known that widely used piston and centrifugal machines for the exploitation and transportation of oil and gas resources are widely used. Investigation and control of the vibration processes of these machines by sensors are current problems for oil and gas transportation. Therefore, in this research, we consider both piston and centrifugal pumps and compressors which are important for practice.

Compressor-pump stations used on the oil and gas pipeline for transport of these mineral resource. Current methods of reciprocating and centrifugal machinery design, see e.g. [1, 2], are based on partial empirical relations and do not take into account many of the factors that are affecting the mechanical and thermodynamic processes in these machines. To date, no applied formulas are available for durability and reliability of the parts, for mechanical losses in the bearings, for energy losses in the kinematical couples of machines used in the oil industry, nor has there been basic research of the impact phenomena in these kinematic couples. On the other hand, very detailed models leading to computer intensive simulations have recently been developed for piston ring dynamics in the context of blow-by estimation [2], for lubrication conditions of the ring/liner contact [3], for dynamically loaded bearings and for stress and failure analysis of assembly parts. However, until now, the insight gained with these models have not found the way into more detailed models for reciprocating machines such as piston and centrifugal compressors. To this end, a modular system modelling approach as described below combined with model reduction is necessary in order to keep the computational efforts in reasonable magnitudes.

Thus, it is required to ensure efficient operation of machines to reduce the substantial losses and possible downtime that may be caused by these reasons. Analysis of the literature shows that the method of measuring parameters of parametric data was used when conducting technical monitoring and signal processing to analyze vibration signals from oscillating machines, etc. [4, 5]. The ultimate goal of the Structural Health Monitoring (Structural Health Monitoring – SHM) system using sensors is to improve the decision-making process in order to ensure reliable protection of the structure in case of failure [6]. The difficulty is associated with an increase in the number of machine parts and their functionality, which hampers and complicates the procedure for monitoring the technical condition. In the assessment procedures of the situation in order to reduce the complexity in recent years, a new method of monitoring the technical condition has been developed, based on a data-driven approach (data-driven methods) [7, 8]. It is known that there are many causes of damage to the units. They are: Misalignment; Unbalance, mechanical looseness; eccentric rotor; curved shaft; resonance; rolling element bearing defects; gears defects; blades and vanes; pumps and compressors, related problems; electrical problems and etc. [9]. Some of these cases, such as misalignment, unbalance, curved shaft, resonance, eccentric rotor, etc., can cause a statistical unbalance [10–12]. Approaching the speed of rotation of the shaft to a critical value is appeared by an increase in the vibration of the shaft. Most shafts and axles work in the subcritical region. To reduce the danger of resonance, the stiffness of the shafts increases, and the frequency of their rotation decreases $n \leq 0{,}7n_{cr}$. To increase the bending stiffness of the shafts and axles, it is recommended to place the parts on them as close as possible to the stops. The criterion for the rigidity of the shaft are the conditions for proper operation of gears and bearings, as well as vibration resistance. Thus, the main criterion for the effectiveness of shafts and axles is their strength and rigidity. For the shaft or axle to work, especially for long shafts it is necessary to ensure: bending stiffness, torsional stiffness.

3 Analytical Analysis of the Transverse Oscillations of Vibrating Parts of Machines

Designed elements, taking into account the provision of static or fatigue strength, sometimes fail due to insufficient rigidity or vibration. The issue of the stability of the shaft of rotating machines is important for its normal operation. The differential equation of transverse oscillations of the machines under the action of general resultant forces N acting on their vibration elements (varying in size and direction depending on time), has the form:

$$\frac{\partial^4 y}{\partial x^4} - \frac{N}{EJ} \cdot \frac{\partial^2 y}{\partial x^2} + \frac{m}{EJ} \cdot \frac{\partial^2 y}{\partial t^2} = 0. \tag{1}$$

Where y is the deflection of machine part, m is where mass is per unit length and EJ is bending stiffness in the plane of oscillation of the shaft.

We will look for a solution to this equation in the following form:

$$y(x,t) = S_1(x)cos\frac{\omega t}{2} + T_1(x)sin\frac{\omega t}{2} + S_2(x)cos\omega t + T_2(x)sin\omega t \tag{2}$$

Where $S_1(x)$, $S_2(x)$, $T_1(x)$, $T_2(x)$ are unknown funktions depend on coordinates x of elements of oscillation parts, ω is the frequancy (angular velocity) of oscillation.

To solve Eq. (1) taking into account (2), we formulated the boundary conditions of the problem and have the analytical solution for different compressor stations [1, 2]. The main idea of the research is to assess the technical condition using a data-driven method for identifying and diagnosing failures and defects by collecting data from sensors into computer memory and processing them. After that, the traditional statistical analysis used a data method based on machine learning, data mining [13–15].

4 Formulation of the Problem

Taking into account everything noticed, the purpose of the study is to conduct experiments, compare the obtained data and results, as well as monitor the tests that implement the system in a real environment. It should be noted that maintenance systems based on the technical condition (CBM –the condition of maintenance), prognostics and health management (PHM - prognostics and health management) were created for different machines, they were introduced and used in industry. Experimental equipment consists of an electric motor with adjustable rotations turnover, a coupling, a cyclone pump, a fluid inlet, outgoing fluid, a fluid supply line, a fluid outlet line, a shut-off valve fluid, several other hydraulic valves, etc. (Fig. 2). The received vibration signals from the observed experiments in equipment that simulated real units of measurement show the possibility of using the presented method.

There are 2 pieces of "DS 18B20 Temperature Sensor MANYEE 5Pcs 1M/39.37" Waterproof Digital Temperature Temp Sensor Thermal Probe with Stainless Steel Tube Probe −55 °C ~ +125 °C for Arduino/Raspberry Pi/Car Refriger" sensors, 2 pieces of "Generic Carbon Steel Alloy Variable Pump Water/Air Pressure Sensor(DC 5V)" pressure sensors, 2 pieces of "Generic DC 5 ~ 24V Water Flow Sensor Nylon + Fiber Glass DN25" flaw rate sensors and 2 pieces "WINGONEER 5PCS High Sensitive vibration sensor module SW −18010P" vibration sensors were mounted to the monitoring units (the angles between the vibration sensors should be 900).

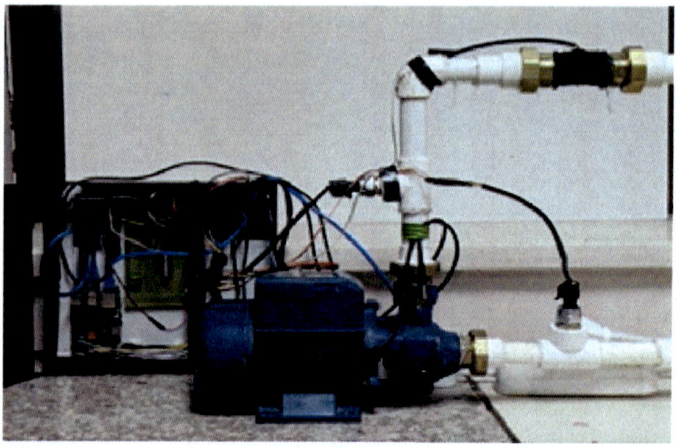

Fig. 2. The experimental testing installation for measurement of vibration signals of a pump system by sensors.

In this case, sensor recordings are transmitted in real time to the hard disk of the Raspberry board. We can write this data into CSV, XLS, etc. files in a computer store or send to other applications or DBMS using web services. The storage of sensor records on the hard disk is carried out by the PuTTY software. PuTTY is an open source, free terminal emulator and file-sharing program. PuTTY is universal software that supports several other network protocols, such as SCP, SSH, Telnet, rlogin, etc. allows you connect to remote computers [13]. After collecting the data received from monitored devices by observers on the Raspberry hard drive using the WinSCP software application, we copy the file with the collected data from the memory of the controller board to the computer store. WinSCP (Windows Secure Copy) is a client protocol for the Microsoft Windows environment that supports other protocols such as SFTP, FTP, WebDAV, Amazon S3, and SCP. Its main goal is to ensure a reliable and secure data transfer between local and remote computers. In the study [14, 15] measured vibration signals using sensors (transducer) and, based on the data obtained, they graphically determined the time dependence function. We investigate the sensor's recording data was recorded. Before processing the data, we need to clear the row data from unnecessary data, and then prepare the data for the next analysis. Figure 3 shows a graphical function of the dependence of vibration signals on the time domain in 20 min. Below is a graph of vibration signals in the time domain.

To control the technical condition of the equipment, general vibration signals were measured, allowing evaluating the current technical condition of the unit. The dynamics of the general vibration signals allows you to monitor the dynamics of the technical condition of the equipment and identify defects at an early stage. Measurement of general vibrations is called the first stage and allows you to monitor various types of machines, providing a simple vibration analysis. The measurements fulfilled in this stage are not enough to detect defects, required the detailed spectral analysis of equipment vibrations to localize of defects. The majority cases the sources of defects are detected in their frequency specters (frequency domain). Required vibration analysators in the 2nd stage to provide vibro diagnosis, which use the Fast Fourier Transformation vibration signals.

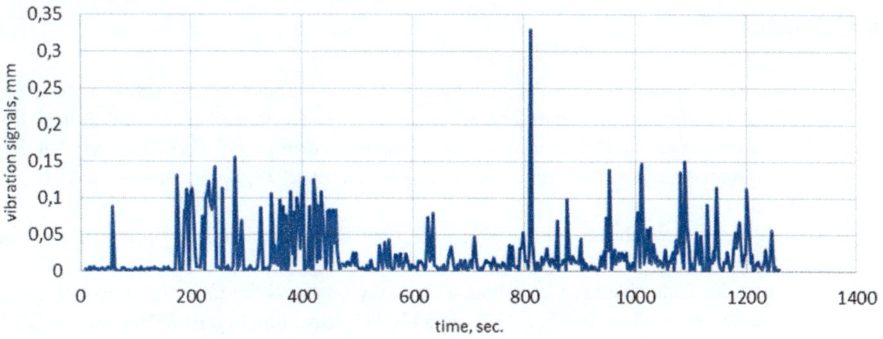

Fig. 3. Graph of vibration signals in the time domain for 20 min.

The Fast Fourier Transformation of time domain relatives based on acquired data from experiments to frequency domain relatives is presented in Fig. 3.

Thus various types of stresses are created in the details of the units that are under the influence of static, dynamic, and alternative loads during operation. These units work in critical processes under the influence of aggressive and abrasive environments. Many units operate at high and variable temperatures. All these facts reduce the service life of the units and cause failures. If we can determine the cause of failures at an early stage, we can eliminate these causes and increase the service life of the units. In this case, in order to reduce the number of machine failures and maintain their normal operation, we need to monitor and evaluate the technical condition of the units.

5 Conclusion

Investigation and control of the vibration processes of the machines used in oil and gas industry by sensors are current problems for oil and gas transportation. Therefore in this research, we consider the piston and centrifugal pumps and compressors which are important for practice. By installing sensors on different parts of the pump-compressor machines, we get the parametric data from the sensors described above. By collecting these data into databases and processing them, we can make a conclusion for decision-making. We calculated the values of general vibration signals and plotted them. The obtained data and graphs make it easy to assess the basic technical conditions of the object, to describe the reality of the technical condition. Comparing the experimental values obtained with the base reference date, we can say what is the technical condition of the installation. At the same time, we can estimate the actual technical condition of the machines and their individual parts using graphical methods, comparing the graphs obtained from experimental data and the graphs from results of theoretical research.

References

1. Bakhshaliev, V.I., Aslan-Zada, F.E., Ismayil, I.A.: Development of innovative methods of fuzzy logic for increase of durability and reliability of piston machines used in oil industry. In: Seventh International Conference on Soft Computing, Computing with Words and Perceptions in System Analysis, Decision and Control–ICSCCW–2013, Izmir, Turkey, pp. 101–110 (2013)
2. Davitashvili, N., Bakhshaliev, V.: Dynamics of Crank-Piston mechanisms. Springer Publisher (2016). https://doi.org/10.1007/978-981-10-0323-3
3. Wannatong, K., Chanchaona, S., Sanitjai, S.: Simulation algorithm for piston ring dynamics. Simul. Model. Pract. Theory **16**(1), 127–146 (2008). https://doi.org/10.1016/j.simpat.2007.11.004
4. Burstein, L., Ingman, D.P.: Pore ensemble statistics in application to lubrication under reciprocate motion. Tribol. Trans. **43**(2), 205–212 (2000). https://doi.org/10.1080/10402000008982330
5. Feiyun, C., Jin, C., Guangming, D., Michael, P.: Vibration model of rolling element bearings in a rotor-bearing system for fault diagnosis. J. Sound. Vib. **332**(8), 2081–2097 (2013). https://doi.org/10.1016/j.jsv.2012.11.029
6. Gu, J., et al.: Manufacturing quality assurance for a rotate vector reducer with vibration technology. J. Mech. Sci. Technol. **33**(5), 1995–2001 (2019). https://doi.org/10.1007/s12206-019-0401-3
7. Hsu, Y., Brennen, C.E.: Fluid flow equations for Rotordynamic flows in seals and leakage paths. Trans. ASME J. Fluids Eng. **124**, 176–181 (2002). https://doi.org/10.1115/1.1436093
8. Erneux, T.: Mechanical Vibrations. Presented at the (2009). https://doi.org/10.1007/978-0-387-74372-1_6
9. Vachtsevanos, G., Lewis, F., Roemer, M., Hess, A., Wu, B.: Intelligent Fault Diagnosis and Prognosis for Engineering Systems. John Wiley & Sons Inc (2006). https://doi.org/10.1002/9780470117842
10. Vachtsevanos, G., Lewis, F., Roemer, M., Hess, A., Wu, B.: Intelligent Fault Diagnosis and Prognosis for Engineering Systems, p. 960. John Wiley & Sons (2006). https://doi.org/10.1002/9780470117842
11. Hong, L., Dhupia, J.S.: A time-domain approach to diagnose gearbox fault based on measured vibration signals. J. Sound Vib. **333**(7), 2164–2180 (2014). https://doi.org/10.1016/j.jsv.2013.11.033
12. Hu, D., Jia, L., Yang, L.: Dimensional analysis on resistance characteristics of labyrinth seals. J. Therm. Sci. **23**(6), 516–522 (2014). https://doi.org/10.1007/s11630-014-0736-0
13. Chaaria, F., Bartelmusb, W., Zimrozb, R., Fakhfakha, T., Haddara, M.: Gearbox vibration signal amplitude and frequency modulation. J. Shock Vibration **19**(4), 635–652 (2012). https://doi.org/10.3233/SAV-2011-0656
14. Soua, S., Van Lieshout, P., Perera, A., Gan, T.-H., Bridge, B.: Determination of the combined vibrational and acoustic emission signature of a wind turbine gearbox and generator shaft in service as a pre-requisite for effective condition monitoring. J. of Renewable Energy. **51**, 175–181. Elsevier (2013). https://doi.org/10.1016/j.renene.2012.07.004
15. Bucinskas, V., Mitrouchev, P., Sutinys, E., Sesok, N., Iljin, I., Morkvenaite-Vilkonciene, I.: Evaluation of comfort level and harvested energy in the vehicle using controlled damping. Energies **10**(11), 1742 (2017). https://doi.org/10.3390/en10111742

Financial Performance Analysis with Intuitive Fuzzy Logic and Entropy-Based Multi-criteria Decision Making Method

Vuqar Salahli(✉) and Abil Suleymanov

Department of Computer Enigieering, Odlar Yurdu University, Baku AZ1072, Azerbaijan
vsalahli@ekin.com, abil@bsb-azeri.org.uk

Abstract. Due to the rapid development of technology, the demand for local furniture is growing day by day. The growing demand for furniture has made the furniture sector vital to the economic development of countries. This study examines the financial performance, heuristic fuzzy logic, and multi-criteria decision-making methods of 8 furniture companies operating in Azerbaijan. The study used real data from 8 furniture companies for 2016–2020. In the decision-making process; 5 main criteria and 13 sub-criteria such as growth, activity, leverage, profitability, and liquidity were used. According to the analysis, Embawood had the best performance among the 8 furniture companies, while Furkanmobilya had the worst performance.

Keywords: Fuzzy logic · Financial performance · Multi-criteria decision making

1 Introduction

The disappearance of borders with the globalization process has caused companies to face increasing competition conditions [1]. Companies that can adapt to these competitive conditions can survive, while companies that cannot adapt first lose their market share and profitability, and then cannot survive [2]. In order to be successful in competition, companies should use their resources correctly and effectively and keep their performance as high as possible. Performance concept; the level of achievement of business objectives [2]. The ability of an enterprise to continue its success and to be able to provide continuity in this success finds its answer with performance measurement.

Companies that can keep their financial performance at a high level can have a competitive advantage in national and international markets. Companies that have a competitive advantage increase their profitability and market share. This naturally ensures the success and survival of the company [3]. The survival of companies is a very important process for the company and there are many factors in order to ensure the continuation of this process. One of the most important factors companies to survival of companies, analyze and interpret their financial performance in the most accurate way.

In the measurement of financial performance, the financial data of the company (balance sheet, profit, and loss statements) are approached from different angles, and

R. A. Aliev et al. (Eds.): ICSCCW 2021, LNNS 362, pp. 711–718, 2022.
https://doi.org/10.1007/978-3-030-92127-9_94

answers are sought for different questions. Indicators such as profitability, the volume of sales, efficient (and profitable) use of resources, indebtedness rate, ability to pay debts, and the company's position in the market, which point to the financial performance of the company, are analyzed with the help of different ratios [4].

The financial performances of the enterprises are calculated with alternative methods in order to see the targeted and achieved status. While making this calculation, many criteria should be taken into consideration and an evaluation should be made accordingly. For this purpose, it is a very common method to use multi-criteria decision-making methods that allow evaluation by considering different criteria in performance analysis [5].

The universe we live in is full of vague, inconsistent, and incomplete information. It is not easy to process this complex information with classical methods. For this reason, Zadeh (1965) defined the theory of fuzzy sets in order to express these complex and uncertain situations in real life [6, 7]. Zadeh (1965) stated that fuzzy logic reflects the mental working processes of people more accurately than classical logic in situations of uncertainty [8, 9]. The success of fuzzy logic in modeling uncertain situations and its propensity to human thought and decision systems make it possible to use fuzzy logic in the field of finance [10].

Especially in recent years, multi-criteria decision-making methods with fuzzy logic have been widely used in decision-making problems, such as choosing the best among the alternatives and ranking the alternatives. At the same time, the use of fuzzy logic in finance is relatively new, but the number of studies is growing day by day [11].

In this study, a multi-criteria decision-making method based on entropy was developed using intuitive fuzzy sets, which are generalizations of Zadeh's fuzzy sets [12]. This method makes it more flexible and more accurate to rank alternatives according to defined criteria. At work; In a decision process where the criteria were not fully known, first of all, the data of 8 furniture companies operating in the furniture sector for 2016–2020 were examined and these data of alternatives were first evaluated in terms of language, then this linguistic assessment was expressed in heuristic fuzzy numbers. The weight data of the criteria were then calculated by the concept of entropy, and finally, using all the data obtained, the alternatives were listed.

2 Fuzzy Set

Lotfi Zadeh (1965) was the first to propose the concept of fuzzy logic [13]. Zadeh has shown that many concepts can be defined linguistically better than traditional mathematics, and fuzzy logic and its expressions in fuzzy sets provide a better model of real life. In the classical set concept, an element with a characteristic function is either an element or not an element of a set. However, in the fuzzy set concept, whether an apple belongs to a cluster or not is defined by a membership function that assigns a membership degree to each object in the range [0,1] [14, 15].

Let X be a non-empty set and $x \in X$. Then, for the function $\mu_A(X) : X \to [0, 1]$, $\mu_A(X)$, x is a fuzzy set A in X to show the membership degree of the x element;

It is expressed as $A = \{(x, \mu_A(x)) : x \in X\}$. In fact, every membership function is a function that returns the elements of a classical universal set to a number in the range [0, 1] [16].

Intuitive Fuzzy Set
In the fuzzy set A, the degree of belonging of an element to the set is μ_A, while the degree of not belonging is 1-μ_A. Therefore, the sum of the degrees of belonging and not belonging is equal to 1. However, this situation is insufficient to explain the uncertainty in some problems. For this reason, intuitive fuzzy set theory, which is the generalization of fuzzy set theory, is proposed [17].

3 An Application of Decision Making Method

In this section, the financial performance of 8 furniture companies traded in Azerbaijan between 2016–2020 is analyzed using a multi-criteria decision-making method based on heuristic fuzzy logic and entropy. The study used data from the balance sheets of 8 furniture companies for 2016–2020. The data were obtained from the Finnet database. Companies in the evaluation process, respectively; Embawood (X_1), Woodpecker (X_2), Damla mebel (X_3), Gumush mobilya (X_4), Hacıoğulları (X_5), Saloglu Mebel (X_6), Kardashlar Mebel (X_7), Furkan Mebel (X_8) Evaluation criteria, Literature 5 main and 13 sub-criteria were identified and these criteria are presented in Table 1.

Table 1. Evaluation criteria for financial performance

Ratio group	Ratios used (Criteria)	Purpose of use
Growth rates	Growth in assets (Y_1) Growth in sales (Y_2) Growth in equity (Y_3)	Ratios that show the growth in the assets of the business
Activity rates	Current asset turnover rate (Y_4) Receivable turnover rate (Y_5) Active RPM (Y_6) Debt turnover rate (Y_7)	Ratios that show how effectively the assets of the business are used
Leverage ratios	Total debt/Total assets (Y_8) Fixed asset/Perpetual capital (Y_9)	Ratios that measure the ability of businesses to meet their long-term obligations
profitability ratios	Return on assets (Y_{10}) Return on equity (Y_{11})	Ratios that measure how efficiently the assets of the business are used
Liquidity ratios	Current ratio (C Y_{12}) Liquid ratio (C Y_{13})	Ratios that measure the ability of a business to pay its short-term debts

The proposed method for evaluating the financial performance of furniture companies using a multi-criteria decision-making method based on intuitive fuzzy logic and entropy consists of the following steps:

First of all, the data of 8 furniture companies traded in 2016–2020 were obtained from the Finnet database. The 5-year averages of these data were then calculated and presented in Table 2.

After the 5-year averages were calculated for each criterion of the alternatives, each alternative was evaluated with the help of linguistic terms, taking into account the sector

Table 2. 5-Year average data of alternatives by criteria

Criteria / Alternatives	Growth rates			Activity rates			
	Y_1	Y_2	Y_3	Y_4	Y_5	Y_6	Y_7
Embawood (X_1)	50.11	50.14	89.68	3.21	12.14	1.16	1.16
Woodpecker (X_2)	26.04	62.88	83.64	1.36	8.25	0.21	0.17
Damla mebel (X_3)	13.54	16.18	13.57	3.58	5.52	0.68	0.71
Gumush mobilya (X_4)	8.04	298.2	14.26	0.71	23.41	0.06	0.08
Hacıoğulları (X_5)	19.25	27.05	10.34	1.52	10.15	0.23	0.21
Saloglu mebel (X_6)	9.51	30.49	−7.25	1.56	9.25	0.08	0.14
Kardashlar mebel (X_7)	16.75	12.46	7.55	1.96	4.40	0.97	0.39
Furkan mebel (X_8)	19.25	27.05	10.34	1.52	10.17	0.23	0.19
Sector average	19.49	64.09	32.62	2.11	11.04	0.51	0.46

Criteria / Alternatives	Leverage ratios		Profitability ratios		Liquidity ratios	
	Y_8	Y_9	Y_{10}	Y_{11}	Y_{12}	Y_{13}
Embawood (X_1)	0.81	1.08	2.17	11.81	0.89	0.64
Woodpecker (X_2)	0.91	1.33	−2.84	−26.9	0.43	0.38
Damla mebel (X_3)	0.70	1.14	2.21	7.02	0.66	0.65
Gumush mobilya (X_4)	0.57	1.34	−10.7	−24.8	0.34	0.17
Hacıoğulları (X_5)	0.81	1.16	−0.85	−4.38	0.60	0.49
Saloglu mebel (X_6)	0.36	0.98	−4.41	−8.12	3.31	3.16
Kardashlar mebel (X_7)	0.74	1.19	−2.04	−11.8	0.66	0.28
Furkan mebel (X_8)	0.81	1.14	−0.85	−4.40	0.60	0.47
Sector average	0.71	1.16	−1.04	−5.20	1.03	0.85

averages. During the evaluation, extreme and extreme values were excluded in order to obtain a healthier result. As a result of this evaluation, the intuitive fuzzy number corresponding to each linguistic term was determined. Intuitive fuzzy numbers corresponding to each linguistic term are obtained. The linguistic terms used in the evaluation of alternatives and the corresponding intuitive fuzzy numbers for these linguistic terms are presented in Table 3.

Table 3. Linguistic terms used in evaluation of criteria and intuitive fuzzy number equivalent

linguistic terms	Intuitive fuzzy numbers
Very Good (VG)	(0,90; 0,10; 0,00)
Good (G)	(0,75; 0,20; 0,15)
Medium (M)	(0,50; 0,45; 0,05)
Bad (B)	(0,35; 0,60; 0,05)
Very Bad (VB)	(0,10; 0,90; 0,00)

After the criterion weights were found, the weighted average values were calculated for each alternative and are presented in Table 4.

Table 4. Weighted average final values of alternatives

Alternatives	μ	ϑ
Embawood (X_1)	0.657	0.335
Woodpecker (X_2)	0.513	0.476
Damla mebel (X_3)	0.454	0.510
Gumush mobilya (X_4)	0.537	0.47
Hacıoğulları (X_5)	0.575	0.403
Saloglu mebel (X_6)	0.366	0.594
Kardashlar mebel (X_7)	0.564	0.395
Furkan mebel (X_8)	0.362	0.612

After calculating the weighted μ and ϑ values for each alternative, the score function of each alternative was calculated and the ranking was made accordingly. The scores of the alternatives are presented in Table 5.

$$\text{Score}(X_1) = 0.657 - 0.335 = 0.322$$
$$\text{Score}(X_2) = 0.513 - 0.476 = 0.037$$
$$\vdots$$
$$\text{Score}(X_8) = 0.362 - 0.612 = -0.25$$

Table 5. Ranking of alternatives

Alternatives	μ-ϑ
Embawood (X_1)	0.322
Woodpecker (X_2)	0.037
Damla mebel (X_3)	−0.056
Gumush mobilya (X_4)	0.067
Hacıoğulları (X_5)	0.172
Saloglu mebel (X_6)	−0.228
Kardashlar mebel (X_7)	0.169
Furkan mebel (X_8)	−0.25

4 Conclusion

The development of technology and the process of globalization have led companies to improve their operations and production processes. New technologies, new needs, and changing production relationships have led manufacturers to focus on different production behaviors in the production process. This situation increased the level of competition and pushed companies to different searches. In a globalizing world, companies that want to increase their profitability must adapt to increasing competition.

The level of competition affects the behavior, profitability, level of production and financial performance of companies. On the other hand, with the development of technology, the growing demand of countries for furniture has led to an increase in the level of competition of every company in the furniture sector. To survive in this competitive environment, companies need to maintain high financial performance. At this point, it is very important for companies to properly analyze their financial performance. For this reason, many methods have been developed to analyze the financial performance of companies. Especially in recent years, multi-criteria decision-making methods are often used to analyze financial performance.

In this study, the financial performance of furniture companies operating in Azerbaijan is analyzed using a multi-criteria decision-making method based on heuristic fuzzy logic and entropy. Five main and 13 sub-criteria were used as criteria, and the main criteria were defined as growth, activity, leverage, profitability and liquidity ratios, respectively. The 5-year averages of the data obtained for each alternative were evaluated in terms of language by comparison with sector averages, and a decision matrix was formed by finding the value of an intuitive fuzzy number corresponding to each linguistic term. After the decision matrix was created, the weight of each criterion was calculated and the score of each alternative was calculated using the weights of this criterion.

According to the results, Embawood, which has the best financial performance among the companies, showed the worst performance as Furkan furniture. Many factors play a role in this result in the financial performance of companies. Criteria in the decision-making process have played an active role in shaping this outcome, and these criteria play an important role in guiding the future activities of companies. Improving the decision-making criteria and performance of companies to maintain their existence and continue to operate will be beneficial both in terms of maintaining their presence and increasing their competitiveness.

In this study, heuristic fuzzy logic was used, along with decision-making methods based on a number of criteria. Given the uncertainty in the financial world, it is hoped that this proposed method will shed some light on the current situation from a different perspective and will help researchers in future research as it is an effective and useful method in uncertainty.

References

1. Atanassovi, K.: Intuitionistic fuzzy sets. Fuzzy Sets Syst. **20**, 87–96 (1986)
2. Garg, H.: Entropy Based Multi-Criteria Decision Making Method Under Fuzzy Environment and Unknown Attribute Weights (June 2015). https://doi.org/10.4172/2229-8711.1000182
3. Asadabadi, M.R.: The Stratified Multi Criteria Decision Making Method (August 2018). https://doi.org/10.1016/j.knosys.2018.07.002
4. Alimohammadlou, M., Bonyani, A.: Fuzzy BWANP multi-criteria decision-making method. Decis. Sci. (January 2019) https://doi.org/10.5267/j.dsl.2018.4.002
5. Kaya, G.K., Öztürk, F.: A Comparison of the Multi-criteria Decision-Making Methods for the Selection of Researchers. (January 2021) https://doi.org/10.1007/978-3-030-76724-2_12
6. Chen, Z., Yang, W.: A new multiple attribute group decision making method in intuitionistic fuzzy setting. Appl. Math. Model **35**, 4424–4437 (2011)
7. Haseli, G., Sheikh, R., Sana, S.S.: Base-criterion on multi-criteria decision-making method and its applications. Int. J. Manag. Sci. Eng. (July 2019) https://doi.org/10.1080/17509653.2019.1633964
8. Imamguluyev, R.: Determination of correct lighting based on fuzzy logic model to reduce electricity in the workplace. In: Conference: International Conference on Eurasian EconomiesAt: Baku, Azerbaijan (September 2020). https://doi.org/10.36880/C12.02456
9. Imamguluyev, R.: Optimal solution of indoor lighting system based on fuzzy logic. In: Conference: 2nd International Baku Conference on Scientific Research at: Baku, Azerbaijan (May 2021)
10. Aliev, R., Tserkovny, A.: Fuzzy logic for incidence geometry. In: Book: Beyond Traditional Probabilistic Data Processing Techniques: Interval, Fuzzy etc. Methods and Their Applications. (February 2020) https://doi.org/10.1007/978-3-030-31041-7_4
11. Zadeh, L.A., Aliev, R.: Fuzzy logic theory and applications: part I and part II. pp. 61(December 2018). https://doi.org/10.1142/10936
12. Zadeh, L.A.: Toward a theory of fuzzy information granulation and its centrality in human reasoning and fuzzy logic. Fuzzy Sets Syst. **90**, 111–127 (1997)
13. Zadeh, L.A.: Fuzzy sets. Inf. Control **8**(3), 338–350 (1965)
14. Valiyev, A., Imamguluyev, R., Ilkin, G.: Application of fuzzy logic model for daylight evaluation in computer aided interior design areas. In: Aliev, R.A., Kacprzyk, J., Pedrycz, W., Jamshidi, Mo., Babanli, M., Sadikoglu, F.M. (eds.) ICAFS 2020. AISC, vol. 1306, pp. 715–722. Springer, Cham (2021). https://doi.org/10.1007/978-3-030-64058-3_89
15. Imamguluyev, R., Suleymanli, T., Umarova, N.: Evaluation of the effectiveness of integration processes in production enterprises based on the fuzzy logic model. In: Aliev, R.A., Kacprzyk, J., Pedrycz, W., Jamshidi, Mo., Babanli, M., Sadikoglu, F.M. (eds.) ICAFS 2020. AISC, vol. 1306, pp. 133–139. Springer, Cham (2021). https://doi.org/10.1007/978-3-030-64058-3_17

16. Imamguluyev, R.: Application of fuzzy logic model for correct lighting in computer aided interior design areas. In: Kahraman, C., Cevik Onar, S., Oztaysi, B., Sari, I.U., Cebi, S., Tolga, A.C. (eds.) INFUS 2020. AISC, vol. 1197, pp. 1644–1651. Springer, Cham (2021). https://doi.org/10.1007/978-3-030-51156-2_192
17. Oglu, A.R.B., Kizi, I.I.T.: A method for forecasting the demand for pharmaceutical products in a distributed pharmacy network based on an integrated approach using fuzzy logic and neural networks. In: Kahraman, C., Cevik Onar, S., Oztaysi, B., Sari, I.U., Cebi, S., Tolga, A.C. (eds.) INFUS 2020. AISC, vol. 1197, pp. 998–1007. Springer, Cham (2021). https://doi.org/10.1007/978-3-030-51156-2_116

Selection of Data Science Software by Using Choquet Integral

V. H. Salimov(✉)

Department of Computer Engineering, Azerbaijan State Oil and Industry University, Azadliq Street 20, Baku AZ1010, Azerbaijan

Abstract. This article is devoted to a software selection problem. Usually, this problem is considered as a multicriteria decision making (MCDM) problem. One of the important problems in MCDM is criteria interaction. An effective approach to solving this problem is the use of fuzzy measure, particularly Sugeno λ-measure. In this paper, we use Choquet integral w.r.t. Sugeno λ-measure in software selection problem with four criteria and 5 alternatives. The comparative analysis of additive and non-additive measures are provided.

Keywords: Software selection · Criteria interaction · Fuzzy measure · λ-Sugeno measure · Choquet integral

1 Introduction

The problems of software selection are very actual [1–3]. Usually, these problems are considered in the framework of multicriteria decision making (MCDM). The major problem of MCDM is the problem of interaction of criteria. For this problem, application of the methods like simple arithmetic mean, geometric mean, weighted arithmetic mean, median and weighted geometric mean is not adequate. One of the methods for solving this problem is the use of the special function - fuzzy (non-additive) measure. Research on non-additive measures was initiated in the works of G. Choquet [4] and M. Sugeno [5]. Later Sugeno [6] proposed a special aggregator, Sugeno integral. This integral w.r.t. fuzzy measure allowed to solve decision making problems with interaction of criteria.

The major problem is the construction of this measure. For solving this problem many methods exist [7–10]. One of them is solving this problem on base of expert's assessments. Another approach relies on an application of the least square method [11].

In this work, we use Choquet integral w.r.t. fuzzy λ-Sugeno measure in solving a decision problem of selection of a data science software.

2 Preliminaries

Definition 1. Fuzzy Measure. Consider a set of criteria $X = \{x_1, x_2, x_3 \ldots\}$. Denote $P(X)$ a power set of X. A fuzzy measure is a mapping $g : P(X) \to [0, 1]$ that satisfies the two conditions:

R. A. Aliev et al. (Eds.): ICSCCW 2021, LNNS 362, pp. 719–723, 2022.
https://doi.org/10.1007/978-3-030-92127-9_95

(i) $g(\emptyset) = 0$ *and* $g(X) = 1$
(ii) *if* $A \subseteq B$ *then* $g(A) \leq g(B), A, B \subseteq X$.

In general, fuzzy measure is non-additive.

A value of fuzzy measure $g(A)$ is the level of importance of A. In general, fuzzy measure may satisfy the following conditions:

$g(A \cup B) = g(A) + g(B)$ (additive measure),
$g(A \cup B) \geq g(A) + g(B)$ (super-additive measure),
$g(A \cup B) \leq g(A) + g(B)$ (sub-additive measure).

Definition 2. λ-Sugeno Measure. Let λ be real number $\lambda \in (-1, \infty)$. λ-Sugeno measure is the function g:P(X) → [0,1] that fulfills the following two conditions:

$$g_\lambda(\mathrm{X}) = 1,$$

$$g_\lambda(A \cup B) = g_\lambda(A) + g_\lambda(B) + \lambda g_\lambda(A) g_\lambda(B), A, B \subseteq X, A \cap B = \emptyset,$$

Parameter λ is found by solving the following equation:

$$\lambda + 1 = \prod\nolimits_{i=1}^{n} (\lambda g_i + 1).$$

Depending on the value of λ as a root of the above equation, the following cases are possible:
if $-1 < \lambda < 0$ then $\sum_{i=1}^{n} g_i > g(X)$ (super-additive measure),
if $\lambda = 0$ then $\sum_{i=1}^{n} g_i = g(X)$ (additive measure),
if $\lambda > 0$ then $\sum_{i=1}^{n} g_i < g(X)$ (sub-additive measure).

Definition 3. Choquet Integral. Assume g is a fuzzy measure defined on X, Choquet integral for this function $f : X \to [0, \infty]$ has the form:

$$\int f dg = \sum\nolimits_{i=1}^{N} (f(x_i) - f(x_{i-1}))g(x_i),$$

subject to the condition:

$$f(x_1) \leq f(x_2) \leq f(x_3) \ldots\ldots f(x_n) f(x_0) = 0.$$

3 Software Selection Problem

Assume we have the software selection problem with 4 benefit criteria $X = (x_1, x_2, x_3, x_4)$, where x_1- functionality, x_2- cost, x_3- reliability, x_4- usability. Five hypothetical alternatives are considered A_1,..., A_5. Experts have presented the assessments of all alternatives by criteria as it is shown in Table 1.

Table 1. The assessments of all alternatives by criteria

	x_1	x_2	x_3	x_4
A_1	0.7	0.5	0.4	0.9
A_2	0.2	0.4	0.7	0.5
A_3	0.7	0.5	0.9	0.4
A_4	0.5	0.4	0.7	0.9
A_5	0.2	0.5	0.9	0.7

The expert assessments of fuzzy measures for any criterion are given below:

$$g_\lambda(x_1) = 0.7, \quad g_\lambda(x_2) = 0.5, \quad g_\lambda(x_3) = 0.5, \quad g_\lambda(x_4) = 0.7.$$

In order to find λ, we need to solve the equation for 4 criteria:

$$\lambda + 1 = \prod_{i=1}^{4} (\lambda g_i + 1),$$

that is

$$\lambda + 1 = (0.7\lambda + 1)(0.5\lambda + 1)(0.5\lambda + 1)(0.7\lambda + 1).$$

The following solutions are obtained:

$$\lambda_1 = 0,$$

$$\lambda_2 = -2.49 + 1.75i,$$

$$\lambda_3 = -2.49 - 1.75i,$$

$$\lambda_4 = -0.97.$$

The root $\lambda = -0.97$ is chosen. The result shows the criteria have negative relationship, i.e. λ-Sugeno measure is sub-additive. Further, we determine values of fuzzy measures for all groups of the criteria

$$g_\lambda(x_1, x_2) = g_\lambda(x_1) + g_\lambda(x_2) + \lambda g_\lambda(x_1) g_\lambda(x_2) = 0.8605,$$

$$g_\lambda(x_1, x_3) = g_\lambda(x_1) + g_\lambda(x_3) + \lambda g_\lambda(x_1) g_\lambda(x_3) = 0.8605,$$

$$g_\lambda(x_1, x_4) = g_\lambda(x_1) + g_\lambda(x_4) + \lambda g_\lambda(x_1) g_\lambda(x_4) = 0.9247,$$

$$g_\lambda(x_2, x_3) = g_\lambda(x_2) + g_\lambda(x_3) + \lambda g_\lambda(x_2) g_\lambda(x_3) = 0.7575,$$

$$g_\lambda(x_2, x_4) = g_\lambda(x_2) + g_\lambda(x_4) + \lambda g_\lambda(x_2)g_\lambda(x_4) = 0.8605,$$

$$g_\lambda(x_3, x_4) = g_\lambda(x_3) + g_\lambda(x_4) + \lambda g_\lambda(x_3)g_\lambda(x_4) = 0.8605,$$

$$g_\lambda(x_1, x_2, x_3) = g_\lambda(x_1, x_2) + g_\lambda(x_3) + \lambda g_\lambda(x_1, x_2)g_\lambda(x_3) = 0.9432,$$

$$g_\lambda(x_1, x_2, x_4) = g_\lambda(x_1, x_2) + g_\lambda(x_4) + \lambda g_\lambda(x_1, x_2)g_\lambda(x_4) = 0.9762,$$

$$g_\lambda(x_1, x_3, x_4) = g_\lambda(x_1, x_3) + g_\lambda(x_4) + \lambda g_\lambda(x_1, x_3)g_\lambda(x_4) = 0.9762,$$

$$g_\lambda(x_2, x_3, x_4) = g_\lambda(x_2, x_3) + g_\lambda(x_4) + \lambda g_\lambda(x_2, x_3)g_\lambda(x_4) = 0.9432,$$

$$g_\lambda(x_1, x_2, x_3, x_4) = 1.$$

The results demonstrate strong interaction between x_1 *and* x_4, weak interaction between x_2 *and* x_3, strong interaction between x_1, x_3, x_4 and weak interaction x_2, x_3, x_4. Finally, we computed values of Choquet integral for all alternatives:

$$\begin{aligned} C_1 = \int fdg &= x_3 g_\lambda(x_1, x_2, x_3, x_4) + (x_2 - x_3)g_\lambda(x_1, x_2, x_4) \\ &+ (x_1 - x_2)g_\lambda(x_1, x_4) + (x_4 - x_1)g_\lambda(x_4) = 0.8225, \end{aligned}$$

$$\begin{aligned} C_2 = \int fdg &= x_1 g_\lambda(x_1, x_2, x_3, x_4) + (x_2 - x_1)g_\lambda(x_2, x_3, x_4) \\ &+ (x_4 - x_2)g_\lambda(x_3, x_4) + (x_4 - x_3)g_\lambda(x_4) = 0.5746, \end{aligned}$$

$$\begin{aligned} C_3 = \int fdg &= x_4 g_\lambda(x_1, x_2, x_3, x_4) + (x_2 - x_4)g_\lambda(x_1, x_2, x_3) \\ &+ (x_1 - x_2)g_\lambda(x_1, x_3) + (x_3 - x_1)g_\lambda(x_3) = 0.7664, \end{aligned}$$

$$\begin{aligned} C_4 = \int fdg &= x_2 g_\lambda(x_1, x_2, x_3, x_4) + (x_1 - x_2)g_\lambda(x_1, x_3, x_4) \\ &+ (x_3 - x_1)g_\lambda(x_3, x_4) + (x_4 - x_3)g_\lambda(x_4) = 0.8097, \end{aligned}$$

$$\begin{aligned} C_5 = \int fdg &= x_1 g_\lambda(x_1, x_2, x_3, x_4) + (x_2 - x_1)g_\lambda(x_2, x_3, x_4) \\ &+ (x_4 - x_2)g_\lambda(x_3, x_4) + (x_3 - x_4)g_\lambda(x_3) = 0.7550. \end{aligned}$$

Thus, the best alternative is A_1 as $C_1 = 0.8225$ is the highest value of Choquet integral. We also considered the case of additive measure ($\lambda = 0$). In this case we have $C_1 = 1.01$ $C_2 = 0.76$ $C_3 = 0.91$ $C_4 = 0.97$ $C_5 = 1.11$. The best alternative is A_5. As we see, the results for additive and non-additive cases are different. This means that for multi-criteria decision making problem with criteria interaction, the application of the additive model is incorrect.

4 Conclusion

The paper is devoted to the problem of software selection. The problem with 4 criteria and 5 alternatives has been considered. The attention is focused at the problem of criteria interaction. As the fuzzy measure, the Sugeno function has been used. The Choquet integral was used for comparison of alternatives. The results for non-additive and additive cases are presented.

References

1. Azadeh, A., Shirkouhi, S., Rezaie, K.: A robust decision-making methodology for evaluation and selection of simulation software package. Int. J. Adv. Manuf. Technol. **47**, 381–393 (2010). https://doi.org/10.1007/s00170-009-2205-6
2. Karaarslan, N.: An application for modular capability-based ERP software selection using AHP method. Int. J. Adv. Manuf. Technol. **42**, 1025–1033 (2009). https://doi.org/10.1007/s00170-008-1522-5
3. Smirlis, Y.G., Zeimpekis, V., Kaimakamis, G.: Data envelopment analysis models to support the selection of vehicle routing software for city logistics operations. Oper. Res. Int. J. **12**, 399–420 (2012). https://doi.org/10.1007/s12351-010-0100-4
4. Choquet, G.: Theory of capacities. Annales de l'Institut Four. **5**, 131–295 (1953)
5. Sugeno, M.: Theory of fuzzy integrals and its applications. Ph.D. Thesis, Tokyo, 237 p (1974). https://doi.org/10.1016/0020-0255(85)90026-X
6. Murofushi, T., Sugeno, M.: An interpretation of fuzzy measures and the Choquet integral as an integral with respect to a fuzzy measure. Fuzzy Sets Syst. **29**, 201–227 (1989). https://doi.org/10.1016/0165-0114(89)90194-2
7. Tan, C., Chen, C.: Intuitionistic fuzzy Choquet integral operator for multi-criteria decision making. Exp. Syst. Appl. **37**, 149–157 (2010). https://doi.org/10.29218/srmsjoms.v4i01.15907
8. Meyer, P., Roubens, M.: On the use of the Choquet integral with fuzzy numbers in multiple criteria decision support. Fuzzy Sets Syst. **157**, 927–938 (2006). https://doi.org/10.1017/CBO9780511521720
9. Ceberi, M., Wang, X.: A hybrid algorithm to extract fuzzy measures for software quality assessment. J. Uncert. Syst. **7**(3), 219–237 (2013)
10. Pasrija, V., Kumar, S.: Assessment of software quality: Choquet integral approach. Procedia Technol. **6**, 153–162 (2012). https://doi.org/10.1016/j.protcy.2012.10.019
11. Grabisch, M.: The application of fuzzy integrals in multicriteria decision making. Eur. J. Oper. Res. **89**, 445–456 (1996). https://doi.org/10.1016/0377-2217(95)00176-X

Artificial Intelligence and Fuzzy Set Theory in the Methodology for Studying the Dynamics of the Financial and Economic State of the IT Industry

N. G. Kuznetsov[1], N. Rodionova[1(✉)], Kh. Abdurahman[1], and M. Vatolina[2]

[1] Rostov State University of Economics, B. Sadovaya street, 69, 344002 Rostov-on-Don, Russia
kuznecov@rsue.ru

[2] Rostov State Transport University, 2, Rostovskogo Strelkovogo Polka Narodnogo Opolcheniya Sq., Rostov-on-Don 344038, Russia

Abstract. We have proposed a fuzzy-multiple methodology for studying the dynamics of the financial and economic state of a given industry based on time series of data from open Internet sources. The technique is as follows. Based on the parsing of the For Fair Business and Audit-IT sites, a database of industry enterprises suitable for analysis is compiled by groups: micro-enterprises, mini-enterprises, small enterprises, medium-sized enterprises, large enterprises (divided by the number of employees). For each enterprise, for each studied year, a data line is formed: financial stability indicators (autonomy coefficients; provision with own circulating assets; investment coverage), solvency indicators (current, quick and absolute liquidity ratios), performance indicators (sales profitability; net profit margin; return on assets). On the basis of the well-known Audit-IT technique, the indicators are assessed, after which the assessments are aggregated by the group based on the previously developed fuzzy-multiple methodology (using fuzzy multilevel [0,1] – classifiers). After that, on the basis of systems of fuzzy-logical conclusions, the calculation of the dynamics of each indicator for the considered time period is made. Then the estimates are aggregated for each of the groups, which results in a final assessment of the dynamics of the group's financial and economic condition. The technique has been tested on the IT industry of the Rostov region for 2017–2019.

Keywords: Financial and economic condition · Fuzzy multilevel classifiers · Linguistic variable · Creditworthiness · Micro-enterprise

1 Introduction

Financial condition is an important characteristic of the financial activity of an enterprise, which determines the potential of the organization, competitiveness and self-sufficiency. The analysis of the financial condition of the enterprise is currently especially relevant because: 1) reflects the success of the enterprise, as well as the possibility of its further functioning; 2) demonstrates which aspects of the organization's economic activities

R. A. Aliev et al. (Eds.): ICSCCW 2021, LNNS 362, pp. 724–731, 2022.
https://doi.org/10.1007/978-3-030-92127-9_96

need to be paid attention to in order to improve its efficiency; 3) allows you to make a choice in favor of ways to improve efficiency that will best contribute to the elimination of deficiencies in the organization.

The purpose of the analysis of the financial condition is a timely assessment by obtaining the required number of the most meaningful indicators that demonstrate a real and reliable representation of the financial condition of the enterprise [1], as well as forecasting its further financial condition and assessing the possibility of increasing efficiency.

Summarizing the existing methods of financial analysis, the following approaches can be conditionally distinguished [2]:

- based on obtaining operational information by calculating a system of relative indicators characterizing the financial condition of the enterprise (J.K. Van Horn and others);
- implemented on the basis of a combination of absolute and relative indicators and allowing to obtain complete information about the ongoing processes in the financial condition of the enterprise (methods of Sheremet A.D., Selezneva I.N., Kovalev V.V., Savitskaya G.V., Podolsky V.I., etc.);
- the purpose of which is to build uniform complex values (ratings) of the financial condition of the enterprise and allow comparisons in time and space (methods of Sheremet A.D., Sukhovoy L.F., Chernova V.A., Kondrakova N.P., Berezhnoy E V., Dontsova L.V. and Nikiforova N.A. and others);
- allowing to identify and quantify the probability of bankruptcy of an enterprise by constructing multifactor models (models of Altman E., Tishaw G., Taffler R., Biver U., Zaitseva OP, Fedotova MA, etc.);
- based on the relationship between the financial condition of the enterprise and the market price of a share (Bliss J., Vinacor A., etc.);
- allowing to predict the level of efficiency of investment in certain securities and the degree of risk associated with it (Foster J. and others), etc.

However, an analysis of the literature did not allow us to find universal methods and models suitable for analyzing the state of entire industries, as well as constructing integral indicators based on the analysis. Studies devoted to the analysis of the financial and economic condition of individual industries, as a rule, are based on a time analysis of individual indicators of either the largest enterprises in the industry, or the sum of such indicators. There are practically no models to quantify the level of development and dynamics of the entire industry as a whole. It was not possible to find articles in which the research would be carried out not on a limited sample of enterprises, but on all information about the industry available from open data sources.

Purpose of the work: development and testing of a methodology that allows you to analyze the dynamics of financial and economic indicators of a given industry within a given period of time in order to optimize its management, based on open Internet data sources, using licensed software, as well as systems of fuzzy-logical conclusions.

Research objectives:

- Using the previously developed methodology for collecting and analyzing data from open Internet sources ("Audit-IT", [3]), analyze the financial and economic condition of IT enterprises in the Rostov region for 2017–2019.
- To develop a fuzzy-multiple methodology for analyzing the dynamics of individual financial and economic indicators of a given industry in the considered time period, as well as an analysis of the dynamics of the financial and economic state of the industry as a whole.
- On the basis of the obtained data and the constructed fuzzy-multiple methodology, analyze the dynamics of such indicators of the IT industry, such as the sum of revenues, expenses and profits, paid taxes, social benefits, etc., as well as analyze the dynamics of the financial and economic state of the industry as a whole.
- Based on the results obtained, draw conclusions that allow assessing the state of the industry as a whole and optimizing its management.

The study was carried out using a specially developed methodology [4] aimed at obtaining an exhaustive analysis of the development of the relevant industry based on open Internet sources, which determines its relevance in the context of limited access to financial statements of enterprises. When developing the methodology, the existing methods for assessing social and economic systems were used by a set of indicators [5].

2 Research Methodology and Material

2.1 The Technique Audit-IT and its Modification, Developed Earlier

The methodology proposed by us is based on the well-known Audit-IT [3] methodology for assessing the financial and economic condition of an enterprise according to three groups of coefficients. The first group – indicators of financial stability: coefficient of autonomy; the coefficient of provision with own circulating assets; investment coverage ratio. The second group – indicators of solvency: current (total) liquidity ratio; quick (intermediate) liquidity ratio; absolute liquidity ratio. Third group: return on sales; net profit rate; return on assets (ROA). In addition, the Final score is used, which is calculated according to a specific algorithm based on a variety of indicators and characterizes the general state of the enterprise. For each of the indicators, on the basis of expert assessments, the intervals of values are set: "excellent", "good", "unsatisfactory", "critical" (for each industry its own). Thus, each enterprise for the reporting year has its own assessment system for each of the indicators under consideration. On the basis of this system, an expert analysis of the financial and economic condition of the enterprise is carried out, conclusions are drawn.

A modification of the methodology [4, 5] is proposed to assess not an individual enterprise, but an entire industry, by groups: micro-enterprises, mini-enterprises, small enterprises, medium-sized enterprises, large enterprises (division by the number of employees). For each enterprise, for each studied year, a data line is formed: three indicators of financial stability, solvency, efficiency, as well as the final score. Based on the Audit-IT methodology, the indicators are assessed, after which the assessments are aggregated by the group using fuzzy multilevel [0,1] – classifiers. Thus, each group of enterprises for the reporting year has its own system of assessments for each of the indicators under consideration. After that, on the basis of the resulting system, a standard expert analysis of the financial and economic condition of a group of enterprises is carried out, and conclusions are drawn.

In this paper, a modification of the above-described methodology is proposed, which makes it possible to assess the dynamics of the development of the financial and economic condition of individual groups and the entire industry as a whole.

2.2 Used Math Tool

To assess the dynamics of the financial and economic state of the industry, the theory of fuzzy sets is used. A linguistic variable is taken into consideration, the term-set of which consists of four terms: 1 – "Stable tendency towards a worsening of the situation, a critical situation"; 2 – "Tendency to deteriorate, unsatisfactory"; 3 – "Tendency to improve, good"; 4 – "Steady tendency to improve, excellent." Membership functions are also trapezoidal. To assess the dynamics of the industry development, for each group. the following algorithm is proposed.

1. The average values of all ten indicators (nine coefficients plus the final score) for each of the groups under consideration are tabulated for the years under study.
2. Indicators are normalized, that is, they are divided by the highest value of the indicator in its time series; after that, the aggregated value of the indicator is calculated, taking into account the temporal significance, in accordance with the Fishburne formula: P = 0,167·P(2017) + 0,333·P(2018) + 0,5·P(2019)
3. A summary table of aggregated values of the studied indicators for five groups is compiled, linguistic recognition of terms is performed in accordance with normative estimates;
4. A comprehensive assessment of the industry is calculated on the basis of a set of studied indicators, as well as a system of fuzzy-logical conclusions, standard four-level [0,1] – classifiers, followed by linguistic recognition of the result.

3 Results

Based on the parsing of the site "For Honest Business" [6] for 2017, data on 880 enterprises of the Rostov region were examined. It was revealed that 634 enterprises have the data required for the analysis.

Accordingly, for 2018 data are available for 633 enterprises. At the same time, for 2019, the website "For Honest Business" contains information about 468 enterprises from the list for the last year (864 enterprises), that is, 54.17% of the headcount of the previous year. At the same time, 337 enterprises have information (against 633 enterprises in the previous year), that is, 53.24% of the headcount in the previous year. Thus, the share of enterprises suitable for analysis in 2019 is 72% of the total, against 73.26% in the previous year. For each year, a detailed analysis was carried out • comparative analysis of the financial condition of IT-industry enterprises by groups.

Tables 1 and 2 show the process of assessing the dynamics of the financial and economic condition of groups of enterprises. Below are the numerical value of complex assessments of the dynamics of the financial and economic state of groups with subsequent linguistic recognition. For each group, on the basis of the numerical values of the indicators of the dynamics of the coefficients, the analysis of indicators of financial stability, indicators of solvency, indicators of the efficiency of the enterprises of the groups, as well as the final score for the groups was carried out.

Micro-Enterprises

1. **Analysis of the dynamics of financial stability indicators** indicates, in general, an increase the degree of solvency (creditworthiness) of the enterprises of the Micro-enterprise group, their availability of funds necessary to maintain a stable and efficient operation.
2. **Analysis of solvency indicators**. Low value of the indicator of the dynamics of the current (total) liquidity ratiospeaks of an increase in the difficulties in repaying the organizations of the group of their currentobligations. Low value of the indicator of the dynamics of the ratio of quick (intermediate) liquidity speaks of deterioration the ability of companies to pay off their short-term liabilities on time using highly liquid assets. Low value of the indicator of dynamics coefficient absolute liquidityspeaks of deterioration the ability of companies to pay off their short-term liabilities at the expense of cash and cash equivalents in the form of marketvaluable papersand deposits, i.e. absolutely liquid assets.
3. **Analysis of performance indicators** showed, in general, an increase the efficiency of using assets, equity capital, borrowed and invested capital, circulating and non-circulating assets, respectively.

Final Score. Micro-enterprises' enterprises meet the rating "good". Consequently, the relevant enterprises belong to the category of borrowers for whom the likelihood of obtaining credit resources is high (good creditworthiness).

Comprehensive group assessment:
$G1 = 0.1250.0 + 0.375\cdot 0{,}4 + 0.625\cdot 0,6 + 0.875\cdot 0.0 = 0.525$ (excellent).

Medium Enterprises

1. **Analysis of the dynamics of indicators of financial stability**. The low value of the indicator of the dynamics of the autonomy ratio shows the strengthening of dependence on creditors; the low value of the indicator of the dynamics of the coefficient of provision with own circulating assets indicates a deterioration in the structure of the balance sheet of the enterprises of the group as a whole; the low value of the indicator of the dynamics of the investment coverage ratio indicates the illiquidity of the assets of companies and their inability to fully pay off their current accounts. Thus, all of the above indicates the presence of a serious tendency

Table 1. Calculation for the group of Micro-enterprises.

	Indicator	Weight	Term1	Term2	Term3	Term4
1	Autonomy ratio	0.1	0	1	0	0
2	Coefficient of provision with own circulating assets	0.1	0	1	0	0
3	Investment coverage ratio	0.1	0	0	1	0
4	Current (total) liquidity ratio	0.1	0	1	0	0
5	Quick (intermediate) liquidity ratio	0.1	0	0	1	0
6	Absolute liquidity ratio	0.1	0	1	0	0
7	Return on sales	0.1	0	0	1	0
8	Net profit margin	0.1	0	0	1	0
9	Return on assets (ROA)	0.1	0	0	1	0
10	Final score	0.1	0	0	1	0
	Term weights		0	0,4	0.6	0
	Coefficients of the terms		0.125	0.375	0.625	0.875

towards a decrease in the degree of solvency (creditworthiness) of enterprises of the Medium Enterprises group, which determines the availability of funds, in order to maintain a stable and efficient operation of enterprises; a high degree of dependence on borrowed capital.

2. **Analysis of indicators of solvency.** The low value of the indicator of the dynamics of the current (total) liquidity ratio indicates an increase in difficulties in repaying the group's organizations of their current liabilities. The low value of the indicator of the dynamics of fast (intermediate) liquidity indicates a deterioration in the ability of companies to pay off their short-term liabilities on time with the help of highly liquid assets. The low value of the indicator of the dynamics of the absolute liquidity ratio indicates a deterioration in the ability of companies to pay off their short-term liabilities at the expense of cash and cash equivalents in the form of market securities and deposits, i.e. absolutely liquid assets.
3. **Analysis of performance indicators** showed, in general, a decrease in the efficiency of using assets, equity capital, borrowed and invested capital, circulating and non-circulating assets, respectively.
4. **Final score**. Medium enterprises are rated "unsatisfactory". Consequently, these entities can be considered as counterparties in the relationship with which a prudent approach to risk management is required. They can apply for credit resources, but the decision largely depends on the analysis of additional specific factors (neutral creditworthiness).

Comprehensive group assessment:
G4 = 0,125·0,0 + 0,375· 1,0 + 0,625·0, 0 + 0,875·0,0 = 0,375 (unsatisfactory.)

Table 2. Calculation for the group of Medium enterprises.

	Indicator	Weight	Term1	Term2	Term3	Term4
1	Autonomy ratio	0.1	0	1	0	0
2	Coefficient of provision with own circulating assets	0.1	0	1	0	0
3	Investment coverage ratio	0.1	0	1	0	0
4	Current (total) liquidity ratio	0.1	0	1	0	0
5	Quick (intermediate) liquidity ratio	0.1	0	1	0	0
6	Absolute liquidity ratio	0.1	0	1	0	0
7	Return on sales	0.1	0	1	0	0
8	Net profit margin	0.1	0	1	0	0
9	Return on assets (ROA)	0.1	0	1	0	0
10	Final score	0.1	0	1	0	0
	Term weights		0	1	0	0
	Coefficients of the terms		0.125	0.375	0.625	0.875

4 Conclusions

The theoretical significance of the work performed consists in the development of a comprehensive analysis methodology aimed at studying the dynamics of the financial and economic development of the industry, using open data sources, artificial intelligence tools and the theory of fuzzy sets. The technique has a clear algorithm that is easily implemented using standard software tools, modifiability, the ability to take into account expert opinions. In addition, it, practically unchanged, can be transferred to any other industry of an arbitrarily chosen region. The practical significance of the work lies in the conclusions regarding the dynamics of the development of the IT industry in the Rostov region.

As a result of the study, it was found that an improvement in financial stability is observed only in the group of micro-enterprises, in other groups there is more or less debt. Large enterprises are characterized by an increase in the efficiency of using assets, equity capital, borrowed and invested capital, circulating and non-circulating assets, respectively; the opposite trend is observed for the rest of the groups. At the same time, small and medium-sized enterprises are characterized by decrease the efficiency of using assets, equity capital, borrowed and invested capital, circulating and non-circulating assets, respectively, the rest of the groups are characterized by an increase. Small and medium-sized enterprises are neutral lenders that should be approached with caution. The rest of the enterprises belong to categories of borrowers for whom the likelihood of obtaining credit resources is high (good creditworthiness).

References

1. Alekseeva, O.A., Gorbachev, A.S.: Financial analysis of the enterprise: essence, problems and prospects. Kant **2**(5) (2012) . https://cyberleninka.ru/article/n/finansovyy-analiz-deyatelnosti-predpriyatiya-suschnost-problemy-i-perspektivy 10 Sept. 2021
2. Vasilishin, T.M.: Classification of methods and models of financial analysis of enterprises. Austrian J. Humanities Social Sci. 3–4. (2014). https://cyberleninka.ru/article/n/klassifikatsiya-metodov-i-modeley-finansovogo-analiza-predpriyatiy (дата обращения: 10.09.2021)
3. Audit-IT/ru: Financial analysis based on reporting data. [Electronic resource]. https://www.audit-it.ru/finanaliz/features/
4. Gorbunov, S.N., Usatyi, R.S., Saplev, M.V.: Research of the financial and economic condition of enterprises in the IT industry based on the analysis of Internet data. Intellectual Resour. Reg. Dev. **2**, 310–316 (2020)
5. Vovchenko, N.G., Sakharova, L.V., Epifanova, T.V., Kokhanova, V.S.: Fuzzy-multiple modification of the spectrum-point methodology for assessing the financial condition of the company (based on the audit-IT). In: Aliev, R.A., Kacprzyk, J., Pedrycz, W., Jamshidi, M., Babanli, M.B., Sadikoglu, F.M. (eds.) ICSCCW 2019. AISC, vol. 1095, pp. 275–283. Springer, Cham (2020). https://doi.org/10.1007/978-3-030-35249-3_35
6. FOR HONEST BUSINESS: Informational portal. [Electronic resource]. https://zachestnyibiznes.ru

Method for Diagnostics and Selection of the Type of Market Structure of Commodity Markets Based on Fuzzy Set Theory

Rena Mikayilova[1,2] and Ramiz Alekperov[2,3](✉)

[1] Digital Technologies and Applied Informatics, Azerbaijan State Economic University, Baku AZ1147, Azerbaijan
rana.mikayilova@unec.edu.az
[2] Azerbaijan State Oil and Industry University, Baku AZ1010, Azerbaijan
[3] Department of Computer Engineering, Odlar Yurdu University, Baku AZ1072, Azerbaijan

Abstract. This work is devoted to the issues of diagnostics and the choice of the type of market structure of the commodity market in the conditions of a free economy and global systemic changes, as a necessary component of economic growth and transformation of commodity markets. The paper concludes that the main and most difficult stage in diagnosing product markets is the process of selecting index indicators and coefficients for the study of product markets. It is proposed to use a complex of indices and coefficients that allow determining the level of the competitive environment in the commodity markets. The method for determining the type of the market structure of the commodity market based on the analysis of the state of competition is presented. The proposed method is based on the use of fuzzy inference capabilities, which is implemented on the basis of the MATLAB tool environment.

Keywords: Commodity market · Structure of the commodity market · Competition · Concentration · Fuzzy set theory · Fuzzy logical conclusion

1 Introduction

The market structure is largely determined by the level of homogeneity, i.e. concentration of its players. Determining the degree of market concentration allows one to assess the level of its competitiveness in the industry market.

The most important characteristic of the type of market structure is the level of competition, that is, the ability of an individual seller to influence the general market situation. The more competitive the market, the lower the opportunity. The competition itself can be both price (change in price) and non-price (change in the quality of goods, design, service and advertising) [1].

So, for example, the market concentration of the production of a good expresses the share of large firms participating in a given market, or the share of large enterprises in an industry participating in the volume of production of a product, as well as in terms of sales in the market. In accordance with the theory of the commodity market, monopoly

R. A. Aliev et al. (Eds.): ICSCCW 2021, LNNS 362, pp. 732–740, 2022.
https://doi.org/10.1007/978-3-030-92127-9_97

power is largely determined by concentration. Along with this, there are a number of factors that are more isolated and operate independently on the commodity market.

It should be noted that the structure of the commodity market includes many boundaries, which are expressed in its various indicators and coefficients, which can be used to determine the level of market concentration [2, 3].

For a deeper study of the commodity market, it's forecasting, and diagnostics, we propose to use a set of indices and coefficients, which are combined and presented in Table 1.

Group I – the Herfindahl-Hirschman coefficient, which includes the sum of the squares of the specific weight of all enterprises operating in the commodity market.

Group II – the Hall-Tideman index, which is determined on the basis of comparing the relative ranks of enterprises and their share.

Group III – the coefficient of relative concentration, which is determined by the ratio of the largest firms in the market and their regulation of the share of the sale of goods.

Group IV – the coefficient of entropy, which is expressed by the average specific weight of firms operating in the market according to the natural logarithm of its reciprocal.

Group V – the variance of the logarithms of the specific weight, containing the distribution of sales volumes between firms.

Group VI – the index of the maximum specific gravity, which is determined by the ratio of the maximum and minimum specific gravity.

Group VII – the Ginny concentration coefficient – characterizes the level of distribution of the population units according to the degree of the trait.

Specific formulas for calculating the above indices and coefficients are described below:

I. Herfindahl-Hirschman coefficient (HHI).

Used to implement antitrust policy that takes into account both the number of enterprises and the level of inequality in the market. It is defined as the sum of the squaress of the shares of all enterprises operating on the product market according to the formula:

$$HHI = \sum_{i=1}^{n} q_i^2 \tag{1}$$

where qi is the volume of production of goods in the given commodity market of the i-th enterprise; n is the number of enterprises. The lower the HHI, the lower the concentration; HHI is aiming for one, for a monopoly market – for 10,000.

II. Hall-Tydeman Index (HT). It is determined on the basis of comparing the relative ranks of enterprises and their share.

$$HT = \frac{1}{2(\sum_{i=1}^{n}\left[(R_i q_i) - 1)\right]} \tag{2}$$

where Ri –1,2,3… the rank of the i-th enterprise in the market;
qi is the share of the i-th enterprise in the product market.

In the commodity market, the coefficient reaches its maximum value (NT (max) = 1). Its minimum value for the market of perfect competition is NT (min) = 0.

III. The coefficient of relative concentration (K) includes the ratio of the number of major firms in the market and the share of goods controlled by them:

$$K = (20 + 3\beta)/\alpha \tag{3}$$

where β is the share of the main firms in the market in the total number of firms,%;
α is the share of sales of these firms in the total volume of goods sold, %.

At K > 1, there is no concentration, the product market is competitive. At K ≤ 1, there is a high level of concentration on the market.

IV. The entropy coefficient (E) is expressed by the average specific weight of firms operating in the market, weighted by the natural logarithm of its reciprocal:

$$E = \frac{1}{n}\sum\nolimits_{i=1}^{n} q_i \ln\left(\frac{1}{q_i}\right) \tag{4}$$

where E is the coefficient of entropy;
qi is the share of production of the i-th enterprise on the commodity market;
n is the number of enterprises on the market.

This coefficient expresses the level of market concentration and helps to specifically investigate the degree and dynamics of concentration: the larger the coefficient of entropy (E), the greater the economic uncertainty, the lower the level of concentration of sellers in the market.

V. The variance of the logarithms of market shares (σ2) includes the distribution of sales volumes between firms:

$$\sigma^2 = \frac{1}{n}\sum\nolimits_{i=1}^{n} (\ln(q_i) - \ln(q_{aver}))^2 \tag{5}$$

where qi is the specific weight of the i-th firm in the commodity market;
q_{aver} is the average market share of one firm, equal to 1/n;
n is the number of firms moving on the market.

Variance expresses the possible market power of firms through the inequality of their sizes.

VI. The index of the maximum share (Idmax) is determined as follows:

$$I_{dmax} = \frac{(Q_{max} - M_q)}{(Q_{max} + M_q)} \tag{6}$$

where M (q) is the arithmetic average market share;
Qmax is the maximum share in this market.

This means that for a competitive market, the share of each firm will be equal to 1/n. The level of deviation of the firm's specific weight from this value will characterize the level of difference from this value.

Limit values:

- index from 1 to 0.75 – monopoly market;
- index from 0.75 to 0.50 – oligopolistic market;
- index from 0.50 to 0.25 – the market of monopolistic competition;
- index from 0.25 to 0 – the market is competitive.

VII. The Gini concentration coefficient (G) includes the level of distribution of population units by degree and takes values from 0 to 1. The larger it is, the greater the level of concentration and uneven distribution in the market. The Gini coefficient is reflected by the Lorenz curve, showing the relationship between the % market share and the number of firms operating in the market.

This indicator characterizes the ratio of the area of the bounded actual Lorentz curve and the Lorentz curve for a stable distribution to the area of a triangle bounded by the Lorentz curve for a stable distribution of shares and the abscissa and ordinate axes.

There are a number of options for calculating the Gini coefficient (G). In this case, we can highlight:

$$G = \left|1 - 2\sum\nolimits_{i=n}^{n} d_{xi}d(cum)_{yi} + \sum\nolimits_{i=n}^{n} d_{xi}d_{yi}\right| \quad (7)$$

where dxi is the share of the firm in each group;

dyi is the share of each group of firms in the total volume of production in a particular market;

d (cum) yi is the cumulative share of production, ranked from smallest to largest firm in the market.

Analysis of the results of all factors for forecasting and diagnosing the commodity market showed that an important stage in determining the concentration of the commodity market is the selection of the necessary index indicators and coefficients, which is mainly of a relative nature. It is the relativity of these indicators that makes it possible to realistically assess the commodity market.

The relativity of these indicators, on the other hand, is expressed in the same way, in that the boundaries of these indicators are expressed subjectively and verbally. For example,

1. the greater the coefficient of entropy (E), the greater the economic uncertainty, the lower the level of concentration of sellers in the market;
2. the lower the HHI, the lower the concentration, etc.

For a better formulation and representation of these factors, the theory of fuzzy sets, which was first proposed by Lutfi Zadeh in 1965 [4], is suitable.

On the other hand, the need to take into account the proposed factors when solving the problem of forecasting and diagnosing the commodity market makes it possible to classify this problem as a decision-making problem based on fuzzy inference [5–12], using the base of fuzzy rules.

2 Formulation of the Problem

The analysis of the proposed indicators for solving the problem of diagnosing commodity markets has shown that their intervals, on the basis of which it is possible to determine the level of concentration of commodity markets, are fuzzy. On the other hand, experts always make decisions about the concentration of commodity markets subjectively, based on experience and intuition, which determines the use of fuzzy inference systems to solve this problem.

To implement systems based on fuzzy rules, many fuzzy inference algorithms have been developed. Fuzzy inference algorithms differ mainly in the type of rules used, logical operations and a kind of defuzzification method. Fuzzy inference models have been developed for Mamdani, Sugeno, Larsen, Tsukamoto, Aliyev (ALI-1, ALI-2, ALI-3, ALI-4) [6, 9].

As you can see from the figure, the fuzzy inference system consists of the following blocks:

1. Knowledge base which includes: 1.1 Rule Base – It contains fuzzy IF-THEN rules. 1.2 Database – It defines the membership functions of fuzzy sets used in fuzzy rules.
2. Fuzzification Interface Unit – It converts the crisp quantities into fuzzy quantities.
3. Decision-making Unit – It performs operation on rules.
4. Defuzzification Interface Unit – It converts the fuzzy quantities into crisp quantities.

We propose to use the Mamdani algorithm using the MATLAB instantiated environment for the implementation of a fuzzy inference system [13–16].

In the database of the inference system, based on the research results, which are given in Table 1, the membership functions of the index indicators are implemented, which are presented in the form of linguistic terms with trapezoidal membership functions.

Using the index indicators HHI, K, E, G, 81 rules were developed for a logical conclusion to determine the level of competition, which make it possible to determine the degree of the possibility of forming a monopoly in the trade markets of the Republic of Azerbaijan. A snippet of the inference rule base is shown below:

R1: If (HHI is H) and (K is H) and (E is M) and (G is L) then (Competitiveness is Middle) (1)

R2: If (HHI is L) and (K is H) and (E is L) and (G is M) then (Competitiveness is Middle) (1)

R3: If (HHI is L) and (K is H) and (E is L) and (G is H) then (Competitiveness is Middle) (1)

R4: If (HHI is L) and (K is M) and (E is H) and (G is H) then (Competitiveness is Middle) (1)

R5: If (HHI is M) and (K is M) and (E is H) and (G is L) then (Competitiveness is High) (1)

Table 1. Fuzzy intervals of index indicators and general rules for determining the possibility of forming a monopoly or oligopoly.

	Rules				
Coefficients		Concentration		Competitiveness	The probability of forming a monopoly or oligopoly
1. Herfindahl-Hirschman Index (HHI)	IF	High	THEN	Low	High
		Middle		Middle	Middle
		Low		High	Low
2. Hall-Tideman Index (HT)		Low		High	Low
		Middle		Middle	Middle
		High		Low	High
3. Relative concentration coefficients (K)		Low		High	Low
		Middle		Middle	Middle
		High		Low	High
4. Entropy coefficient (E)		High		Low	High
		Middle		Middle	Middle
		Low		High	Low
5. Index of the maximum share (Imax)		High		Low	High
		Middle		Middle	Middle
		Low		High	Low
6. Gini coefficient		Low		High	Low
		Middle		Middle	Middle
		High		Low	High

R6: If (HHI is H) and (K is H) and (E is H) and (G is L) then (Competitiveness is High) (1) etc.

According to the developed rules, the received responses of index indicators and their influence on the levels of competition in the commodity markets, which are presented in Figs. 1, 2, 3 and 4.

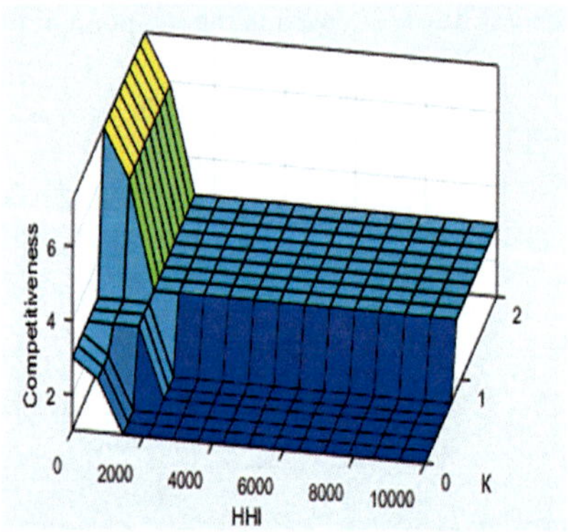

Fig. 1. Surface HHI-K-Competitiveness

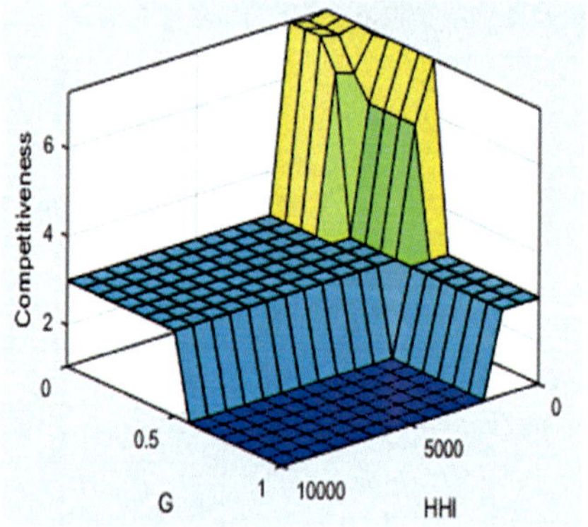

Fig. 2. Surface HHI-G-Competitiveness

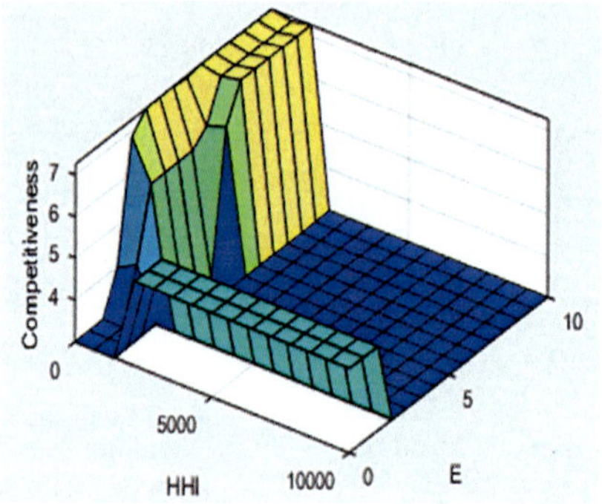

Fig. 3. Surface HHI-E-Competitiveness

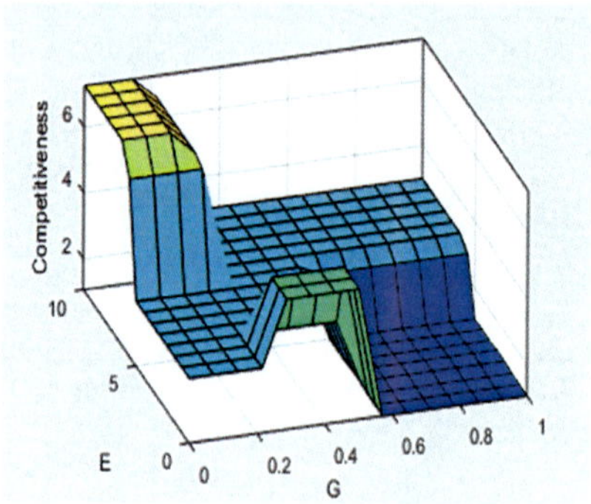

Fig. 4. Surface G-E-Competitiveness

3 Determining the Level of Concentration of Commodity Markets

To determine the level of concentration of the commodity markets of the Republic of Azerbaijan, we used the results of calculating index indicators: Herfindahl-Hirschman coefficient (HHI), relative concentration coefficient, entropy coefficient, Gini concentration coefficient for industrial, agricultural, and consumer goods markets in 2013, 2017 and 2020. The results are given for determining the levels of competition for the specified product markets are shown in Table 2.

The table shows that the level of competition for the industrial product market in 2013 was low, that is, there was a certain monopoly. But since 2017, the situation in this product market has changed for the better and moved to a level of high competition, which indicates that there is no monopoly on this product market today.

The level of competition in the consumer product market in 2013 was low, and in 2017 it moved to the level of medium competition and today there is high competition in this product market.

Table 2. The results of the inference system

	Index indicators	2013	2017	2020
Industrial commodity market	HHI	1681	2809	3320
	K	0.232	0.288	0.289
	E	4.3	3.5	3.97
	G	4.123	5.204	4.99
Result of Fuzzy Inference System	Competitiveness	Low	Low	Low
Agricultural commodity market	Index indicators	2013	2017	2020
	HHI	324	169	171.4
	K	0.194	0.105	0.285
	E	5.1	9.5	9.66
	G	2.278	2.059	2.153
Result of Fuzzy Inference System	Competitiveness	Middle	High	High
Consumer product market	Index indicators	2013	2017	2020
	HHI	484	225	809
	K	0.008	0.003	0.888
	E	1.25	3.3333	5.195
	G	1.725	1.878	2.042
Result of Fuzzy Inference System	Competitiveness	Low	Middle	High

However, the level of competition for the industrial product market turned out to be low and did not change over the years under study, which suggests that there is a certain monopoly on this product market.

4 Conclusion

Diagnostics of the level of competition of the republic's commodity markets (industrial commodity market, agrarian commodity market, consumer commodity market), carried out on the basis of a fuzzy logical conclusion, showed that positive dynamics is observed in the agrarian and consumer goods markets in terms of organizing pure competition. This suggests that in these markets, commodity producers and consumers will find their rightful place not only in the domestic but also in foreign markets.

The studies carried out have confirmed that the country's commodity markets are mainly developing and expanding steadily, and the existing shortcomings are eliminated in accordance with the adopted state programs for the socio-economic development of the republic in all important areas.

The studies carried out and the results obtained confirm the correctness and expediency of applying the theory of fuzzy sets to solve the problems of diagnosing and choosing the type of market structure of commodity markets, using the complex of indices and coefficients for assessing the level of concentration proposed in this work.

References

1. Galyautdinov, R.R.: Industry markets in the Russian economy: features of the structure and characteristics of leadership Types of market structures: perfect competition, monopolistic competition, oligopoly and monopoly, Website of an economics teacher (2015). http://galyautdinov.ru/post/tipy-rynochnyx-struktur. Accessed 18 July 2021
2. Yusupova, A.T.: Industry markets in the Russian economy: features of the structure and characteristics of leadership. Vestnik NQU **8**(3), 53–61 (2008)
3. Smirnova, Y.B.: Methodology for determining the type of structure of the commodity market. https://cyberleninka.ru/article/n/metodika-opredeleniya-tipa-struktury-tovarnogo-rynka/. Accessed 18 July 2021
4. Zadeh, L.A.: Fuzzy sets. Inf. Control **8**(3), 338–353 (1965). https://doi.org/10.1016/s0019-9958(65)90241-x
5. Zadeh, L.A.: Fuzzy logic. Computer **21**(4), 83–93 (1988). https://doi.org/10.1109/2.53
6. Aliev, R.A., Fazlollahi, B., Aliev, R.R.: Soft Computing and its Applications in Business and Economics. Springer, Berlin, Heidelberg (2004). https://doi.org/10.1007/978-3-540-44429-9
7. Aliev, R.A., Pedrycz, W.: Fundamentals of a fuzzy-logic-based generalized theory of stability. IEEE Trans. Syst. Man Cybern. Part B Cybern. **39**(4), 971–988 (2009). https://doi.org/10.1109/TSMCB.2008.2010523
8. Aliev, R.A., Aliev, R.R.: Soft Computing and Its Applications. WORLD SCIENTIFIC (2001). https://doi.org/10.1142/4766
9. Piegat, A.: Methods of fuzzy modeling. Stud. Fuzziness Soft Comput. (2001). https://doi.org/10.1007/978-3-7908-1824-6_6
10. Oglu, A.R.B., Kizi, I.I.T.: A method for forecasting the demand for pharmaceutical products in a distributed pharmacy network based on an integrated approach using fuzzy logic and neural networks. In: Kahraman, C., Cevik Onar, S., Oztaysi, B., Sari, I.U., Cebi, S., Tolga, A.C. (eds.) INFUS 2020. AISC, vol. 1197, pp. 998–1007. Springer, Cham (2021). https://doi.org/10.1007/978-3-030-51156-2_116
11. Oglu, A.R.B., Oglu, S.V.M.: Estimation of potential locations of trade objects on the basis of fuzzy set theory. In: Kahraman, C., Cevik Onar, S., Oztaysi, B., Sari, I.U., Cebi, S., Tolga, A.C. (eds.) INFUS 2020. AISC, vol. 1197, pp. 228–237. Springer, Cham (2021). https://doi.org/10.1007/978-3-030-51156-2_28
12. Kizi, M.R.N.: Comprehensive research of the commodity market: theoretical, methodological and modern approaches using neural networks with fuzzy logic. In: Kahraman, C., Cevik Onar, S., Oztaysi, B., Sari, I.U., Cebi, S., Tolga, A.C. (eds.) INFUS 2020. AISC, vol. 1197, pp. 973–980. Springer, Cham (2021). https://doi.org/10.1007/978-3-030-51156-2_113
13. D'yakonov, V., Kruglov, V.: MATLAB Mathematical Extension Packages, p 480. Piter (2001) (in Russian)
14. Fuzzy Logic Toolbox: https://www.mathworks.com/products/fuzzy-logic.html
15. Imamguluyev, R.: Application of fuzzy logic model for correct lighting in computer aided interior design areas. In: Kahraman, C., Cevik Onar, S., Oztaysi, B., Sari, I.U., Cebi, S., Tolga, A.C. (eds.) INFUS 2020. AISC, vol. 1197, pp. 1644–1651. Springer, Cham (2021). https://doi.org/10.1007/978-3-030-51156-2_192
16. Valiyev, A., Imamguluyev, R., Ilkin, G.: Application of fuzzy logic model for daylight evaluation in computer aided interior design areas. In: Aliev, R.A., Kacprzyk, J., Pedrycz, W., Jamshidi, M.O., Babanli, M., Sadikoglu, F.M. (eds.) ICAFS 2020. AISC, vol. 1306, pp. 715–722. Springer, Cham (2021). https://doi.org/10.1007/978-3-030-64058-3_89

Decision Making on Tourism by Using Natural Language Processing

Nigar F. Huseynova(✉)

Joint MBA Program, Azerbaijan State Oil and Industry University, 20 Azadlig Avenue, Baku AZ1010, Azerbaijan
huseynova.nigar@asoiu.edu.az

Abstract. The paper deals with measuring the effectiveness of Natural Language in decision-making to have comprehensive results. It's more convenient for people to comprehend the meaning of linguistic terms rather than quantitative data. A lot of studies have shown that people don't understand or they are reluctant to act upon numerical data. We surveyed people presenting two different types of information: Natural Language Generation (NLG) and graphical representation to measure the predicted number of tourist arrivals to three countries in four quarters. Fuzzy Inference analysis was applied on the base of the collected data. As a result of the survey, it became clear that people are more inclined in making decisions on NLG rather than on state-of-the-art graphics. Only 27% of the people made decision on graphical based information, the others chose NLG conveyed by "IF…Then" rules. The validity of the decisions made on NLG is 78%, while the other one is less than 62%.

Keywords: Natural Language Processing · Natural Language Generation · Decision making under uncertainty · Fuzzy If-Then rules · Fuzzy Inference Systems · Deep learning

1 Introduction

The main purpose of Natural Language Processing which is one of the main subsets of Artificial Intelligence is to support computers to learn and process human languages. It makes computers be close to a human-level intelligence, comprehension, intuition [1]. It is an area which incorporates computer science, artificial intelligence, and linguistics. But fully understanding and reproducing the meaning of language is an extremely difficult task. Computers are unable to convert the same semantic understanding, meaning as human. They don't possess the same intuitive comprehension of natural languages [2]. Recent advances in Artificial Intelligence support computers to present innovations in natural language. Deep Learning makes possible to write programs to enhance translation, semantic understanding and summarization [3]. etc. All the mentioned issues enable people to comprehend and process computations on large texts.

Natural Language Generation is one of the subsets of Natural Language Processing which enables computers to write, to produce a human language text response based

R. A. Aliev et al. (Eds.): ICSCCW 2021, LNNS 362, pp. 741–747, 2022.
https://doi.org/10.1007/978-3-030-92127-9_98

on some data input [4]. Natural Language Generation (NLG) technology enables us to get better results than traditionally used quantitative data representation techniques for supporting exact decision-making [5]. NLG has the potential to automatically produce linguistic descriptions of uncertainty to support decision-making. It can be used as a successful method to support decision-making.

A lot of NLG researchers have investigated the conversion of numbers into linguistic units [6]. We analyze the importance of converting numerical representations into Natural Language to get better choices in decision-making [7]. We consider the exemplar task of measuring number of tourist arrivals during four quarters of a year for 3 selected countries. Section 2 gives introductory information about criteria of fuzzy If-Then rules, Sect. 3 discusses the main goal of the paper and some introductory information about the research. Section 4 explains the implementation of the process, gives results derived from Fuzzy logic-based analysis. Section 5 comprises the conclusion of the work.

2 Preliminaries

Definition 1. Fuzzy number [8, 9]: A fuzzy number is a set A on R which has the following features: a) A is an ordinary fuzzy set; b) A is a convex fuzzy set; c) $\alpha-$ cut of A, A^{α} is a closed interval for each $\alpha \in [0, 1]$; d) the support of A, A^{+0} is bounded.

Definition 2. Trapezoidal fuzzy number: A Trapezoidal fuzzy number (TrFN) denoted by $\tilde{A}$ is defined as (a_1, a_2, a_3, a_4) where the membership function

$$\mu_{\tilde{A}}(x) = \begin{Bmatrix} 0, \ x \le a_1 \\ \dfrac{x - a_1}{a_2 - a_1}, \ a_1 \le x \le a_2 \\ 1, \ a_2 \le x \le a_3 \\ \dfrac{a_4 - x}{a_4 - a_3}, \ a_3 \le x \le a_4 \\ 0, \ x \ge a_4 \end{Bmatrix}$$

Definition 3. Fuzzy If-Then rules [8, 9]: Fuzzy if-then rule representations are generally used to indicate the conventional statements that fuzzy logic possess.

A common fuzzy if-then rule is expressed as follows:

If x is A then y is B

where A and B denote linguistic values conveyed by fuzzy sets.

Multi-input multi-output fuzzy system is formulated in the following form

$$\text{If } X_1 \ is \ A_1 \text{ and } X_2 \text{ is } A_2 \text{ and} \ldots \text{ and } X_n \text{ is } A_n$$
$$\text{Then}$$
$$Y_1 \text{ is } B_1 \text{ and } Y_2 \text{ is } B_2 \text{ and} \ldots \text{ and } Y_m \text{ is } B_m$$

where A_i and B_i are data pieces.

3 Statement of the Problem

The paper analyzes which knowledge is more appropriate for human comprehension, how decision-making process is held by using NLP, what conveniences it contributes to decision making. We measure different types of knowledge to compare. Firstly, we gathered statistical data on the number of tourist arrivals to the selected countries (A, B, C countries) in 4 quarters. The statistical data on tourist arrivals have been taken from official sites of the countries. Through graphical representations constructed on statistics and textual forecasts conveyed by If-Then rules the predicted number of tourist arrivals is predicted. Survey participants are asked to initially choose the best touristic destination and time to visit. They must choose graphics or If-Then rules to make decision. Inputs are seasons which don't have universal boundaries, they're based on weather conditions. Output (low, medium, high) is the number of tourist arrivals. Fuzzy Inference analysis was applied to formulate mapping of inputs and outputs. The seasons are described by trapezoidal fuzzy numbers and the parameters of trapezoidal fuzzy numbers are months. Winter is the union of two fuzzy numbers. If-Then rules are constructed as follows:

IF winter (1,1,2,4) ∪ (10,12,12,12) THEN low
IF spring (2,4,5,7) THEN low
IF summer (5,7,8,10) THEN high
IF autumn (8,10,10.5,12) THEN low

For B country
IF winter (1,1,2,4) ∪ (10,12,12,12) THEN low
IF spring (2,4,5,7) THEN low
IF summer (5,7,8,10) THEN high
IF autumn (8,10,10.5,12) THEN medium

For C country
IF winter (1,1,2,4) ∪ (10,12,12,12) THEN low
IF spring (2,4,5,7) THEN medium
IF summer (5,7,8,10) THEN high
IF autumn (8,10,10.5,12) THEN low

4 Solution of the Problem

As an example, we are presenting calculations for A country:

1. Statistical data of tourist arrivals for 3 countries was gathered, the figures were normalized, standard deviation was found. According to the obtained figures graphics were set up [Figs. 1, 2 and 3]:

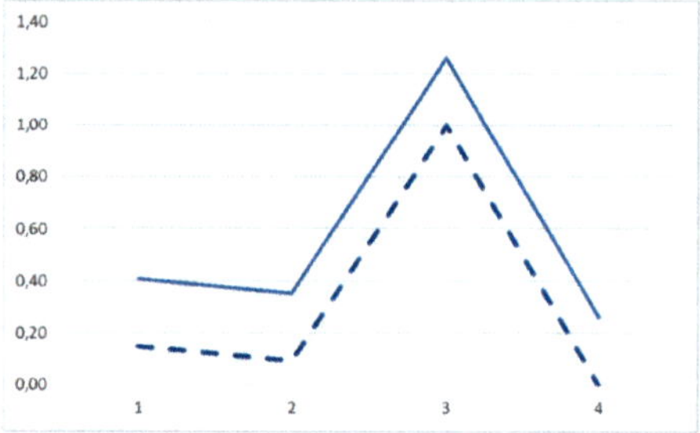

Fig. 1. Graphical representation of tourist statistics under uncertainty for A country

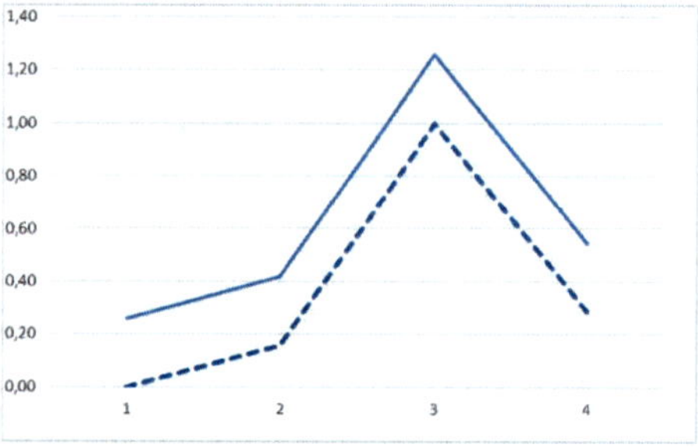

Fig. 2. Graphical representation of tourist statistics under uncertainty for B country

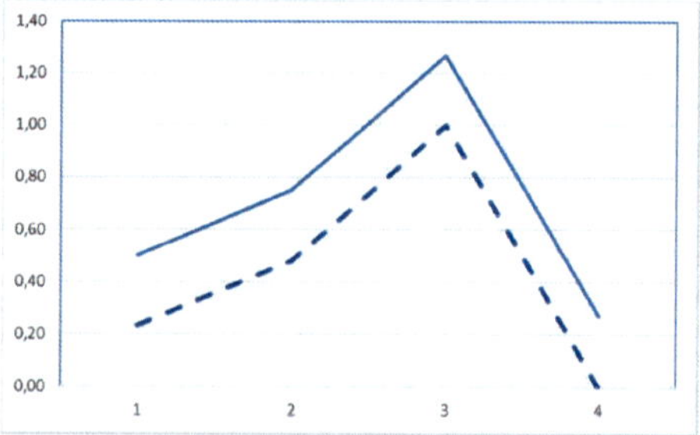

Fig. 3. Graphical representation of tourist statistics under uncertainty for C country

2. Survey for choosing appropriate place and time to visit was held among 15 adults. Only 4 of them made decisions through graphical representations, the others did through if-then rules
3. Fuzzy Inference Systems analysis based on the statistical data and if then rules were applied [Figs. 4 and 5].

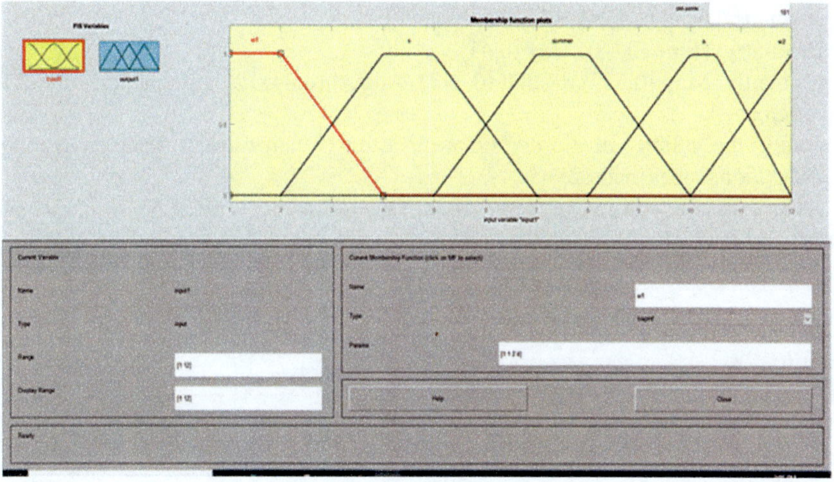

Fig. 4. Fuzzy partition of input

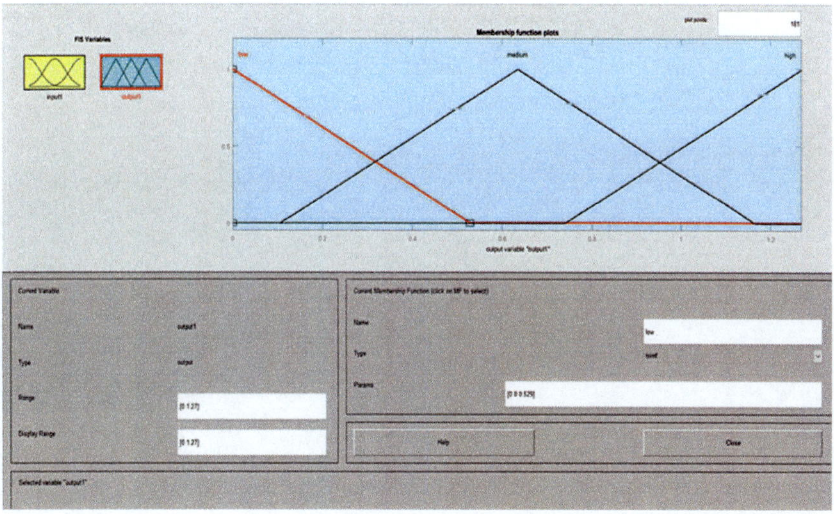

Fig. 5. Fuzzy partition of output

4. To analyze which type of data support people achieve more accurate results we measured both the sides decisions. Fuzzy Inference Systems was applied to measure the validity of decisions. First, we measured decisions based on graphics.

Two participants of the survey chose to visit A country in March and in August (0.635 and 1.1), the second participant chose B country in December (0.27) and the other one selected C country in July (0.84).

The validity analysis of results of If-Then rules-based decisions are as follows:

A country: May (1.1), April (1.1), June (1.07),

B country: (June (0.68), May (0.41),

C country: May (0.507), March (0.377), August (0.84) (2 participants), July (0.84) (2 participants).

Totally the validity of the decisions made on If-then rules is 78%, on graphical representation is less than 62%.

As an example, you can see screenshot of the FIS analysis made on the choice of A country in March [Fig. 6].

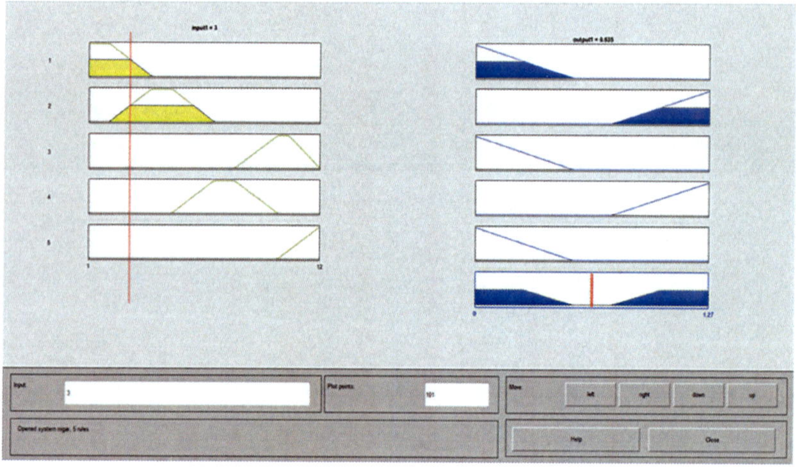

Fig. 6. FIS analysis on the validity of the decisions

5 Conclusion

We introduce results of linguistic data generation which is an important issue and so far has become uninvestigated by data-to-text community. We discovered that there are comparable differences between NLG and graphical based choices of people. NLG results in a 73% increase in task success over graphics and the total validity of the decisions made on textual information are higher than the other one. It's much more convenient and reliable for people to make decisions on NLG rather than graphical representations. These are significant findings, as traditional decision-making systems commonly rely on quantitative representations.

A substantial area for future work is adaptive NLG approaches that can create text-based systems in other areas of life.

References

1. Natural language processing: From Wikipedia, the free encyclopedia

2. Huseynova, F.: Computing with words in natural language processing. In: Aliev, R.A., Kacprzyk, J., Pedrycz, W., Jamshidi, M.O., Babanli, M.B., Sadikoglu, F.M. (eds.) ICSCCW 2019. AISC, vol. 1095, pp. 621–625. Springer, Cham (2020). https://doi.org/10.1007/978-3-030-35249-3_79
3. Bateman, J., Zock, M.: Natural Language Generation, 2nd edn. The Oxford Handbook of Computational Linguistics (2016). https://doi.org/10.1093/9780199573691.013.01
4. Zadeh, L.A.: What computing with words means to me. IEEE Comput. Intell. Mag. (2010). https://doi.org/10.1109/MCI.2009.934561
5. Ivaschenko, A., Milutkin, M.: HR Decision-Making Support Based on Natural Language Processing. In: Kravets, A.G., Groumpos, P.P., Shcherbakov, M., Kultsova, M. (eds.) CIT&DS 2019. CCIS, vol. 1083, pp. 152–161. Springer, Cham (2019). https://doi.org/10.1007/978-3-030-29743-5_12
6. Zadeh, L.A.: From computing with numbers to Computing with words- from Manipulation of Measurements to Manipulation of Perception. In: Wang, P. (eds.) Computing with Word. John Wiley & Sons. Inc., pp 35–67 (2001). https://doi.org/10.1109/81.739259
7. Gkatzia, D., Lemon, O., Rieser, V.: Natural language generation enhances human decision-making with uncertain information. In: Proceedings of the 54th Annual Meeting of the Association for Computational Linguistics, vol. 2, Short Papers (2016). https://doi.org/10.18653/v1/p16-2043
8. Aliev, R.A.: Fundamentals of the Fuzzy Logic-based Generalized Theory of Decisions. Springer-Verlag, Berlin, Heidelberg (2013). https://doi.org/10.1007/978-3-642-34895-2
9. Adilova, N.E.: Quality criteria of fuzzy if-then rules and their calculations. In: Aliev, R.A., Kacprzyk, J., Pedrycz, W., Jamshidi, M.O., Babanli, M., Sadikoglu, F.M. (eds.) ICAFS 2020. AISC, vol. 1306, pp. 55–62. Springer, Cham (2021). https://doi.org/10.1007/978-3-030-64058-3_7

Solving the Problem of Building an Automatic Control System for the Process of Water Chemical Treatment Using Fuzzy Logic

K. A. Mammadova(✉) and E. N. Aliyeva

Department of Computer Engineering, Azerbaijan State Oil and Industry University, Baku, Azerbaijan

Abstract. This article solves the problem of building an automatic control system for the process of chemical water treatment using fuzzy logic. The problem of automation of chemical water purification processes at thermal power plants is very relevant. Technological process of chemical preparation of water at the local level is the determinant for the effective operation of the main equipment of heating networks. In particular, it testifies to the high social significance of this technological process. The quality of chemical preparation of water is one of the most important factors, directly forming the basic characteristics of the system of heat supply of the population and industrial enterprises. The aim of the work is to solve the scientific and technical problem of developing algorithms for fuzzy control of a number of parallel objects to control the filters of the heating network feeder.

Keywords: Chemical preparation of water · Mode of operation of filters · Consumption vector of filters · Fuzzy logic · Rule base · Fuzzy regulator

1 Introduction

Currently, despite the rather high level of automation of technological processes at thermal power plants, there are a number of processes where the influence of the human factor. However, these processes are decisive because of the technological features of the object and the difficulties arising from the application of classical methods of control theory. As a consequence, it is impossible to avoid the influence of erroneous actions of the operator on the process of chemical water treatment. This can lead to overconsumption of chemicals and make-up water, as well as increase the risk of emergency conditions [1]. When solving the problem of automatic control of the heating network replenishment, it is necessary to be based on both the technological requirements for the processes of chemical water purification and operating modes, and on the experience of operators who have been operating the facility for a long time and have the necessary knowledge about possible uncertainties that arise during the operation of the facility.

This work solves the problem of constructing an automatic control system for the chemical water purification process using fuzzy logic. The use of fuzzy logic allows, in

R. A. Aliev et al. (Eds.): ICSCCW 2021, LNNS 362, pp. 748–756, 2022.
https://doi.org/10.1007/978-3-030-92127-9_99

conditions of limited and uncertain information about the characteristics of the control object, to use the knowledge of the operating personnel in the control process most fully. This knowledge is used in the form of logical-linguistic approximations of formal models of control objects [2–4].

In accordance with the set goal, the following main tasks are solved in the case: Structural synthesis of a fuzzy controller for group control of some parallel objects; Development of a methodology to form the basis of production rules for the management of a group of facilities; Development and research of methodology of parametric synthesis of fuzzy control system based on passive experience; Development of experimental-industrial tests and fuzzy control system for operation of the filter group of the drinking water device of the heating network.

A method of structural-parametric synthesis of a fuzzy system is proposed to control the operation of the filter unit of the heating system drinking water block. This system is based on the use of configurable membership functions that ensure that management effects are consistent with benchmarks [4].

The operator, in a situation of such uncertainty, going out of the current state of the object and experience of operation, in essence, intuitively makes decisions on the management of their work.

2 Structure of the Work

The device, which performs the process of chemical purification of water to supply the heating network with drinking water, consists of several N-cation exchange filters designed to soften the water. Opens when the filter is finished. The water is cleaned during operation and before the filters are applied. The purpose of water purification is to eliminate compression of the cation exchange layer to reduce the hydraulic resistance of the filter. It is also to ensure that the aqueous solution enters the cation exchange layers freely. The load of the block consisting of six filters is determined by the current water consumption. Therefore, the number of filters operating in parallel is determined by the nominal load requirement depending on the required capacity of the block. The decision moving the filter from one state to another is currently made by the operator. That is, the load distribution in the filter block is controlled manually. This reduces the efficiency and accuracy of the decision. The task of automatic control is to determine the moment when the filter switches to one of the possible states (operation, backup, cleaning). Thus, the problem of load distribution between the filters of the blocks must be solved in order to obtain a high water flow to each block and to perform the required mode in each special filter of the block [3, 4].

2.1 Solution of the Work

The problem is non-standard due to the complexity of the regulatory object. When solving a problem, object identification is performed. Due to the lack of clear rules in an uncertain environment, it is difficult to apply the classical approach when describing the mathematical model of an object and changing filters. The operator relies on the current state of the enterprise and its own experience in such uncertainty. The system considers it expedient to apply the method of intellectual management as a tool that allows the use of knowledge about the object in the form of logical-linguistic interpretation. Intellectual management, knowledge of the object in the solution of the problem is effectively used in the form of logical linguistic rules. It is a tool that allows you to use all the accumulated experience in the preparation of the object when building a decision-making algorithm. Management systems based on the use of fuzzy logic in solving the problem are developed. Of all the intelligent management systems used, the most appropriate is selected for the following reasons:

1. This method is more suitable for modeling the activities of an experienced operator.
2. Ability to use rich experience in the operation of the management facility and management decisions based on a basic knowledge's base for real situations.

2.2 Solution to the Work

Practical solution of the problem is as follows: In this work, the solution of the problem of establishing a classical fuzzy regulator for the management of parallel facilities was considered. A methodology for establishing a universal framework for block management of parallel filters has been developed. The operating modes characterizing the water consumption through the filters were analyzed and the following terms of the linguistic variables "operating modes of the filters" were determined on the basis of expert evaluation [5]:

A1: "Minimum lower than nominal consumption" (MLNC) – 200 m^3/h;
A2: "Consumption below the nominal" (CBN) – 150 $\div$ 300 m^3/h;
A3: "Nominal consumption" (NC) – 250–400 m^3/h;
A4: "Maximum consumption above Nominal" (MCAN) – 350 m^3/h;

The operating modes of the filters included in the block were analyzed and on the basis of expert assessment the following terms of the linguistic variables "operating modes of the filters connected to the system" were determined:

F1: "Filters not joined" (FNJ) – 0 number;
F2: "Little joined" (LJ) – 0 $\div$ 3 number;
F3: "Joined the required number" (JRN) – 3 $\div$ 5;
F4: "All joined" (AJ) – 6.

The operating modes characterizing the percentage of purity of water consumption through the filters were analyzed. Based on the expert assessment, the following terms of the linguistic variables "purity percentage of water consumption of filters" were determined:

B1: "Much lower than normal" (TLN) – 0% $\div$ 25%;
B2: "Below the normal" (BN) – 35% $\div$ 50%;
B3: "Normal" (N) – 50% $\div$ 75%;
B4: "More than normal" (MN) – 75% $\div$ 100%.

Where A = {x, μ (x)} fuzzy input set - water entering the filter. B = {y, μ (y)} is a fuzzy result, i.e. the percentage of water purity. F = {z, μ (z)} is a fuzzy input, i.e. the number of working filters. $A \subset X$, $B \subset Y$, $F \subset Z$. X∊ {0; 500}, Y∊ {0; 100}, Z ∊ {0; 6} are content of water, filters, and water purity (inlet and outlet). R (x, y, z) is a fuzzy attitude matrix defined by an expert in the scheme of fuzzy language rules:

The fuzzy rules for control:

İF X = A1 and Z = F1 Then Y = B1
or
İF X = "Minimum lower than nominal consumption" and Z = "Filters not joined" Then Y = "Much lower than normal"
İF X = A2 and Z = F2 Then Y = B2
or
İF X = "Minimum lower than nominal consumption" and Z = "Filters little joined" Then Y = "Below the normal"
....

Using standard trapezoidal membership functions [4, 7], *X* fuzzy model of the control object input is formed with (1)–(4):

$$\mu_1(x) = \min(\max(2 - 0.02 \cdot (x - 100), 0), 1) \tag{1}$$

$$\mu_{A_2}(x) = \max\left(1 - \frac{|x - 100|}{50}, 0\right) \tag{2}$$

$$\mu_{A_3}(x) = \min(\max(2 - 0.02 \cdot (x - 400), 0), 1) \tag{3}$$

$$\mu_{A_4}(x) = \min(\max(2 - 0.02 \cdot (500 - x), 0), 1) \tag{4}$$

Using standard trapezoidal belonging functions, *A* fuzzy model of the control object's output is formed with (5)–(7):

$$\mu_{F_1}(y) = \min(\max(2 - 0.02 \cdot (z - 2), 0), 1) \tag{5}$$

$$\mu_{F_2}(y) = \max(1 - |z - 2|, 0) \tag{6}$$

$$\mu_{F_3}(x) = \min(\max(2 - 0.02 \cdot (z - 3), 0), 1)$$

$$\mu_{F_4}(y) = \min(\max(2 - 0.02 \cdot (6 - z), 0), 1) \tag{7}$$

Using standard trampezoidal belonging functions, *F* fuzzy model of the control object's output is formed with (8)–(10):

$$\mu_{B_1}(y) = \min(\max(2 - 0.02 \cdot (y - 37), 0), 1) \tag{8}$$

$$\mu_{B_2}(y) = \max\left(1 - \frac{|y - 75|}{25}, 0\right) \tag{9}$$

$$\mu_{B_3}(x) = \min(\max(2 - 0.02 \cdot (x - 75), 0), 1)$$

$$\mu_{B_4}(y) = \min(\max(2 - 0.02 \cdot (100 - y), 0), 1) \tag{10}$$

The set of functions describes (1)–(4) a model of the mode of operation of the filter based on fuzzy logic (Fig. 1) [5, 6]:

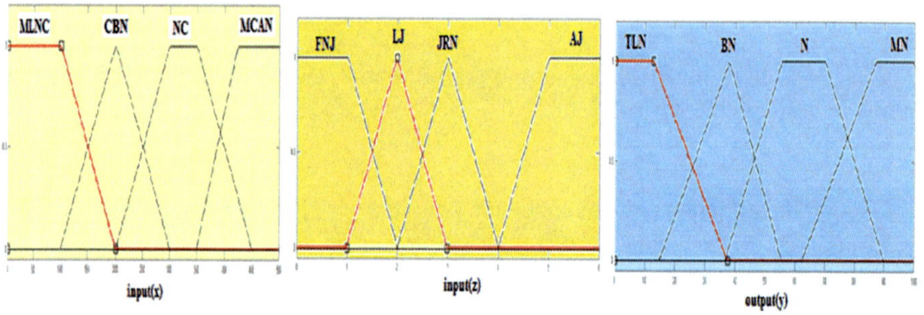

Fig. 1. Fuzzy model of filter input (x) and (Z), output (y) operation mode

For these models, two main types have been developed: Mamdani type of fuzzy regulators for working filters with a known number. This is sufficient to simulate the maximum number of combination models of different filter operating modes. Combinations of modes in accordance with (1) and (4) are not used because they are not applied in practice. The decision to change the filter is described in the production rules based on the combination of modes of all currently operating block filters. Each production rule is written in the form "If… Then…". For example, "If all the filters work in one mode, then the amount of clean water at the outlet will be minimal." 46 production rules have been developed for the Mamdani type system. Deflation in the Mamdani system was performed using the "maximum maximum" method.

When developing a methodology for forming a universal base of rules for fuzzy control of a group of parallel operating objects, a study of the control object was carried out. Based on the results of the study, it was concluded that all possible combinations of filter operation modes can be conditionally divided into three groups, differing in the number of distinguishable filter operation modes [6, 7]:

- the first group - all filters work in one mode;
- the second group - two different filter modes are distinguished;
- the third group - three different modes are distinguished.

Based on this division, a universal base is formulated, consisting of 11 rules. These rules cover all possible combinations of filter modes and regardless of the number of filters in operation (here c1, c2,... represent all possible combinations of filter operating modes) (11)–(21):

$$\mu_{C_1}(x) = \min(\mu_{A_1}(x_i)), \tag{11}$$

$$\mu_{C_1}(x) = \min(\mu_{A_1}(x_1), \mu_{A_1}(x_2), \mu_{A_1}(x_3), \mu_{A_1}(x_4), \mu_{A_1}(x_5), \mu_{A_1}(x_6))$$

$$\mu_{C_2}(x) = \min(\mu_{A_2}(x_i)), \tag{12}$$

$$\mu_{C_2}(x) = \min(\mu_{A_2}(x_1), \mu_{A_2}(x_2), \mu_{A_2}(x_3), \mu_{A_2}(x_4), \mu_{A_2}(x_5), \mu_{A_2}(x_6))$$

$$\mu_{C_3}(x) = \min(\mu_{A_3}(x_i)), \tag{13}$$

$$\mu_{C_3}(x) = \min(\mu_{A_3}(x_1), \mu_{A_3}(x_2), \mu_{A_3}(x_3), \mu_{A_3}(x_4), \mu_{A_3}(x_5), \mu_{A_3}(x_6))$$

$$\mu_{C_4}(x) = \min(\mu_{A_4}(x_i)), \tag{14}$$

$$\mu_{C_4}(x) = \min(\mu_{A_4}(x_1), \mu_{A_4}(x_2), \mu_{A_4}(x_3), \mu_{A_4}(x_4), \mu_{A_3}(x_5), \mu_{A_2}(x_6))$$

$$\mu_{C_5}(x) = \min(\min_j\left(\max_i(\mu_{A_1}(x_i), \mu_{A_2}(x_i))\right), \max_i(\mu_{A_1}(x_i)), \max_i(\mu_{A_2}(x_i)) \tag{15}$$

$$\mu_{C_6}(x) = \min\left(\min_j\left(\max_i(\mu_{A_1}(x_i), \mu_{A_3}(x_i))\right)\right), \max_i(\mu_{A_1}(x_i)), \max_i(\mu_{A_3}(x_i)) \tag{16}$$

$$\mu_{C_7}(x) = \min\left(\min_j\left(\max_i(\mu_{A_2}(x_i), \mu_{A_3}(x_i))\right)\right), \max_i(\mu_{A_2}(x_i)), \max_i(\mu_{A_3}(x_i)) \tag{17}$$

$$\mu_{C_8}(x) = \min\left(\min_j\left(\max_i(\mu_{A_2}(x_i), \mu_{A_4}(x_i))\right)\right), \max_i\left(\mu_{A_2}(x_i), \max_i(\mu_{A_4}(x_i))\right) \tag{18}$$

$$\mu_{C_9}(x) = \min\left(\min_j\left(\max_i(\mu_{A_3}(x_i), \mu_{A_4}(x_i))\right)\right), \max_i\left(\mu_{A_3}(x_i), \max_i(\mu_{A_4}(x_i))\right) \tag{19}$$

$$\mu_{C_{10}}(x) = \min\left(\min_j\left(\max_{i,j=3}(\mu_{A_j}(x_i))\right)\right), \max_i(\mu_{A_1}(x_i)), \max_i(\mu_{A_2}(x_i)), \max_i(\mu_{A_3}(x_i)) \tag{20}$$

$$\mu_{C_{11}}(x) = \min\left(\min_j\left(\max_{i,j=3}(\mu_{A_j}(x_i))\right)\right), \max_i(\mu_{A_2}(x_i)), \max_i(\mu_{A_3}(x_i)), \max_i(\mu_{A_4}(x_i)) \tag{21}$$

Here the parameter $x = (x_1, x_2, x_3, \ldots x_W)$ is the vector of consumption of working filters, *W* is the number of working filters, *N* is the number of members of the linguistic variable "filter operation mode", $\bigcap$ is the fuzzy *AND* operation, $\bigcup$ is the fuzzy *OR* operation.

The formulated algorithm can be written using a matrix of coefficients, where the line denotes the rule number, and the column denotes the tested mode of operation (the first column is "Below the minimum possible", the second is "below the nominal", the third is "Nominal", the fourth is "Above the maximum allowable"). The described matrix looks like this:

$$
P= \begin{matrix}
1 & 0 & 0 & 0 \\
0 & 1 & 0 & 0 \\
0 & 0 & 1 & 0 \\
0 & 0 & 0 & 1 \\
1 & 1 & 0 & 0 \\
1 & 0 & 1 & 0 \\
1 & 0 & 0 & 1 \\
1 & 1 & 0 & 0 \\
0 & 0 & 1 & 1 \\
0 & 1 & 1 & 1 \\
1 & 1 & 1 & 0
\end{matrix} \tag{22}
$$

Using the matrix p, the condition for finding the filter in one of the checked states is written as follows:

$$
\bigcap_{i=1}^{W}\left(\bigcup_{j=1}^{N}\left(p_{j,m}\cap\mu_{A_j}(x_i)\right)\right)=\min\left(\max\left(\min\left(p_{j,m},\mu_{A_j}(x_i)\right)\right)\right), \tag{23}
$$
$$
i=\overline{1,6},\ j=\overline{1,4},\ m=\overline{1,11}
$$

where ***i, m*** is the matrix element located in the ***i***-th row, the ***m***-th column. In (23), the checked states are included in the sum over ***j*** with a factor of one, and undesirable states - with a factor of zero. If the filter is only in an undesirable state, then the sum over ***i***, and, therefore, the product over ***i*** will be equal to zero and the rule being checked will not affect the value of the membership function $\mu_{c,A}(x_i)$.

Checking that in each of the checked states there is at least one filter using the matrix p can be written as an expression:

$$
\bigcap_{j=1}^{N}\left[\bigcup_{i=1}^{W}\mu_{A_j}(x_i)(1-p_{j,m})\right]=\min_{j}\left[\max_{i}(\max(\mu_{A_j}(x_i))),\left(1-p_{j,m}\right)\right] \tag{24}
$$

In order not to check the conditions that are not checked by this rule, the signal "there are filters in this situation" is deliberately given.

$$
\left(1-p_{j,m}\right) \tag{25}
$$

For them, the bracket (19) will be equal to one. This means the sum of ***i***. For tested situations (19) it will disappear. The result of ***i*** will depend on whether at least one filter is in the tested condition. It had to be found. At the next step, as a result of performing the fuzzy operation AND, we get a universal rule describing on the basis of fuzzy rules from expressions (22) and (23):

$$\bigcap_{i=1}^{W}\left(\bigcup_{j=1}^{N} p_{j,m} \cap \mu_{A_j}(x_i)\right) \cap \bigcap_{j=1}^{N}\left(\bigcup_{i=1}^{W} \mu_{A_j}(x_i) \cup (1 - p_{j,m})\right)$$
$$= \min\left[\min_i\left(\max_j(p_{j,m})\right), \mu_{A_j}(x_i); \min_i\left(\max_j(\mu_{A_j}(x_i)), (1 - p_{j,m})\right)\right] \tag{26}$$

Rates are then assigned to each of the four possible classes ("exclude", "weaken", "ignore", "enter the system"). The highest value for each class is calculated by the formula (26).

The term B_1 - "softening" of operations - corresponds to the rules (11), (12), (15), (16), (20).

In the (27) calculates the membership function of the term B1:

$$\mu_{B_1}(x) = \mu_{C_1}(x) \cup \mu_{C_2}(x) \cup \mu_{C_5}(x) \cup \mu_{C_6}(x) \cup \mu_{C_{10}}(x)$$
$$= \max(\mu_{C_1}(x), \mu_{C_2}(x), \mu_{C_5}(x), \mu_{C_6}(x), \mu_{C_{10}}(x)). \tag{27}$$

The term B_2 - "opening from the system" - corresponds to the rules (17), (18), (21). In the (28) calculates the membership function of the term B2:

$$\mu_{B_2}(x) = \mu_{C_7}(x) \cup \mu_{C_8}(x) \cup \mu_{C_{11}}(x) = \max(\mu_{C_7}(x), \mu_{C_8}(x), \mu_{C_{11}}(x)) \tag{28}$$

The term B_3 - "not taken into account" - corresponds to the rule (17).
In the (29) calculates the membership function of the term B3:

$$\mu_{B_3}(x) = \mu_{C_7}(x) \tag{29}$$

The term B_4 - "inclusion in the system" - corresponds to the rules (14), (19).
In the (30) calculates the membership function of the term B4:

$$\mu_{B_4}(x) = \mu_{C_4}(x) \cup \mu_{C_9}(x) = \max(\mu_{C_4}(x), \mu_{C_9}(x)) \tag{30}$$

In fuzzy management, the final operation is the procedure of converting a fuzzy result into a physical variable - a fuzzy price:

$$Z = \underset{(B_1,B_2,B_3,B_4)}{\arg} \max\{\mu_{B_1}(x), \mu_{B_2}(x), \mu_{B_3}(x), \mu_{B_4}(x)\} \tag{31}$$

Thus, the solution is chosen according to the degree of belonging to which of the terms B (27)–(30).

In this study, a fuzzy regulatory system was established through passive practice.

3 Conclusion

The parameters of the fuzzy regulator were determined using the membership functions of the original model of the operating modes of the filters. Decisions made by a fuzzy regulator should be as consistent as possible with some standard decisions (training example). The training form was developed on the basis of archived information about the boiler house of the heating system. A training sample of key situations was developed using archived data. The main situation is determined by the combination of current consumption of block filters and the decision of the operator according to this consumption.

References

1. Ahmed, A.S.E.D., Elhosseini, M.A., Arafat Ali, H.: Modelling and Practical Studying of Heat Recovery Steam Generator (HRSG) drum dynamics and approach point effect on control valves. Ain Shams Eng. J. **9**(4), 3187–3196 (2018). https://doi.org/10.1016/j.asej.2018.06.004
2. Singhala, P., Shah, D.N., Patel, B.: Temperature control using fuzzy logic. Int. J. Instrument. Control Syst. **4**(1), 1–10 (2014). https://doi.org/10.5121/ijics.2014.4101
3. Martelli, E., Alobaid, F., Elsido, C.: Design optimization and dynamic simulation of steam cycle power plants: a review. Front. Energy Res. **9**, 1–31 (2021). https://doi.org/10.3389/fenrg.2021.676969
4. Piyali, G., Akhtar, K., Aladin, Z.: Fuzzy logic-based energy management system of stand-alone renewable energy system for a remote area power system. Aust. J. Elect. Electron. Eng. **16**(4), 1–12 (2019). https://doi.org/10.1080/1448837X.2019.1588091
5. Zadeh, L.A.: A note on Z-numbers. Inf. Sci. **181**(14), 2923–2932 (2011). https://doi.org/10.1016/j.ins.2011.02.022
6. Aliev, R.R., Mraiziq, D.A.T., Huseynov, O.H.: Expected utility based decision making under Z-information and its application. Comput. Intell. Neurosci. **2015**, 1–11 (2015). https://doi.org/10.1155/2015/364512
7. Mammadova, K.A., Sultanova, A.B., Aliyeva, Y.N., Huseynova, A.N.: Problem solving based on Z-numbers. In: 14th International Conference on Applications of Fuzzy Systems, Soft Computing, and Artificial Intelligence Tools – ICAFS-2020, pp. 555–564 (2020). https://doi.org/10.1007/978-3-030-64058-3, ISBN: 978-3-030-64057-6

Fuzzy Multi-criteria Decision Making Approach for Choosing the Best Subcontractor Company

Rashad M. Suleymanli(✉)

Azerbaijan State Oil and Industry University, Azadlıq 20, AZ1010 Baku, Azerbaijan
rashad_suleymanli@yahoo.com

Abstract. In the cutting edge of the business world it is impossible to overcome all of the challenges without contribution of subcontractors which means outsourcing activity. As a business practice outsourcing is hiring a party outside a company to do services or make goods that were conventionally performed by the company's own employees and staff. It is usually undertaken as cost-cutting measure that can contain various types of occupations from manufacturing to the head office. Another point is about focusing on main points of business that outsourcing can handle this issues by spinning off less important operations to outside companies. The main objective of this paperwork is exhibiting how fuzzy analytical hierarchy process works to select appropriate subcontractor company. In this study three alternatives are compared over four criteria which characterize these companies based on their performance requirements respectively. In the aim of eliminating ambiguity of information and to calculate relative weights of selecting method, Fuzzy AHP has been chosen that can be effective tool to deal with the given problem.

Keywords: Outsourcing · Fuzzy analytical hierarchy process · Triangular fuzzy number

1 Introduction

Nowadays outsourcing has become an integral part of business operations that is "Buying more goods and services from outside domestic or foreign vendors" [1]. It mostly considers the contracting of a business process, operational, and non-core functions, such as manufacturing, facility management, call center. Outsourcing has grown much due to the general changes in business philosophy. It is a business strategy by which an organization entrusts major tasks to more specialized providers, that ultimately contributes to the companies' business rather than performing within its own facilities. Outsourcing companies are third parties who execute any task for a certain period of time, activity, job or process that was traditionally performed by employees of the company.

The reasons why a company may decide to outsource a part of their business are resource shortages, cost savings, business flexibility, reduced overhead costs and others. In order to decide which company is appropriate to select various decision making methods are applied however all classical multi-attribute decision making methods are

R. A. Aliev et al. (Eds.): ICSCCW 2021, LNNS 362, pp. 757–764, 2022.
https://doi.org/10.1007/978-3-030-92127-9_100

of an assumption that as criteria and weights are always given in crisp numbers, criteria ratings and ranking degree of their importance can be easily and precisely implemented. However, in real life of the decision making issues face constraints from imprecise and vague criteria [2]. In order to deal with complex engineering problems MCDM is one of the methodological tools [3].

An array of problems with vague information in MCDM are found by decision makers since the properties of these problems often require such kind of information. When we need human knowledge modelling and human evaluations, fuzzy set approaches can be applied. More than past 40 years this study has been studied extensively and is considered as important tool for problem modelling and solution. Lots of the early interest in this theory pertained to represent uncertainty in human cognitive processes. Many definitions are available for fuzzy logic and derivations. Definition of Type–1 fuzzy sets is: If X denotes a classic set of objects, named the universe, in which elements are consist of x [4]. Membership in a classical subset A of X is viewed as a characteristic function μ_A from [0, 1] to {0, 1} where {0, 1} is a valuation set; While 1 indicates membership 0 is non-membership.

To deal with problems in various areas like engineering, business, health and medical sciences fuzzy set approach has grand value. Integrated approach of MCDM with fuzzy set has been applied successfully over the years. MCDM as a branch in which fuzzy set theory found a wide application area. The fuzzy set was first brought into the MCDM science by Bellman and Zadeh. As a branch of mathematics, fuzzy logic allows a computer to model the real world the same way that people do. Their suggestion is to define decision as the conflux between the constraints and the goals. Here maximizing decision is considered a point where fuzzy decision attains its maximum value $\mu A(X)$: R = [0, 1] is equal to (1) (Fig. 1)

$$\mu_A(X) = \begin{cases} \frac{x-a1}{a2-a1}, & \text{If } a_1 \le x \le a_2 \\ \frac{a3-x}{a3-a2} r & \text{if } a_2 \le x \le a_3 \\ 0 & \text{otherwise} \end{cases} \tag{1}$$

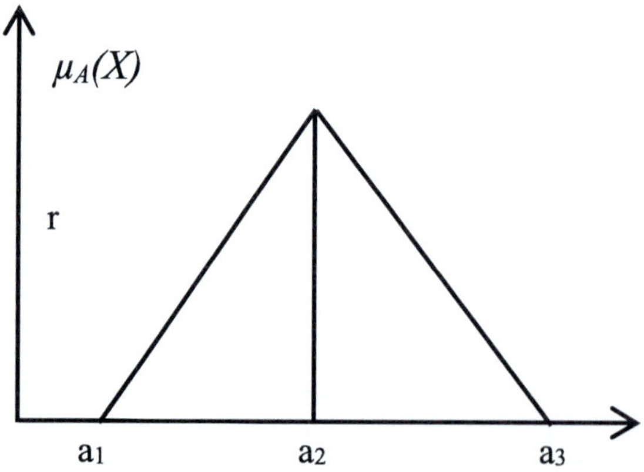

Fig. 1. Membership function of elements of Set A

Analytical hierarchy process is a kind of additive weighting tool for analyzing and solving issues based on the synthesis of math and psychology. It was developed by T. L. Saaty in the 1970s and has been deeply investigated and studied. It has been widely used in group decision making, and its use is supported in several fields such as government, business, industry, healthcare and education. The AHP method's advantage is not only in finding "correct" decision, but also compared to other methods with its nice mathematical properties which help decision makers to define the solution that best suits their target as it is difficult for them to determine importance weights accurately. Final step of the AHP application is calculation of numerical priorities for each alternative. Various works have been dedicated to development of Fuzzy AHP methodology since its proposal. C. Kahraman et al. [5] have used this approach to compare and select supplier firms which is a multi-criterion problem to decide which one to select as a complex problem. F. Zhu et al. [6] have proposed innovative framework in order to give solution to stochastic multi-criteria problems when there are vagueness in criteria performance values and criteria weights at the same time over the problem of reservoir flood control operations. F. Junior et al. [7] have compared Fuzzy AHP and Fuzzy TOPSIS methods in order to contribute to the selection problem of suppliers which encompasses enough information about both approaches. S. Patil and R. Kant [8] attempted to combine a Fuzzy AHP and Fuzzy TOPSIS methods in the aim of identify and prioritize the solutions of Knowledge Management adoption in Supply Chain to eliminate barriers.

2 Statement of Problem

The case is to select optimum subcontractor company by using fuzzy AHP method. One of the three alternatives denoted as Company A, Company B and Company C has to win the tender and to be rent by Client Company based on performance requirements which are taken as criteria. These criteria are determined by conducting a survey among some client companies what they prefer and want from subcontractor companies to handle work packages to them. Following this method, the subcontractor companies are ranked and the best one is determined. The hierarchical structure has shown in Fig. 2.

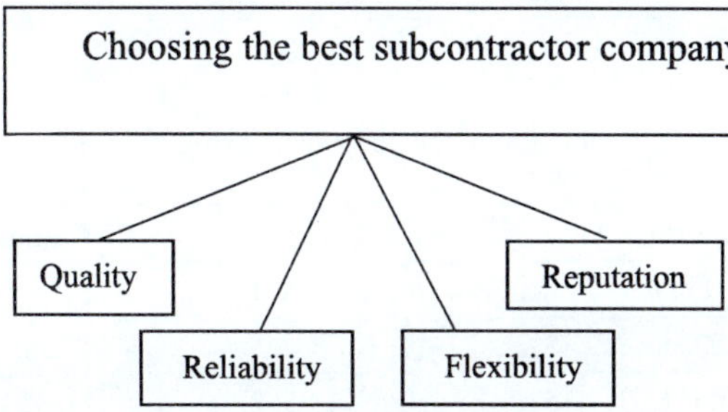

Fig. 2. Research hierarchical structure

Pair-wise comparisons value is exhibited in triangular fuzzy numbers (TFN). The linguistic scale is translated to fuzzy scale through triangular fuzzy conversion scale which is described in Table 1. Different values of criteria are shown from Tables 2, 3, 4, 5, and 6 in triangular fuzzy numbers.

Table 1. Triangular fuzzy conversion scale [3]

Linguistic description	TFS	TFR
Just equal	(1, 1, 1)	(1, 1, 1)
Equally important	(1/2, 1, 3/2)	(2/3, 1, 2)
Weakly important	(1, 3/2, 2)	(1/2, 2/3, 1)
Strongly more important	(3/2, 2, 5/2)	(2/5, 1/2, 2/3)
Very strongly more important	(2, 5/2, 3)	(1/3, 2/5, 1/2)
Absolutely more important	(5/2, 3, 7/2)	(2/7, 1/3, 2/5)

Table 2. Evaluation of the companies to the quality

Quality	Company: A	Company: B	Company: C
Company: A	(1, 1, 1)	(1.28, 1.45, 1.78)	(1.38, 1.85, 2.05)
Company: B	(1.32, 1.96, 2.35)	(1, 1, 1)	(1.63, 2.02, 2.48)
Company: C	(2.36, 2.85, 3.48)	(2.95, 3.37, 3.95)	(1, 1, 1)

Table 3. Evaluation of the companies to the reliability

Reliability	Company: A	Company: B	Company: C
Company: A	(1, 1, 1)	(0.92, 1.36, 1.98)	(1.10, 1.56, 2.14)
Company: B	(0.56, 1.06, 1.84)	(1, 1, 1)	(1.35, 1.95, 2.31)
Company: C	(2.15, 2.85, 3.24)	(2.70, 3.12, 3.85)	(1, 1, 1)

Table 4. Evaluation of the companies to the flexibility

Flexibility	Company: A	Company: B	Company: C
Company: A	(1, 1, 1)	(0.52, 0.95, 1.25)	(1.21, 1.84, 2.84)
Company: B	(0.92, 1.54, 2.15)	(1, 1, 1)	(1.56, 1.98, 2.35)
Company: C	(1.05, 1.67, 2.15)	(0.68, 1.35, 2.01)	(1, 1, 1)

Table 5. Evaluation of the companies to the reputation

Reputation	Company: A	Company: B	Company: C
Company: A	(1, 1, 1)	(0.95, 1.85, 2.61)	(1.10, 2.07, 2.75)
Company: B	(1.26, 2.14, 2.85)	(1, 1, 1)	(1.96, 2.45, 2.98)
Company: C	(1.23, 1.59, 2.12)	(1.29, 1.95, 2.15)	(1, 1, 1)

Table 6. Evaluation matrix to the goal

Criteria	Company: A	Company: B	Company: C	Company:D
Quality	(1, 1, 1)	(0.49, 1.01, 1.95)	(0.48, 1.48, 2.15)	(0.59, 0.77, 1.26)
Reliability	(2.05, 2.36, 2.65)	(1, 1, 1)	(1.99, 2.26, 2.86)	(1.36, 1.95, 2.36)
Flexibility	(0.85, 1.45, 2.26)	(0.89, 1.66, 2.16)	(1, 1, 1)	(0.94, 1.68, 2.28)
Reputation	(0.80, 1.48, 1.72)	(0.89, 1.36, 1.96)	(0.78, 1.52, 1.96)	(1, 1, 1)

3 Solution Methodology

Method of extent analysis which has been proposed by Chang is applied in which each object is extent to each goal, respectively. So with the following signs each extent values for each object can be got.

$$\mathrm{M_{gi1}},\ \mathrm{M_{gi2}},\ \ldots,\ \mathrm{M_{gi}^{m}},\ \mathrm{i} = 1, 2, \ldots, \mathrm{n}$$

Here Triangular Fuzzy Numbers are $\mathrm{M_{gi}^{j}}(\mathrm{j} = 1, 2 \ldots,\ \mathrm{m})$.

Step 1. Value of fuzzy synthetic extent with respect to the ith object is defined:

$$\mathrm{S_k} = \sum\nolimits_{j=1}^{n} M \mathop{=}^{\mathrm{j}}_{\mathrm{gi}} \otimes \left[\sum\nolimits_{i=1}^{n} \sum\nolimits_{j=1}^{m} M_{gi}^{j}\right]^{-1} \tag{2}$$

Here, (3) perform the fuzzy addition operation

$$\sum\nolimits_{j=1}^{n} M_{gi}^{\mathrm{j}} = \left[\sum\nolimits_{i=1}^{n} l_j, \sum\nolimits_{i=1}^{n} m_j, \sum\nolimits_{i=1}^{n} u_j\right] \tag{3}$$

$$\left[\sum\nolimits_{i=1}^{n} \sum\nolimits_{j=1}^{m} M_{gi}^{j}\right]^{-1} = \frac{1}{\sum_{i=1}^{n} u}, \frac{1}{\sum_{i=1}^{n} m}, \frac{1}{\sum_{i=1}^{n} l} \tag{4}$$

Step 2. The degree of possibility of two fuzzy numbers $\mathrm{V(M_2 \geq M_1)} = \sup\left[\min(\mu_{\mathrm{m1}}(\mathrm{x})\mu_{\mathrm{m2}}(\mathrm{y}))\right]$, (4) can be shown as hgt (M1M2) $= \mu_{\mathrm{m2}}$

$$\mu_{\mathrm{m_2}}(\mathrm{d}) = \begin{cases} 1, & \mathrm{m_2 \geq m_1} \\ 0, & \mathrm{l_1 \geq u_2} \\ \frac{l_1 - u_2}{(m_2 - u_2) - (m_1 - l_1)}, & \text{otherwise} \end{cases} \tag{5}$$

To compare $\mathrm{M_1}$ and $\mathrm{M_2}$ both the values $\mathrm{V(M_1 \geq M_2)}$ and $\mathrm{V(M_2 \geq M_1)}$ are required.

Step 3. The possibility degree for convex fuzzy number being greater than k convex fuzzy Mi(i = 1,2,..,k) numbers can be determined by

$$V(M \geq M_1, M_2, \ldots, M_k) = V[(M \geq M_1) \text{ and } (M \geq M_2) \text{ and } \ldots (M \geq M_k)] = \min V(M \geq M_i), \quad (6)$$

$$\text{If } d_i(A_i) = \min V(S_i \geq S_k) \quad (7)$$

For $k = 1, 2..., n;\ k \neq i$ the weight vector is given as

$$W' = (d'(A_1),\ d'(A_2, \ldots, d'(A_n))T \quad (8)$$

$A_i = (i = 1, 2, \ldots n)$ are n elements.

Step 4. With normalization the normalized weight vectors as below:

$$W = (d(A_1),\ d(A_2, \ldots, d(A_n))T \quad \text{where W is a non - fuzzy number} \quad (9)$$

4 Application and Results

As an example all the companies are compared on the criteria of Quality and other calculations will be done analogically. So the figures from Table 2 are taken and following calculations are conducted. According to (2)

$$\sum_{j=1}^{n} M_{g1} = (3.66, 4.3, 4.83)$$

$$\sum_{j=1}^{n} M_{g2} = (3.95, 4.98, 5.83)$$

$$\sum_{j=1}^{n} M_{g3} = (6.31, 7.22, 8.43)$$

According to (3) their reciprocal values are calculated

$$\left[\sum_{i=1}^{n} \sum_{j=1}^{m} Mg_{ij}\right]^{-1} = (1/19.09, 1/16.5, 1/13.92) = (0.05, 0.06, 0.07)$$

Then according to the equation of (1)

$$S_1 = (3.66, 4.30, 4.83) \otimes (0.05, 0.06, 0.07) = (0.19, 0.26, 0.35)$$

$$S_2 = (3.95, 4.98, 5.83) \otimes (0.05, 0.06, 0.07) = (0.21, 0.30, 0.42)$$

$$S_3 = (6.31, 7.22, 8.43) \otimes (0.05, 0.06, 0.07) = (0.33, 0.44, 0.61)$$

Once the calculations of Step 1 completed then we can swift to find degree of possibility by applying (5)

$$V(S_1 \geq S_2) = 0.78;\ V(S_2 \geq S_1) = 1;\ V(S_1 \geq S_3) = 0.1$$

$$V(S_3 \geq S_1) = 1;\; V(S_2 \geq S_3) = 0.39;\; V(S_3 \geq S_2) = 1$$

Then according to the procedure of Step 3 we have to find weight vector by comparison of obtained values

$$V(S_1 \geq S_2, S_3) = 0.1;\; V(S_2 \geq S_1, S_3) = 0.39;\; V(S_3 \geq S_1, S_2) = 1$$

So the weight vector (8)

$$W_{Qlt'} = (0.1; 0.39; 1)$$

After normalization (9)

$$W_{Qlt} = (0.07, 0.26, 0.67)$$

Using the similar calculations obtained results of weight vectors after normalization are as following

$$W_{Rel} = (0.20, 0.22, 0.58)$$

$$W_{Flx} = (0.31, 0.36, 0.33)$$

$$W_{Rep} = (0.33, 0.37, 0.29)$$

If we figure out the results for the main attributes of the goal which characterize all four criteria from Table 6,

$$\sum_{j=1}^{n} M_{g1} = (2.56, 4.26, 6.36)$$

$$\sum_{j=1}^{n} M_{g2} = (6.40, 7.57, 8.87)$$

$$\sum_{j=1}^{n} M_{g3} = (3.68, 5.79, 7.70)$$

$$\sum_{j=1}^{n} M_{g4} = (3.47, 5.36, 6.64)$$

The weight vector is

$$W' = (0.55, 1, 0.76, 0.66)$$

After normazization

$$W = (0.18, 0.34, 0.26, 0.22)$$

We have to calculate alternative priority weights (APW) in order to determine the best subcontractor company (Table 7)

$$APW_A = 0.18 \times 0.07 + 0.34 \times 0.20 + 0.26 \times 0.31 + 0.22 \times 0.33 = 0.23$$

$$APW_B = 0.18 \times 0.26 + 0.34 \times 0.22 + 0.26 \times 0.36 + 0.22 \times 0.37 = 0.30$$

$$APW_C = 0.18 \times 0.67 + 0.34 \times 0.58 + 0.26 \times 0.33 + 0.22 \times 0.29 = 0.47$$

Referring to the results of calculations Company C will be hired by client company.

Table 7. Final evaluation matrix

Criteria/Alternatives	Quality 0.18	Reliability 0.34	Flexibility 0.26	Reputation 0.22	Alternative priority weight
Company: A	0.07	0.20	0.31	0.33	0.23
Company: B	0.26	0.22	0.36	0.37	0.30
Company: C	0.67	0.58	0.33	0.29	0.47

5 Conclusion and Discussion

The purpose of this research paper was to define the best subcontractor company by applying Fuzzy AHP methodology and assign the work packages to this company. The top three companies were selected with four performance requirements and comparison was drawn to decide which one would be priority to be selected. In order to get a solution fuzzy AHP method was applied with respect to ambiguity of human decision-making processes. Among all the criteria reliability dominates over its peers. In terms of companies' works company C exceeds other ones with high quality and reliable performance.

References

1. Kotler, P.: Kevin Lane Keller, Marketing Management, p. 622 (2014)
2. Bellmann, R.E., Zadeh, L.A.: Decision making in fuzzy environment. Manag. Sci. **17**(4), 141–164 (1970). https://doi.org/10.1287/mnsc.17.4.B141
3. Kahraman, C.: Fuzzy multi-criteria decision making (2008). https://doi.org/10.1007/978-0-387-76813-7
4. Aliyev, R.A., Huseynov, O.H.: Decision theory with imperfect information, pp. 43–52 (2014). https://doi.org/10.1142/9186
5. Kahraman, C., Cebeci, U., Ulukan, Z.: Multi-criteria supplier selection using fuzzy AHP. Logist. Inf. Manag. **16**(6), 382–394 (2003). https://doi.org/10.1108/09576050310503367
6. Zhu, F., Zhong, P., Sun, Y.: Multi-criteria group decision making under uncertainty: application in reservoir flood control operation. Environ. Model. Softw. **100**, 236–251 (2018). https://doi.org/10.1016/j.envsoft.2017.11.032
7. Junior, F., Osiro, L., Carpinetti, L.: A comparison between fuzzy AHP and fuzzy TOPSIS methods to supplier selection. Appl. Soft Comput. **21**, 194–209 (2014). https://doi.org/10.1016/j.asoc.2014.03.014
8. Patil, S., Kant, R.: A fuzzy AHP-TOPSIS framework for ranking the solutions of knowledge management adoption in supply chain to overcome its barriers. Exp. Syst. Appl. **41**(2), 679–693 (2014). https://doi.org/10.1016/j.eswa.2013.07.093

Determination of Radial Strains in Sealing Elements with Rubber Matrix Based on Fuzzy Sets

J. N. Aslanov[1(✉)], A. B. Sultanova[2], Z. S. Huseynli[1], and F. F. Mustafayev[1]

[1] Azerbaijan State University of Oil and Industry, Baku AZ1010, Azerbaijan
camaladdin.aslanov@asoiu.edu.az

[2] Azerbaijan State University of Oil and Industry, Institute of Control Systems of ANAS, Baku AZ1010, Azerbaijan

Abstract. The research work is devoted to the study of the characteristics of the stress distribution in seals with a rubber matrix, depending on the degree of deformation. Although the determination of the relative deformation of seals by traditional methods allows us to obtain a fairly accurate result in the cross-sectional area under consideration, depending on the load distribution, the results obtained during the distribution do not fully reflect reality. In this regard, it is advisable to use modern information technologies that allow obtaining the necessary analytical forecasting data. One of the most promising areas today is artificial intelligence and fuzzy logic technologies. For this purpose, the research paper considered the issue of determining the radial strains caused by the relative deformation of the seal on the basis of fuzzy logic. The article proposes an approach based on the theory of fuzzy sets to solving deformation problems in rubber matrix humidifiers. The results of the study were used as initial data.

Keywords: Fuzzy logic · Rubber sealer · Deformation · Internal strains · Radial strains · Saddle · Algorithm Mamdani · Sealing elements

1 Introduction

In modern conditions, the process of oil refining is complex and ambiguous and requires accurate knowledge of all internal and external factors. However, it is impossible to get complete information in many cases. Therefore, the process of oil production takes place in the conditions of uncertainty that accompanies various situations [1]. The normal operation of any technological line under the pressure of the working medium is largely determined by the tightness of the detachable connections of the components of this line (apparatuses, pipelines, machines) and the strength of the parts that make up these components, including the parts of these connections. Violation of tightness reduces the efficiency of production due to losses of raw materials and the final product, leads to environmental pollution and reduces the safety of equipment operation in case of leaks of toxic, explosive and fire-hazardous working environments. The tightness of the split connection is provided by a sealing element placed between the sealing parts of the

R. A. Aliev et al. (Eds.): ICSCCW 2021, LNNS 362, pp. 765–773, 2022.
https://doi.org/10.1007/978-3-030-92127-9_101

connection [2, 3]. Rubber is widely used as a sealing element material. Therefore, it is advisable to use rubber seals during the study.

2 Solution Method

Rubber seals used in industrial machines and equipment are indispensable elements during filling metal-to-metal gaps, creating tightness in any nodes, and in breaking the connection between the environments.

One of the important issues is the resistance of rubber sealing elements to wear due to climatic conditions and strains due to deformation, depending on the load on them. These elements operate at high pressures, loads and aggressive environments in machine and equipment nodes. Their failure in working nodes leads to failure of machinery and equipment. These elements play an important role in equipment used in the production of oil and gas in high pressure flowing wells. The seal assembly of a set of equipment used in oil and gas production in the oil and gas industry is equipped with rubber elements. Rubber cups and gaskets, which are the main sealing elements of fountain equipment, operate under high pressure, load and aggressive environment. Failure of these units leads to the termination of the operation of the entire fountain complex. U-shaped, O-shaped, triangular, rectangular and trapezoidal cross-section sealing elements are used in high pressure gate structures and packers of fountain and drilling manifolds. They are exposed to the corrosive environment, working in aggressive environments and under pressure. Their failure leads to the shutdown of the shutter structures and environmental pollution. These failures are directly related to the strain-strain state in the seat in which the sealing element is seated and from its material. From this point of view, the determination by modern methods of the strain-strain state of rubber sealing elements of gate structures and the application of the results obtained in their design is an important task [4, 5].

Determination of the relative deformation of seals by traditional methods does not allow obtaining a sufficiently accurate result. In this regard, it is advisable to use modern information technologies that allow obtaining the necessary analytical forecasting data.

One of the most promising areas today is artificial intelligence and fuzzy logic technologies. For this, the research work considered the issue of determining the relative deformation of the seal on the basis of fuzzy logic [6].

Experiments were carried out to determine the relative deformation of the sealing element of the valve structure operating under high pressure (Fig. 1, Fig. 2), proposed by us. The parameters of the rubber sealing element are shown in Table 1.

Table 1. Table of experiments to determine the relative deformation of seals

P_W, MPa	Time, day	Type of material	Seal diameter
$65 \div 70$	30	Rubber B14 matrix panel	Outer diameter – 60 mm Inner diameter – 50 mm

The experimental values for the original database are shown in Table 2.

In theory, the radial strain in the sealing element is determined as follows. The radial strain is determined as follows:

$$\sigma_r = -\frac{2F}{\pi\delta} \cdot \frac{cos\theta}{r} \tag{1}$$

where: F – is the dispersed force acting on the seal;
δ – thickness of the seal;
θ – the angle at point M, formed by the radius vector with the Z axis, $\theta = const.$

Table 2. Experimental values for the original database

F;$10^6 N$	δ; mm	r; mm	Radial strain
0	8	60	$\sigma_r = -\frac{2F}{\pi\delta} \cdot \frac{cos\theta}{r}$
1	7,74	60,1	
1,5	7,4	60,3	
2	7,15	60,6	
2,5	6,88	61,00	
3	6,5	61,2	
3,5	6,3	61,5	
4	6,2	62,00	
4,5	0,00	Saturation	

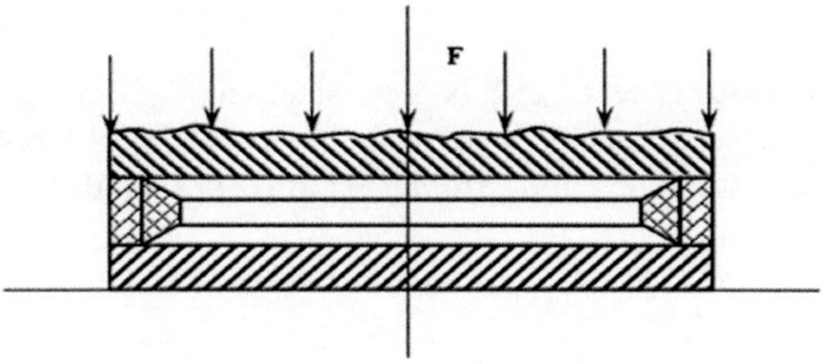

Fig. 1. General view of the rubber sealant

The input parameters adopted for solving the problem are shown in Table 3.

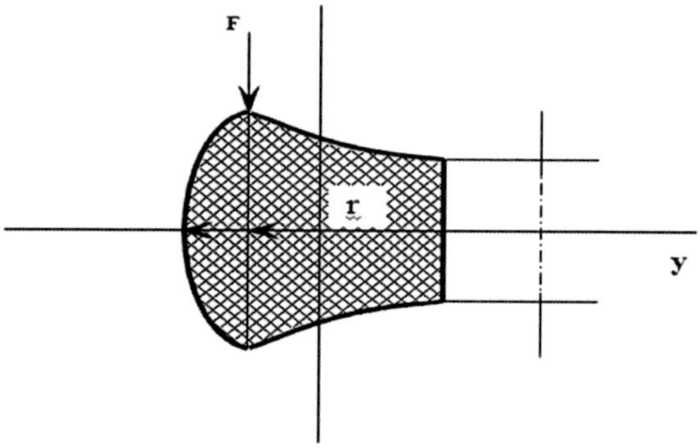

Fig. 2. Load diagram of the cross-section of the rubber seal

Table 3. Nominal value of input variables

Variables	F (diffused force)	δ (seal thickness)	R (seal diameter)	σ (radial strain)
Un.ofmeasure	$(10^6 N)$	(mm)	(mm)	10^3 (Pa)
Change interval	0–4.5	6.2–8	60–62	$100 * 10^3 - 200 * 10^3$

3 Preliminaries

Fuzzy control is widely used in applications such as servomotor and process control. [1–3].

Definition 1. Type-1 fuzzy sets. Let X be a classical set of objects, called the universe, whose generic elements are denoted x. Membership in a classical subset A of X is often viewed as a characteristic function μ_A from [0,1] to {0,1} such that

$$\mu_A(x) = \begin{cases} 1 \ if \ x \in A \\ 0 \ if \ x \notin A \end{cases}$$

where {0, 1} is called a valuation set; 1 indicates membership while 0—non membership. If the valuation set is allowed to be in the real interval [0, 1], then A is called a fuzzy set, μ_A is the grade of membership of x in A: $\mu_A(x) : \mathrm{X} \rightarrow [0, 1]$.

A fuzzy set A is a subset of X and is defined as

$$A = \{(x, \mu_A(x), |x \in X)\}$$

Definition 2. A triangular fuzzy number $\breve{A}$ can be defined by a triplet (a_1, a_2, a_3) where the membership can be determined as the following equation:

$$\mu_{\tilde{A}}(x) = \begin{cases} 0, & x \in [-\infty, a_1] \\ \frac{x-a_1}{a_2-a_1}, & x \in [a_1, a_2] \\ \frac{a_3-x}{a_3-a_2}, & x \in [a_2, a_3] \\ 0, & x \in [a_3, +\infty] \end{cases}$$

The Mamdani algorithm: Let us first note that the mechanism of fuzzy conclusions used in various expert and control systems is based on a knowledge base formed by specialists of the subject area in the form of a set of fuzzy predicate rules of the form:

R1: İF X is A1, THEN Z is B1,

R2: İF X is A2, THEN Z is B2,

....

RN: İF X is AN, THEN Z is BN.

where x is an input variable (a name for known data values), z is an output variable (a name for the data value to be calculated); Ai and Vi are fuzzy sets defined on X and Z, respectively, using the membership and (z) functions [7, 8].

4 Statement of the Problem and Solution

The main goal of the work is to determine the degree of deformation in seals with a rubber matrix using fuzzy logic.

We will use the Mamdani model as the fuzzy logic inference model. Fuzzy Toolbox Matlab package was chosen as a simulation tool.

Fuzzy rules for Mamdani extraction are based on specific input and output variables. Three input models and one output model are offered, some of the rules are shown below. For this task (6) "IF-THEN" will be defined as a set of production rules. Fuzzy inference is based on the Mamdani model [9, 10], based on the following fuzzy knowledge base.

Rule 1. IF the diffused force acting on the seal is small and the thickness of the seal does not change AND the diameter of the seal does not change, THEN the radial strain is close to 0.

Rule 2. IF the diffused force acting on the seal is below normal AND the thickness of the seal is below normal AND the diameter of the seal is above normal, THEN the radial strain is low.

Rule 3. IF the diffused force acting on the seal is close to normal AND the thickness of the seal is much lower than normal AND the diameter of the seal is much larger than normal, THEN the radial strain is slightly lower.

Rule 4. If the diffused force acting on the seal is normal AND the thickness of the seal is too small AND the diameter of the seal is too large, THEN the radial strain is normal.

Rule 5. If the diffused force acting on the seal is more than normal, AND the thickness of the seal is much less than normal, AND the diameter of the seal is much larger than the norm, then the radial strain is greater than normal.

Rule 6. If the diffused force acting on the seal is too large, AND the thickness of the seal is too small, AND the diameter of the seal is too large, then the radial strain is dangerously high.

To solve the problem, a fuzzy model was built with 3 input and 1 output variables. During solving a problem, input and output variables are represented as linguistic variables, and each linguistic variable is defined as the following set of terms: The input variables are shown in Table 4, output variable are shown in Table 5 (Fig. 3, Fig. 4, Fig. 5).

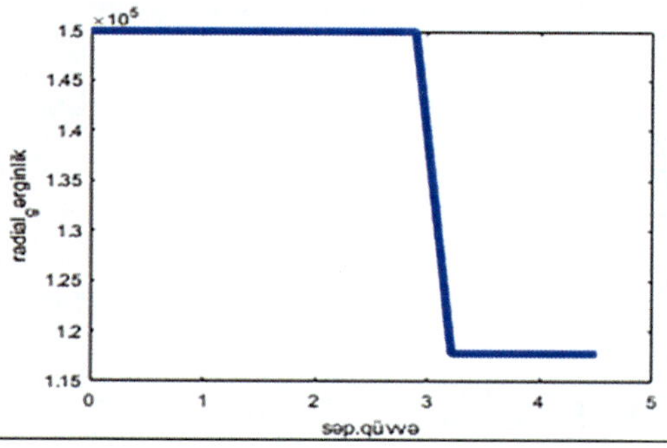

Fig. 3. Radial strain dependence on diffused force

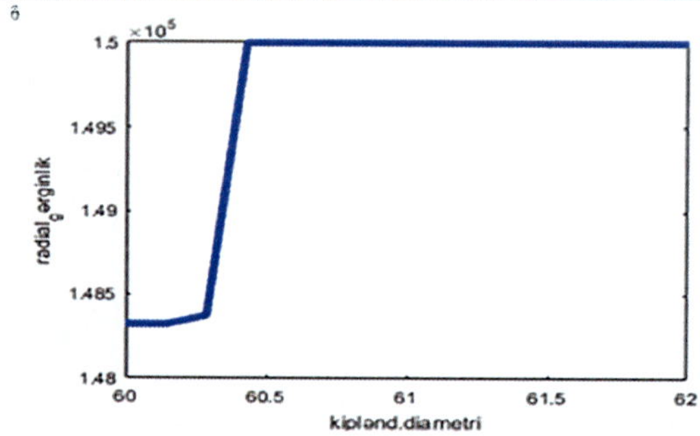

Fig. 4. Radial strain dependence on diameter of the seal

The Mamdani-type extraction model was used as the main fuzzy logical result model. Matlab\Fuzzy Toolbox was chosen as the simulation tool. Experimental researches have shown that the calculation results are acceptable for assessing the radial deformation strain during compression by the axial load of a circular seal. Researches have found the existence of permanent deformation only on the outer contour of the seal. The rate of change of the radial strain depending on the given diffused force – F, δ – thickness and diameter of the r-seal are given in Table 6.

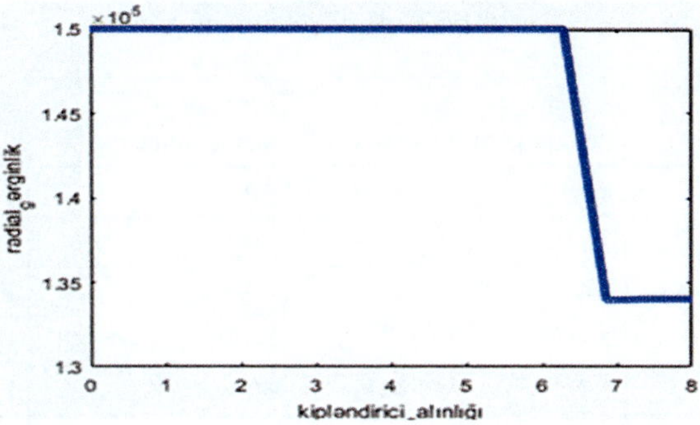

Fig. 5. Radial strain dependence on thickness of the seal

Table 4. Input variables

№	Input variables	Input variable values		
		Less	Normal	More
1	F (diffused force)	[0 1.5 2.0]	[2 2.5 3.0]	[3.0 3.5 4.0 4.5]
2	δ (seal thickness)	[0 5 6.3]	[6.0 6.9 7]	[7.0 7. 25 7.75]
3	r (seal diameter)	[60.0 60.2 60.5]	[60.5 61.0 61.3]	[61.2 61.5 62.5]

Table 5. Output variable

№	Output variable	Output variable values		
		Low	Normal	Tall
1	Radial strain	[100000 110000 130000]	[130000 140000 160000]	[160000 180000 200000]

Table 6. Research results

F; diffused force	δ; mm thickness of the seal	r; mm seal diameter	Radial strain
0	8	60	1.5e+05
1	7,74	60,1	1.87e+05
1,5	7,4	60,3	1.84e+05
1.6	7.35	60.35	1.84e+05
1.7	7.32	60.4	1.47e+05
1.8	7.3	60.4	1.47e+05
2	7,15	60,6	1.34e+05
2,5	6,88	61,00	1.34e+05
3	6,5	61,2	1.18e+05
3,5	6,3	61,5	1.16e+05
4	6,2	62.00	1.5 e+05
4,5	0,00	Saturation	–

5 Conclusion

The characteristic of the radial deformation of the seal by its volume shows that the internal strain of the seal must be loaded until the deformation is saturated. This state corresponds to the maximum value of the internal strain of the seal. Overloading will damage the seal. Researches have found permanent deformation only on the outer contour of the seal. This means that during preparing the seal, the structural dimensions must be selected according to the degree of deformation change. Taking this change into account, the outer diameter of its seat should be 12% larger than the outer diameter of the seal.

Acknowledgments. This work was supported by the Science Development Foundation under the President of the Republic of Azerbaijan – Grant no. EIF-MQM-ETS-2020-1(35)-08/04/1-M-04.

References

1. Berry, K.J., Mielke, P. W., Johnston, J. E.: Permutation Statistical Methods (2016). https://doi.org/10.1007/978-3-319-28770-6
2. Nuraddin, A.J., Baxman, S.A., Abulfas, H.I.: Model design for predicting the efficiency of improved valve constructions during statistical data based exploitation. IFAC-PapersOnLine **52**(25), 547–550 (2019). https://doi.org/10.1016/j.ifacol.2019.12.603
3. Aliev, R.: Fuzzy-genetic approach to aggregate production-distribution planning in supply chain management. Inf. Sci. **177**(20), 4241–4255. https://doi.org/10.1016/j.ins.2007.04.012
4. Aliev, R.A., Pedrycz, W., Huseynov, O.H., Eyupoglu, S.Z.: Approximate reasoning on a basis of Z number-valued if–then rules. IEEE Trans. Fuzzy Syst. **25**(6), 1589–1600 (2017). https://doi.org/10.1109/tfuzz.2016.2612303

5. Коротка, Л.I.: Fuzzy modeling of conduct elements of chemical equipment. East. Eur. J. Enterp. Technol. **2**(4(62)), 12–15. https://doi.org/10.15587/1729-4061.2013.12363
6. Aliev, R.A., Fazlollahi, B., Aliev, R.R.: Soft Computing and Its Application in Business and Economics, Springer Verlag, 446 p. (2004). https://doi.org/10.1007/978-3-540-44429-9
7. Mamedov, V.T., Mamedov, G.A., Aslanov, J.N.: Stress–strain state of sealing rubber membranes at large deformations. J. Appl. Mech. Tech. Phys. **61**(2), 286–291 (2020). https://doi.org/10.1134/s0021894420020157
8. Nuraddin, A.J., Baxman, S.A.: Forecasting of improved straightforward valves technical condition using fuzzy inference models. IFAC-PapersOnLine **51**(30), 12–14 (2018). https://doi.org/10.1016/j.ifacol.2018.11.236
9. Bahn, S.: Workplace hazard identification and management: the case of an underground mining operation. Saf. Sci. **57**, 129–137 (2013). https://doi.org/10.1016/j.ssci.2013.01.010
10. Shtovba, S.D.: Fuzzy model tuning based on a training set with fuzzy model output values. Cybern. Syst. Anal. **43**(3), 334–340 (2007). https://doi.org/10.1007/s10559-007-0054-6

Mobile Robot Navigation with Preference-Based Fuzzy Behaviors

L. M. Zeinalova(✉) and B. O. Jafarov

Azerbaijan State University of Oil and Industry, Baku, Azerbaijan
L.zeynalova@asoiu.edu.az

Abstract. A behavior based fuzzy control for two-wheeled mobile robot with navigation is considered. Solving the problem of avoiding obstacles inevitably becomes an important task of robotics in case of impossibility to calculate the trajectory before the start of its movement, or unforeseen changes in the working environment, or in the absence of information on the exact location of obstacles. A problem can be interpreted as a requirement for robot's movement from a starting point to the given target avoiding the obstacles, reaching the target in the shortest time, avoiding trapping. Getting from an initial position and cruising to the next position is under closed-loop control. A mobile robot should stop at the target within a very small position error. For coordination of these behaviors a multi-objective techniques was applied. A selection from a set of behaviors was represented as vector optimization problem. Behaviors are selected as Pareto optimal solutions, using lexicographic method.

Keywords: Fuzzy logic · Mobile robot · Behavior-based navigation · Pareto optimal solution · Multi-objective optimization

1 Introduction

A modern mobile robot needs to estimate the working environment, to forecast a trajectory to escape the possible obstacles on its way. One of the control architectures is behavior-based or in other words a reactive architecture. It doesn't need a complete model of environment [1]. Prior knowledge about the environment is uncertain and unreliable. It should be noted the unpredictable character of environment's dynamic. Behavior-based navigation identify one or several control outputs (behaviors) to sensor inputs. The considered behaviors then are "mixed" [2]. An adequate actions should be generated to avoid collisions with simple obstacles, if one or more objects are detected in the vicinity of the mobile robot using perceptual means. Also an adequate actions are necessary in the case when the surrounding the mobile robot environment may contain obstacles to personal shape (polygons, walls, ellipses and etc.). In this case, the mobile robot should be able to efficient avoidance of these obstacles without collisions. Fuzzy logic is applied to overcome these difficulties. Fuzzy control has shown its effectiveness for navigation problems, but the synthesis of fuzzy controllers, which allows one to

R. A. Aliev et al. (Eds.): ICSCCW 2021, LNNS 362, pp. 774–782, 2022.
https://doi.org/10.1007/978-3-030-92127-9_102

obtain the desired robot's behavior, is not always trivial. The main disadvantage of the concept of using fuzzy logic is the lack of a generalized methodology in the presence of a significant number of determined parameters (parameters of membership functions, rules of fuzzy inference, etc.). Fuzzy control is used to implement different behaviors [3–5, etc.]. Several models using fuzzy logic are used for behavior coordination, for example, hierarchical fuzzy control, game strategy selection models for soccer robots. A knowledge in fuzzy logic controllers is represented in the form of IF…,THEN… linguistic rules. The computer and practical simulations show good robustness characteristics [3]. In the recent years, the application of intelligent control methods in the navigation of mobile robots has been increasingly used. This, in addition to fuzzy logic, includes neural networks [6, 7], neuro-fuzzy [8, 9], genetic algorithms [10] and other methods. In this paper we use fuzzy logic in mobile robot navigation in uncertain environment. The behavior based architecture is used. Behaviors are the following: target reaching, obstacles avoiding, and local minimum avoidance. Multi-objective optimization method is used to control the robot.

The paper is organized as follows: Sect. 2 presents the architecture of the behavior based navigation. Section 3 shows simulation results.

2 Behavior Based Navigation

In behavior based navigation the navigation task is subdivided into small subtasks. It allows to add new behaviors to the system [2]. In this problem we split the task of robot navigation into three behaviors: target reaching, obstacles avoiding, and local minimum avoidance [4]. In this study we consider a robot with two drive wheels. This is a non-holonomic control system [5]. Here x and y are coordinates of the mobile robot, θ_m represents the angle orientation of the robot, u and ω are linear and angular velocities of the robots, r is the wheels radius, and D is the distance between two wheels. The distance to the target and the angular error between the actual orientation and the target orientation are denoted, correspondingly as position error (EP) and actual orientation error (EA). which are the inputs of the fuzzy logic controller [5]:

$$\begin{aligned} EP &= \sqrt{D_x^2 + D_y^2}, \\ EA &= \theta_m - \theta \\ D_x &= x_g - x \\ D_y &= y_g - y \\ \theta &= \arctan(D_y, D_x) \end{aligned} \tag{1}$$

Movement of the robot with the target can be found as

$$E = EP_i - EP_{i-1} \tag{2}$$

The architecture of given mobile robot navigator is shown in Fig. 1. The computation module compares the real coordinates of the mobile robot with the coordinates of the target to calculate EP, EA, E and θ, according to Eq. (1).The navigation task is realized by multi-objective optimization of the behavior's objective for command fusion of the considered behaviors. Fuzzy controller realizes all behaviors and contains the fuzzification and inference procedures. The sensors around the robot are collected in three modules. We reduce the dimensions of inputs by grouping the robot's sensors reading into three options which are left, front and right [6]. The module 1 is for the right sensors and their inputs are d_1, d_2, d_3. The module 2 is for the front sensors. d_4, d_5 are their inputs. The module 3 is for left sensors. The inputs are the distances $d_6,\ d_7,\ d_8$. We determine the minimal value of all sensors in the corresponding groups $d_l,\ d_r,\ d_r$. For the fuzzification of the input and output variables we use trapezoidal type of membership functions. The knowledge base of fuzzy controller consists of IF…, THEN… linguistic rules. Each module has different interpretation to the linguistic terms. For example, for sensors ranging from 0.3 to 6.6 the linguistic term is Near.

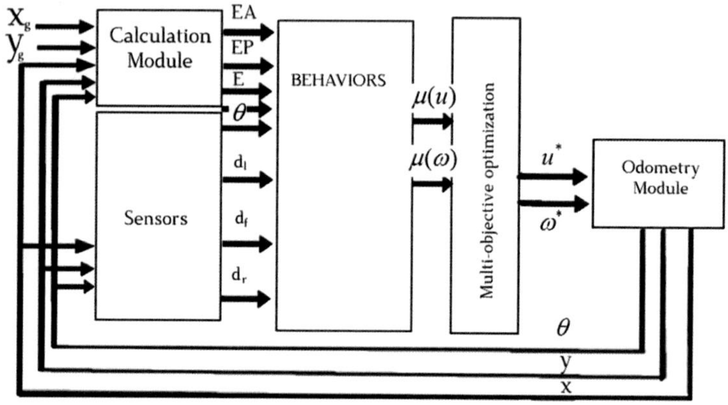

Fig. 1. Fuzzy behavior control system

The linguistic rules for obstacle avoidance behavior are as follows:

IF d_l is A_{1i} AND d_f is A_{2i} AND d_r is A_{3i} AND EA is A_{4i} THEN u is B_{1k} AND ω is B_{2k}.

For goal reaching behavior:

IF EP is A_{1i} AND EA is A_{2i} THEN u is B_{1k} AND ω is B_{2k}.

For local minimum avoidance:

IF d_l is A_{1i} AND d_f is A_{2i} AND d_r is A_{3i} AND EA is A_{4i} AND EP is A_{5i} THEN u is B_{1k} AND ω is B_{2k}.

For logic inference of control variables we apply max-min compositional rule for each behavior and determine $\mu_{ob.av.}(u^*)$, $\mu_{goal.reach}(u^*)$, $\mu_{loc.\min.avoid}(u^*)$, $\mu_{ob.av.}(\omega^*)$, $\mu_{goal.reach}(\omega^*)$, $\mu_{loc.\min.avoid}(\omega^*)$ taking into account all rules in linguistic tables

$$\begin{aligned}
\mu_{ob.av.}(u^*) &= \max(\mu_{ob.av.1}(u), \mu_{ob.av.2}(u), \ldots, \mu_{ob.av.k}(u)) \\
\mu_{ob.av.}(\omega^*) &= \max(\mu_{ob.av.1}(\omega), \mu_{ob.av.2}(\omega), \ldots, \mu_{ob.av.k}(\omega)) \\
\mu_{goal.reach.}(u^*) &= \max(\mu_{goal.reach.1}(u), \mu_{goal.reach.2}(u), \ldots, \mu_{goal.reach.m}(u)) \\
\mu_{goal.reach.}(\omega^*) &= \max(\mu_{goal.reach.1}(\omega), \mu_{goal.reach.2}(\omega), \ldots, \mu_{goal.reach.m}(\omega)) \\
\mu_{loc.\min.avoid}(u^*) &= \max(\mu_{loc.\min.avoid_1}(u), \mu_{loc.\min.avoid_2}(u), \ldots, \mu_{loc.\min.avoid_n}(u))
\end{aligned} \quad (3)$$

where l, m, n means the number of rule in linguistic table.

According to definition of multi-objective decision problem the optimal control values are determined as

$$\begin{aligned}
u^* &= \arg\max(\mu_{ob.av.}(u^*), \mu_{goal.reach}(u^*), \mu_{loc.\min.avoid}(u^*)) \\
\omega^* &= \arg\max(\mu_{ob.av.}(\omega^*), \mu_{goal.reach}(\omega^*), \mu_{loc.\min.avoid}(\omega^*))
\end{aligned} \quad (4)$$

The solution of multi-objective optimization problem must be Pareto optimal one as there cannot be the best solution for all objectives. In order to find a Pareto optimal solution we used a lexicographic method. According to this method, preferences are specified by ordering the objective functions according to their importance. Here the descending order of behaviors is the following: 1) local minimum avoidance, 2) obstacle avoidance behavior, 3) goal reaching. Then the local avoidance problem is solved as a single objective problem. If the optimum solution is not unique, we maximize the second objective function for obstacle avoidance behavior. If the solution is unique and maximal then process stops. There are different types and shapes of obstacles a robot can face on its path and sometimes it is difficult to predict the presence of a local minimum in a given situation. In this case we take into account that if the robot's direction of travel is more than 90° off target, then it is very likely about to get trapped. In the literature we can find many examples of using fuzzy PID controllers in mobile robots navigation. And it‘s important to have tuning of the fuzzy PID controllers parameters to avoid an impact on each fbehavior. There are various methods of fuzzy controller parameters tuning. In our work we used fuzzy adaptive control system with PID controller in motion control part [11].

The used kinematic model for simulation is represented by equation [5]:

$$\begin{bmatrix} \dot{x} \\ \dot{y} \\ \dot{\theta} \end{bmatrix} = \begin{bmatrix} \cos\theta & 0 \\ \sin\theta & 0 \\ 0 & 1 \end{bmatrix} \cdot \begin{bmatrix} u \\ \omega \end{bmatrix} \quad (5)$$

3 Simulation Results

Let consider an example of mobile robot behavior navigation to confirm the effectiveness of the control system. We decompose the robot navigation task into three small independent behaviors. The first behavior focuses on obstacle avoidance. The second behavior focuses on goal reaching. And the third behavior is a kind of local minimum avoidance. Each behavior consists of a set of fuzzy linguistic IF…THEN … rules. Our goal is to support a robot navigation with no collisions, trapping and reaching the target in the shortest time. The goal reaching behavior drives the robot to the target direction at the fastest speed. Only in the presence of obstacles, as well as approaching the target, the robot should reduce its speed. In the case of a change in position and a long distance to the target, the angle EA takes the maximum value, while the movement speed should be low. If the robot has the same direction towards the target, the angle is reduced to zero and the speed is maximized. As robot approaches the target, the speed decreases to zero. The outputs of all behaviors are linear and steering velocities. As result of fuzzification using trapezoidal membership function type the obstacle distance of each option d (m), EA (rad), linear u (m/s) and angular ω (rad/s) velocities, EP and the movement of the robot with the target E are represented by fuzzy sets {Very Near, Near, Far}, {Negative Large, Negative, Zero, Positive, Positive Large}, {{Small, Medium, Large}, {Negative Large, Negative, Zero, Positive, Positive Large}, {Very Near, Near, Far}, {Negative, Zero, Positive},correspondingly, and are shown in Fig. 2.

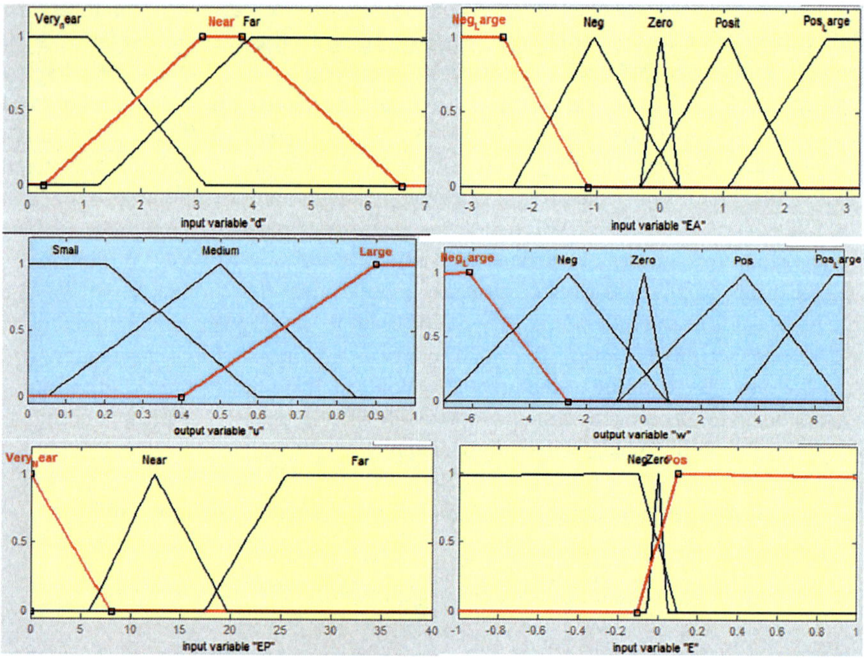

Fig. 2. Fuzzy sets for input and output variables

Tables 1–3 presents control rules defined for three behaviors.

Table 1. Linguistic rules for obstacle avoidance

Rule	Inputs				Outputs	
	d_l	d_f	d_r	EA	U	ω
1	Very Near	Very Near	Far	Positive	Small	Positive
2	Far	Very Near	Very Near	Positive	Medium	Positive
3	Near	Very Near	Very Near	Positive	Medium	Positive
4	Near	Very Near	Near	Pos.Large	Large	Pos.Large
5	Very Near	Very Near	Far	Positive	Medium	Pos.Large
6	Very Near	Very Near	Near	Pos.Large	Medium	Negative
7	Far	Very Near	Far	Pos.Large	Medium	Pos.Large
8	Near	Very Near	Far	Zero	Medium	Negative
9	Far	Very Near	Near	Pos.Large	Medium	Positive
10	Very Near	Near	Very Near	Pos.Large	Medium	Positive
11	Very Near	Very Near	Very Near	Negative	Small	Positive
12	Very Near	Very Near	Near	Positive	Small	Positive
13	Very Near	Near	Near	Zero	Medium	Negative
14	Very Near	Near	Far	Zero	Medium	Negative
15	Very Near	Far	Near	Zero	Medium	Negative
16	Very Near	Far	Far	Neg.Large	Small	Neg.Large
17	Very Near	Far	Far	Negative	Small	Negative
18	Very Near	Far	Far	Zero	Large	Zero
19	Very Near	Very Near	Very Near	Pos. Large	Large	Zero
20	Very Near	Near	Far	Positive	Large	Zero
21	Near	Near	Far	Pos.Large	Medium	Positive
22	Far	Near	Very Near	Pos.Large	Medium	Positive
23	Near	Near	Very Near	Pos.Large	Medium	Positive
24	Far	Far	Very Near	Neg.Large	Large	Zero
25	Far	Far	Very Near	Negative	Large	Zero
26	Far	Far	Very Near	Zero	Large	Zero
27	Far	Far	Very Near	Pos.Large	Small	Pos.Large
28	Far	Far	Very Near	Positive	Small	Pos.Large

Table 2. Linguistic rules for goals reaching

Rule	Input		Output	
	EP	EA	u	ω
1	Very Near	Zero	Small	Zero
2	Very Near	Negative	Small	Negative
3	Very Near	Neg.Large	Small	Neg.Large
4	Very Near	Positive	Small	Positive
5	Very Near	Posit. Large	Small	Posit. Large
6	Near	Zero	Medium	Zero
7	Near	Negative	Medium	Negative
8	Near	Neg.Large	Medium	Neg.Large
9	Near	Positive	Medium	Positive
10	Near	Posit. Large	Medium	Posit. Large
11	Far	Zero	Large	Zero
12	Far	Negative	Large	Negative
13	Far	Neg.Large	Large	Neg.Large
14	Far	Positive	Large	Positive
15	Far	Posit. Large	Large	Posit. Large

Table 3. Linguistic rules for local minimum avoidance

Rule	Input					Output	
	d_l	d_f	d_r	E	EA	u	ω
1	Very Near	Very Near	Very Near	Zero	Zero	Small	Positive
2	Far	Very Near	Very Near	Positive	Positive	Medium	Positive
3	Far	Very Near	Very Near	Negative	Pos.Large	Small	Pos Large
4	Far	Far	Very Near	Positive	Positive	Medium	Zero
5	Far	Far	Far	Positive	Pos.Large	Medium	Zero
6	Far	Far	Far	Positive	Positive	Medium	Zero
7	Far	Far	Far	Positive	Pos.Large	Medium	NegLarge

Let the start position will be $(x_s, y_s, \theta_s = (2, 1.8, 180°))$ and the target position is $(x_g, y_g, \theta_g = (10, 2, 0°))$. Initially the distance from the robot to goal is equal to 4 and the angle between robot and target orientation is equal to −1.6 rad according to (1).

$$EP = \sqrt{D_x^2 + D_y^2} = \sqrt{(10-2)^2 + (2-1.8)^2} \approx 4$$

$$EA = \theta_s - \theta = \arctan(0.2, 8) - 3,14 = arctg(8/0.2) - 3,14 = 1,54 - 3,14 = -1,6$$

For example, $d_r = 0.725$, $d_f = 1{,}04$, $d_l = 0.836$. It is obvious that at the starting moment value of the movement E = EP = 5. We determine the optimal value of each output control signal applying multi-objective optimization at the initial moment according to (3) and (4):

For obstacle avoidance behavior: $u^* = 0.48$, $\omega^* = -0.077$.

For goal reaching $u^* = 0.792$, $\omega^* = -3.13$.

For local minimum avoidance $u^* = 0.5$, $\omega^* = 0.01$.

We choose the maximal value for membership functions of control variables in case of local minimum avoidance behavior according to (3) and lexicographic method. But in this case these values are true for 3 rules simultaneously. As they are not unique we consider the control values for obstacle avoidance behavior. As they are unique, but not greater than control values for local minimum avoidance behavior then we choose $u^* = 0.5$, $\omega^* = 0.01$. After this we set new values to x_s and y_s and calculate the new values of EP and EA and so on. The mobile robot will move according to these position and orientation. And these paths will be saved in the robot as a consequent number of (x, y) points. After 15 runs we have determined an error between the last position and the target as equal to 0.07.

The Fig. 3 presents the robot trajectory with start position (2, 1.8, 180°) and target position (10, 2, 0°).

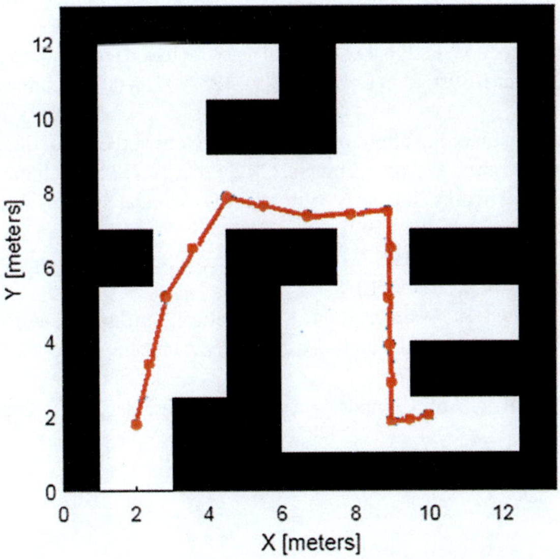

Fig. 3. Navigation of robot in unknown environment

4 Conclusion

In this paper we considered a behavior of the mobile robot. We applied a multi-objective optimization for optimal choice of control variables. Three behaviors: target reaching, obstacles avoiding, and local minimum avoidance are implemented using fuzzy logic. To verify the method a simulation was performed. It has been established that fuzzy logic devices should make decisions not taking into account the presence of any one obstacle, but taking into account the nearest obstacles in front, to the left and to the right of the link. An error between the last position of robot and the target equal to 0.07 that shows an effectiveness of this methodology.

References

1. Nakhaeinia, S., Tang, H., Noor, S.B.M., Motlagh, O.: A review of control architectures for autonomous navigation of mobile robot. Int. J. Phys. Sci. **6**, 169–174 (2011). https://doi.org/10.5897/IJPS10.540
2. Yahmedi, S.A., El-Tahir El-Dirdiri, A.B, Pervez, T.: Behavior based control of a robotic based navigation aid for the blind. In: Control and Applications Conference, p. 287–309, St. Petersburg, Russia (2009). https://doi.org/10.5772/25621
3. Saffiotti, A.: The uses of fuzzy logic in autonomous robot navigation. Soft Comp. **1**, 180–197 (1997). https://doi.org/10.1007/s005000050020
4. Nguyen, T.T.V., Phung, M.D., Vinh, T.Q.: Behavior-based navigation of mobile robot in unknown environments using fuzzy logic and multi-objective optimization. Int. J. Control and Autom. **10**, 349–364 (2017). https://doi.org/10.14257/ijca.2017.10.2.29
5. Mo, H., Tang, Q., Meng, L.: Behavior-based fuzzy control for mobile robot navigation. Math. Probl. Eng. **2013**, 1–10 (2013). https://doi.org/10.1155/2013/561451
6. Carelli, R., Camacho, E.F., Patino, D.: A neural network based feedforward adaptive controller for robots. IEEE Trans. Syst. Man Cybern. **25**(9), 1281–1288 (1995). https://doi.org/10.1109/21.400506
7. Lewis, F.L., Jagannathan, S., Yeşildirek, A.: Neural Network Control of Robot Manipulator and Non-Linear Systems. Taylor & Francis, CRC Press LLC Philadelphia (1999)
8. Kosko, B.: Neural Networks and Fuzzy Systems: A Dynamical Systems Approach to Machine Intelligence. Prentice-Hall, Inc. (1991)
9. Kundu, S., Parhi, R., Deepak, B.B.V.L.: Fuzzy-neuro based navigational strategy for mobile robot. J. Sci. Eng. Res. **3**, 1–6 (2012)
10. Manikas, T., Ashenayi, K., Wainwright, R.L.: Genetic algorithms for autonomous robot navigation. In: IEEE Instrumentation and Measurement Magazine, pp.26–31 (2008). https://doi.org/10.1109/MIM.2007.4428579
11. Aliev, R.A., Aliev, R.R.: Soft Computing and Its Applications. World Scientific (2001)

Author Index

A

Abbasov, Mikael H., 246
Abdalla, Mohamed Almojtba Hamid Ali, 316
Abdukaxxarov, Inomjon, 654
Abdullahi, Jazuli, 165
Abdullayev, Tarlan, 384
Abdurahman, Kh., 724
Abiyev, Rahib H., 347, 375
Abizada, Sanan, 347, 375
Adilov, Farukh, 505
Adilova, Nigar E., 28, 237
Agammedova, S. A., 567
Aghayev, Farid H., 537
Ahmadov, Shamil A., 43
Akdağ, Rojda Kılınçaslan, 183
Akdağ, Şahin, 183, 630
Akhundov, V. J., 545
Akkaya, Nurullah, 347, 375
Akperov Imran, G., 689
Alasgarova, F., 156
Alekperov, Ramiz, 732
Alekseychik, Tamara, 474
Alexeev Anton, S., 365
Ali, Adila, 671
Aliev, Araz R., 229
Aliev, R. A., 1, 3
Aliyev R. R., 420
Aliyev, A., 59
Aliyev, Elchin, 671
Aliyev, I. R., 124
Aliyev, Nazim K., 599
Aliyev, Rafig R., 14
Aliyeva, A. S., 496
Aliyeva, Aysel, 95
Aliyeva, E. N., 748
Aliyeva, Kamala, 371, 445
Alizada, Parvin, 662
Alizadeh, Akif V., 636
Almezhghwi, Khaled, 116
Altıparmak, Hamit, 261
Anene, Promise Isioma, 237
Aslanov, J. N., 765
Averkin, Alexey, 3, 102
Aytac, Ersin, 347, 375
Azizov, R. Z., 140

B

Babaei, Ebrahim, 19
Babanli, J. M., 205
Babanli, K. M., 401
Babanli, M. B., 7, 459
Babanly, M. B., 513
Babayev, A. I., 580
Bağlama, Başak, 183
Bakhshali, Valeh I., 704
Batishcheva Galina, A., 365, 488
Bayramov, I. Y., 522
Baysal, Ahmet Bahadir, 85
Bekirova, Aygun A., 704
Bektas, Senol, 19
Bilgehan, Bülent, 148
Bogachev, Taras, 474
Bozhenyuk, Alexander, 51
Bratishchev Alexander, V., 488

C

Çakır, Esra, 34
Çamur, Hüseyin, 316

R. A. Aliev et al. (Eds.): ICSCCW 2021, LNNS 362, pp. 783–785, 2022.
https://doi.org/10.1007/978-3-030-92127-9

D
Dadasheva, Aygul, 324
Davudova, Revana I., 680
Dovlatova, Khatira J., 173

E
Efendiyev, G. M., 331
Elkiran, Gozen, 165
Emamalipour, Soheil, 110

F
Farkhadov, Z. I., 140
Fathi, Jamal, 132
Fedorova, I. V., 621

G
Gahramanov, Ilkin, 437
Gardashova, Latafat A., 75, 302
Gashamoglu, Ahmad, 615
Gasimzade, T., 156
Gazanfarli, M. Kh., 278
Guliyev, A., 156
Guliyev, Javanshir, 420
Gulyamov, Shuhrat, 412
Gunsel, Irfan, 347
Gurbanov, A. N., 522

H
Habibov, I. A., 567
Hasanov, Adalat B., 340
Hasanov, G. S., 580
Hasanov, Mehman H., 246
Hasanov, Vasif, 645
Hasanova, Leyla, 308
Hasan-zade, D. S. Mamed, 580
Huseynli, Z. S., 765
Huseynov, O. H., 420
Huseynova, Farida, 560
Huseynova, Nigar F., 741

I
Ibrahimli, Mubarize S., 340
Igamberdiev, Husan, 654
Imamguluyev, Rahib, 384, 437
Imanov, Elbrus, 552
Imanov, G., 59
Imanova, Gunay E., 465
Imanova, Gunel, 465
Inan, Yucel, 67
Islamov, Islam J., 246
Ismayil, Ismayil A., 704
Ismayilova, K. H., 459
Ismayilova, Nigar T., 229
Ivanyan, Arsen, 505

J
Jabbarova, Aynur I., 214
Jabbarova, K. I., 214, 286
Jabiyeva, Aynur J., 270
Jafarov, B. O., 774
Jafarov, S. M., 496
Jamalov, Zeynal, 591

K
Kaba, Şerife, 261
Kacprzyk, Janusz, 2, 51
Kalantarly, Naila, 645
Karazhanova, M. K., 331
Karimov, R. I., 190
Kassem, Youssef, 316
Kazimov, Mirali S., 599
Kerimov, Djahid A., 355, 607
Khalilov, Rashad, 630
Khodkar, Kasra, 294
Khoshimov, Jamshid, 412
Khramov, Vladimir V., 529
Kılınç, Çağla, 183
Kiyaschenko Tatyana, A., 365
Knyazeva, Margarita, 51
Koilybayev, B. N., 331
Kurinova, Yana, 474
Kuzminov, Alexander N., 392
Kuznetsov, N. G., 724

L
Lozina, E. N., 621
Lukyanova Galina, V., 365
Lyudmila, Sakharova, 689

M
Madatova, Shahzada G., 599
Mamirov, Uktam, 654
Mammadov, A. N., 513
Mammadov, J. F., 124
Mammadova, K. A., 748
Mardanov, Misir, 591
Mardanov, Nail T., 704
Mehdiyev, Nihad, 480
Memmedova, Konul, 221, 450
Mikayilova, Rena, 732
Mirzayev, O. M., 522
Muradov, Nusrat, 697
Murtuzaeva, M., 59
Musayev, A. F., 278
Mustafa, Ansari, 392
Mustafayev, F. F., 765

N
Nourani, Vahid, 110, 294
Nuriyev, Aziz, 85

O
Ogly, Akperov Imran Gurru, 575
Orujlu, E. N., 513
Ovsyannikov Victor, A., 365

P
Paknezhad, Nardin Jabbarian, 110
Pedrycz, Witold, 6
Popova, L. K., 621

R
Rakhimov, Sh. R., 124
Reshetnikov, A. G., 196
Rodionova, N., 724
Ruso, Kemal, 347
Rzayev, Ramin, 591, 671
Rzayeva, Inara, 645

S
Sadikoglu, Fahreddin, 110, 148, 294
Sadikoglu, Gunay, 237
Sadıkoğlu, Saide, 221, 630
Safarov, Rza S., 537
Salahli, M. A., 156
Salahli, Vuqar, 156, 711
Salimov, V. H., 719
Salmanov, Salman, 302
Sardarova, I. Z., 522
Say, Gorkem, 347
Sebnem, Guldal Kan, 450
Seidzade, A. E., 513
Sergei, Rogozhin, 488
Shah, Zubair, 552
Shahir, Farzad Mohammadzadeh, 19
Shahli, Arash, 294
Sharypova, T. N., 621
Shevchenko, A. A., 196
Shevchenko, A. V., 196
Strekov, A. S., 331
Stryukov, Michael B., 392
Suleymanli, Rashad M., 757
Suleymanli, Tural, 254
Suleymanov, Abil, 711
Sultanova, A. B., 428, 765

T
Tagiev, D. B., 513
Taş, Mehmet Ali, 34
Ternovsky, Oleg A., 392
Tezer, Murat, 630
Tuzunkan, Murat, 347

U
Ulukan, Ziya, 34
Ulyanov, S. V., 196
Umarova, Niyar, 384
Usmanova, Nargiza, 412

V
Valiyev, Ahmad, 437
Vatolina, M., 724
Vovchenko, Natalia, 474

Y
Yarushev, Sergey, 102
Yirtici, Tolga, 347
Yücesoy, Yücehan, 183
Yusifov, Salahaddin I., 537
Yusupbekov, Azizbek, 654
Yusupbekov, Nadirbek, 412
Yusupbekov, Nodirbek, 505
Yuvalı, Meliz, 261

Z
Zeinalova, L. M., 774
Zhetekova, L. B., 331
Zhuravleva Maria, I., 488